AF712937

International Series in Operations Research & Management Science

Founding Editor
Frederick S. Hillier

Volume 356

Series Editor
Camille C. Price, Department of Computer Science, Stephen F. Austin State University, Nacogdoches, TX, USA

Editorial Board Members

Emanuele Borgonovo, Department of Decision Sciences, Bocconi University, Milan, Italy

Barry L. Nelson, Department of Industrial Engineering and Management Sciences, Northwestern University, Evanston, IL, USA

Bruce W. Patty, Veritec Solutions, Mill Valley, CA, USA

Michael Pinedo, Stern School of Business, New York University, New York, NY, USA

Robert J. Vanderbei, Princeton University, Princeton, NJ, USA

Associate Editor

Joe Zhu, Business School, Worcester Polytechnic Institute, Worcester, MA, USA

The book series **International Series in Operations Research and Management Science** encompasses the various areas of operations research and management science. Both theoretical and applied books are included. It describes current advances anywhere in the world that are at the cutting edge of the field. The series is aimed especially at researchers, advanced graduate students, and sophisticated practitioners.

The series features three types of books:

• Advanced expository books that extend and unify our understanding of particular areas.

• Research monographs that make substantial contributions to knowledge.

• Handbooks that define the new state of the art in particular areas. Each handbook will be edited by a leading authority in the area who will organize a team of experts on various aspects of the topic to write individual chapters. A handbook may emphasize expository surveys or completely new advances (either research or applications) or a combination of both.

The series emphasizes the following four areas:

Mathematical Programming: Including linear programming, integer programming, nonlinear programming, interior point methods, game theory, network optimization models, combinatorics, equilibrium programming, complementarity theory, multiobjective optimization, dynamic programming, stochastic programming, complexity theory, etc.

Applied Probability: Including queuing theory, simulation, renewal theory, Brownian motion and diffusion processes, decision analysis, Markov decision processes, reliability theory, forecasting, other stochastic processes motivated by applications, etc.

Production and Operations Management: Including inventory theory, production scheduling, capacity planning, facility location, supply chain management, distribution systems, materials requirements planning, just-in-time systems, flexible manufacturing systems, design of production lines, logistical planning, strategic issues, etc.

Applications of Operations Research and Management Science: Including telecommunications, health care, capital budgeting and finance, economics, marketing, public policy, military operations research, humanitarian relief and disaster mitigation, service operations, transportation systems, etc.

This book series is indexed in Scopus.

Francisco Saldanha-da-Gama • Shuming Wang

Facility Location Under Uncertainty

Models, Algorithms and Applications

Springer

Francisco Saldanha-da-Gama
Sheffield University Management School
Sheffield, UK

Shuming Wang
University of Chinese Academy of Science
Beijing, China

ISSN 0884-8289 ISSN 2214-7934 (electronic)
International Series in Operations Research & Management Science
ISBN 978-3-031-55926-6 ISBN 978-3-031-55927-3 (eBook)
https://doi.org/10.1007/978-3-031-55927-3

© The Editor(s) (if applicable) and The Author(s), under exclusive license to Springer Nature Switzerland AG 2024
This work is subject to copyright. All rights are solely and exclusively licensed by the Publisher, whether the whole or part of the material is concerned, specifically the rights of translation, reprinting, reuse of illustrations, recitation, broadcasting, reproduction on microfilms or in any other physical way, and transmission or information storage and retrieval, electronic adaptation, computer software, or by similar or dissimilar methodology now known or hereafter developed.
The use of general descriptive names, registered names, trademarks, service marks, etc. in this publication does not imply, even in the absence of a specific statement, that such names are exempt from the relevant protective laws and regulations and therefore free for general use.
The publisher, the authors, and the editors are safe to assume that the advice and information in this book are believed to be true and accurate at the date of publication. Neither the publisher nor the authors or the editors give a warranty, expressed or implied, with respect to the material contained herein or for any errors or omissions that may have been made. The publisher remains neutral with regard to jurisdictional claims in published maps and institutional affiliations.

This Springer imprint is published by the registered company Springer Nature Switzerland AG
The registered company address is: Gewerbestrasse 11, 6330 Cham, Switzerland

If disposing of this product, please recycle the paper.

To all who have contributed to our growth as researchers and as persons. This volume would not be the same without their inspiration and guidance in many stages of our lives.

Preface

Facility location under uncertainty is not a new topic. Nevertheless, the theoretical and technological advances observed in recent decades have led to many developments in terms of models, solution techniques, and applications for problems in this field. Such developments have been mostly triggered by the size and complexity of many real-world applications fed by the amount of data currently available and by the involved decision-making processes inherent to a globalized economy. Other aspects have also come into play such as sustainability and resilience.

Different sources of uncertainty in facility location and related problems have become prevalent such as those related to costs and demand levels and can hardly be ignored in the decision-making processes. Moreover, the surge of unexpected events such as floods or earthquakes' major impact at different levels of the enterprises has made the decision-makers more aware of the need to anticipate uncertainty when making decisions. In facility location, this is particularly relevant due to the long-term nature of many such decisions that are strategic and involve relevant investments.

Quantitative decision-making in areas such as supply chain planning, telecommunications network design, marketing, and humanitarian logistics planning, to name a few, has strongly benefited from considering facility location models capturing uncertainty and thus seeking solutions that can better anticipate unexpected changes in the underlying settings. The increased complexity observed in terms of modeling has been followed by significant developments in solution algorithms.

Although many procedures developed for tackling facility location problems under uncertainty are rooted in well-known optimization areas (robust optimization, stochastic programming, etc.), facility location problems often call for special tailoring when considering such procedures. Depending on the exact problem being studied, not only is it important to understand what is the adequate modeling paradigm but also how a model can be solved efficiently. This is even more relevant nowadays given that the amount of available data and the dimension of the problems to solve may easily become cumbersome. The development of progressively more comprehensive models as well as efficient special-tailored solution algorithms is of major relevance.

This volume aims to provide researchers, postgraduate students, and practitioners with systemic information of the models and algorithms currently available for hedging against uncertainty when making facility location decisions. Successful application areas are also presented. The book was written in a way that allows the reader to take advantage of its contents irrespective of their experience in the area. A "beginner" can easily grasp the necessary knowledge to successfully deal with new applications for existing and new problems. An experienced researcher or practitioner can directly use the most advanced models and algorithms thus directly overcoming major challenges.

The need for producing a structured overview of facility location under uncertainty emerged naturally from the experience of the authors working on many related topics. The idea was to gather in a single volume much of the existing knowledge that so far was scattered in different books and scholarly journals. This was a long journey requiring many hours of devoted work that the authors believe can be useful for a large audience.

The authors are deeply thankful to Ms. Jialin Yan and all the other staff of Springer Nature for all the help and support in this project.

Sheffield, UK Francisco Saldanha-da-Gama
Beijing, China Shuming Wang

Contents

Acronyms and Symbols

AD	Anaerobic digestion
ALNS	Adaptive large neighborhood search
ARP	Arc routing problem
BFS	Basic feasible solution
BoM	Bill-of-materials
CDF	Cumulative distribution function
CFLP	Capacitated facility location problem
CLRP	Capacitated location-routing problem
CVaR	Conditional value-at-risk
CVRP	Capacitated vehicle routing problem
DRO	Distributionally robust optimization
DVSS	Dynamic value of the stochastic solution
ERN	Estimated random noise
EVPI	Expected value of perfect information
FGLS	Feasible generalized least squares
FTL	Full-truck-load
FCCP	Fuzzy chance-constrained programming
GDRC	Globalized distributionally robust counterpart
UHLP	Uncapacitated hub location problem
LARP	Location-arc routing problem
LIRP	Location-inventory-routing problem
LRP	Location-routing problem
MHL	Multi-period hub location
MILP	Mixed-integer linear programming
MIP	Mixed-integer programming
NPV	Net present value
PSO	Particle swarm optimization
PoR	Price-of-robustness
RLRP	Reliable location-routing problem
ROI	Return on investment
SAA	Sample average approximation

SCM	Supply chain management
SOCP	Second-order conic programming
SOS2	Special ordered set of type 2
SUR	Seemingly unrelated regression
UFLP	Uncapacitated facility location problem
VaR	Value-at-risk
VNS	Variable neighborhood search
VRP	Vehicle routing problem
VSS	Value of the stochastic solution
WS	Wait-and-see (Wait-and-See solution value)
WTS	Wasserstein-ball-and-time-series-model-based
α-RAVPI	α-risk aversion value of perfect information
α-RAVSS	α-risk aversion value of the stochastic solution
$\mathbb{R}$	Set of real numbers
$\mathbb{R}^n$	Set of real vectors with n components
$\mathbb{R}^n_-$	Set of real vectors with n non-positive components
$\mathbb{R}^n_+$	Set of real vectors with n non-negative components
$\mathbb{R}^n_{++}$	Set of real vectors with n positive components
$\mathbb{Z}$	Set of integer numbers
$\mathbb{Z}^n$	Set of integer vectors with n components
$\mathbb{Z}^n_-$	Set of integer vectors with n non-positive components
$\mathbb{Z}^n_+$	Set of integer vectors with n non-negative components
$(a)^+,\ [a]^+$	$\max\{0, a\}$
$\|\cdot\|_l$	l-norm, $l = 1, 2, \ldots, \infty$
$\|\cdot\|_l^*$	Dual norm, $\|\boldsymbol{y}\|_l^* = \sup_{\|\boldsymbol{z}\|_l \le 1} \boldsymbol{z}^\top \boldsymbol{y}$
$\mathrm{dom}(\cdot)$	Domain of the argument function
$\mathrm{ri}(\cdot)$	Relative interior of the argument set
$\partial f(\boldsymbol{x})$	Subdifferential of the convex function f at $\boldsymbol{x}$
$\mathbb{E}_{\mathbb{P}}[\boldsymbol{g}(\boldsymbol{\xi})]$	Expected value of random vector $\boldsymbol{g}(\boldsymbol{\xi})$ with probability measure $\mathbb{P}$, where $\boldsymbol{\xi}$ denotes an n-dimensional random vector, $\mathbb{P}$ its distribution, and $\boldsymbol{g} : \mathbb{R}^n \to \mathbb{R}^m$
$\mathbb{I}(\cdot)$	Indicator function equal to one if the argument is true and zero otherwise
$[S]$	Set $\{1, \ldots, S\}$
$\boldsymbol{A} \succ \boldsymbol{0}$	Positive definite matrix $\boldsymbol{A}$
$\mathrm{Diag}(\boldsymbol{a})$	Square null matrix apart from the diagonal whose elements are the components of a vector $\boldsymbol{a} = (a_1, \ldots, a_n)$

Chapter 1
Introduction

Abstract This is the introductory chapter of the book. General aspects related to Location Science are reviewed. The role that uncertainty plays in the field is then discussed. The structure of the volume is explained, namely the purpose of each of the four parts. Several hints are provided regarding the better way to take advantage of the contents of this volume. Finally, some terminology and notation used throughout this book are briefly summarized.

Keywords Location science · Uncertainty in facility location problems

1.1 Location Science

Facility location problems lie at the core of a very active research area known as Location Science [326]. This is the area devoted to the study and development of models, algorithms, and applications involving the location of a set of facilities or equipment to serve a group of demand points in the best possible way. The meaning of "best" depends on the specific problem being considered, namely the objectives and constraints underlying it.

The terms "facility" and "equipment" should be looked at in a broad sense. They may refer to factories, distribution centers, warehouses, sensors, stores, teams mastering some skill, shelves in a warehouse, gates in an airport, ambulances, other emergency vehicles, *et cetera*. These examples also indicate that a facility can be mobile or immobile.

We face a unit-demand facility location problem when some "unitary" commodity (e.g., an inspection activity or some other service request) is to be provided. Alternatively, demand may refer to some quantity of a commodity. When it comes to supplying the customers, we often distinguish between single allocation (or single sourcing) and multiple allocation. In the first case, each demand node or customer must be fully served by a single facility, and in the second case, several facilities may jointly serve or supply the same customer. Additionally, we may be dealing with a single or multiple commodities.

© The Author(s), under exclusive license to Springer Nature Switzerland AG 2024
F. Saldanha-da-Gama, S. Wang, *Facility Location Under Uncertainty*,
International Series in Operations Research & Management Science 356,
https://doi.org/10.1007/978-3-031-55927-3_1

A feature of relevance in facility location is the pattern adopted for a facility to serve the demand nodes. Different possibilities exist. In some cases, we are dealing with the demand for some commodity or service that originates at demand nodes and must be supplied from existing or to-be-located facilities. In this setting, the demand nodes may be directly allocated to one or more such facilities (direct allocation). Alternatively, they may be included in one or several routes originated at and destined for the facilities. Finally, we may have a set of origin–destination pairs with some flow to be shipped between them. In this case, even if possible, shipping the flow directly between the nodes may render the resulting system too expensive or inefficient. Instead, several nodes can be set up to act as transshipment nodes (concentrators or hubs) to consolidate and redistribute the flow. Such nodes may also originate or be the destination of some flow. The above paradigms lead to different facility location problems that we briefly revisit in Chap. 2.

Facility location problems can be so diverse that they have been categorized according to different features. In addition to aspects already mentioned (e.g., allocation and service pattern), the problems are often distinguished according to the location space. We face a continuous facility location problem when the facilities can be located anywhere in the space underlying the problem. A network facility location problem is such that the location of the facilities is restricted to a given network with demand emerging in the same network (nodes or edges). Finally, we have a discrete facility location problem when a finite set of potential locations for the facilities is given.

An alternative for classifying facility location problems was proposed by Eiselt and Marianov [185], which considers the type of objective function: minsum, covering, and minmax. In any case, depending on the application or problem investigated, we find a wide variety of objectives in facility location problems such as cost, revenue, number of facilities, demand captured, and return on investment, to name a few.

Location Science interacts with several different areas, among which we find geography, economics, logistics, transportation, telecommunications, industrial engineering, *et cetera* [326]. Applications of facility location emerge in those and other related areas. Many such applications call for the use of discrete facility location models. This may explain why discrete facility location has progressively become predominant—namely in the recent decades—in the facility location literature. The same holds when uncertainty comes into play. For this reason, we also give a predominant role to discrete facility location problems. Nevertheless, when of relevance, we refer the reader to other settings.

1.2 The Role of Uncertainty in Facility Location

The location of facilities is often a strategic long-term, decision. Often, it involves significant costs. Moreover, the installed facilities should remain operating for a considerable time, during which changes may occur. If they are predictable, then time-dependent decisions can be considered to better account for the occurring changes. If that is not the case, one must find a way to hedge against uncertainty. This is a feature of relevance in this volume. Note, however, that time-dependent decision-making may also require uncertainty to be hedged against. Depending on the problem being investigated, different sources of uncertainty may emerge. These include but are not limited to demand levels, transportation costs, location of the demand, capacity of the facilities, supply amount, production yields, availability of the facilities, and availability of the transportation channels.

Facility location under uncertainty has attracted much attention from the scientific community. Significant knowledge has been accumulated in the field. This can be observed in different overviews: [145, 353, 510]. In those articles and the recent book [186], we can observe different types of problems that have been investigated. In some cases, we are dealing with so-called congestion. This applies when the customers' requests for service have a probabilistic behavior. If a facility is busy when a new service request arrives, we say that congestion occurs [66]. Another perspective is triggered by settings where disruption can occur at the facilities or in the underlying transportation or distribution network. We may immunize the system against such unpredictable events or identify its vulnerabilities. When the (unpredictable) disruptions cannot be described probabilistically (or that is not relevant), we fall in the area of facility location under disaster events [479]. Three types of models are usually considered: interdiction, protection, and design models. In the first case, we identify vulnerabilities and quantify the impact of a disruption. In the second case, the system is hardened by establishing "protective" components. Finally, we have models used for planning for service and supply systems, which are resilient to disruption.

Nowadays, more than ever, many challenges need to be faced, which makes the problems progressively more complex, not only in terms of the decisions to integrate but also in terms of the uncertainty and the way it can affect those decisions. Moreover, significant computational challenges are often faced just because of the size of the problems to solve. Thus, capturing uncertainty in facility location problems has become progressively more necessary and challenging. The economic globalization, the challenges posed by Industry 4.0, the upcoming challenges triggered by Industry 5.0, the political turmoils faced in many regions around the globe, *et cetera*, give more relevance to embedding uncertainty when making facility location decisions.

Looking into solutions that hedge against uncertainty calls for developing adequate models and algorithms for tackling them. Throughout this volume, step by step, we discuss modeling aspects and efficient solution techniques for accomplishing this. As we will see, often, we consider discrete optimization models when formulating the problems of interest. We usually distinguish among different paradigms for capturing uncertainty since each requires a different analysis. The exact paradigm to use depends on two aspects: (i) the knowledge we have about the uncertainty (e.g., is there a known CDF underlying it? If so, do we know it? Is this relevant?) and (ii) several preferences related to the decision-maker (e.g., constraints hardness and level of risk aversion). The main paradigms discussed in this volume are robust and distributionally robust optimization, two- and multi-stage stochastic programming, and chance-constrained programming. Occasionally, we may refer to other possibilities (e.g., uncertainty theory).

As we will, some families of facility location problems and some paradigms for capturing uncertainty in those problems have been studied more frequently than others. In the former case (e.g., stochastic and distributionally robust hub location), we can extend the discussion, which is not the case in the latter (e.g., location-inventory problems).

Several aspects will be focused throughout the book, although not having a chapter specifically dedicated to them. These include online optimization, congestion, interdiction, and fortification. The reason is that these topics appear very often related to specific problems or modeling paradigms that are already discussed in the volume.

Regarding online optimization—optimization problems having no or incomplete knowledge of the future—we face them, among other cases, when a single-stage decision-making process under uncertainty is adopted. This occurs often throughout the volume. Therefore, although not explicitly stating the problems in that context, we are considering them when adopting a single-stage decision-making process under uncertainty.

1.3 Structure of the Book

This volume is divided into four parts: principles and background, modeling paradigms and solution techniques, extended models, and comprehensive uncertainty with applications.

I. Principles and background

   Chapter 2: Discrete facility location problems
   Chapter 3: Decision-making under uncertainty: ingredients for modeling
   Chapter 4: Optimization under uncertainty

   This first part presents the background for the topics discussed later in the volume. A summary of different classes of discrete facility location problems and some usual models adopted for them is presented. Different aspects related to uncertainty are discussed: its nature, its representation, the nature of the decision-maker, sources of uncertainty, *et cetera*. In this part, different paradigms in terms of optimization under uncertainty are also revisited with an emphasis on general modeling aspects and solution techniques.

II. Modeling paradigms and solution techniques

   Chapter 5: Robust facility location
   Chapter 6: Stochastic facility location
   Chapter 7: Facility location with chance constraints
   Chapter 8: Distributionally robust facility location

This part focuses on different modeling paradigms and solution techniques that can be considered for facility location under uncertainty. Fundamental classes of problems are used in this part, namely those reviewed in Chap. 2. Often, the decisions to make are still limited to location and transportation. This allows for a better focus on mechanisms for hedging against uncertainty in different settings. The four chapters in this part lay the ground for considering more comprehensive models and uncertainty, thus preparing the reader for Parts III and IV.

III. Extended models

Chapter 9: Location-inventory problems
Chapter 10: Facility location with routing decisions
Chapter 11: Hub location
Chapter 12: Logistics and supply chain management
Chapter 13: Territory design

In this part, different families of facility location problems are extended by considering additional decisions or settings, such as inventory, routing, multiple facility layers, multiple objectives, time-dependent decisions, specific types of chance constraints, *et cetera*. In this part, we also start discussing different types of applications.

IV. Comprehensive uncertainty with applications

Chapter 14: Bilevel recycling facility location
Chapter 15: Multi-period hub location with time series models
Chapter 16: Data-driven robust production planning

This final part discusses several problems embedding more comprehensive uncertainty, thus leading to more involved models. The presented discussion of models and solution techniques is complemented with computational results, allowing a deeper analysis of the problems analyzed. The first application emerges in strategic reverse logistics planning and is thus aligned with sustainable supply chain planning. The second application focuses on data-driven hub location. Time series modeling is embedded in the ambiguity set definition. Finally, the third application emerges in the context of servitization strategic planning, which is becoming pervasive in many industries.

As the above contents make clear, this book does not seek to contain encyclopedic knowledge on facility location under uncertainty. That would require a much larger volume. Instead, we seek to discuss a representative set of models and techniques for hedging against uncertainty while making facility location decisions. Thus, despite the extensive and up-to-date references list considered, this volume does not dispense further reading. For instance, in the volume edited by Eiselt and Marianov [186], we can find additional applications of relevance in Location Science that are not considered in this volume.

1.4 Information Flow Throughout the Book

The first two parts of this volume are segmented in terms of the contents presented; neither is this the case with Part III nor Part IV. In Chaps. 9–16, we find problems gathering different aspects treated separately in Chaps. 2–8. A reader not well-acquainted with facility location and optimization under uncertainty should pay particular attention to Parts I and II before going into Parts III and IV.

Although the chapters are connected, someone targeting a specific problem and uncertainty paradigm can easily find a flow through the book without reading them all. For instance, a reader interested in distributionally robust hub location problems should focus on Chaps. 5, 6, 11, and 15. Someone focusing on humanitarian logistics should target Chaps. 5, 6, and 12. For a reader working in strategic supply chain planning, Chaps. 5–8, 9, 10, and 12 should be the target.

In Fig. 1.1, a relation between the different parts and chapters of the book is depicted.

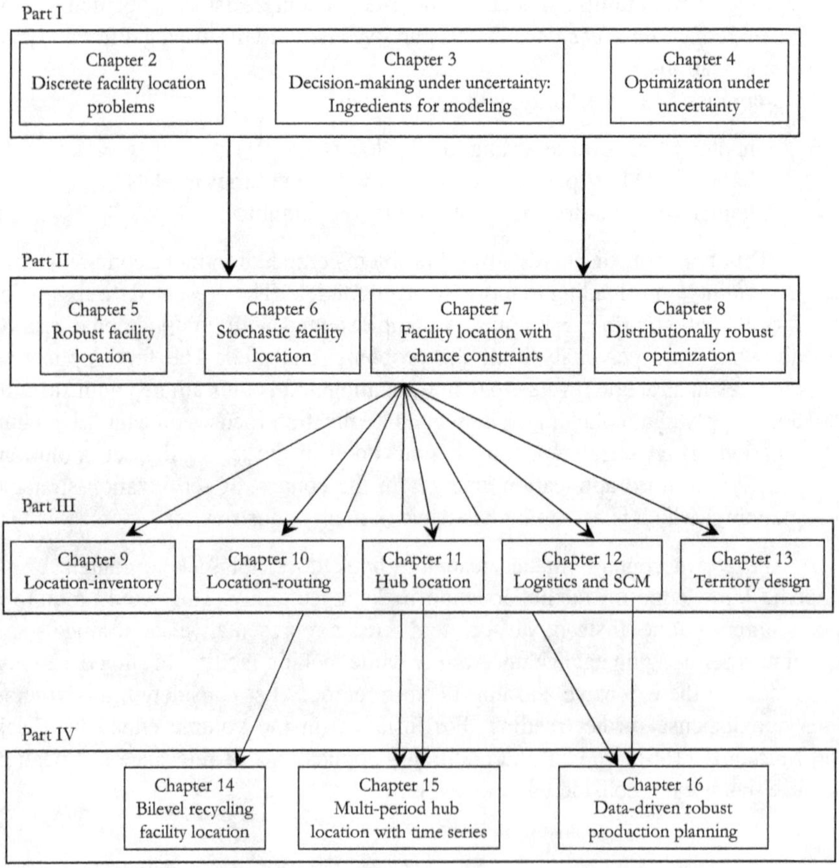

Fig. 1.1 Information flow throughout the book

This figure makes it clear that Parts III and IV have their roots in Parts I and II. Nevertheless, all chapters in Part III can be read independently like the chapters in Part IV (assuming that the background from Parts I and II has been acquired).

All chapters contain a short abstract, allowing a reader to grasp the related content. A single list of references is provided at the end of the volume. Given the relation between the content of different chapters, presenting a chapter-related reference list would require some references to be listed several times. A single list avoids this and helps the reader, who can look for some specific reference in a single numbered list.

A significant effort was made to ensure that the notation and terminology used in the volume are consistent. Unless unavoidable, the same notation is always used with the same meaning. The same happens with the terms. A list of acronyms and symbols used throughout is available after the table of contents.

Part I
Principles and Background

Chapter 2
Discrete Facility Location Problems

Abstract This chapter reviews several well-known classes of facility location problems. These include problems with direct allocation (a demand node is served directly from a facility), problems with routing decisions (a demand node is served as part of a route), and hub location (there is a commodity to be shipped among different pairs of demand nodes). The emphasis is on modeling aspects, but references are provided regarding state-of-the-art solution algorithms. The chapter also covers models for handling unexpected facility disruptions, which include interdiction models and fortification.

Keywords Direct allocation · Routing · Median objectives · Fixed-charge problems · Covering objectives · Hub location · Interdiction · Fortification

2.1 Facility Location with Direct Allocation

Discrete facility location is a central element in this volume. In this section, we revisit some of the most well-known facility location problems assuming direct allocation. This means that the facilities serve each customer directly (e.g., the customers go to the facility to be served, or single-customer routes are defined for supplying the demand). Depending on the problem, different costs may be incurred.

The p-median problem is possibly one of the most famous problems in Location Science. The seminal work by Hakimi [254] would lay the ground for much work done. Let J be a finite set of nodes. For each $j \in J$, let d_j be the demand of node j w.r.t. some commodity. Suppose we wish to install p facilities to serve those customers and have identified a finite set I of possible locations. Let c_{ij} denote the unit cost for supplying demand node j from facility $i \in I$. The problem can be formulated mathematically by considering two sets of decision variables: binary

© The Author(s), under exclusive license to Springer Nature Switzerland AG 2024

F. Saldanha-da-Gama, S. Wang, *Facility Location Under Uncertainty*,
International Series in Operations Research & Management Science 356,
https://doi.org/10.1007/978-3-031-55927-3_2

location variables y_i indicating whether a location $i \in I$ is selected and binary allocation variables x_{ij} equal to one if and only if node $j \in J$ is supplied from a facility located at $i \in I$. Mathematically we consider [373]

$$\text{minimize} \quad \sum_{i \in I} \sum_{j \in J} c_{ij} d_j x_{ij} \tag{2.1}$$

$$\text{subject to} \quad \sum_{i \in I} x_{ij} = 1 \qquad \forall j \in J, \tag{2.2}$$

$$\sum_{i \in I} y_i = p, \tag{2.3}$$

$$x_{ij} \le y_i \qquad \forall i \in I,\ j \in J, \tag{2.4}$$

$$x_{ij} \in \{0, 1\} \qquad \forall i \in I,\ j \in J, \tag{2.5}$$

$$y_i \in \{0, 1\} \qquad \forall i \in I. \tag{2.6}$$

The objective function accounts for the total cost incurred to serve the customers. Constraints (2.2) ensure that each demand node is allocated to one and only one facility, and Constraint (2.3) imposes that exactly p facilities are installed. Constraints (2.4) guarantee that all allocations are made to operating facilities. Finally, we find the domain constraints.

A particular case of interest much considered in the literature regards $I = J$. This allows using only the x-variables since the role of y_j can be played by x_{jj}.

The p-median problem has been extensively investigated. Despite being NP-hard, efficient exact and approximate algorithms are nowadays available that tackle large instances within a reasonable computing time. The interested reader can refer to [373] and the references therein for further details and references.

The above problem does not account for fixed costs associated with the network structures, namely the facilities. If such costs are relevant, we face a fixed-charge facility location problem. Let f_i be the fixed cost for a facility installed at $i \in I$. We keep using the notation already presented. To make the modeling framework more general, we assume that a facility operating at $i \in I$ has an operating capacity q_i. In this case, we face a capacitated facility location problem. If single allocation is assumed, then the problem can be formulated as follows:

$$\text{minimize} \quad \sum_{i \in I} f_i y_i + \sum_{i \in I} \sum_{j \in J} c_{ij} d_j x_{ij}, \tag{2.7}$$

$$\text{subject to} \quad \sum_{i \in I} x_{ij} = 1 \qquad \forall j \in J, \tag{2.2}$$

$$\sum_{j \in J} d_j x_{ij} \le q_i y_i \qquad \forall i \in I, \tag{2.8}$$

$$x_{ij} \in \{0, 1\} \qquad \forall i \in I,\ j \in J, \tag{2.5}$$

$$y_i \in \{0, 1\} \qquad \forall i \in I. \tag{2.6}$$

The objective function constrains fixed costs associated with the facilities and those for supplying the demand. Again, we have constraints imposing that each customer or demand node is allocated to one and only one facility, Constraints (2.2). Inequalities (2.8) ensure that the total amount supplied by each facility is not above its capacity. These constraints also play the role of (2.4).

The uncapacitated version of the above problem leads to the well-known uncapacitated facility location problem. In this case, Constraint (2.8) is replaced by

$$x_{ij} \le y_i \quad \forall i \in I,\ j \in J, \tag{2.9}$$

and also relaxing the integrality of the x-variables because there is always an optimal solution that follows a single-allocation pattern.

Under a multiple-allocation setting, the x-variables are usually redefined to represent the amount shipped from location $i \in I$ to demand node $j \in J$. The model becomes

$$\text{minimize} \quad \sum_{i \in I} f_i y_i + \sum_{i \in I} \sum_{j \in J} c_{ij} x_{ij} \tag{2.10}$$

$$\text{subject to} \quad \sum_{i \in I} x_{ij} = d_j \qquad \forall j \in J, \tag{2.11}$$

$$\sum_{j \in J} x_{ij} \le q_i y_i \qquad \forall i \in I, \tag{2.12}$$

$$x_{ij} \ge 0 \qquad \forall i \in I,\ j \in J, \tag{2.13}$$

$$y_i \in \{0, 1\} \qquad \forall i \in I. \tag{2.6}$$

The meaning of the constraints can be easily understood from those already described for the previous models. Fixed-charge discrete facility location problems are at the core of much existing work in Location Science. In particular, many variants and extensions have been proposed, some of which are mentioned in this chapter. For other details and algorithmic developments, the reader should refer to [209].

One aspect related to the above fixed-charge facility location problems regards the combination in a single objective function of costs associated with the facilities and those related to transportation or distribution. The interested reader can refer to [470] for a discussion on the relevance of considering such a combination.

The above models are valid when the parameters are static, i.e., no changes are foreseen, and the location decisions can and should be implemented in a single step. In practice, this is not always realistic. First, we must often deal with time-dependent parameters such as costs and demands. Additionally, setting up the facilities may not be possible in a single step, which calls for a progressive facility phase-in. In both cases, time becomes a dimension of relevance that must be considered. For different reasons (see, e.g., [181]), time is often discretized, thus determining specific moments for implementing the decisions: A multi-period facility location problem emerges.

Let us assume that the changes in the underlying parameters can be predicted accurately for the planning horizon of interest. In particular, we denote by T the planning horizon, which is assumed to be divided into a finite number of periods. Additionally, let d_{jt} be the demand of customer $j \in J$ in period $t \in T$, we consider c_{ijt} as the unit transportation cost from facility $i \in I$ to customer $j \in J$ in period $t \in T$, and finally, f_{it} denotes the cost for having a facility operating at $i \in I$ in period $t \in T$. The decision variables are now indexed in the periods: y_{it} is equal to one if a facility is operating at $i \in I$ in period $t \in T$ and zero otherwise, and x_{ijt} is the amount shipped from location $i \in I$ to customer $j \in J$ in period $t \in T$. We can adopt the following mathematical model for representing the multi-period fixed-charge facility location problem:

$$\text{minimize} \quad \sum_{t \in T} \left(\sum_{i \in I} f_{it} y_{it} + \sum_{i \in I} \sum_{j \in J} c_{ijt} x_{ijt} \right) \tag{2.14}$$

$$\text{subject to} \quad \sum_{i \in I} x_{ijt} = d_{jt} \qquad \forall j \in J,\ t \in T, \tag{2.15}$$

$$\sum_{j \in J} x_{ijt} \leq q_i y_{it} \qquad \forall i \in I,\ t \in T, \tag{2.16}$$

$$y_{it} \leq y_{i,t+1} \qquad \forall i \in I,\ t \in T \setminus \{|T|\}, \tag{2.17}$$

$$x_{ijt} \geq 0 \qquad \forall i \in I,\ j \in J,\ t \in T, \tag{2.18}$$

$$y_{it} \in \{0, 1\} \qquad \forall i \in I,\ t \in T. \tag{2.19}$$

One particular case of the above problem is the so-called implicit multi-period facility location problem, following the terminology of [146]. In this case, the selected facilities are installed at the beginning of the planning horizon and remain so throughout it. A model can be obtained straightforwardly from the above by removing the time index from the y-variables and ignoring Constraint (2.17).

The inclusion of time as an extra dimension in facility location problems can be triggered by the need to make additional decisions such as inventory management [141, 475], capacity reduction or expansion [286, 287], *et cetera*. The above type of model can be explored for modeling a system already in operation. In this case, closing existing facilities is also a decision that may be considered. For an overview of multi-period facility location problems and many related details, the reader can refer to [403] and the references therein.

A different class of facility location problems stems from a setting in which a demand node can be supplied or served by a facility only if it lies within a certain distance or travel time from the facility. These define the so-called covering location problems. Two types of problems are usually distinguished: set covering and maximal covering problems. In the former, the goal is to install facilities to cover all demand nodes while minimizing the total cost associated with the facilities [525]. In the latter, a limit is imposed on the number of facilities to install (e.g., due to a limited budget). The goal is to maximize the total demand covered [129].

A notable case in line with the set covering problem is the p-center problem in which the coverage radius is endogenous, thus resulting from the decision to make [255]. The discrete version of this problem can be formulated using the binary location variables y_i already introduced and the binary allocation variables x_{ij}. Now, c_{ij} stands for the distance, cost, travel time, *et cetera*, if demand node $j \in J$ is allocated to a facility operating at $i \in I$. A mathematical model inheriting much from the one presented above for the p-median problems is the following [154]:

$$\text{minimize} \quad z \tag{2.20}$$

$$\text{subject to} \quad \sum_{i \in I} c_{ij} x_{ij} \leq z \quad \forall j \in J, \tag{2.21}$$

$$\sum_{i \in I} x_{ij} = 1 \qquad \forall j \in J, \tag{2.2}$$

$$\sum_{i \in I} y_i \leq p, \tag{2.22}$$

$$x_{ij} \leq y_i \qquad \forall i \in I,\ j \in J, \tag{2.4}$$

$$y_i \in \{0, 1\} \qquad \forall i \in I, \tag{2.6}$$

$$x_{ij} \in \{0, 1\} \qquad \forall i \in I,\ j \in J. \tag{2.5}$$

Several other formulations have been proposed in the literature for the p-center problem. The interested reader can refer to [13, 98, 99, 133, 188, 215]. State-of-the-art algorithms can also be found in those works.

The set covering problem can be formulated using only the location variables y_i as far as some preprocessing is performed. In particular, for every demand node $j \in J$, we define the set $I_j \subseteq I$ of locations that cover the node (e.g., the distance is below the maximum given threshold). Assuming that opening a facility at $i \in I$ incurs a cost f_i, the model becomes

$$\text{minimize} \quad \sum_{i \in I} f_i y_i \tag{2.23}$$

$$\text{subject to} \quad \sum_{i \in I_j} y_i \geq 1 \quad \forall j \in J, \tag{2.24}$$

$$y_i \in \{0, 1\} \quad \forall i \in I. \tag{2.6}$$

If all costs f_i are equal, then the problem reduces to minimizing the number of facilities selected.

Finally, we recall the maximal covering location problem. In this case, a maximum number of p facilities can be installed, covering as much demand as possible. Again consider sets I_j, $j \in J$. However, now we need to define covering variables: u_j is a binary variable equal to one if and only if demand node $j \in J$ is covered by at least one of the selected facilities. Note that we only need to account for the demand covered (to maximize it), so it is not relevant to have a two-index allocation variable indicating the facility that ensures some node coverage. The problem can be formulated as follows:

$$\text{maximize} \quad \sum_{j \in J} d_j u_j \tag{2.25}$$

$$\text{subject to} \quad \sum_{i \in I_j} y_i \geq u_j \quad \forall j \in J, \tag{2.26}$$

$$\sum_{i \in I} y_i \leq p, \tag{2.27}$$

$$u_j \in \{0, 1\} \quad \forall j \in J, \tag{2.28}$$

$$y_i \in \{0, 1\} \quad \forall i \in I. \tag{2.6}$$

In [225], the authors propose a general modeling framework capturing the two covering location problems above and several others. The authors also provide an overview of exact and approximate algorithms for these problems. An important algorithmic contribution is provided in [138]. The interested reader can refer to that reference and to [131] that focuses on the history, applications, and recent advances in the field.

2.2 Facility Location with Routing Decisions

As an alternative to directly supplying the demand nodes from the open facilities (or having the customers travel to a facility to be served), some applications require the design of routes originated and destined at the facilities to the demand nodes as part of one or several such routes. The best way to accomplish that becomes embedded in the decision-making process. We distinguish between problems in which demand emerges at the nodes (like we have been considering so far) and those in which it originates at the arcs of the underlying transportation network. The former setting leads to a location-routing problem, whereas the latter leads to a location-arc routing problem.

For now, we keep denoting by I the potential locations for the facilities and by J the demand nodes. As before, q_i represents the capacity of a facility installed at $i \in I$ and d_j the demand of customer $j \in J$, and f_i is the fixed setup cost for facility $i \in I$. A directed connected graph $G = (V, E)$ with $V = I \cup J$ is assumed to represent the underlying transportation network. A unit transportation cost is associated with each arc. We consider a finite set M of homogeneous vehicles available, each with capacity Q. Thus, we allow determining up to $|M|$ routes for supplying the customers. Additionally, we incur a fixed cost, F, for every vehicle used.

We continue with the location variables y_i $(i \in I)$. The binary allocation variables, x_{ij}, can also be considered to indicate whether demand node $j \in J$ is supplied as part of a route originated at depot $i \in I$. Additionally, for $m \in M$ and $\ell, \ell' \in V$, $\ell \neq \ell'$, $v_{\ell\ell' m}$ is a decision variable equal to one if route m serves node ℓ' immediately after serving node ℓ and zero otherwise. Finally we consider costs $c_{\ell\ell'}$ representing the shortest unit transportation cost in the network G between nodes ℓ and ℓ' $(\ell, \ell' \in V, \ell \neq \ell')$.

Assuming a route to start and end in the same depot, the problem can be formulated mathematically as follows:

$$\text{minimize} \quad \sum_{i \in I} f_i y_i + \sum_{m \in M} \sum_{\ell, \ell' \in V, \ell' \neq \ell} c_{\ell\ell'} v_{\ell\ell' m} + F \sum_{m \in M} \sum_{\ell \in I} \sum_{\ell' \in J} v_{\ell\ell' m}, \tag{2.29}$$

$$\text{subject to} \quad \sum_{j \in J} d_j x_{ij} \leq q_i y_i \qquad \forall i \in I, \tag{2.8}$$

$$\sum_{m \in M} \sum_{\ell \in V, \ell \neq j} v_{\ell j m} = 1 \qquad \forall j \in J, \tag{2.30}$$

$$\sum_{j \in J} \left(d_j \sum_{\ell \in V, \ell \neq j} v_{\ell j m} \right) \leq Q \qquad \forall m \in M, \tag{2.31}$$

$$\sum_{\ell' \in V, \ell' \neq \ell} v_{\ell\ell' m} - \sum_{\ell' \in V, \ell' \neq \ell} v_{\ell' \ell m} = 0 \qquad \forall m \in M,\ \ell \in V, \tag{2.32}$$

$$\sum_{\ell \in I} \sum_{\ell' \in J} v_{\ell\ell' m} \leq 1 \qquad \forall m \in M, \tag{2.33}$$

$$\sum_{m \in M} \sum_{\ell \in S} \sum_{\ell' \in V \setminus S} v_{\ell\ell' m} \geq 1 \qquad \forall I \subseteq S \subset V, \tag{2.34}$$

$$\sum_{\ell' \in J} v_{i\ell' m} + \sum_{\ell' \in V, \ell' \neq j} v_{j\ell' m} \leq 1 + x_{ij} \quad \forall i \in I,\ j \in J,\ m \in M, \tag{2.35}$$

$$v_{\ell\ell' m} \in \{0, 1\} \qquad \forall \ell, \ell' \in V,\ \ell \neq \ell',\ m \in M, \tag{2.36}$$

$$x_{ij} \in \{0, 1\} \qquad \forall i \in I,\ j \in J, \tag{2.5}$$

$$y_i \in \{0, 1\} \qquad \forall i \in I. \tag{2.6}$$

The objective function (2.29) represents the total cost incurred, which includes the facility setup costs (first term), transportation (second term), and vehicle usage (third term). Inequalities (2.8) stands for the capacity constraints for the facilities (depots). They also ensure that nodes can only be supplied from operating facilities. Equalities (2.30) ensure that every demand node is included in one and only one route. Inequalities (2.31) guarantees that no route exceeds the vehicle capacity. Equalities (2.32) states that a set of routes is created. Inequalities (2.33) ensure that each route uses (at most) a single facility. Constraints (2.34) ensure that every route involves a facility. Constraints (2.35) guarantee that a route including depot i and demand node j forces x_{ij} to be one. In other words, j is served by a route originated at (and destined to) depot i. Conversely, if x_{ij} is equal to zero, no route can involve simultaneously depot i and demand node j. Finally, we specify the domain of the decision variables.

The above model was introduced in [437] and [438], and describes the well-known capacitated location-routing problem (CLRP) [327]. The interested reader can refer to [12] and [364] for recent overviews on this class of problems.

A recent research trend in the context of location-routing problems seeks to minimize the latency, i.e., the sum of the arrival times at the demand nodes. The latency location-routing problem extends the cumulative capacitated vehicle routing problem [401]. In turn, it can be seen as an extension of the traveling repairman problem [210]. In the cumulative capacitated vehicle routing problem, there is a single depot where a fleet of capacitated vehicles is dispatched. The traveling repairman problem involves a single depot and a single uncapacitated vehicle.

A formulation for the latency location-routing problem can be easily obtained by adapting the above cost-oriented location-routing problem. Let $b_{\ell\ell'}$ be the travel time between nodes ℓ and ℓ' with $\ell, \ell' \in V$, $\ell \neq \ell'$. Consider additional decision variables accounting for the arrival times at the demand nodes. In particular, we denote by t_{jm} the arrival time at demand node $j \in J$ if visited by vehicle $m \in M$. The problem can be formulated as follows:

$$\text{minimize} \quad \sum_{m \in M} \sum_{j \in J} t_{jm}, \tag{2.37}$$

$$\text{subject to} \quad (2.5),\ (2.6),\ (2.8),\ (2.30)\text{–}(2.33),\ (2.35),$$

$$t_{jm} \geq t_{\ell m} + b_{\ell j} - M(1 - v_{\ell jm}) \qquad \forall j \in J,\ \ell \in V,\ \ell \neq j,\ m \in M, \tag{2.38}$$

$$t_{jm} \geq 0 \qquad \forall j \in J,\ m \in M. \tag{2.39}$$

A closely related model was introduced by Moshref-Javadi and Lee [389]. Constraints (2.38) ensure that if an arc is used, the arrival time at the end node is at least equal to the arrival time at the origin node plus the travel time between the nodes. Constraints (2.39) define the domain of the t-variables.

Recently, [412] introduced improved models and algorithms for the problem. For completeness, we review the state-of-the-art model that is model 2 in that paper. The idea is to consider a multi-layered network to represent the problem. Each layer represents the backward position of a node in a route. For instance, layer 1 denotes the last position, layer 2 is the second last, *et cetera*. The layers are indexed in a set we denote by R, whose cardinality is an upper bound on the maximum number of demand nodes we may observe in a route (which can be easily obtained from the demands and capacities of the vehicles). Each layer from 2 to $|R|$ contains $|I|+|J|$ nodes. The first $|I|$ corresponds to the potential locations for the depots, and the remaining $|J|$ corresponds to the demand nodes. The first layer only contains a copy of the demand nodes.

The following decision variables are considered. For $r \in R$, $j \in J$, $\ell \in V$, $\ell \neq j$, $v_{\ell jr}$ is a binary variable that is equal to one if and only if demand node $j \in J$ is visited at layer r immediately after the route has visited node ℓ. For $i \in I$, y_i keeps indicating whether location i is selected for installing a depot. z_{jr} is a binary variable equal to one if and only if demand node $j \in J$ is visited at layer $r \in R$. For $i \in I$ and $j \in J$, w_{ij} denotes the load of a vehicle when leaving depot i to visit demand node j next. Finally, $u_{jj'}$ is the load of a vehicle when it is departing from demand node $j \in J$ to immediately visit demand node $j' \in J$, $j' \neq j$.

The problem can be formulated as follows:

$$\text{minimize} \quad \sum_{i\in I}\sum_{j\in J}\sum_{r\in R} r b_{ij} v_{ijr} + \sum_{\substack{j,j'\in J\\ j'\neq j}}\sum_{r\in R} r b_{jj'} v_{jj'r}, \tag{2.40}$$

$$\text{subject to} \quad \sum_{r\in R} z_{jr} = 1 \qquad \forall j \in J, \tag{2.41}$$

$$\sum_{r\in R}\sum_{i\in I}\sum_{j\in J} v_{ijr} \leq |M|, \tag{2.42}$$

$$\sum_{j\in J} z_{j1} = \sum_{r\in R}\sum_{i\in I}\sum_{j\in J} v_{ijr}, \tag{2.43}$$

$$\sum_{j'\in J: j'\neq j} v_{jj'r} = z_{j,r+1} \qquad \forall j \in J,\ r \in \{1,\ldots,|R|-1\}, \tag{2.44}$$

$$\sum_{\ell\in V:\ell\neq j} v_{\ell jr} = z_{jr} \qquad \forall j \in J,\ r \in \{1,\ldots,|R|-1\}, \tag{2.45}$$

$$v_{ij,|R|} = z_{j,|R|} \qquad \forall i \in I,\ j \in J, \tag{2.46}$$

$$v_{ijr} \leq y_i \qquad \forall i \in I,\ j \in J,\ r \in R, \tag{2.47}$$

$$\sum_{i\in I} w_{ij} + \sum_{j'\in J:j\neq j} u_{j'j} - \sum_{j'\in J:j'\neq j} u_{jj'} = d_j \qquad \forall j \in J, \tag{2.48}$$

$$d_j \sum_{r\in R} v_{ijr} \leq w_{ij} \qquad \forall i \in I,\ j \in J, \tag{2.49}$$

$$d_j \sum_{r\in R} v_{j'jr} \leq u_{j'j} \qquad \forall j, j' \in J,\ j \neq j', \tag{2.50}$$

$$w_{ij} \leq Q \sum_{r\in R} v_{ijr} \qquad \forall i \in I,\ j \in J, \tag{2.51}$$

$$u_{jj'} \leq (Q - d_j) \sum_{r\in R} v_{jj'r} \qquad \forall j, j' \in J,\ j \neq j', \tag{2.52}$$

$$u_{jj'} \geq 0 \qquad \forall j, j' \in J,\ j \neq j', \tag{2.53}$$

$$v_{\ell jr} \in \{0,1\} \qquad \forall j \in J,\ \ell \in V,\ \ell \neq j,\ r \in R, \tag{2.54}$$

$$w_{ij} \geq 0 \qquad \forall i \in I,\ j \in J, \tag{2.55}$$

$$y_i \in \{0,1\} \qquad \forall i \in I, \tag{2.6}$$

$$z_{jr} \in \{0,1\} \qquad \forall j \in J,\ r \in R. \tag{2.56}$$

The objective function (2.40) accounts for the total latency. Constraints (2.41) ensure that each demand node is visited exactly once. Inequality (2.42) stands for the maximum number of vehicles that can be used. Equalities (2.43) ensure that the number of routes created is equal to the number of vehicles used. Constraints (2.44)–(2.46) ensure a consistent choice for the values of v- and z-variables. Constraints (2.47) guarantee that the routes only start at open depots. Equalities (2.48) tracks the load of a vehicle at each customer. Constraints (2.49) and (2.50) establish the minimum values for the loads when leaving the depots and when leaving the demand nodes, respectively. Constraints (2.51) and (2.52) ensure that the capacity of the vehicles is satisfied. Finally, we have the domain constraints.

We note that [412] consider a maximum number of depots to open. The corresponding constraints are straightforward. Thus, we omit them.

As we have already observed, we may be facing demand along the arcs or edges of the underlying network instead of having it originate at the nodes. Popular applications of this situation include garbage collection and road maintenance, to mention a few. This setting leads to the class of so-called arc-routing problems (ARPs). By including the location of the depots in the decision-making process, we obtain the capacitated location-arc routing problems (LARPs). It is well-known that one can transform a VRP problem into an arc-routing problem and vice versa (see, e.g., [240, 351]). Nevertheless, solving a capacitated arc-routing problem as a capacitated VRP is often impractical. This explains why specially tailored models and algorithms have been proposed in the literature. The same analysis can be made to the multi-depot versions of both problems and consequently to problems in which the location of the depots is also a decision to make.

Again, we consider a transportation network $G = (V, E)$ with V containing I—the set of given potential locations for the depots. E contains the arcs. M is again the set of (homogeneous) vehicles available, each with capacity Q. We assume enough vehicles to allow a feasible partition of the demand arcs into routes. We denote by $d_{(\ell,\ell')}$ the demand in arc $(\ell, \ell') \in E$. Let E_d be the set of arcs in E that have demand; all the other arcs can still be used as part of the routes to design, but no service will be provided. Let us define a complete graph, $\tilde{G} = (\tilde{V}, \tilde{E})$ such that $\tilde{V}$ contains all nodes in I as well as all nodes that correspond to extremes of an arc in E_d. The unit cost associated with an arc $(\ell, \ell') \in \tilde{E}$ that we denote by $\tilde{c}_{\ell\ell'}$ is equal to $c_{\ell\ell'}$ if $(\ell, \ell') \in E_d$, and it is equal to the smallest transportation cost between ℓ and ℓ' in $G = (V, E)$ otherwise. If both $\ell, \ell' \in I$, then we set $\tilde{c}_{\ell\ell'} = \infty$.

To formulate the problem, we need to redefine the allocation variables so that they refer to arcs: $x_{i(\ell,\ell')}$ is equal to 1 if and only if the demand in arc $(\ell, \ell') \in E_d$ is supplied from facility $i \in I$. Regarding the routing variables v, they are now defined only in $\tilde{E}$ as follows: $v_{\ell\ell'm}$ is equal to one if route m uses arc (ℓ, ℓ') and zero otherwise, $\ell, \ell' \in \tilde{E}$. By including costs associated with the origins to set operating, we are led to a mathematical model for the LARP (see [84]):

$$\text{minimize} \quad \sum_{i\in I} f_i y_i + \sum_{m\in M} \sum_{(\ell,\ell')\in\tilde{E}} \tilde{c}_{\ell\ell'} v_{\ell\ell'm} + F \sum_{m\in M} \sum_{\ell\in I} \sum_{\substack{\ell'\in\tilde{V}\setminus I \\ (\ell,\ell')\in\tilde{E}}} v_{\ell\ell'm} \tag{2.57}$$

$$\text{subject to} \quad \sum_{(\ell,\ell')\in E_d} d_{(\ell,\ell')} x_{i(\ell,\ell')} \le q_i y_i \qquad \forall i \in I, \tag{2.58}$$

$$\sum_{m\in M} v_{\ell\ell'm} = 1 \qquad \forall (\ell, \ell') \in E_d, \tag{2.59}$$

$$\sum_{(\ell,\ell')\in E_d} d_{(\ell,\ell')} v_{\ell\ell'm} \le Q \qquad \forall m \in M, \tag{2.60}$$

$$\sum_{\ell'\in\tilde{V}} v_{\ell\ell'm} - \sum_{\ell'\in\tilde{V}} v_{\ell'\ell m} = 0 \qquad \forall m \in M,\ \ell \in \tilde{V}, \tag{2.61}$$

$$\sum_{\ell\in I} \sum_{\ell'\in\tilde{V}} v_{\ell\ell'm} \le 1 \qquad \forall m \in M, \tag{2.62}$$

$$\sum_{m\in M} \sum_{\ell\in S} \sum_{\ell'\in\tilde{V}\setminus S} v_{\ell\ell'm} \ge 1 \qquad \forall I \subseteq S \subset \tilde{V}, \tag{2.63}$$

$$\sum_{\ell'\in\tilde{V}\setminus I} v_{i\ell'm} + v_{\ell\ell'm} \le 1 + x_{i(\ell,\ell')} \quad \forall i \in I,\ (\ell, \ell') \in E_d,\ m \in M, \tag{2.64}$$

$$y_i \in \{0, 1\} \qquad \forall i \in I, \tag{2.6}$$

$$x_{i(\ell,\ell')} \in \{0, 1\} \qquad \forall i \in I,\ (\ell, \ell') \in E_d, \tag{2.65}$$

$$v_{\ell\ell' m} \in \{0, 1\} \qquad \forall (\ell, \ell') \in E_d,\ m \in M. \tag{2.66}$$

Both the objective function and the constraints have a straightforward interpretation after the discussion for the CVRP. This is a quite challenging problem. Borges Lopes et al. [84] propose heuristic procedures for finding feasible solutions and introduce a new set of test instances. This is a problem still scarcely addressed in the literature.

2.3 Hub Location

When some commodity is to be shipped among many origin–destination pairs, we often look into the design of a network that allows significant savings in transportation costs while improving the overall shipment efficiency. This calls for the location of so-called hubs, i.e., nodes that act as transshipment points consolidating and redistributing the flow.

We consider a set of nodes, say V, such that some flow d_{ij} must be sent from i to j, both in V. This flow is usually referred to as "demand". Thus, we keep denoting it using the notation d. Let $K \subseteq V$ be the set of nodes that can be selected as hubs, i.e., we assume that the potential hubs correspond to nodes that also originate or are destination of some flow. Thus, they also belong to V. The decisions to make include the network design (hubs to open and hub edges to activate) and the routes for the flow. If a hub is set up at node k, then a capacity q_k is also associated.

Regarding the costs, we define f_k as a fixed cost for setting up a hub at node $k \in K$ and $g_{k\ell}$ the cost for setting up an edge between hubs k and ℓ ($k, \ell \in K$, $k < \ell$). In hub location problems, all flow is usually charged in all path segments from the origin until the destination. Nevertheless, the inter-hub flow is discounted (e.g., due to economies of scale). Hence, we define $c_{\ell\ell'}$ as the unit cost for the flow sent directly from ℓ to ℓ' ($\ell, \ell' \in V$). Additionally, we consider a discount-cost factor, $\alpha \in]0, 1]$ for the inter-hub flow. If needed, we use O_i (D_i) to represent all the flow originated at (destined to) $i \in V$.

The single-allocation hub location problem emerges when each node is allocated to one and only one hub. It can be formulated using three sets of decision variables: (i) node allocation to hubs, (ii) inter-hub edges, and (iii) flow. We continue adopting the allocation binary variables x_{ik} equal to one if node $i \in V$ is allocated to hub $k \in K$ and zero otherwise. For $k, \ell \in K$ with $k < \ell$, $z_{k\ell}$ is a binary variable indicating whether hub edge $\{k, \ell\}$ is installed. Finally, for every $i \in V$ and $k, \ell \in K$, we consider flow variables $u_{ik\ell}$ indicating the amount of flow originated at node i that is routed via hubs k directly to hub ℓ (thus using hub edge $\{k, \ell\}$). The problem can be formulated mathematically as follows:

$$\text{minimize} \quad \sum_{k \in K} f_k x_{kk} + \sum_{\substack{k,\ell \in K \\ k<\ell}} g_{k\ell} z_{k\ell} + \sum_{i \in V} \sum_{k \in K} c_{ik}(O_i + D_i) x_{ik}$$

$$+\alpha \sum_{i \in V} \sum_{k,\ell \in K} c_{k\ell} u_{ik\ell} \tag{2.67}$$

$$\text{subject to} \quad \sum_{k \in K} x_{ik} = 1 \qquad \forall i \in V, \tag{2.68}$$

$$\sum_{i \in V} O_i x_{ik} + \sum_{i \in V} \sum_{\ell \in K} u_{i\ell k} \leq q_k x_{kk} \qquad \forall k \in K, \tag{2.69}$$

$$\sum_{\substack{\ell \in K \\ \ell \neq k}} u_{ik\ell} - \sum_{\substack{\ell \in K \\ \ell \neq k}} u_{i\ell k} = O_i x_{ik} - \sum_{j \in V} d_{ij} x_{jk} \qquad \forall i \in V, k \in K, \tag{2.70}$$

$$z_{k\ell} \leq x_{kk} \qquad \forall k, \ell \in K, \ k < \ell, \tag{2.71}$$

$$z_{k\ell} \leq x_{\ell\ell} \qquad \forall k, \ell \in K, \ k < \ell, \tag{2.72}$$

$$u_{ik\ell} + u_{i\ell k} \leq O_i z_{k\ell} \qquad \forall i \in V, \ k, \ell \in K, \ k < \ell, \tag{2.73}$$

$$u_{ik\ell} \geq 0 \qquad \forall i \in V, \ k, \ell \in K, \tag{2.74}$$

$$x_{ik} \in \{0, 1\} \qquad \forall i \in V, \ k \in K, \tag{2.75}$$

$$z_{k\ell} \in \{0, 1\} \qquad \forall k, \ell \in K, \ k < \ell. \tag{2.76}$$

The objective function represents the total cost for designing the network (setup cost for hubs and hub edges) plus the total transportation cost. Constraints (2.68) ensure that all nodes are allocated to some hub (to themselves in case of hubs). Constraints (2.69) refer to the capacity at the hubs. Note that these constraints also ensure that flow can only be sent to hubs that have been set up. Equalities (2.70) regard the usual flow divergence constraint ensuring a consistent flow routing through the network. Constraints (2.71) and (2.72) guarantee that when a hub edge is installed, both extremes correspond to hubs. Constraints (2.73) impose that the inter-hub flow is routed exclusively using the installed hub edges. Finally, (2.75) and (2.76) specify the domain of the decision variables.

The multiple-allocation version of the above problem allows a node to interact with more than one hub. For this reason, we need additional information to track the flow through the network. We keep using variables z and u. However, now we redefine x_{ik} as the amount of flow originated at node i that is consolidated at hub k being redistributed from there ($i \in V$, $k \in K$). Additionally, we need to consider binary hub location variables that we denote by y_k according to the terminology we have been using, indicating whether node k is selected as a hub ($k \in K$). Finally, we consider flow distribution variables, $u'_{i\ell j}$, that for $i, j \in V$ and $\ell \in K$ represent the amount of flow originated at node i that reaches node j directly from hub ℓ. The multiple-allocation hub location problem can be formulated as follows:

$$\text{minimize} \quad \sum_{k\in K} f_k y_k + \sum_{\substack{k,\ell\in K\\ k<\ell}} g_{k\ell} z_{k\ell} + \sum_{i\in V}\sum_{k\in K} c_{ik} x_{ik}$$

$$+ \alpha \sum_{i\in V}\sum_{k,\ell\in K} c_{k\ell} u_{ik\ell} + \sum_{i,j\in V}\sum_{\ell\in K} c_{\ell j} u'_{i\ell j} \tag{2.77}$$

$$\text{subject to} \quad \sum_{k\in K} x_{ik} = O_i \qquad \forall i \in V, \tag{2.78}$$

$$\sum_{\ell\in K} u'_{i\ell j} = d_{ij} \qquad \forall i, j \in V, \tag{2.79}$$

$$\sum_{i\in V} x_{ik} + \sum_{i\in V}\sum_{\ell\in K} u_{i\ell k} \le q_k y_k \qquad \forall k \in K, \tag{2.80}$$

$$\sum_{\substack{\ell\in K\\ \ell\ne k}} u_{ik\ell} - \sum_{\substack{\ell\in K\\ \ell\ne k}} u_{i\ell k} = x_{ik} - \sum_{j\in V} u'_{ikj} \qquad \forall i \in V,\ k \in K, \tag{2.81}$$

$$z_{k\ell} \le y_k \qquad \forall k, \ell \in K,\ k < \ell, \tag{2.82}$$

$$z_{k\ell} \le y_\ell \qquad \forall k, \ell \in K,\ k < \ell, \tag{2.83}$$

$$u_{ik\ell} + u_{i\ell k} \le O_i z_{k\ell} \qquad \forall i \in V,\ k, \ell \in K,\ k < \ell, \tag{2.84}$$

$$y_k \in \{0, 1\} \qquad \forall k \in K, \tag{2.85}$$

$$z_{k\ell} \in \{0, 1\} \qquad \forall k, \ell \in K,\ k < \ell, \tag{2.86}$$

$$x_{ik} \ge 0 \qquad \forall i \in V,\ k \in K, \tag{2.87}$$

$$u_{ik\ell} \ge 0 \qquad \forall i \in V,\ k, \ell \in K, \tag{2.88}$$

$$u'_{i\ell j} \ge 0 \qquad \forall i, j \in V,\ \ell \in K. \tag{2.89}$$

Again, the objective function accounts for the total cost (network design and transportation). Constraints (2.78) ensure that all flow enters the hub network. Equalities (2.79) impose that, for all pairs of nodes (i, j), all the flow from the origin to the destination arrives at the latter. Inequalities (2.80) refer to the capacity of the open hubs. These constraints also ensure that flow can only be routed through hubs that have been set up. Constraints (2.81) ensure the correct flow routing through the network. Together with (2.80) they also guarantee that $\sum_{i\in V} u'_{i\ell j} \le D_j y_\ell$, $j \in V$, $\ell \in K$, i.e., the consistency between positive flows u' and the operationality of the hubs distributing that flow is ensured. Constraints (2.82) and (2.83) impose that the extremes of a hub edge are indeed hubs, whereas Constraints (2.84) state that all the flow shipped between hubs is routed via hub edges. Finally, (2.85)–(2.89) state the domain of the decision variables.

In some applications, a node selected as a hub cannot send its outbound flow to another hub, i.e., all hubs process their outbound flow. Accordingly, all the flow routed through the hub network is discounted in terms of cost. Also, in that case, all the flow destined to a hub is distributed via that hub, i.e., all hubs distribute their inbound flow. These constraints can be easily accommodated by considering

$$x_{ik} \leq O_i(1 - y_i) \qquad \forall i, k \in K,\ i \neq k, \tag{2.90}$$

$$u'_{i\ell j} \leq D_j(1 - y_j) \qquad \forall j, \ell \in K,\ j \neq \ell. \tag{2.91}$$

The objective functions of the hub location problems discussed so far are cost-oriented. Recently, in the context of distribution logistics, some authors have investigated profit maximization hub location problems (see, e.g., [14, 515]).

The above models consider hub network design decisions. This is relevant in many practical applications. Nevertheless, much hub location literature still assumes a specific topology [92]. A fully inter-hub connectivity has been considered in many papers in the area [134, 137, 150, 554]. The hub location problems initially investigated in the literature [415, 416] relied on that assumption. The motivation stems from the airline Industry, given that world hubs or major hubs within a country are often fully interconnected. This assumption simplifies the above models considerably and even further when hub capacities are ignored.

Under a complete hub-level network, assuming that the unit transportation costs are proportional to the distances and that the latter satisfy the triangle inequality, it is immediate to conclude that the flow between two nodes traverses at least one but at most two hubs. This motivated the development of four-index-variable MILP models. This was first accomplished for p-hub median problems (single and multiple allocation) by Skorin-Kapov et al. [508]: The new variables are used to define the full path to be used from origin to destination leading to a so-called multi-commodity formulation. In the case of single-allocation, we keep using the x-variables to represent the allocation decisions. Nevertheless, define $w_{ijk\ell}$ as the fraction of the flow from i to j shipped via hubs k and ℓ. If the number of hubs to locate is set equal to p, the problem can be formulated as follows:

$$\text{minimize} \quad \sum_{i,j \in V} \sum_{k,\ell \in K} d_{ij} C_{ijk\ell} w_{ijk\ell}, \tag{2.92}$$

$$\text{subject to} \quad \sum_{k \in K} x_{ik} = 1 \qquad \forall i \in V, \tag{2.93}$$

$$\sum_{k \in K} x_{kk} = p, \tag{2.94}$$

$$x_{ik} \leq x_{kk} \qquad \forall i \in V,\ k \in K, \tag{2.95}$$

$$\sum_{\ell \in K} w_{ijk\ell} = x_{ik} \qquad \forall i, j \in V,\ k \in K, \tag{2.96}$$

$$\sum_{k \in K} w_{ijk\ell} = x_{j\ell} \qquad \forall i, j \in V,\ \ell \in K, \tag{2.97}$$

$$x_{ik} \in \{0, 1\} \qquad \forall i \in V,\ k \in K, \tag{2.98}$$

$$w_{ijk\ell} \geq 0 \qquad \forall i, j \in V,\ \forall k, \ell \in K, \tag{2.99}$$

In this model, $C_{ijk\ell}$ is the unit cost for shipping flow from i to j via hubs k (as the first hub) and ℓ (as the second). This cost is easily computed: $C_{ijk\ell} = c_{ik} + \alpha c_{k\ell} + c_{\ell j}$, $i, j \in V$, $k, \ell \in K$. Some authors also consider multiplication factors for the collected and distributed flows. Given that adapting the exposition to that case is straightforward, we omit it here. We also note that given that each node is allocated to a single hub, and assuming that the unit costs satisfy the triangle inequality, then we can half the routing variables considering only $w_{ijk\ell}$ with $i < j$ since this is enough to account to all the flow from i to j and from j to i. When relevant, we take advantage of this.

The multiple-allocation version can be formulated using the location variables y and the w-variables above defined:

$$\text{minimize} \quad \sum_{i,j \in V} \sum_{k\ell \in K} d_{ij} C_{ijk\ell} w_{ijk\ell}, \tag{2.100}$$

$$\text{subject to} \quad \sum_{k,\ell \in K} w_{ijk\ell} = 1 \qquad \forall i, j \in V, \tag{2.101}$$

$$\sum_{k \in K} y_k = p, \tag{2.102}$$

$$\sum_{\ell \in K} w_{ijk\ell} \leq y_k \qquad \forall i, j \in V,\ k \in K, \tag{2.103}$$

$$\sum_{k \in K} w_{ijk\ell} \leq y_\ell \qquad \forall i, j \in V,\ \ell \in K, \tag{2.104}$$

$$w_{ijk\ell} \geq 0 \qquad \forall i, j \in V,\ \forall k\ell \in K, \tag{2.105}$$

$$y_k \in \{0, 1\} \qquad \forall k \in K, \tag{2.106}$$

Despite having a tight linear programming relaxation, those models rapidly become intractable due to the number of variables needed. For this reason, a trade-off between tightness and model-tractability was sought in [192] for single allocation and [193] for multiple allocation. Those authors introduced the three-indexed variables u and u' considered above. The single-allocation p-hub median problem investigated in [192] can be looked at as a particular case of (2.67)–(2.76) by removing all the information related to the network design decision (hub edges) and making some additional adjustments:

$$\text{minimize} \quad \sum_{i \in V} \sum_{k \in K} c_{ik}(O_i + D_i)x_{ik} + \alpha \sum_{i \in V} \sum_{k \in K} \sum_{\ell \in K} c_{k\ell} u_{ik\ell} \tag{2.107}$$

$$\text{subject to} \quad \sum_{k \in K} x_{ik} = 1 \qquad \forall i \in V, \tag{2.68}$$

$$\sum_{k \in K} x_{kk} = p, \tag{2.108}$$

$$\sum_{\substack{\ell \in K \\ \ell \neq k}} u_{ik\ell} - \sum_{\substack{\ell \in K \\ \ell \neq k}} u_{i\ell k} = O_i x_{ik} - \sum_{j \in V} d_{ij} x_{jk} \quad \forall i \in V,\ k \in K, \tag{2.70}$$

$$u_{ik\ell} \geq 0 \qquad \forall i \in V,\ k, \ell \in K, \tag{2.74}$$

$$x_{ik} \in \{0, 1\} \qquad \forall i \in V,\ k \in K. \tag{2.75}$$

Fixed costs for the hubs can be easily introduced with the consequent removal Constraint (2.108). If we add capacity constraints, we obtain the model introduced in [194].

Note, however, that the first attempts to formulate the p-hub median problem were accomplished in [415] and [416]. In particular, in [416] the author formulated the problem as a binary quadratic optimization problem. A very successful use of that formulation is explored in [462]. The model makes use of the allocation variables only:

$$\text{minimize} \quad \sum_{i \in V} \sum_{k \in K} c_{ik}(O_i + D_i)x_{ik} + \alpha \sum_{i,j \in V} \sum_{k,\ell \in K} d_{ij} c_{k\ell} x_{ik} x_{j\ell} \tag{2.109}$$

$$\text{subject to} \quad \sum_{k \in K} x_{ik} = 1 \qquad \forall i \in V, \tag{2.68}$$

$$\sum_{k \in K} x_{kk} = p, \tag{2.108}$$

$$x_{ik} \leq x_{kk} \qquad \forall i \in V,\ k \in K, \tag{2.110}$$

$$x_{ik} \in \{0, 1\} \qquad \forall i \in V,\ k \in K. \tag{2.75}$$

The multiple-allocation p-hub median problem can be easily obtained from (2.77)–(2.89) again by ignoring the network design decisions, ignoring Constraints (2.90) and (2.91) and making minor adjustments, leading to

$$\text{minimize} \quad \sum_{i\in V}\sum_{k\in K} c_{ik}x_{ik} + \alpha \sum_{i\in V}\sum_{k\in K}\sum_{\ell\in K} c_{k\ell}u_{ik\ell} + \sum_{i\in V}\sum_{\ell\in K}\sum_{j\in V} c_{\ell j}u'_{i\ell j} \tag{2.111}$$

$$\text{subject to} \quad \sum_{k\in K} y_k = p \qquad \forall k \in K, \tag{2.112}$$

$$\sum_{k\in K} x_{ik} = O_i \qquad \forall i \in V, \tag{2.78}$$

$$\sum_{\ell\in K} u'_{i\ell j} = d_{ij} \qquad \forall i, j \in V, \tag{2.79}$$

$$\sum_{\substack{\ell\in K\\ \ell\neq k}} u_{ik\ell} - \sum_{\substack{\ell\in K\\ \ell\neq k}} u_{i\ell k} = x_{ik} - \sum_{j\in V} u'_{ikj} \qquad \forall i \in V,\ k \in K, \tag{2.81}$$

$$\sum_{\ell\in K} u_{ik\ell} \le x_{ik} \qquad \forall i \in V,\ k \in K, \tag{2.113}$$

$$x_{ik} \le O_i y_k \qquad \forall k \in K, \tag{2.114}$$

$$u'_{i\ell j} \le d_{ij} y_\ell \qquad \forall i, j \in V,\ \ell \in K, \tag{2.115}$$

$$u_{ik\ell} \ge 0 \qquad \forall i \in V,\ k, \ell \in K, \tag{2.88}$$

$$u'_{i\ell j} \ge 0 \qquad \forall i, j \in V,\ \ell \in K. \tag{2.89}$$

$$x_{ik} \ge 0 \qquad \forall i \in V,\ k \in K, \tag{2.87}$$

$$y_k \in \{0, 1\} \qquad \forall k \in K. \tag{2.85}$$

Fixed costs for the hubs can be easily introduced with the consequent removal of Constraints (2.112). If we add capacity constraints, we obtain one of the models introduced in [184].

For the uncapacitated multiple-allocation hub location problem, [256] present a variant for the four-indexed model presented above. Again, we have a multi-commodity formulation. We continue adopting the location variables y_k and the four indexed w-variables. We directly consider the version with fixed costs:

$$\text{minimize} \quad \sum_{k\in K} f_k y_k + \sum_{i,j\in V}\sum_{k,\ell\in K} d_{ij}C_{ijk\ell}w_{ijk\ell}, \tag{2.116}$$

$$\text{subject to} \quad \sum_{k,\ell\in K} w_{ijk\ell} = 1 \qquad \forall i, j \in V, \tag{2.117}$$

$$\sum_{\ell \in K} w_{ijk\ell} + \sum_{\ell \in K \setminus \{k\}} w_{ij\ell k} \le y_k \quad \forall k \in K,\ i, j \in V, \tag{2.118}$$

$$w_{ijk\ell} \ge 0 \quad \forall i, j \in V,\ k, \ell \in K, \tag{2.119}$$

$$y_k \in \{0, 1\} \quad \forall k \in K. \tag{2.120}$$

In the above model, $C_{ijk\ell}$ is again the unit cost for shipping flow from i to j ($i, j \in V$) through hubs k (as the first hub) and ℓ (as the second). Constraints (2.118) guarantee that if a node is selected as a hub, its fixed cost is paid. The other elements are straightforward, considering all the explanations already provided in this section. Since the hubs are uncapacitated, we can relax the integrality constraints on the x-variables because there is always an optimal solution such that it occurs. By removing the fixed costs in the above formulation and inserting a constraint on the number of hubs to select, we obtain a formulation for the p-hub median problem.

We now focus on a relevant family of hub location problems in which the key performance indicator is the maximum of some observable measure. This leads to a so-called hub center problem. As discussed in [18], different cases fit this setting. We consider a common one in which we wish to minimize the maximum cost among all origin–destination pairs. Although we keep invoking a transportation cost, it can relate, for instance, to transportation time. In this case, the so-called p-hub center problem consists of selecting p nodes for becoming hubs, minimizing the maximum transportation costs between any origin–destination pair. For the single-allocation case, several models have been proposed in the literature (see [18] for an overview). Next, we consider a model that is useful when the time comes to capturing uncertainty. We adopt the notation underlying model (2.92)–(2.99). The demands are no longer relevant, so we discard them. We keep assuming that the potential locations for the hubs are gathered in a set $K \subseteq V$. The p-hub center problem can be formulated as follows:

$$\text{minimize} \quad B, \tag{2.121}$$

$$\text{subject to} \quad B \ge C_{ijk\ell} w_{ijk\ell} \quad \forall i, j \in V,\ k, \ell \in K, \tag{2.122}$$

$$\sum_{k \in K} x_{ik} = 1 \quad \forall i \in V, \tag{2.123}$$

$$\sum_{k \in K} x_{kk} = p, \tag{2.124}$$

$$x_{ik} \le x_{kk} \quad \forall i \in v,\ k \in K, \tag{2.125}$$

$$w_{ijk\ell} \ge x_{ik} + x_{j\ell} - 1 \quad \forall i, j \in V,\ k, \ell \in K, \tag{2.126}$$

$$w_{ijk\ell} \in \{0, 1\} \quad \forall i, j \in V,\ k, \ell \in K, \tag{2.127}$$

$$x_{ik} \in \{0, 1\} \quad \forall i \in V,\ k \in K. \tag{2.128}$$

The objective function and Constraints (2.122) guarantee that the maximum cost among origin–destination pairs is minimized. The meaning of the other constraints is straightforward from the discussions already made in this section.

In the above problem, we ensure that all demands are covered while minimizing the maximum coverage (cost). Another class of hub location problems emerges when the coverage radius is exogenous. In this case, we seek to locate the hubs and route the flows in a way that captures as much demand as possible. We are led to a p-hub maximal covering problem. It can be easily formulated from the previous problem as follows:

$$\text{maximize} \quad \sum_{i,j \in V} \sum_{k,\ell \in K} d_{ij} a_{ijk\ell} w_{ijkl}, \tag{2.129}$$

$$\text{subject to} \quad (2.123)–(2.128).$$

In this model, $a_{ijk\ell}$ is a binary parameter that for each $i, j \in V, k, \ell \in K$ is equal to one if and only if the cost for shipping demand d_{ij} is smaller than or equal to a given acceptable threshold.

2.3.1 Hub Location Routing

In the above-discussed hub location models, nodes are served directly from the hubs. In practice, as for the CLRP and LARP, we may face a setting that calls for the nodes to be served as part of routes, each of which has its origin and destination at a hub.

To simplify the exposition, we directly extend what we have discussed to the CLRP. Again, assume a fleet of vehicles M, each having capacity Q. We keep assuming capacitated hubs, with q_k denoting the capacity of a hub operating at $k \in K$.

A central aspect of hub location routing is that the route originating at a depot (hub) must visit the nodes to deliver or connect the commodity. Hence, we must design routes with simultaneous pickup and delivery [165]. In other words, when it comes to embedding routing decisions, a hub location problem induces a multi-depot pickup and delivery CVRP combined with hub location and hub-network design. However, unlike the CLRP, when assessing the load carried in a vehicle, it is not enough to know the nodes that are part of a route. We need to assess the amount carried in every arc of a route. A vehicle can visit some node next in its route if the amount to collect minus the amount to distribute at the node does not raise the load beyond its capacity.

To intertwine hub location with location routing, we assume that two types of vehicles exist: those visiting the nodes to collect and distribute the commodity, whose set is denoted by M, and those that move between hubs, represented by set M^H. We keep defining by Q the capacity of the vehicles that visit the nodes. We denote by Q^H the capacity of the inter-hub vehicles. Typically, $Q^H > Q$. The setup cost for a route (vehicle) originated at and destined for the hub is again denoted by F. The other notation is similar to what we have already presented.

We assume that all nodes can be served with a single visit. Thus, they are included in a single route. Accordingly, our base hub location model is the single allocation. Additionally, we consider that the route starts by distributing the flow at the hub where it starts but collects the flow originating at the hub only at its end. This prevents the vehicle from carrying unnecessary load throughout an entire route.

We continue adopting the x-variables for allocation (like in the CLRP and single-allocation hub location problems), the z-variables for the hub edges to install, and the u-variables for the flow routing. Concerning the routing decisions for collecting and distributing the commodity at the nodes, we keep using the v-variables presented when we discussed the CLRP.

Due to the pickup and delivery nature of the underlying routing problem, we must consider two additional sets of decision variables: w^D_{ijm} represents the amount of flow still to be delivered that is on-board vehicle $m \in M$ when it is departing from i to visit j $(i, j \in V)$, and w^C_{ijm} is the amount of collected flow on-board vehicle $m \in M$ when it departs from i to visit j $(i, j \in V)$. To account for the inter-hub costs, we assess the number of vehicles (freights) made, which is represented by another decision variable, $\zeta_{k\ell}$—the number of vehicles that traverse hub-arc (k, ℓ). Finally, we denote by $c_{k\ell}$ the freight cost from hub k to hub ℓ corresponding to one trip. The total inter-hub cost depends on the number of freights made. In turn, the latter depends on the total amount transported. This cost already embeds economies of scale when that is the case. The hub location-routing problem with pickup and delivery that we have just described can now be formulated as follows:

$$\text{minimize} \quad \sum_{k\in K} f_k x_{kk} + \sum_{k\in K}\sum_{\ell\in K,\, k<\ell} g_{k\ell} z_{k\ell} + F\sum_{m\in M}\sum_{k\in K}\sum_{\ell\in V} v_{k\ell m}$$

$$+ \sum_{m\in M}\sum_{\ell\in V}\sum_{\ell'\in V} c_{\ell\ell'} v_{\ell\ell' m} + \sum_{k\in K}\sum_{\ell\in K} c_{k\ell}\,\zeta_{k\ell} \tag{2.130}$$

subject to (2.68)–(2.72),

$$\sum_{i\in V} u_{ik\ell} \le Q^H \zeta_{k\ell} \qquad \forall i \in V,\ k, \ell \in K, \tag{2.131}$$

$$\sum_{i\in V} u_{i\ell k} \le Q^H \zeta_{\ell k} \qquad \forall i \in V,\ k, \ell \in K, \tag{2.132}$$

$$\zeta_{k\ell} + \zeta_{\ell k} \le |M^H| z_{k\ell} \qquad \forall k, \ell \in K,\ k < \ell, \tag{2.133}$$

$$\sum_{m\in M}\sum_{\ell\in V} v_{\ell j m} = 1 \qquad \forall j \in V, \tag{2.134}$$

$$\sum_{\ell'\in V} v_{\ell\ell' m} - \sum_{\ell'\in V} v_{\ell'\ell m} = 0 \qquad \forall m \in M,\ \ell \in V, \tag{2.135}$$

$$\sum_{k \in K} \sum_{j \in V} v_{kjm} \leq 1 \qquad \forall m \in M, \tag{2.136}$$

$$\sum_{m \in M} \sum_{\ell \in S} \sum_{\ell' \in V \setminus S} v_{\ell\ell'm} \geq 1 \qquad \forall K \subseteq S \subset V, \tag{2.137}$$

$$\sum_{\ell' \in J} v_{i\ell'm} + \sum_{\ell' \in V} v_{k\ell'm} \leq 1 + x_{ik} \qquad \forall i \in V,\ k \in K,\ m \in M, \tag{2.138}$$

$$w^C_{ijm} + W^D_{ijm} \leq Q v_{ijm} \qquad \forall i, j \in V,\ m \in M, \tag{2.139}$$

$$\sum_{m \in M} \sum_{i \in V} w^D_{ijm} - D_j = \sum_{m \in M} \sum_{i \in V} w^D_{jim} \qquad \forall j \in V, \tag{2.140}$$

$$\sum_{m \in M} \sum_{i \in V} w^C_{ijm} + O_j = \sum_{m \in M} \sum_{i \in V} w^C_{jim} \qquad \forall j \in V, \tag{2.141}$$

$$\sum_{i \in V} w^D_{ikm} = \sum_{i \in V} w^C_{kim} = 0 \qquad \forall k \in K,\ m \in M, \tag{2.142}$$

(2.75)–(2.76),

$$v_{ijm} \in \{0, 1\} \qquad \forall i, j \in V,\ m \in M, \tag{2.143}$$

$$w^D_{ijm}, w^C_{ijm} \geq 0 \qquad \forall i, j \in V,\ m \in M, \tag{2.144}$$

$$\zeta_{k\ell} \geq 0 \text{ and integer} \qquad \forall k, \ell \in K. \tag{2.145}$$

The objective function accounts for all costs: hub location and hub network design (first two terms in the first row), vehicle usage (first row, third term), and transportation (second row). Constraints (2.131)–(2.133) account for the number of times the hub edges are traversed in each direction and also that such transportation can occur only in hub edges. Constraints (2.134)–(2.138) are inherited directly from the CLRP and thus have already been explained. Inequalities (2.139) impose that some load can be carried out in arc (i, j) and in route m if and only if the arc is included in that route. In that case, the capacity of the vehicle must be respected. Equalities (2.140) and (2.141) update the amounts loaded in the vehicles in the different routes. Constraints (2.142) guarantee that all the flow is distributed. They also ensure that no collected flow is carried by a vehicle when departing from a hub. Finally, we have the domain of the decision variables.

The above problem has the flavor of both the single-allocation hub location problem with hub network design decisions and the CLRP with simultaneous pickup and delivery. In the above model, we followed the "spirit" of location-routing problems, so the transportation costs are not multiplied by the amounts transported. Nevertheless, it is easy to adapt the model to account for those.

2.4 Interdiction and Fortification

In this section, we review some modeling possibilities for hedging against uncertainty when it cannot be quantified in any way or when the decision-maker does not find that quantification of relevance.

Unpredictable disruptions can occur irrespective of having an intentional or casual reason. This can easily jeopardize the operational capability of a system. Three main types of optimization models have been proposed in the literature in the context of a system relying on a set of operating facilities: (i) interdiction models, (ii) fortification models, and (iii) design models. In the first case, the goal is to find vulnerable components in the system and quantify the impact of losing some (e.g., a facility). The second case is used to ensure additional protection for the network structures and thus better prepare them for some unexpected event. Finally, the third family of models targets the design of resilient systems. Scaparra and Church [479] provide a detailed overview of this topic. In this section, we review several aspects concerning the first two types of models. The third type is explored in different chapters in this volume.

Suppose we have an operating system using a set of uncapacitated facilities providing some commodity or service.

An r-interdiction median problem seeks to find the facilities (r out of those currently operating) whose removal leads to the highest deterioration of the objective function of interest. Let I be the set of facilities currently operating. For $i \in I$ and $j \in J$, let K_{ij} denote the set of facilities in I that serve j at a cost higher than i: $K_{ij} = \{k \in I \mid d_{kj} > d_{ij}\}$. We keep using the customary notation in this chapter: c_{ij} denotes the unit cost for supplying demand node $j \in J$ from facility $i \in I$, and d_j stands for the demand of node $j \in J$. The problem can be formulated as follows:

$$\text{maximize} \quad \sum_{i \in I} \sum_{j \in J} c_{ij} d_j x_{ij} \tag{2.146}$$

$$\text{subject to} \quad \sum_{i \in I} x_{ij} = 1 \qquad \forall j \in J, \tag{2.147}$$

$$\sum_{i \in I} s_i = r, \tag{2.148}$$

$$\sum_{k \in K_{ij}} x_{kj} \leq s_i \qquad \forall i \in I,\ j \in J, \tag{2.149}$$

$$s_i \in \{0, 1\} \qquad \forall i \in I, \tag{2.150}$$

$$x_{ij} \in \{0, 1\} \qquad \forall i \in I,\ j \in J. \tag{2.151}$$

In this model, for $i \in I$, s_i is a binary variable equal to one if and only if operating facility $i \in I$ is disrupted (thus interdicted). The variable x_{ij} is one if demand node $j \in J$ is allocated to facility $i \in I$ and zero otherwise. The objective function (2.146) seeks to maximize the total cost after the interdiction of r facilities. Constraints (2.147) ensure that each customer is assigned to a facility after a disruption. Equality (2.148) states that exactly r facilities are interdicted. Inequalities (2.149) ensure that a demand node is assigned to a facility serving it at a higher cost when better options are interdicted. Finally, Constraints (2.150) and (2.151) define the domain constraints.

The structure of the problem guarantees that there is always one optimal solution in which the x-variables are binary. Thus, the integrality constraints on these variables can be relaxed. Extensions of the above problem to account for capacitated facilities were investigated in [6, 478, 570]. The interested reader can refer to those references for further details.

In the context of covering location, an r-interdiction problem seeks to find the r facilities out of those currently operating whose removal leads to the minimum demand covered (from the remaining facilities). In another way, we seek to find the most critical facilities out of those currently operating. Define I as the set of operating facilities and $I_j \subseteq I$ those covering node $j \in J$. Let v_j be a binary variable equal to one if demand node $j \in J$ is covered after disruption and zero otherwise. Additionally, we let s_i be a binary variable equal to one if and only if facility location $i \in I$ is interdicted. The problem can be formulated as follows:

$$\text{minimize} \quad \sum_{j \in J} d_j v_j, \tag{2.152}$$

$$\text{subject to} \quad v_j \geq 1 - s_i \qquad \forall j \in J,\ i \in I_j, \tag{2.153}$$

$$\sum_{i \in I} s_i = r, \tag{2.154}$$

$$v_j \in \{0, 1\} \qquad \forall j \in J, \tag{2.155}$$

$$s_i \in \{0, 1\} \qquad \forall i \in I. \tag{2.156}$$

The objective function (2.152) accounts for the total demand covered after interdiction. Inequalities (2.153) ensure that a node is covered unless all facilities that can cover it are interdicted. Equality (2.154) sets the number of facilities supposedly disrupted equal to r. Finally, we have the domain constraints.

Interdiction problems seek to assess worst-case scenario losses upon the surge of a disruption. A parametric analysis on p allows assessing how different disruption levels may affect the system. Nevertheless, in practice, it is more relevant to use interdiction for assessing worst-case scenarios when we seek to implement a location or protection plan. In this case, the extension of an interdiction problem leads to a so-called interdiction problem with fortification. This possibility is better cast in the context of bi-level optimization.

Let us consider the r-interdiction median problem above. Suppose we can protect (fortify) a set of operating facilities to ensure that some unpredictable event does not prevent us from using such facilities. In this case, we wish to identify the facilities out of the operating ones to fortify. We keep using I for the set of facilities currently operating. Let z_i be a binary variable equal to one if and only if facility $i \in I$ is fortified. Following [479], if exactly b facilities are to be fortified, we can formulate the problem as follows:

$$\text{minimize} \quad H(z), \tag{2.157}$$

$$\text{subject to} \quad \sum_{i \in I} z_i = b, \tag{2.158}$$

$$z_i \in \{0, 1\} \qquad \forall i \in I, \tag{2.159}$$

with

$$H(z) = \max \quad (2.146), \tag{2.160}$$

$$\text{subject to} \quad s_i \leq 1 - z_i \qquad \forall i \in I, \tag{2.161}$$

$$(2.147)\text{–}(2.151). \tag{2.162}$$

The upper-level problem seeks to minimize the highest possible cost after r facilities are disrupted. This is ensured by fortifying b facilities ($b + r \leq |I|$). Constraints (2.161) ensure that a fortified facility is not disrupted. As we can observe, the lower-level problem consists of the r-interdiction median problem with the additional protective constraints (2.161).

The above classes of problems emerge in the context of a system that is in operation. However, when the decision-maker wishes to design a network to better hedge against disruption, she/he seeks to find facility location strategies for coping in the best possible way with loss or disruption—the system becomes more resilient to unexpected events. The mathematical models adopted in this case usually embed an interdiction model in a multi-level optimization model such that the location decision is made at the upper level but accounting for the consequences, which are accounted for at the lower levels.

We illustrate this case in the context of covering facility location. In particular, we take again the r-interdiction covering problem and see how to embed location decisions. Now, I denotes the finite set of potential locations for facilities. The problem consists of selecting p locations for installing the facilities to maximize the demand covered before and after interdiction. Considering again binary location variables y_i that for each location $i \in I$ indicate whether location i is selected for installing a facility and considering binary variables, say u_j, that for each $j \in J$ indicate whether a customer is covered before disruption, the problem can be cast as the following bi-level optimization problem:

$$\text{maximizing} \quad \alpha \sum_{j \in J} d_j u_j + (1 - \alpha) H(\mathbf{y}), \tag{2.163}$$

$$\text{subject to} \quad \sum_{i \in I} y_i = p, \tag{2.164}$$

$$\sum_{i \in I_j} y_i \geq u_j \qquad \forall j \in J, \tag{2.165}$$

$$y_i \in \{0, 1\} \qquad \forall i \in I, \tag{2.166}$$

$$u_j \in \{0, 1\} \qquad \forall j \in J, \tag{2.167}$$

with

$$H(\boldsymbol{y}) = \min \quad (2.152),$$

$$v_j \geq y_i - s_i \qquad \forall i \in I_j,\ j \in J, \tag{2.168}$$

$$(2.154)\text{–}(2.156).$$

The upper-level problem seeks to locate p facilities maximizing a weighted sum of the demand covered before and after the disruption. Note that in the lower-level problem, we see constraints (2.168) instead of (2.153) because now not all facilities in I will necessarily be operating.

None of the above problems explicitly account for uncertainty in the optimization models. Moreover, they often assume a complete risk-averse attitude of the decision-maker toward risk, which means that a decision is sought to better hedge against the worst possible scenario.

2.5 Conclusions and Further Reading

This chapter provided an overview of some widely used facility location problems and popular optimization models designed for them. In terms of discrete facility location, much more could be presented. Neither is this the purpose of this chapter nor this volume. The interested reader can refer to several references that can help complement the information provided in the previous sections. Some suggestions include [131, 154, 179, 180, 185, 326, 382, 402].

Chapter 3
Decision-Making Under Uncertainty: Ingredients for Modeling

Abstract This chapter discusses several aspects of relevance when embedding uncertainty in optimization models aiming at supporting decision-making. These include uncertainty nature and representation, scenario generation and reduction, risk measurement, robustness measures, the relevance of hedging against uncertainty and ways for assessing it, sources of uncertainty in particular facility location problems, and adaptive/non-adaptive decisions in facility location.

Keywords Uncertainty representation · Uncertainty sources · Scenario generation and reduction · Risk assessment · Modeling relevance assessment · Adaptive decisions

3.1 Introduction

In a deterministic decision-making problem, all the parameters are given beforehand and assumed immune to uncertainty. Nevertheless, in many problems, we seek to plan here and now for a future that we do not know precisely. This means that some (possibly all) parameters underlying the problem vary in some sets, and we do not know the exact values we will face. Usual examples in the context of facility location include transportation time and demand levels. In such a case, it becomes desirable that the decisions to make are those that better hedge against such uncertainty.

When the parameters of a decision-making problem are not deterministic, we can distinguish among two situations: (1) We can identify some uncertainty set for the unknown parameters although we cannot (or it is irrelevant to do it) describe their behavior using some probability distribution function, and (2) uncertainty can be measured probabilistically, and we wish to take advantage of that when making decisions.

Decision-making under uncertainty calls for analyzing a few aspects in detail, namely those impacting the modeling paradigm to adopt and the tools for tackling the models. In this chapter, we discuss several such aspects: the nature of the uncertainty and its representation, scenario generation and reduction, risk assess-

© The Author(s), under exclusive license to Springer Nature Switzerland AG 2024

F. Saldanha-da-Gama, S. Wang, *Facility Location Under Uncertainty*,
International Series in Operations Research & Management Science 356,
https://doi.org/10.1007/978-3-031-55927-3_3

ment, robustness measures, the relevance of hedging against uncertainty, uncertainty sources, and, finally, adaptive/non-adaptive decisions.

3.2 Uncertainty Nature and Representation

When dealing with uncertainty in a decision-making problem, it is relevant to understand how much we know about the future and its relevance to the problem. This knowledge impacts the modeling paradigm we can (or should) use and, consequently, the techniques to consider for tackling the models. This is particularly relevant when optimization models are involved.

Uncertain parameters in a decision-making problem can either be discrete, continuous or mixed. If uncertainty can be cast as a random vector, say $\boldsymbol{\xi}$, to which some CDF (or families of CDFs) can be associated, we can resort to stochastic programming models and techniques. This includes two- or multi-stage stochastic programming, chance-constrained programming, distributionally robust optimization, *et cetera.* If no probabilistic representation exists for the unknowns, but we can identify the set they belong to—uncertainty set—then we can resort to robust optimization models and techniques. Chapter 4 provides details related to these paradigms.

Irrespective of the case, a scenario is a possible future outcome of the random experiment underlying the problem. In other words, a scenario fully specifies all uncertain parameters. We may have a finite or an infinite number of scenarios. The same holds if uncertainty is captured simply by an infinite uncertainty set.

When uncertainty is represented by a random vector with a given CDF, some probability (or probability density) can be associated with each scenario. In this case, special attention must be paid to the objective function to be used in an optimization model since its correct shape may depend on the answer to some questions: Are all scenarios relevant? If not, which scenarios should we use? Implicitly? Explicitly? We will further discuss this later in the chapter.

When uncertainty cannot be measured probabilistically (or it is irrelevant to do it), we may resort to uncertainty sets. If they are infinite, then we must characterize them. Some popular types of such sets include box uncertainty, its generalization to polyhedral uncertainty, and ellipsoidal uncertainty [55]. In Chap. 4, we present further details.

An important aspect to discuss at this stage regards the mechanism(s) for fully describing the uncertainty underlying some problem, e.g., the uncertainty set of relevance or the CDF if that is the case.

In some situations, uncertainty is related to driving forces (political conditions in a specific region, economic trends, technological developments, *et cetera*) that influence the data of the problem. In this case, it is up to the decision-maker to understand these driving forces and how they influence that input. This understanding leads to a complete definition of the relevant scenarios for the problem. The decision-maker may need to ask experts about plausible scenarios or ambiguity sets. In some cases, by using subjective probabilities using eliciting probability distributions [107, 413, 414], it is possible to associate a probability with each scenario. Accordingly, we end up with a CDF for the underlying random vector. In other situations, historical data exist that allows fitting a CDF to the underlying data: A theoretical probability distribution is found. A current trend in line with the search for the domain of the unknowns is that of data-driven decision-making. The goal is to embed (historical) data (e.g., using an econometric model) into the modeling to condition the feasibility or ambiguity sets.

Eventually, we end up with a fully defined optimization problem under uncertainty, i.e., we have identified the decision variables, the objective function, the constraints, and the relevant data (possibly uncertain).

3.3 Scenario Generation and Reduction

When making decisions under uncertainty, one often has to deal with a scenario set (finite or infinite) that should be embedded into a model. In robust optimization problems, we typically use the entire set of scenarios to derive a so-called robust counterpart. In stochastic programming models, the number of scenarios is often too large (possibly infinite) to allow finding a tractable deterministic counterpart. In this case, we may need to resort to scenario generation or scenario reduction techniques. This is one possibility for instantiating large deterministic equivalent models or restricting the set of scenarios in a sampling scheme used within a solution procedure. The idea is to consider a sample of scenarios to formulate a tractable deterministic counterpart. Of course, such a sample should lead to a model (or a set of models) that allows finding approximate solutions to the whole problem.

Let us consider a general optimization problem under uncertainty as follows:

$$V^* = \min_{\boldsymbol{x}_1 \in \mathcal{X}_1} \mathcal{F}_1(\boldsymbol{x}_1, \boldsymbol{\xi}). \tag{3.1}$$

In this problem, $\boldsymbol{x}_1$ represents a here-and-now decision to make, $\mathcal{X}_1$ is the corresponding feasibility set, $\boldsymbol{\xi}$ is the uncertainty vector, whose support we denote by Ξ, and $\mathcal{F}_1$ is a function that maps the uncertainty $\boldsymbol{\xi}$ to some real value given a decision $\mathcal{X}_1$, i.e., $\mathcal{F}_1(\boldsymbol{x}_1, \boldsymbol{\xi}) \in \mathbb{R}$. Note that function $\mathcal{F}_1$ can be explicitly given (in the case of a single-stage problem) or may depend on another optimization problem. For instance, in a two-stage problem, we may have

$$\mathcal{F}_1(\boldsymbol{x}_1, \boldsymbol{\xi}) = \boldsymbol{c}_1 \boldsymbol{x}_1 + \min_{\boldsymbol{x}_2 \in \mathcal{X}_2} \mathcal{F}_2(\boldsymbol{x}_1, \boldsymbol{x}_2, \boldsymbol{\xi}). \tag{3.2}$$

This can be easily generalized to multiple stages. Nevertheless, the relevant aspect to discuss concerns the possibility of having a support Ξ of a cardinality so large (possibly infinite) that we need to instantiate the deterministic equivalent model by using a subset of scenarios. The goal in this case should be to find a sample of scenarios, say $\boldsymbol{\xi}_1, \ldots, \boldsymbol{\xi}_S$, such that the problem

$$V^S = \min_{\substack{x_1 \in \mathcal{X}_1 \\ \boldsymbol{\xi} \in \{\boldsymbol{\xi}_1, \ldots, \boldsymbol{\xi}_S\}}} \mathcal{F}_1(\boldsymbol{x}_1, \boldsymbol{\xi}) \tag{3.3}$$

is tractable. The optimal solution to (3.3) provides an approximate solution to (3.1). We would like to find a set $\boldsymbol{\xi}_1, \ldots, \boldsymbol{\xi}_S$ such that $V^S = V^*$ although, again, this is elusive. At most, we may seek to have the non-negative difference $V^S - V^*$ as small as possible. It does not come as a surprise that the size of the sample is a relevant aspect. On the one hand, it should be large enough so that (3.3) provides an acceptable approximation to (3.1); on the other hand, it should be small enough to make the approximate model manageable using optimization techniques. Some experimentation may be required at this stage. For the moment, we assume that the size S of the sample has been previously decided.

Although the following discussion targets stochastic programming models, in some cases, the same principles can be used to reduce the number of working scenarios in other settings, namely in robust optimization, if desirable. Several possibilities have been proposed in the literature for reducing the number of scenarios to use, either by selecting a sample of the entire set or by generating a sample of new scenarios that represent well the original ones.

One possibility is to consider a standard Monte Carlo sampling in which an S-dimensional sample is randomly drawn from Ξ according to the CDF of $\boldsymbol{\xi}$. Sample average approximation is a well-known technique in stochastic programming proposed in [312] that makes use of several such samples and the corresponding restricted models to approximate the optimal solution of a stochastic programming problem.

Another possibility is to determine a scenario subset of a prescribed cardinality and a probability measure based on this set, which is close to the initial distribution in terms of some natural probability metric. The reader should refer to [166, 182, 277], and the references therein for further details.

When the number of scenarios is finite but too large to be fully considered, we can consider some similarity measure among scenarios and use a clustering method to split Ξ into S groups such that the scenarios that are close to each other according to that measure end up in the same group. The final sample results from selecting a representative in each group. The clustering approaches differ in two aspects: (i) the metric used to compare scenarios and (ii) the request (or its absence) that the representatives should be part of the original set Ξ. Two well-known clustering algorithms are the S-means [4, 82] and the S-medoids [303], with S standing for the number of groups to create. The former seeks to minimize the sum of the squares of the ℓ_2 distances within each cluster (cluster variance). The centroid in each cluster is the weighted average scenario in that cluster using the probabilities as weights. When the underlying CDF is unknown, the standard average can be used. The latter differs from the former by imposing that, in each cluster, the representative should be a scenario belonging to the original set. The exact number of groups to consider is also a matter of analysis and may depend on the specific problem or setting. Recent references complementing the information provided can be found in [268] and [306].

3.4 Risk Assessment

Uncertainty underlying a decision-making problem that can be quantified probabilistically leads to what is often called risk. Informally, (financial) risk corresponds to a level of uncertainty in terms of the relation between the actual outcome obtained by implementing some decision and the investment that needs to be made. Risk is usually seen as something negative that depends on a loss (e.g., financial or reputational) and an occurrence probability. Depending on the context, this view may need to be expanded. The discussion presented in [263] for supply chain management is a clear example.

It is not the purpose of this volume to discuss such concepts in detail but to highlight their use in optimization models seeking their application to facility location problems. We often distinguish among three attitudes of a decision-maker toward risk: risk-neutral, risk-averse, and risk-seeker. This is particularly relevant if we consider financial assets and thus focus on financial risk. To better assess the attitude of a decision-maker toward (financial) risk, we can convert monetary into utility values. The latter are values in an appropriate scale that reflect preferences. That scale is called the utility function for money. No decision-maker is indifferent to financial risk. In Fig. 3.1, we can observe an example of a utility function characterizing a specific decision-maker. In this case, the decision-maker values a financial amount of 40 monetary units, double that of 10.

A decision-maker is said to be risk averse when s/he/it prefers a guaranteed outcome to a chance outcome with the same expected utility. Graphically, this corresponds to a utility function such as the one depicted in the above figure: the

Fig. 3.1 Utility function

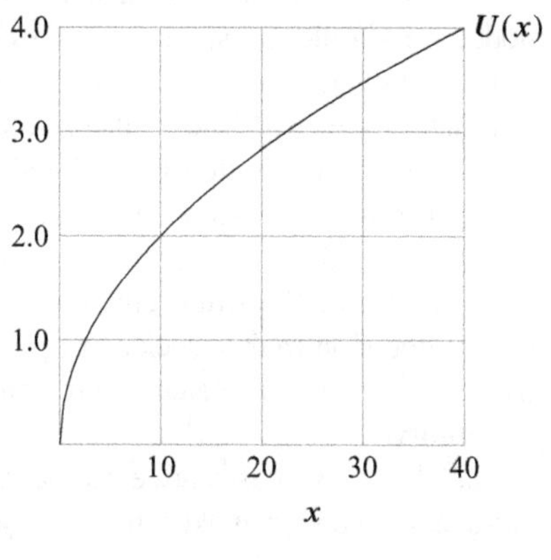

marginal utility for money is decreasing in this case. A risk-neutral decision-maker is indifferent between a chance outcome with some expected value and a guaranteed gain having the same value. In this case, the utility function is well described by a line segment: the marginal utility for money is constant. Finally, a risk-taking or risk-seeking decision-maker prefers a chance outcome to a guaranteed gain having the same expected value. In this case, we observe a convex and non-decreasing utility function, i.e., an increasing marginal utility for money.

In practice, we must find the exact utility function associated with a decision-maker. The process for accomplishing this is called utility elicitation [25]. The reader can find such a mechanism discussed in detail in [317]. In the same book, the authors draw attention to several functional forms adopted in practice for a utility function. These include but are not limited to

- the quadratic utility, $U(x) = \lambda x - x^2$, where $\lambda > 0$ controls the level of risk aversion;
- the exponential utility, $U(x) = 1 - e^{-\lambda x}$, $\lambda > 0$;
- the power utility, $U(x) = (x^{1-\lambda} - 1)/(1 - \lambda)$, $\lambda \geq 0$, $\lambda \neq 1$;
- the logarithmic utility, $U(x) = \log x$, which is a particular case of the power utility.

For additional details, the interested reader should refer to [317] and the references therein.

Since the attitude of a decision-maker toward risk depends on the financial values involved in the practice, we may observe a decision-maker that hybridizes the above attitudes, depending on the amount involved: for a small amount, risk-seeking may be adequate; at some point, the neutral attitude emerges; finally, above a certain threshold, risk aversion may start being observed. In this volume, we focus mainly on risk-neutral and risk-averse decision-makers.

3.5 On the Objective Functions

The attitude of a decision-maker toward risk strongly impacts the optimization model we should consider for a specific problem. A particular component much influenced by such attitude is the objective function. Assume *w.l.o.g.* that we seek to minimize some cost. Due to uncertainty, there is often a component of the total cost that cannot be computed beforehand. In this case, one may look into a function that incorporates in some way joint information for all or at least for a set of scenarios. A fully risk-averse decision-maker typically plans considering the worst future scenario. We often end up with a minmax optimization problem. A risk-neutral decision-maker seeks to optimize the expected (or conditional expected) loss or utility, i.e., all scenarios are considered, each being weighted according to its probability.

The above two cases are extreme. One may ask whether they are adequate for a risk-averse decision-maker that is not complete risk-averse. Several measures have been proposed in the literature for this case, which include the minmax regret, the value-at-risk (VaR), and the conditional value-at-risk (CVaR). The latter has become progressively more popular, which justifies some considerations in this text.

Let Ξ be a set of real-valued random variables defined on a probability space (Ω, $\mathcal{E}$, $\mathbb{P}$), where Ω is a set of outcomes, $\mathcal{E}$ is a set of events, and $\mathbb{P}$ is a function that assigns probabilities to the events.

Definition 3.1 (Risk Measure) A real-valued function $\rho : \Xi \to \mathbb{R}$ is called a risk measure if it satisfies the following conditions:

1. (Normalization) $0 \leq \rho(X) \leq 1 \ \forall X \in \Xi$.
2. (Monotonicity) $\rho(X') \leq \rho(X'') \forall X', X'' \in \Xi$ with $X' \preceq X''$, where $X' \preceq X''$ indicates that $X'(\omega) \geq X''(\omega) \forall \omega \in \Omega$.
3. (Translational invariance) $\rho(X + a) = \rho(X) + a, \forall X \in \Xi, a \in \mathbb{R}$.

Definition 3.2 (Coherent Risk Measure) A real-valued function $\rho : \Xi \to \mathbb{R}$ is called a coherent risk measure if it is a risk measure that also satisfies the following conditions:

1. (Subadditivity) $\rho(X' + X'') \leq \rho(X') + \rho(X''), \forall X', X'' \in \Xi$.
2. (Positive Homogeneity) $\rho(\lambda X) = \lambda \rho(X) \forall X \in \Xi, \lambda > 0$.

Coherent risk measures, along with their properties, have been introduced in [33]. Further extensions would be proposed in [161]. Nowadays, coherence gathers widely accepted properties for a risk measure due to its advantages, including: (i) A coherent risk measure helps determine the risk involved in a business and estimate the profitability more accurately, (ii) a coherent risk measure contributes to a decrease in the risk of loss by boosting the return for all future sales, and (iii) a coherent risk measure stimulates risk diversification by separately categorizing the assets. Note, however, that in some cases, the properties of coherent risk measures may not be appropriate (e.g., the scale of potential losses is an issue). Note also that despite being quite popular nowadays, coherent risk measures do not guarantee the total accuracy.

Thinking ahead to the contents of this volume, let $F(\boldsymbol{y}, \boldsymbol{\xi})$ denote a cost function that depends on a decision vector $\boldsymbol{y} \in Y$ as well as on a random vector $\boldsymbol{\xi}$. For some $\boldsymbol{y}$, let $\mathbb{P}[F(\boldsymbol{y}, \boldsymbol{\xi}) \leq \eta]$ denote the probability that $F(\boldsymbol{y}, \boldsymbol{\xi})$ does not exceed a value η.

Definition 3.3 (Value-at-Risk) For a given confidence level α, the Value-at-Risk (VaR) of $\boldsymbol{y} \in Y$ is denoted by $\eta_\alpha(\boldsymbol{y})$ and is given by the minimum threshold η for which the probability that $F(\boldsymbol{y}, \boldsymbol{\xi})$ does not exceed η is at least α, i.e.,

$$\eta_\alpha(\boldsymbol{y}) = \min\{\eta : \mathbb{P}[F(\boldsymbol{y}, \boldsymbol{\xi}) \leq \eta] \geq \alpha\}.$$

In simple words, the VaR is a quantile of the loss distribution. By using the α-VaR as an objective function, the decision-maker seeks to minimize the threshold for the risk level $1 - \alpha$. Figure 3.2 illustrates this concept.

The VaR is a risk measure because it is (i) normalized, (ii) monotonous, and (iii) translation-invariant [474]. Nevertheless, there are some drawbacks when using

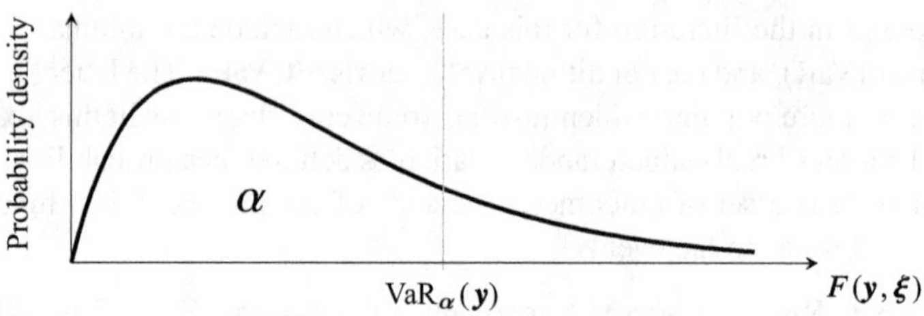

Fig. 3.2 The value-at-risk

VaR as an objective function. First, when we consider a solution $\mathbf{y} \in Y$ using this objective, the value $F(\mathbf{y}, \boldsymbol{\xi})$ may be too high for some scenarios whose cost exceeds this minimum threshold even if their joint probability is low ($<1-\alpha$). Second, the VaR often leads to a non-convex objective function. Overall, we see that VaR is not a coherent risk measure.

An alternative to VaR is to minimize the expected cost only for the scenarios whose costs exceed the threshold for the confidence level α. This leads to the so-called α-Conditional Value-at-Risk (α-CVaR).

Definition 3.4 (Conditional Value-at-Risk) The α-Conditional Value-at-Risk of a solution $\mathbf{y} \in Y$, which we denote by $\Psi_\alpha(\mathbf{y})$, is given by the expected value of the total cost considering only the $(1-\alpha) \times 100\%$ worst scenarios for solution $\mathbf{y}$, i.e., it is the expected cost/loss conditional to the scenarios whose value exceeds $\eta_\alpha(\mathbf{y})$.

The α-CVaR is an interesting measure to embed in an optimization model. First, it is a coherent risk measure [33]. Second, it has the expected cost (for a risk-neutral decision-maker) as a particular case. When α is zero, all scenarios become involved in the expectation to be computed. Third, although hardly a closed form can be derived for a solution of minimum α-CVaR (hardly an analytical representation of $\eta_\alpha(\mathbf{y})$ can be found), we know how to write CVaR explicitly and embed it in an optimization model. Rockafellar and Uryasev [459] suggest using the function

$$\Phi_\alpha(\mathbf{y}, \eta) = \eta + \frac{1}{1-\alpha}\mathbb{E}\left[(F(\mathbf{y}, \boldsymbol{\xi}) - \eta) \mid F(\mathbf{y}, \boldsymbol{\xi}) > \eta\right],$$

which, in case the set of scenarios, say Ω, is finite, can be written as

$$\Phi_\alpha(\mathbf{y}, \eta) = \eta + \frac{1}{1-\alpha}\sum_{\omega \in \Omega}\left(F(\mathbf{y}, \boldsymbol{\xi}_\omega) - \eta\right)^+ \pi_\omega.$$

In this expression, $\boldsymbol{\xi}_\omega$ stands for the observation of $\boldsymbol{\xi}$ under scenario $\omega \in \Omega$ and $a^+ = \max\{0, a\}$. Now the α-CVaR of a solution $\mathbf{y} \in Y$ can be computed as

$$\Psi_\alpha(\mathbf{y}) = \Phi_\alpha(\mathbf{y}, \eta(\mathbf{y})) = \min_{\eta} \Phi_\alpha(\mathbf{y}, \eta).$$

Hence, the problem of finding the feasible vector $y \in Y$ with the smallest α-CVaR can be expressed as

$$\min_{y \in Y} \Psi_\alpha(y) = \min_{y \in Y} \Phi_\alpha(y, \eta(y)) = \min_{y \in Y, \eta} \Phi_\alpha(y, \eta).$$

Finally, we can reformulate the problem as

$$\text{minimize} \quad \eta + \frac{1}{1-\alpha} \sum_{\omega \in \Omega} \pi_\omega \zeta_\omega, \tag{3.4}$$

$$\text{subject to} \quad y \in Y, \tag{3.5}$$

$$\zeta_\omega \geq F(y, \xi_\omega) - \eta \qquad \forall \omega \in \Omega, \tag{3.6}$$

$$\zeta_\omega \geq 0, \qquad \forall \omega \in \Omega. \tag{3.7}$$

The following concept introduced in [320] generalizes that of α-CVaR.

Definition 3.5 (Higher Moment Coherent Risk Measures) The family of the higher moment coherent risk measures, $\text{HMCR}_{\alpha,p}(\xi)$ for the random outcome X with order $p \geq 1$, is defined as follows:

$$\text{HMCR}_{\alpha,p}(\xi) = \inf_{\eta \in \mathbb{R}} \left\{ \eta + \frac{1}{1-\alpha} \|(\xi - \eta)^+\|_p \right\}, \tag{3.8}$$

where $\|\varpi\|_p = \sqrt[p]{\mathbb{E}[|\varpi|^p]}$.

In the context of this definition, α-CVaR $= \text{HMCR}_{\alpha,1}$. When $p > 1$, the risk is evaluated via the higher tail moments of the loss distributions. It can better handle heavy-tailed distributions. The exact values of p and α to consider depend on the decision-maker, namely on his/her/its risk attitude. The value of p controls the loss risk and the extreme events.

In addition to the above coherent risk measures, others may be of interest, such as the weighted average of CVaR and expected value, whose definition is now straightforward. Other possibilities exist but will not be explored in this book.

3.6 The Relevance of Hedging Against Uncertainty

Optimization problems involving uncertain parameters are often more involved and complex to solve than those in which all the information is perfectly known beforehand.

For decision-makers who need to plan for an uncertain future, some questions emerge: (i) Is it relevant to hedge against uncertainty? and (ii) if yes, is there any simplified optimization model one can use to approximate the model we are considering for capturing uncertainty?

Depending on the setting we are studying, these questions have different answers. In this section, we resort to the concepts, skipping notation and specific problems. The application of these concepts will then be detailed later in different chapters and contexts.

We start by assuming we are dealing with a risk-neutral decision-maker. When uncertainty is represented by a random vector with some given CDF, we can use the so-called expected value of perfect information to assess the relevance of hedging against uncertainty.

Definition 3.6 (Expected Value of Perfect Information (EVPI)) The expected value of perfect information (EVPI) is the maximum amount a decision-maker accepts paying to get complete knowledge about the future.

This concept was first introduced in the context of Decision Analysis and was later extended to other types of stochastic optimization problems. Suppose we can find the optimal solution for each single-scenario problem. Considering the underlying probability distribution, we can specify the distribution of the optimal values.

Definition 3.7 (Wait-and-See Solution Value) The wait-and-see solution (WS) value is the expected value of the distribution of the optimal values (when this distribution can be found).

Denote by SP the value associated with the best decision that can be made before the uncertainty is revealed. The expected value of perfect information is computed as $\text{EVPI} = \text{SP} - \text{WS}$.

Regarding the possibility of using a simplified model, a natural choice is the (single-scenario) problem we obtain when replacing the random variables with their expectation. After solving that simplified problem, we can use the corresponding solution and evaluate the objective function of the stochastic problem. We obtain the expected cost of the optimal solution of the expected value problem. Denote this result by EEV.

Definition 3.8 (Value of the Stochastic Solution) The value of the stochastic solution (VSS) is the difference between the value of the objective function of the stochastic problem evaluated in the optimal solution of the expected value problem and the optimal value of the stochastic problem.

In other words, we have $\text{VSS} = \text{EEV} - \text{SP}$.

The above measures are popular in the literature despite being tailored for a risk-neutral decision-maker. Noyan [408] considered the necessary changes to adapt to a risk-averse decision-maker.

As in the case of the EVPI, we can start by solving all single-scenario problems separately, thus characterizing the distributional problem. Now we can compute the α-CVaR for any α. Let us call it the α-RAWS.

Definition 3.9 (Risk Aversion Value of Perfect Information) The α-Risk Aversion Value of Perfect Information (α-RAVPI) is the difference between the optimal value of the stochastic problem and the α-CVaR using the optimal values of the single scenarios.

In other words, α-RAVPI = SP − α-RAWS. It is straightforward to conclude that we always have α-RAVPI ≥ 0 because the optimal solution to the stochastic problem is a feasible solution to each single-scenario problem.

Regarding a risk-aware VSS, we can proceed as follows. We first solve the single-scenario problem induced by replacing the random variables with their expectations. Let us consider the resulting first-stage decision. We solve the single-scenario problems for all scenarios, setting that first-stage decision. By doing so, we obtain a distribution of objective function values whose α-CVaR we can compute. Denote by α-RAEV the value obtained.

Definition 3.10 (Risk Aversion Value of the Stochastic Solution) The α-Risk Aversion Value of the Stochastic Solution (α-RAVSS) is the difference between the α-CVaR obtained by solving the single-scenario problem for all scenarios by fixing the fist-stage solution according to the expected value problem and the optimal value of the stochastic problem.

In other words, α-RAVSS = α-RAEV − SP. Again, it is easy to conclude that we always have α-RAVSS ≥ 0.

Finally, we note that all values VSS, $EVPI$, α-VPI, and α-VSS can be normalized and multiplied by 100, giving rise to percentage values w.r.t. the optimal value of the stochastic problem.

Suppose now that no probabilistic information exists that we can explore. In this case, to assess the relevance of hedging against uncertainty, we can directly compare the optimal value of the robust optimization problem with that of the nominal problem. Nevertheless, a more insightful concept was introduced by Bertsimas and Sim [75], which is the Price-of-Robustness. An uncertain parameter comes with a nominal value and a maximum deviation one may observe w.r.t. that value. In practice, it is not reasonable to consider that all parameters will be observed at the highest distance from the nominal value. Suppose we limit the sum for all uncertain parameters of their (positive) deviations w.r.t. the nominal values. In this case, we will be working with so-called budget uncertainty.

Definition 3.11 (Price-of-Robustness) For a given budget, the Price-of-Robustness (PoR) is the difference between the worst-case objective value of the robust solution and the nominal (objective) value of the nominal solution.

The PoR quantifies the trade-off between the objective value and constraint violations. It is useful when the objective function is not uncertain. In this case, the PoR is the amount that has to be paid for being robust against constraint violations.

We should note that the PoR can also be used when the objective is uncertain. Nevertheless, this is discouraged because it compares the nominal solution in the case of certainty with the robust solution where the worst case occurs, i.e., two completely different situations are being compared.

In the context of specific topics, other possibilities may emerge to quantify the relevance of using a more involved model instead of a simplified one. The same holds to check whether hedging against uncertainty is of relevance. For instance, in Chap. 12, we will discuss the value of a risk-aware supply chain solution. In Chap. 13, we refer to the compactness value of the stochastic solution. These specific ideas will be discussed in the related chapters.

3.7 Sources of Uncertainty

Depending on the problem being investigated, one may face different sources of uncertainty. To narrow our discussion aligning it to the contents of this volume, we directly consider problems in Location Science.

When studying a facility location problem, it is customary to start by building a model, i.e., an abstraction of the reality in which we focus on the features that we consider relevant for finding a viable solution to the problem. As pointed out in [392], there is inherent uncertainty when we build a model, feed it with data, solve it, and implement the obtained solutions. In particular, that author distinguishes uncontrollable errors (thus inducing uncertainty) related to model abstraction and specification, attribute specification, data creation process, spatial model properties, model solution, and implementation.

Although acknowledging all aspects raised in [392], throughout this volume, we assume that some model type has been decided upon and that we wish to hedge against the uncertainty that is quantifiable in some way. This includes aspects gathered in [392] such as attribute specifications and data creation processes. In any case, our starting point is that the adopted modeling framework is an acceptable abstraction of the actual problem to solve.

Base data underlying a facility location problem include the demand levels or, more generally, the weights of the nodes to serve. Thus, in many problems investigated in the literature, the uncertainty stems from those. Wagner et al. [533] discuss demand uncertainty in facility location by distinguishing between systematic and unsystematic risk factors. The former may result from diverse aspects such as economic changes, the emergence of new companies as competitors, changes in policy, and large-scale demographic changes over the area that affect the demand across the entire system. Unsystematic sources of uncertainty refer to elements causing demand perturbations at a local level, e.g., changes in the population of a specific area or demand fluctuations over time.

In addition to the demand levels, other parameters of relevance to a facility location problem may be subject to uncertainty: (i) the distribution costs (dependent, e.g., on the energy prices), (ii) traveling time (depending on the traffic), (iii) location of the customers (e.g., in emergency facility location), and (iv) the revenues for the commodities shipped (e.g., reacting to economic conditions).

Depending on the application area, we may find other parameters conveying uncertainty. We detail some possibilities.

When the p-median problem is at stake, the number of facilities to locate may be an uncertainty source (see, e.g., [65]). Such uncertainty may stem from some budget that has still not been observed or specified.

In supply chain management, upon the occurrence of a disruption, the production or distribution capacity may change unexpectedly (see, e.g., [264, 265]) the same holding with the distribution network (e.g., affected roads or train lines). In reverse logistics planning, the uncertainty often stems from the amount of returned products [212].

In humanitarian logistics, uncertainty may stem from the demand requiring some first aid commodity, and from the availability of that commodity (e.g., dependent on donations) or, from the network connectivity (see [50, 177]).

In covering location problems, the coverage capability of a facility/equipment can be a source of uncertainty (see [367]) depending, for instance, on the irregularity of the terrain or the weather conditions.

Given the large variety of uncertainty sources in a facility location problem, a classification is attempted in [503]. The authors distinguish between provider-side, receiver-side, and in-between uncertainty. The first is related to the facilities, e.g., capacity or availability. The second is related to the demand nodes, e.g., demand levels or position. The third conveys aspects involving the transportation network, e.g., transportation costs or available transportation channels.

Above all, when facing a specific problem, we must precisely identify all the involved parameters and those subject to uncertainty. Only after the sources of uncertainty in a problem have been identified, can we start working out possibilities for accounting for that uncertainty. This can be done in different ways, as already discussed in this chapter and as extensively detailed throughout this book.

3.8 *Ex Ante* Versus *Ex Post* Decisions

Several types of decisions are involved in a facility location problem. Some can be seen as strategic (e.g., the location of a large production plant), and some hold at a tactical or operational level (e.g., the exact amount to ship to each customer).

Often, strategic decisions have a long-lasting effect, in the sense that once implemented, they can hardly be reversed in the short term. On the other hand, operational decisions can be more easily adjusted since they usually hold for a shorter time. When different types of decisions are involved in a problem, we always seek to integrate them.

When the time comes to implement the obtained solutions, several components may be considered at separate moments. Under uncertainty, this aspect is particularly relevant because some decisions may need to be implemented before uncertainty is revealed (e.g., the location of a manufacturing facility); others can wait for the uncertainty to be disclosed, adapting to it. In the former case, we talk about *ex ante* or non-adaptive decisions, and the latter are usually called *ex post*, adaptive, or recourse decisions.

Interestingly, much flexibility exists in the context of facility location for considering the same type of decisions being *ex ante* or *ex post*. It is a specific application that determines that. Even the location decisions that in the majority of the literature on facility location under uncertainty are tackled as *ex ante*, can in some cases be cast at an operational decision level and thus become *ex post* (see, e.g., [208, 269]).

In other applications, we see decisions acting simultaneously as *ex ante* and *ex post*. A classical case is that involving endogenous capacity decisions. Before uncertainty is revealed, some base capacities can be decided for the facilities which are then adapted according to how uncertainty is disclosed (e.g., using extra labor or outsourcing—see [9, 10, 171]).

Another setting in which the same type of decision can be simultaneously *ex ante* and *ex post* is that of a multi-period planning problem. If uncertainty is progressively revealed over time, a decision made in some period other than the first is adaptive w.r.t. the past periods and non-adaptive regarding the forthcoming periods.

3.9 Conclusions and Further Reading

A model is always an abstraction of the reality in which we seek to capture the features of relevance for solving some problem. In simple wording, a model is always an approximation of the reality. Capturing uncertainty is another element that can contribute to making the abstraction richer.

In this chapter, we sought to provide the reader with a few thoughts and ideas that should be clarified before detailing a specific model for abstracting a decision-making problem under uncertainty. Many aspects of relevance in this context are discussed in the specialized literature, e.g., [316].

In the context of facility location, the interested reader can complement the information provided in this chapter with that presented in [365]. A detailed discussion is presented on the sources of customer-related uncertainty. These include unplanned purchases, the taste for variety—given product heterogeneity, imperfect information provided to the customers about product and store features, and imperfect information on customers available to decision-makers. The authors also provide examples of how to model such uncertainty embedding it in a facility location problem.

Chapter 4
Optimization Under Uncertainty

Abstract This chapter discusses several aspects of relevance in optimization under uncertainty. Different paradigms are reviewed, such as Robust Optimization, Stochastic Programming, Chance-Constrained Programming, and Distributionally Robust Optimization. The reviewed concepts, models, and techniques seek to provide the reader with the background for better capturing the contents related to facility location to be presented in the remainder of the volume.

Keywords Modeling paradigms · Properties · Solution techniques · Discrete optimization under uncertainty · Multi-stage decision-making

4.1 Introduction

Many decision-making problems are formulated mathematically as optimization problems. For this reason, it is important to discuss different mechanisms that can be adopted for embedding uncertainty in such models. They are strongly related to the knowledge about the future and the hardness (or softness) of the constraints involving uncertain elements. The following pages review different paradigms for embedding uncertainty in an optimization model. We start with Robust Optimization, assuming uncertainty captured by uncertainty sets. We proceed to Stochastic Programming models, thus assuming that uncertainty can be represented by a random vector with some given CDF. In this case, when the decision-maker is satisfied if some constraints hold with some (typically high) probability level, we fall into Chance-Constrained Programming, which we also revisit. Finally, we discuss Distributionally Robust Optimization models assuming that a random vector captures uncertainty although we do not know its exact probability distribution; instead, an ambiguity set of distributions is adopted. The chapter ends with some conclusions and directions for further reading.

© The Author(s), under exclusive license to Springer Nature Switzerland AG 2024
F. Saldanha-da-Gama, S. Wang, *Facility Location Under Uncertainty*,
International Series in Operations Research & Management Science 356,
https://doi.org/10.1007/978-3-031-55927-3_4

4.2 Robust Optimization

As we have already explained, Robust Optimization is an adequate paradigm for abstracting a decision-making problem when the underlying data is uncertain and can be captured using uncertainty sets [55, 57, 58, 60, 75].

To better convey several aspects of relevance in the topic, we start by considering a standard linear optimization problem given by

$$\text{minimize} \quad \boldsymbol{c}^{\top}\boldsymbol{x}, \tag{4.1}$$

$$\text{subject to} \quad \boldsymbol{A}\boldsymbol{x} \geq \boldsymbol{b}, \tag{4.2}$$

$$\boldsymbol{x} \geq \boldsymbol{0}, \tag{4.3}$$

with $\boldsymbol{x} \in \mathbb{R}^n$ standing for the vector of decision variables, $\boldsymbol{A} = [\boldsymbol{a}_1, \dots, \boldsymbol{a}_n] \in \mathbb{R}^{m \times n}$, $\boldsymbol{b} \in \mathbb{R}^m$, and $\boldsymbol{c} \in \mathbb{R}^n$ denoting parameters possibly uncertain. Equivalently, the model can be written as

$$\text{minimize} \quad w, \tag{4.4}$$

$$\text{subject to} \quad (4.2),\ (4.3),$$

$$\boldsymbol{c}^{\top}\boldsymbol{x} \leq w, \tag{4.5}$$

and thus, w.l.o.g. we can focus on uncertainty stemming from the constraints.

Often, the data of real-world linear optimization problems is not precisely known. The most common reasons for data uncertainty have been discussed in Chap. 3. They include prediction, measurement, or implementation errors. In this section, we consider uncertainty in parameters involved in the linear constraints (4.2). Let us assume that both $\boldsymbol{A}$ and $\boldsymbol{b}$ are uncertain and that the corresponding uncertainty set $\mathcal{U} \in \mathbb{R}^{m \times (n+1)}$ was identified. Hence, the definition of the uncertain linear optimization problem is as follows:

Definition 4.1 (Uncertain Linear Optimization Problem) An uncertain linear optimization problem with a certain objective is a collection

$$\left\{ \underset{\boldsymbol{x} \geq \boldsymbol{0}}{\text{minimize}} \left\{ \boldsymbol{c}^{\top}\boldsymbol{x} : \boldsymbol{A}\boldsymbol{x} \geq \boldsymbol{b} \right\} \right\}_{(\boldsymbol{A},\boldsymbol{b}) \in \mathcal{U}} \tag{4.6}$$

of linear optimization problems, $\text{minimize}_{\boldsymbol{x} \geq \boldsymbol{0}} \left\{ \boldsymbol{c}^{\top}\boldsymbol{x} : \boldsymbol{A}\boldsymbol{x} \geq \boldsymbol{b} \right\}$, sharing a common structure namely, having a common number of constraints and variables but with the data varying in a given uncertainty set $\mathcal{U}$.

Unlike a single optimization problem, the concepts of feasible/optimal solution and optimal value do not apply to a family of optimization problems like (4.6). How to (re)define/adapt these concepts depends on the underlying "decision environment." Next, we assume the following setting [55]:

1. All decision variables in (4.6) represent "here and now" decisions; they must be assigned specific numerical values as a result of solving the problem before the actual data is revealed.
2. The decision-maker is fully responsible for the consequences of the decisions to be made when, and only when, the actual data is within the prespecified uncertainty set $\mathcal{U}$.
3. The constraints in (4.6) are "hard"—we cannot tolerate their violation when the data is in $\mathcal{U}$.

The above assumptions determine, in a more or less unique fashion, the meaningful (feasible) solutions to the uncertain problem (4.6). The following definition of a robust feasible solution is due.

Definition 4.2 (Robust Feasible Solution) A vector $\boldsymbol{x} \in \mathbb{R}^n$ is a robust feasible solution to (4.6) if it satisfies all realizations of the constraints from the uncertainty set, that is,

$$\boldsymbol{A}\boldsymbol{x} \geq \boldsymbol{b} \quad \forall (\boldsymbol{A}, \boldsymbol{b}) \in \mathcal{U}.$$

After we agree on what meaningful candidate solutions to the uncertain problem (4.6) are, we must establish a way to quantify their quality. Finally, we can seek the best robust value of the objective function among all robust (feasible) solutions to the problem. This brings us to the central concept of this section, the Robust Counterpart of an uncertain optimization problem, which is defined as follows:

Definition 4.3 (Robust Counterpart) The Robust Counterpart of the uncertain linear optimization problem (4.6) with certain objective is the optimization problem

$$\text{minimize} \quad \boldsymbol{c}^\top \boldsymbol{x}, \tag{4.1}$$

$$\text{subject to} \quad \boldsymbol{A}\boldsymbol{x} \geq \boldsymbol{b} \quad \forall (\boldsymbol{A}, \boldsymbol{b}) \in \mathcal{U}, \tag{4.7}$$

$$\boldsymbol{x} \geq \boldsymbol{0}, \tag{4.8}$$

consisting of minimizing the robust value of the objective over the robust feasible solutions to the uncertain problem.

An optimal solution to the Robust Counterpart is called a robust optimal solution to (4.6), and the optimal value of the Robust Counterpart is called the robust optimal value of (4.6). If $\mathcal{U}$ is infinite, this so-called robust counterpart is a semi-infinite-dimensional optimization problem that, in general, cannot be tackled directly by efficient linear optimization algorithms. For this reason, in practice, we seek to replace (4.7) with a finite set of constraints in a way that the latter implies the former. The idea is to leverage the convexity of the uncertainty set $\mathcal{U}$ and the associated duality arguments. We also note that the complicating part of the above robust counterpart is the one that involves the vector of decision variables. For this reason, to better convey the most significant aspects of relevance in this chapter, we concentrate on uncertainty stemming from the coefficients of the decision variables.

Furthermore, w.l.o.g. we focus on the single-constraint robust counterpart by noting that the "column-wise" uncertainty in (4.7) can be transformed in the "row-wise" uncertainty [59]. Denoting $\boldsymbol{a} = [a_1, \ldots, a_n]^\top \in \mathbb{R}^n$, we have

$$\text{minimize} \quad \boldsymbol{c}^\top \boldsymbol{x}, \tag{4.1}$$

$$\text{subject to} \quad \boldsymbol{a}^\top \boldsymbol{x} \geq b \quad \forall \boldsymbol{a} \in \mathcal{U}, \tag{4.9}$$

$$\boldsymbol{x} \geq \boldsymbol{0}, \tag{4.10}$$

i.e., the constraint $\boldsymbol{a}^\top \boldsymbol{x} \geq b$ should be satisfied for all $\boldsymbol{a} \in \mathcal{U}$ or, equivalently, the worst-case value of $\boldsymbol{a}^\top \boldsymbol{x}$ should be larger than or equal to b:

$$\min_{\boldsymbol{a} \in \mathcal{U}} \boldsymbol{a}^\top \boldsymbol{x} \geq b.$$

4.2.1 Robust Counterparts Under Different Uncertainty Sets

In this section, we investigate the computational tractability of the robust counterpart of an uncertain linear optimization problem. The situation here turns out to be as good as it could be: We shall see, essentially, that the Robust Counterpart of the uncertain linear optimization problem with uncertainty set $\mathcal{U}$ is computationally tractable whenever the convex uncertainty set $\mathcal{U}$ itself is computationally tractable. We present three examples.

Example 4.1 (Interval Uncertainty) Under interval uncertainty, $a_i \in [a_i^-, a_i^+]$, $i = 1, 2, \ldots, n$, i.e., the uncertainty set can be represented by $\mathcal{U}_I = [a_1^-, a_1^+] \times [a_2^-, a_2^+] \times \cdots \times [a_n^-, a_n^+]$. It is not difficult to see that under the interval uncertainty set $\mathcal{U}_I$, Constraint (4.9) is equivalent to the following inequality:

$$\sum_{i=1}^{n} \mu_i x_i - \sigma_i |x_i| \geq b, \tag{4.11}$$

where $\mu_i = \frac{a_i^- + a_i^+}{2}$, $\sigma_i = \frac{a_i^+ - a_i^-}{2}$, $\forall i = 1, \ldots, n$.

Accordingly, by introducing a set of auxiliary variables λ_i, $i = 1, \ldots, n$, the robust optimization problem that consists of minimizing (4.1) subject to (4.9) and (4.10) can be written as the following linear optimization problem:

$$\text{minimize} \quad \boldsymbol{c}^\top \boldsymbol{x},$$

$$\text{subject to} \quad \sum_{i=1}^{n} \mu_i x_i - \sigma_i \lambda_i \geq b,$$

$$\begin{aligned}
\lambda_i &\geq -x_i && \forall i = 1, 2, \ldots, n, \\
\lambda_i &\geq x_i && \forall i = 1, 2, \ldots, n, \\
x_i &\geq 0 && \forall i = 1, \ldots, n.
\end{aligned}$$

Under interval uncertainty $\mathcal{U}_I$, we have an uncertainty set defined by a deviation around a central or nominal point. Looking at the robust counterpart (4.11), we see in the left-hand side again a central point (the first term) corrected using a penalty (the second term).

Example 4.2 (Polyhedral-Based Uncertainty) Under polyhedral-based uncertainty, the set $\mathcal{U}$ can be written as

$$\mathcal{U}_{\mathrm{P}} = \{\boldsymbol{a} \in \mathbb{R}^n \;:\; \boldsymbol{B}\boldsymbol{a} \geq \boldsymbol{r}\},$$

where $\boldsymbol{B} \in \mathbb{R}^{m \times n}$ and $\boldsymbol{r} \in \mathbb{R}^m$. Using this uncertainty set, we can reformulate the constraints of the robust counterpart (4.9) as follows:

$$\min_{\boldsymbol{a}\,:\,\boldsymbol{B}\boldsymbol{a} \geq \boldsymbol{r}} \; \boldsymbol{a}^\top \boldsymbol{x} \geq b.$$

By duality, we can derive the following equivalent constraints:

$$\max_{\boldsymbol{p}\,:\,\boldsymbol{p}^\top \boldsymbol{B} = \boldsymbol{x}^\top,\, \boldsymbol{p} \geq \boldsymbol{0}} \; \boldsymbol{p}^\top r \geq b,$$

where $\boldsymbol{p}$ is the dual vector with $\boldsymbol{p} \in \mathbb{R}^m$. Accordingly, the robust optimization problem that consists of minimizing (4.1) subject to (4.9) and (4.10) can be written as the following linear optimization problem:

$$\begin{aligned}
\text{minimize} \quad & \boldsymbol{c}^\top \boldsymbol{x}, \\
\text{subject to} \quad & \boldsymbol{p}^\top \boldsymbol{r} \geq b, \\
& \boldsymbol{p}^\top \boldsymbol{B} = \boldsymbol{x}^\top, \\
& \boldsymbol{p} \geq \boldsymbol{0}, \\
& \boldsymbol{x} \geq \boldsymbol{0}.
\end{aligned}$$

Example 4.3 (Norm-Based Uncertainty) Under a norm-based uncertainty, the set $\mathcal{U}$ can be written as

$$\mathcal{U}_\ell^\gamma = \{\boldsymbol{a} \in \mathbb{R}^n \;:\; \|\boldsymbol{M}(\boldsymbol{a} - \boldsymbol{\mu})\|_\ell \leq \gamma\},$$

where $\boldsymbol{M} \in \mathbb{R}^{n \times n}$ is a non-singular matrix. Such an uncertainty set can be looked at as the variation regions defined by the deviation of $\boldsymbol{a}$ from a nominal vector $\boldsymbol{\mu}$

rotated using the matrix $\boldsymbol{M}$, measured by a distance metric $\|\cdot\|_\ell$, $\ell = 1, 2, \ldots, \infty$. Parameter γ represents a predefined uncertainty threshold.

Using the above uncertainty set, we can reformulate the constraints of the robust counterpart (4.9) as follows:

$$- \max_{\boldsymbol{a}:\|\boldsymbol{M}(\boldsymbol{a}-\boldsymbol{\mu})/\gamma\|_\ell \le 1} -\boldsymbol{a}^\top \boldsymbol{x} \ge b. \tag{4.12}$$

Recalling the definition of dual norm, $\|\boldsymbol{y}\|_\ell^* = \sup_{\|\boldsymbol{z}\|_\ell \le 1} \boldsymbol{z}^\top \boldsymbol{y}$, we can reformulate the Constraint (4.12) as

$$\boldsymbol{x}^\top \boldsymbol{\mu} - \gamma \left\| (\boldsymbol{M}^{-1})^\top \boldsymbol{x} \right\|_\ell^* \ge b. \tag{4.13}$$

Under norm-based uncertainty, we observe that the robust counterpart (4.13) can be decomposed as a nominal value (the first term) corrected by a penalty (the second term). In this case, the parameter γ controls the penalty level to consider.

Three popular norms considered in the literature are ℓ_1, ℓ_2, and ℓ_∞. For an n-dimensional vector $\boldsymbol{w}$, recall that

$\|\boldsymbol{w}\|_1 = \sum_{i=1}^n |w_i|$,
$\|\boldsymbol{w}\|_2 = (w_1^2 + w_2^2 + \cdots + w_n^2)^{\frac{1}{2}}$, and
$\|\boldsymbol{w}\|_\infty = \max_{i=1,\ldots,n} |w_i|$.

Under an uncertainty set $\mathcal{U}_1^\gamma$, the reformulation of the robust linear optimization problem consisting of minimizing (4.1) subject to (4.10) and (4.9) can be written as

$$\begin{aligned}
\text{minimize} \quad & \boldsymbol{c}^\top \boldsymbol{x}, \\
\text{subject to} \quad & \boldsymbol{x}^\top \boldsymbol{\mu} - \gamma\lambda \ge b, \\
& \lambda \ge (\boldsymbol{M}^{-1})_i^\top \boldsymbol{x} \quad \forall i = 1, 2, \ldots, n, \\
& \lambda \ge -(\boldsymbol{M}^{-1})_i^\top \boldsymbol{x} \quad \forall i = 1, 2, \ldots, n, \\
& \boldsymbol{x} \ge \boldsymbol{0},
\end{aligned}$$

where $(\boldsymbol{M}^{-1})_i$ stands for the i-th row of matrix $\boldsymbol{M}^{-1}$. Recall that $\|\boldsymbol{w}\|_\infty^* = \|\boldsymbol{w}\|_1$. The above model shows that under an uncertainty set $\mathcal{U}_1^\gamma$, the robust counterpart can be reformulated as a linear optimization problem.

Under an uncertainty set $\mathcal{U}_\infty^\gamma$, the problem can be reformulated as the following linear optimization problem:

$$\begin{aligned}
\text{minimize} \quad & \boldsymbol{c}^\top \boldsymbol{x}, \\
\text{subject to} \quad & \boldsymbol{x}^\top \boldsymbol{\mu} - \gamma \sum_{i=1}^n \lambda_i \ge b,
\end{aligned}$$

$$\lambda_i \geq (M^{-1})_i^\top x \quad \forall i = 1, 2, \ldots, n,$$
$$\lambda_i \geq -(M^{-1})_i^\top x \quad \forall i = 1, 2, \ldots, n,$$
$$x \geq 0.$$

Under an uncertainty set $\mathcal{U}_2^\gamma$, the problem can be written as

$$\begin{aligned} \text{minimize} \quad & c^\top x, \\ \text{subject to} \quad & \left\| (M^{-1})^\top x \right\|_2 \leq (x^\top \mu - b)/\gamma, \\ & x \geq 0. \end{aligned}$$

Recall that $\|w\|_2^* = \|w\|_2$. The robust counterpart can be reformulated as a second-order conic programming problem.

4.2.2 Robust Counterparts Using Fenchel Duality

So far, we have considered uncertainty affecting a single linear constraint leading to the robust counterpart (4.9). In this section, we provide a systematic way to construct the robust nonlinear uncertainty inequality that is concave in the uncertain parameters. We use convex analysis (support functions, conjugate functions, and Fenchel duality) to convert the robust counterpart into an explicit and computationally tractable set of constraints.

We start by recalling some concepts and theorems (see [52, 458]). In what follows, dom(·) stands for the domain of the argument function, and ri(·) stands for the relative interior of the argument set. Let $f : \mathbb{R}^n \to \mathbb{R}$ be a closed convex function with $\mathrm{dom}(f) = \{x \mid f(x) < \infty\}$. Let $g : \mathbb{R}^n \to \mathbb{R}$ be a closed concave function with $\mathrm{dom}(g) = \{x \mid g(x) > -\infty\}$.

Definition 4.4 (Conjugate Function) The convex conjugate of f is defined as

$$f^*(y) = \sup_{x \in \mathrm{dom}(f)} \left\{ y^\top x - f(x) \right\}.$$

The concave conjugate of g is defined as

$$g_*(y) = \inf_{x \in \mathrm{dom}(g)} \left\{ y^\top x - g(x) \right\}. \tag{4.14}$$

For a function $g(\cdot,\cdot)$ of two vector variables, $g_*(\cdot,\cdot)$ will denote the partial concave conjugate concerning the first variable. Next, we introduce the *support function*. This is a particular conjugate function of great relevance.

Definition 4.5 (Support Function) The support function is the conjugate of the indicator function. The indicator function on the set S is defined as

$$\delta(\boldsymbol{x} \mid S) = \begin{cases} 0, & \text{if } \boldsymbol{x} \in S, \\ \infty, & \text{otherwise.} \end{cases} \tag{4.15}$$

Then the conjugate function of $\delta(\boldsymbol{x} \mid S)$,

$$\delta^*(\boldsymbol{y} \mid S) = \sup_{\boldsymbol{x} \in S} \boldsymbol{y}^\top \boldsymbol{x}, \tag{4.16}$$

is the so-called support function of the set S.

Next, we revisit the well-known Fenchel duality theorem. First, we define the primal problem:

$$\inf\{f(\boldsymbol{x}) - g(\boldsymbol{x}) \mid \boldsymbol{x} \in \text{dom}(f) \cap \text{dom}(g)\} \quad \text{(P)}.$$

The Fenchel dual of (P) is given by

$$\sup\{g_*(\boldsymbol{y}) - f^*(\boldsymbol{y}) \mid \boldsymbol{y} \in \text{dom}(g_*) \cap \text{dom}(f^*)\} \quad \text{(D)}.$$

The Fenchel duality theorem holds:

Theorem 4.1 (Fenchel Duality) *If* $\text{ri}(\text{dom}(f)) \cap \text{ri}(\text{dom}(g)) \neq \emptyset$, *then the optimal value of* (P) *coincides with the optimal value of* (D), *and the latter is attainable. Likewise, if* $\text{ri}(\text{dom}\, g_*) \cap \text{ri}(\text{dom}\, f^*) \neq \emptyset$, *then the optimal value of* (P) *coincides with the optimal value of* (D), *and the former is attainable.*

Since $f^{**} = f$ and $g_{**} = g$, we have the dual of (D) given by (P).

We now focus on a more general robust optimization problem with a nonlinear constraint:

$$\text{minimize} \quad \boldsymbol{c}^\top \boldsymbol{x}, \tag{4.17}$$

$$\text{subject to} \quad f(\boldsymbol{a}, \boldsymbol{x}) \leq 0, \tag{4.18}$$

$$\boldsymbol{x} \in \mathbb{R}^n. \tag{4.19}$$

In this model, $f(\boldsymbol{a}, \boldsymbol{x})$ is a closed proper concave function in the vector of uncertainty parameters $\boldsymbol{a} \in \mathbb{R}^m$ for all $\boldsymbol{x} \in \mathbb{R}^n$. We assume the uncertainty set $\mathcal{U}$ to be given by

$$\mathcal{U} = \left\{\boldsymbol{a} \in \mathbb{R}^n \; : \; \boldsymbol{a} = \bar{\boldsymbol{a}} + \boldsymbol{U}\boldsymbol{\zeta}, \; \boldsymbol{\zeta} \in \mathcal{W}\right\},$$

where $\bar{\boldsymbol{a}}$ represents a nominal vector, $\boldsymbol{U} = [\boldsymbol{u}_1, \ldots, \boldsymbol{u}_r] \in \mathbb{R}^{n\times r}$ is a perturbation matrix, and $\mathcal{W} \subset \mathbb{R}^r$ is a nonempty, convex, and compact set whose relative interior contains the null vector.

The robust counterpart of (4.17)–(4.19) is then

$$f(\boldsymbol{a}, \boldsymbol{x}) \leq 0, \ \forall \boldsymbol{a} \in \mathcal{U}, \tag{4.20}$$

which can be rewritten as

$$\underset{\boldsymbol{a}\in\mathcal{U}}{\text{maximize}}\, f(\boldsymbol{a}, \boldsymbol{x}) \leq 0. \tag{4.21}$$

Using the definition of indicator functions (4.15) and using Theorem 4.1, we can derive an equivalent reformulation for (4.20), which is used to derive tractable robust counterparts.

Theorem 4.2 (Fenchel Robust Counterpart, [53]) *Let* $\bar{\boldsymbol{a}} \in \text{ri}(\text{dom}(f(\cdot, \boldsymbol{x})))$ $\forall \boldsymbol{x} \in \mathbb{R}^n$. *Then the vector* $\boldsymbol{x} \in \mathbb{R}^n$ *satisfies* (4.20) *if and only if* $\boldsymbol{x} \in \mathbb{R}^n$ *and* $\boldsymbol{v} \in \mathbb{R}^m$ *satisfy the single inequality*

$$\boldsymbol{v}^\top \bar{\boldsymbol{a}} + \delta^* \left(\boldsymbol{A}^\top \boldsymbol{v} \mid \mathcal{W}\right) - f_*(\boldsymbol{v}, \boldsymbol{x}) \leq 0, \tag{4.22}$$

in which the support function δ^* *and the partial concave conjugate function* f_* *are defined in* (4.16) *and* (4.14), *respectively.*

Ben-Tal et al. [53] study the tractable reformulations of (4.22) under different support functions $\delta^*(\boldsymbol{A}^\top \boldsymbol{v} \mid \mathcal{W})$ and functions $f_*(\boldsymbol{v}, \boldsymbol{x})$. The reader can refer to [53, 55] for more detailed discussions on Robust Optimization techniques.

Applications of Robust Optimization include energy [151, 400], project scheduling [89, 237], water management [243, 433], inventory management [28, 56], and other areas namely facility location, which will be discussed in depth throughout this volume.

In addition, the interested reader can find excellent reviews on Robust Optimization in [60, 68, 217, 242].

4.3 Stochastic Programming

Like in the previous section, to better focus our discussion, we start by considering a general linear optimization problem:

$$\text{minimize} \quad \boldsymbol{c}^\top \boldsymbol{x}, \tag{4.1}$$

$$\text{subject to} \quad \boldsymbol{A}\boldsymbol{x} \geq \boldsymbol{b}, \tag{4.2}$$

$$\boldsymbol{x} \geq \boldsymbol{0}, \tag{4.3}$$

with $\boldsymbol{x} \in \mathbb{R}^{n_1}$ standing for the vector of decision variables, $\boldsymbol{A} = [\boldsymbol{a}_1, \ldots, \boldsymbol{a}_{n_1}] \in \mathbb{R}^{m_1 \times n_1}$, $\boldsymbol{b} \in \mathbb{R}^{m_1}$, and $\boldsymbol{c} \in \mathbb{R}^{n_1}$.

In general terms, stochastic linear programming deals with optimization problems like the above one involving parameters whose behavior is described by some probability distribution. In some problems, a decision must be made and evaluated here and now. The decision is made before the uncertainty is disclosed, and no correction is allowed to this decision.

Often, a definite evaluation of the performance of a solution must wait for uncertainty to be revealed. When this is the case, we can think of a two-stage process: In the first stage, the here-and-now solution is implemented. Then, at some point in time, uncertainty is revealed. This allows measuring accurately the initial decision made. In many cases, corrective decisions are possible—adaptive or recourse decisions are made. In any case, the above setting leads to the family of two-stage stochastic programming problems. These are very important in decision-making under uncertainty.

The above two-stage process can be extended when multiple decision stages are allowed (e.g., in a multi-period decision-making problem). Given the relevance of two- and multi-stage stochastic programming in Location Science, we directly discuss them. Nevertheless, the interested reader can refer to [295, 529] for well-established concepts in stochastic linear optimization.

4.3.1 Two-Stage Stochastic Linear Programming

A stochastic linear programming problem can be generically formulated as follows:

$$\text{minimize} \quad \boldsymbol{c}^\top \boldsymbol{x} + \mathcal{Q}(\boldsymbol{x}), \tag{4.23}$$

$$\text{subject to} \quad \boldsymbol{A}\boldsymbol{x} = \boldsymbol{b}, \tag{4.24}$$

$$\boldsymbol{x} \geq \boldsymbol{0}. \tag{4.25}$$

The function $\mathcal{Q}(\boldsymbol{x})$ is called the recourse function and seeks to provide a performance measure of the first-stage decision $\boldsymbol{x}$. We denote by $\boldsymbol{\xi}$ the random vector underlying the problem. Typically, the function provides an aggregate measure of the future assets considering a subset of realizations (possibly all) of $\boldsymbol{\xi}$. Since uncertainty is not explicitly stated in the above model, we call it the deterministic equivalent model.

Assuming a neutral attitude of the decision-maker toward risk, a natural choice for the recourse function is the expected value of the "consequence" faced by choosing the first-stage solution $\boldsymbol{x}$. In this case, we take

$$\mathcal{Q}(\boldsymbol{x}) = \mathbb{E}_{\mathbb{P}}[Q(\boldsymbol{x}, \boldsymbol{\xi})],$$

with $\mathbb{P}$ standing for the probability distribution of $\boldsymbol{\xi}$ and $Q(\boldsymbol{x}, \boldsymbol{\xi})$ denoting the optimal value of a second-stage optimization problem:

$$Q(\boldsymbol{x}, \boldsymbol{\xi}) = \min \quad \boldsymbol{q}^\top \boldsymbol{y}, \tag{4.26}$$

$$\text{s. t.} \quad \boldsymbol{T}\boldsymbol{x} + \boldsymbol{W}\boldsymbol{y} = \boldsymbol{h}, \tag{4.27}$$

$$\boldsymbol{y} \geq \boldsymbol{0}, \tag{4.28}$$

with $\boldsymbol{q} \in \mathbb{R}^{n_2}$, $\boldsymbol{T} \in \mathbb{R}^{m_2 \times n_1}$, $\boldsymbol{W} \in \mathbb{R}^{m_2 \times n_2}$, and $\boldsymbol{h} \in \mathbb{R}^{m_2}$. $\boldsymbol{y} \in \mathbb{R}^{n_2}$ is the vector of second-stage decision variables. This is a single-stage stochastic programming problem with $\boldsymbol{\xi} = (\boldsymbol{q}, \boldsymbol{h}, \boldsymbol{W}, \boldsymbol{T})$. Each observation of $\boldsymbol{\xi}$ is called a scenario and fully determines all the second-stage data and thus fully defines the second-stage problem.

Summing up, at the first stage, we optimize an immediate cost $\boldsymbol{c}^\top \boldsymbol{x}$ plus the expected cost of the optimal second-stage decision. In turn, the second-stage problem can be seen as a problem ensuring the best decision when uncertainty is revealed. Also, we can consider the solution to the second-stage problem as a recourse action, thus using the term $\boldsymbol{W}\boldsymbol{y}$ to compensate for any inconsistency of the system $\boldsymbol{T}\boldsymbol{x} = \boldsymbol{h}$. For this reason, $\boldsymbol{W}$ is called the recourse matrix. When this matrix is deterministic, we say that we have fixed recourse. A fixed recourse helps to characterize the feasibility region, which is good from a computational perspective. Moreover, a fixed recourse is often found in practical problems. Unless stated otherwise, we assume this case hereafter.

One important feature of stochastic programming is the non-anticipativity principle [460] stating that a decision should depend only on the information available when it is to be implemented. In the above model, like in many models considered in the literature, this principle is implicit since the first-stage solution represented by variables $\boldsymbol{x}$ is not adaptable to the occurring scenario.

One particular aspect of interest when modeling a two-stage stochastic programming problem regards the first-stage feasibility. The interested reader should refer to [78, 259] for all details. The question is whether a first-stage solution can lead to second-stage infeasibility or unboundedness. Some related concepts have been developed accordingly.

The problem has complete recourse when the second-stage feasibility holds, irrespective of the first-stage choice. In this case, $\forall \boldsymbol{\eta} \in \mathbb{R}^{m_2}$, $\exists \boldsymbol{y} \geq \boldsymbol{0}$ such that $\boldsymbol{W}\boldsymbol{y} = \boldsymbol{\eta}$, i.e., the polyhedron

$$P(\boldsymbol{\eta}) = \left\{ \boldsymbol{y} : \boldsymbol{W}\boldsymbol{y} = \boldsymbol{\eta}, \boldsymbol{y} \geq \boldsymbol{0} \right\}$$

is nonempty $\forall \boldsymbol{\eta} \in \mathbb{R}^{m_2}$. A more relevant concept for us is that of relatively complete recourse. It holds when every feasible first-stage solution has a second-stage feasible completion, irrespective of the scenario. It can be proved that this holds if

$$\mathbb{P}[Q(\boldsymbol{x}, \boldsymbol{\xi}) < \infty] = 1, \ \forall \boldsymbol{x} \in \mathbb{R}^{n_1} \text{ with } \boldsymbol{A}\boldsymbol{x} = \boldsymbol{b}, \ \boldsymbol{x} \geq 0.$$

Overall, a two-stage stochastic programming problem is considered to be well-defined when

$$\min\left\{\mathbb{E}_{\mathbb{P}}\left(\left[Q(\boldsymbol{x},\boldsymbol{\xi})\right]^{+}\right),\mathbb{E}_{\mathbb{P}}\left(\left[Q(\boldsymbol{x},\boldsymbol{\xi})\right]^{-}\right)\right\}<\infty,\ \forall\boldsymbol{x}\in\mathbb{R}^{n_1}\text{ with }\boldsymbol{A}\boldsymbol{x}=\boldsymbol{b},\ \boldsymbol{x}\geq\boldsymbol{0}.$$

One particular case of complete recourse is that of simple recourse in which $\boldsymbol{W} = [\boldsymbol{I} \mid -\boldsymbol{I}]$.

In practice, the general problem is difficult to solve because computing $\mathbb{E}_{\mathbb{P}}\left[Q(\boldsymbol{x},\boldsymbol{\xi})\right]$ requires multiple integration.

Consider the case where the support of $\boldsymbol{\xi}$ is finite. We can index all scenarios (observations of $\boldsymbol{\xi}$) in a finite set, say Ω with each $\omega \in \Omega$ specifying an observation $\boldsymbol{\xi}_\omega = (\boldsymbol{q}_\omega, \boldsymbol{h}_\omega, \boldsymbol{W}_\omega, \boldsymbol{T}_\omega)$ of $\boldsymbol{\xi}$.

For every $\omega \in \Omega$, let π_ω be the probability associated with scenario ω. We can go further by formulating the so-called extensive form of the deterministic equivalent. Denoting by $\boldsymbol{y}_\omega$ the second-stage decision in case scenario $\omega \in \Omega$ occurs, we have the following model:

$$\text{minimize} \quad \boldsymbol{c}^\top\boldsymbol{x} + \sum_{\omega\in\Omega}\pi_\omega\boldsymbol{q}_\omega^\top\boldsymbol{y}_\omega, \tag{4.29}$$

$$\text{subject to} \quad \boldsymbol{A}\boldsymbol{x} = \boldsymbol{b}, \tag{4.24}$$

$$\boldsymbol{T}_\omega\boldsymbol{x} + \boldsymbol{W}_\omega\boldsymbol{y}_\omega = \boldsymbol{h}_\omega \quad \forall\omega\in\Omega, \tag{4.30}$$

$$\boldsymbol{x} \geq \boldsymbol{0}, \tag{4.25}$$

$$\boldsymbol{y}_\omega \geq \boldsymbol{0} \quad \forall\omega\in\Omega. \tag{4.31}$$

Even this model may be difficult to solve if the number of scenarios increases. Popular techniques for handling it include the Sample Average Approximation and the L-shaped method, which we revisit next due to their relevance.

4.3.1.1 Sample Average Approximation (SAA)

This is a Monte Carlo technique introduced by Kleywegt et al. [312] for reducing the number of scenarios to a manageable size either when we have a finite although large number of scenarios or when the support of the underlying random vector is infinite. Suppose that the support of $\boldsymbol{\xi}$ has a very large cardinality (possibly infinity). In this case, we can define an S-dimensional random sample $(\boldsymbol{\xi}_1, \ldots, \boldsymbol{\xi}_S)$ with $\boldsymbol{\xi}_s$ i.i.d. $\boldsymbol{\xi}$. The recourse function $Q(\boldsymbol{x}) = \mathbb{E}_{\mathbb{P}}[Q(\boldsymbol{x},\boldsymbol{\xi})]$ can be approximated by the average

$$\hat{Q}(\boldsymbol{x}) = \frac{1}{S}\sum_{s=1}^{S} Q(\boldsymbol{x},\boldsymbol{\xi}_s),$$

with

$$Q(\boldsymbol{x}, \boldsymbol{\xi}_s) = \min \quad \boldsymbol{q}_s^\top \boldsymbol{y}_s, \tag{4.32}$$

$$\text{s. t.} \quad \boldsymbol{T}_s\boldsymbol{x} + \boldsymbol{W}_s\boldsymbol{y}_s = \boldsymbol{h}_s, \tag{4.33}$$

$$\boldsymbol{y}_s \geq \boldsymbol{0}. \tag{4.34}$$

Consequently, the two-stage stochastic problems can be approximated by a finite scenario set problem

$$\text{minimize} \quad \boldsymbol{c}^\top \boldsymbol{x} + \frac{1}{S}\sum_{s=1}^{S} Q(\boldsymbol{x}, \boldsymbol{\xi}_s), \tag{4.35}$$

$$\text{subject to} \quad \boldsymbol{A}\boldsymbol{x} = \boldsymbol{b}, \tag{4.24}$$

$$\boldsymbol{x} \geq \boldsymbol{0}. \tag{4.25}$$

This so-called SAA problem is similar to the above extensive form of the deterministic equivalent but using the sampled scenarios and setting their probability equal to $1/N$.

Note that the SAA mechanism described is not an algorithm for approximating the two-stage stochastic programming problem; we still have to solve this problem by an appropriate procedure.

Let $\hat{\boldsymbol{x}}$ be a feasible solution to the problem (4.23)–(4.25), found by solving an SAA problem or using another approximate algorithm. Denote by $\nu(\boldsymbol{x})$ the objective function (4.23). The optimality gap quality of the solution $\hat{\boldsymbol{x}}$ is given by $\nu(\hat{\boldsymbol{x}}) - \nu^*$, with ν^* denoting the optimal value of the problem. This optimality gap can be estimated using SAA. We focus separately on each term of the above gap.

The exact value $\nu(\hat{\boldsymbol{x}})$ can be estimated by Monte Carlo sampling. Considering a random sample $\boldsymbol{\xi}_1, \ldots, \boldsymbol{\xi}_{S'}$ i.i.d. $\boldsymbol{\xi}$, the exact value of $\nu(\hat{\boldsymbol{x}})$ can be estimated using the sample average $\hat{\nu}_{S'}(\hat{\boldsymbol{x}}) = \boldsymbol{c}^\top \hat{\boldsymbol{x}} + \frac{1}{N}\sum_{s=1}^{S'} Q(\hat{\boldsymbol{x}}, \boldsymbol{\xi}_s)$. Furthermore, the sample variance of $\hat{\nu}_{S'}(\hat{\boldsymbol{x}})$ is given by

$$\hat{\sigma}^2_{S'}(\hat{\boldsymbol{x}}) = \frac{1}{S'(S'-1)}\sum_{s=1}^{S'} \left(Q(\hat{\boldsymbol{x}}, \xi_s) - \hat{\nu}_{S'}(\hat{\boldsymbol{x}})\right)^2.$$

Assuming that S' is high (which is a reasonable assumption in practice), the central limit theorem can be invoked, leading to an approximate confidence upper bound for $\nu(\hat{\boldsymbol{x}})$ at a level $(1-\alpha) \times 100\%$:

$$UB_{S'}(\hat{\boldsymbol{x}}) = \hat{\nu}_{S'}(\hat{\boldsymbol{x}}) + z_{1-\alpha}\hat{\sigma}_{S'}(\hat{\boldsymbol{x}}),$$

with $z_{1-\alpha}$ standing for the $(1-\alpha)$-quantile of a standardized Gaussian distribution.

We now focus on finding a lower bound on ν^*. Let ν_S denote the random variable representing the optimal value of the SAA problem based upon an S-dimensional random sample (ν_N is a random variable due to the randomness of the sample). It is easy to see that $\nu^* \geq \mathbb{E}[\nu_S]$ [312]. On the other hand, $\mathbb{E}[\nu_S]$ can be estimated by solving a series of, say, R, SAA problems, each based upon a sample of size S. Denote by $\nu_S^1, \ldots, \nu_S^R$ the optimal values of those SAA problems. Hence, $\bar{\nu}_{S,R} = \frac{1}{R}\sum_{r=1}^{R} \nu_S^r$ is an unbiased estimator of $\mathbb{E}[\nu_S]$. Furthermore, since the R samples are independent (and so are the values $\nu_S^1, \ldots, \nu_S^R$), we can estimate the variance of the average $\bar{\nu}_{S,R}$ by

$$\hat{\sigma}_{S,R}^2 = \frac{1}{R(R-1)} \sum_{r=1}^{R} \left(\nu_S^r - \bar{\nu}_{S,R}\right)^2.$$

Consequently, a lower bound on $\mathbb{E}[\nu_S]$ at a $100(1-\alpha)\%$ confidence level is given by

$$LB_{S,R} = \bar{\nu}_{S,R} - t_{\alpha,R-1}\hat{\sigma}_{S,R},$$

with $t_{\alpha,R-1}$ standing for the $1-\alpha$ quantile of a t-distribution with $R-1$ degrees of freedom. In practice, we often consider small values of R, which justifies using the t-distribution. Given that $\nu^* \geq \mathbb{E}[\nu_S]$, $LB_{S,R}$ gives a statistical lower bound on ν^*. Finally, we can estimate the gap $(\nu(\hat{\boldsymbol{x}}) - \nu^*)$ by

$$UB_{N'}(\hat{\boldsymbol{x}}) - LB_{S,R}.$$

4.3.1.2 Benders Decomposition (L-Shaped Method)

Despite relying on linear programming, an SAA problem or, more generally, its use for tackling a two-stage stochastic linear program with a finite number of scenarios may become computationally unaffordable when their number is high. Next, we revisit a popular decomposition approach that can be considered: the L-shaped method. This is a specialized procedure for a problem of the form:

$$\begin{aligned}
\text{minimize} \quad & \boldsymbol{c}^\top \boldsymbol{x} + \sum_{\omega\in\Omega} \pi_\omega \boldsymbol{q}_\omega^\top \boldsymbol{y}_\omega, && && (4.29)\\
\text{subject to} \quad & \boldsymbol{A}\boldsymbol{x} = \boldsymbol{b}, && && (4.24)\\
& \boldsymbol{T}_\omega \boldsymbol{x} + \boldsymbol{W}_\omega \boldsymbol{y}_\omega = \boldsymbol{h}_\omega && \forall \omega \in \Omega, && (4.30)\\
& \boldsymbol{x} \geq \boldsymbol{0}, && && (4.25)\\
& \boldsymbol{y}_\omega \geq \boldsymbol{0} && \forall \omega \in \Omega, && (4.31)
\end{aligned}$$

where Ω is a finite set indexing the scenarios. Recall that the second term of the objective function is the recourse function, i.e., $\mathcal{Q}(\boldsymbol{x}) = \sum_{\omega \in \Omega} \pi_\omega \boldsymbol{q}_\omega^\top \boldsymbol{y}_\omega$. The L-shaped method is an iterative procedure that involves the resolution of small-scale problems in each iteration.

We assume that we have fixed recourse, i.e., matrix $\boldsymbol{W}$ is deterministic (scenario invariant). We also assume that for all feasible $\boldsymbol{x}$, the second-stage optimal value is finite, i.e., $Q(\boldsymbol{x}, \omega) < \infty, \forall \omega \in \Omega$ with

$$Q(\boldsymbol{x}, \omega) = \min \quad \boldsymbol{q}_\omega^\top \boldsymbol{y}, \tag{4.36}$$

$$\text{s. t.} \quad \boldsymbol{T}_\omega \boldsymbol{x} + \boldsymbol{W} \boldsymbol{y} = \boldsymbol{h}_\omega, \tag{4.37}$$

$$\boldsymbol{y} \geq \boldsymbol{0}. \tag{4.38}$$

Hereafter, we omit the scenario index in $\boldsymbol{y}$ whenever no confusion arises. By the strong duality holding for linear programming, for every $\omega \in \Omega$, we can equivalently write

$$Q(\boldsymbol{x}, \omega) = \max_{\boldsymbol{p}} \left\{ \boldsymbol{p}^\top (\boldsymbol{h}_\omega^\top - \boldsymbol{T}_\omega \boldsymbol{x}) \mid \boldsymbol{W}^\top \boldsymbol{p} \leq \boldsymbol{q}_\omega \right\}.$$

Denoting by $\boldsymbol{p}_\omega^*$ an optimal solution of the above problem, we have $Q(\boldsymbol{x}, \omega) = \boldsymbol{p}_\omega^* (\boldsymbol{h}_\omega^\top - \boldsymbol{T}_\omega \boldsymbol{x})$. Given that $Q(\boldsymbol{x}, \omega)$ is convex function in $\boldsymbol{x}$, for a fixed $\hat{\boldsymbol{x}}$ we obtain an optimal $\hat{\boldsymbol{p}}_\omega^*$ and conclude that $(-\boldsymbol{T}_\omega^\top \hat{\boldsymbol{p}}_\omega^*)$ is a subgradient of $Q(\boldsymbol{x}, \omega)$ at $\hat{\boldsymbol{x}}$. Accordingly, for the same $\hat{\boldsymbol{x}}$, we can find a subgradient of $\mathcal{Q}(\boldsymbol{x})$ at $\hat{\boldsymbol{x}}$:

$$\boldsymbol{s} = -\sum_{\omega \in \Omega} \pi_\omega \boldsymbol{T}_\omega^\top \hat{\boldsymbol{p}}_\omega^*.$$

In our two-stage stochastic problem (4.23)–(4.25), we can now replace $\mathcal{Q}(\boldsymbol{x})$ with $\mathcal{Q}(\hat{\boldsymbol{x}}) + \boldsymbol{s}^\top (\boldsymbol{x} - \hat{\boldsymbol{x}})$ thus obtaining a lower bound problem

$$\min_{x \in \mathcal{X}} \boldsymbol{c}^\top \boldsymbol{x} + \left[\mathcal{Q}(\hat{\boldsymbol{x}}) + \boldsymbol{s}^\top (\boldsymbol{x} - \hat{\boldsymbol{x}}) \right] \leq \min_{x \in \mathcal{X}} \boldsymbol{c}^\top \boldsymbol{x} + \mathcal{Q}(\boldsymbol{x}),$$

with $\mathcal{X} = \{\boldsymbol{x} \in \mathbb{R}^n : \boldsymbol{A}\boldsymbol{x} = \boldsymbol{b}, \boldsymbol{x} \geq \boldsymbol{0}\}$.

Suppose now that we consider a set of solutions $\hat{\boldsymbol{x}}_\ell$, $\ell = 1, \ldots, L$, and a set of corresponding subgradients of $\mathcal{Q}(\boldsymbol{x})$ at those points, $\boldsymbol{s}_\ell$, $\ell = 1, \ldots, L$. In this case, we can solve the following linear programming problem:

$$\text{minimize} \quad \theta + \boldsymbol{c}^\top \boldsymbol{x}, \tag{4.39}$$

$$\text{subject to} \quad \theta \geq \mathcal{Q}(\hat{\boldsymbol{x}}_\ell) + \boldsymbol{s}_\ell^\top (\boldsymbol{x} - \hat{\boldsymbol{x}}_\ell), \qquad \ell = 1, \ldots, L, \tag{4.40}$$

$$\boldsymbol{A}\boldsymbol{x} = \boldsymbol{b}, \tag{4.24}$$

$$\boldsymbol{x} \geq \boldsymbol{0}. \tag{4.25}$$

This problem is called a master lower bound problem.

When a new subgradient $\hat{s}$ is obtained for a new $\hat{x}$, the corresponding cut can be added. The optimal solution $(\theta^*, \boldsymbol{x}^*)$ to the above problem defines a lower bound on the optimal value of our two-stage stochastic linear programming problem, which is given by $\theta^* + \boldsymbol{c}^\top \boldsymbol{x}^*$. On the other hand, given that $\boldsymbol{x}^*$ is feasible to our problem (it always satisfies (4.24) and (4.25)), then $\boldsymbol{c}^\top \boldsymbol{x}^* + \mathcal{Q}(\boldsymbol{x}^*)$ is an upper bound on the optimal solution of the stochastic problem. Recall that $\mathcal{Q}(\boldsymbol{x}^*) = \sum_{\omega\in\Omega} \pi_\omega Q(\boldsymbol{x}^*, \omega)$. Hence, we can devise an iterative process that starts with no cut and then iterates between solving the above problem, thus finding an optimal θ and $\boldsymbol{x}$ and using the latter to derive a new cut. The process stops when the gap between the lower and upper bounds is small enough: $\mathcal{Q}(\boldsymbol{x}^*) - \theta^* \le \epsilon$ with ϵ standing for a desired tolerance.

By construction, we see that the sequence of lower bounds provided by the master lower bound problems is non-decreasing. In fact, from one iteration to the following, a new cut is added. The above iterative procedure is a Benders decomposition algorithm (L-shaped method without feasibility constraints). Computationally, it involves three tasks in each iteration: (i) solving the master lower bound problem, (ii) evaluating $\mathcal{Q}(\boldsymbol{x}^*)$ by solving $|\Omega|$ linear optimization problems to find $Q(\boldsymbol{x}^*, \omega)$, $\omega \in \Omega$, and (iii) computing the subgradient $\boldsymbol{s}$ by solving a dual subproblem.

For further details and particular cases related to the L-shaped method, the interested reader should refer to the specialized literature, e.g., [78, 259].

4.3.2 Two-Stage Stochastic Mixed-Integer Programming

Discrete facility location problems are the focus of this volume. These problems are usually formulated as mixed-integer linear programming problems. For this reason, we revisit some concepts and tools related to stochastic mixed-integer programming. To make the discussion more focused we consider a risk-neutral decision-maker.

A two-stage stochastic mixed-integer program for a risk-neutral decision-maker can be formulated as follows:

$$\text{minimize} \quad \boldsymbol{c}^\top \boldsymbol{x} + \mathcal{Q}(\boldsymbol{x}), \tag{4.41}$$

$$\text{subject to} \quad \boldsymbol{A}\boldsymbol{x} = \boldsymbol{b}, \tag{4.42}$$

$$\boldsymbol{x} \in \mathbb{R}_+^{n_1} \times \mathbb{Z}_+^{p_1}, \tag{4.43}$$

with $\mathcal{Q}(\boldsymbol{x}) = \mathbb{E}_{\mathbb{P}}[Q(\boldsymbol{x}, \boldsymbol{\xi})]$, $\boldsymbol{\xi} = (\boldsymbol{q}, \boldsymbol{h}, \boldsymbol{T}, \boldsymbol{W})$, and

$$Q(\boldsymbol{x}, \boldsymbol{\xi}) = \min \quad \boldsymbol{q}^\top \boldsymbol{y}, \tag{4.44}$$

$$\text{s. t.} \quad \boldsymbol{T}\boldsymbol{x} + \boldsymbol{W}\boldsymbol{y} = \boldsymbol{h}, \tag{4.45}$$

$$\boldsymbol{y} \in \mathbb{R}_+^{n_2} \times \mathbb{Z}_+^{p_2}. \tag{4.46}$$

We assume that the matrices and vectors of parameters have adequate dimensions. We use similar terminology as introduced for stochastic linear problems, so we do not repeat the meaning of the elements involved.

As for the linear case, if the support of the random vector $\boldsymbol{\xi}$ is finite, we can index the possible realizations of the random vector in a set Ω that is also finite. In this case, we can again go further in terms of modeling by indexing in Ω the second-stage variables and parameters and thus formulating the extensive form of the deterministic equivalent:

$$\text{minimize} \quad \boldsymbol{c}^\top \boldsymbol{x} + \sum_{\omega \in \Omega} \pi_\omega \boldsymbol{q}_\omega^\top \boldsymbol{y}_\omega, \tag{4.29}$$

$$\text{subject to} \quad \boldsymbol{A}\boldsymbol{x} = \boldsymbol{b}, \tag{4.24}$$

$$\boldsymbol{T}_\omega \boldsymbol{x} + \boldsymbol{W}_\omega \boldsymbol{y}_\omega = \boldsymbol{h}_\omega \qquad \forall \omega \in \Omega, \tag{4.30}$$

$$\boldsymbol{x} \in \mathbb{R}_+^{n_1} \times \mathbb{Z}_+^{p_1}, \tag{4.43}$$

$$\boldsymbol{y} \in \mathbb{R}_+^{n_2} \times \mathbb{Z}_+^{p_2}. \tag{4.46}$$

Stochastic integer or mixed-integer programming gathers challenges from mixed-integer and stochastic programming. For a specific scenario, $Q(\boldsymbol{x}, \omega)$ is the objective function of a mixed-integer program, and thus it is non-convex and discontinuous. Moreover, the evaluation of the second-stage function $Q(\boldsymbol{x}, \omega)$ for each first-stage decision and scenario may be challenging (possibly NP-hard). To add to those, the evaluation of the recourse function $\mathcal{Q}(\boldsymbol{x})$ for a first-stage decision may become too challenging. Even if the underlying random vector has a discrete probability distribution (we can avoid integrating the optimal value of an MILP), the number of scenarios may be high. This may render the computation of the (many) second-stage MILP problems unaffordable.

Next, we briefly discuss some cases for which efficient tools have been developed. In the context of this book, we must necessarily avoid many details. For those, the interested reader should refer to the specialized references (e.g., [78, 259]).

4.3.2.1 Continuous Recourse

In case we have $\boldsymbol{y} \in \mathbb{R}^{n_2}$, the analysis presented in Sect. 4.3.1.2 can be applied in the same way. The difference is that in the lower bounding problem (minimizing (4.39) subject to (4.40), (4.24), and (4.25)), we must replace (4.25) with (4.43). We are assuming the matrices and vectors of parameters with adequate dimensions. Like in Sect. 4.3.1.2, the algorithm converges after a finite number of iterations. Nevertheless, the master problem is now a mixed-integer program, which may call for a specially tailored procedure to be efficiently solved. For instance, instead of solving one MILP master problem in each iteration, we can think of using a so-called branch-and-Benders-cut algorithm that involves a single branch-and-cut tree with the Benders cuts being derived and added while exploring that tree (see, e.g., [227, 423]).

4.3.2.2 Mixed-Integer Recourse

Our two-stage stochastic mixed-integer program can be written as follows:

$$\text{minimize} \quad \boldsymbol{c}^\top \boldsymbol{x} + \sum_{\omega \in \Omega} \pi_\omega \theta_\omega, \tag{4.47}$$

$$\text{subject to} \quad \boldsymbol{A}\boldsymbol{x} = \boldsymbol{b}, \tag{4.42}$$

$$\theta_\omega \geq Q(\boldsymbol{x}, \omega) \qquad \forall \omega \in \Omega, \tag{4.48}$$

$$\boldsymbol{x} \in \mathbb{R}_+^{n_1} \times \mathbb{Z}_+^{p_1}, \tag{4.43}$$

and

$$Q(\boldsymbol{x}, \omega) = \min \quad \boldsymbol{q}_\omega^\top \boldsymbol{y}, \tag{4.49}$$

$$\text{s. t.} \quad \boldsymbol{W}_\omega \boldsymbol{y} = \boldsymbol{h}_\omega - \boldsymbol{T}_\omega \boldsymbol{x}, \tag{4.50}$$

$$\boldsymbol{y} \in \mathbb{R}_+^{n_2} \times \mathbb{Z}_+^{p_2}. \tag{4.46}$$

As mentioned above, $Q(\boldsymbol{x}, \omega)$ is the optimal value of an MILP problem. The Benders cuts as above derived are still valid. However, the master problem constraints (4.48) cannot be ensured by Benders cuts only. Next, we will see how to overcome this.

Integer L-Shaped Method

If we have pure binary first-stage variables, i.e., $\boldsymbol{x} \in \{0, 1\}^{n_1}$, then we can apply the integer L-shaped method proposed by Laporte and Louveaux [324]. The algorithm is a branch-and-cut procedure. Each iteration includes continuous and integer optimality cuts, used as in a standard Benders decomposition. The continuous cuts can be Benders cuts generated from the linear programming relaxation (e.g., [29]). The integer cuts may consist of conditions imposed to exclude the current solution or result from binary optimality cuts derived using a lower bound on the recourse function. Given a first-stage feasible solution $\hat{\boldsymbol{x}}$, let $J^1 = \{j : \hat{x}_j = 1\}$ and $J^0 = \{j : \hat{x}_j = 0\}$. This solution can be excluded by imposing

$$\sum_{j \in J^1} x_j - \sum_{j \in J^0} x_j \leq |J^1| - 1. \tag{4.51}$$

Additionally, the following binary cuts are valid:

$$\theta_\omega \geq Q(\hat{x}, \omega) - \left[Q(\hat{x}, \omega) - L\right] \left(\sum_{j \in J^1} (1 - x_j) + \sum_{j \in J^0} x_j \right). \tag{4.52}$$

L is a lower bound on the second-stage optimal objective function for every first-stage feasible solution and every scenario. A branch-and-cut algorithm can now be devised. In each node of the branching tree, cuts are added. When necessary, branching is performed on factional variables; when an integer solution is found, binary optimality cuts are added. The interested reader can refer to [29, 324, 357] for further details.

Dual Decomposition

As we emphasized when presenting the first two-stage stochastic linear programming problem, the so-called non-anticipativity principle is implicit since the first-stage variables are not indexed in the scenarios and thus do not adapt to each occurring one. A frequent challenge when tackling stochastic programming problems (even with a finite number of scenarios) is that the models tend to be too large. One family of algorithms that seek to overcome this aspect are those exploring dual decomposition [106]. The idea is to impose the non-anticipativity principle explicitly. This can be accomplished by considering a so-called variable splitting representation. Afterward, the non-anticipativity constraints are relaxed in a Lagrangian fashion.

Let us consider the extensive form of the deterministic equivalent problem consisting of minimizing (4.29) subject to (4.24), (4.30), (4.43), and (4.46). The corresponding variable splitting representation can be written as follows:

$$\text{minimize} \quad \sum_{\omega \in \Omega} \pi_\omega \left(\boldsymbol{c}^\top \boldsymbol{x}_\omega + \boldsymbol{q}_\omega^\top \boldsymbol{y}_\omega \right), \tag{4.53}$$

$$\text{subject to} \quad \boldsymbol{A}\boldsymbol{x}_\omega = \boldsymbol{b}, \qquad \forall \omega \in \Omega, \tag{4.54}$$

$$\boldsymbol{T}_\omega \boldsymbol{x}_\omega + \boldsymbol{W}_\omega \boldsymbol{y}_\omega = \boldsymbol{h}_\omega \qquad \forall \omega \in \Omega, \tag{4.55}$$

$$\boldsymbol{x}_\omega = \sum_{\tau \in \Omega} \pi_\tau \boldsymbol{x}_\tau \qquad \forall \omega \in \Omega, \tag{4.56}$$

$$\boldsymbol{x}_\omega \in \mathbb{R}_+^{n_1} \times \mathbb{Z}_+^{p_1}, \qquad \forall \omega \in \Omega, \tag{4.57}$$

$$\boldsymbol{y} \in \mathbb{R}_+^{n_2} \times \mathbb{Z}_+^{p_2}. \tag{4.46}$$

Constraints (4.56) ensure the non-anticipativity principle. Thus are called the non-anticipativity constraints. By dualizing such constraints in a Lagrangian fashion, we obtain

$$\mathcal{L}(\boldsymbol{\mu}) = \min \quad \sum_{\omega \in \Omega} \pi_\omega \left(\left[\boldsymbol{c} + (\boldsymbol{\mu}_\omega - \hat{\boldsymbol{\mu}} \right]^\top \boldsymbol{x}_\omega + \boldsymbol{q}_\omega^\top \boldsymbol{y}_\omega \right), \tag{4.58}$$

$$\text{s. t.} \quad (4.46),\ (4.54),\ (4.55), (4.57),$$

with $\hat{\boldsymbol{\mu}} = \sum_{\omega \in \Omega} \pi_\omega \boldsymbol{\mu}_\omega$. The relaxed problem can now be decomposed in the scenarios:

$$\mathcal{L}(\boldsymbol{\mu}) = \sum_{\omega \in \Omega} \pi_\omega \mathcal{L}_\omega(\boldsymbol{\mu}_\omega),$$

with

$$\mathcal{L}_\omega(\boldsymbol{\mu}_\omega) = \min \quad (\boldsymbol{c}\boldsymbol{\mu}_\omega)^\top + \boldsymbol{q}_\omega^\top \boldsymbol{y}_\omega, \tag{4.59}$$

$$\text{s. t.} \quad (4.42),\ (4.43),\ (4.46),\ (4.50).$$

Each subproblem is a mixed-integer linear program. The multipliers $\boldsymbol{\mu}_\omega$, $\omega \in \Omega$ are assumed to have been normalized so that $\hat{\boldsymbol{\mu}} = 0$. The Lagrangian dual consists of

$$\max_{\boldsymbol{\mu}} \left\{ \mathcal{L}(\boldsymbol{\mu}) : \sum_{\omega \in \Omega} \pi_\omega \boldsymbol{\mu}_\omega = 0 \right\}.$$

The well-known subgradient algorithm is one possibility for tackling the problem. However, it may be slow in practice and very sensitive to the choices made for the step sizes. An alternative can be to formulate a master problem, namely,

$$\max_{\boldsymbol{\mu}, \boldsymbol{\theta}} \left\{ \sum_{\omega \in \Omega} \pi_\omega \theta_\omega : \theta_\omega \leq \mathcal{L}_\omega(\boldsymbol{\mu}_\omega),\ \sum_{\omega \in \Omega} \pi_\omega \boldsymbol{\mu}_\omega = 0 \right\}.$$

Now, subgradient cuts can be used to approximate the constraints $\theta_\omega \leq \mathcal{L}_\omega(\boldsymbol{\mu}_\omega)$, $\omega \in \Omega$, making use of the upper bound problem (relaxed master problem):

$$\min \quad \sum_{\omega \in \Omega} \pi_\omega \theta_\omega, \tag{4.60}$$

$$\text{s. t.} \quad \theta_\omega \leq \mathcal{L}_\omega(\boldsymbol{\mu}_\omega) + (\boldsymbol{x}_\omega^\ell - \bar{\boldsymbol{x}}^\ell)^\top (\boldsymbol{\mu}_\omega - \boldsymbol{\mu}_\omega^\ell), \quad \ell = 1, \ldots, L, \tag{4.61}$$

$$\sum_{\omega \in \Omega} \pi_\omega \boldsymbol{\mu}_\omega = 0, \tag{4.62}$$

with $\bar{x}^\ell = \sum_{\omega \in \Omega} \pi_\omega x_\omega^\ell$. Following this approach, the relaxed master problem is solved. Then we solve the MILP subproblems to find the subgradient cuts and add them to the master problem; this process repeats until a termination criterion is met.

Other possibilities exist for tackling two-stage stochastic mixed-integer programming with mixed-integer recourse both exact and approximate: branch-and-fix coordination (e.g., [16]), scenario cluster decomposition (e.g., [197]), progressive hedging (e.g., [541]), *et cetera.* The interested reader should refer to the specialized literature for further algorithmic developments. Some further readings include [83, 259].

4.3.3 Multi-stage Stochastic Mixed-Integer Programming

The two-stage stochastic programming framework discussed so far can be extended to a multi-stage setting. There is a decision to be implemented now. Then, uncertainty is partially realized. Afterward, another decision is implemented (the second-stage decision), and then, again, more uncertainty is revealed. A subsequent third-stage decision is implemented. *Et cetera.* In any case, a finite number of stages are assumed. We consider directly mixed-integer variables in all stages.

Assuming a finite support for the stochastic process underlying the problem, the uncertainty can be modeled using a so-called scenario tree. Such a tree is depicted in Fig. 4.1. One scenario is a complete branch of the tree from the root node to a leaf. Accordingly, each leaf fully characterizes one scenario. In Fig. 4.1, we find six scenarios. Note that the tree does not need to be symmetric. Associated with each

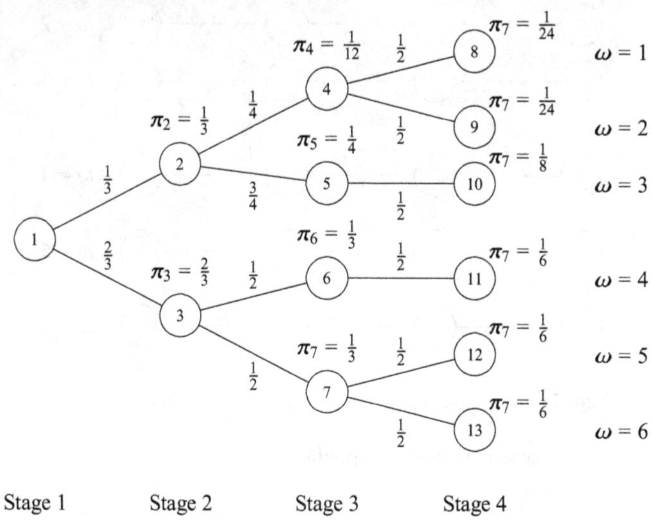

Fig. 4.1 A four-stage scenario tree

node, we have the probability that the sequence of events leads us to that node. In the particular case of the leaves, the probabilities are those associated with the corresponding scenarios. Let $\mathcal{N}$ stand for the set of all nodes in the scenario tree. For each $n \in \mathcal{N} \setminus \{1\}$, let $\Gamma(n)$ be the set that includes n and all its predecessors back to the root node. We can now formulate a general multi-stage (mixed-integer) stochastic programming model:

$$\text{minimize} \quad \sum_{n \in \mathcal{N}} \pi_n \boldsymbol{c}_n^\top \boldsymbol{x}_n, \tag{4.63}$$

$$\text{subject to} \quad \sum_{m \in \Gamma(n)} \boldsymbol{A}_{mn} \boldsymbol{x}_m = \boldsymbol{h}_n \qquad \forall n \in \mathcal{N}, \tag{4.64}$$

$$\boldsymbol{x}_n \in \mathcal{X}_n \subseteq \mathbb{R}_+^{n_1 - p_1} \times \mathbb{Z}_+^{p_1}, \quad \forall n \in \mathcal{N}. \tag{4.65}$$

In this model, $\boldsymbol{x}_n$ represents the decision to make at node $n \in \mathcal{N}$. Above all, the model makes it clear that a decision made at a node depends on the decisions made in all nodes in the (unique) path originated at the root node.

Like in the two-stage case, a successful way to deal with such models is to consider a variable splitting representation. In the multi-stage case, this becomes more involved.

In Fig. 4.2, the scenarios of Fig. 4.1 are represented using replicas of the nodes. Let T be the number of stages. Denote by $\boldsymbol{x}_{\omega,t}$ the decision in stage t of scenario ω.

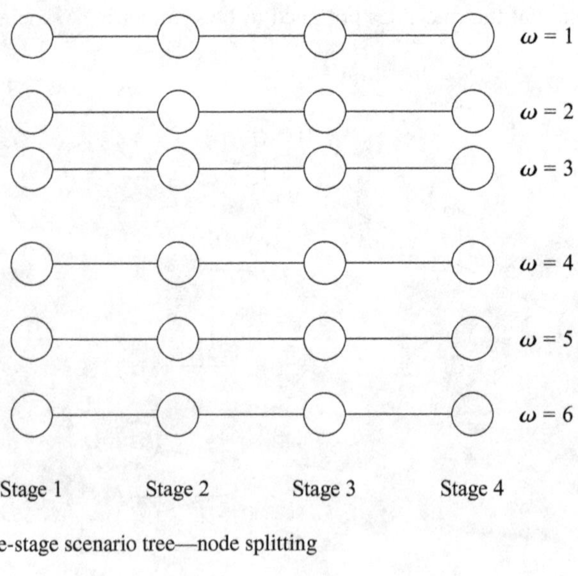

Fig. 4.2 A three-stage scenario tree—node splitting

Likewise, we consider $\boldsymbol{c}_{\omega t}$ and $\boldsymbol{h}_{\omega t}$. We can now reformulate the previous model according to

$$\text{minimize} \quad \sum_{\omega \in \Omega} \sum_{\tau=1}^{T} \pi_\omega \boldsymbol{c}_{\omega t}^\top \boldsymbol{x}_{\omega t}, \tag{4.66}$$

$$\text{subject to} \quad \sum_{\tau=1}^{t} \boldsymbol{A}_{\omega \tau t} \boldsymbol{x}_{\omega t} = \boldsymbol{h}_{\omega t} \qquad \forall t = 1, \ldots, T,\ \forall \omega \in \Omega, \tag{4.67}$$

$$\boldsymbol{x}_{\omega t} \in \mathcal{X}_{\omega t} \subseteq \mathbb{R}_+^{n_1 - p_1} \times \mathbb{Z}_+^{p_1} \qquad \forall t = 1, \ldots, T,\ \forall \omega \in \Omega, \tag{4.68}$$

$$\boldsymbol{B}\boldsymbol{x} = 0. \tag{4.69}$$

Constraint (4.69) is the non-anticipativity constraint. For a scenario tree like the one depicted in Fig. 4.2, these constraints must ensure that for all nodes in stage 1, the decision is the same. Likewise, for the top three nodes in stage 2, the decision again must coincide, the same holding for the bottom three nodes in stage 2, *et cetera.*

If we relax the non-anticipativity constraints, the problem can be decomposed into scenarios. The Lagrangian dual becomes

$$\max_{\boldsymbol{\mu}} \mathcal{L}(\boldsymbol{\mu}) = \sum_{\omega \in \Omega} \min \quad \sum_{t=1}^{T} \pi_\omega \boldsymbol{c}_{\omega t}(\boldsymbol{\mu})^\top \boldsymbol{x}_{\omega, t}, \tag{4.70}$$

$$\text{s. t.} \quad \sum_{\tau=1}^{t} \boldsymbol{A}_{\omega \tau t} \boldsymbol{x}_{\omega t} = \boldsymbol{h}_{\omega t} \qquad \forall t = 1, \ldots, T, \tag{4.71}$$

$$\boldsymbol{x}_{\omega t} \in \mathcal{X}_{\omega t} \subseteq \mathbb{R}_+^{n_1 - p_1} \times \mathbb{Z}_+^{p_1} \quad \forall t = 1, \ldots, T. \tag{4.72}$$

The dual decomposition approach proposed by Carrøe and Schultz [106] that we have already mentioned in the context of two-stage problems can be considered: We embed the Lagrangian dual in a branch-and-bound procedure. Branching is used to ensure non-anticipativity.

One type of heuristic that has proven successful for dealing with multi-stage systems design is the so-called relax-and-fix matheuristic. Its basic version solves the linear relaxation of a large MILP model. Then, some variables are fixed, leading to a restricted MILP model whose linear relaxation is solved again. In turn, this leads to more variables to be fixed. The algorithm proceeds until all the integer variables are fixed to some integer value in their domain. This matheuristic was introduced in [174] and further investigated in [201], where different strategies for relaxing and fixing the integer variables are compared. More recently, this matheuristic has been successfully applied to rather challenging multi-stage decision-making processes, namely those underlying a multi-stage stochastic programming problem. The reader can refer to [7, 198, 200, 291] for several breakthroughs related with the procedure.

As far as multi-stage mixed-integer stochastic programming is concerned, several other references of interest include [196, 248, 259].

More recently, [199] explored the concept of strategic multi-stage operational two-stage stochastic scenario trees in the context of multi-stage stochastic programming. The purpose is to better integrate strategic and operational decisions in multi-stage decision-making processes. There is a strategic multi-stage scenario tree such that in each node an operational two-stage scenario tree is considered. In Sect. 12.2.1 we provide additional details in the context of a capacitated multi-echelon facility location.

One important aspect when considering a multi-stage stochastic model concerns its need. As discussed in Sect. 3.6, in the two-stage stochastic programming, assuming a risk-neutral decision maker, this is often assessed by the so-called value of the stochastic solution. This value can be extended to the multi-stage case in different ways (see, e.g., [195, 578]), leading to the value of the multi-stage stochastic solution. The methodology is often the same: to compute an expected value solution as an approximation to the stochastic problem. This can be accomplished by using a conditional expected value approach. It corresponds to using a dynamic solution to the average scenario as proposed in the above works. The process can be briefly summarized as follows. Compute the expected values for all random variables, and find the corresponding deterministic solution. Fix the first-stage decision found for the deterministic problem. For each node in the subsequent stage of the scenario tree, compute conditional expected values for all random variables in the sub-tree and solve the (deterministic) problem thus induced from that node. Use the solutions to fix the decisions for the current stage. Proceed to the next stage, and iterate likewise until the leaf nodes of the scenario tree are reached. In Chap. 6, we provided further details for a specific problem.

Multi-stage stochastic programming for a risk-averse decision-maker is an important topic, although not much explored in the literature. Homem-de-Mello and Pagnoncelli [273] is a notable exception. Shortly, we should expect significant developments related to this.

4.4 Chance-Constrained Programming

Among the existing approaches for dealing with uncertainty in optimization, we find chance constraints, introduced in [112]. In this case, a solution is acceptable if some constraints are satisfied with a probability greater than or equal to a given threshold. Generically, the problem can be formulated as follows:

$$\text{minimize} \quad \boldsymbol{c}^\top \boldsymbol{x}, \tag{4.73}$$

$$\text{subject to} \quad \mathbb{P}[\boldsymbol{A}\boldsymbol{x} \geq \boldsymbol{b}] \geq 1 - \alpha, \tag{4.74}$$

$$\boldsymbol{x} \in \mathcal{X}, \tag{4.75}$$

where $\boldsymbol{A} \in \mathbb{R}^{m\times n}$ or $\boldsymbol{b} \in \mathbb{R}^m$ contains random components. The vector $\boldsymbol{c}$ may also be stochastic.

For $i = 1, \ldots, m$, denote by $\boldsymbol{a}_i$ the i-th row of $\boldsymbol{A}$. The above model can be equivalently written as

$$\text{minimize} \quad \boldsymbol{c}^\top \boldsymbol{x}, \tag{4.73}$$

$$\text{subject to} \quad \mathbb{P}\left[\boldsymbol{a}_i^\top \boldsymbol{x} \geq b_i,\ i = 1, \ldots, m\right] \geq 1 - \alpha, \tag{4.76}$$

$$\boldsymbol{x} \in \mathcal{X}. \tag{4.75}$$

This is what is called a model with joint chance constraints. Note that (4.76) is equivalent to

$$\mathbb{P}\left[\min_{i=1,\ldots,m} \left\{\boldsymbol{a}_i^\top \boldsymbol{x} - b_i\right\} \geq 0\right] \geq 1 - \alpha, \tag{4.77}$$

which emphasizes the nonlinearity of the joint constraints.

In case $\boldsymbol{c}$ is stochastic, the objective function can itself be expressed as a probability, thus leading to a so-called probabilistic optimization model [95, 528]. To clarify the idea, assume $\boldsymbol{A}$ and $\boldsymbol{b}$ deterministic. The problem to solve is the following:

$$\text{minimize} \quad \mathbb{P}\left[\boldsymbol{c}^\top \boldsymbol{x} \geq \zeta\right], \tag{4.78}$$

$$\text{subject to} \quad \boldsymbol{A}\boldsymbol{x} \geq \boldsymbol{b}, \tag{4.79}$$

$$\boldsymbol{x} \in \mathcal{X}. \tag{4.75}$$

In the above model, ζ is a given cost target or tolerance level. For this reason, the problem is also termed an aspirational optimization problem or a target-based optimization problem [117, 170].

4.4.1 Chance Constraints and Risk Measures

An alternative to the above probabilistic optimization problem (4.78), (4.79), and (4.75) is to minimize the cost tolerance ζ given a probability level $1 - \alpha$. This leads to

$$\text{minimize} \quad \zeta, \tag{4.80}$$

$$\text{subject to} \quad \mathbb{P}\left[\boldsymbol{c}^\top \boldsymbol{x} \leq \zeta\right] \geq 1 - \alpha, \tag{4.81}$$

$$(4.75),\ (4.79),$$

which is closely related to the value-at-risk (VaR), i.e., a percentile of the underlying probability distribution. Sarykalin et al. [474] point out this close relationship. Recall the terminology introduced in Sect. 3.5 with $\mathcal{F}(\boldsymbol{y}, \boldsymbol{\xi})$ denoting a cost function that depends on a decision vector $\boldsymbol{x} \in \mathcal{X}$ as well as on a random vector $\boldsymbol{\xi}$. Sarykalin et al. [474] note that

$$\mathbb{P}\left[\mathcal{F}(\boldsymbol{x}, \boldsymbol{\xi}) \leq \eta\right] \geq 1 - \alpha \Leftrightarrow \mathrm{VaR}_{1-\alpha}(\boldsymbol{x}) \leq \eta.$$

In our case, we conclude that

$$\mathbb{P}\left[\boldsymbol{c}^\top \boldsymbol{x} \leq \zeta\right] \geq 1 - \alpha \Leftrightarrow \mathrm{VaR}_{1-\alpha}(\boldsymbol{x}) \leq \zeta.$$

The non-convexity of VaR w.r.t. $\boldsymbol{x}$ together with the above equivalence shows that the chance constraints in general are non-convex [397, 398]. In practice, the conditional value-at-risk (CVaR) can be employed to achieve a tractable safeguarding approximation for the chance constraints (see again [474]):

$$\mathrm{CVaR}_{1-\alpha}(F(\boldsymbol{x}, \boldsymbol{\xi})) \leq \eta \Rightarrow \mathbb{P}\left[F(\boldsymbol{x}, \boldsymbol{\xi}) \leq \eta\right] \geq 1 - \alpha, \tag{4.82}$$

where

$$\mathrm{CVaR}_{1-\alpha}(F(\boldsymbol{x}, \boldsymbol{\xi})) = \mathbb{E}\left[F(\boldsymbol{x}, \boldsymbol{\xi}) \mid F(\boldsymbol{x}, \boldsymbol{\xi}) > \mathrm{VaR}_{1-\alpha}(F(\boldsymbol{x}, \boldsymbol{\xi}))\right].$$

Chance-constrained programming poses some significant challenges in practice. From a modeling perspective, we must know the underlying probability distribution. This may not be available. Overall, computing the probability function in general is NP-hard. In practice, the viability of tackling the problem often relies on the possibility of finding a tractable deterministic equivalent. We proceed by discussing some cases of relevance in this volume.

4.4.2 Deriving a Deterministic Equivalent

We focus on the situation in which a single chance constraint is involved in the problem, i.e., $m = 1$,

$$\mathbb{P}\left[\boldsymbol{a}^\top \boldsymbol{x} \geq b\right] \geq 1 - \alpha.$$

Assume additionally that b is deterministic. In this case, if $\boldsymbol{a} \sim \mathcal{N}(\boldsymbol{\mu}, \boldsymbol{\Sigma})$ and $\alpha < 0.5$, then the chance constraint has the following deterministic equivalent [259]:

$$\boldsymbol{\mu}'\boldsymbol{x} - \Phi^{-1}(1 - \alpha)\sqrt{\boldsymbol{x}'\boldsymbol{\Sigma}\boldsymbol{x}} \geq b.$$

This means that the problem reduces to a tractable second-order conic programming problem. Unfortunately, when $\boldsymbol{a}$ has a general distribution, multiple integration would be needed to compute the chance constraints, which prevents their practical use.

Consider the more general case in which b is also stochastic. In case a sample $(\boldsymbol{a}^1, b^1), \ldots, (\boldsymbol{a}^S, b^S)$ is available, we can go further in terms of deriving a potentially tractable optimization model. First, note that

$$\mathbb{P}\left[\boldsymbol{a}^\top \boldsymbol{x} \geq b\right] = \mathbb{E}\left[\mathbb{I}\left(\boldsymbol{a}^\top \boldsymbol{x} \geq b\right)\right],$$

where $\mathbb{I}(\cdot)$ is the indicator function. In turn, the expected value can be estimated using the above sample by computing the average

$$\frac{1}{S}\sum_{s=1}^{S} z_s, \quad \text{with} \quad z_s = \begin{cases} 1, & b^s - \boldsymbol{x}^\top \boldsymbol{a}^s \leq 0, \\ 0, & \text{otherwise.} \end{cases} \qquad s = 1, \ldots, S.$$

Hence, the single-constrained version of chance-constrained problem (4.73), (4.76), (4.75) can be approximated by the MILP

$$\text{minimize} \quad \boldsymbol{c}^\top \boldsymbol{x}, \tag{4.73}$$

$$\text{subject to} \quad \frac{1}{S}\sum_{s=1}^{S} z_s \geq 1 - \alpha \tag{4.83}$$

$$b^s - \boldsymbol{x}^\top \boldsymbol{a}^s \leq \Theta(1 - z_s) \quad \forall s = 1, \ldots, S, \tag{4.84}$$

$$z_s \in \{0, 1\} \quad \forall s = 1, \ldots, S, \tag{4.85}$$

$$\boldsymbol{x} \in \mathcal{X}, \tag{4.75}$$

with Θ denoting a sufficiently large number. Naturally, the size of the above MILP problem grows fast with the sample size.

The reasoning can be extended to the case with multiple joint chance constraints (4.74). For each $i = 1, \ldots, m$, let us assume that a sample $(\boldsymbol{a}_i^1, b_i^1), \ldots, (\boldsymbol{a}_i^S, b_i^S)$ is available for $(\boldsymbol{a}_i, b_i)$. Given the natural growth of the above MILP model, we explore an approximation alternative as mentioned in (4.82).

Recall that (4.76) and (4.77) are equivalent. In turn, the latter—leveraging the relation (4.82)—can be approximated by the corresponding CVaR constraint:

$$\text{CVaR}_{1-\alpha}\left(\max_{i=1,\ldots,m}\left\{b_i - \boldsymbol{x}^\top \boldsymbol{a}_i\right\}\right) \leq 0.$$

Given the CVaR reformulation (3.4)–(3.7), we can rewrite (also considering the sample) the CVaR constraint

$$\min_{\eta}\left\{\eta+\frac{1}{\alpha}\mathbb{E}\left[\max_{i=1,\ldots,m}\left\{b_i-\boldsymbol{x}^\top\boldsymbol{a}_i\right\}-\eta\right]^+\right\}\leq 0$$

as a linear system:

$$\eta+\frac{1}{\alpha}\frac{1}{S}\sum_{s=1}^{S}\gamma_s\leq 0, \tag{4.86}$$

$$\gamma_s\geq 0 \qquad \forall s=1,\ldots,S, \tag{4.87}$$

$$\gamma_s\geq\max_{i=1,\ldots,m}\left\{b_i^s-\boldsymbol{x}^\top\boldsymbol{a}_i^s\right\}-\eta \qquad \forall s=1,\ldots,S, \tag{4.88}$$

$$\eta\in\mathbb{R}. \tag{4.89}$$

The approximate model that is a linear programming now becomes straightforward:

$$\begin{aligned}\text{minimize}\quad & (4.73),\\ \text{subject to}\quad & (4.75),\ (4.86)\text{–}(4.89).\end{aligned}$$

More detailed discussions on approximations of chance-constrained optimization can be found in [5, 96, 120, 289, 397, 398]. Recent advances also consider the exact and approximate reformulations for distributionally robust chance constraints, which can be found in [97, 119, 290, 321, 498, 546, 547, 581].

4.5 Distributionally Robust Optimization

In Sects. 4.3 and 4.4, we assumed that uncertainty can be quantified using some specific probability distribution. In practice, one often has limited information about the stochastic parameters.

To address these issues, a robust formulation for stochastic programming was first proposed by Scarf [480]. The idea is to define a so-called ambiguity set $\mathcal{F}$ of possible probability distributions that are assumed to include the true one. The objective function is then reformulated concerning the worst-case expected cost over the choice of distributions in this set. A Distributionally Robust Optimization (DRO) program emerges, i.e.,

$$\underset{\boldsymbol{x}\in\mathcal{X}}{\text{minimize}}\left(\sup_{\mathbb{P}\in\mathcal{F}}\mathbb{E}_{\mathbb{P}}[h(\boldsymbol{x},\boldsymbol{\xi})]\right), \tag{4.90}$$

where $h(\boldsymbol{x},\boldsymbol{\xi})$ stands for a convex cost function in $\boldsymbol{x}$.

In this section, we briefly introduce several common types of ambiguity sets $\mathcal{F}$ for DRO models: (i) moment-based, (ii) Wasserstein, and (iii) event-wise ambiguity sets. We also introduce several other modeling frameworks for optimization under ambiguity, including robust satisficing and globalized DRO. In particular, we discuss how to derive tractable reformulations for these models. For more details, the interested reader should refer to the literature which includes [76, 160, 224, 238, 322, 340, 350, 385, 444, 543].

4.5.1 Moment-Based Ambiguity

One possibility for dealing with ambiguity in the distribution driving the uncertain parameters in an optimization problem consists of relying on empirical estimates of the mean $\boldsymbol{\mu}_0$ and covariance matrix $\boldsymbol{\Sigma}_0$ of the distribution. However, even in such cases, hardly can one be entirely confident in these estimates. For this reason, one can represent uncertainty using a parameterized set [160]:

$$(\mathbb{E}[\boldsymbol{\xi}] - \mu_0)^\top \boldsymbol{\Sigma}_0^{-1} (\mathbb{E}[\boldsymbol{\xi}] - \mu_0) \leq \gamma_1, \tag{4.91}$$

$$\mathbb{E}\left[(\boldsymbol{\xi} - \boldsymbol{\mu}_0)(\boldsymbol{\xi} - \boldsymbol{\mu}_0)^\top\right] \preceq \gamma_2 \Sigma_0. \tag{4.92}$$

Inequality (4.91) establishes that the mean of $\boldsymbol{\xi}$ lies within an ellipsoid centered at the estimate μ_0, whereas (4.92) imposes that the covariance matrix $\mathbb{E}\left[(\boldsymbol{\xi} - \boldsymbol{\mu}_0)(\boldsymbol{\xi} - \boldsymbol{\mu}_0)^\top\right]$ lies in a bounded positive semi-definite cone. In other words, it describes how likely $\boldsymbol{\xi}$ is close to $\boldsymbol{\mu}_0$ in terms of the correlations expressed in $\boldsymbol{\Sigma}_0$. The parameters γ_1 and γ_2 provide natural means of quantifying the confidence one has in $\boldsymbol{\mu}_0$ and $\boldsymbol{\Sigma}_0$, respectively.

The above conditions induce the ambiguity set that we can detail as follows:

$$\mathcal{F} = \left\{ \mathbb{P} \in \mathcal{P}(\mathbb{R}^m) \left| \begin{array}{l} \mathbb{P}(\xi \in \mathcal{S}) = 1 \\ \left(\mathbb{E}_{\mathbb{P}}[\boldsymbol{\xi}] - \boldsymbol{\mu}_0\right)^\top \boldsymbol{\Sigma}_0^{-1} \left(\mathbb{E}_{\mathbb{P}}[\boldsymbol{\xi}] - \boldsymbol{\mu}_0\right) \leq \gamma_1 \\ \mathbb{E}_{\mathbb{P}}\left[(\boldsymbol{\xi} - \boldsymbol{\mu}_0)(\boldsymbol{\xi} - \boldsymbol{\mu}_0)^\top\right] \preceq \gamma_2 \Sigma_0 \end{array} \right. \right\}. \tag{4.93}$$

In this set, $\mathcal{P}$ is the space of probability distributions on $\mathbb{R}^m$ and $\mathcal{S} \subseteq \mathbb{R}^m$ stands for a closed convex set known to contain the support of $\mathbb{P}$. As pointed out by Delage and Ye [160], the above set can be seen as a generalization of many previously proposed sets (e.g., exact mean and support constraints or mean and covariance matrix).

A major issue when it comes to tackling the DRO problem regards the inner problem

$$\underset{\mathbb{P} \in \mathcal{F}}{\text{maximize}} \ \mathbb{E}_{\mathbb{P}}[h(\boldsymbol{x}, \boldsymbol{\xi})], \tag{4.94}$$

for a fixed $\boldsymbol{x}$. This problem can be formulated as a conic linear problem. Assuming $h(\boldsymbol{x}, \boldsymbol{\xi})$ $\mathbb{P}$-integrable for all $\mathbb{P} \in \mathcal{F}, \gamma_1, \gamma_2 \geq 0$, and $\boldsymbol{\Sigma}_0 \succ 0$, it can be proved that the optimal value to the above problem is equal to the optimal value of the following problem:

$$\underset{\boldsymbol{Q},\boldsymbol{q},r,t}{\text{minimize}} \quad r + t \tag{4.95}$$

$$\text{subject to} \quad r \geq h(\boldsymbol{x}, \boldsymbol{\xi}) - \boldsymbol{\xi}^\top \boldsymbol{Q}\boldsymbol{\xi} - \boldsymbol{\xi}^\top \boldsymbol{q} \quad \forall \boldsymbol{\xi} \in \mathcal{S} \tag{4.96}$$

$$t \geq \left(\gamma_2 \boldsymbol{\Sigma}_0 + \boldsymbol{\mu}_0 \boldsymbol{\mu}_0^\top\right) \bullet \boldsymbol{Q} + \boldsymbol{\mu}_0^\top \boldsymbol{q} + \sqrt{\gamma_1} \left\| \boldsymbol{\Sigma}_0^{1/2} \left(\boldsymbol{q} + 2\boldsymbol{Q}\boldsymbol{\mu}_0\right) \right\|, \tag{4.97}$$

$$\boldsymbol{Q} \succeq 0, \tag{4.98}$$

where $(A \bullet B)$ refers to the Frobenius inner product between matrices, $\boldsymbol{Q} \in \mathbb{R}^{m \times m}$ is a symmetric matrix, the vector $\boldsymbol{q} \in \mathbb{R}^m$, and $r, t \in \mathbb{R}$. In addition, there exists a set of real-valued assignments for $(\boldsymbol{Q}, \boldsymbol{q}, r, t)$ that achieves optimality for problem (4.95)–(4.98). A complete proof can be found in [160]. The same authors show that under some mild assumptions regarding the support set $\mathcal{S}$ and function $h(\boldsymbol{x}, \boldsymbol{\xi})$ the above problem is a convex optimization problem that can be solved up to any precision ϵ in a time that is polynomial in $\log(1/\epsilon)$ and in the size of the problem. The ellipsoid method is an algorithm ensuring the complexity.

Having found a way to deal with the inner problem, we turn our attention to the actual DRO problem of interest, namely

$$\underset{x \in \mathcal{X}}{\text{minimize}} \left(\max_{\mathbb{P} \in \mathcal{F}} \mathbb{E}_{\boldsymbol{\xi}}[h(\boldsymbol{x}, \boldsymbol{\xi})] \right). \tag{4.99}$$

We assume that set $\mathcal{X}$ is convex and compact (closed and bounded). Additionally, we assume that function $h(\boldsymbol{x}, \boldsymbol{\xi})$ is convex in $\boldsymbol{x}$. Under a set of additional mild conditions (see [160]), the above DRO problem is equivalent to

$$\underset{\boldsymbol{x},\boldsymbol{Q},\boldsymbol{q},r,t}{\text{minimize}} \quad r + t \tag{4.100}$$

$$\text{subject to} \quad r \geq h_k(\boldsymbol{x}, \boldsymbol{\xi}) - \boldsymbol{\xi}^\top \boldsymbol{Q}\boldsymbol{\xi} - \boldsymbol{\xi}^\top \boldsymbol{q} \quad \forall \boldsymbol{\xi} \in \mathcal{S},\ k = 1, \ldots, K, \tag{4.101}$$

$$t \geq \left(\gamma_2 \Sigma_0 + \mu_0 \mu_0^\top\right) \bullet \boldsymbol{Q} + \mu_0^\top \boldsymbol{q} + \sqrt{\gamma_1} \left\| \Sigma_0^{1/2} \left(\boldsymbol{q} + 2\boldsymbol{Q}\mu_0\right) \right\|, \tag{4.102}$$

$$\boldsymbol{Q} \succeq 0, \tag{4.103}$$

$$x \in \mathcal{X}, \tag{4.104}$$

which is a convex optimization problem that can be solved up to any precision ϵ in a time that is polynomial in $\log(1/\epsilon)$ and in $\boldsymbol{x}$ and $\boldsymbol{\xi}$.

Furthermore, we introduce the DRO model under moment-based ambiguity set with lifting proposed by Wiesemann et al. [543]. Accordingly, we consider an ambiguity set $\mathcal{F}$ in the following standard form:

$$\mathcal{F} = \left\{ \mathbb{P} \in \mathcal{P}\left(\mathbb{R}^P \times \mathbb{R}^Q\right) \middle| \begin{array}{l} \mathbb{E}_{\mathbb{P}}[\boldsymbol{A}\boldsymbol{\xi} + \boldsymbol{B}\boldsymbol{v}] = \boldsymbol{b}, \\ \mathbb{P}\left[(\boldsymbol{\xi}, \boldsymbol{v}) \in \mathcal{C}_\ell\right] \in [\underline{\boldsymbol{p}}_\ell, \overline{\boldsymbol{p}}_\ell] \quad \forall \ell \in \mathcal{L} \end{array} \right\}. \tag{4.105}$$

In this set, $\mathcal{P}(\mathbb{R}^P \times \mathbb{R}^Q)$ stands for the space of non-negative measures and probability distributions on $\mathbb{R}^P \times \mathbb{R}^Q$ and $\mathbb{P}$ represents a joint probability distribution of the random vector $\boldsymbol{\xi} \in \mathbb{R}^P$ appearing in function $h(\boldsymbol{x}, \boldsymbol{\xi})$ and some auxiliary random vector $\boldsymbol{v} \in \mathbb{R}^Q$. We assume that $\boldsymbol{A} \in \mathbb{R}^{K \times P}$, $\boldsymbol{B} \in \mathbb{R}^{K \times Q}$, $\boldsymbol{b} \in \mathbb{R}^K$, and $\mathcal{L} = \{1, \ldots, L\}$. The so-called confidence sets $\mathcal{C}_\ell$ are defined as

$$\mathcal{C}_\ell = \left\{ (\hat{\boldsymbol{\xi}}, \hat{\boldsymbol{v}}) \in \mathbb{R}^P \times \mathbb{R}^Q : \boldsymbol{C}_\ell \hat{\boldsymbol{\xi}} + \boldsymbol{D}_\ell \hat{\boldsymbol{v}} \preccurlyeq_{\mathcal{K}_\ell} \boldsymbol{c}_\ell \right\} \tag{4.106}$$

with $\boldsymbol{C}_\ell \in \mathbb{R}^{L_\ell \times P}$, $\boldsymbol{D}_\ell \in \mathbb{R}^{L_\ell \times Q}$, $\boldsymbol{c}_\ell \in \mathbb{R}^{L_\ell}$, and $\mathcal{K}_\ell$ being proper cones. Additionally, it is assumed that $\underline{\boldsymbol{p}}_\ell, \overline{\boldsymbol{p}}_\ell \in [0, 1]$ and $\underline{\boldsymbol{p}}_\ell \leq \overline{\boldsymbol{p}}_\ell$ for all $\ell \in \mathcal{L}$.

The following regularity conditions are assumed in [543]: (1) the confidence set $\mathcal{C}_L$ is bounded and has probability 1, that is, $\underline{\boldsymbol{p}}_L = \overline{\boldsymbol{p}}_L = 1$, and (2) there is a distribution $\mathbb{P} \in \mathcal{F}$ such that $\mathbb{P}\left[(\boldsymbol{\xi}, \boldsymbol{v}) \in \mathcal{C}_\ell\right] \in (\underline{\boldsymbol{p}}_\ell, \overline{\boldsymbol{p}}_\ell)$ whenever $\underline{\boldsymbol{p}}_\ell < \overline{\boldsymbol{p}}_\ell$, $\ell \in \mathcal{L}$.

The first condition ensures that the confidence set with the highest index, $\mathcal{C}_L$, contains the support of the joint random vector $(\boldsymbol{\xi}, \boldsymbol{v})$. The second condition stipulates that there is a probability distribution $\mathbb{P} \in \mathcal{F}$ satisfying the bounds in (4.105) as strict inequalities whenever the corresponding interval $[\underline{\boldsymbol{p}}_\ell, \overline{\boldsymbol{p}}_\ell]$ is non-degenerate. This assumption allows exploiting the strong duality results from [281] and [490] to reformulate a distributionally robust counterpart.

The ambiguity set $\mathcal{F}$ in (4.105) specifies joint probability distributions for the uncertain problem parameters $\boldsymbol{\xi}$ and an auxiliary random vector $\boldsymbol{v}$ that does not explicitly appear in (4.90). The inclusion of such an auxiliary random vector allows for capturing a rich set of structural information about the marginal distribution of $\boldsymbol{\xi}$ in a unified manner.

When designing the ambiguity set, one should embed all the available information about the correct marginal distribution of $\boldsymbol{\xi}$. In other words, $\mathcal{F}$ should be chosen as the tightest ambiguity set for which one knows with certainty that such marginal distribution is in $\Pi_{\boldsymbol{\xi}} \mathcal{F} = \cup_{\mathbb{P} \in \mathcal{F}} \left\{ \Pi_{\boldsymbol{\xi}} \mathbb{P} \right\}$, where $\Pi_{\boldsymbol{\xi}} \mathbb{P}$ stands for the marginal distribution of $\boldsymbol{\xi}$ under probability distribution $\mathbb{P}$.

Regarding function $h(\boldsymbol{x}, \boldsymbol{\xi})$, [543] take it as being convex and piecewise affine in the decision variables $\boldsymbol{x}$ and the random vector $\boldsymbol{\xi}$:

$$h(\boldsymbol{x}, \boldsymbol{\xi}) = \max_{j \in \mathcal{J}} h_j(\boldsymbol{x}, \boldsymbol{\xi}),$$

with $\mathcal{J} = \{1, \ldots, n\}$, and the auxiliary functions $h_j : \mathbb{R}^N \times \mathbb{R}^P \to \mathbb{R}$ are of the form

$$h_j(\boldsymbol{x}, \boldsymbol{\xi}) = \boldsymbol{s}_j(\boldsymbol{\xi})^\top \boldsymbol{x} + t_j(\boldsymbol{\xi})$$

with $\boldsymbol{s}_j(\boldsymbol{\xi}) = \boldsymbol{S}_j\boldsymbol{\xi} + \boldsymbol{s}_j$ $\boldsymbol{S}_j \in \mathbb{R}^{N\times P}$ and $\boldsymbol{s}_j \in \mathbb{R}^N$, $t_j(\boldsymbol{\xi}) = \boldsymbol{t}_j^\top \boldsymbol{\xi} + t_j$, $\boldsymbol{t}_j \in \mathbb{R}^P$ and $t_j \in \mathbb{R}$.

The same authors show that under the above condition as well as under the two regularity conditions mentioned, it becomes possible to efficiently solve optimization problems involving distributionally robust counterpart of the type (4.90) if and only if the ambiguity set $\mathcal{F}$ satisfies a nesting condition, namely the following one:

$$\forall \ell, \ell' \in \mathcal{L}, \ell \neq \ell', \text{ we have either } C_\ell \Subset C_{\ell'},\ C_{\ell'} \Subset C_\ell \text{ or } C_i \cap C_{i'} = \emptyset,$$

with "$\Subset$" standing for strict inclusion. This condition induces a strict partial order on the confidence sets C_ℓ concerning the $\Subset$-relation with the additional requirement that incomparable sets must be disjoint. Recall that the relation $C_\ell \Subset C_{\ell'}$ can be checked efficiently if (i) C_ℓ is an ellipsoid and $C_{\ell'}$ is the intersection of finitely many half-spaces or ellipsoids or (ii) both C_ℓ and $C_{\ell'}$ are polyhedral. If both C_ℓ and $C_{\ell'}$ are described by intersections of finitely many half-spaces or ellipsoids, then we efficiently verify a sufficient condition for $C_\ell \Subset C_{\ell'}$ (see [543]).

Let $\mathcal{A}(\ell) = \{\ell\} \cup \left\{\ell' \in \mathcal{L} : C_\ell \Subset C_{\ell'}\right\}$ and $\mathcal{D}(\ell) = \left\{\ell' \in \mathcal{L} : C_{\ell'} \Subset C_\ell\right\}$ be the index sets of all supersets (antecedents) and all strict subsets (descendants) of C_ℓ, respectively. Under the two regularity conditions above presented together with the detailed condition for $h(\boldsymbol{x}, \boldsymbol{\xi})$, the problem (4.90) has a tractable reformulation if the above nesting condition holds. In particular, it is equivalent to the following MILP:

$$\text{minimize} \quad \boldsymbol{b}^\top \boldsymbol{\beta} + \sum_{\ell\in\mathcal{L}} \left[\overline{p}_\ell \kappa_\ell - \underline{p}_\ell \lambda_\ell\right], \tag{4.107}$$

$$\text{subject to} \quad \boldsymbol{c}_\ell^\top \boldsymbol{\phi}_{\ell j} + \boldsymbol{s}_j^\top \boldsymbol{x} + t_j \leq \sum_{\ell'\in\mathcal{A}(\ell)} [\kappa_{\ell'} - \lambda_{\ell'}] \qquad \forall \ell \in \mathcal{L}, \forall j \in \mathcal{J}, \tag{4.108}$$

$$\boldsymbol{C}_\ell^\top \boldsymbol{\phi}_{\ell j} + \boldsymbol{A}^\top \boldsymbol{\beta} = \boldsymbol{S}_j^\top \boldsymbol{x} + \boldsymbol{t}_j \qquad \forall \ell \in \mathcal{L}, \forall j \in \mathcal{J}, \tag{4.109}$$

$$\boldsymbol{D}_\ell^\top \boldsymbol{\phi}_{\ell j} + \boldsymbol{B}^\top \boldsymbol{\beta} = 0 \qquad \forall \ell \in \mathcal{L}, \forall j \in \mathcal{J}, \tag{4.110}$$

$$\boldsymbol{\beta} \in \mathbb{R}^K, \boldsymbol{\kappa}, \boldsymbol{\lambda} \in \mathbb{R}_+^L, \boldsymbol{\phi}_{\ell j} \in \mathcal{K}_\ell^* \qquad \forall \ell \in \mathcal{L}, \forall j \in \mathcal{J}, \tag{4.111}$$

$$\boldsymbol{x} \in \mathcal{X}, \tag{4.112}$$

where $\mathcal{K}_\ell^*$ is the dual cone of $\mathcal{K}_\ell$, $\forall \ell \in \mathcal{L}$.

If the confidence set C_ℓ is described by linear, conic-quadratic, or semidefinite inequalities, then the above model provides a linear, conic-quadratic, or semidefinite reformulation of the DRO model (4.90), respectively. Hence, the nesting condition discussed is sufficient for the tractability of the DRO model (4.90). Unfortunately, if that condition is violated, model (4.90) can be strongly NP-hard.

Despite the apparent simplicity of expectation conditions and probability bounds, ambiguity sets of the type (4.105) are powerful because they generalize many uncertainty sets used in classical robust optimization.

The expected value of $\boldsymbol{\xi}$ can easily be set equal to any value considered of interest through the adequate instantiation of the expectation condition in (4.105). Nevertheless, the ambiguity set we are considering allows capturing (even if just partially) information about certain higher order moments of $\boldsymbol{\xi}$. This can be accomplished by a lifting technique that we recall next.

Following the terminology introduced by Wiesemann et al. [543] we say that the *$\mathcal{K}$-epigraph* of a function $\boldsymbol{f} : \mathbb{R}^M \to \mathbb{R}^N$ and a proper cone $\mathcal{K}$ is *conic representable* if the set $\left\{(\boldsymbol{x}, \boldsymbol{y}) \in \mathbb{R}^M \times \mathbb{R}^N : \boldsymbol{f}(\boldsymbol{x}) \preccurlyeq_{\mathcal{K}} \boldsymbol{y}\right\}$ can be written using conic inequalities, possibly involving a cone different from $\mathcal{K}$ and additional auxiliary variables. We have the following lifting theorem.

Theorem 4.3 *Let $\boldsymbol{f} \in \mathbb{R}^M$ and $\boldsymbol{g} : \mathbb{R}^P \to \mathbb{R}^M$ be a function with a conic representable $\mathcal{K}$-epigraph, and consider the ambiguity set*

$$\mathcal{G} = \left\{ \mathbb{P} \in \mathcal{P}\left(\mathbb{R}^P\right) \middle| \begin{array}{l} \mathbb{E}_{\mathbb{P}}[\boldsymbol{g}(\boldsymbol{\xi})] \preccurlyeq_{\mathcal{K}} \boldsymbol{f} \\ \mathbb{P}\left[\boldsymbol{\xi} \in C_\ell\right] \in [\underline{p}_\ell, \overline{p}_\ell] \quad \forall \ell \in \mathcal{L} \end{array} \right\}, \tag{4.113}$$

as well as the lifted ambiguity set

$$\mathcal{F} = \left\{ \mathbb{P} \in \mathcal{P}\left(\mathbb{R}^P \times \mathbb{R}^M\right) \middle| \begin{array}{l} \mathbb{E}_{\mathbb{P}}[\boldsymbol{v}] = \boldsymbol{f}, \\ \mathbb{P}[\boldsymbol{g}(\boldsymbol{\xi}) \preccurlyeq_{\mathcal{K}} \boldsymbol{v}] = 1, \\ \mathbb{P}\left[\boldsymbol{\xi} \in C_\ell\right] \in [\underline{p}_\ell, \overline{p}_\ell] \quad \forall \ell \in \mathcal{L} \end{array} \right\},$$

which involves the auxiliary random vector $\boldsymbol{v} \in \mathbb{R}^M$.

The detailed proof can be found in [543]. Theorem 4.3 shows that (i) $\mathcal{G} = \Pi_{\boldsymbol{\xi}}\mathcal{F}$ and (ii) $\mathcal{F}$ can be reformulated as an instance of the standardized moment-based ambiguity set (4.105).

This relevant result allows identifying several ambiguity sets from the literature as particular cases of the ambiguity set (4.105). This includes mean-based, variance-based, and high-order moment-based ambiguity sets. For instance, by defining the function $\boldsymbol{g}(\boldsymbol{\xi})$ to be linear, it becomes possible to specify ambiguity sets that impose conic constraints on the mean value of $\boldsymbol{\xi}$.

4.5.2 Wasserstein Metric Ambiguity Set

We focus now on a type of ambiguity set of great relevance in practice based on the Wasserstein metric. Next, we closely follow [385].

Probability metrics are a means to quantify distances in the space of probability distributions. A popular one is the Wasserstein metric, which is defined on the space $\mathcal{P}(\Xi)$ of all probability distributions $\mathbb{P}$ supported on set Ξ with $\mathbb{E}_{\mathbb{P}}[\|\xi\|] = \int_{\Xi} \|\boldsymbol{\xi}\| \mathrm{d}\mathbb{P}(\boldsymbol{\xi}) < \infty$. To make this text self-contained, we review its definition [299].

Definition 4.6 The Wasserstein metric $d_{\mathrm{W}} : \mathcal{P}(\Xi) \times \mathcal{P}(\Xi) \to \mathbb{R}_+$ is defined as

$$d_{\mathrm{W}}(\mathbb{P}_1, \mathbb{P}_2) = \inf \left\{ \int_{\Xi^2} \|\boldsymbol{\xi}_1 - \boldsymbol{\xi}_2\| \, \mathbb{Q}\left(\mathrm{d}\boldsymbol{\xi}_1, \mathrm{d}\boldsymbol{\xi}_2\right) \middle| \begin{array}{l} \mathbb{Q} \text{ is a joint distribution of } \boldsymbol{\xi}_1 \text{ and } \boldsymbol{\xi}_2 \\ \text{with marginals } \mathbb{P}_1 \text{ and } \mathbb{P}_2, \text{ respectively} \end{array} \right\}$$

for all distributions $\mathbb{P}_1, \mathbb{P}_2 \in \mathcal{P}(\Xi)$, where $\|\cdot\|$ represents an arbitrary norm on $\mathbb{R}^m$.

A generalized p-Wasserstein metric for $p \geq 1$ is obtained by taking $\|\boldsymbol{\xi}_1 - \boldsymbol{\xi}_2\|^p$, instead of $\|\boldsymbol{\xi}_1 - \boldsymbol{\xi}_2\|$. In the context of this volume, unless otherwise said, we focus exclusively on the 1-Wasserstein metrics defined above. Note that it is sometimes also referred to as the Kantorovich–Rubinstein metric.

The Wasserstein metric d_{W} can be used to construct an ambiguity set in the context of DRO. Let us consider a set of the past realizations of the underlying random vector $\hat{\Xi}_N = \left\{\hat{\boldsymbol{\xi}}_1, \ldots, \hat{\boldsymbol{\xi}}_N\right\} \subseteq \Xi$. The induced empirical probability distribution is given by

$$\widehat{\mathbb{P}}_N = \frac{1}{N} \sum_{i=1}^{N} \delta_{\hat{\boldsymbol{\xi}}_i},$$

where $\delta_{\hat{\boldsymbol{\xi}}_i}$ is the Dirac distribution taken on $\hat{\boldsymbol{\xi}}_i$ with mass one ($i = 1, \ldots, N$). Hence, the empirical distribution reduces to the uniform distribution on $\widehat{\Xi}_N$. An ambiguity set can be defined as

$$\mathcal{F}(\varepsilon) = \left\{\mathbb{P} \in \mathcal{P}(\Xi) : d_{\mathrm{W}}\left(\widehat{\mathbb{P}}_N, \mathbb{P}\right) \leq \varepsilon\right\}, \tag{4.114}$$

which can be viewed as the Wasserstein ball of radius ε centered at the empirical distribution $\widehat{\mathbb{P}}_N$.

Next, we recall that the inner worst-case expectation problem in the DRO model (4.90) over this ambiguity set can be reformulated as a finite convex program for many functions $h(\boldsymbol{x}, \boldsymbol{\xi})$ of practical interest. Let us consider that $\boldsymbol{x}$ is fixed. The inner problem is defined by

$$\sup_{\mathbb{P} \in \mathcal{F}(\varepsilon)} \mathbb{E}_{\mathbb{P}}[h(\boldsymbol{x}, \boldsymbol{\xi})]. \tag{4.115}$$

Given that $\boldsymbol{x}$ is fixed, we are dealing with a cost function $g(\boldsymbol{\xi})$. Let us assume that $g(\boldsymbol{\xi})$ is defined as the pointwise maximum of more elementary measurable functions $g_\ell : \mathbb{R}^m \to \overline{\mathbb{R}}$, i.e., $g(\boldsymbol{\xi}) = \max_{\ell=1,\dots,L} \{g_\ell(\boldsymbol{\xi})\}$.

The worst-case expectation problem (4.115) constitutes an infinite-dimensional optimization problem over probability distributions and thus appears to be intractable. However, under the assumption that $g(\boldsymbol{\xi})$ can be written as the maximum of concave functions, [385] show that it is equal to the optimal value of the following finite convex program:

$$
\begin{aligned}
\inf_{\lambda, s, z, v} \quad & \lambda \varepsilon + \frac{1}{N} \sum_{j=1}^{N} s_j, \\
\text{s. t.} \quad & [-g_\ell]^* \left(z_{j\ell} - v_{j\ell}\right) + \sigma_\Xi \left(v_{j\ell}\right) - \left\langle z_{j\ell}, \hat{\boldsymbol{\xi}}_j \right\rangle \le s_j \\
& \qquad\qquad\qquad\qquad \forall j = 1, \dots, N,\ \ell = 1, \dots, L, \\
& \left\| z_{j\ell} \right\|_* \le \lambda, \qquad\qquad \forall j = 1, \dots, N,\ \ell = 1, \dots, L.
\end{aligned}
$$

In the above model, $[-g_\ell]^* \left(z_{j\ell} - v_{j\ell}\right)$ denotes the conjugate of $(-g_\ell)$ evaluated at $z_{j\ell} - v_{j\ell}$[1] and $\left\| z_{j\ell} \right\|_*$ the dual norm of $z_{l\ell}$.[2] Moreover, σ_Ξ is the conjugate of the characteristic function of Ξ, i.e., the support function of Ξ.

4.5.3 Event-wise Ambiguity Set

Next, we briefly review event-wise robust distributionally optimization [121]. The event-wise robust stochastic optimization (RSO) framework extends the moment-based and distance-based distributionally robust optimization frameworks [54, 76, 160, 224, 385, 543] mentioned in Sects. 4.5.1 and 4.5.2. It combines scenario-tree-based stochastic linear optimization and distributionally robust optimization in a unified framework. Moreover, the uncertainty associated with the event-wise RSO model comprises discrete and continuous random variables.

Let s denote a discrete random scenario taking values in set $\{1, \dots, S\}$ and $\boldsymbol{\xi}$ represent a continuous random vector with outcomes in $\mathbb{R}^I$.

Conditioning on a scenario $s \in \{1, \dots, S\}$, the support set of the random vector $\boldsymbol{\xi}$ is tractable and conic representable. We denote it by Ξ_s. The joint probability distribution of $(\boldsymbol{\xi}, s)$ is denoted by $\mathbb{P} \in \mathcal{F}$, whereas $\mathcal{F}$ is the ambiguity set of

---

[1] $g^*(\boldsymbol{\zeta}) = \sup_{\boldsymbol{\xi}} \langle \boldsymbol{\zeta}, \boldsymbol{\xi} \rangle - g(\boldsymbol{\xi})$.

[2] $\|\boldsymbol{\zeta}\|_* = \sup_{\|\boldsymbol{\xi}\| \le 1} \langle \boldsymbol{\zeta}, \boldsymbol{\xi} \rangle$.

probability distributions that share some identical distributional information. The so-called event-wise ambiguity is stated as follows:

$$\mathcal{F} = \left\{ \mathbb{P} \in \mathcal{P}\left(\mathbb{R}^I \times \{1, \ldots, S\}\right) \;\middle|\; \begin{array}{ll} (\tilde{\boldsymbol{\xi}}, \tilde{s}) \sim \mathbb{P} & \\ \mathbb{E}_{\mathbb{P}}\left[\tilde{\boldsymbol{\xi}} \mid \tilde{s} \in \mathcal{S}_k\right] \in \mathcal{L}_k, & \forall k = 1, \ldots, K \\ \mathbb{P}\left[\tilde{\boldsymbol{\xi}} \in \Xi_s \mid \tilde{s} = s\right] = 1, \; \forall s = 1, \ldots, S & \\ \mathbb{P}[\tilde{s} = s] = p_s, & \forall s = 1, \ldots, S \end{array} \right\} \tag{4.116}$$

for given events $\mathcal{S}_k$, $k = 1, \ldots, K$, and given closed and convex sets Ξ_s, $s = 1, \ldots, S$, and $\mathcal{P} \subseteq \left\{ \boldsymbol{p} \in \mathbb{R}^S_{++} \mid \sum_{s=1}^S p_s = 1 \right\}$. The random variable $\tilde{s}$ represents a set of scenarios whose realization probabilities may be uncertain. For each scenario, the support of the random variable $\tilde{\boldsymbol{\xi}}$ may differ. The expectation of $\tilde{\boldsymbol{\xi}}$ conditional on the event realization may also vary from one event to another.

Consider an objective function $h(\boldsymbol{x}, \tilde{\boldsymbol{\xi}}, \tilde{s})$ and an event-wise ambiguity set (4.116), and we can formulate the associated event-wise RSO model as

$$\sup_{\mathbb{P} \in \mathcal{F}} \mathbb{E}_{\mathbb{P}}\left[h(\boldsymbol{x}, \tilde{\boldsymbol{\xi}}, \tilde{s})\right]. \tag{4.117}$$

We can effectively determine the worst-case expectation over the event-wise ambiguity set $\mathcal{F}$ by solving a classical robust optimization problem. Assuming that Slater's condition holds, the worst-case expectation (4.117) can be reformulated as the following classical robust optimization problem:

$$\inf \quad \gamma, \tag{4.118}$$

$$\text{s. t.} \quad \gamma \geq \boldsymbol{\alpha}^\top \boldsymbol{p} + \sum_{k=1}^K \boldsymbol{\beta}_k^\top \boldsymbol{\mu}_k \quad \forall \boldsymbol{p} \in \mathcal{P}, \; \frac{\boldsymbol{\mu}_k}{\sum_{s \in \mathcal{S}_k} p_s} \in \mathcal{L}_k, \; k = 1, \ldots, K, \tag{4.119}$$

$$\alpha_s + \sum_{k \in \mathcal{K}_s} \boldsymbol{\beta}_k^\top \boldsymbol{\xi} \geq h(\boldsymbol{x}, \boldsymbol{\xi}, s) \qquad \forall \boldsymbol{\xi} \in \Xi_s, \; s = 1, \ldots, S, \tag{4.120}$$

$$\gamma \in \mathbb{R}, \boldsymbol{\alpha} \in \mathbb{R}^S, \boldsymbol{\beta}_k \in \mathbb{R}^I \qquad \forall k = 1, \ldots, K, \tag{4.121}$$

where $\mathcal{K}_s = \{k \in \{1, \ldots, K\} \mid s \in \mathcal{S}_k\}$, for each $s = 1, \ldots, S$.

The detailed proof can be found in [121]. The tractability of the above problem depends on the uncertainty sets Ξ_s, $s \in \{1, \ldots, S\}$. For practicability, these sets are confined to tractable conic representable ones such as polyhedral or second-order conic ones.

4.5.4 Robust Satisficing

A so-called robust satisficing framework was introduced in [350] leveragin [87, 88]. It favors solutions for which a risk-aware objective function would best attain an acceptable target even when the actual probability distribution deviates from the empirical distribution. Unlike robust optimization, the robust satisficing model does not restrict distributions to an ambiguity set. Instead, it allows the Nature to take its course and, as much as possible, limit the impact of ambiguity on the risk-aware objective should the actual distribution deviate from the empirical distribution. The satisficing decision-maker specifies an acceptable target, or loss of optimality compared with the empirical optimization model, as a trade-off for the model's ability to withstand "more" uncertainty.

Before presenting the robust satisficing model, we first provide the following definition of the probability distance function.

Definition 4.7 (Probability Distance Function) A probability distance function $\Delta(\mathbb{P}, \hat{\mathbb{P}})$ is a non-negative function on the domain of probability distributions such that $\Delta(\mathbb{P}, \hat{\mathbb{P}}) = 0$ if $\mathbb{P} = \hat{\mathbb{P}}$.

Under this definition, the probability distance-based ambiguity sets would include the Wasserstein metric discussed in Sect. 4.5.2. We now present the robust satisficing model:

$$\kappa_\tau = \min \quad k \tag{4.122}$$

$$\text{s. t.} \quad \mathbb{E}_{\mathbb{P}}[h(\boldsymbol{x}, \boldsymbol{\xi})] - \tau \leq k\Delta(\mathbb{P}, \hat{\mathbb{P}}) \quad \forall \mathbb{P} \in \mathcal{P}(\Xi) \tag{4.123}$$

$$\boldsymbol{x} \in \mathcal{X}, \tag{4.124}$$

$$k \geq 0, \tag{4.125}$$

where $h(\boldsymbol{x}, \boldsymbol{\xi}) : \mathcal{X} \times \Xi \longmapsto \mathbb{R}$. Furthermore, τ represents a targeted cost level or an acceptable loss of optimality relative to the empirical optimization model. Defining $Z_0 = \min_{\boldsymbol{x}\in\mathcal{X}} \mathbb{E}_{\hat{\mathbb{P}}}[h(\boldsymbol{x}, \tilde{\xi})]$, we should choose $\tau \geq Z_0$.

Since the actual distribution is unknown and may deviate from $\hat{\mathbb{P}}$, we do not expect the actual objective Z_0 to be achievable. Having a target higher than Z_0 provides more leeway to improve the robustness of the model. A violation from the desired target level occurs whenever $\mathbb{E}_{\mathbb{P}}[h(\boldsymbol{x}, \boldsymbol{\xi})] > \tau$.

Like in robust optimization, the choice of the probability distance function depends, among other things, on how we can efficiently optimize the robust counterparts and, more importantly, how well the solutions would perform in empirical tests.

Intuitively, the robust satisficing model minimizes the fragility of the system, which is measured by the worst-case magnitude, over all possible distributions $\mathbb{P} \in \mathcal{P}(\Xi)$, of the expected target violation under a distribution normalized by the statistical distance of this distribution from $\hat{\mathbb{P}}$. The optimal proportionality factor

κ_τ effectively describes the fragility level of the model. As κ_τ decreases in value, a lower magnitude of the expected target violation could occur under any distribution.

We mainly focus on using the Wasserstein distance discussed in Sect. 4.5.2. The above robust satisficing problem under the Wasserstein metric d_W can be equivalently written as

$$\kappa_\tau = \min \quad k, \tag{4.126}$$

$$\text{s. t.} \quad \mathbb{E}_{\mathbb{P}}[h(\boldsymbol{x}, \boldsymbol{\xi})] - \tau \le k\mathbb{E}_{\mathbb{Q}}\left[\|\boldsymbol{\xi} - \boldsymbol{u}\|\right], \ \forall \mathbb{Q} \in \mathcal{F}, \tag{4.127}$$

$$\boldsymbol{x} \in \mathcal{X}, k \ge 0, \tag{4.128}$$

where the ambiguity set of the joint distribution is defined as

$$\mathcal{F} = \left\{ \mathbb{Q} \in \mathcal{P}(\Xi^2) \left| \begin{array}{l} (\boldsymbol{\xi}, \boldsymbol{u}) \sim \mathbb{Q} \\ \mathbb{Q}\left[\boldsymbol{u} = \hat{\boldsymbol{\xi}}_s\right] = \frac{1}{S}, \ \forall s \in \{1, \ldots, S\} \end{array} \right. \right\}. \tag{4.129}$$

To derive the tractable reformulation of problem (4.126)–(4.128), we have the following (see [350] for further details).

Suppose that the support set Ξ is polyhedral, i.e., $\Xi = \{\boldsymbol{\xi} \in \mathbb{R}^N \mid \boldsymbol{C}\boldsymbol{\xi} \le \boldsymbol{l}\}$. Then the robust satisficing model (4.126)–(4.128) with $h(\boldsymbol{x}, \boldsymbol{\xi}) = \max_{j=1,\ldots,n} \left\{a_j \boldsymbol{x}^\top \boldsymbol{\xi} + b_j\right\}$ can be reformulated as follows:

$$\text{minimize} \quad k, \tag{4.130}$$

$$\text{subject to} \quad \frac{1}{S}\sum_{s=1}^{S} y_s \le \tau, \tag{4.131}$$

$$y_s \ge a_j \boldsymbol{x}^\top \hat{\boldsymbol{\xi}}_s + b_j + \boldsymbol{\eta}_{js}^\top \boldsymbol{l} - \boldsymbol{\eta}_{js}^\top \boldsymbol{C} \hat{\boldsymbol{\xi}}_s \quad \forall s = 1, \ldots, S, \ j = 1, \ldots, n, \tag{4.132}$$

$$k \ge \left\| a_j \boldsymbol{x} - \boldsymbol{C}^\top \boldsymbol{\eta}_{js} \right\|_* \quad \forall s = 1, \ldots, S, \ i = 1, \ldots, n, \tag{4.133}$$

$$\boldsymbol{\eta}_{is} \ge \boldsymbol{0} \quad \forall s = 1, \ldots, S, \ i = 1, \ldots, n, \tag{4.134}$$

$$\boldsymbol{x} \in \mathcal{X}, \tag{4.135}$$

$$k \ge 0. \tag{4.136}$$

The interested reader can also refer to several works leveraging the robust satisficing framework, including [116, 446, 506].

4.5.5 Globalized Distributionally Robust Counterpart

The globalized distributionally robust counterpart (GDRC) paradigm is proposed in [340]. This paradigm minimizes the violation of constraints involving uncertainty when the underlying distribution does not reside in the prespecified ambiguity set. Rooted in the domain of optimization under uncertainty, this paradigm unifies the frameworks of globalized robust optimization [52] and robust satisficing [350]. The GDRC paradigm focuses on parameter uncertainty embedded in a (possibly nonlinear) constraint as follows:

$$f(\boldsymbol{x}, \boldsymbol{\xi}) \leq 0.$$

The vector $\boldsymbol{x} \in \mathcal{X} \subseteq \mathbb{R}^K$ represents the decision variable, and $\boldsymbol{\xi} \in \Xi \subseteq \mathbb{R}^L$ is the vector of uncertain parameters.

For any fixed decision $\boldsymbol{x}$, the cost function $f(\boldsymbol{x}, \boldsymbol{\xi})$ is proper concerning the uncertainty $\boldsymbol{\xi}$ that follows an unknown probability distribution. In particular, the globalized distributionally robust counterpart (GDRC) is as follows:

$$\mathbb{E}_{\mathbb{P}}[f(\boldsymbol{x}, \boldsymbol{\xi})] \leq \gamma \cdot \min_{\mathbb{Q} \in \mathcal{F}} d(\mathbb{P}, \mathbb{Q}) \quad \forall \mathbb{P} \in \mathcal{P}(\Xi), \tag{4.137}$$

which involves the following inputs: (i) a tolerance level $\gamma \geq 0$, (ii) a non-negative statistical metric d that measures the proximity between probability distributions and satisfies $d(\mathbb{P}, \mathbb{Q}) = 0$ if $\mathbb{P} = \mathbb{Q}$, and (iii) a prescribed ambiguity set $\mathcal{F} \subseteq \mathcal{P}(\Xi)$. Here, the ambiguity set $\mathcal{F}$, when properly defined, can be sized with a parameter $\theta \geq 0$ to achieve adjustable robustness. By decomposing the globalized distributionally robust counterpart (4.137) into

$$\begin{cases} \mathbb{E}_{\mathbb{P}}[f(\boldsymbol{x}, \boldsymbol{\xi})] \leq 0, & \forall \mathbb{P} \in \mathcal{F}, \\ \mathbb{E}_{\mathbb{P}}[f(\boldsymbol{x}, \boldsymbol{\xi})] \leq \gamma \cdot \min_{\mathbb{Q} \in \mathcal{F}} d(\mathbb{P}, \mathbb{Q}), & \forall \mathbb{P} \in \mathcal{P}(\Xi) \backslash \mathcal{F}, \end{cases}$$

it becomes clear that, for any distribution within the ambiguity set $\mathcal{F}$, the model guarantees strict inner robustness to no constraint violation; on the other hand, for distributions not residing in $\mathcal{F}$, the tolerance level γ ensures soft outer robustness that limits possible constraint violation to a proportion of the probability distribution's distance to the ambiguity set $\mathcal{F}$. Integrating these two, the globalized distributionally robust counterpart (4.137) provides, for all distribution $\mathbb{P} \in \mathcal{P}(\Xi)$, globalized robustness in terms of limited constraint violation.

By varying its inputs, the GDRC (4.137) recovers several models mentioned in the previous sections when facing parameter uncertainty, summarized in the following:

1. When $\gamma = 0$, Constraint (4.137) reduces to the traditional robust counterpart:

$$f(\boldsymbol{x}, \boldsymbol{\xi}) \leq 0 \quad \forall \boldsymbol{\xi} \in \Xi.$$

2. When $\gamma \to \infty$, Constraint (4.137) becomes the distributionally robust counterpart:

$$\mathbb{E}_{\mathbb{P}}[f(\boldsymbol{x}, \boldsymbol{\xi})] \leq 0 \quad \forall \mathbb{P} \in \mathcal{F}.$$

3. When $\gamma \to \infty$ and $\mathcal{F} = \{\hat{\mathbb{P}}\}$, Constraint (4.137) reduces to the stochastic counterpart under the reference distribution $\hat{\mathbb{P}}$:

$$\mathbb{E}_{\hat{\mathbb{P}}}[f(\boldsymbol{x}, \boldsymbol{\xi})] \leq 0.$$

4. When $\mathcal{F} = \{\hat{\mathbb{P}}\}$, Constraint (4.137) coincides with the robust satisficing counterpart mentioned in Sect. 4.5.4:

$$\mathbb{E}_{\mathbb{P}}[f(\boldsymbol{x}, \boldsymbol{\xi})] \leq \gamma \cdot d(\mathbb{P}, \hat{\mathbb{P}}) \quad \forall \mathbb{P} \in \mathcal{P}(\Xi).$$

5. With proper specifications of its inputs, Constraint (4.137) recovers the original globalized robust counterpart advocated by Ben-Tal et al. [52]:

$$f(\boldsymbol{x}, \boldsymbol{\xi}) \leq \gamma \cdot \min_{\boldsymbol{\zeta} \in \mathcal{U}_1} c_1(\boldsymbol{\xi}, \boldsymbol{\zeta}) \quad \forall \boldsymbol{\xi} \in \mathcal{U}_2,$$

where $\mathcal{U}_1 \subseteq \mathcal{U}_2$ and $c_1(\boldsymbol{\xi}, \boldsymbol{\zeta})$ measures the non-negative distance between the parameters $\boldsymbol{\xi}$ and $\boldsymbol{\zeta}$.

These results showcase the rich modeling flexibility of the globalized distributionally robust counterpart (4.137). Apart from such expressiveness, the globalized distributionally robust counterpart, with proper choices of the statistical metric d and ambiguity set $\mathcal{F}$, also admits an attractive tractable reformulation.

Next, we focus on the 1-Wasserstein distance given in Definition 4.6 as the statistical metric d and consider the ambiguity set $\mathcal{F}$ as the Wasserstein ambiguity set $\mathcal{F}(\varepsilon)$ given in (4.114). The corresponding globalized distributionally robust counterpart (4.137) can then be re-expressed as

$$\sup_{\mathbb{P} \in \mathcal{P}(\Xi), \mathbb{Q} \in \mathcal{F}(\varepsilon)} \{\mathbb{E}_{\mathbb{P}}[f(\boldsymbol{x}, \boldsymbol{\xi})] - \gamma \cdot d_{\mathrm{W}}(\mathbb{P}, \mathbb{Q})\} \leq 0. \tag{4.138}$$

To derive the tractable formulation of the GDRC (4.138), we define the convex conjugate of a proper function $f : \mathbb{R}^L \mapsto \mathbb{R}$ by $f^*(\boldsymbol{x}) =$

$\sup\{x^\top \boldsymbol{\xi} - f(\boldsymbol{\xi}) \mid \boldsymbol{\xi} \in \mathbb{R}^L\}$. We use $f^{\bar{*}}(\boldsymbol{x}, \boldsymbol{\xi})$ with a bar sign to emphasize the partial convex conjugate concerning the second part of arguments, $\boldsymbol{\xi}$. We also define the indicator function of a set Ξ through $\delta(\boldsymbol{\xi} \mid \Xi) = 0$ if $\boldsymbol{\xi} \in \Xi$; $= \infty$ otherwise. Furthermore, the support function is defined as $\delta^*(\boldsymbol{x} \mid \Xi) = \sup\{\boldsymbol{x}^\top \boldsymbol{\xi} \mid \boldsymbol{\xi} \in \Xi\}$. The following reformulation holds.

Suppose that the support set Ξ is convex and closed, and for any fixed $\boldsymbol{x} \in \mathcal{X}$, the cost function takes the form $f(\cdot, \boldsymbol{\xi}) = \max_{k=1,\ldots,K} f_k(\cdot, \boldsymbol{\xi})$ with constituent functions $f_k(\cdot, \boldsymbol{\xi})$ being proper, upper semicontinuous, and concave in $\boldsymbol{\xi}$. Then, for any fixed decision $x \in \mathcal{X}$, given the Wasserstein ambiguity set $\mathcal{F}(\varepsilon)$ in (4.114), we have that

$$\sup_{\mathbb{P} \in \mathcal{P}(\Xi), \mathbb{Q} \in \mathcal{F}(\varepsilon)} \{\mathbb{E}_{\mathbb{P}}[f(\boldsymbol{x}, \boldsymbol{\xi})] - \gamma \cdot d_{\mathrm{W}}(\mathbb{P}, \mathbb{Q})\}$$

equals the optimal value of the finite-dimensional convex program

$$\inf \quad \varepsilon t + \frac{1}{N} \sum_{n=1}^{N} s_n \tag{4.139}$$

$$\text{s. t.} \quad [-f_k]^{\bar{*}}(\boldsymbol{x}, \boldsymbol{u}_{nk} - \boldsymbol{w}_{nk}) + \delta^*(\boldsymbol{u}_{nk} \mid \Xi) - \hat{\boldsymbol{\xi}}_n^\top \boldsymbol{w}_{nk} \le s_n$$

$$\forall n = 1, \ldots, N,\ k = 1, \ldots, K, \tag{4.140}$$

$$\|\boldsymbol{w}_{nk}\|_* \le t \qquad \forall n = 1, \ldots, N,\ k = 1, \ldots, K, \tag{4.141}$$

$$\boldsymbol{u}_{nk}, \boldsymbol{w}_{nk} \in \mathbb{R}^I \qquad \forall n = 1, \ldots, N,\ k = 1, \ldots, K, \tag{4.142}$$

$$\boldsymbol{s} \in \mathbb{R}^N, \tag{4.143}$$

$$t \in [0, \gamma]. \tag{4.144}$$

Recall that N is the number of past realizations used to build the empirical probability distribution $\widehat{\mathbb{P}}_N$ that in turn is used for defining the ambiguity set $\mathcal{F}(\varepsilon)$.

The detailed proof of the above model can be found in [340], by which we can equivalently reformulate the globalized distributionally robust optimization model and globalized robust satisficing model as finite-dimensional convex programs. The globalized distributionally robust chance-constraint model can be reformulated as a deterministic mixed-integer program. Furthermore, the tractable formulations of the globalized distributionally robust counterpart (4.137) for other ambiguity sets based on probability metrics, optimal transport, ϕ-divergences, or moment conditions, as well as the event-wise ambiguity set can be found in [340].

4.6 Conclusion and Further Reading

In this chapter, we have briefly introduced four types of optimization-under-uncertainty models. These include robust optimization, stochastic program, chance-constrained program, and distributional robust optimization. Different approaches deal with the uncertainty from different perspectives and data conditions. For instance, stochastic optimization models work better when the probability distribution of the parameters is given. Moreover, the amount of available data should be sufficiently high. Robust optimization models are more suitable under data insufficiency, where the probability distribution of the parameters cannot be well calibrated. Deeper discussions and further advances can be found in [60, 68, 217, 242] (for robust optimization), [5, 96, 120, 289, 397, 398] (for chance-constrained optimization), [29, 78, 224, 259, 324, 357] (for stochastic optimization), and [76, 160, 238, 322, 340, 350, 385, 444, 543] (for distributionally robust optimization).

In particular, as generalizations to the two-stage/multi-stage stochastic programming models as discussed in Sect. 4.3, the two-stage/multi-stage distributionally robust optimization can be found in [32, 74, 76, 257, 428, 430]. Also, the adaptive optimization models under ambiguity are discussed in the framework of robust satisficing and globalized distributionally robust optimization. See more details in [340, 350].

Furthermore, in the era of Industry 4.0, disruptive technologies, such as IoT, 5G, and digital twins, have significantly enhanced the data accessibility for production planning and control, which enables several statistical and machine learning methods to harness the side information or covariates for predicting and estimating the demand [43, 70, 72]. Meanwhile, optimization-under-uncertainty methodologies go more and more closely with data science, and many data-driven or machine learning-based optimization approaches are developed under uncertainty, for instance, predict-then-optimize framework [343, 349], estimation-fortified robust satisficing model [506], and residuals-based robust optimization paradigm [297, 298, 439]. Along this direction, the data-driven optimization-under-uncertainty methodologies can be applied to the facility location problems, such as to be discussed in Chaps. 8, 15, and 16 of this book.

Finally, we refer to [434] for a comprehensive and unified framework of sequential decisions under uncertainty. The area can be split into many sub-fields (15 are mentioned). These have been studied under names such as dynamic programming, stochastic programming, optimal control, simulation optimization, optimal learning, and multi-armed bandit problems. That volume seeks a unified analysis of the entire area.

Part II
Modeling Paradigms and Solution Techniques

Chapter 5
Robust Facility Location

Abstract This chapter discusses different possibilities for modeling and solving robust discrete facility location problems. Initially, a finite uncertainty set is assumed. Several concepts are discussed, leading to different objective functions. Then, facility location problems considering interval and ellipsoidal uncertainty are analyzed. Adaptive robust optimization in the context of facility location is also considered. The content is extended by considering multi-period facility location problems, laying the ground for more comprehensive models to be covered later in the book.

Keywords Finite uncertainty sets · Infinite uncertainty sets · Adaptive decisions · Time-dependent decisions

5.1 Introduction

A core assumption in this chapter is that the uncertainty underlying a facility location problem is not quantified using a probability distribution. This may be due to the lack of data allowing estimating the underlying cumulative distribution function. Alternatively, we may seek a feasible solution irrespective of the observed scenario. In this case, as discussed in Chap. 4, we consider an uncertainty set containing all the viable scenarios for the uncertain vector of parameters.

When no probabilistic information is available for the underlying parameters, a robustness measure can be considered. There are several notable examples in the literature. Two popular measures are the minmax cost and the minmax regret. Moreover, an uncertainty set can be finite or infinite. This influences the models and solution techniques one can or should use. As in the previous chapter, we use well-known facility location problems throughout the text.

© The Author(s), under exclusive license to Springer Nature Switzerland AG 2024
F. Saldanha-da-Gama, S. Wang, *Facility Location Under Uncertainty*,
International Series in Operations Research & Management Science 356,
https://doi.org/10.1007/978-3-031-55927-3_5

The remainder of this chapter is organized as follows. In Sect. 5.2, uncertainty is captured by a finite set of scenarios. Section 5.3 focuses on modeling alternatives triggered by an infinite uncertainty set. In both sections, a subsection is devoted to problems gathering adaptive decisions. This aspect is extended in Sect. 5.4 when time is discussed as an extra dimension of relevance. The chapter ends with some conclusions and directions for further reading.

5.2 Finite Uncertainty Sets

We start with a fundamental problem in Location Science: the discrete p-median problem. We recall a model presented in Chap. 2:

$$\text{minimize} \quad \sum_{i \in J} \sum_{j \in J} c_{ij} d_j x_{ij}, \tag{5.1}$$

$$\text{subject to} \quad \sum_{i \in J} x_{ij} = 1 \qquad \forall j \in J, \tag{5.2}$$

$$x_{ij} \leq x_{ii} \qquad \forall i \in I,\ j \in J, \tag{5.3}$$

$$\sum_{i \in J} x_{ii} = p, \tag{5.4}$$

$$x_{ij} \in \{0, 1\} \qquad \forall i \in I,\ j \in J. \tag{5.5}$$

In this model, J is the set of demand nodes that coincides with the potential locations for the facilities; d_j is the demand of node $j \in J$, and c_{ij} is the unit cost or travel time between nodes i and j $(i, j \in J)$. The decision variables define the allocation of the nodes to the facilities with $x_{ii} = 1$ "proxying" the location variables, thus indicating that node $i \in J$ is selected for locating a facility. The objective function accounts for the total cost, Constraints (5.2) ensure that each demand node is allocated to one and only one facility, and Constraints (5.3) guarantee that allocations involve facilities that are installed. Constraint (5.4) imposes that exactly p facilities are installed. Finally, we have domain constraints.

In the p-median problem, uncertainty can stem from the demands d_j or the costs c_{ij}. With this in mind, let Ω denote a finite set of scenarios previously identified for those parameters. Recall that one scenario fully specifies the uncertainty underlying the problem. Let us assume *ex ante* location decisions, i.e., the facilities must be located here and now and thus before uncertainty is revealed. In this case, it helps consider specific binary location decision variables, y_i, equal to one if and only if a facility is located at $i \in J$.

A minmax cost p-median problem can be formulated. Let $d_{j\omega}$ denote the demand of node $j \in J$ under scenario $\omega \in \Omega$; likewise, we take $c_{ij\omega}$ as the unit cost or travel time between nodes i and j $(i, j \in J)$ under scenario $\omega \in \Omega$. The problem becomes

$$\begin{aligned}
\text{minimize} \quad & v, && && (5.6)\\
\text{subject to} \quad & \sum_{i \in J} \sum_{j \in J} d_{j\omega} c_{ij\omega} x_{ij\omega} \leq v && \forall \omega \in \Omega, && (5.7)\\
& \sum_{i \in J} x_{ij\omega} = 1 && \forall j \in J,\ \omega \in \Omega, && (5.8)\\
& x_{ij\omega} \leq y_i && \forall i, j \in J,\ \omega \in \Omega, && (5.9)\\
& \sum_{i \in J} y_i = p, && && (5.10)\\
& x_{ij\omega} \in \{0, 1\} && \forall i, j \in J,\ \omega \in \Omega, && (5.11)\\
& y_i \in \{0, 1\} && \forall i \in J. && (5.12)
\end{aligned}$$

The objective function (5.6) together with (5.7) ensures that the maximum cost across the scenarios is computed and minimized. Constraints (5.8) impose that in every scenario, all demand nodes are allocated to one facility (which can change from one scenario to another). Inequalities (5.9) guarantee that in every scenario, allocations are made only to open facilities. Inequality (5.10) ensures that exactly p facilities are located here and now, irrespective of the observed scenario. Finally, we have the domain of the decision variables.

The above minmax perspective indicates a full risk-averse decision-maker. Nevertheless, by planning for the worst-case scenario, we are planning for a future that may be very unlikely. This may render the solution overly conservative. A better compromise is often sought, which consists of minimizing the maximum regret across all scenarios.

The regret of a solution in a scenario is measured by the difference between the cost of that solution if the scenario occurs and the optimal cost that can be achieved under that scenario. The reader can refer to [319] for further details.

Denote by v^*_ω the optimal value of the single-scenario problem induced by scenario ω. In other words, this would be the problem one would solve if we knew in advance the occurring scenario. The minmax regret p-median problem can be formulated replacing (5.7) with

$$\sum_{i \in J} \sum_{j \in J} d_{j\omega} c_{ij\omega} x_{ij\omega} - v^*_\omega \leq v \quad \forall \omega \in \Omega. \tag{5.13}$$

Serra and Marianov [485] consider a minmax regret p-median problem as above although initially scaling the demands: In each scenario, the demands are divided by the total demand under that scenario. The authors also point out that when the optimal objective function differs significantly across the scenarios, the relative regret is a more appropriate robustness measure (see also [319]). In this case, it is advisable to replace (5.13) with

$$\frac{\sum_{i \in J} \sum_{j \in J} d_{j\omega} a_{ij\omega} x_{ij\omega} - v_{\omega}^*}{v_{\omega}^*} \leq v, \quad \omega \in \Omega. \tag{5.14}$$

Serra and Marianov [485] develop a heuristic for this problem and for the minmax p-median problem by noticing that both constraints (5.7) and (5.14) are not integer-friendly. The heuristic adapts the two-phase procedure introduced in [519] to the new problems. In the first phase, a set of initial p facilities is found, and in the second, the location decision obtained in phase 1 is changed. The new objective function value is computed. If this leads to an improvement, it is saved as the best solution so far; otherwise, the change is reverted, and another is attempted. The initial operating facilities are obtained by solving a p-median problem for each scenario. The best p facilities in one scenario can be assessed in all other scenarios regarding the average weighted cost. This allows building a matrix with each scenario associated with one row and a p facility solution associated with each column. The diagonal contains the optimal weighted costs for each scenario. The regret of each column (selection of p facilities) in each scenario can be easily computed using this matrix, leading to a regret matrix. The selection of the initial p facilities for the robust optimization problem depends on the objective: In the minmax p-median problem, [485] take the first matrix, compute the maximum in each row, select the scenario that minimizes that maximum, and take for the p initial facilities the best for that scenario. For minmax regret, similar reasoning is followed but using the regret matrix.

If the allocation of customers to facilities is also an *ex ante* decision, the models above can be easily adapted. In this case, the scenario index should be removed from the allocation variables, i.e., the allocation variables become those introduced in model (5.1)–(5.5). Furthermore, the location variables y_i are no longer necessary, since the variables x_{ii} $(i \in J)$ can play their role.

A different but related robust facility location problem is studied in [486]. The authors consider a firm that seeks to locate p facilities in a competitive environment, i.e., another firm is already in the market, with a certain number of facilities and serving the customers in J. The new firm captures the demand of a node if a new facility is installed in a location closer to the node than an existing one. In case of a tie, the new firm and the existing one share the demand of the node. The location of the facilities (existing and new) does not necessarily coincide with the location of the demand nodes. The authors start by considering uncertainty only for the demand. Afterward, this is extended to the location of the facilities owned by the firm already in the market. It is reasonable to assume that those locations can change. We directly consider the most comprehensive uncertainty setting. Moreover, we assume that a finite set of scenarios Ω has been identified for capturing uncertainty.

In addition to the notation already introduced, we consider $k_{j\omega}$ standing for the closest existing facility to node $j \in J$ under scenario $\omega \in \Omega$. Let us assume that a set I of potential facility locations is identified by the firm entering the market. In this case, for every scenario $\omega \in \Omega$ and for every demand node $j \in J$, we can identify the locations in I that are closer to j than the closest existing facility (w.r.t. some adequate distance): $I^{<}_{j\omega} = \{i \in I : \ell_{ij} < \ell_{k_{j\omega}j}\}$. We can also find the locations in I that are at a similar distance to the closest existing facility: $I^{=}_{j\omega} = \{i \in I : \ell_{ij} = \ell_{k_{j\omega}j}\}$. In the above sets, ℓ represents the distance between nodes. The goal is to maximize the demand captured by the new firm.

Due to the underlying uncertainty, two relevant objectives emerge: maximizing the minimum demand captured and minimizing the maximum regret across the different scenarios. We adopt the location decision variables y_i already used above for the location of the new facilities. Two additional sets of binary variables are also considered. For $j \in J$ and $\omega \in \Omega$, variable $w_{j\omega}$ equals one if and only if node j is captured by the new firm, under scenario ω. Additionally, $z_{j\omega}$ is a variable equal to one if and only if the demand of node $j \in J$ is shared between the existing and the new firm under scenario ω. Considering the first objective, the problem can be formulated as follows:

$$\text{maximize} \quad v, \tag{5.15}$$

$$\text{subject to} \quad \sum_{j\in J} d_{j\omega} w_{j\omega} + \sum_{j\in J} \frac{d_{j\omega}}{2} z_{j\omega} \geq v \qquad \forall \omega \in \Omega, \tag{5.16}$$

$$w_{j\omega} \leq \sum_{i\in I^{<}_{j\omega}} y_i \qquad \forall j \in J,\ \omega \in \Omega, \tag{5.17}$$

$$z_{j\omega} \leq \sum_{i\in I^{=}_{j\omega}} y_i \qquad \forall j \in J,\ \omega \in \Omega, \tag{5.18}$$

$$w_{j\omega} + z_{j\omega} \leq 1 \qquad \forall j \in J,\ \omega \in \Omega, \tag{5.19}$$

$$\sum_{i\in J} y_i = p, \tag{5.20}$$

$$y_i \in \{0, 1\} \qquad \forall i \in I, \tag{5.21}$$

$$w_{j\omega} \in \{0, 1\} \qquad \forall j \in J,\ \omega \in \Omega, \tag{5.22}$$

$$z_{j\omega} \in \{0, 1\} \qquad \forall j \in J,\ \omega \in \Omega. \tag{5.23}$$

The objective function seeks to maximize the minimum demand captured, which is accounted for in (5.16). Constraints (5.17) and (5.18) ensure the consistency between demand captured and shared and the new facilities located. Constraints (5.19) ensure that the demand of a node cannot be simultaneously captured in full and shared (in each scenario). Constraint (5.20) guarantees that p new facilities are located. Finally, we have the domain constraints.

Denote by v^*_ω the maximum demand captured under scenario ω. This value is obtained by solving the single-scenario version of the above maxmin problem induced by scenario ω. The minmax regret maximum capture facility location problem can be formulated by replacing (5.15) with

$$\text{minimize} \quad v \tag{5.24}$$

and replacing (5.16) with

$$v^*_\omega - \left(\sum_{j \in J} d_{j\omega} w_{j\omega} + \sum_{j \in J} \frac{d_{j\omega}}{2} z_{j\omega} \right) \leq v \quad \forall \omega \in \Omega. \tag{5.25}$$

Serra et al. [486] propose a heuristic procedure for this problem similar to the one proposed in [485]. The same reasoning can be followed in both versions of the problem.

Vatsa and Jayaswal [531] investigate a minmax regret multi-period covering location problems. Uncertainty regards the number of servers that can be installed in each period. A pure phase-in problem is considered, i.e., once installed, a server remains operating until the end of the planning horizon. Uncertainty is related to the available servers. A finite set of scenarios is assumed. The authors developed a Benders decomposition algorithm. Vatsa and Jayaswal [532] extend that work by considering capacities associated with the facilities.

5.2.1 *Adaptive Decisions*

As we discussed in Part I of this volume, under uncertainty, it may be possible to correct an initial solution by making another decision that adapts to the occurring scenario. As we will see in Chap. 6, this is pervasive in stochastic programming models and methods for facility location problems. Nevertheless, in robust optimization, we can also consider that possibility. This idea was formalized in [337] and named recoverable robustness.

To expand the analysis we have been presenting, consider now another classical facility location: the uncapacitated facility location problem (UFLP). As we reviewed in Chap. 2, the problem can be formulated as follows:

$$\text{minimize} \quad \sum_{i \in I} f_i y_i + \sum_{i \in I} \sum_{j \in J} c_{ij} d_j x_{ij}, \tag{5.26}$$

$$\text{subject to} \quad \sum_{i \in I} x_{ij} = 1 \qquad \forall j \in J, \tag{5.27}$$

$$x_{ij} \leq y_i \qquad \forall i \in I,\ j \in J, \tag{5.28}$$

$$y_i \in \{0, 1\} \qquad \forall i \in I, \tag{5.29}$$

$$x_{ij} \geq 0 \qquad \forall i \in I,\ j \in J. \tag{5.30}$$

In this model, I denotes the set of potential locations for the facilities, J is the set of customers, f_i represents the setup cost for facility $i \in I$, c_{ij} corresponds to the unit cost for supplying the demand of customer $j \in J$ from facility $i \in I$, and d_j is the demand of customer $j \in J$. The binary variable y_i indicates whether a facility is installed at $i \in I$, and the continuous variable x_{ij} represents the fraction of the demand of customer $j \in J$ that is supplied from facility $i \in I$. Note that there is always an optimal solution with all demand notes allocated to a single facility. The above model can be looked at as valid for the single-allocation problem.

Let us assume that not all allocation channels are viable. Accordingly, define a set A containing the pairs (i, j), $(i \in I,\ j \in J)$ that can be used. The problem is written as

$$\text{minimize} \quad \sum_{i \in I} f_i y_i + \sum_{(i,j) \in A} c_{ij} d_j x_{ij}, \tag{5.31}$$

$$\text{subject to} \quad \sum_{i \in I:(i,j) \in A} x_{ij} = 1 \qquad \forall j \in J, \tag{5.32}$$

$$x_{ij} \leq y_i \qquad \forall (i, j) \in A, \tag{5.33}$$

$$(5.29),\ (5.30). \tag{5.34}$$

Let us consider that a tentative location-allocation decision is to be made without knowing the actual data we will face. In particular, assume that we do not know in advance the exact potential location for the facilities, the same holding with the demand nodes, and the allocation channels available. Once uncertainty is disclosed, an adaptive decision is made to recover feasibility when the first-stage decision is infeasible. The second-stage decision includes (i) opening new facilities, (ii) allocating customers to open facilities (either opened in the first or the second stage), and (iii) reallocating customers that were initially allocated to facilities that turn out to be unavailable.

Assume uncertainty captured by a finite set of scenarios Ω. Each scenario $\omega \in \Omega$ fully specifies the actual set of potential locations for the facilities I^ω, the set of customers J^ω, and the set of transportation channels A^ω available. To distinguish the initial setting, we denote by I^0, J^0, and A^0 the initial assumed conditions with $I^0 = \bigcup_{\omega \in \Omega} I^\omega$, $J^0 = \bigcup_{\omega \in \Omega} J^\omega$, and $A^0 = \bigcup_{\omega \in \Omega} A^\omega$. We also assume that there is at least one feasible first-stage solution. Additionally, for every $\omega \in \Omega$, each customer $j \in J^\omega$ has at least one allocation channel available in A^ω.

To formulate the problem, we denote by y_i^0 the first-stage location decisions, $i \in I^0$, and x_{ij}^0 the allocation variables, $(i, j) \in A^0$. The first-stage costs are denoted by f_i^0, $i \in I^0$, and c_{ij}^0, $(i, j \in A^0)$. For each scenario $\omega \in \Omega$, we define likewise, y_i^ω, $i \in I^\omega$, x_{ij}^ω, $(i, j) \in A^\omega$, f_i^ω, $i \in I^\omega$, and c_{ij}^ω, $(i, j) \in A^\omega$. We assume a single allocation of demand nodes to facilities irrespective of the stage in which the decision is made. The adaptive robust optimization problem can be formulated as follows:

$$\text{minimize} \quad \sum_{i \in I^0} f_i^0 y_i^0 + \sum_{(i,j) \in A^0} c_{ij}^0 x_{ij}^0 + \mathcal{R}(\boldsymbol{x}^0, \boldsymbol{y}^0), \tag{5.35}$$

$$\text{subject to} \quad x_{ij}^0 \leq y_i^0 \qquad \forall i \in I^0,\ (i,j) \in A^0, \tag{5.36}$$

$$\sum_{(i,j) \in A^0} x_{ij}^0 \qquad \forall j \in J^0, \tag{5.37}$$

$$x_{ij}^0 \in \{0,1\} \qquad \forall (i,j) \in I^0, \tag{5.38}$$

$$y_i^0 \in \{0,1\} \qquad \forall i \in I^0, \tag{5.39}$$

with $\mathcal{R}(\boldsymbol{x}^0, \boldsymbol{y}^0) = \max_{\omega \in \Omega} g(\boldsymbol{x}^0, \boldsymbol{y}^0, \omega)$ and

$$g(\boldsymbol{x}^0, \boldsymbol{y}^0, \omega) = \min \quad \sum_{i \in I^\omega} f^\omega (y_i^\omega - y_i^0) + \sum_{(i,j) \in A^\omega} c_{ij}^\omega x_{ij}^\omega + \sum_{(\ell,j) \in A^\omega} r_{\ell j}^\omega z_{\ell j}^\omega,$$

$$+ \sum_{i \in I^0 \setminus I^\omega} p_i^\omega y_i^0, \tag{5.40}$$

$$\text{s. t.} \quad \sum_{(i,j) \in A^0} x_{ij}^0 + \sum_{(i,j) \in A^\omega} x_{ij}^\omega = 1 \qquad \forall j \in J^\omega, \tag{5.41}$$

$$\sum_{(i,j) \in A^0 \setminus A^\omega} x_{ij}^0 \leq \sum_{(i,j) \in A^\omega} z_{\ell j}^\omega \qquad \forall j \in J^\omega, \tag{5.42}$$

$$x_{ij}^\omega + z_{ij}^\omega \leq y_i^\omega \qquad \forall i \in I^\omega,\ (i,j) \in A^\omega, \tag{5.43}$$

$$y_i^0 \leq y_i^\omega \qquad \forall i \in I^\omega, \tag{5.44}$$

$$x_{ij}^\omega \{0,1\} \qquad \forall (i,j) \in A^\omega, \tag{5.45}$$

$$y_i^\omega \in \{0,1\} \qquad \forall i \in I^\omega, \tag{5.46}$$

$$z_{ij}^\omega \in \{0,1\} \qquad \forall (i,j) \in A^\omega. \tag{5.47}$$

In the second-stage problem, $r_{\ell j}^\omega$ stands for the cost incurred for reallocating customer $j \in J^\omega$ to facility $\ell \in I^\omega$ under scenario $\omega \in \Omega$. Additionally, p_i^ω is a penalty incurred under scenario $\omega \in \Omega$ due to not having available in the second stage a facility $i \in I^0$ that was opened in the first one. The first-stage objective function accounts for the initial location and allocation costs and the maximum cost we may need to pay for recovering feasibility in the second stage.

The second-stage objective function accounts for the location of facilities, the allocation costs in the second stage, the reallocation costs, and the penalty costs.

Constraints (5.41) ensure that all customers are allocated to a facility either in the first or in the second stage. Constraints (5.42) ensures that if a customer was allocated to a facility in the first stage, it must be so to an open facility in the second stage. Inequality (5.43) ensures when a customer is allocated or reallocated to a facility, the latter must be open in the second stage. Constraint (5.44) guarantees that if for some $\omega \in \Omega$ a facility in I^{ω} is opened in the first stage, then it must be kept open in the second stage. Finally, we have the domain constraints.

The above model assumes that each customer must be allocated or reallocated to exactly one facility in the second stage. This assumption may be too strong in practice. Nevertheless, the above model can still be used by considering a dummy facility with a setup cost equal to zero and with a first-stage allocation cost equal to the reallocation in the second stage and equal to the cost of not allocating a customer.

Álvarez-Miranda et al. [21] study the above problem. The authors also consider a more conservative perspective in which no new facilities can be opened later. Thus, no reallocation is allowed. This leads to a robust optimization problem that treats the UFLP like we treated the p-median problem at the beginning of this section.

The above model is compact and thus can be tackled by a general-purpose solver for small- or medium-sized instances. For large-scale instances, specially tailored procedures may be needed. Álvarez-Miranda et al. [21] propose transforming a Benders decomposition into a pure branch-and-cut approach. In this way, it is possible to use powerful general-purpose solvers. In particular, they propose generating Benders cuts and adding them on the fly as lower bounds on the second-stage objective function. This is done whenever a fractional solution to the master problem is obtained when solving the linear relaxation. Further details regarding separation mechanisms and enhancements can be found in [21].

5.3 Infinite Uncertainty Sets

The above models fit settings with uncertainty captured by a finite set of scenarios. In practice, however, this is not always a correct representation of the uncertainty: An uncertain parameter may lie in an infinite set.

Let us directly consider the well-known UFLP formulated as (5.26)–(5.30). A common source of uncertainty in that problem is the demand. A simple way of capturing such uncertainty is via intervals, i.e., considering box uncertainty. Under box uncertainty, demand d_j ($j \in J$) is assumed to lie in an interval

$$\mathcal{U}_j^B = [\overline{d}_j - \epsilon\Delta_j, \overline{d}_j + \epsilon\Delta_j],\ 0 \leq \epsilon \leq 1.$$

The parameter ϵ measures the uncertainty "magnitude," $\overline{d}_j$ denotes a reference value for the demand of customer $j \in J$, which is commonly referred to as the nominal value, and Δ_j is a scaling factor.

A particular case of box uncertainty stems from having $\Delta_j = \overline{d}_j$ ($j \in J$), which leads to the intervals $\mathcal{U}_j^B = [\overline{d}_j(1-\epsilon), \overline{d}_j(1+\epsilon)]$, $j \in J$. Denote $\mathcal{U}^B = \mathcal{U}_1^B \times \cdots \times \mathcal{U}_{|J|}^B$ and $\boldsymbol{d}$ the vector of demands, $\boldsymbol{d} = (d_1, \ldots, d_{|J|})^\top$. We have

$$\mathcal{U}^B = \left\{ \boldsymbol{d} \in \mathbb{R}^{|J|} \mid -1 \leq \frac{d_j - \overline{d}_j}{\epsilon \overline{d}_j} \leq 1,\ \forall j \in J \right\},$$

i.e., the multi-dimensional unit box is given by the absolute normalized deviations [46]. We can now formulate the so-called robust counterpart of model (5.26)–(5.30). This is accomplished by considering an auxiliary variable v, which allows writing the objective function of the problem as

$$\text{Minimize} \quad v \tag{5.48}$$

with the following constraint included in the model

$$\sum_{i \in I} f_i y_i + \sum_{i \in I} \sum_{j \in J} c_{ij} d_j x_{ij} \leq v. \tag{5.49}$$

By considering a so-called augmented version of these constraints, namely

$$\sum_{i \in I} f_i y_i + \max_{d \in \mathcal{U}^B} \left\{ \sum_{i \in I} \sum_{j \in J} c_{ij} d_j x_{ij} \right\} \leq v, \tag{5.50}$$

the robust counterpart of (5.49) becomes

$$\sum_{i \in I} f_i y_i + \sum_{i \in I} \sum_{j \in J} c_{ij} \left[\overline{d}_j (1+\epsilon) \right] x_{ij} \leq v. \tag{5.51}$$

The robust counterpart of (5.26)–(5.30) consists of minimizing (5.48) subject to (5.27)–(5.30) and (5.51).

A drawback of box uncertainty is the possibility of having all the uncertain parameters taking their worst values simultaneously. This is often not realistic. Nikoofal and Sadjadi [406] avoid the resulting overly conservative solutions by setting a maximum total scaled variation for the uncertain parameters. In their work, the authors consider a discrete p-median problem with interval uncertainty associated with the distances (or travel times). In particular, for each pair (i, j), $i, j \in J$, they assume that ℓ_{ij} (the distance or travel time between i and j) can take any value within a given $[\underline{\ell}_{ij}, \overline{\ell}_{ij}]$. Additionally, the values ℓ_{ij} are assumed to satisfy

$$\sum_{i,j \in J,\, i<j} (\ell_{ij} - \underline{\ell}_{ij})/(\overline{\ell}_{ij} - \underline{\ell}_{ij}) \leq L,$$

where L denotes a maximum level previously imposed for the total scaled variation.

The disadvantage of box uncertainty mentioned above can also be overcome by considering ellipsoidal uncertainty, which we represent as follows:

$$\mathcal{U}^E = \left\{ \boldsymbol{d} \in \mathbb{R}^{|J|} \mid \sum_{j \in J} \left[\frac{d_j - \overline{d}_j}{\epsilon \overline{d}_j} \right]^2 \leq L^2 \right\}$$
$$= \left\{ \boldsymbol{d} \in \mathbb{R}^{|J|} \mid (\boldsymbol{d} - \overline{\boldsymbol{d}})^\top \boldsymbol{\Lambda}^{-1} (\boldsymbol{d} - \overline{\boldsymbol{d}}) \leq L^2 \right\},$$

with $\boldsymbol{d}$ standing for the demand vector, L denoting a parameter, and $\boldsymbol{\Lambda}_{|J| \times |J|}$ a diagonal matrix whose generic entry is $\sigma_j = \epsilon \overline{d}_j$. Since $\boldsymbol{\Lambda}$ is a positive definite matrix, the set $\mathcal{U}^E$ defines an ellipsoid. As pointed out in [46], the set induced by $L = 1$ is the largest ellipsoid contained in $\mathcal{U}^B$, while the one stemming from $L = \sqrt{|J|}$ is the smallest ellipsoid containing $\mathcal{U}^B$.

Under ellipsoidal uncertainty, the augmented constraint for (5.49) is similar to (5.50) but replacing $\mathcal{U}^B$ with $\mathcal{U}^E$. If we denote $V_j = \sum_{i \in I} c_{ij} x_{ij}$ and $\boldsymbol{V} = (V_1, \ldots, V_{|J|})^\top$, then the augmented constraint can be written as

$$\sum_{i \in I} f_i y_i + \max_{d \in \mathcal{U}^E} \boldsymbol{V}^\top \boldsymbol{d} \leq v. \tag{5.52}$$

The problem that consists of finding a point $\boldsymbol{d} \in \mathcal{U}^E$ maximizing $\boldsymbol{V}^\top \boldsymbol{d}$ can be easily solved by standard optimization techniques. The optimal solution is

$$\boldsymbol{d} = \overline{\boldsymbol{d}} + L \frac{1}{\sqrt{\boldsymbol{V}' \boldsymbol{\Lambda} \boldsymbol{V}}} \boldsymbol{\Lambda} \boldsymbol{V},$$

and the optimal value of the problem is

$$\boldsymbol{V}^\top \overline{\boldsymbol{d}} + L \sqrt{\boldsymbol{V}^\top \boldsymbol{\Lambda} \boldsymbol{V}}.$$

This leads to the following robust counterpart of (5.49):

$$\sum_{i \in I} f_i y_i + \sum_{j \in J} \overline{d}_j V_j + L \sqrt{\sum_{j \in J} \sigma_j^2 V_j^2} \leq v. \tag{5.53}$$

The nonlinearity in the above expression is typically handled by introducing a new variable, $W = \sqrt{\sum_{j \in J} \sigma_j^2 V_j^2}$, which allows casting the problem as a conic programming problem. We detail this aspect using a fixed-charge multi-period facility location problem under uncertainty, namely that investigated in [46].

We recall some notation introduced in Chap. 2 when introducing the base multi-period facility location problem (2.14)–(2.18). Set T represents a planning horizon partitioned into several periods; d_{jt} is the demand of customer $j \in J$ in period $t \in T$.

Additionally, we consider a unit profit p_{ij} for satisfying customer $j \in J$ from facility $i \in I$. This cost results, for instance, from revenue after subtracting a delivery or transportation cost. Let r_{it} be the unit production cost at a facility located at $i \in I$ in period $t \in T$. Parameter f_i keeps representing a fixed cost for setting up a facility at $i \in I$. This cost may include, for instance, capacity activation.

Three types of decisions are considered: location, transportation, and production. For $i \in I$, y_{it} keeps denoting a binary variable equal to 1 if and only if a facility is operating at $i \in I$ in period $t \in T$, x_{ijt} represents now the proportion of the demand of customer $j \in J$ in period $t \in T$ that is supplied from location $i \in I$, and z_{it} is the production level at facility $i \in I$ in period $t \in T$. Model (2.14)–(2.18) can now be easily adapted to the new setting:

$$\text{maximize} \quad \sum_{i\in I}\sum_{j\in J}\left(p_{ij}\sum_{t\in T} d_{jt}x_{ijt}\right) - \sum_{i\in I}\sum_{t\in T} r_{it}z_{it} - \sum_{i\in I} f_i y_i, \tag{5.54}$$

$$\text{subject to} \quad \sum_{j\in J} d_{jt}x_{ijt} \le z_{it} \qquad \forall i \in I,\ t \in T, \tag{5.55}$$

$$\sum_{i\in I} x_{ijt} \le 1 \qquad \forall j \in J,\ t \in T, \tag{5.56}$$

$$z_{it} \le \Theta y_i \qquad \forall i \in I,\ t \in T, \tag{5.57}$$

$$x_{ijt} \ge 0 \qquad \forall i \in I,\ j \in J,\ t \in T, \tag{5.58}$$

$$y_i \in \{0, 1\} \qquad \forall i \in I, \tag{5.59}$$

$$z_{it} \ge 0 \qquad \forall i \in I,\ t \in T. \tag{5.60}$$

In this model, the objective function assesses the total profit. Constraint (5.55) accounts for the total production at the facilities, Constraints (5.56) ensure that no customer receives more than the respective demand, and Constraints (5.57) ensure that production occurs at operating facilities. In these inequalities, Θ denotes a sufficiently large constant. Finally, we have the domain of the decision variables.

We assume that demand is uncertain. In particular, we consider an uncertainty set for the demand of each customer in each period: $d_{jt} \in \mathcal{U}_{jt}$, $j \in J, t \in T$. Again, we can replace the objective function with a variable, say v. The model becomes

$$\text{maximize} \quad v, \tag{5.61}$$

$$\text{subject to} \quad \sum_{i\in I}\sum_{j\in J}\left(p_{ij}\sum_{t\in T} d_{jt}x_{ijt}\right) - \sum_{i\in I}\sum_{t\in T} r_{it}z_{it} - \sum_{i\in I} f_i y_i \ge v, \tag{5.62}$$

$$(5.55)\text{–}(5.60).$$

Hence, under the uncertainty sets considered above, the augmented constraint for (5.62) is

$$\min_{d_{jt}\in\mathcal{U}_{jt}}\left\{\sum_{i\in I}\sum_{j\in J}\left(p_{ij}\sum_{t\in T}d_{jt}x_{ijt}\right)\right\}-\sum_{i\in I}\sum_{t\in T}r_{it}z_{it}-\sum_{i\in I}f_iy_i\geq v. \tag{5.63}$$

The augmented constraint for (5.55) becomes

$$\max_{d_{jt}\in\mathcal{U}_{jt}}\left\{\sum_{j\in J}d_{jt}x_{ijt}\right\}\leq z_{it},\quad \forall i\in I,\ t\in T. \tag{5.64}$$

The type of inequality determines the way to write the augmented constraints.

Under box uncertainty, let us assume

$$\mathcal{U}_{jt}=[\bar{d}_{jt}(1-\epsilon_t),\bar{d}_{jt}(1-\epsilon_t)],\quad \forall i\in,\ t\in T.$$

In this case, the robust counterpart is straightforward:

$$\text{maximize}\quad v, \tag{5.61}$$

$$\text{subject to}\quad \sum_{i\in I}\sum_{j\in J}\left(p_{ij}\sum_{t\in T}(\bar{d}_{jt}-\epsilon_t)x_{ijt}\right)-\sum_{i\in I}\sum_{t\in T}r_{it}z_{it}-\sum_{i\in I}f_iy_i\geq v, \tag{5.65}$$

$$\sum_{j\in J}(\bar{d}_{jt}+\epsilon_t)x_{ijt}\leq z_{it},\quad \forall i\in I,\ t\in T. \tag{5.66}$$

(5.56)–(5.60).

Also, Constraints (5.57) can be fine-tuned considering

$$z_{it}\leq\left(\sum_{j\in J}(\bar{d}_{jt}+\epsilon_t)\right)y_i\qquad \forall i\in I,\ t\in T. \tag{5.67}$$

We are facing a compact model. Hence, we can, in principle, solve it for small- or medium-sized instances using a general-purpose solver.

The problem becomes more involved under ellipsoidal uncertainty:

$$\begin{aligned}\mathcal{U} &= \left\{ \boldsymbol{d} \in \mathbb{R}^{|J|\times|T|} \mid \sum_{j\in J}\sum_{t\in T}\left[\frac{d_{jt}-\bar{d}_{jt}}{\epsilon_t \bar{d}_{jt}}\right]^2 \le L^2 \right\} \\ &= \left\{ \boldsymbol{d} \in \mathbb{R}^{|J|\times|T|} \mid (\boldsymbol{d}-\bar{\boldsymbol{d}})^\top \Sigma^{-1}(\boldsymbol{d}-\bar{\boldsymbol{d}}) \le L^2 \right\}, \end{aligned} \tag{5.68}$$

where Σ is a $(|J|\times|T|)\times(|J|\times|T|)$ diagonal matrix whose entries are $\sigma_{jt} = \epsilon_t \bar{d}_{jt}$, $j \in J, t \in T$. Under this uncertainty set, the augmented constraint for (5.62) is now

$$\min_{d\in\mathcal{U}} \left\{ \sum_{i\in I}\sum_{j\in J}\left(p_{ij}\sum_{t\in T} d_{jt}x_{ijt}\right)\right\} - \sum_{i\in I}\sum_{t\in T} r_{it}z_{it} - \sum_{i\in I} f_i y_i \ge v. \tag{5.69}$$

The augmented constraint for (5.55) becomes

$$\max_{d\in\mathcal{U}} \left\{ \sum_{j\in J} d_{jt}x_{ijt} \right\} \le z_{it}, \quad i \in I,\ t \in T. \tag{5.70}$$

Let us analyze the first one. To do so we denote $V_{jt} = \sum_{i\in I} p_{ij}x_{ijt}$, $j \in J, t \in T$. Thus (5.69) becomes

$$\min_{d\in\mathcal{U}} \boldsymbol{V}^\top \boldsymbol{d} - \sum_{i\in I}\sum_{t\in T} r_{it}z_{it} - \sum_{i\in I} f_i y_i \ge v. \tag{5.71}$$

The problem that consists of finding a point $\boldsymbol{d} \in \mathcal{U}$ minimizing $\boldsymbol{V}^\top \boldsymbol{d}$ can again be easily solved by standard optimization techniques. The optimal solution is

$$\boldsymbol{d} = \bar{\boldsymbol{d}} - L\frac{1}{\sqrt{\boldsymbol{V}^\top \boldsymbol{\Lambda} \boldsymbol{V}}}\boldsymbol{\Lambda}\boldsymbol{V},$$

and the optimal value of the problem is

$$\boldsymbol{V}^\top \bar{\boldsymbol{d}} - L\sqrt{\boldsymbol{V}^\top \boldsymbol{\Lambda} \boldsymbol{V}}.$$

This leads to the following robust counterpart for (5.62):

$$\sum_{j\in J}\sum_{t\in T} \bar{d}_{jt}V_{jt} - L\sqrt{\sum_{j\in J}\sum_{t\in T}\sigma_{jt}^2 V_{jt}^2} - \sum_{i\in I}\sum_{t\in T} r_{it}z_{it} - \sum_{i\in I} f_i y_i \ge v. \tag{5.72}$$

We can linearize the above inequality by considering $W = \sqrt{\sum_{j\in J}\sum_{t\in T}\sigma_{jt}^2 V_{jt}^2}$. Now, the robust counterpart of (5.62) is jointly defined by

$$\sum_{j\in J}\sum_{t\in T}\bar{d}_{jt}V_{jt} - LW - \sum_{i\in I}\sum_{t\in T} r_{it}z_{it} - \sum_{i\in I} f_i y_i \geq v, \tag{5.73}$$

$$V_{jt} = \sum_{i\in I} p_{ij}x_{ijt} \qquad \forall j \in J,\ t \in T, \tag{5.74}$$

$$W = \sqrt{\sum_{j\in J}\sum_{t\in T}\sigma_{jt}^2 V_{jt}^2}. \tag{5.75}$$

Given that uncertainty is associated with the demands, we can assume that the ellipsoid capturing the uncertainty is in the positive orthant of $\mathbb{R}^{|J|\times|T|}$. This fact together with the maximization goal for v allows relaxing equalities (5.74) and (5.75):

$$V_{jt} \leq \sum_{i\in I} p_{ij}x_{ijt} \qquad \forall j \in J,\ t \in T, \tag{5.74$'$}$$

$$W \geq \sqrt{\sum_{j\in J}\sum_{t\in T}\sigma_{jt}^2 V_{jt}^2}. \tag{5.75$'$}$$

Constraints (5.70) can be handled easily considering the discussion made in the context of Constraint (5.49). Now, the robust counterpart becomes

$$\sum_{j\in J}\bar{d}_{jt}x_{ijt} + L\sqrt{\sum_{j\in J}\sigma_{jt}^2 x_{ijt}^2} \leq z_{it} \quad \forall i \in I,\ t \in T. \tag{5.76}$$

Consider $S_{it} = \sqrt{\sum_{j\in J}\sigma_{jt}^2 x_{ijt}^2}$, $i \in I, t \in T$. Assuming that the unit costs r_{it} are non-negative (and thus the production is to be minimized), we conclude that the robust counterpart of (5.55) becomes jointly defined by

$$\sum_{j\in J}\bar{d}_{jt}x_{ijt} + LS_{it} \leq z_{it} \qquad \forall i \in I,\ t \in T. \tag{5.77}$$

$$S_{it} \geq \sqrt{\sum_{j\in J}\sigma_{jt}^2 x_{ijt}^2} \qquad \forall i \in I,\ t \in T. \tag{5.78}$$

Note that parameter L represents a safety level for the uncertainty. Nothing forbids adapting that parameter to the constraints, thus considering different values for (5.62) and (5.55). In the latter case, we can even consider a time-dependent parameter. We also observe that inequalities (5.75′) and (5.78) define second-order conic constraints. However, variables x_{ijt} appear in the right-hand side in both cases. We can consider auxiliary variables $w_{jt} \geq \sigma_{jt} V_{jt}$ $(j \in J, t \in T)$ and $s_{ijt} \geq \sigma_{jt} x_{ijt}$, $i \in I$, $j \in J$, $t \in T$. Finally, we obtain the robust counterpart of our multi-period capacitated facility location problem (5.54)–(5.60):

$$\text{maximize} \quad v, \tag{5.61}$$

$$\text{subject to} \quad (5.73), (5.77), (5.74'),$$

$$w_{jt} \geq \sigma_{jt} V_{jt} \qquad \forall j \in J,\ t \in T, \tag{5.79}$$

$$s_{ijt} \geq \sigma_{jt} x_{ijt} \qquad \forall i \in I,\ j \in J,\ t \in T, \tag{5.80}$$

$$W \geq \sqrt{\sum_{j \in J} \sum_{t \in T} w_{jt}^2}, \tag{5.81}$$

$$S_{it} \geq \sqrt{\sum_{j \in J} s_{ijt}^2} \qquad \forall i \in I,\ t \in T, \tag{5.82}$$

$$(5.56)\text{–}(5.60).$$

This problem is a second-order conic programming problem that nowadays can be tackled efficiently by several off-the-shelf solvers.

The above problem assumes that all decisions are made here and now. Averbakh and Berman [38, 39] assume a similar framework but in the context of the p-center problem. In the first work, the authors assume a problem with uncertain node weights. The goal is to minimize the maximum regret. In the second article, box uncertainty is assumed for the node weights and edge lengths. A single center is to be located. In this case, a polynomial algorithm is proposed assuming the underlying network to be a tree.

In a multi-period setting as described above, it is natural that the decisions progressively adapt to the demand observed. Next, we discuss two-stage stochastic robust optimization facility location problems. Afterward, in Sect. 5.4, we consider a more general adaptive setting. Before that, we mention a different robustness concept introduced in [105] within the context of continuous facility location. The concept can be extended to network or discrete problems. In that paper, nominal values are assumed to have been estimated for the (uncertain) weights of a set of nodes. A maximum value is preset for the weighted distance between a single facility to be located and the demand nodes. The robustness of a location is then defined as the minimum deviation of the vector of weights regarding the nominal vector that renders that location infeasible. The goal of the problem is to find the most robust location. This yields a nonlinear fractional model that the authors tackle by existing methods and *ad hoc* procedures they propose in the paper.

5.3.1 Adaptive Decisions

We proceed the discussion with problems in which some decisions are adaptive, i.e., adjust to the revealed uncertainty. To enrich the discussion and provide the reader with a broader set of tools, we extend the analysis of Sect. 5.2 to a capacitated facility location problem and a more involved type of uncertainty sets—budget uncertainty sets—initially proposed in [75].

Consider the capacitated facility location problem formulated as follows:

$$\text{minimize} \quad \sum_{i \in I} f_i y_i + \sum_{i \in I} \sum_{j \in J} c_{ij} d_j x_{ij}, \tag{5.26}$$

$$\text{subject to} \quad \sum_{i \in I} x_{ij} = 1 \qquad \forall j \in J, \tag{5.27}$$

$$\sum_{j \in J} x_{ij} \leq q_i y_i \qquad \forall i \in I, \tag{5.83}$$

$$y_i \in \{0, 1\} \qquad \forall i \in I, \tag{5.29}$$

$$x_{ij} \geq 0 \qquad \forall i \in I,\ j \in J. \tag{5.30}$$

In the above model, recall that q_i is the capacity of a facility operating at $i \in I$.

Let us assume demand uncertainty characterized by the following budget-uncertainty set:

$$\mathcal{U}^{\mathrm{D}} = \left\{ \boldsymbol{d} \in \mathbb{R}^{|J|} \mid d_j = \overline{d}_j + \epsilon_j \Delta_j,\ 0 \leq \epsilon_j \leq 1,\ \sum_{j \in J} \epsilon_j \leq \Gamma^{\mathrm{D}} \right\}.$$

For $j \in J$, ϵ_j measures the uncertainty "magnitude," and $\overline{d}_j$ is the nominal value of the demand and Δ_j being a scaling factor for demand node j. In addition to demand uncertainty, [123, 124] also propose capturing facility disruption risk. In particular, the authors consider a related uncertainty set defined as

$$\mathcal{U}^{\mathrm{F}} = \left\{ \boldsymbol{z} \in \{0, 1\}^{|I|} \mid \sum_{i \in J} z_i \leq \Gamma^{\mathrm{F}} \right\},$$

with $z_i = 1$ indicating that facility $i \in I$ is disrupted. The above set limits to Γ^{F} the number of facilities that can be disrupted.

The above uncertainty sets can be combined as follows:

$$\mathcal{U} = \left\{ (\boldsymbol{d}, \boldsymbol{z}) \in \mathbb{R}^{|J|} \times \{0, 1\}^{|I|} \mid \boldsymbol{d} \in \mathcal{U}^{\mathrm{D}},\ \boldsymbol{z} \in \mathcal{U}^{\mathrm{F}} \right\}.$$

An adaptive robust optimization model can now be formulated:

$$\text{minimize} \quad \sum_{i \in I} y_i + \max_{(\boldsymbol{d}, \boldsymbol{z}) \in \mathcal{U}} g(\boldsymbol{y}, \boldsymbol{d}, \boldsymbol{z}), \tag{5.84}$$

$$\text{subject to} \quad y_i \in \{0, 1\} \qquad \forall i \in I, \tag{5.85}$$

where $g(\boldsymbol{y}, \boldsymbol{d}, \boldsymbol{z})$ is the optimal value of the second-stage problem:

$$\text{minimize} \quad \sum_{i \in I} \sum_{j \in J} c_{ij} x_{ij}, \tag{5.86}$$

$$\text{subject to} \quad \sum_{i \in I} x_{ij} = d_j \qquad \forall j \in J, \tag{5.87}$$

$$\sum_{i \in I} x_{ij} \leq q_i y_i (1 - z_i) \qquad \forall i \in I, \tag{5.88}$$

$$x_{ij} \geq 0 \qquad \forall i \in I, \; j \in J. \tag{5.89}$$

The meaning of the different components of the model should be immediate, given the discussion presented so far. Note that we can go further by relaxing the integrality of the z-variables, thus allowing a facility to be only partially disrupted.

One important aspect related to the above model is that we may not have a feasible second-stage completion for a first-stage decision. In fact, given a set of open facilities, it may happen that for some realization of the uncertainty, demand cannot be supplied. Nevertheless, this case can be easily accommodated in the above model. This can be accomplished by considering a dummy facility with unlimited capacity, whose unit service cost equals the unit cost for unsupplied demand. The setup cost of this "facility" is zero. Moreover, it cannot be disrupted (the corresponding z-variable is fixed to zero). Without loss of generality, we keep working with the above model.

The described problem can be solved using a column-and-constraint generation algorithm as proposed in [123, 124] that builds on the one proposed in [564] for the two-stage robust optimization problems. The master problem can be formulated as follows:

$$\text{minimize} \quad w, \tag{5.90}$$

$$\text{subject to} \quad w \geq \sum_{i \in I} f_i y_i + \sum_{i \in I} \sum_{j \in J} c_{ij} x_{ij}^{\ell} \quad \forall \ell = 1, \ldots, L, \tag{5.91}$$

$$\sum_{i \in I} x_{ij}^{\ell} = d_j^{\ell} \qquad \forall \ell = 1, \ldots, L, \; j \in J, \tag{5.92}$$

$$\sum_{j \in J} x_{ij}^{\ell} \leq q_i y_i (1 - z_i^{\ell}) \qquad \forall \ell = 1, \ldots, L, \; i \in I, \tag{5.93}$$

$$x_{ij}^{\ell} \geq 0 \qquad \forall \ell = 1, \ldots, L,\ i \in I,\ j \in J, \tag{5.94}$$

$$y_i \in \{0, 1\} \qquad \forall i \in I. \tag{5.95}$$

The above problem seeks to find the best location decision given that a certain set of L worst-case scenarios has been previously identified for the subproblems. The index ℓ indicates information added to the master problem after solving the ℓ-th subproblem. Additionally, d_j^{ℓ} and z_i^{ℓ} regard the worst-case scenario after the ℓ-th solved subproblem. Thus, the master problem provides a lower bound of the optimal value of the two-stage robust optimization problem.

The identification of the relevant scenarios to embed in the master problem is done by solving a subproblem: $\max_{(\boldsymbol{d},z)\in\mathcal{U}} g(\hat{\boldsymbol{y}}, \boldsymbol{d}, z)$, with $\hat{\boldsymbol{y}}$ standing for a solution obtained by solving a master problem. In particular, for a given first-stage solution $\hat{\boldsymbol{y}}$, $\max_{(\boldsymbol{d},z)\in\mathcal{U}} g(\hat{\boldsymbol{y}}, \boldsymbol{d}, z)$ can be obtained by solving the following MILP:

$$\text{maximize} \sum_{i\in I}\sum_{j\in J} c_{ij}x_{ij}, \tag{5.96}$$

$$\text{subject to} \sum_{i\in I} x_{ij} \geq \overline{d}_j + \epsilon_j \Delta_j \qquad \forall j \in J, \tag{5.97}$$

$$\sum_{j\in J} x_{ij} \leq q_i \hat{y}_i (1 - z_i) \qquad \forall i \in I, \tag{5.98}$$

$$\alpha_j - \beta_i \leq c_{ij} \qquad \forall i \in I,\ j \in J, \tag{5.99}$$

$$\sum_{i\in I} x_{ij} \leq \overline{d}_j + \epsilon_j \Delta_j + \Theta_j^I (1 - s_j^I) \qquad \forall j \in j, \tag{5.100}$$

$$\alpha_j \leq \Theta_j^I s_j^I \qquad \forall j \in J, \tag{5.101}$$

$$\sum_{j\in J} x_{ij} \geq q_i \hat{y}_i (1 - z_i) + \Theta_i^{II} (s_i^{II} - 1) \qquad \forall i \in I, \tag{5.102}$$

$$\beta_i \leq \Theta_i^{II} s_i^{II} \qquad \forall i \in I, \tag{5.103}$$

$$\alpha_j - \beta_i \geq c_{ij} + \Theta_{ij}^{III} (s_{ij}^{III} - 1) \qquad \forall i \in I,\ j \in J, \tag{5.104}$$

$$x_{ij} \leq \Theta_{ij}^{III} s_{ij}^{III} \qquad \forall i \in I,\ j \in J, \tag{5.105}$$

$$\epsilon_j \leq 1 \qquad \forall j \in J, \tag{5.106}$$

$$\sum_{j\in J} \epsilon_j \leq \Gamma, \tag{5.107}$$

$$\sum_{i\in I} z_i \leq \Gamma^{\mathrm{F}}, \tag{5.108}$$

$$x_{ij} \geq 0 \qquad \forall i \in I,\ j \in J, \tag{5.109}$$

$$\alpha_j \geq 0 \qquad \forall j \in J, \tag{5.110}$$

$$\beta_i \geq 0 \qquad \forall i \in I, \tag{5.111}$$

$$s_j^I, s_i^{II}, s_{ij}^{III} \in \{0, 1\} \qquad \forall i \in I,\ j \in J, \tag{5.112}$$

with $\Theta_j^I, \Theta_i^{II}, \Theta_{ij}^{III}$ standing for large constants that, nonetheless, can be easily fine-tuned setting, $\Theta_j^I = \overline{d}_j + \Delta_j$, $\Theta_i^{II} = \max\left\{q_i, \max_{j \in J}\{c_{ij}(\overline{d}_j + \Delta_j)\}\right\}$, $\Theta_{ij}^{III} = \max\left\{q_i, c_{ij}(\overline{d}_j + \Delta_j)\right\} + c_{ij}$.

The column-and-constraint generation procedure for the two-stage robust optimization problem is summarized in Algorithm 5.1 (see also [123, 124]).

Algorithm 5.1: Column-and-constraint generation algorithm

LB $\leftarrow -\infty$; UB $\leftarrow \infty$: $L = 0$;
Solve the deterministic model with $d_j = \overline{d}_j$, $j \in J$, to obtain an initial location solution $\hat{\boldsymbol{y}}$.
Set LB equal to the objective function value of the deterministic model.
repeat
  Solve the subproblem using $\hat{\boldsymbol{y}}$ to find the worst-case scenario $(\hat{\boldsymbol{d}}, \hat{z})$ and denote by ψ its optimal value;
  UB $\leftarrow \min\{\text{UB}, \psi\}$;
  if *(UB-LB)/UB* $> \varepsilon$ **then**
    Solve the master problem to get a location decision $\hat{\boldsymbol{y}}$ and its optimal value ϕ;
    Set LB $\leftarrow \phi$;
    $L \leftarrow L + 1$;
until *The given optimality gap* ε *is reached*;

For a problem like the one described above, [124] investigate a so-called linear decision rule assuming that uncertainty is associated only with facility disruptions. We recall that a linear decision rule restricts the second-stage variables to an affine function of the uncertain parameters [57].

In our case, this entails setting $x_{ij} = \boldsymbol{A}_{ij}^\top z + b_{ij}$, $i \in I$, $j \in J$, with $\boldsymbol{A}_{ij} \in \mathbb{R}^{|I|}$, $b_{ij} \in \mathbb{R}$. In [124] it is also noted that in the adjustable robust counterparts, the uncertainty set $\mathcal{U}^F$ has the integrality property, i.e., we can relax the integrality constraints imposed on the z-variables.

The problem stemming from assuming an affine decision rule can be formulated as an MILP problem by dualizing the robust constraints. Alternatively, a Benders decomposition procedure can be inspired in [31].

We conclude this section by discussing adaptive robust optimization in the context of another classical facility location problem: the p-center problem.We consider the model introduced in [154] revisited in Sect. 2.1:

$$\text{minimize} \quad z \tag{5.113}$$

$$\text{subject to} \quad \sum_{i \in I} c_{ij} x_{ij} \leq z \quad \forall j \in J, \tag{5.114}$$

$$\sum_{i \in I} x_{ij} = 1 \quad \forall j \in J, \tag{5.115}$$

$$\sum_{i \in I} y_i \leq p, \tag{5.116}$$

$$x_{ij} \leq y_i \quad \forall i \in I, \ j \in J, \tag{5.117}$$

$$y_i \in \{0, 1\} \quad \forall i \in I, \tag{5.118}$$

$$x_{ij} \in \{0, 1\} \quad \forall i \in I, \ j \in J. \tag{5.119}$$

Assuming uncertainty associated with parameters c_{ij} (e.g., travel times) and also that the facilities are to be installed here and now, a two-stage problem emerges in case we allow the allocation of the demand nodes to the facilities to be implemented only after uncertainty is revealed.

Let us consider box uncertainty:

$$\mathcal{U} = \left\{ \boldsymbol{c} \in \mathbb{R}^{|I| \times |J|} \mid c_{ij}^{-} \leq c_{ij} \leq c_{ij}^{+} \ \forall i \in I, \ j \in J \right\}.$$

The above uncertainty set entails an infinite number of scenarios. Nevertheless, [183, 355] identify a finite dominating set of scenarios that allows finding the optimal solution. We directly consider the former work due to the reduced cardinality of the dominating set.

Let $\hat{\boldsymbol{y}}$ be a feasible location decision for the p-center problem. Note that any subset of p locations is feasible. For $j' \in J$, consider the cost scenario $\omega_{j'}(\hat{\boldsymbol{y}})$ defined by

$$c_{ij}(\hat{\boldsymbol{y}}) = \begin{cases} c_{ij}^{+} & \text{if } j = j' \text{ and } \hat{y}_i = 1, \\ c_{ij}^{-}, & \text{otherwise,} \end{cases}$$

$\forall i \in I, \ j \in J$. In [183], it is shown that given a feasible location decision $\hat{\boldsymbol{y}}$ for the p-center problem, there exists $j \in J$ such that scenario $\omega_j(\hat{\boldsymbol{y}})$ leads to the maximal deviation for $\hat{\boldsymbol{y}}$. This result justifies restricting uncertainty to a finite set of scenarios, namely

$$\Omega = \left\{ \omega_j(\boldsymbol{y}) \mid j \in J, \ \boldsymbol{y} \in \mathcal{Y} \right\},$$

with $\mathcal{Y} = \left\{ \boldsymbol{y} \in \mathbb{R}^{|I|} \mid \sum_{i \in I} y_i = p \right\}$.

Considering this finite set of scenarios, we can reformulate the two-stage robust p-center problem. Seeking to avoid the usually overly conservative minmax solution (in our case, min-max-min), we can directly think of a minmax regret objective. For $\omega \in \Omega$, let v^*_ω be the optimal value of the deterministic problem induced by that single scenario. A minmax regret p-center problem can be formulated as follows:

$$\text{minimize} \quad \zeta, \tag{5.120}$$

$$\text{subject to} \quad \zeta \geq \sum_{i \in I} c_{ij\omega} x_{ij\omega} - v^*_\omega \quad \forall j \in J,\ \omega \in \Omega, \tag{5.121}$$

$$\sum_{i \in I} x_{ij\omega} = 1 \quad \forall j \in J,\ \omega \in \Omega, \tag{5.122}$$

$$\sum_{i \in I} y_i \leq p, \tag{5.123}$$

$$x_{ij\omega} \leq y_i \quad \forall i \in I,\ j \in J,\ \omega \in \Omega, \tag{5.124}$$

$$y_i \in \{0, 1\} \quad \forall i \in I, \tag{5.125}$$

$$x_{ij\omega} \in \{0, 1\} \quad \forall i \in I,\ j \in J,\ \omega \in \Omega. \tag{5.126}$$

Duran-Mateluna et al. [183] extend this analysis in different ways. First, the authors consider that a weight (e.g., demand) is also associated with each node in $j \in J$. Second, they explore different formulations for the problem stemming from existing optimization models for the classical p-center problem. Additionally, after observing that the number of constraints involved in the problem can grow too fast (Ω is combinatorial), they propose a column and constraint generation algorithm. The latter can be summarized as follows:

A restricted set of scenarios is considered (possibly empty). In each iteration, the algorithm finds a solution satisfying all currently in the scenario set. If the solution does not satisfy some scenario in the whole set Ω, the most violated scenario is included in the restricted set. Again, a solution is found. When no violated scenario is found, an optimal solution is returned. Note this algorithm can be executed considering any base formulation for the p-center problem. Duran-Mateluna et al. [183] explore this fact by comparing different versions of the algorithm. The authors also develop a branch-and-cut algorithm based on two specific base models for the p-center problem, namely those introduced in [13, 215].

Finally, we note that [183] also extends the works by [355, 356] by considering adaptive decisions and by developing exact algorithms.

5.4 Multi-period Robust Facility Location

We now focus on a multi-stage robust optimization facility location problem extending the settings just discussed. We assume an underlying multi-period facility location. We start by formulating the (deterministic) problem in a way more convenient for capturing uncertainty.

We assume most of the notation and terminology in use so far. Hence, we keep denoting by d_{jt} the demand of node $j \in J$ in period $t \in T$. We assume a pure phase-in problem. This means that when a facility is installed, it remains operating until the end of the planning horizon. In this case, we can redefine the location variables: y_{it} is equal to one if and only if a facility is installed at $i \in I$ at the beginning of period $t \in T$ remaining operating until the end of the planning horizon. A fixed cost f_i is paid for installing a facility at $i \in I$. Variables x_{ijt} keep representing the amount shipped from facility $i \in I$ to demand node $j \in J$ in period $t \in T$. We consider a unit revenue r_{ij} for the commodity shipped from facility $i \in I$ to demand node $j \in J$. We assume a maximization-oriented objective driven by the total revenue net present value (NPV). Under this objective, supplying the entire demand is no longer a goal. Instead, we consider a unit penalty, denoted by s, for the unsupplied demand and define an additional set of decision variables. For $j \in J$ and $t \in T$, z_{jt} is the unsupplied amount of customer j in period t. Finally, we denote by β_t the present value of one monetary unit t periods ahead—discount factor related to period $t \in T$. The multi-period facility location problem can be formulated as follows:

$$\text{maximize} \quad \phi(\boldsymbol{x}, \boldsymbol{y}, \boldsymbol{z}), \tag{5.127}$$

$$\text{subject to} \quad \sum_{i \in I} x_{ijt} + z_{jt} = d_{jt} \qquad \forall j \in J,\ t \in T, \tag{5.128}$$

$$\sum_{j \in J} x_{ijt} \leq q_i \sum_{\tau=0}^{t} y_{it} \qquad \forall i \in I,\ t \in T, \tag{5.129}$$

$$x_{ijt} \geq 0 \qquad \forall i \in I,\ j \in J,\ t \in T, \tag{5.130}$$

$$\boldsymbol{y} \in \mathcal{Y}, \tag{5.131}$$

$$y_{it} \in \{0, 1\} \qquad \forall i \in I,\ t \in T, \tag{5.132}$$

$$z_{jt} \geq 0 \qquad \forall j \in J,\ t \in T, \tag{5.133}$$

with $\phi(\boldsymbol{x}, \boldsymbol{y}, \boldsymbol{z}) = \sum_{t \in T} \phi_t(\boldsymbol{x}_t, \boldsymbol{y}_t, \boldsymbol{z}_t)$ denoting the total NPV. For $t \in T$,

$$\phi_t(\boldsymbol{x}_t, \boldsymbol{y}_t, \boldsymbol{z}_t) = \beta_t \sum_{j \in J} \left[\sum_{i \in I} r_{ij} x_{ijt} - s\, z_{jt} \right] - \beta_{t-1} \sum_{i \in I} f_i x_{it}. \tag{5.134}$$

In the above expression, the location costs for facilities installed at the beginning of period t are corrected $t-1$ periods. The revenues and penalties for unsupplied demands are corrected t periods (we refer them to the end of period T), $t \in T$.

In the above model, Constraints (5.128) ensure that the demand of each customer in each period is fully supplied or the unsupplied demand is accounted for to be penalized. Constraints (5.129) regard the capacity of the facilities. In each period, location i can supply up to q_i units only if a facility is established in that location at the beginning of a period in $\{1, \ldots, t\}$. If not, then no unit can be shipped from that location. Constraints (5.132) accommodate additional requirements on the location decisions through a polyhedral set denoted by $\mathcal{Y}$. The remaining constraints refer to the domain of the decision variables.

In what follows, we assume the following general model for describing the demand over time:

$$d_{jt} = d_{jt-1} + \eta_{it} \quad \forall i \in I, t \in T,$$

where η_{it} is the variation observed in the demand of customer $j \in J$ in period t. Moreover, $d_{j0} = \hat{d}_{j0}$ is the demand of customer $j \in J$ observed in the period before the beginning of the planning horizon.

For $j \in J$ and $k = 1, \ldots, K$, let $\hat{d}_{j,-k}$ denote the demand of customer j observed k periods before the beginning of the planning horizon. This allows computing the observed variations:

$$\hat{\eta}_{j,-k} = \hat{d}_{j,-k} - \hat{d}_{j,-(k+1)} \quad \forall j \in J,\ k = 1, \ldots, K-1.$$

For $j \in J, t \in T$, denote by $\mathcal{F}_t(\hat{\boldsymbol{\eta}}_j)$ the t-year ahead forecast of the variation η_{jt} made at the beginning of the planning horizon where

$$\hat{\boldsymbol{\eta}}_j = \left(\hat{\eta}_{j,0}, \hat{\eta}_{j,-1}, \ldots, \hat{\eta}_{j,-(K-1)}\right)^\top.$$

Making use of the data $\hat{\boldsymbol{\eta}}_j$ and the forecast model $\mathcal{F}_t(\hat{\boldsymbol{\eta}}_j)$, we can determine spread statistics $\mathcal{S}_t(\hat{\boldsymbol{\eta}}_j)$ for the forecasting error $\eta_{jt} - \mathcal{F}_t(\hat{\boldsymbol{\eta}}_j)$, such that the possibility of

$$\left|\eta_{jt} - \mathcal{F}_t(\hat{\boldsymbol{\eta}}_j)\right| \leq \mathcal{S}_t(\hat{\boldsymbol{\eta}}_j)$$

is high, for $t \in T$.

Without loss of generality, both $\mathcal{F}_t(\hat{\boldsymbol{\eta}}_j)$ and $\mathcal{S}_t(\hat{\boldsymbol{\eta}}_j)$ are presented here in the general sense, and they are specified by choosing the suitable models (statistics) based on the characteristics of the given data.

Furthermore, to mitigate conservativeness, we incorporate budget uncertainty and consider the following budget model to (dynamically) limit the variation of future demand variation:

$$\eta_{jt} = \mathcal{F}_t(\hat{\boldsymbol{\eta}}_j) + \varrho_{jt}\mathcal{S}_t(\hat{\boldsymbol{\eta}}_j) \qquad \forall j \in J,\ t \in T \tag{5.135}$$

$$\sum_{t\in T} |\varrho_{jt}| \le \Gamma_j^1 \qquad \forall i \in I, \tag{5.136}$$

$$\sum_{j\in J} |\varrho_{jt}| \le \Gamma_t^2 \qquad \forall t \in T, \tag{5.137}$$

$$\varrho_{jt} \in [-1, 1] \qquad \forall j \in J,\ t \in T, \tag{5.138}$$

where $\Gamma_j^1 \le |T|$ $(j \in J)$ are the values controlling the total variation of demand variation for customer j over the entire horizon T. In turn, $\Gamma_t^2 \le |I|$ $(t \in T)$ are the values limiting the total demand variation over all the demand nodes. Finally, ϱ_{it}, $i \in I$, and $t \in T$ are the budget-uncertainty variables. In the above budget model, each spread statistic $\mathcal{S}_t(\hat{\boldsymbol{\eta}}_j)$ can be interpreted as a budget unit to be assigned through the budget variable ϱ_{jt}.

From the above increment model, it is easy to conclude that

$$d_{jt} = \hat{d}_{j0} + \sum_{\tau=1}^{t} \mathcal{F}_\tau(\hat{\boldsymbol{\eta}}_j) + \sum_{\tau=1}^{t} \varrho_{j\tau}\mathcal{S}_\tau(\hat{\boldsymbol{\eta}}_j) \quad \forall j \in J,\ t \in T.$$

This implies that each d_{jt} has one more free budget unit $\mathcal{S}_t(\hat{\boldsymbol{\eta}}_j)$ than d_{jt-1}, which captures the uncertainty propagation effect in future demand over time.

Gathering all the above ingredients, we can consider a so-called budget-incorporated prediction set for demands $\boldsymbol{d}_t$, based on the demands $\hat{\boldsymbol{d}}_0, \boldsymbol{d}_1, \ldots, \boldsymbol{d}_{t-1}$, $t \in T$.

In practice, demands are observed period by period, i.e., $\boldsymbol{d}_1, \boldsymbol{d}_2, \ldots, \boldsymbol{d}_{|T|}$. Therefore, we can construct the budget-incorporated prediction set in a nested fashion to realize the ambiguity-averse evaluation process period by period.

For each $t \in T$, we denote by $\boldsymbol{d}_{[t-1]} = (\hat{\boldsymbol{d}}_0, \boldsymbol{d}_1, \boldsymbol{d}_2, \ldots, \boldsymbol{d}_{t-1})$ the scenario path of demands up to time period $t-1$, with $\boldsymbol{d}_{[0]} = \hat{\boldsymbol{d}}_0$.

Given the scenario path $\boldsymbol{d}_{[t-1]}$, by the construction of the budget-based model for each $\boldsymbol{d}_t$, we actually also know the budget variables $\varrho_{j\tau} \in [-1, 1]$, $j \in J$, $\tau = 1, \ldots, t-1$ that determine each $d_{j\tau}$. Therefore, the prediction set for $\boldsymbol{d}_t$ can be expressed as follows:

$$\mathcal{U}_t^\Gamma(\boldsymbol{d}_{[t-1]}) = \left\{ \boldsymbol{d}_t \in \mathbb{R}^{|J|} \;\middle|\; \begin{array}{l} d_{it} = d_{jt-1} + \mathcal{F}_t(\hat{\boldsymbol{\eta}}_j) + \varrho_{jt}\mathcal{S}_t(\hat{\boldsymbol{\eta}}_j),\ j \in J \\ \sum_{j\in J} |\varrho_{jt}| \le \Gamma_t^2 \\ |\varrho_{jt}| \le \Gamma_j^1 - \mathbb{I}(t \ge 2)\left[\sum_{\tau=1}^{t-1} |\varrho_{j\tau}|\right],\ j \in J \\ \varrho_{jt} \in [-1, 1],\ j \in J \end{array} \right\}, \tag{5.139}$$

where $\mathbb{I}(t \geq 2)$ is an indicator function taking the value of 1 when $t \geq 2$ and 0 otherwise ($t = 1$), which is used to distinguish the cases $t = 1$ and $t \geq 2$.

Based on the budget-incorporated prediction set (5.139), we can build the multi-period robust facility location model. Given any (feasible) location decision $\boldsymbol{y} = \left(\boldsymbol{y}_1, \boldsymbol{x}_2, \ldots, \boldsymbol{y}_{|T|}\right)$ and the uncertainty budget Γ, we now evaluate the "Ambiguity-Averse Performance" of design $\boldsymbol{y}$.

Denote by $\boldsymbol{y}_{[t]} = (\boldsymbol{y}_1, \ldots, \boldsymbol{y}_t)$ the given location designs over all periods up to the beginning of period $t \in T$. At each time period $t \in T$ with any given demand forecast $\boldsymbol{d}_t$, we denote by

$$\mathcal{X}_t\left(\boldsymbol{y}_{[t]}, \boldsymbol{d}_t\right) = \left\{(\boldsymbol{x}_t, z_t) \in \mathbb{R}_+^{|I|\times|J|} \times \mathbb{R}_+^{|J|} \mid (5.128),\ (5.129)\right\}$$

the feasible set for decisions $(\boldsymbol{x}_t, z_t)$ at time period t given $\boldsymbol{y}_{[t]}$ and $\boldsymbol{d}_t$.

Therefore, by the structure of the multi-period facility location problem—point-wise forecasting model (5.127)–(5.133)—the decisions $\boldsymbol{x}_t$ and z_t are made by maximizing the NPV at the period t, which is only affected by $\boldsymbol{y}_{[t]}$ and the demands $\boldsymbol{d}_t$ at that period:

$$\max_{(\boldsymbol{x}_t, z_t)\in\mathcal{X}_t(\boldsymbol{y}_{[t]}, \boldsymbol{d}_t)} \phi_t\left(\boldsymbol{y}_t, \boldsymbol{x}_t, z_t\right),$$

where $\phi_t\left(\boldsymbol{y}_t, \boldsymbol{x}_t, z_t\right)$ is the NPV function at the period t as defined in (5.134). Furthermore, we denote by

$$\Upsilon_t(\boldsymbol{y}_{[t]}, \boldsymbol{d}_t) = \max_{(\boldsymbol{x}_t, z_t)\in\mathcal{X}_t(\boldsymbol{y}_{[t]}, \boldsymbol{d}_t)} \phi_t\left(\boldsymbol{y}_t, \boldsymbol{y}_t, z_t\right) \tag{5.140}$$

the optimal NPV at time period t given $\boldsymbol{y}_{[t]}$ and $\boldsymbol{d}_t$.

Recall that in each period t, the forecast $\boldsymbol{d}_t$ cannot be perfect; we therefore evaluate the ambiguity-averse performance of the feedstock allocation at period t that hedges the ambiguity within $\mathcal{U}_t^{\Gamma}(\boldsymbol{d}_{[t-1]})$, which solves the following minmax optimization problem:

$$\min_{\boldsymbol{d}_t\in\mathcal{U}_t^{\Gamma}(\boldsymbol{d}_{[t-1]})} \Upsilon_t(\boldsymbol{y}_{[t]}, \boldsymbol{d}_t) = \min_{\boldsymbol{d}_t\in\mathcal{U}_t^{\Gamma}(\boldsymbol{d}_{[t-1]})} \max_{(\boldsymbol{x}_t, z_t)\in\mathcal{X}_t(\boldsymbol{y}_{[t]}, \boldsymbol{d}_t)} \phi_t\left(\boldsymbol{y}_{[t]}, \boldsymbol{y}_t, z_t\right). \tag{5.141}$$

The above ambiguity-averse performance (5.141) depends on the demands $\boldsymbol{d}_{[t-1]}$ up to period $t-1$. We also note that the ambiguity-averse demand $\boldsymbol{d}_t^{\dagger}$ solved from (5.141) at period t as the input of $\mathcal{U}_t^{\Gamma}(\boldsymbol{d}_{[t]})$ will affect sequentially the future periods' ambiguity-averse evaluations.

Embedding the above ambiguity-averse evaluation components of all periods $t \in T$, we can formulate the overall ambiguity-averse NPV performance, denoted by $\mathcal{Z}_{\text{NPV}}(\boldsymbol{y})$, which solves the following nested minmax dynamic optimization problem:

$$\mathcal{Z}_{\text{NPV}}(\boldsymbol{y}) = \min_{\boldsymbol{d}_1 \in \mathcal{U}_1^{\Gamma}(\boldsymbol{d}_{[0]})} \left[\Upsilon_1(\boldsymbol{y}_{[1]}, \boldsymbol{d}_1) + \min_{\boldsymbol{d}_2 \in \mathcal{U}_2^{\Gamma}(\boldsymbol{d}_{[1]})} \left[\Upsilon_2(\boldsymbol{y}_{[2]}, \boldsymbol{d}_2) + \dots \right.\right.$$
$$\left.\left. + \min_{\boldsymbol{d}_{|T|} \in \mathcal{U}_{|T|}^{\Gamma}(\boldsymbol{d}_{[T-1]})} \left[\Upsilon_T(\boldsymbol{y}_{[|T|]}, \boldsymbol{d}_T) \right] \dots \right] \right], \tag{5.142}$$

where $\mathcal{U}_t^{\Gamma}(\boldsymbol{d}_{t-1})$, $t \in T$, are the scenario path-dependent prediction sets given by (5.139).

We have just established in (5.142) the evaluation problem of ambiguity-averse NPV performance given location design $\boldsymbol{y}$, from the realistic information realization perspective. However, problem (5.142) is challenging to handle directly. Wang et al. [537] provide a tractable reformulation and develop an efficient cutting plane algorithm to solve it. For more details, the readers are referred to that work.

5.5 Conclusion and Further Reading

In this section, we have introduced robust facility location problems under finite and infinite uncertainty sets. We highlighted the differences and related challenges. Adaptive decisions were part of the discussion throughout the chapter that culminated in a multi-period robust facility location problem under a prediction uncertainty set. The latter integrates a set of forecasts and is constructed based on user-specified levels of forecasting errors.

Different types of uncertainty sets apply to different situations. A finite uncertainty set is suitable when a certain number of scenarios can be accurately identified. An infinite uncertainty set as a natural extension (e.g., convex full) of the former applies to situations with limited information acquired on the uncertain parameters. Also, when forecasting information is available, an uncertainty budget-incorporated prediction set can be constructed as done in (5.139) of Sect. 5.4. This can effectively describe the underlying uncertainty.

Additional insights on the topic can be found in different sources. Snyder [510] provides an overview, [24, 71, 216, 536, 564, 577] consider two-stage (adaptive) robust facility location, and [31] and [445] investigate multi-period robust facility location.

An interesting perspective focused on in [24] regards the fact that under disruption, demand may also change, which impacts the network design.

Chapter 6
Stochastic Facility Location

Abstract This chapter discusses stochastic facility location problems. It starts with location models considering reliability objectives and then focuses on two-stage decision-making processes. Different settings are analyzed, including *ex ante* location or allocation decisions, unit-demand customers, multi-period facility location, risk-averse models, and covering location problems. Together with the modeling aspects, we also discuss algorithmic developments.

Keywords Reliability objectives · Recourse decisions · Unit-demand problems · Multi-stage problems · Risk aversion · Congestion

6.1 Introduction

In this chapter, we assume the uncertainty underlying a facility location problem to be represented by a random vector with a given cumulative distribution function (CDF). This raises different challenges from modeling and resolution perspectives. Using well-known facility location problems, the existing knowledge is progressively explored and discussed throughout the chapter. The problems at the core of the discussion include the p-median problem, the capacitated facility location problem, and the maximal covering location problem. Several variants and extensions are also considered. We start by focusing on reliability objectives. Afterward, we concentrate on problems with recourse decisions (two and multiple stage). A specific section is then devoted to models and solution techniques for risk-averse decision-makers. The chapter proceeds by considering covering location problems and ends with an overview and suggestions for further reading.

© The Author(s), under exclusive license to Springer Nature Switzerland AG 2024
F. Saldanha-da-Gama, S. Wang, *Facility Location Under Uncertainty*,
International Series in Operations Research & Management Science 356,
https://doi.org/10.1007/978-3-031-55927-3_6

6.2 Reliability Objectives

When probabilities can be associated with the observations of the random vector underlying an optimization problem, different paradigms can be considered for formulating and analyzing it. As we have done in the previous chapter, we consider a well-known discrete facility location problem to start our discussion: the discrete p-median problem. Recalling the synthesis presented in Chap. 2, the problem can be formulated as follows:

$$\text{minimize} \quad \sum_{i,j \in J} c_{ij} d_j x_{ij}, \tag{6.1}$$

$$\text{subject to} \quad \sum_{i \in J} x_{ij} = 1 \qquad \forall j \in J, \tag{6.2}$$

$$x_{ij} \leq x_{ii} \qquad \forall i, j \in J, \tag{6.3}$$

$$\sum_{i \in J} x_{ii} = p, \tag{6.4}$$

$$x_{ij} \in \{0, 1\} \qquad \forall i, j \in J. \tag{6.5}$$

In the model, J is the set of demand nodes that also defines the potential locations for the facilities. The binary variable x_{ij} equals one if and only if demand node $j \in J$ is allocated to facility $i \in J$. The problem seeks to select p facilities and allocate the demand nodes to them. The objective is to minimize the total weighted cost for supplying the demand.

Assume that both the demands d_j and the costs c_{ij} (e.g., distances or travel times) are sources of uncertainty. Suppose that such uncertainty can be captured by a finite set of scenarios, Ω. Every scenario $\omega \in \Omega$ determines both the demands, denoted by $d_{j\omega}$, and the costs, denoted by $c_{ij\omega}$, $i, j \in J$. Let π_ω be the probability of scenario $\omega \in \Omega$ that we assume to be given. One possibility for edging against uncertainty is to adopt the concept of α-robustness introduced in [509]. In this case, we seek a solution minimizing the expected cost but such that the relative regret in each scenario is less than or equal to some given tolerance α. Recall the minmax regret formulation presented in Chap. 5 for the discrete p-median problem. Let v^*_ω be the optimal value of the single-scenario p-median problem induced by scenario ω. The α-robust discrete p-median problem becomes

$$\text{minimize} \quad \sum_{\omega \in \Omega} \sum_{i,j \in J} \pi_\omega d_{j\omega} c_{ij\omega} x_{ij\omega}, \tag{6.6}$$

$$\text{subject to} \quad \sum_{i,j \in J} d_{j\omega} c_{ij\omega} x_{ij\omega} \leq (1 + \alpha) v^*_\omega \quad \forall \omega \in \Omega, \tag{6.7}$$

$$\sum_{i \in J} x_{ij\omega} = 1 \qquad \forall j \in J,\ \omega \in \Omega, \tag{6.8}$$

$$x_{ij\omega} \leq y_i \qquad \forall i, j \in J,\ \omega \in \Omega, \tag{6.9}$$

$$\sum_{i \in J} y_i = p, \tag{6.10}$$

$$x_{ij\omega} \in \{0, 1\} \qquad \forall i, j \in J, \tag{6.11}$$

$$y_i \in \{0, 1\} \qquad \forall i \in J. \tag{6.12}$$

As pointed out in [509], this model generalizes the well-known models proposed in [383] and [542] who seek to minimize the expected total cost in a p-median problem (Constraints (6.9) are omitted in those works). The former considers here-and-now location decisions and adapts the allocation to how uncertainty is revealed. The latter assumes that all decisions are to be implemented here and now.

Snyder and Daskin [509] also apply these ideas to the UFLP. The extension is straightforward. We omit it here. Those authors show that the α-robust discrete p-median problem is NP-hard (it has the p-median problem as a particular case). Even the feasibility problem is NP-complete if we have at least two scenarios. For this reason and algorithmic purposes, the authors prove that

$$\sum_{\omega \in \Omega} \pi_\omega (1 + \alpha) v^*_\omega$$

is a valid surrogate upper bound. Furthermore, an efficient Lagrangian-based procedure can be developed for the problem by replicating the x variables and reformulating the problem as follows:

$$\text{minimize} \quad \gamma \sum_{\omega \in \Omega} \sum_{i,j \in J} \pi_\omega d_{j\omega} c_{ij\omega} x_{ij\omega} + (1 - \gamma) \sum_{\omega \in \Omega} \sum_{i,j \in J} \pi_\omega d_{j\omega} c_{ij\omega} z_{ij\omega}, \tag{6.13}$$

$$\text{subject to} \quad \sum_{i,j \in J} d_{j\omega} c_{ij\omega} z_{ij\omega} \leq (1 + \alpha) v^*_\omega \qquad \forall \omega \in \Omega, \tag{6.14}$$

$$\sum_{i \in J} z_{ij\omega} = 1 \qquad \forall j \in J,\ \omega \in \Omega, \tag{6.15}$$

$$z_{ij\omega} = x_{ij\omega} \qquad \forall i, j \in J,\ \omega, \in \Omega, \tag{6.16}$$

$$z_{ij\omega} \in \{0, 1\} \qquad \forall i, j \in J, \tag{6.17}$$

$$(6.9) - (6.12).$$

By relaxing the coupling constraints (6.16) using Lagrangian multipliers $\lambda_{ij\omega}$, the relaxed problem can be decomposed into two problems: $\mathcal{L}(\boldsymbol{\lambda}, \boldsymbol{x}, \boldsymbol{y})$ and $\mathcal{L}(\boldsymbol{\lambda}, \boldsymbol{z})$:

$$\mathcal{L}(\boldsymbol{\lambda}, \boldsymbol{x}, \boldsymbol{y}) = \min \sum_{\omega\in\Omega} \sum_{i,j\in J} \left(\gamma\pi_\omega d_{j\omega} c_{ij\omega} - \lambda_{ij\omega}\right) x_{ij\omega}, \tag{6.18}$$

$$\text{s. t.} \quad (6.9) - (6.12),$$

$$\mathcal{L}(\boldsymbol{\lambda}, \boldsymbol{z}) = \min \sum_{\omega\in\Omega} \sum_{i,j\in J} \left((1-\gamma)\pi_\omega d_{j\omega} c_{ij\omega} + \lambda_{ij\omega}\right) z_{ij\omega}, \tag{6.19}$$

$$\text{s. t.} \quad (6.14),\ (6.15),\ (6.17).$$

The first subproblem above can be solved by computing the values

$$\beta_i = \sum_{\omega\in\Omega} \sum_{j\in J} \min\{0, \gamma\pi_\omega d_{j\omega} c_{ij\omega} - \lambda_{ij\omega}\},\ i \in J,$$

and setting $y_i = 1$ for the locations yielding the p smallest such values. Each $x_{ij\omega}$ is then set equal to one if $y_i = 1$ and $\gamma\pi_\omega d_{j\omega} c_{ij\omega} - \lambda_{ij\omega} < 0$.

The second subproblem can be decomposed into $|\Omega|$ instances of the multi-choice knapsack problem [93]. The instance for scenario $\omega \in \Omega$ has $|J|$ classes, each determined by one demand node $j \in J$. Each class contains $|I|$ elements, one for each potential facility. The "item" $i \in I$ in class $j \in J$ has a weight $d_{js} c_{ij\omega}$ and a value $(1-\gamma)\pi_\omega d_{j\omega} c_{ij\omega} + \lambda_{ij\omega}$. Including this "item" in the knapsack indicates that demand node j is assigned to facility i. The capacity of the knapsack is set equal to $(1+\alpha)v^*_\omega$. This problem can be solved efficiently (see, e.g., [305]). As noted in [509], if a heuristic is used, then it should be one producing a lower bound so that the Lagrangian problem provides a valid lower bound.

Once a Lagrangian problem is solved, we can easily find an upper bound by opening the facilities such that $y_i = 1$ and assigning the demand nodes in each scenario to the lowest-cost assignment facility.

We seek to solve the Lagrangian dual:

$$\max_{\boldsymbol{\lambda}} \mathcal{L}(\boldsymbol{\lambda}, \boldsymbol{x}, \boldsymbol{y}) + \mathcal{L}(\boldsymbol{\lambda}, \boldsymbol{z}).$$

This can be accomplished using, for instance, subgradient optimization. If needed, the final gap can be closed using a branch-and-bound procedure.

When the number of scenarios is too large, the large-scale models obtained easily become intractable. Moreover, in practice, focusing explicitly on all scenarios may not be that relevant. In this case, restricting the scenario set is desirable, if not unavoidable. Daskin et al. [156] introduce the concept of α-reliability set:

Definition 6.1 An α-reliability set is a subset of scenarios that have a joint probability at least equal to α $(0 < \alpha < 1)$.

Two reliability measures have emerged that explore this concept: the maximum regret over an α-reliability set and the α-reliability mean excess regret.

In the first case, one seeks to minimize the maximum regret over a subset of scenarios with the condition that the total probability of the selected scenarios is at least α. The problem can be formulated considering one additional decision binary variable, u_ω, equal to one if and only if scenario $\omega \in \Omega$ is considered. The minmax regret p-median problem over an α-reliability set can now be formulated as follows:

$$\text{minimize} \quad v \tag{6.20}$$

$$\text{subject to} \quad \left(\sum_{i,j \in J} d_{j\omega} c_{ij\omega} x_{ij\omega} - v_\omega^* \right) - M_\omega (1 - u_\omega) \leq v \quad \forall \omega \in \Omega, \tag{6.21}$$

$$\sum_{\omega \in \Omega} \pi_\omega u_\omega \geq \alpha, \tag{6.22}$$

$$(6.8) - (6.12),$$

$$u_\omega \in \{0, 1\} \qquad \forall \omega \in \Omega. \tag{6.23}$$

Daskin et al. [156] solve this model using a general-purpose solver.

As pointed out in [114], the α-reliable minmax regret has some drawbacks. First, it does not assess the regret for the scenarios not included in the reliability set. Second, it does not distinguish between the cases in which (i) the regrets in the tail are only slightly worse than the minimum α-reliable maximum regret and (ii) the regrets in the tail are significantly larger. Finally, the α-reliable maximum regret is non-smooth, non-convex, and multi-extreme as a function of the decision variables. These facts led the authors to adopt the so-called α-reliable mean excess regret. This measure weights the maximum regret over the reliability set and the conditional regret expectation over the scenarios not included in the reliability set. This is equivalent to considering the conditional value-at-risk using the regret as the loss function. Hence, we seek to minimize the expected regret regarding the worst-case scenarios, namely those whose collective probability of occurrence is at most $1-\alpha$. Therefore, we can easily adapt the above models (also recalling some contents of Sect. 3.5):

$$\text{minimize} \quad \eta + \frac{1}{1-\alpha} \sum_{\omega \in \Omega} \pi_\omega \zeta_\omega \tag{6.24}$$

$$\text{subject to} \quad \zeta_\omega \geq \left(\sum_{i,j \in J} d_{j\omega} c_{ij\omega} x_{ij\omega} - v_\omega^* \right) - \eta \qquad \forall \omega \in \Omega, \tag{6.25}$$

$$(6.8) - (6.12),$$

$$\zeta_\omega \geq 0 \quad \forall \omega \in \Omega. \tag{6.26}$$

This model can be solved using a general-purpose solver (see [114]).

6.3 Recourse Decisions

The previous section focused on a feature we often observe in facility location under uncertainty: Some decisions adapt to the different scenarios observed (e.g., allocation decisions), but others do not (e.g., the location of the facilities). The former are often called recourse decisions (or adaptive if we adopt the terminology used in Chap. 5). In this section, we further explore this aspect.

6.3.1 *Uncapacitated Facilities*

We start with the p-median problem and the setting investigated in [65]. In that problem, an additional set of up to q facilities can be added to the initial p. However, the exact value of q is uncertain. The problem consists of finding the initial set of p facilities so that the expected cost in the future for supplying the demand is minimized.

Let $x_{ij}^{p+\omega}$ be a binary allocation decision variable equal to one if demand node $j \in J$ is assigned to a facility located at $i \in J$ when $\omega \in \Omega$ additional facilities are located ($\Omega = \{0, 1, \ldots, q\}$). Likewise, $x_{ii}^{p+\omega}$ is a binary variable equal to one if a facility is located at node $i \in J$ when $\omega \in \Omega$ new facilities are to be installed and 0 otherwise. Denoting by π_ω the probability of scenario $\omega \in \Omega$ and using the other customary notation, the problem can now be formulated as follows:

$$\text{minimize} \quad \sum_{\omega \in \Omega} \left(\pi_\omega \sum_{i,j \in J} c_{ij} x_{ij}^{p+\omega} \right), \tag{6.27}$$

$$\text{subject to} \quad \sum_{i \in J} x_{ii}^{p} = p, \tag{6.28}$$

$$\sum_{i \in J} x_{ii}^{p+\omega} = \omega \qquad \forall \omega \in \Omega \setminus \{0\}, \tag{6.29}$$

$$x_{ij}^{p} \leq x_{ii}^{p} \qquad \forall i, j \in J, \tag{6.30}$$

$$x_{ij}^{p+\omega} \leq x_{ii}^{p} + x_{ii}^{p+\omega} \qquad \forall i, j \in J,\ \omega \in \Omega \setminus \{0\}, \tag{6.31}$$

$$\sum_{j \in J} x_{ij}^{p+\omega} = 1 \qquad \forall i \in J,\ \omega \in \Omega, \tag{6.32}$$

$$x_{ij}^{p+\omega} \in \{0, 1\} \qquad \forall i, j \in J,\ \omega \in \Omega. \tag{6.33}$$

In this model, the objective function (6.27) accounts for the total expected allocation cost. Constraints (6.28) and (6.29) ensure that the proper number of facilities is installed according to the occurring scenario. Constraints (6.30) and (6.31) guarantee that, in each scenario, allocations can only be made to installed facilities. Constraints (6.32) state that all nodes must be allocated to a facility in every scenario. Finally, we have the domain constraints.

Berman and Drezner [65] tackle the above compact model using a general-purpose solver. When at most one facility can be installed in the future, advantages can be taken from the resulting objective function. This motivated the authors to develop an exchange approach (the one introduced in [519]) and a tabu search algorithm. We refer the interested reader to that work for further details.

We proceed with another classical problem, namely the UFLP. We retrieve the model reviewed in Chap. 2:

$$\text{minimize} \quad \sum_{i \in I} f_i y_i + \sum_{i \in I} \sum_{j \in J} c_{ij} d_j x_{ij}, \tag{6.34}$$

$$\text{subject to} \quad \sum_{i \in I} x_{ij} = 1 \qquad \forall j \in J, \tag{6.35}$$

$$x_{ij} \leq y_i \qquad \forall i \in I, \tag{6.36}$$

$$y_i \in \{0, 1\} \qquad \forall i \in I, \tag{6.37}$$

$$x_{ij} \geq 0, \qquad \forall i \in I,\ j \in J. \tag{6.38}$$

Let us assume uncertainty to be associated with the demands and distribution costs. In particular, consider it captured by a random vector $\boldsymbol{\xi} = \big((c_{ij})_{i \in I,\, j \in J}, (d_j)_{j \in J}\big)$ that has a given cumulative distribution function.

Assuming *ex ante* location decisions, the model to be adopted depends on the *ex post* decisions and also on the attitude of the decision-maker toward risk. When dealing with *ex post* allocation decisions, the following stochastic uncapacitated facility location problem with recourse can be considered:

$$\text{minimize} \quad \sum_{i \in I} f_i y_i + \mathcal{Q}(\boldsymbol{y}), \tag{6.39}$$

$$\text{subject to} \quad \sum_{i \in I} y_i \geq 1, \tag{6.40}$$

$$y_i \in \{0, 1\} \quad \forall i \in I, \tag{6.37}$$

with $\mathrm{Q}(\boldsymbol{y}) = \mathbb{E}\left[\mathcal{Q}(\boldsymbol{y}, \boldsymbol{\xi})\right]$, and $\mathcal{Q}(\boldsymbol{y}, \boldsymbol{\xi})$ denoting the optimal value of the problem

$$\text{minimize} \quad \sum_{i \in I} \sum_{j \in J} c_{ij} d_j x_{ij}, \tag{6.41}$$

$$\text{subject to} \quad \sum_{i \in I} x_{ij} = 1 \qquad \forall j \in J, \tag{6.42}$$

$$x_{ij} \leq y_i \qquad \forall i \in I,\ j \in J, \tag{6.43}$$

$$x_{ij} \geq 0 \qquad \forall i \in I,\ j \in J. \tag{6.44}$$

A model (6.41)–(6.44) is induced by every realization of $\boldsymbol{\xi}$. The allocation decisions x_{ij} $(i \in I,\ j \in J)$ eventually adapt to each such realization. Regarding the location variables y_i, $i \in I$, they correspond to *ex ante* (first-stage) decisions. Hence, they must hold for all possible realizations of the random variables. The expectation defining the recourse function $\mathrm{Q}(\boldsymbol{y})$ conveys a neutral attitude of the decision-maker toward risk. Later, we discuss other possibilities and the consequences from a modeling perspective.

Due to Constraint (6.40), the problem has relatively complete recourse, i.e., for every first-stage feasible solution, there is always a second-stage completion irrespective of the observation of the random vector.

If uncertainty is captured by a finite set of scenarios or a finite-dimensional sample of scenarios can be drawn from the underlying distribution, we can further specify the optimization model. In particular, we can define scenario-indexed parameters and variables as done in the previous section. Let Ω denote the set of scenarios.

Let $c_{ij\omega}$ be the cost for supplying customer $j \in J$ from facility $i \in I$ under scenario $\omega \in \Omega$; $d_{j\omega}$ stands for the demand of customer $j \in J$ under scenario $\omega \in \Omega$. We consider a decision variable $x_{ij\omega}$ representing the fraction of the demand of customer $j \in J$ that is satisfied from facility $i \in I$ under scenario $\omega \in \Omega$. We can adopt the following extensive form of the deterministic equivalent (i.e., all scenarios are explicitly considered in the model):

$$\text{minimize} \quad \sum_{i \in I} f_i y_i + \sum_{\omega \in \Omega} \pi_\omega \left(\sum_{i \in I} \sum_{j \in J} c_{ij\omega} d_{j\omega} x_{ij\omega} \right), \tag{6.45}$$

$$\text{subject to} \quad (6.37),$$

$$\sum_{i \in I} x_{ij\omega} = 1 \quad \forall j \in J,\ \omega \in \Omega, \tag{6.46}$$

$$x_{ij\omega} \leq y_i \qquad \forall i \in I,\ j \in J,\ \omega \in \Omega, \tag{6.47}$$

$$x_{ij\omega} \geq 0 \qquad \forall i \in I,\ j \in J,\ \omega \in \Omega. \tag{6.48}$$

In the above model, the non-anticipativity nature of the first-stage decisions[1] is implicit because each first-stage decision variable has the same value across all scenarios. Note also that in this model, Constraint (6.40) becomes redundant and thus is omitted.

A direct extension of the above model regards a time-changing demand. If we discretize time, then a planning horizon T is considered that is divided into several periods. In this case, assuming that the facilities are to be opened at the beginning of the planning horizon and do not change throughout it, then we are led to an "implicit" multi-period facility location problem following the terminology of [146]. Assuming that each scenario fully determines the demand for all periods (e.g., a scenario specifies a demand trend), then we denote by $d_{jt\omega}$ the demand of customer $j \in J$ in period $t \in T$ under scenario $\omega \in \Omega$. Likewise, $c_{ijt\omega}$ is the unit transportation cost from location $i \in I$ to customer $j \in J$ in period $t \in T$ under scenario $\omega \in \Omega$. The transportation decisions can be represented by decision variables $x_{ijt\omega}$ denoting the proportion of the demand of customer $j \in J$ in period $t \in T$ satisfied by a facility open at $i \in I$ under scenario $\omega \in \Omega$. A mathematical model for the problem is the following:

$$\text{minimize} \quad \sum_{i \in I} f_i y_i + \sum_{\omega \in \Omega} \pi_\omega \left(\sum_{t \in T} \sum_{i \in I} \sum_{j \in J} c_{ijt\omega} d_{jt\omega} x_{ijt\omega} \right), \tag{6.49}$$

$$\text{subject to} \quad (6.37),$$

$$\sum_{i \in I} x_{ijt\omega} = 1 \quad \forall j \in J,\ t \in T,\ \omega \in \Omega, \tag{6.50}$$

$$x_{ijt\omega} \leq y_i \quad \forall i \in I,\ j \in J,\ t \in T,\ \omega \in \Omega, \tag{6.51}$$

$$x_{ijt\omega} \geq 0 \quad \forall i \in I,\ j \in J,\ t \in T,\ \omega \in \Omega. \tag{6.52}$$

Later in this chapter, we discuss the consequences of location decisions changing over the planning horizon.

The stochastic uncapacitated facility location problem just presented is a two-stage stochastic mixed-integer program with continuous recourse. Therefore, it can be solved by a Benders decomposition-based procedure (see Sect. 4.3.2). This can be accomplished by writing the problem as follows:

$$\text{minimize} \quad \sum_{i \in I} f_i y_i + \sum_{\omega \in \Omega} \pi_\omega \theta_\omega, \tag{6.53}$$

$$\text{subject to} \quad (6.37),\ (6.40),$$

$$\theta_\omega \geq \mathcal{Q}(y, \omega) \quad \forall \omega \in \Omega. \tag{6.54}$$

[1] A decision should depend only on the information available at the time it is implemented (see [460]).

For each $\omega \in \Omega$,

$$Q(\boldsymbol{y}, \omega) = \min \sum_{i \in I} \sum_{j \in J} \sum_{t \in T} c_{ijt\omega} d_{jt\omega} x_{ijt\omega}, \tag{6.55}$$

$$\text{s. t.} \quad \sum_{i \in I} x_{ijt\omega} = 1 \qquad \forall j \in J,\ t \in T, \tag{6.56}$$

$$x_{ijt\omega} \leq y_i \qquad \forall i \in I,\ j \in J,\ t \in T, \tag{6.57}$$

$$x_{ijt\omega} \geq 0 \qquad \forall i \in I,\ j \in J,\ t \in T. \tag{6.58}$$

Dualizing the above, we obtain

$$Q(\boldsymbol{y}, \omega) = \max \sum_{j \in J} \sum_{t \in T} \lambda_{jt\omega} - \sum_{i \in I} \sum_{j \in J} \sum_{t \in T} y_i \nu_{ijt\omega}, \tag{6.59}$$

$$\text{s. t.} \quad \lambda_{jt\omega} - \nu_{ijt\omega} \leq c_{ijt\omega} d_{jt\omega} \qquad \forall i \in I,\ j \in J,\ t \in T, \tag{6.60}$$

$$\nu_{ijt\omega} \geq 0 \qquad \forall i \in I,\ j \in J,\ t \in T. \tag{6.61}$$

A standard Benders decomposition algorithm can now be devised. We start by solving a relaxed master problem consisting of minimizing (6.53) subject to (6.37) and (6.40). Thus, we obtain a solution $(\boldsymbol{y}^1, \boldsymbol{\theta}^1)$. For every $\omega \in \Omega$, we check whether $\theta^1_\omega \geq Q(\boldsymbol{y}^1, \omega)$. If not, the cut

$$\theta_\omega \geq \sum_{j \in J} \sum_{t \in T} \lambda^*_{jt\omega} - \sum_{i \in I} \sum_{j \in J} \sum_{t \in T} y_i \nu^*_{ijt\omega}$$

is added, where $\boldsymbol{\lambda}^*_\omega$ and ν^*_ω are the optimal values of the dual variables associated with (6.56) and (6.57). The enhanced relaxed Master problem is solved, leading to an optimal solution $(\boldsymbol{y}^2, \boldsymbol{\theta}^2)$, which is again checked. The process proceeds until it eventually converges.

Improvements and variants can be considered for this procedure. An improvement consists of using stabilization techniques (see, e.g., [211]). A variation consists of aggregating the Benders cuts as often done in the literature (see, e.g., [78]). Unfortunately, the relaxed master problem may still be too much time-consuming. In this case, one alternative is to add Benders cuts per needed within a branch-and-cut algorithm. We sketch the main steps in Algorithm 6.1.

Algorithm 6.1: A branch-and-Benders-cut algorithm

Solve the linear programming relaxation of the master problem to obtain a solution $(\hat{\mathbf{y}}, \hat{\boldsymbol{\theta}})$;
if *the LP bound exceeds the incumbent* **then**
    prune
if $\hat{\mathbf{y}}$ *is integer* **then**
    Solve the scenario-induced subproblems to generate Benders cuts;
    if *some* $(\hat{\mathbf{y}}, \theta_\omega)$ *violates a Benders cut* **then**
        add the the cut to the LP;
        reoptimize the LP;
else
    Branch;

Algorithm 6.1 can be improved in different ways. For instance, we can start by solving the linear relaxation of the problem using a plain Benders decomposition and keeping the cuts obtained. Later, in the algorithm, when we observe that $\hat{\mathbf{y}}$ is not an integer, we can still solve the scenario subproblems and add the violated Benders cuts.

6.3.2 Capacitated Facilities

We consider again the single-period problem to discuss another relevant setting: the existence of capacitated facilities. This is important in many applications (e.g., in logistics and supply chain management). However, some difficulties emerge. Denote by q_i the capacity of a facility established at $i \in I$. A model for the capacitated stochastic facility location problem is obtained from (6.34)–(6.38) by replacing (6.36) with

$$\sum_{j \in J} d_j x_{ij} \leq q_i y_i, \quad \forall i \in I. \tag{6.62}$$

With the inclusion of these constraints, the problem no longer has relatively complete recourse. In fact, given a first-stage solution, a feasible completion may not exist in the second stage for one or several observations of the underlying random vector.

Second-stage infeasibility is often an indication of an undesirable first-stage solution. A natural way of dealing with this issue is to penalize the non-satisfied demand. In practice, such a penalty may correspond, for example, to a loss-opportunity cost or some outsourcing course of action. Denote by ψ_j the demand of customer $j \in J$ not supplied from the open facilities and denote by μ_j the corresponding unit penalty cost. Note that ψ_j depends on the occurring realization of the random vector $\boldsymbol{\xi}$. We can still consider the first-stage problem that consists of minimizing (6.39) subject to (6.37) and (6.40). However, the second-stage problem must be rewritten as follows:

$$\text{minimize} \quad \sum_{i \in I} \sum_{j \in J} c_{ij} d_j x_{ij} + \sum_{j \in J} \mu_j \psi_j, \tag{6.63}$$

$$\text{subject to} \quad \sum_{i \in I} x_{ij} + \frac{\psi_j}{d_j} = 1 \quad \forall j \in J, \tag{6.64}$$

$$(6.44), \ (6.62),$$

$$\psi_j \geq 0 \qquad \forall j \in J. \tag{6.65}$$

Again, if a finite set of scenarios exist, we can consider scenario-indexed recourse variables, $x_{ij\omega}$, $\psi_{j\omega}$, and parameters, $c_{ij\omega}$, $d_{j\omega}$, $\mu_{j\omega}$ $(i \in I, \ j \in J, \ \omega \in \Omega)$. This allows writing the deterministic equivalent in its extensive form that again can be tackled using a Benders decomposition with a mixed-integer (relaxed) master problem or a branch-and-Benders-cut as presented in the previous section. Now, the Benders cuts are of the form

$$\theta_\omega \geq \sum_{j \in J} \lambda^*_{j\omega} - \sum_{i \in I} q_i \, y_i \, \nu^*_{i\omega}, \tag{6.66}$$

where $\lambda^*_{j\omega}$ and $\nu^*_{i\omega}$ are the optimal values of the dual variables associated with scenario-indexed constraints (6.62) and (6.64), respectively, considering the dual of the scenario subproblems induced by a solution to the (relaxed) master problem.

In the capacitated model just described, capacities are exogenous. Louveaux [352] considers a stochastic facility location problem with endogenous capacities. In particular, capacities must be decided in advance, i.e., before uncertainty is disclosed—they correspond to *ex ante* decisions. A unit cost g_i is assumed for operating capacity to be installed at location $i \in I$. The author also considers variable production costs and revenues associated with demand satisfaction. A new decision variable z_i $(i \in I)$ must be introduced, representing the capacity to be installed at location $i \in I$. With revenues, it is no longer necessary to consider Constraint (6.40). Furthermore, it may not be profitable to satisfy all the demand. The trade-off between revenues and costs will determine the best service level for each customer. The capacitated model formulated above can be easily adapted to the new conditions, leading to the model proposed in [352]:

$$\text{minimize} \quad \sum_{i \in I} f_i y_i + \sum_{i \in I} g_i z_i + \mathcal{Q}(y, z), \tag{6.67}$$

$$\text{subject to} \quad (6.37),$$

$$z_i \geq 0 \quad \forall i \in I, \tag{6.68}$$

with $\mathrm{Q}(\boldsymbol{y}, \boldsymbol{z}) = \mathbb{E}\,[Q(\boldsymbol{y}, \boldsymbol{z}, \boldsymbol{\xi})]$, and

$$Q(\boldsymbol{y}, \boldsymbol{z}, \boldsymbol{\xi}) = \min \quad \sum_{i \in I} \sum_{j \in J} c_{ij} d_j x_{ij}, \tag{6.69}$$

$$\text{s. t.} \quad \sum_{i \in I} x_{ij} \leq 1 \qquad \forall j \in J, \tag{6.70}$$

$$(6.43),\ (6.44),$$

$$\sum_{j \in J} d_j x_{ij} \leq z_i \qquad \forall i \in I. \tag{6.71}$$

In this model, the parameters c_{ij} embed the production costs and the revenues above mentioned, i.e., one expects many such costs to be negative, thus indicating a positive profit for supplying a customer from a facility. Considering this problem, [354] assume that stochasticity is captured by a finite number of scenarios, Ω, and develop a dual-based procedure for tackling the extensive form of the deterministic equivalent. Extending the notation as done already before, we obtain

$$\text{minimize} \quad \sum_{i \in I} f_i y_i + \sum_{i \in I} g_i z_i - \sum_{\omega \in \Omega} \sum_{i \in I} \sum_{j \in J} p_{ij\omega} x_{ij\omega}, \tag{6.72}$$

$$\text{subject to} \quad (6.37),\ (6.68),$$

$$-\sum_{i \in I} x_{ij\omega} \geq -1 \qquad \forall j \in J,\ \omega \in \Omega, \tag{6.73}$$

$$-\sum_{j \in J} d_{j\omega} x_{ij\omega} + z_i \geq 0 \qquad \forall i \in I, \omega \in \Omega, \tag{6.74}$$

$$-x_{ij\omega} + y_i \geq 0, \qquad \forall i \in I,\ j \in J, \omega \in \Omega, \tag{6.75}$$

$$x_{ij\omega} \geq 0, \qquad \forall i \in I,\ j \in J,\ \omega \in \Omega, \tag{6.76}$$

with $p_{ij\omega} = -\pi_\omega c_{ij\omega} d_{j\omega}$ denoting the (weighted) profit for supplying all the demand of customer $j \in J$ from facility $i \in I$. Relaxing the integrality conditions on the y variables, the dual of the resulting linear programming problem can be written as

$$-\,(\text{minimize} \quad \sum_{\omega \in \Omega} \sum_{i \in I} v_{i\omega}), \tag{6.77}$$

$$\text{subject to} \quad -v_{j\omega} - d_{j\omega} u_{i\omega} - w_{ij\omega} \leq -p_{ij\omega} \quad \forall i \in I,\ j \in J,\ \omega \in \Omega, \tag{6.78}$$

$$\sum_{j \in J} \sum_{\omega \in \Omega} w_{ij\omega} \leq f_i \qquad \forall i \in I, \tag{6.79}$$

$$\sum_{\omega \in \Omega} u_{i\omega} \leq g_i \qquad \forall i \in I, \tag{6.80}$$

$$u_{i\omega} \geq 0 \qquad \forall i \in I,\ \omega \in \Omega, \tag{6.81}$$

$$v_{j\omega} \geq 0 \qquad \forall j \in J,\ \omega \in \Omega, \tag{6.82}$$

$$w_{ij\omega} \geq 0 \qquad \forall i \in I,\ j \in J,\ \omega \in \Omega. \tag{6.83}$$

Given that $w_{ij\omega} \geq \max\{0, p_{ij\omega} - v_{j\omega} - d_{j\omega}u_{i\omega}\}$, we can set it equal to that lowest limit and formulate a condensed dual:

$$\text{minimize} \quad \sum_{\omega \in \Omega} \sum_{i \in I} v_{i\omega}, \tag{6.84}$$

$$\text{subject to} \quad \sum_{j \in J} \sum_{\omega \in \Omega} \max\{0, p_{ij\omega} - v_{j\omega} - d_{j\omega}u_{i\omega}\} \leq f_i \quad \forall i \in I, \tag{6.85}$$

$$(6.80) - (6.82).$$

The complementary slackness conditions are the following:

$$v_{j\omega}\left(1 - \sum_{i \in I} x_{ij\omega}\right) \qquad \forall j \in J,\ \omega \in \Omega,$$

$$u_{i\omega}\left(z_i - \sum_{j \in J} d_{j\omega}x_{ij\omega}\right) \qquad \forall i \in I,\ \omega \in \Omega,$$

$$y_i\left(f_i - \sum_{j \in J} \sum_{\omega \in \Omega} \max\{0, p_{ij\omega} - v_{j\omega} - d_{j\omega}u_{i\omega}\}\right) \qquad \forall i \in I,$$

$$z_i\left(g_i - \sum_{\omega \in \Omega} u_{i\omega}\right) \qquad \forall i \in I,$$

$$w_{ij\omega}\left(y_i - x_{ij\omega}\right) \qquad \forall i \in I,\ j \in J,\ \omega \in \Omega.$$

Inspired by the algorithm of [191], a four-step procedure can be devised, sketched in Algorithm 6.2.

Algorithm 6.2: Major phases of a dual-based algorithm for stochastic facility location with endogenous capacities

1 A dual-ascent heuristic is applied to the condensed dual.
2 A primal solution is constructed based on the above complementary slackness conditions.
3 A primal–dual adjustment procedure adjusts primal and dual solutions to decrease the total violation in the complementary slackness conditions.
4 A branch-and-bound algorithm is applied if the complementary violation remains.

The dual-ascent heuristic consists of a sequence of $\boldsymbol{v}$- and $\boldsymbol{u}$-ascent steps. The former seeks to increase the value of the objective function; the latter targets the increase of the slacks in Constraints (6.85), thus allowing the $\boldsymbol{v}$ variables to increase further (in the following $\boldsymbol{v}$-step). For the other details of the algorithm, the interested reader can refer to [191, 354].

A different problem emerges when the distribution decisions become *ex ante*. In this case, penalties are assumed for inventory shortage or surplus in the second stage. In addition to the notation already introduced, we denote by ϕ_j the inventory surplus at customer $j \in J$ and by λ_j the corresponding unit cost. Assuming deterministic distribution costs (they are now associated with an *ex ante* decision), we can formulate the stochastic facility location problem as follows:

$$\text{minimize} \quad \sum_{i \in I} f_i y_i + \sum_{i \in I} \sum_{j \in J} c_{ij} x_{ij} \mathrm{Q}(x), \tag{6.86}$$

$$\text{subject to} \quad (6.37),\ (6.43),\ (6.44),$$

with $\mathrm{Q}(\boldsymbol{x}) = \mathbb{E}\left[Q(\boldsymbol{x}, \boldsymbol{\xi})\right]$, and

$$Q(\boldsymbol{x}, \boldsymbol{\xi}) = \min \quad \sum_{j \in J} \lambda_j \phi_j + \sum_{j \in J} \mu_j \psi_j, \tag{6.87}$$

$$\text{s. t.} \quad \psi_j - \phi_j = d_j \left(1 - \sum_{i \in I} x_{ij}\right) \qquad \forall j \in J, \tag{6.88}$$

$$\psi_j, \phi_j \geq 0 \qquad \forall j \in J. \tag{6.89}$$

Capacities can be easily included in the above model leading to the so-called stochastic transportation location problem, which has been investigated by several authors (e.g., [213, 272]).

6.3.3 Detaching Allocation and Distribution Decisions

So far, we have assumed that the allocation and distribution decisions are made simultaneously (with the former implicit in the latter), either after or before uncertainty is disclosed. However, in some settings, these decisions are made separately. This is the case in logistics applications when a contract is previously signed, determining a priori the distribution channels but leaving the shipping quantities to be decided later depending upon the observed values of the stochastic parameters. The same situation occurs in service-providing companies that need to segment the customers a priori by allocating each customer to a server or facility. These are cases in which we must explicitly consider allocation decision variables.

We keep using variables $\boldsymbol{x}$ for allocation. In particular, we consider x_{ij} equal to 1 if and only if customer $j \in J$ is allocated to facility $i \in I$. The single-allocation version of the problem was introduced in [325], who proposed the following optimization model:

$$\text{minimize} \quad \sum_{i \in I} f_i y_i + \sum_{i \in I} \sum_{j \in J} b_{ij} x_{ij} - \mathrm{Q}(\boldsymbol{x}), \tag{6.90}$$

$$\text{subject to} \quad x_{ij} \le y_i \qquad \forall i \in I,\ j \in J, \tag{6.91}$$

$$\sum_{i \in I} x_{ij} \le 1 \qquad \forall j \in J, \tag{6.92}$$

$$x_{ij}, y_i \in \{0, 1\} \qquad \forall i \in I,\ j \in J, \tag{6.93}$$

with $\mathrm{Q}(\boldsymbol{x}) = \mathbb{E}[Q(\boldsymbol{x}, \boldsymbol{\xi})]$ and

$$Q(\boldsymbol{x}, \boldsymbol{\xi}) = \max \quad \sum_{i \in I} \sum_{j \in J} r_{ij} d_j w_{ij}, \tag{6.94}$$

$$\text{s. t.} \quad w_{ij} \le x_{ij} \qquad \forall i \in I,\ j \in J, \tag{6.95}$$

$$\sum_{j \in J} d_j w_{ij} \le q_i \qquad \forall i \in I, \tag{6.96}$$

$$w_{ij} \ge 0 \qquad \forall i \in I,\ j \in J. \tag{6.97}$$

In the above model, b_{ij} is a fixed cost for allocating customer $j \in J$ to facility $i \in I$ and r_{it} is the unit net revenue when supplying customer $j \in J$ from facility $i \in I$. $\boldsymbol{\xi} = [d_1, \ldots, d_{|J|}]$ is the underlying random vector. The transportation amounts between facilities and customers are represented by variables w_{ij}, $i \in I$, $j \in J$. The other notation was already introduced before. In the above problem, facilities are capacitated. Furthermore, a service level of 100% is not imposed.

From a stochastic programming point of view, we are facing a two-stage problem with continuous recourse. Thus, it can be tackled by a plain Benders decomposition (with an integer relaxed master problem) or using a branch-and-Benders-cut algorithm, as discussed in other cases. Laporte et al. [325] consider the second possibility, enriching it with some additional features. The linear relaxation of the master problem can be written as follows:

$$\text{minimize} \quad V(\boldsymbol{y}, \boldsymbol{x}, \theta) = \sum_{i \in I} f_i y_i + \sum_{i \in I} \sum_{j \in J} b_{ij} x_{ij} + \theta, \tag{6.98}$$

$$\text{subject to} \quad (6.91),\ (6.92),$$

$$\theta \geq -\mathrm{Q}(\boldsymbol{x}), \tag{6.99}$$

$$0 \leq y_i \leq 1 \qquad \forall i \in I, \tag{6.100}$$

$$0 \leq x_{ij} \leq 1 \qquad \forall i \in I,\ j \in J. \tag{6.101}$$

This problem can be enhanced using the following inequalities:

$$\theta \geq -\sum_{i \in I} \max_{j \in J}\{r_{ij}\} q_i y_i, \tag{6.102}$$

and

$$\theta \geq \max \left\{ -\sum_{i \in I} \max_{j \in J}\{r_{ij}\} q_i,\ -\sum_{j \in J} \max_{i \in I}\{r_{ij}\} \mathbb{E}[d_j] \right\}, \tag{6.103}$$

which define lower bounds on the maximum revenue that can be obtained.

As customary in a Benders decomposition algorithm, Constraint (6.99) is initially relaxed. This yields a problem that can be enhanced using optimality cuts. Considering a location-allocation solution $\{\hat{\boldsymbol{y}}, \hat{\boldsymbol{x}}\}$ to the relaxed master problem, such cuts can be derived following the methodology proposed in [324] for two-stage stochastic problems with integer recourse:

$$\theta + \left(\mathrm{Q}(\hat{\boldsymbol{x}}) + L\right) \left(\sum_{(i,j) \in \hat{S}} x_{ij} - \sum_{(i,j) \notin \hat{S}} x_{ij} \right) \geq \left(\mathrm{Q}(\hat{\boldsymbol{x}}) + L\right) \left(|\hat{S}| - 1 \right) + L, \tag{6.104}$$

with L denoting the lower bound stated in (6.103), and

$$\hat{S} = \{(i, j) \mid i \in I,\ j \in J,\ \hat{x}_{ij} = 1\}.$$

If we define $\hat{J}^1 = \{j \in J \mid \sum_{i \in I} \hat{x}_{ij} = 1\}$ and $\hat{J}^0 = \{j \in J \mid \sum_{i \in I} \hat{x}_{ij} = 0\}$ and denote by $\hat{\imath}(j)$ the facility that $j \in \hat{J}^1$ is allocated to, then we can simplify the above cut:

$$\theta + \left(\mathrm{Q}(\hat{\boldsymbol{x}}) + L\right) \left(\sum_{j \in \hat{J}^1} x_{\hat{\imath}(j),j} - \sum_{j \in \hat{J}^0} \sum_{i \in I} x_{ij} \right) \geq \left(\mathrm{Q}(\hat{\boldsymbol{x}}) + L\right) \left(|\hat{J}^1| - 1 \right) + L. \tag{6.105}$$

Finally, a component of the procedure is the computation of $\mathrm{Q}(\hat{\boldsymbol{x}})$. For a specific observation of the demands, we can find a closed form for that value:

$$\mathrm{Q}(\hat{\boldsymbol{x}}) = \sum_{i \in I, \hat{y}_i = 1} \sum_{j \in \hat{J}^1, \hat{\imath}(j) = i} r_{ij} \mathbb{E} \left[\min \left\{ d_j, \max \left(0, q_i - \sum_{\substack{k \in \hat{J}^1 \mid \hat{\imath}(k) = i \\ r_{ik} \geq r_{ij}}} d_k \right) \right\} \right]. \tag{6.106}$$

Laporte et al. [325] show that the above expression can be easily computed for several discrete probability distributions and different continuous ones. The overall procedure proposed by those authors is summarized in Algorithm 6.3.

In the problem investigated in [325], some demand may remain unsupplied depending on the capacities and the revenues.

We focus now on the case in which demand should be fully supplied. We can accomplish this by considering transportation costs in the second stage (no longer revenues) and assuming some recourse action in case the operating capacities cannot supply the demand.

To make the problem more general, we assume a minimum number of allocations ℓ_i to a facility $i \in I$ to justify opening it. We continue considering a fixed cost f_i associated with each open facility $i \in I$. The problem can be formulated generically as follows:

$$\text{minimize} \quad \sum_{i \in I} f_i y_i + \mathrm{Q}(\boldsymbol{y}, \boldsymbol{x}), \tag{6.107}$$

$$\text{subject to} \quad \sum_{i \in I} x_{ij} = 1 \qquad \forall j \in J, \tag{6.108}$$

$$x_{ij} \leq y_i \qquad \forall i \in I,\ j \in J, \tag{6.109}$$

$$\ell_i y_i \leq \sum_{j \in J} x_{ij} \qquad \forall i \in I, \tag{6.110}$$

$$x_{ij} \in \{0, 1\} \qquad \forall i \in I,\ j \in J. \tag{6.111}$$

$$y_i \in \{0, 1\} \qquad \forall i \in I. \tag{6.112}$$

Algorithm 6.3: A branch-and-Benders-cut algorithm for a stochastic facility location with a priori allocation

$\overline{V} \leftarrow \infty$ // incumbent best upper bound;
Insert problem (6.91), (6.92), (6.98), (6.100), (6.101) in the list of the pending nodes;
while *list of pending nodes is non-empty* **do**
  Select a node in the list;
  Solve the problem induced by the selected node and denote the solution by $\{\hat{y}, \hat{x}, \hat{\theta}\}$;
  if $V(\hat{y}, \hat{x}, \hat{\theta}) > \overline{V}$ **then**
    phathom the current problem;
  else
    check for integrality constraints;
    if *one is violated* **then**
      create two new nodes (a fractional variable is set to 0 and 1);
    else
      Compute $\mathrm{Q}(\hat{x})$;
      Compute $\hat{V} = V(\hat{y}, \hat{x}, -\mathrm{Q}(\hat{x}))$;
      if $\hat{V} < \overline{V}$ **then**
        $\overline{V} \leftarrow \hat{V}$;
        Store $\{\hat{y}, \hat{x}\}$;
      if $\hat{\theta} \geq -\mathrm{Q}(\hat{x})$ **then**
        Fathom the current node;
      else
        Add an optimality cut (6.105);

The recourse function contains two components: one related to the transportation of the amounts made available by the operational capacity and another related to the penalty for capacity shortage. The former will be termed service cost. As for the latter, given that all demand must be supplied, we assume that an extra cost is incurred for acquiring the missing amount, which, somehow, indicates an outsourcing of the activity. For this reason, the second component is termed the outsourcing cost. Thus, assuming a risk-neutral decision-maker, we set

$$\mathrm{Q}(\boldsymbol{y}, \boldsymbol{x}) = \mathbb{E}\left[\text{service cost} + \text{outsourcing cost}\right].$$

For each open facility $i \in I$, we keep assuming a service capacity q_i. Let ξ_j represent the demand of customer $j \in J$. We assume that ξ_j, $j \in J$ are i.i.d. random variables with a finite mean μ. We assume that 0 is part of the support of the underlying distribution. This means that a customer may turn out to have no demand. For each first-stage solution, denote by z_i the number of customers assigned to facility $i \in I$, i.e., $z_i = \sum_{j \in J} x_{ij}$.

The major challenge is to find the recourse function value for an a priori solution. This problem has been studied in different settings in [9] and Bieniek [77].

Let us denote by N_i the number of customers with non-null demand (called demand customers in the above works). It is straightforward to conclude that for each $i \in I$, N_i is distributed according to a binomial distribution with parameters z_i

and p, with $p = 1 - \mathbb{P}[\xi_j = 0]$ for any $j \in J$, i.e., $\mathbb{P}[N_i = s] = \binom{z_i}{s} p^s (1-p)^{z_i - s}$, $s = 0, \ldots, z_i$. Denote by ζ_{tps} the probability that a binomial random variable with parameters t and p takes the value s. Accordingly, the above probability is $\zeta_{z_i ps}$.

For this problem, [9] introduce two recourse actions: facility outsourcing and customer outsourcing. In the first case, each facility keeps supplying the customers allocated to it. An extra cost is incurred for obtaining the missing capacity. This corresponds to a temporary increase in the capacity of a facility due, for instance, to an external rent or purchase of the missing resources needed to satisfy the occurring demand. For each facility $i \in I$, we denote by g_i the cost per demand unit beyond its capacity. We keep adopting c_{ij} as the unit cost for serving customer $j \in J$ from facility $i \in I$.

In the case of customer outsourcing, when the demand of the customers assigned to some facility $i \in I$, say η_i, exceeds q_i, an amount of $\eta_i - q_i$ is served directly from an external source. A FIFO policy is assumed to decide the demand to serve from the facility and to outsource. The cost for supplying each outsourced unit is denoted by $\hat{g}_i$. It remains dependent on the facility a customer is initially assigned to.

The following result holds.

Proposition 6.1 *Let $F^{*m}(.)$ be the m-fold convolution of the distribution functions of the random variables ξ_r, $r = 1, \ldots, m$:*

1. *Under a facility outsourcing recourse action,*

$$\mathcal{Q}(\boldsymbol{y}, \boldsymbol{x}) = \mathcal{Q}(\boldsymbol{x}) = \sum_{i \in I} \sum_{j \in J} c_{ij} x_{ij} \mu + \sum_{i \in I} g_i \mu \left(z_i - \frac{1}{p} \sum_{s=1}^{z_i} \zeta_{z_i ps} \sum_{m=1}^{s} F^{*m}(q_i) \right). \tag{6.113}$$

2. *Under a customer outsourcing recourse action,*

$$\mathcal{Q}(\boldsymbol{y}, \boldsymbol{x}) = \mathcal{Q}(\boldsymbol{x}) = \sum_{i \in I} \sum_{j \in J} \frac{c_{ij} \mu}{z_i p} x_{ij} \sum_{s=1}^{z_i} \zeta_{z_i ps} \sum_{m=1}^{s} F^{*m}(q_i) + \sum_{i \in I} \hat{g}_i \mu \left(z_i - \frac{1}{p} \sum_{s=1}^{z_i} \zeta_{z_i ps} \sum_{m=1}^{s} F^{*m}(q_i) \right). \tag{6.114}$$

Proof See [77]. □

The above expressions for the recourse functions involve z_i, $i \in I$ as an upper limit of a sum and thus depend on the decision to make. This challenge can be overcome by discretizing the first-stage decisions [245] because the above expressions are linear if we know the number of customers assigned to a facility.

We start by considering the case of facility outsourcing and by defining y_i^t a binary variable equal to one if and only if a facility is established at $i \in I$ and t customers are assigned to it, $t \in \{\ell_i, \ldots, |J|\}$. It should be $\sum_{t=\ell_i}^{|J|} y_i^t \leq 1, i \in I$. Furthermore, $y_i^t = 1$ if and only if $z_i = t$. Thus, (6.113) can be linearized as follows:

$$\sum_{i\in I}\sum_{j\in J} c_{ij}x_{ij}\mu + \sum_{i\in I}\sum_{t=\ell_i}^{|J|} g_i\mu y_i^t\left(t - \frac{1}{p}\sum_{s=1}^{t}\zeta_{tps}\sum_{m=1}^{s}F^{*m}(q_i)\right).$$

Hence, the objective function of the problem becomes

$$v_1(\boldsymbol{y}, \boldsymbol{x}) = \sum_{i\in I}\sum_{t=\ell_i}^{|J|} a_i^t y_i^t + \sum_{i\in I}\sum_{j\in J} b_{ij}x_{ij}, \tag{6.115}$$

with

$$a_i^t = f_i + g_i\mu\left(t - \frac{1}{p}\sum_{s=1}^{t}\zeta_{tps}\sum_{m=1}^{s}F^{*m}(q_i)\right) \qquad \forall i \in i,\ t = \ell_i, \ldots, |J|,$$

$$b_{ij} = c_{ij}\,\mu \qquad \forall i \in I,\ j \in J.$$

A compact formulation can now be presented for the deterministic equivalent problem under facility outsourcing:

$$\text{minimize} \quad v_1(\boldsymbol{y}, \boldsymbol{x}), \tag{6.116}$$

$$\text{subject to} \quad (6.108),\ (6.111),$$

$$\sum_{j\in J} x_{ij} = \sum_{t=\ell_i}^{|J|} t y_i^t \qquad \forall i \in I, \tag{6.117}$$

$$\sum_{t=\ell_i}^{|J|} y_i^t \leq 1 \qquad \forall i \in I, \tag{6.118}$$

$$y_i^t \in \{0, 1\} \qquad \forall i \in I,\ \ell_i \leq t \leq |J|. \tag{6.119}$$

In this model, Equalities (6.117) relate the assignment variables and the new location variables. Inequalities (6.118) ensure the consistency in the choice of the y_i^t, i.e., at most one $i \in I$ can be equal to 1. Moreover, together with Constraints (6.117), Inequalities (6.118) guarantee that if a facility is open, then the minimum number of assigned customers is satisfied.

We now consider the customer outsourcing recourse action. A deterministic equivalent can be obtained by discretizing the location and allocation variables accounting for the number of customers allocated to a facility. As for facility outsourcing, we define y_i^t as a binary variable equal to 1 if a facility is located at $i \in I$ and t customers in total are allocated to it ($t = \ell_i, \ldots, |J|$) and 0 otherwise. Likewise, we define x_{ij}^t as a binary variable equal to 1 if and only if customer $j \in J$ is allocated to facility $i \in I$ that has t customers assigned to it ($t = \ell_i, \ldots, |J|$). Using the new variables, (6.114) can be linearized as follows:

$$\sum_{i \in I} \sum_{j \in J} c_{ij} \mu \sum_{t=\ell_i}^{|J|} \frac{1}{tp} x_{ij}^t \sum_{s=1}^{t} \zeta_{tpr} \sum_{m=1}^{s} F^{*m}(q_i)$$

$$+ \sum_{i \in I} \sum_{t=\ell_i}^{|J|} \hat{g}_i \mu y_i^t \left(t - \frac{1}{p} \sum_{s=1}^{t} \zeta_{tps} \sum_{m=1}^{s} F^{*m}(q_i) \right).$$

Hence, the objective function of the problem becomes

$$v_2(\boldsymbol{y}, \boldsymbol{x}) = \sum_{i \in I} \sum_{t=\ell_i}^{|J|} \hat{a}_i^t y_i^t + \sum_{i \in I} \sum_{j \in J} \sum_{t=\ell_i}^{|J|} \hat{b}_{ij}^t x_{ij}^t, \tag{6.120}$$

with

$$\hat{a}_i^t = f_i + \hat{g}_i \mu \left(t - \frac{1}{p} \sum_{s=1}^{t} \zeta_{tps} \sum_{m=1}^{s} F^{*m}(q_i) \right) \qquad \forall i \in I,\ t = \ell_i, \ldots, |J|,$$

$$\hat{b}_{ij}^t = c_{ij} \mu \frac{1}{tp} \sum_{s=1}^{t} \zeta_{tps} \sum_{m=1}^{s} F^{*m}(q_i) \qquad \forall i \in I,\ j \in J,\ t = \ell_i, \ldots, |J|.$$

The compact formulation for the deterministic equivalent problem under customer outsourcing is the following:

$$\text{minimize} \quad v_2(\boldsymbol{y}, \boldsymbol{x}), \tag{6.121}$$

$$\text{subject to} \quad \sum_{i \in I} \sum_{t=\ell_i}^{|J|} x_{ij}^t = 1 \qquad \forall j \in J, \tag{6.122}$$

$$\sum_{j \in J} x_{ij}^t = t y_i^t \qquad \forall i \in I, \tag{6.123}$$

$$\sum_{t=\ell_i}^{|J|} y_i^t \leq 1 \qquad \forall i \in I, \tag{6.124}$$

$$x_{ij}^t \in \{0, 1\} \qquad \forall i \in I,\ j \in J,\ t = \ell_i, \ldots, |J|. \tag{6.125}$$

$$y_i^t \in \{0, 1\} \qquad \forall i \in I,\ t = \ell_i, \ldots, |J|. \tag{6.126}$$

The meaning of the expressions is straightforward, given the discussion already presented.

Bieniek [77] shows how the above models can be particularized to some probability distributions. Nevertheless, it should be pointed out that such developments aimed at extending a first work [9] for unit-demand problems that we detail next.

6.3.4 Unit-Demand Problems

In service planning systems, the demands often consist of a request for a service. Under uncertainty, this means that the demands are represented by Bernoulli random variables. Such a problem was introduced in [9]. The setting is the same that would be later adopted in [77] and thus already detailed. We keep using the same notation and terminology.

When the demands of the customers, ξ_j, $j \in J$, are described by homogeneous Bernoulli distributions with some parameter p, the total demand at a facility is equal to the number of demand customers (customers that request the service). In this case, the model that consists of minimizing (6.116) subject to (6.108), (6.111), (6.117)–(6.119) is valid using

$$a_i^t = f_i + g_i \sum_{s=q_i+1}^{t} \zeta_{tps}(s - q_i) \quad \forall i \in I,\ t = \ell_i, \ldots, |J|,$$

$$b_{ij} = pc_{ij} \quad \forall i \in I,\ j \in J$$

in the definition of (6.116).

In the case of customer outsourcing, the model that consists of minimizing (6.121) subject to (6.122)–(6.125) using

$$\hat{a}_i^t = f_i + \hat{g}_i \sum_{s=q_i+1}^{t} \zeta_{tps}(s - q_i) \quad \forall i \in I,\ t = \ell_i, \ldots, |J|,$$

$$\hat{b}_{ij}^t = c_{ij} \left[\sum_{s=0}^{t} \zeta_{tps} \frac{\min\{q_i, s\}}{t} \right] \quad \forall i \in I,\ j \in J,\ t = \ell_i, \ldots, |J|$$

in the definition of (6.121).

In [9] the authors also introduce the problem when demand probabilities are general. In this case, the demand ξ_j of customer $j \in J$ is now assumed to be a random variable following a Bernoulli distribution with parameter p_j. For each first-stage solution $\{\boldsymbol{y}, \boldsymbol{x}\}$, we keep denoting by z_i the number of customers assigned to facility $i \in I$. Additionally, we denote by η_i the random variable representing the number of customers who request the service (again referred to as demand customers) among those assigned to facility $i \in I$ (i.e., $\eta_i = \sum_{j \in J} \xi_j x_{ij}$). The probability distribution of η_i is quite involved since it depends on the actual values of x_{ij} $(j \in J)$. Denote by $\mathbb{P}_x(\eta_i = s)$ the probability that η_i is equal to s $(s = 0, \ldots, z_i)$.

Under facility outsourcing, a customer allocated to a facility keeps being served by the facility. Hence, the customer is served if it has demand (calls for the service). Thus, the expected service cost can be written as

$$\mathbb{E}[\text{service cost}] = \sum_{i \in I} \sum_{j \in J} c_{ij} p_j x_{ij}. \tag{6.127}$$

Regarding the expected outsourcing cost at a facility, we can specify the different possibilities for η_i, which leads to

$$\begin{aligned} \mathbb{E}[\text{Outsourcing cost}] &= \sum_{i \in I} \mathbb{P}(\eta_i = s) \times \mathbb{E}[\text{outsourcing cost} | \eta_i = s] \\ &= \sum_{i \in I} g_i \left(\sum_{s = q_i + 1}^{z_i} \mathbb{P}_x(\eta_i = s)(s - q_i) \right). \end{aligned} \tag{6.128}$$

Under customer outsourcing, when the number of customers allocated to some facility $i \in I$ requesting the service (demand customers) exceeds q_i, $\eta_i - q_i$ customers must be served directly from an external source. Again, a FIFO policy is assumed for deciding which customers to serve from the facility and which ones to outsource. The cost for supplying each outsourced customer remains denoted by g_i $(i \in I)$. Denote by $\mathbb{P}_i(s)$ the conditional probability of serving a demand customer assigned to facility $i \in I$ given that the total number of demand customers allocated to that facility is s (i.e., $\eta_i = s$). We have $\mathbb{P}_i(s) = (1/s) \times \min\{q_i, s\}$.

The outsourcing cost is computed similarly to facility outsourcing, leading again to (6.128). The service cost is now computed differently because the outsourced customers are no longer served by the facility they are allocated to

$$\begin{aligned} \mathbb{E}[\text{service cost}] &= \sum_{i \in I} \sum_{s=0}^{z_i} \mathbb{P}(\eta_i = s) \times \mathbb{E}[\text{service cost} | \eta_i = s] \\ &= \sum_{i \in I} \sum_{s=0}^{z_i} \left[\mathbb{P}(\eta_i = s) \sum_{j \in J} \mathbb{P}(\xi_j = 1 | \eta_i = s) \mathbb{P}_i(s) c_{ij} x_{ij} \right]. \end{aligned} \tag{6.129}$$

The above expressions use the probability distribution of η_i, $i \in I$. For a given a priori solution $\{\hat{y}, \hat{x}\}$, this probability is given by

$$\mathbb{P}(\eta_i = s) = \sum_{S \subseteq J_i : |S| = s} \left(\prod_{j \in S} p_j \right) \left(\prod_{j \in J_i \setminus S} (1 - p_j) \right)$$

with $J_i = \{j \in J \mid \hat{x}_{ij} = 1\}$.

This general setting would be investigated in [10], who proposed a heuristic algorithm consisting of two phases. First, a GRASP algorithm is used for building two pools of solutions—one based on quality and another on diversity. Second, a path relinking procedure is devised for connecting solutions from both pools, hoping that better feasible solutions can be found during the process.

The greedy randomized procedure consists of two phases: construction and improvement. The former uses an evaluation function that estimates the increase in the objective function due to incorporating some element in the current (partial) solution. In our problem, this materializes by selecting a new facility to open. In each iteration, a restricted candidate list (RCL) is built using the set of the lowest incremental costs. Denote by δ_i the (estimate in the) incremental cost by setting facility $i \in I$ open. Set $\delta_{\min}$ ($\delta_{\max}$) equal to the lowest (highest) incremental cost. The RCL in each iteration becomes

$$\text{RCL} \leftarrow \{i \in I \mid y_i = 0 \wedge \delta_{\min} \leq \delta_i \leq \delta_{\min} + \alpha(\delta_{\max} - \delta_{\min})\},$$

with α being a user-defined parameter that controls the size of the list.

In the first iteration, all facilities are closed. In [10] the authors suggest using a different formula for estimating the incremental costs:

$$\delta_i \leftarrow \frac{1}{u_i} \left(f_i + \sum_{j \in J} c_{ij} \right)$$

with u_i standing for an extended (artificial) capacity defined for each facility as

$$u_i = \max \left\{ \ell_i, \left\lfloor \frac{n}{\overline{p}} \frac{q_i}{\sum_{\iota \in I} q_\iota} \right\rfloor \right\}.$$

In the above expression, $\overline{p}$ is the average of the values p_j, $j \in J$.

Once at least one facility has been included in the partial solution, the incremental cost δ_i for a facility $i \in I$ not in the solution is estimated by summing the facility setup cost and the variation in the service cost when reassigning customers from their current assignments to the candidate facility. For each customer $j \in J$, we denote by $i(j)$ the facility it is currently assigned to. The above variation can be computed as

$$\sigma_{ij} = c_{ij} - c_{i(j)j} - p_j\, g(i(j)) \frac{[z_{i(j)} - q_{i(j)}]^+}{z_{i(j)}}.$$

These values can be sorted non-decreasingly leading to $\sigma_{ij_{[1]}}, \ldots, \sigma_{ij_{[|J|]}}$. Finally, the incremental cost δ_i can be estimated as

$$\delta_i \leftarrow f_i + \sum_{\iota=1}^{r_i} \sigma_{ij_{[\iota]}},$$

with $r_i = \min\{u_i, \max\{q \mid \sigma_{ij_{[q]}}\}\}$ chosen in a way that the lower throughout ℓ_i is satisfied and also customers are reassigned as much as possible among those with a negative estimate for the reassignment cost.

Once the RCL is built, a facility is randomly chosen; a new iteration begins.

Algorithm 6.4 structures the whole randomized greedy. The construction phase proceeds iterating according to a user-defined probability γ, and as far as at least one negative incremental cost is found.

Algorithm 6.4: GRASP for stochastic facility location with unit-demand customers

---

1 Build the first RCL;
2 Select a facility randomly in RCL and assign all customers to it;
3 **while** *there are negative incremental costs* **do**
4   generate $u \sim U[0, 1]$;
5   **if** $u > \gamma$ **then**
6   exit the construction phase
7   **else**
8   Build the RCL;
9   Select a facility i^* randomly in RCL and open it;
10   Reassign to i^* the r_{i^*} customers with the smallest reallocation costs;
11 **if** *solution is not feasible* **then**
12   Restore feasibility;
13 Perform a local search to improve the solution;

After the construction phase, it may happen that for some facility $i \in I$, the lower limit ℓ_i is violated (line 11). In this case, a facility i^* is chosen to be closed according to

$$i^* \in \arg \max_{i \in I \mid y_i = 1 \wedge (\ell_i - z_i) > 0} \left\{ f_i + \sum_{j \in J, i(j) = i} c_{ij} + g(i)[\overline{p}_i\, z_i - q_i]^+ \right\},$$

where $\overline{p}_i$ denotes the average demand probability of all customers assigned to facility i. Once the facility to close is chosen, the customers are reassigned greedily, i.e., the reassignments contribute the least to the (estimate) increase in the solution cost. Initially, reassignments are made only to infeasible facilities (that need more customers assigned to them). If feasibility is reached, then the reassignment is made considering the estimates

$$c_{ij} + [z_i + 1 - q_i]^+ p_j \frac{g_i\, q_i}{z_i\,(z_i + 1)}.$$

Otherwise, we need to reassign customers currently assigned to facilities i' such that $z_i > \ell_i$ to a facility i with $z_i < \ell_i$. This can be done greedily using the following estimates for the reassignment costs:

$$p_j(c_{ij} - c_{i'j}) + [z_i + 1 - q_i]^+ p_j \frac{g_i\, q_i}{z_i\,(z_i + 1)} - [z_{i'} - q_{i'}]^+ p_j \frac{g_{i'}\, q_{i'}}{z_{i'}\,(z_{i'} + 1)}.$$

Finally, when a feasible solution is obtained, a local search is performed to improve it. Two moves can be considered. While an improvement is achieved:

- Customer reassignment. In this case, the following estimate can be used for the variation in the cost when reassigning a customer j from facility $i_1 = i(j)$ to i_2:

$$\sigma_{j,i_2} = p_j(c_{i_2 j} - c_{i_1 j}) + [z_{i_2} + 1 - q_{i_2}]^+ g_{i_2} p_j \sum_{s=q_{i_2}}^{z_{i_2}} \zeta_{z_{i_2}, \bar{p}_{i_2}, s}$$
$$- [z_{i_1} - q_{i_1}]^+ g_{i_1} p_j \sum_{s=q_{i_1}+1}^{z_{i_1}} \zeta_{z_{i_1}, \bar{p}_{i_1}, s}.$$

- Customers exchange. In this case, the following estimate can be used for the variation in the cost when switching customers j_1 and j_2:

$$\sigma_{j_1 j_2} = c_{i_1 j_2} + c_{i_2 j_1} - c_{i_1 j_1} - c_{i_2 j_2}$$
$$+ g_{i_2} \max\{p_{j_1} - p_{j_2}, p_{j_2} - p_{j_1}\} \left[\frac{[z_{i_1} - K_{i_1}]^+}{z_{i_1}} - \frac{[z_{i_2} - K_{i_2}]^+}{z_{i_2}} \right].$$

Each iteration of the above procedure leads to a solution to the problem. Using the sequence of solutions obtained, two lists of solutions are built in [10], one driven by quality (the best ones) and another driven by diversity (those whose 0/1 solution vectors are more distinct from all the solutions found so far). From here, a path relinking procedure can be executed. For each solution in the quality list (target solution), a solution is selected at random in the diverse list. Then, progressively, the following moves are performed until the latter solution is converted into the former: exchange facilities, close facilities, open facilities, and reassign customers. Next, we briefly summarize these moves to make these contents self-contained.

Closed facilities in the target solution and open in the current solution are candidates for being closed. When a facility is closed, each of the customers assigned to it is taken: (i) if the facility it is assigned to in the target solution is open, the customer is assigned to it; otherwise, (ii) the customer is assigned greedily to the open facilities.

Open facilities in the target solution and closed in the current solution are the candidates for opening. When a facility is opened, we automatically assign to it all customers assigned to it in the target solution. If necessary, feasibility should be restored. A criterion for selecting the facility to open is the minimum incremental cost.

When one facility i_1 is open, and another one i_2 is closed: (i) all customers initially assigned to i_2 are assigned to i_1; (ii) if $\ell_{i_1} > \ell_{i_2}$ and there are less than ℓ_{i_1} customers initially assigned to i_1, then feasibility needs to be restored (as already explained). A criterion for selecting a pair of facilities to exchange is that it leads to the smallest increment in the cost.

Finally, customers can also be reassigned. In this case, the following estimate in the cost can be considered when reassigning customers j from facility $i_1 = i(j)$ to i_2:

$$\sigma_{j,i_2} = c_{i_2 j} - c_{i_1 j} + g_{i_2}\frac{p_j}{z_{i_2}+1}[z_{i_2}+1-q_{i_2}]^+ - g_{i_1}\frac{p_j}{z_{i_1}}[z_{i_1}-q_{i_1}]^+.$$

One final element in this procedure regards the computation of the objective function for an a priori solution. Recall that such an objective function is intractable. Thus, simulation remains a possibility. In this case, convergence can be accelerated by using the so-called antithetic estimates (see, e.g., [214]). The idea is simple: When generating a sequence of pseudo-random numbers, say $\rho_1, \ldots, \rho_T$, we also consider the sequence $1-\rho_1, \ldots, 1-\rho_T$. The antithetic estimate is the average of the outcomes using both sequences. The simulation process for approximating the objective function value of a given a priori solution terminates when the average cost (overall simulations so far) stabilizes according to some user-defined precision.

Finally, we cite [11], who include additional outsourcing policies. The authors assume uncertainty captured by a finite set of scenarios and propose compact MILP models that can be tackled up to proven optimality using an off-the-shelf solver.

6.4 Multi-stage Stochastic Facility Location

In this section, we go beyond the above contents by discussing the need to capture the multi-stage nature of uncertainty in optimization models for facility location problems.

To better focus on the topic, let us consider a planning horizon T divided into several periods. Assume that we seek to make decisions for an already operating system. Therefore, in addition to installing new facilities over time, we can consider closing some existing ones. This is a way to model facility/capacity relocation [377]. Accordingly, we partition the set of locations I into two subsets. I^c contains the locations where facilities are initially operating—that can be removed during the planning horizon. I^o includes the locations where new facilities can be opened during the planning horizon. As customary in this volume, we denote by f_{it} the cost for having facility $i \in I$ operating in period $t \in T$. It may include fixed setup costs for the facilities paid only once and when the facility is established or removed (see [403]). Let d_{jt} be the demand of customer $j \in J$ in period $t \in T$, q_{it} the capacity of a facility operating at $i \in I$ in period $t \in T$, and c_{ijt} the unit transportation cost from location $i \in I$ to demand node $j \in J$ in period $t \in T$. Note that we are also considering time-dependent capacities, which may be relevant, for instance, in logistics applications.

Two sets of time-dependent decision variables are considered: y_{it} is a binary variable indicating where a facility is operating at $i \in I$ in period $t \in T$, and x_{ijt} is the amount to send from location $i \in I$ to demand node $j \in J$ in period $t \in T$. The problem can be formulated as follows:

$$\text{minimize} \quad \sum_{t\in T}\sum_{i\in I} f_{it} y_{it} + \sum_{t\in T}\sum_{i\in I}\sum_{j\in J} c_{ijt} x_{ijt}, \tag{6.130}$$

$$\text{subject to} \quad \sum_{i\in I} x_{ijt} = d_{jt} \qquad \forall j \in J,\ t \in T, \tag{6.131}$$

$$\sum_{j\in J} x_{ijt} \leq q_{it} y_{it} \qquad \forall i \in I,\ t \in T, \tag{6.132}$$

$$y_{it} \leq y_{it-1} \qquad \forall i \in I^c,\ t \in T \setminus \{1\}, \tag{6.133}$$

$$y_{it} \geq y_{it-1} \qquad \forall i \in I^o,\ t \in T \setminus \{1\}, \tag{6.134}$$

$$x_{ijt} \geq 0 \qquad \forall i \in I,\ j \in J,\ t \in T. \tag{6.135}$$

$$y_{it} \in \{0, 1\} \qquad \forall i \in I,\ t \in T. \tag{6.136}$$

The meaning of most components in this model is clear from all models already discussed. Regarding Constraints (6.133) and (6.134), they guarantee that once an existing facility is closed, it is no longer established (until the end of the planning horizon) and vice versa: once installed, a new facility remains operating until the end of the planning horizon.

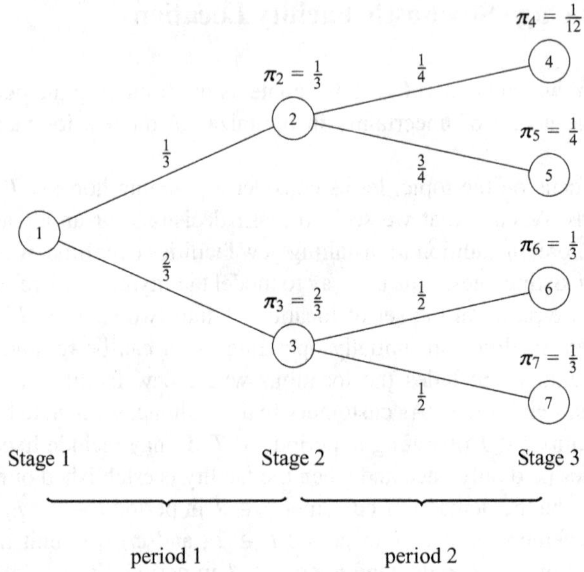

Fig. 6.1 A three-stage scenario tree—two periods in the planning horizon

Let us assume that all parameters are potential sources of uncertainty. In particular, for each $t \in T$, let $\boldsymbol{\xi}_t = (\boldsymbol{d}_t, \boldsymbol{f}_t, \boldsymbol{c}_t, \boldsymbol{q}_t)$ with $\boldsymbol{d}_t = (d_{jt})_{j \in J}$, $\boldsymbol{f}_t = (f_{it})_{i \in I}$, $\boldsymbol{c}_t = (c_{ijt})_{i \in I,\, j \in J}$, and $\boldsymbol{q}_t = (q_i t)_{i \in I,\, t \in T}$. We assume that $\{\boldsymbol{\xi}_1, \boldsymbol{\xi}_2, \ldots\}$ is a discrete-time stochastic process. As discussed in Sect. 4.3, in this case, a possible approach for embedding uncertainty in an optimization model consists of discretizing the distribution of the stochastic process (e.g., by sampling), thus constructing a scenario tree. In what follows, we assume that a scenario tree has been identified or sampled beforehand that captures a meaningful set of future scenarios—possible evolution of the above stochastic process. In Fig. 6.1, we illustrate one such tree for a two-period planning horizon and thus with three decision stages. The first stage corresponds to the beginning of period 1. The second one refers to the end of period 1 ($\boldsymbol{\xi}_1$ has been observed) and to the beginning of period 2. Finally, stage 3 refers to the end of period 2 ($\boldsymbol{\xi}_2$ has been observed).

One scenario corresponds to a sequence of events from the beginning of the first period until the end of the last one, i.e., one scenario fully determines all the information for the entire planning horizon. Thus, it induces a deterministic multi-period that can be formulated according to (6.130)–(6.135). In Fig. 6.1, we have four scenarios corresponding to the leaf nodes 4–7. We adopt the following notation inspired by some existing literature (see, e.g., [94]):

$\mathcal{N}$, set of nodes in the scenario tree.
M, the number of stages ($M = |T| + 1$).
Ω, a set of scenarios—leaf nodes.
$\gamma(n)$, immediate predecessor of node $n \in \mathcal{N} \setminus \{1\}$.

$\sigma(n)$, stage to which node n belongs to, $n \in \mathcal{N}$.
$\rho(n)$, period ending at node $n \in \mathcal{N} \setminus \{1\}$. Due to the relation between stages and periods, we have $\rho(n) = \sigma(n) - 1$.
π^n, probability that node $n \in \mathcal{N}$ occurs. Given that in each stage $m \in \{1, \ldots, M\}$ some node certainly occurs, the probabilities of the nodes satisfy $\sum_{n\in\mathcal{N}|\sigma(n)=m} \pi^n = 1$. The probability associated with node 1 is one—the *status quo*.

A node $n \in \mathcal{N} \setminus \{1\}$ in the scenario tree represents a possible sequence of events from the beginning of period one until the end of period $\rho(n)$. Assuming independent events, the probability associated with each node n is the product of the probabilities associated with the events whose sequence leads to node n. In Fig. 6.1, we also depict these values. The probabilities of the events are presented next to the edges. The probability associated with the sequence of events leading to a node is presented near the node.

We can now formally introduce the stochastic parameters associated with the problem:

f_i^n, cost for operating a facility operating at $i \in I$ in period $\rho(n)$ if node $n \in \mathcal{N} \setminus \{1\}$ occurs. This is one observation of the random variable $f_{i\rho(n)}$.
d_j^n, demand of customer $j \in J$ in period $\rho(n)$ if node $n \in \mathcal{N} \setminus \{1\}$ occurs. This is one observation of the random variable $d_{j\rho(n)}$.
q_i^n, capacity of a facility operating at $i \in I$ in period $\rho(n)$ if node $n \in \mathcal{N} \setminus \{1\}$ occurs. This is one observation of the random variable $q_{i\rho(n)}$.
c_{ij}^n, unit transportation cost from location $i \in I$ to customer $j \in J$ in period $\rho(n)$ if node $n \in \mathcal{N} \setminus \{1\}$ occurs. This is one observation of the random variable $c_{ij\rho(n)}$.
α^n, discount factor to apply to the costs in period $\rho(n)$, $n \in \mathcal{N} \setminus \{1\}$.

The decisions to make in a stage depend on the decisions made in the previous stages and also on the evolution of the underlying stochastic process. Furthermore, the decision about the open facilities in some period must be implemented before knowing the exact operation costs that will be incurred as well as the transportation costs and the demands since uncertainty will be revealed only during the period. Formally, we define:

$$y_i^n = \begin{cases} 1, & \text{if facility } i \in I \text{ is operating in period } \sigma(n) \text{ and the node} \\ & n \in \mathcal{N} \setminus \{\Omega\} \text{ occurs;} \\ 0, & \text{otherwise.} \end{cases}$$

x_{ij}^n = amount to send from facility $i \in I$ to customer $j \in J$ in period $\rho(n)$ if node $n \in \mathcal{N} \setminus \{1\}$ occurs.

Using the above notation and terminology, we derive the following multi-stage stochastic programming model:

$$\text{minimize} \quad \sum_{n\in\mathcal{N}\setminus\{1\}} \pi^n\alpha^n \left(\sum_{i\in I} f_i^n y_i^{\gamma(n)} + \sum_{i\in I}\sum_{j\in J} c_{ij}^n x_{ij}^n\right), \tag{6.137}$$

$$\text{subject to} \quad \sum_{i\in I} x_{ij}^n = d_j^n \qquad \forall j \in J,\ n \in \mathcal{N}\setminus\{1\}, \tag{6.138}$$

$$\sum_{j\in J} x_{ij}^n \leq q_i^n y_i^{\gamma(n)} \qquad \forall i \in I,\ n \in \mathcal{N}\setminus\{1\}, \tag{6.139}$$

$$y_i^n \leq y_i^{\gamma(n)} \qquad \forall i \in I^c,\ n \in \mathcal{N}\setminus\{\Omega\cup\{1\}\}, \tag{6.140}$$

$$y_i^n \geq y_i^{\gamma(n)} \qquad \forall i \in I^o,\ n \in \mathcal{N}\setminus\{\Omega\cup\{1\}\}, \tag{6.141}$$

$$x_{ij}^n \geq 0 \qquad \forall i \in I,\ j \in J,\ n \in \mathcal{N}\setminus\{1\}. \tag{6.142}$$

$$y_i^n \in \{0,1\} \qquad \forall i \in I,\ n \in \mathcal{N}\setminus\{\Omega\}. \tag{6.143}$$

The first component of the objective function can be written differently, thus better emphasizing its reasoning:

$$\begin{aligned}\sum_{n\in\mathcal{N}\setminus\{1\}} \pi^n\alpha^n \sum_{i\in I} f_i^n y_i^{\gamma(n)} &= \sum_{n\in\mathcal{N}\setminus\Omega}\left(\sum_{i\in I} y_i^n \sum_{\nu\in\mathcal{N}|\gamma(\nu)=n} \pi^\nu\alpha^\nu f_i^\nu\right)\\ = \sum_{n\in\mathcal{N}\setminus\Omega}\left(\pi^n \sum_{i\in I} y_i^n \sum_{\nu\in\mathcal{N}|\gamma(\nu)=n} \frac{\pi^\nu}{\pi^n}\alpha^\nu f_i^\nu\right) &= \sum_{n\in\mathcal{N}\setminus\Omega}\left(\pi^n\sum_{i\in I} y_i^n\, \mathbb{E}[\boldsymbol{F}_i^n]\right),\end{aligned} \tag{6.144}$$

where $\boldsymbol{F}_i^n$ is the random variable representing the (discounted) operational cost for facility $i \in I$ when we decide to operate it in the period starting at node n. This expected cost is a weighted sum of the total operational cost observed at the successors of n, using the probability of observing those successors conditional on starting from node n. Like in the deterministic model, the operational costs can easily accommodate fixed setup costs.

In the above model, the objective function (6.137) accounts for the total expected cost for the entire planning horizon. Recall that when we decide that a facility should be operating at some period, we still do not know the corresponding operational cost, which explains the expected value observed in (6.144). On the other hand, demand satisfaction is accomplished at the end of the period. Thus, we already know the corresponding costs and values for the demand. This explains the second term of the objective function (6.137).

Constraints (6.138) ensure demand satisfaction in every period and scenario. Inequalities (6.139) regard the capacity of the facilities in each period and scenario. Constraints (6.140) ensure the phasing-out for initially existing facilities (once closed, they cannot be open again), whereas Constraints (6.141) guarantee that once a new facility is opened, it should remain open until the end of the planning horizon. Finally, (6.142) and (6.143) define the domain of the decision variables.

In the above problem, we allow a previously existing facility to never operate during the planning horizon. Additionally, no facility removal is considered at the end of the last period since that would not impact the cost optimality.

The uncertainty associated with the capacity of the facilities, as considered in the above model, has raised an important branch of research that concerns the need to build the so-called resilient supply chain networks. Resilience is the capability of a system to hedge against disruptions by anticipating them or, if this is not possible, to recover from them in the most cost- and time-efficient way. The interested reader can refer to [265, 431] as well as to the references therein.

The above model can, in principle, be tackled by techniques such as those discussed in Sect. 4.3.3. However, in practice, the number of scenarios can be large, which is quite likely given that we are considering a multi-period stochastic problem—a scenario tree can quickly grow beyond tractability. Furthermore, the above problem stems from a simple setting. It will likely become more complex when dealing with real-world applications. For this reason, the development of approximate algorithms may be unavoidable. Next, we illustrate one possibility that has proven successfully for approximating large-scale multi-stage stochastic facility location problems (see, e.g., [7, 198, 432]): a relax-and-fix procedure already discussed in Sect. 4.3.3.

The reasoning underlying a relax-and-fix algorithm is to solve a sequence of problems not as challenging as the original one. In our case, a difficulty stems from the integrality of the y variables. Let us denote problem (6.137)–(6.143) by P. We start by partitioning the y variables into $M-1$ sets as follows: For $m = 1, \ldots, M-1$, V_m contains the y variables corresponding to decisions to make in stage m, i.e., V_m contains the variables y_i^n with $i \in I$ and $n \in \mathcal{N} \mid \sigma(n) = m$.

For each $m = 1, \ldots, M-1$, we define a problem P^m such that (i) the y variables in $V_1, \ldots, V_{m-1}$ are fixed according to the solutions to problems $P^1, \ldots, P^{m-1}$, and (ii) the integrality condition on the y variables in $V_{m+1}, \ldots, V_{M-1}$ is relaxed. Thus, in problem P^m, the only binary variables are those in V_m.

We start in stage 1 by solving the problem by relaxing the integrality of all the y variables in all stages apart from stage 1. In stage 2, we fix that first-stage decision according to the finding in stage 1 and relax the integrality of all y variables from stage 3 onward. Then, we obtain the solution for stage 2. The procedure proceeds until all stages have been considered (apart from the last one since the decisions on the y variables are made only up to stage $M-1$).

Denote by $V(.)$ the optimal value of a problem, $\underline{V}(.)$ a valid lower bound, and $\overline{V}(.)$ a valid upper bound. The procedure is summarized in Algorithm 6.5. This algorithm is an enhancement proposed by Escudero and Salmeron [201] for the original relax-and-fix procedure. Lines 8–13 seek to aggregate stages if, at some

Algorithm 6.5: A relax-and-fix algorithm for multi-stage stochastic facility location

```
m ← 1;
repeat
    Solve P^m;
    if P^m is feasible then
        V̲(P) ← V(P^m);
        m ← m + 1;
    else
        if m > 1 then                                // Aggregation of stages
            V_{m-1} ← V_{m-1} ∪ V_m;
            for k=m,...,M-2 do
                V_k ← V_{k+1}
            M ← M − 1;
            m ← m − 1;
until m = M − 1 or P^1 is infeasible;
if P^1 is infeasible then
    P is infeasible;
else
    V̄(P) ← V(P^{M-1});
```

point, infeasibility is reached. Infeasibility in one of the original stages does not indicate that the entire problem is infeasible. In the worst case, the procedure ends up solving the original problem. Nevertheless, as pointed out by the same authors, this is unlikely in practice. Finally, an optimality gap can be estimated using the lower and upper bounds returned by the procedure (if it ends successfully).

As mentioned in Sect. 4.3.3, when considering a multi-stage stochastic model (as we have just done), we should look into its actual need. We detail the process summarized in that section but now in the context of the multi-stage stochastic facility location problem we have discussed in this section. Consider a scenario tree as discussed above and illustrated in Fig. 6.1. Consider the following additional notation:

Ω^n, a set of scenarios induced by node $n \in \mathcal{N}$ or, in other words, those scenarios that include the node
S^n, a set of all successors of node $n \in \mathcal{N} \setminus \{\Omega\}$
S_1^n, a set of immediate successors of node $n \in \mathcal{N} \setminus \{\Omega\}$

The parameters capturing uncertainty in our problem are f_i^n, d_j^n, q_i^n, c_{ij}^n, $n \in \mathcal{N} \setminus \{1\}$.

A possible use of the expected value solution consists of solving a model, say, EV^n in every node of the scenario tree. This should be done sequentially for stages $1, 2, \ldots, M-1$. In a specific stage, for every node in that stage, we can consider the sub-tree rooted at that node and compute the corresponding conditional expected solution considering all the parameters' values according to the path from the root node to the node and aggregating in a single "expected" scenario the information for the subsequent stages. The process can be formalized as follows:

(i) Solve problem EV^1 denoting its optimal value by $V(EV^1)$ and save the optimal values of the first-stage variables $\{\hat{y}^1\}$. Problem EV^1 is a single-scenario problem rooted at node 1, i.e., we consider a single node in every stage. For each stage $m = 2, \ldots, M$, we consider the following data defining the unique scenario in that stage:

$$\overline{f}_{im} = \sum_{n \in \mathcal{N} | \sigma(n)=m} \pi^n f_i^n; \qquad \overline{d}_{jm} = \sum_{n \in \mathcal{N} | \sigma(n)=m} \pi^n d_j^n;$$

$$\overline{q}_{im} = \sum_{n \in \mathcal{N} | \sigma(n)=m} \pi^n q_i^n; \qquad \overline{c}_{ijm} = \sum_{n \in \mathcal{N} | sigma(n)=m} \pi^n c_{ij}^n.$$

After solving problem EV^1, we obtain a solution that we denote by $\{\hat{y}_m^1\}_{m=1,\ldots,M-1}$ and $\{\hat{x}_m^1\}_{m=2,\ldots,M}$. For the following stages, we fix the first-stage solution according to the values $\{\hat{y}_1\}$.

(ii) For every stage $m \in \{2, \ldots M - 1\}$ we proceed as follows: For every node $n \in \mathcal{N}$ belonging to stage m (i.e., such that $\sigma(n) = m$), solve model EV^n denoting its optimal value by $V(EV^n)$. Model EV^n corresponds to a single-scenario (deterministic) problem, which has the decisions corresponding to the previous stages fixed according to the computations already done. For each of the remaining stages, a single node is considered. In particular, for every remaining stage $\kappa = \sigma(n) + 1, \ldots, M$, the following data are considered (defining a single (sub)scenario until the end):

$$\overline{f}_{i\kappa} = \sum_{\nu \in S^n | \sigma(\nu)=\kappa} \frac{\pi^\nu}{\pi^n} f_i^\nu; \qquad \overline{d}_{j\kappa} = \sum_{\nu \in S^n | \sigma(\nu)=\kappa} \frac{\pi^\nu}{\pi^n} d_j^\nu;$$

$$\overline{q}_{i\kappa} = \sum_{\nu \in S^n | \sigma(\nu)=\kappa} \frac{\pi^\nu}{\pi^n} q_i^\nu; \qquad \overline{c}_{ij\kappa} = \sum_{\nu \in S^n | \sigma(\nu)=\kappa} \frac{\pi^\nu}{\pi^n} c_{ij}^\nu.$$

After solving problem EV^n using the above data, we obtain a solution that we denote by $\{\hat{y}_\kappa^n\}_{\kappa=\sigma(n),\ldots,M-1}$ and $\{\hat{x}_\kappa^n\}_{\kappa=\sigma(n)+1,\ldots,M}$. When considering problems EV^ν for the nodes ν that are successors of n (i.e., $\nu \in S_1^n$), we fix the solution corresponding to node n according to the component of the above solution corresponding to stage $\sigma(n)$ (the stage to which node n belongs to).

Remark 6.1 Every model EV^n "aggregates" the costs from the beginning of period $\sigma(n)$ until the end of the planning horizon. In particular, when we consider a problem EV^n in stage $M - 1$, we are dealing with a problem at the beginning of period T. The decisions that remain to be made regard the transportation in the last period. In other words, a model EV^n in stage $M - 1$ is a two-stage stochastic programming problem determining the location decisions in the first stage (beginning of the last period) and determining the (expected cost with) transportation in the second stage (end of the planning horizon).

Following [195], we define the expected value in stage m of using the above dynamic solution of the average scenario, and we denote it by EDEV_m ($m = 1, \ldots, M-1$) as the expected value of the optimal values of problems EV^n problems, with $n \in \mathcal{N}$, such that $\sigma(n) = m$. In other words, we define:

$$\text{EDEV}_m = \sum_{n \in \mathcal{N} \mid \sigma(n)=m} \pi^n V(EV^n), \quad m = 1, \ldots, M-1.$$

Finally, we compute the dynamic value of the stochastic solution as

$$DVSS = EDEV_{M-1} - SP,$$

or, in relative terms:

$$\%DVSS = \frac{DVSS}{SP} \times 100,$$

where SP is the optimal value of the multi-stage stochastic model.

The DVSS indicates the extent to which we can obtain a good approximation to the multi-stage stochastic problem by solving a sequence of single-scenario problems (and thus hopefully easier-to-solve models). Nevertheless, it does not assess the relevance of capturing uncertainty in our problem. For this reason, we consider a well-known indicator: the Expected Value of Perfect Information (EVPI). The EVPI measures the amount the decision-maker would be willing to pay to get access to perfect information about the future. The higher the EVPI, the higher the role of uncertainty in the problem (and, thus, the more relevant it is to capture it). This is a value computed easily: For each scenario $\omega \in \Omega$, we find the optimal value of the corresponding multi-period single-scenario problem. Denote that value by $V(\omega)$. The value of the wait-and-see solution value is computed as:

$$WS = \sum_{\omega \in \Omega} \pi^\omega V(\omega),$$

and, finally

$$EVPI = SP - WS,$$

or, in relative terms:

$$\%EVPI = \frac{EVPI}{WS} \times 100.$$

6.5 Dealing with Risk Aversion

In all the problems discussed in this chapter, the recourse function has been represented by an expectation. As mentioned before, this conveys a neutral attitude of the decision-maker toward risk. Location decisions are often strategic and involve significant investments. Accordingly, a risk-averse attitude toward risk cannot be disregarded as a possibility to be considered. If uncertainty is described probabilistically, we can consider a mean–variance objective [370] expanding the recourse function to account for variance. Taking, as an example, model (6.39)–(6.44), this consists of defining

$$\mathrm{Q}(\boldsymbol{y}) = \mathbb{E}\left[\mathcal{Q}(\boldsymbol{y}, \boldsymbol{\xi})\right] - \lambda \mathrm{Var}\left[\mathcal{Q}(\boldsymbol{y}, \boldsymbol{\xi})\right]. \tag{6.145}$$

Such a modeling framework in facility location is not new (see [293]). Nevertheless, in the context of facility location problems, such a model has a clear disadvantage: It often results in a non-linear large-scale mixed-integer model. Next, we discuss an alternative.

Consider again the setting underlying problem (6.39)–(6.44). Assume a risk-averse decision-maker. A measure of risk quite popular in this case is the Conditional Value-at-Risk. We keep assuming that uncertainty is represented by a finite set of scenarios Ω.

For a first-stage decision $\boldsymbol{y}$ and for a scenario $\omega \in \Omega$, we can formulate the second-stage problem as follows ($\boldsymbol{\xi}_\omega$ stands for the observation of the underlying random vector under scenario $\omega \in \Omega$):

$$\mathcal{Q}(\boldsymbol{y}, \boldsymbol{\xi}_\omega) = \min \sum_{i \in I} \sum_{j \in J} c_{ij\omega} d_{j\omega} x_{ij\omega}, \tag{6.146}$$

$$\text{s. t.} \quad \sum_{i \in I} x_{ij\omega} = 1, \qquad \forall j \in J, \tag{6.147}$$

$$x_{ij\omega} \leq y_i, \qquad \forall i \in I,\ j \in J, \tag{6.148}$$

$$x_{ij\omega} \geq 0, \qquad \forall i \in I,\ j \in J. \tag{6.149}$$

For each first-stage decision $\boldsymbol{y}$, $\mathcal{Q}(\boldsymbol{y}, \boldsymbol{\xi})$ is a random variable described by some CDF. As discussed in Sect. 3.5, a risk-averse decision-maker is not worried about the "favorable" or "good" scenarios. The worrisome scenarios are the worst. A complete risk-averse decision-maker focuses on the most extreme case. However, in practice, a decision-maker may be risk averse but up to some extent. Thus, the focus can be put on a certain proportion of the worst-case scenarios for the decision to make. This gives rise to the Conditional Value-at-Risk. Assume uncertainty captured by a finite set of scenarios. Model (3.4)–(3.7) can be adopted. Considering a confidence level

α, i.e., assume that the decision-maker focuses on the $(1-\alpha)100\%$ worst scenarios for a first-stage decision. In our case, this materializes as follows:

$$\text{minimize} \quad \eta + \frac{1}{1-\alpha} \sum_{\omega \in \Omega} \pi_\omega \zeta_\omega, \tag{6.150}$$

$$\text{subject to} \quad \zeta_\omega \geq \sum_{i \in I} \sum_{j \in J} c_{ij\omega} d_{j\omega} x_{ij\omega} - \eta \qquad \forall \omega \in \Omega, \tag{6.151}$$

$$\zeta_\omega \geq 0 \qquad \forall \omega \in \Omega, \tag{6.152}$$

$$(6.37),\ (6.46) - (6.48).$$

Due to the strategic nature of many location decisions, these are often assumed to be implemented here and now, as in the above problem as well as in the other two-stage problems discussed so far. However, in some applications, facilities may be seen as elements in a system requiring decisions made at an operational level, thus adapting to how uncertainty is revealed. This is the case, for instance, when renting temporary small-scale facilities, such as a small depot. Next, we revisit one such problem for a risk-averse decision-maker. In particular, we consider a fixed-charge stochastic transportation problem investigated in [208] that extends the risk-neutral version previously introduced in [269].

Let I and J denote a set of facilities/origins and customers/destinations, respectively. A commodity must be shipped from the former to satisfy the demand at the latter. Each facility has a limited capacity, which is known beforehand.

Transportation incurs two costs: a fixed setup cost for activating the connection (link) and a variable cost that depends on the amount of flow sent through the link. Activating a link originated at origin $i \in I$ incurs an additional fixed cost due to the setup of handling operations at the origin. The following deterministic parameters are known:

g_{ij}, setup cost for arc (i, j), $i \in I$, $j \in J$.
c_{ij}, unit transportation cost in arc (i, j), $i \in I$, $j \in J$.
q_i, capacity of facility $i \in I$.
f_i, fixed handling cost at facility $i \in I$. At least one such parameter is strictly positive.

We assume demand to be the source of uncertainty. Hence, $\boldsymbol{\xi} = (d_1, \ldots, d_{|J|})$ is the random vector underlying the problem. We seek to decide about the set of links to activate. The cost of such an a priori solution is the total setup cost for the selected links. After the demand is revealed, a decision is made regarding the origins that will operate (facilities to activate) and the amount to ship from each such origin to each demand destination, using the links activated a priori.

Shortages at the destinations are accepted, incurring a cost. This can be easily modeled mathematically by assuming that shortages correspond to shipments from a fictitious facility with enough capacity and a setup cost equal to 0. The unit transportation costs from such fictitious origin are the unit penalty at the destination.

Hence, w.l.o.g., we use I to denote the set of origins (that may include a fictitious one). Finally, we assume that demand can be captured by a finite set of scenarios Ω.

Thinking of an optimization model and our first-stage decision can be captured by the following set of decision variables:

$$x_{ij} = \begin{cases} 1 & \text{if link } (i, j) \text{ is contracted,} \\ 0 & \text{otherwise,} \end{cases} \qquad i \in I,\ j \in J.$$

The cost of an a priori solution can be written as

$$\sum_{i \in I} \sum_{j \in J} g_{ij} x_{ij}.$$

Given a first-stage decision $\boldsymbol{x}$, the problem of finding a minimum cost second-stage decision for each possible scenario $\omega \in \Omega$ (equivalently, for each possible realization of the demand $\boldsymbol{\xi}$) can be formulated using the following sets of decision variables:

$$y_{i\omega} = \begin{cases} 1, & \text{if origin } i \in I \text{ is used under scenario } \omega \in \Omega, \\ 0, & \text{otherwise.} \end{cases}$$

$w_{ij\omega}$ = flow to be shipped from facility $i \in I$ to customer $j \in J$ under scenario $\omega \in \Omega$.

We also define $d_{j\omega}$, the demand of customer $j \in J$ under scenario $\omega \in \Omega$.

A minimum cost recourse action for the a priori solution $\boldsymbol{x}$ when the realization of the stochastic demand $\boldsymbol{\xi}$ corresponds to scenario $\omega \in \Omega$ can be formulated as follows:

$$Q(\boldsymbol{x}, \boldsymbol{\xi}_\omega) = \min \quad \sum_{i \in I} f_i y_{i\omega} + \sum_{i \in I} \sum_{j \in J} c_{ij} w_{ij\omega}, \tag{6.153}$$

$$\text{s. t.} \quad w_{ij\omega} \le d_{j\omega} x_{ij} \qquad \forall i \in I,\ j \in J, \tag{6.154}$$

$$\sum_{i \in I} w_{ij\omega} \ge d_{j\omega} \qquad \forall j \in J, \tag{6.155}$$

$$\sum_{j \in J} w_{ij\omega} \le q_i y_{i\omega} \qquad \forall i \in I, \tag{6.156}$$

$$w_{ij\omega} \ge 0 \qquad \forall i \in I,\ j \in J, \tag{6.157}$$

$$y_{i\omega} \in \{0, 1\} \qquad \forall i \in I. \tag{6.158}$$

Constraints (6.154) ensure that flows are shipped using distribution links previously activated. Given that I includes a fictitious facility (to account for shortages at the customers), the correct evaluation of the missing amount at a destination requires imposing that the overall flow arriving at each destination $j \in J$ is at least its demand. This is ensured by Constraints (6.155). The origins to activate and their capacity limitations are ensured by (6.156). We are assuming that the fictitious facility has a scenario-independent capacity. This does not need to be the case. Given the scenario, we know the maximum shortage we can observe. Hence, we can better specify the capacity of the fictitious origin. The domain of the decision variables is defined in (6.157) and (6.158). The objective function (6.153) computes the cost of a recourse action for the a priori solution $\boldsymbol{x}$ under scenario $\omega \in \Omega$, as the setup costs of the handling operations at the origins that are used plus the overall routing costs, which include the penalty costs for unmet demand.

The above model can be enhanced in different ways. First, Constraints (6.154) can be replaced by the set of tighter constraints:

$$w_{ij\omega} \leq \overline{D}_{ij\omega} x_{ij} \quad \forall i \in I,\ j \in J,\ \omega \in \Omega, \tag{6.159}$$

where $\overline{D}_{ij\omega} = \min\{q_i, d_{j\omega}\}$. Second, we can explicitly state that the overall capacity of the activated origins must be enough to supply the total demand in each scenario $\omega \in \Omega$:

$$\sum_{i \in I} q_i y_{i\omega} \geq \sum_{j \in J} d_{j\omega} \quad \forall \omega \in \Omega. \tag{6.160}$$

Thinking again of a risk-averse decision-maker focusing on a certain proportion of the worst-case scenarios, we can adapt model (3.4)–(3.7) to our case. This leads to

$$\text{minimize} \quad \eta + \frac{1}{1-\alpha} \sum_{\omega \in \Omega} \pi_\omega \zeta_\omega, \tag{6.161}$$

$$\text{subject to} \quad \zeta_\omega \geq \left(\sum_{i \in I} \sum_{j \in J} g_{ij} x_{ij} + \sum_{i \in I} f_i y_{i\omega} + \sum_{i \in I} \sum_{j \in J} c_{ij} w_{ij\omega} \right) - \eta \ \forall \omega \in \Omega, \tag{6.162}$$

$$\zeta_\omega \geq 0 \quad \forall \omega \in \Omega, \tag{6.163}$$

(6.159), (6.160),

$$\sum_{i \in I} w_{ij\omega} \geq d_{j\omega} \quad \forall j \in J,\ \omega \in \Omega, \tag{6.164}$$

$$\sum_{j \in J} w_{ij\omega} \leq q_i y_{i\omega} \quad \forall i \in I,\ \omega \in \Omega, \tag{6.165}$$

$$w_{ij\omega} \geq 0 \quad \forall i \in I,\ j \in J,\ \omega \in \Omega, \tag{6.166}$$

$$x_{ij} \in \{0, 1\} \quad \forall i \in I,\ j \in J. \tag{6.167}$$

$$y_{i\omega} \in \{0, 1\} \quad \forall i \in I,\ \omega \in \Omega. \tag{6.168}$$

The above formulation is a mixed-integer linear program and thus, in principle, can be tackled by a general-purpose solver for small- to medium-sized instances.

Fernández et al. [208] devise an algorithm that seeks to find lower and upper bounds on an MILP problem seeking to optimize CVaR and apply it to the above problem. Next, we explain the algorithm and directly apply it to the problem.

For a fixed a priori solution $\boldsymbol{x}$, we can solve the second-stage problem (6.153)–(6.158) for every $\omega \in \Omega$, and then we can sort the scenarios by a non-increasingly sorting of the values $\mathcal{Q}(\boldsymbol{x}, \boldsymbol{\xi}_\omega)$, $\omega \in \Omega$. We can re-index the scenarios accordingly: $\omega_1(\boldsymbol{x}), \ldots, \omega_{|\Omega|}(\boldsymbol{x})$. Using $\boldsymbol{\xi}_{r(\boldsymbol{x})}$ instead of $\boldsymbol{\xi}_{\omega_r(\boldsymbol{x})}$ and $\pi_{r(\boldsymbol{x})}$ instead of $\pi_{\omega_r(\boldsymbol{x})}$ to make the notation lighter, we can denote the above sorting by $\boldsymbol{\xi}_{1(\boldsymbol{x})}, \ldots, \boldsymbol{\xi}_{|\Omega|(\boldsymbol{x})}$ with probabilities $\pi_{1(\boldsymbol{x})}, \ldots, \pi_{|\Omega|(\boldsymbol{x})}$, respectively. In other words,

$$\mathcal{Q}(\boldsymbol{x}, \boldsymbol{\xi}_{1(\boldsymbol{x})}) \geq \cdots \geq \mathcal{Q}(\boldsymbol{x}, \boldsymbol{\xi}_{|\Omega|(\boldsymbol{x})}).$$

Let $\ell(\boldsymbol{x})$ be the index such that

$$\pi_{1(\boldsymbol{x})} + \ldots + \pi_{\ell(\boldsymbol{x})-1} \leq 1 - \alpha, \tag{6.169}$$

$$\pi_{1(\boldsymbol{x})} + \ldots + \pi_{\ell(\boldsymbol{x})} > 1 - \alpha. \tag{6.170}$$

Following [459], we conclude that the α-VaR for a first-stage decision $\boldsymbol{x}$ is given by $\eta_\alpha(\boldsymbol{x}) = \mathcal{Q}(\boldsymbol{x}; \boldsymbol{\xi}_{\ell(\boldsymbol{x})})$.

Let $\Omega(\boldsymbol{x}) = \{\omega_r(\boldsymbol{x}) \in \Omega \mid 1 \leq r \leq \ell(\boldsymbol{x})\}$ and let $\boldsymbol{\beta}(\boldsymbol{x})$ be a weight vector induced by $\Omega(\boldsymbol{x})$, whose r-th component is given by

$$\beta_{r(\boldsymbol{x})} = \begin{cases} \frac{\pi_{r(\boldsymbol{x})}}{1-\alpha} & \text{if } \ \omega_r(\boldsymbol{x}) \in \Omega(\boldsymbol{x}) \setminus \{\omega_{\ell(\boldsymbol{x})}\}, \\ 1 - \frac{\sum_{\omega \in \Omega(\boldsymbol{x}) \setminus \{\omega_{\ell(\boldsymbol{x})}\}} \pi_\omega}{1-\alpha} & \text{if } \ \omega_r(\boldsymbol{x}) = \omega_{\ell(\boldsymbol{x})}, \\ 0 & \text{if } \ \omega_r(\boldsymbol{x}) \in \Omega \setminus \Omega(\boldsymbol{x}). \end{cases} \tag{6.171}$$

Given that the non-zero components of $\boldsymbol{\beta}(\boldsymbol{x})$ determine the conditional probabilities for the scenarios in $\Omega(\boldsymbol{x})$, the α-CVaR of $\boldsymbol{x}$ can be written as

$$\vartheta(\boldsymbol{\beta}(\boldsymbol{x}), \boldsymbol{x}) = \sum_{i \in I} \sum_{j \in J} g_{ij} x_{ij} + \sum_{r=1}^{|\Omega|} \beta_{r(\boldsymbol{x})} \mathcal{Q}(\boldsymbol{x}, \boldsymbol{\xi}_{r(\boldsymbol{x})}). \tag{6.172}$$

Unfortunately, we have no closed expression for the weight vector $\boldsymbol{\beta}(\boldsymbol{x})$, which depends on the first-stage decision to make. Therefore, one possibility to partially overcome this difficulty is to focus on the subproblem $\mathcal{H}(\boldsymbol{\beta})$ that we obtain when the weights are fixed to some constant vector $\boldsymbol{\beta} \in \mathbb{R}^{|\Omega|}$ suitably selected. Such a problem can be formulated as follows:

$$\mathcal{H}(\boldsymbol{\beta}) = \min \quad \sum_{i \in I} \sum_{j \in J} c_{ij} y_{ij} + \sum_{\omega \in \Omega} \beta_\omega \left(\sum_{i \in I} \sum_{j \in J} d_{ij} x_{ij\omega} + \sum_{i \in I} h_i z_{i\omega} \right), \tag{6.173}$$

$$\text{s. t.} \quad (6.159),\ (6.160),\ (6.164) - (6.167).$$

Algorithm 6.6 can be used to find lower and upper bounds on the problem.

Algorithm 6.6: Obtaining lower and upper bounds on problem (6.161)–(6.167), (6.159), (6.160)

1 $\boldsymbol{x}^0 \leftarrow$ Any feasible vector $\boldsymbol{x}$;
2 $\boldsymbol{\beta}^0 \leftarrow \boldsymbol{\beta}(\boldsymbol{y}^0)$;
3 UB $\leftarrow \infty$; LB $\leftarrow 0$;
4 $k = 1$
5 **repeat**
6 | Determine $\boldsymbol{y}^k$ by computing $\mathcal{H}(\boldsymbol{\beta}^{k-1})$;
7 | update UB and LB;
8 | $\boldsymbol{\beta}^k \leftarrow \boldsymbol{\beta}(\boldsymbol{y}^k)$;
9 | $k \leftarrow k + 1$;
10 **until** *stopping criterion is met*;

Some aspects of this algorithm require clarification.

Initially (line 1), we can start by sorting the scenarios by decreasing values of their total demand $\sum_{j \in J} d_{j\omega}$. Thus, the initial constant vector $\boldsymbol{\beta}^0$ is computed as the ℓ-weight vector induced by the ℓ-tail obtained according to this sorting.

In each iteration $k \geq 1$ we compute $\mathcal{H}(\boldsymbol{\beta}^{k-1})$, which directly gives a lower bound. Then, we update the scenario vector as $\boldsymbol{\beta}^k = \boldsymbol{\beta}(\boldsymbol{x}^k)$, and we calculate an upper bound using the expression of $\mathcal{H}$ considering the new weights $\boldsymbol{\beta}^k$ and fixing the solution according to the solution $\{\boldsymbol{x}^k, \boldsymbol{y}^k, \boldsymbol{w}^k\}$.

If the updated scenario vector $\boldsymbol{\beta}^k$ coincides with $\boldsymbol{\beta}^{k'}$ for some $k' < k - 1$, we define the new weight vector as $\frac{\boldsymbol{\beta}^k + \boldsymbol{\beta}^{k-1}}{2}$.

The algorithm terminates either when an optimal solution to the problem is found, which happens when $\boldsymbol{\beta}^k = \boldsymbol{\beta}^{k-1}$, or after a maximum number of iterations previously specified. In this last case, we obtain a lower and an upper bound of the optimal solution of the problem.

6.6 Covering Location Problems

As discussed in Chap. 2, covering location problems emerge when a demand node can be served only if it lies within some given coverage measure (e.g., distance or travel time). In the maximal covering location problem, p facilities are selected to cover as much demand as possible. In this case, for each demand node $j \in J$, we define the set I_j of potential locations for the facility that covers the demand node. The problem can be formulated considering the usual location y variables and a set of binary variables, x_j, indicating whether or not demand node $j \in J$ is covered by the selected facilities. Recall the optimization model presented in Chap. 2:

$$\text{maximize} \quad \sum_{j \in J} d_j x_j, \tag{6.174}$$

$$\text{subject to} \quad \sum_{i \in I_j} y_i \geq x_j \quad \forall j \in J, \tag{6.175}$$

$$\sum_{i \in I} y_i \leq p, \tag{6.176}$$

$$x_j \in \{0, 1\} \quad \forall j \in J, \tag{6.177}$$

$$y_i \in \{0, 1\} \quad \forall i \in I. \tag{6.178}$$

The discrete set covering location problem, also presented in Chap. 2, seeks to minimize the number of facilities that cover all demand nodes. Again, we recall an optimization model, where f_i is a fixed cost for locating a facility at $i \in I$:

$$\text{minimize} \quad \sum_{i \in I} f_i y_i \tag{6.179}$$

$$\text{subject to} \quad \sum_{i \in I_j} y_i \geq 1 \quad \forall j \in J, \tag{6.180}$$

$$y_i \in \{0, 1\} \quad \forall i \in I. \tag{6.181}$$

Different sources of uncertainty can emerge in covering location problems, such as the availability of the facilities when the demand nodes call for being served, the response/travel time, the demands, the coverage radius, and possibly others. In the following sections, we discuss these possibilities, considering the covering problems in which they are more relevant or insightful.

6.6.1 Stochasticity in the Availability of the Facilities

Uncertainty in the availability of the facilities has attracted much attention in the specialized literature due to its natural applicability to emergency facilities. In [153] the author introduces one such setting in the context of the maximal covering location problem, assuming that each facility is unavailable (e.g., busy or non-operational) with some probability, say π. Moreover, the availability of each facility is assumed to be independent of that of the others. Hence, if L facilities cover a demand node $j \in J$, then the probability that at least one such facility is available is $1 - \pi^L$. Consequently, the expected demand covered is $\mathbb{E}(D_{j,L}) = (1 - \pi^L)d_j$.

The increase in the expected coverage at node $j \in J$ resulting from increasing the number of facilities covering the node from $L - 1$ to L is equal to

$$\Delta\mathbb{E}(D_{j,L}) = \mathbb{E}(D_{j,L}) - \mathbb{E}(D_{j,L-1}) = \pi^{L-1}(1 - \pi)d_j.$$

The problem that consists of maximizing the expected coverage can be formulated mathematically by redefining the variables $\boldsymbol{x}$: For $j \in J$ and $L \in \{1, \dots, p\}$, x_{jL} is equal to 1 if demand node j is covered by at least L facilities and 0 otherwise. The problem can be formulated as follows:

$$\text{maximize} \quad \sum_{j \in J} \sum_{L=1}^{p} \pi^{L-1}(1 - \pi)d_j x_{jL}, \tag{6.182}$$

$$\text{subject to} \quad \sum_{L=1}^{p} x_{jL} \le \sum_{i \in I_j} y_i \quad \forall j \in J, \tag{6.183}$$

$$x_{jL} \in \{0, 1\} \qquad \forall j \in J,\ L = 1, \dots, p, \tag{6.184}$$

$$(6.176),\ (6.178).$$

If $\pi = 0$, the problem reduces to the deterministic maximal covering location problem.

Proposition 6.2 *Assuming* $0 < \pi < 1$*, an optimal solution to the above problem satisfies Constraints* (6.183) *and* (6.176) *as an equality.*

Proof See [153]. □

Unlike the deterministic version of the problem, locating more than one facility at a location is now a possible optimal policy. Furthermore, for values of π close to 1, it becomes optimal to locate all p facilities in a single location: the one covering the highest demand (if a single one can be identified).

For a fixed value of π, the model can, in principle, be tackled by a general-purpose solver. However, in large-scale instances, this option may be too time-consuming or unsuccessful due to short memory. Daskin [153] propose a heuristic to find high-quality feasible solutions for different values of that probability. This calls for a parametric analysis of π starting at $\pi = 1$ and thus with the solution that installs all servers in the location covering the highest demand. From a current solution, single-node substitutions are tested (moving one server from its current location to an alternative one), and the resulting improvement is considered as a function of π. Then, the value of π for which both solutions are equally good is computed—the root of the improvement function. The substitution leading to the highest such root is adopted, leading to a new current solution; π is set equal to that highest root. The algorithm proceeds until π becomes equal to 0—the entire range of interest of π was explored. An application of the above model to an emergency medical service system design can be found in [152].

A crucial assumption in the setting under discussion is the independence of the different facilities (in terms of availability). This may be too restrictive in practice. Batta et al. [49] relax it by considering a modified objective, namely

$$\text{maximize} \quad \sum_{j\in J}\sum_{L=1}^{p} C(p,\pi,L-1)\pi^{L-1}(1-\pi)d_j x_{jL}, \tag{6.185}$$

with

$$C(p,\pi,L) = \frac{\displaystyle\sum_{j=L}^{p-1} \frac{(p-L-1)!(p-j)}{(j-L)!}\frac{p^j}{p!}\pi^{j-L}}{\left[(1-\pi)\displaystyle\sum_{k=0}^{p-1}\frac{p^k}{k!}\pi^k\right] + \dfrac{p^p\pi^p}{p!}}, \quad L = 0,\ldots,p-1.$$

The factors $C(p, \pi, L)$ are approximate. They were derived in [328] by casting the system as an $M/M/p$ queue and assuming that when a demand node calls for being served, the decision-maker selects an available server randomly without replacement. The modified model reduces to the one proposed in [153] if $C(p, \pi, L) = 1$ for $L = 0, 1, \ldots, p - 1$.

The procedure proposed in [153] can still be adopted. However, unlike the original version, the starting solution (for $\pi = 1$) may not be optimal. The reader can refer to [49] for all details.

6.6.2 Stochasticity in the Demand

We focus again on the maximal covering location problem since this is the one in which demand uncertainty may pose some challenges. Our starting point is the problem (6.174)–(6.178).

For $j \in J$, let ξ_j be the random variable representing the demand of node j. Under uncertainty, a natural objective associated with a risk-neutral decision-maker is the expected value of the demand covered. However, in this case, embedding uncertainty in the model is irrelevant because the value of the stochastic solution is zero. In fact, for any feasible location solution, say $\{\hat{\mathbf{y}}\}$, we can identify the covered demand, say $\{\hat{\mathbf{x}}\}$. Hence,

$$\mathbb{E}_{\xi}\left[\sum_{j\in J}\xi_j\hat{x}_j\right] = \sum_{j\in J}\mathbb{E}_{\xi}\left[\xi_j\hat{x}_j\right] = \sum_{j\in J}\mathbb{E}_{\xi}\left[\xi_j\right]\hat{x}_j,$$

with $\boldsymbol{\xi}$ standing for the random vector of demands. The above equality indicates that if we replace the random variables with their expectations, we obtain a simplified model that is exact for the stochastic problem.

A not-so-straightforward case emerges if one considers a risk-averse decision-maker. Given that uncertainty is measured probabilistically, we can resort to maximizing CVaR. Since we seek to capture as much demand as possible, a bad scenario is one in which the demand served is small. In other words, for a confidence level α, following the contents of Sect. 3.5 (adapted to a maximization objective), we define the function

$$\Phi_\alpha(\boldsymbol{y},\eta) = \eta - \frac{1}{1-\alpha}\mathbb{E}\left[(\eta - \mathcal{F}(\boldsymbol{y},\boldsymbol{\xi})) \mid \mathcal{F}(\boldsymbol{y},\boldsymbol{\xi}) < \eta\right],$$

with $\mathcal{F}(\boldsymbol{y},\boldsymbol{\xi})$ denoting the (random) demand covered for a location solution $\boldsymbol{y}$. The α-CVaR of a solution $\boldsymbol{y} \in Y$ can be computed as

$$\Psi_\alpha(\boldsymbol{y}) = \Phi_\alpha(\boldsymbol{y},\eta(\boldsymbol{y})) = \max_{\eta}\Phi_\alpha(\boldsymbol{y},\eta),$$

and thus the problem of finding the feasible vector $\boldsymbol{y} \in Y$ with the highest α-CVaR can be expressed as

$$\max_{\boldsymbol{y}\in Y}\Psi_\alpha(\boldsymbol{y}) = \max_{\boldsymbol{y}\in Y}\Phi_\alpha(\boldsymbol{y},\eta(\boldsymbol{y})) = \max_{\boldsymbol{y}\in Y,\eta}\Phi_\alpha(\boldsymbol{y},\eta).$$

We are not considering the decision vector $\boldsymbol{x}$ since it is an immediate consequence of the location decision $\boldsymbol{y}$, thus making the notation lighter.

In case uncertainty is represented by a finite set of scenarios Ω or a finite sample of scenarios is drawn from the underlying distribution, then we can consider the function

$$\Phi_\alpha(\boldsymbol{y},\eta) = \eta - \frac{1}{1-\alpha}\sum_{\omega\in\Omega}\left[\eta - \mathcal{F}(\boldsymbol{y},\boldsymbol{\xi}_\omega)\right]^+\pi_\omega.$$

As customary in this volume, $\boldsymbol{\xi}_\omega$ stands for the observation of $\boldsymbol{\xi}$ under scenario $\omega \in \Omega$. Also in this case, denoting by $d_{j\omega}$ the demand of $j \in J$ under scenario

$\omega \in \Omega$, we can reformulate the stochastic maximum covering location problem as

$$\text{maximize} \quad \eta - \frac{1}{1-\alpha}\sum_{\omega \in \Omega} \pi_\omega \zeta_\omega, \tag{6.186}$$

$$\text{subject to} \quad (6.175) - (6.178),$$

$$\zeta_\omega \geq \eta - \sum_{j \in J} d_{j\omega} x_j \qquad \forall \omega \in \Omega, \tag{6.187}$$

$$\zeta_\omega \geq 0, \qquad \forall \omega \in \Omega. \tag{6.188}$$

6.6.3 Stochasticity in Travel Times and Demands

If response or travel times are uncertain, then the sets I_j of potential facilities $i \in I$ that cover node $j \in J$ become ill-defined. In this case, we can set a tolerance, say γ, and include in I_j those locations $i \in I$ such that $\mathbb{P}_{\ell}[\ell_{ij} \leq \tau] \geq \gamma$, with ℓ_{ij} denoting a random variable representing the travel time between $i \in I$ and $j \in J$, and $\mathbb{P}_{\ell}$ standing for the joint probability distribution of the travel times. In other words, we include in I_j the locations that provide some given guarantee of covering node j. We denote this set by $I_j(\gamma)$.

If the random variables ℓ_{ij} are independent ($i \in I$, $j \in J$) and if Φ_{ij} is the cumulative distribution function of ℓ_{ij}, then we can explicitly write the sets $I_j(\gamma)$ as

$$I_j(\gamma) = \{i \in I \mid \tau \geq \chi_\gamma(\Phi_{ij})\}, \tag{6.189}$$

with $\chi_\gamma(.)$ denoting the γ-quantile of the argument distribution. In case the CDF Φ_{ij} is an increasing function, we can just write

$$I_j(\gamma) = \{i \in I \mid \tau \geq \Phi_{ij}^{-1}(\gamma)\}. \tag{6.190}$$

We can directly adopt model (6.179)–(6.181) by replacing I_j by $I_j(\gamma)$ thus obtaining a formulation for the stochastic problem.

It is a simple result to prove that the optimal number of facilities to locate in a non-decreasing function of the tolerance γ. Zhang et al. [567] use this model to cast a set covering facility location problem (on a network). A general-purpose solver is considered for tackling the integer problem. The authors also adapt the setting to the maximal covering location problem. This is straightforward. We omit it here.

The combination of uncertainty in the demands and travel times may be relevant in the maximal covering location problem.

In case uncertainty can be represented by a finite set of scenarios, Ω, each with an associated probability π_ω, we can consider scenario-indexed demands and travel times. Additionally, for $j \in J$ and $\omega \in \Omega$, we consider $x_{j\omega}$ as a binary variable equal to one if and only if demand node $j \in J$ is covered under scenario $\omega \in \Omega$. Note also the set of potential locations covering a demand node becomes scenario dependent as well, i.e., we define $I_{j\omega} = \{i \in I \mid \ell_{ij\omega} \leq \tau\}$ with $\ell_{ij\omega}$ standing for the travel time between $i \in I$ and $j \in J$ under scenario $\omega \in \Omega$. We can write an extensive form of the deterministic equivalent. Considering a risk-neutral decision-maker, we obtain

$$\text{maximize} \quad \sum_{\omega \in \Omega} \pi_\omega \sum_{j \in J} d_{j\omega} x_{j\omega}, \tag{6.191}$$

$$\text{subject to} \quad \sum_{i \in I_{j\omega}} y_i \geq x_{j\omega} \qquad \forall j \in J,\ \omega \in \Omega, \tag{6.192}$$

$$(6.176),\ (6.178),$$

$$x_{j\omega} \in \{0, 1\} \qquad \forall j \in J,\ \omega \in \Omega. \tag{6.193}$$

Adapting this model to maximize the conditional value-at-risk is now a trivial exercise.

A more complex setting emerges if demand is described by a very large (possibly infinite) set of scenarios. One possibility is to embed the above model into a sampling procedure. Next, we discuss another possibility that may be of relevance.

Let us assume that the travel times ℓ_{ij} ($i \in I$, $j \in J$) are independent random variables, each described by a continuous and increasing cumulative distribution function Φ_{ij}. We keep considering a given tolerance γ, defining the minimum covering probability acceptable for each pair (i, j), $i \in I$, $j \in J$, thus defining set $I_j(\gamma)$ accordingly.

We assume now that the demand vector, $\boldsymbol{\xi} = (\xi_j)_{j \in J}$, is described by a continuous and increasing CDF.

For a fixed $\gamma \in (0, 1)$, let $\Xi(\boldsymbol{y}, \boldsymbol{x}|\gamma) = \sum_{j \in J} x_j \xi_j$ represent the demand covered by solution $\{\boldsymbol{y}, \boldsymbol{x}\}$. Note that $\Xi(\boldsymbol{y}, \boldsymbol{x}|\gamma)$ is itself a random variable, which makes an objective function using it directly to be ill-defined. Following the terminology of [567], we consider confidence levels $\gamma, \epsilon \in (0, 1)$ and define an (γ, ϵ)-stochastic maximal covering location problem as follows:

$$\text{maximize} \quad d, \tag{6.194}$$

$$\text{subject to} \quad \mathbb{P}\left[\sum_{j \in J} x_j \xi_j \geq d\right] \geq \epsilon, \tag{6.195}$$

$$\sum_{i \in I_j(\gamma)} y_i \geq x_j \qquad \forall j \in J, \tag{6.196}$$

$$(6.176) - (6.178).$$

In other words, we seek to maximize the minimum threshold such that the probability of covering a demand at least equal to that threshold is above a given tolerance.

The viability of the above model depends on the possibility of writing the quantiles of the distribution function of $\Xi(\boldsymbol{y}, \boldsymbol{x}|\gamma)$ in a tractable way as a function of the decision variables. This allows deriving a deterministic equivalent for Constraint (6.195). We will focus on this aspect in Chap. 7 in general and in Sect. 7.4 for the specific case of that constraint.

Zhang et al. [567] investigate uncertainty in both the demands and travel times in the context of uncertainty theory [339]. Under regular conditions, the authors derive a tractable reformulation for an (γ, ϵ)-maximal covering location problem.

6.6.4 A Unified Model

Marín et al. [367] introduce a general model for covering location problems, capturing uncertainty and time-dependent aspects. The authors consider unit-demand customers and assume uncertainty associated with the covering capability of the facilities, in terms of both proximity and amount. The problem consists of deciding on a finite-length planning horizon divided into periods. We keep denoting by T the planning horizon. Like in the previous sections, a set of locations for the facilities I and a set of demand points J are given. To consider a more general setting, it is assumed that at each location, it is possible to install more than one facility/equipment up to the given limits. In some locations, we may have facilities already operating.

Opening or closing a facility during the planning horizon and operating it incur costs. As customary in multi-period facility location problems (see Chap. 2), opening a facility is planned for the beginning of a period, and closing is planned for the end of a period.

As mentioned above, uncertainty is associated with the covering capability of each pair (i, j), $i \in I$, $j \in J$, and with the minimum threshold for the coverage of each demand point in each period. A (uncertain) benefit is associated with coverage surplus, i.e., the number of facilities/equipment covering a demand point above a minimum threshold. An uncertain penalty is also associated with coverage shortage, i.e., the number of facilities/equipment in shortage for a demand point concerning the minimum threshold. Uncertainty is captured by a finite set of scenarios, Ω each having a probability π_ω, $\omega \in \Omega$, assumed known in advance.

The first-stage decisions regard the opening/closing plan for all locations throughout the planning horizon; in the second stage, we account for the total benefit and penalty induced by the covering plan. Next, we introduce the relevant parameters and decision variables.

Deterministic parameters:

o_{it}, cost of opening a facility at $i \in I$ at the beginning of time period $t \in T$.
c_{it}, cost of closing a facility at $i \in I$ at the end of time period $t \in T \setminus \{|T|\}$.
f_{it}, cost of operating a facility at $i \in I$ during period $t \in T$.
e_i, maximum number of facilities/equipment that can operate at location $i \in I$.
p_t, a maximum number of facilities/equipment that can operate in period $t \in T$.
$\bar{y}_{i0}$, the number of facilities that are open at location $i \in I$ before the beginning of the planning horizon.

Stochastic (scenario-dependent) parameters:

$a_{ijt\omega}$, (binary) parameter indicating whether location $i \in I$ can cover demand point $j \in J$ in time period $t \in T$ under scenario $\omega \in \Omega$.
$b_{jt\omega}$, minimum number of facilities/equipment requested to cover demand point $j \in J$ in period $t \in T$ under scenario $\omega \in \Omega$.
$d_{jkt\omega}$, marginal benefit for covering demand point $j \in J$ with a surplus of at least k facilities in period $t \in T$ under scenario $\omega \in \Omega$.
$h_{jkt\omega}$, marginal penalty for a shortage of at least k facilities in the coverage of demand point $j \in J$ in period $t \in T$ under scenario $\omega \in \Omega$.

In addition to the above notation, we define two auxiliary sets:

$K_{jt\omega} = \{1, \ldots, p_t - b_{jt\omega}\}$, with $p_t - b_{jt\omega}$ representing the maximum surplus in terms of coverage that we may observe in demand point $j \in J$ in period $t \in T$ under scenario $\omega \in \Omega$.
$K'_{jt\omega} = \{1, \ldots, b_{jt\omega}\}$, with $b_{jt\omega}$ standing for the maximum shortage in terms of coverage that we may observe in demand point $j \in J$ in period $t \in T$ under scenario $\omega \in \Omega$.

In the model that follows, we assume a risk-neutral decision-maker. The benefit from coverage surplus is treated as a negative cost. For this reason, in addition to the above notation, we consider:

$g_{jkt\omega} = -d_{jkt\omega}$, the negative marginal cost represents the (marginal) benefit accounted for by $d_{jkt\omega}$.

The values g and h are seen as marginal variations in the cost of having one extra surplus or shortage of equipment. For instance, suppose that at some demand point j in some period t and scenario $\omega \in \Omega$, we have a surplus of two facilities/equipment. In this case, we assume the corresponding "cost" to be $g_{j1t\omega} + g_{j2t\omega}$. This means that $g_{j2t\omega}$ can be interpreted as the increase in the "cost" for going from one to two surplus facilities. If the surplus becomes three, then the "cost" would be $g_{j1t\omega} + g_{j2t\omega} + g_{j3t\omega}$. Hence, an increase of one surplus facility, given that we already have two surplus facilities, yields an increase in the benefit of $g_{j3t\omega}$.

We assume that for each demand point $j \in J$, each period $t \in T$, and scenario $\omega \in \Omega$

$$g_{j1t\omega} \leq g_{j2t\omega} \leq \ldots \leq g_{j|K_{jt\omega}|t\omega} \leq 0$$

and

$$0 \leq h_{j1t\omega} \leq h_{j2t\omega} \leq \ldots \leq h_{j|K'_{jt\omega}|t\omega}.$$

In other words, we assume a non-increasing marginal benefit (for surplus) and a non-decreasing marginal penalty (for shortage). This seems realistic in practice: The marginal benefit of having a large surplus decreases when the number of surplus facilities increases; the marginal penalty for having a shortage of facilities increases with the number of facilities in shortage.

An integer programming model can be presented for the problem considering the following first-stage decision variables:

z_{it}, the number of facilities opened in site $i \in I$ at the beginning of period $t \in T$.
z'_{it}, the number of facilities closed in location $i \in I$ at the end of period $t \in T \setminus \{|T|\}$.
y_{it}, the number of facilities operating in location $i \in I$ in period $t \in T$.

Regarding the second-stage decision, we consider:

$$w_{jkt\omega} = \begin{cases} 1, & \text{if demand point } j \text{ is covered by at least} b_{jt\omega} + k \text{ facilities in} \\ & \text{period } t \text{ and scenario } \omega, \\ 0, & \text{otherwise.} \end{cases}$$
$(j \in J, k \in K_{jt\omega}, t \in T, \omega \in \Omega)$

$$v^s_{jkt} = \begin{cases} 1, & \text{if demand point } j \text{ is covered by at most } b_{jt\omega} - k \text{ facilities in} \\ & \text{period } t \text{ and scenario } \omega, \\ 0, & \text{otherwise.} \end{cases}$$
$(j \in J, k \in K'_{jt\omega}, t \in T, \omega \in \Omega.)$

An optimization model for the general problem is the following:

$$\min \sum_{i\in I}\sum_{t\in T} o_{it} z_{it} + \sum_{i\in I}\sum_{t\in T\setminus\{|T|\}} c_{it} z'_{it} + \sum_{i\in I}\sum_{t\in T} f_{it} y_{it}$$

$$+ \sum_{\omega\in\Omega} \pi_\omega \left[\sum_{t\in T}\sum_{j\in J}\sum_{k\in K_{jt\omega}} g_{jkt\omega} w_{jkt\omega} + \sum_{t\in T}\sum_{j\in J}\sum_{k\in K'_{jt\omega}} h_{jkt\omega} v_{jkt\omega} \right] \tag{6.197}$$

$$\text{s. t.} \sum_{i\in I} y_{it} \leq p_t \qquad \forall t \in T, \tag{6.198}$$

$$y_{i1} = z_{i1} + \bar{y}_{i0} \qquad \forall i \in I, \tag{6.199}$$

$$y_{it} = y_{i,t-1} + z_{it} - z'_{i,t-1} \qquad \forall i \in I,\ t \in T \setminus \{1\}, \tag{6.200}$$

$$\sum_{i \in I} a_{ijt\omega} y_{it} = b_{jt\omega} + \sum_{k \in K_{jt\omega}} w_{jkt\omega} - \sum_{k \in K'_{jt\omega}} v_{jkt\omega} \qquad \forall j \in J,\ t \in T,\ \omega \in \Omega, \tag{6.201}$$

$$w_{j1t\omega} + v_{j1t\omega} \le 1 \qquad \forall j \in J,\ t \in T,\ \omega \in \Omega, \tag{6.202}$$

$$w_{jkt\omega} \le w_{j1t\omega} \qquad \forall j \in J,\ t \in T,\ \omega \in \Omega,\ k = 2, \ldots, |K_{jt\omega}|, \tag{6.203}$$

$$v_{jkt\omega} \le v_{j1t\omega} \qquad \forall j \in J,\ t \in T,\ \omega \in \Omega,\ k = 2, \ldots, |K'_{jt\omega}|, \tag{6.204}$$

$$y_{it} \in \{0, \ldots, e_i\} \qquad \forall i \in I,\ t \in T, \tag{6.205}$$

$$z_{it} \in \{0, \ldots, e_i\} \qquad \forall i \in I,\ t \in T, \tag{6.206}$$

$$z'_{it} \in \{0, \ldots, e_i\} \qquad \forall i \in I,\ t \in T \setminus \{|T|\}, \tag{6.207}$$

$$w_{jkt\omega} \in \{0, 1\} \qquad \forall j \in J,\ k \in K_{jt\omega},\ t \in T,\ \omega \in \Omega, \tag{6.208}$$

$$v_{jkt\omega} \in \{0, 1\} \qquad \forall j \in J,\ k \in K'_{jt\omega},\ t \in T,\ \omega \in \Omega. \tag{6.209}$$

In this model, the objective function (6.197) quantifies the total cost for opening, closing, and operating the facilities (first line) plus the expected benefit and penalty associated with the actual coverage of the demand points. Constraints (6.198) restrict the number of operating centers to p_t, for each $t \in T$, and Constraints (6.199)–(6.200) establish the relationship between y, z, and z' variables. From a stochastic programming point of view, the above constraints together with the domain constraints (6.205)–(6.207) are the so-called first-stage constraints in the sense that they only relate the first-stage decision variables.

The left-hand side of each constraint (6.201) accounts for the number of open facilities that cover a given customer j in a given scenario ω and period t; then the right-hand side of the equality aims to compare this amount with the minimum request $b_{jt\omega}$, to ensure that both the v and w variables take their correct values. Additional constraints (6.202)–(6.204) are required to make either all the v variables or all the w variables (or all of them) take value 0. The fact that the costs $g_{jkt\omega}$ (respectively, $h_{jkt\omega}$) are increasingly sorted guarantees the correctness of the solution. Finally, Constraints (6.205)–(6.209) define the domain of the decision variables.

The above model reduces to the model (COV) from [225] if we set $|T| = \{1\}$, $\Omega = \{1\}$, $\pi_1 = 1$, and $\bar{y}_{i0} = 0$ for $i \in I$. In this case, the z and z' variables are not necessary; only y variables are needed to decide whether some facilities are opened in location i. Specifically, the parameters that adapt the above general model to (COV) are:

$o_{i1} = 0, c_{i1} = 0, f_{i1} = f_i$, for $i \in I$; $p_1 = p$
$b_{j11} = b_j, K_{j11} = K_j$
$g_{jk11} = g_{jk}$ for $j \in J, k \in K_{j11} = K_j$ for $j \in J$
$K'_{j11} = \emptyset$ for $j \in J$

Note that model (COV), presented in [225], does not capture coverage shortage (a demand point j being covered by less than b_j facilities). Thus, we have $K'_j = \emptyset$ for $j \in J$ and vector $\boldsymbol{h}$ is not needed.

The above-presented formulation also generalizes to a stochastic setting the models investigated in [250]. Since all the problems studied in that work are deterministic, they can be regarded as single-scenario problems, i.e., we have $\Omega = \{1\}$ with $\pi_1 = 1$. For simplicity of notation next, index ω is omitted.

First we note that if we set $K_{jt} = K'_{jt} = \emptyset$, then we can replace the equalities in Constraint (6.201) by

$$\sum_{i \in I} a_{ijt} y_{it} \geq b_{jt}, \ \forall j \in J, \ t \in T. \tag{6.210}$$

Note also that in some models considered in [250], facilities cannot be closed during the planning horizon. This can be easily accounted for by setting the closing costs (c_{it}) equal to a large enough value Θ. This ensures that facilities can only be opened during the planning horizon. Similarly, when centers cannot be opened, the same is done with opening costs (o_{it}).

One problem introduced in [250] is the General Dynamic Set Covering Location Problem. In some (empty) potential locations $(I^o \subset I)$, we can open facilities during the planning horizon; in the other locations $(I^c \subset I)$, we have a facility operating that can be closed during the planning horizon. These two sets do not overlap. For each $t \in T$, a subset of customers $J_t \subset J$ must be covered.

We can easily adjust the general model above presented to this case. First we remove w and v variables. Constraints (6.201) are replaced by (6.210). For the remaining parameters, we set: $K_{jt} = K'_{jt} = \emptyset$, $f_{it} = 1$, $p_t = m$, $e_i = 1$, for $i \in I, j \in J, t \in T$. Besides, $o_{it} = M$ for $i \in I^c$ and $o_{it} = 0$ otherwise for $t \in T$. Similarly, $c_{it} = \Theta$ for $i \in I^o$ and $c_{it} = 0$ otherwise. Finally, $b_{jt} = 1$ for $j \in J_t$, $t \in T$ and $b_{jt} = 0$ for $j \notin J_t, t \in T$.

A second model considered in [250] is the so-called Dynamic Maximal Covering Location Problem. This is a multi-period maximum covering location problem, which seeks to minimize the weighted sum of demand points not covered in each period plus the opening and closing costs in each period. Each weight ρ_{jt} is the cost of demand point j if not covered in period t. In this case, the facilities can be opened or closed during the planning horizon.

To adjust the above general model to this problem, we set $K_{jt} = \{1\}$ and $g_{j1t} = \rho_{jt}$ for $j \in J, t \in T$. Constraints (6.201) are replaced by

$$\sum_{i \in I} a_{ijt} y_{it} \geq w_{j1t}, \quad j \in J, t \in \mathcal{T}. \tag{6.211}$$

For the remaining parameters, $o_{it} = o_t$, $c_{it} = c_t$, $f_{it} = 0$, $K'_{jt} = \emptyset$, $p_t = p_t$, $b_{jt} = 0$, and $e_i = 1$ for $i \in I$, $j \in J$, $t \in T$. To obtain the same objective value as in the original model, constant $-\sum_{j \in J} \sum_{t \in \mathcal{T}} g_{j1t}$ must be added to the objective value.

Considering the general modeling framework above described, the authors in [367] note that variables $\boldsymbol{y}$ can be written as function of the variables $\boldsymbol{z}$ and $\boldsymbol{z}'$ namely,

$$y_{it} = \bar{y}_{i0} + \sum_{\tau=1}^{t} z_{i\tau} - \sum_{\tau=1}^{t-1} z'_{i\tau} \qquad \forall i \in I,\ t \in T,$$

which leads to a reformulation as follows:

$$\begin{aligned}
\min \quad & \sum_{i \in I} \sum_{t \in T} \left(o_{it} + \sum_{\tau=t}^{|T|} f_{i\tau} \right) z_{it} + \sum_{i \in I} \sum_{t \in T \setminus \{|T|\}} \left(c_{it} - \sum_{\tau=t+1}^{|T|} f_{i\tau} \right) z'_{it} \\
& + \sum_{i \in I} \sum_{t \in T} f_{it} \bar{y}_{i0} \\
& + \sum_{\omega \in \Omega} \pi_\omega \left(\sum_{t \in T} \sum_{j \in J} \sum_{k \in K_{jt\omega}} g_{jkt\omega} w_{jkt\omega} + \sum_{t \in T} \sum_{j \in J} \sum_{k \in K'_{jt\omega}} h_{jkt\omega} v_{jkt\omega} \right)
\end{aligned} \tag{6.212}$$

s. t. (6.202) – (6.204), (6.206) – (6.209),

$$\sum_{i \in I} \left(\sum_{\tau=1}^{t} z_{i\tau} - \sum_{\tau=1}^{t-1} z'_{i\tau} \right) \leq p_t - \sum_{i \in I} \bar{y}_{i0} \qquad \forall t \in T, \tag{6.213}$$

$$\begin{aligned}
& \sum_{i \in I} a_{ijt\omega} \left(\bar{y}_{i0} + \sum_{\tau=1}^{t} z_{i\tau} - \sum_{\tau=1}^{t-1} z'_{i\tau} \right) \\
& = b_{jt\omega} + \sum_{k \in K_{jt\omega}} w_{jkt\omega} - \sum_{k \in K'_{jt\omega}} v_{jkt\omega} \quad \forall j \in J,\ t \in T,\ \omega \in \Omega,
\end{aligned} \tag{6.214}$$

$$\bar{y}_{i0} + \sum_{\tau=1}^{t} z_{i\tau} - \sum_{\tau=1}^{t-1} z'_{i\tau} \leq e_i \qquad \forall i \in I,\ t \in T, \tag{6.215}$$

$$\bar{y}_{i0} + \sum_{\tau=1}^{t} z_{i\tau} - \sum_{\tau=1}^{t-1} z'_{i\tau} \geq 0 \qquad \forall i \in I,\ t \in T. \tag{6.216}$$

Based on the above reformulation, a Lagrangian relaxation-based procedure can be proposed. Constraints (6.214) can be relaxed in a Lagrangian fashion—associating Lagrangian multipliers $\theta_{jt\omega}$—leading to a problem that can be decoupled into two subproblems:

$$(\text{LR1}_\theta)\ \min \sum_{i\in I}\sum_{t\in T}\left[o_{it} + \sum_{\tau=t}^{|T|}(f_{i\tau} + \sum_{j\in J}\sum_{\omega\in\Omega}\alpha_{j\tau\omega}a_{ij\tau\omega})\right] z_{it}$$

$$+\sum_{i\in I}\sum_{t\in T\setminus\{|T|\}}\left[c_{it} - \sum_{\tau=t+1}^{|T|}(f_{i\tau} + \sum_{j\in J}\sum_{\omega\in\Omega}\theta_{j\tau\omega}a_{ij\tau\omega})\right] z'_{it}$$

$$+\sum_{i\in I}\sum_{t\in T}(f_{it} + \sum_{j\in J}\sum_{\omega\in\Omega}\theta_{jt\omega}a_{ijt\omega})\bar{y}_{i0}$$

s. t. (6.206), (6.207), (6.213), (6.215), (6.216).

$$(\text{LR2}_\theta)\ \min \sum_{j\in J}\sum_{t\in T}\sum_{\omega\in\Omega}\sum_{k\in K_{jt\omega}}(\pi_\omega g_{jkt\omega} - \theta_{jt\omega})w_{jkt\omega} +$$

$$+\sum_{j\in J}\sum_{t\in T}\sum_{\omega\in\Omega}\sum_{k\in K'_{jt\omega}}(\pi_\omega h_{jkt\omega} + \theta_{jt\omega})v_{jkt\omega} - \sum_{j\in J}\sum_{t\in T}\sum_{\omega\in\Omega}\theta_{jt\omega}b_{jt\omega}$$

s. t. (6.202) – (6.204), (6.208), (6.209).

For each vector of multipliers $\boldsymbol{\theta}$, both subproblems can be solved efficiently. The first has the integrality property—we can directly consider its linear relaxation. The second can be decomposed in $j \in J$, $t \in T$, and $\omega \in \Omega$. Each such subproblem can be solved by inspection after relaxing the integrality conditions on the decision variables $\boldsymbol{w}$ and $\boldsymbol{v}$. The above-relaxed problem indeed has the integrality property. Accordingly, the lower bound provided by Lagrangian relaxation will not improve the lower bound provided by the linear relaxation of the general reformulation on variables $\boldsymbol{z}$ and $\boldsymbol{z}'$. However, we can use the solutions provided by subproblems (LR1_θ) and (LR2_θ) to obtain sharp upper bounds to the problem.

Feasible solutions to the problem can be easily derived from an integer optimal solution to (LR1_θ) for some θ (obtained, for instance, using the simplex method). Denote such a solution by $\{\mathbf{z}^*, \mathbf{z}'^*\}$.

If we fix these variables in the reformulated model, we can decompose them according to the different demand nodes, periods, and scenarios. Therefore, for every $j \in J$, $t \in T$, and $\omega \in \Omega$, we can formulate the problem $\mathrm{UB}(\mathbf{z}^*, \mathbf{z}'^*)_{jt\omega}$ that consists of finding the optimal values of variables $\boldsymbol{w}$ and $\boldsymbol{v}$ as follows:

$$
\begin{aligned}
\text{minimize} \quad & \sum_{k \in K_{jt\omega}} \pi_\omega g_{jkt\omega} w_{jkt\omega} + \sum_{k \in K'_{jt\omega}} \pi_\omega h_{jkt\omega} v_{jkt\omega}, \\
\text{subject to} \quad & \sum_{k \in K_{jt\omega}} w_{jkt\omega} - \sum_{k \in K'_{jt\omega}} v_{jkt\omega} = M_{jt\omega}, \\
& w_{j1t\omega} + v_{j1t\omega} \leq 1, \\
& w_{jkt\omega} \leq w_{j1t\omega} && \forall k \in K_{jt\omega} \\
& v_{jkt\omega} \leq v_{j1t\omega} && \forall k \in K'_{jt\omega}, \\
& w_{jkt\omega} \in \{0, 1\} && \forall k \in K_{jt\omega}, \\
& v_{jkt\omega} \in \{0, 1\} && \forall k \in K'_{jt\omega}
\end{aligned}
$$

with $M_{jt\omega} = \sum_{i \in I} a_{ijt\omega} \left(\bar{y}_{i0} + \sum_{\tau=1}^{t} z^*_{i\tau} - \sum_{\tau=1}^{t-1} z'^*_{i\tau} \right) - b_{jt\omega}$.

It is easy to see that the above problem is always feasible. Moreover, for every $j \in J, t \in T, \omega \in \Omega$:

If $M_{jt\omega} < 0$, the optimal solution to the problem $\mathrm{UB}(\mathbf{z}^*, \mathbf{z}'^*)_{jt\omega}$ is:

$\widetilde{w}_{jkt\omega} = 0, \quad k \in K_{jt\omega}$
$\widetilde{v}_{jkt\omega} = 1, \quad k \leq -M_{jt\omega}$
$\widetilde{v}_{jkt\omega} = 0, \quad k > -M_{jt\omega}$

If $M_{jt\omega} > 0$, then the optimal solution becomes:

$\widetilde{v}_{jkt\omega} = 0, \quad k \in K'_{jt\omega}$
$\widetilde{w}_{jkt\omega} = 1, \quad k \leq M_{jt\omega}$
$\widetilde{w}_{jkt\omega} = 0, \quad k > M_{jt\omega}$

If $M_{jt\omega} = 0$, then:

$\widetilde{w}_{jkt\omega} = 0, \quad k \in K_{jt\omega}$
$\widetilde{v}_{jkt\omega} = 0, \quad k \in K'_{jt\omega}$

Hence, the optimal value of the problem $\mathrm{UB}(\mathbf{z}^*, \mathbf{z}'^*)_{jt\omega}$ is

$$
\sum_{k=1}^{M_{jt\omega}} \pi_\omega h_{jkt\omega}
$$

if $M_{jt\omega} < 0$ and

$$\sum_{k=1}^{M_{jt\omega}} \pi_\omega g_{jkt\omega},$$

otherwise.

Combining the optimal solutions to the above problems, we end up with an upper bound on the optimal value of the general model, which is equal to

$$\sum_{i\in I}\sum_{t\in T}\left(o_{it}+\sum_{\tau=t}^{T} f_{i\tau}\right) z_{it}^{*}+\sum_{i\in I}\sum_{t\in T\setminus\{|T|\}}\left(c_{it}-\sum_{\tau=t+1}^{T} f_{i\tau}\right) z_{it}^{\prime *}$$
$$+\sum_{i\in I}\sum_{t\in T} f_{it}\bar{y}_{i0}+\sum_{j\in J}\sum_{t\in T}\sum_{\omega\in\Omega} \mathcal{V}(\mathrm{UB}(\mathbf{z}^{*},\mathbf{z}^{\prime *})_{jt\omega}), \tag{6.217}$$

where $\mathcal{V}(\mathrm{UB}(\mathbf{z}^{*},\mathbf{z}^{\prime *})_{jt\omega})$ denotes the optimal value of subproblem $\mathrm{UB}(\mathbf{z}^{*},\mathbf{z}^{\prime *})_{jt\omega}$.

6.7 Conclusions and Further Reading

Stochastic programming approaches for discrete facility location problems have attracted much attention. More work exists than that summarized in the previous sections. In this chapter, we gathered some base knowledge that may guide an interested reader in building and solving models for stochastic facility location. The contents of this chapter are extended in different ways throughout this volume, namely in the context of particular families of facility location problems.

A work related to this chapter and the previous one is [371]. The authors study a multi-period facility location problem combining two-stage stochastic programming and robust optimization. This is a trend that will mentioned later in this volume. Uncertainty is associated with (i) fixed and assignment costs and (ii) the set of customers in each period. The goal is to minimize the total expected cost. A constraint is imposed on the maximum regret allowed in each scenario.

Chapter 7
Facility Location with Chance Constraints

Abstract This chapter focuses on facility location problems with chance constraints. The analysis starts with chance-constrained capacitated fixed-charge facility location problems. Afterward, the probabilistic set covering and maximal covering location problems are presented. This establishes a link with congestion in facility location problems. Uncertainty captured by a finite set of scenarios is also discussed. Throughout the chapter, emphasis is put on modeling aspects due to their relevance for deriving tractable deterministic equivalent models. When relevant algorithmic contributions exist, they are also detailed.

Keywords Joint chance constraints · Independent chance constraints · Congestion · Probabilistic covering

7.1 Introduction

An important class of stochastic optimization problems includes those with chance constraints. In this case, one or several constraints are not required to hold strictly. In turn, the decision-maker accepts a solution that ensures their satisfaction with some probability.

Chance-constrained programming is of relevance when dealing with reliability issues. As this chapter shows, this may also be sought when deciding where to locate a set of facilities for serving a given set of demand nodes. Furthermore, service level is often related to congestion, i.e., a facility or equipment may be busy when a customer calls for service. Different uncertainty sources may be associated with a problem. In this chapter, we focus on the most commonly considered in the literature, namely, demand and availability of the facilities.

The remainder of this chapter is organized as follows. In Sect. 7.2, joint and independent chance constraints are analyzed in the context of capacitated facility location problems. The following section covers the set covering location problem, whereas Sect. 7.4 is devoted to maximal covering location. Finally, we discuss

© The Author(s), under exclusive license to Springer Nature Switzerland AG 2024
F. Saldanha-da-Gama, S. Wang, *Facility Location Under Uncertainty*,
International Series in Operations Research & Management Science 356,
https://doi.org/10.1007/978-3-031-55927-3_7

settings where a finite set of scenarios is considered (e.g., a scenario reduction technique was applied to the problem).

7.2 Capacitated Facility Location with Chance Constraints

In facility location under demand uncertainty, if the decision-maker wants to plan for satisfying all demand irrespective of its actual level, then a solution may be obtained involving an operational capacity much above the demand level that turns out to be observed. In such a situation, may the setting allow, a suitable alternative may be to seek a solution ensuring that demand does not exceed the capacity of the operating facilities with some given probability.

To describe this paradigm, we consider the classical single-allocation capacitated facility location problem [169, 271] that was reviewed in Chap. 2:

$$\text{minimize} \quad \sum_{i \in I} f_i y_i + \sum_{i \in I} \sum_{j \in J} c_{ij} x_{ij}, \tag{7.1}$$

$$\text{subject to} \quad \sum_{i \in I} x_{ij} = 1 \qquad \forall j \in J, \tag{7.2}$$

$$\sum_{j \in J} d_j x_{ij} \leq q_i y_i \qquad \forall i \in I, \tag{7.3}$$

$$y_i \in \{0, 1\} \qquad \forall i \in I, \tag{7.4}$$

$$x_{ij} \in \{0, 1\} \qquad \forall i \in I,\ j \in J. \tag{7.5}$$

We recall that I is a finite set of potential locations for the facilities and J is a set of demand nodes. Furthermore, y_i is a binary variable equal to one if and only if a facility is established at $i \in I$; x_{ij} is also a binary variable that indicates whether demand node $j \in J$ is allocated to facility $i \in I$. Regarding the parameters, f_i is the fixed cost for setting up a facility at $i \in I$, c_{ij} is the cost for supplying demand node $j \in J$ from facility $i \in J$, d_j is the demand of node $j \in J$, and q_i is the capacity of location $i \in I$. The objective function accounts for the total cost; Constraints (7.2) ensure that each demand node is allocated to one and only one facility; Constraints (7.3) are the capacity constraint that also guarantees that demand nodes can be allocated only to operating facilities. Finally, we have the specification of the domain of the decision variables.

If the decision-maker is satisfied with a solution ensuring demand satisfaction with a given probability, then the "hard" capacity constraints (7.3) can be replaced by

$$\mathbb{P}_{\xi}\left[\sum_{j \in J} d_j x_{ij} \geq q_i y_i\right] \leq \gamma_i \quad \forall i \in I. \tag{7.6}$$

In the above expression and following the terminology used in this volume, $\boldsymbol{\xi}$ is the random vector representing the demand, which is assumed to have a given joint CDF; γ_i a user-defined upper bound on the probability of exceeding the capacity of location $i \in I$ once the plants to be opened and the allocations are determined.

The discussion can be easily extended to other settings of practical relevance. It is often the case that a decision-maker does not wish to install a facility irrespective of its throughput. Again, planning for the worst-case scenario may not be the best. A conservative solution ensuring the minimum throughout may easily violate the capacity constraints (when the actual demand is observed) and vice versa. Thus, one can again consider chance constraints as follows:

$$\mathbb{P}_{\boldsymbol{\xi}}\left[\sum_{j\in J} d_j x_{ij} \le \beta q_i y_i\right] \le \zeta_i \quad \forall i \in I. \tag{7.7}$$

In the above expression, ζ_i stands for the probability—given in advance—for having the facility $i \in I$ violating the minimum utilization rate represented by β.

A desirable feature for chance-constrained optimization is the possibility of finding a deterministic equivalent formulation. This is a model having the probabilistic constraints replaced by deterministic (equivalent) ones. Unfortunately, this is not always doable with the existing knowledge. Nevertheless, there are some successful cases reported in the location analysis literature. Next, we revisit two cases: one using a discrete CDF associated with the demand and another considering a continuous one.

Lin [338] explores the above chance constraints (7.6) assuming that demands d_j, $j \in J$, are independent with each being described by a Poisson random variable $P(\lambda_j)$.

Under this assumption, the total demand assigned to facility $i \in I$, i.e., $\sum_{j\in J} d_j x_{ij}$, is described by Poisson distribution $P(\Lambda_i)$ with $\Lambda_i = \sum_{j\in J} \lambda_j x_{ij}$. Thus, (7.6) can be directly formulated as follows:

$$\sum_{\ell=0}^{q_i y_i} e^{-\Lambda_i} \frac{\Lambda_i^\ell}{\ell!} \ge \gamma_i \quad \forall i \in I.$$

In turn, this can be written as

$$\sum_{j\in J} \lambda_j x_{ij} \le \nu_i y_i \quad \forall i \in I,$$

where ν_i is the largest expected value of the Poisson random variable, ensuring that the cumulative probability up to q_i is equal to at least γ_i if facility i is open. As detailed by the same author, the value ν_i can easily be obtained by a binary search. After replacing the probabilistic constraints (7.6) with their linear deterministic equivalent, the resulting problem becomes a single-allocation capacitated facility location problem that can be tackled by any appropriate method (see [209]).

Let us keep assuming independent demands but now without specifying a particular probability distribution. Nevertheless, let us take $\mu_j = \mathbb{E}[d_j]$, $\sigma_j^2 = \mathbb{V}[d_j]$, $j \in J$.

In practice, if we have many demand nodes, we can expect that each selected facility has many such nodes allocated to it. In this case, considering the random variable $D_i = \sum_{j \in J} d_j x_{ij}$, we can invoke the Central Limit Theorem and assume that

$$\frac{D_i - \mathbb{E}[D_i]}{\sqrt{\mathbb{V}[D_i]}} \approx \text{Gaussian}(0, 1).$$

Hence, the chance constraints (7.6) state that $q_i y_i$ must be larger than or equal to the $(1 - \gamma_i)$-quantile, say $z_{1-\gamma_i}$, of a Gaussian distribution with parameters $\mathbb{E}[D_i]$ and $\mathbb{V}[D_i]$ in which case we can write

$$\frac{q_i y_i - \sum_{j \in J} \mu_j x_{ij}}{\sqrt{\sum_{j \in J} \sigma_j^2 x_{ij}}} \geq z_{1-\gamma_i} \quad \forall i \in I. \tag{7.8}$$

The nonlinearity of the above constraints can be overcome by defining a set of auxiliary variables:

$$\nu_i = \frac{\sqrt{\sum_{j \in J} \sigma_j^2 x_{ij}}}{\sqrt{\sum_{j \in J} \sigma_j^2}} \quad \forall i \in I.$$

Embedding these variables in (7.8) and noting that we always have $0 \leq \nu_i \leq 1$ $\forall i \in I$, we can approximate (7.6) by

$$z_{1-\gamma_i} \nu_i + \sum_{j \in J} \frac{\mu_j}{\sqrt{\sum_{k \in J} \sigma_k^2}} x_{ij} \leq \frac{q_i}{\sqrt{\sum_{k \in J} \sigma_k^2}} y_i \quad \forall i \in I, \tag{7.9}$$

$$\nu_i^2 = \sum_{j \in J} \frac{\sigma_j^2}{\sum_{k \in J} \sigma_k^2} x_{ij} \quad \forall i \in I, \tag{7.10}$$

$$0 \leq \nu_i \leq 1 \quad \forall i \in I. \tag{7.11}$$

Two possibilities emerge for tackling the above (approximate) model. One is to cast it as a second-order conic programming problem and use an off-the-shelf solver to solve it. If this becomes computationally unaffordable, then one can think of leveraging the fact that for every $i \in I$, ν_i^2 can be approximated by a piecewise linear function in [0, 1]. Let us consider $b_1 = 0, b_2, \ldots, b_{n-1}, b_n = 1$ a set of n

distinct points in that interval. They create a partition of that interval into $n-1$ sub-intervals. Furthermore, we have

$$\nu_i = \sum_{m=1}^{n} \lambda_{im} b_m \quad \text{and} \quad \nu_i^2 \approx \sum_{m=1}^{n} \lambda_{im} b_m^2$$

as far as $\lambda_{i1}, \ldots, \lambda_{in} \geq 0$, $\sum_{m=1}^{n} \lambda_{im} = 1$, and $(\lambda_{i1}, \ldots, \lambda_{in})$ is a special ordered set of type 2 (SOS2), i.e., at most two such variables can be different from zero; if two are so, then they must be consecutive. Finally, the problem that consists of minimizing (7.1) subject to (7.2), (7.4)–(7.7), can be approximated by replacing (7.6) and (7.7) with

$$\sum_{j\in J} \frac{\mu_j}{\sqrt{\sum_{k\in J} \sigma_k^2}} x_{ij} + z_{1-\gamma_i} \sum_{m=1}^{n} \lambda_{im} b_m \leq \frac{q_i}{\sqrt{\sum_{k\in J} \sigma_k^2}} y_i \qquad \forall i \in I,$$

$$\sum_{j\in J} \frac{\mu_j}{\sqrt{\sum_{k\in J} \sigma_k^2}} x_{ij} + z_{\zeta_i} \sum_{m=1}^{n} \lambda_{im} b_m \geq \frac{\beta q_i}{\sqrt{\sum_{k\in J} \sigma_k^2}} y_i \qquad \forall i \in I,$$

$$\sum_{j\in J} \frac{\sigma_j^2}{\sum_{k\in J} \sigma_k^2} x_{ij} = \sum_{m=1}^{n} \lambda_{im} b_m^2 \qquad \forall i \in I,$$

$$\sum_{m=1}^{n} \lambda_{im} = y_i \qquad \forall i \in I,$$

$$\lambda_{i1}, \ldots, \lambda_{in} \geq 0 \qquad \forall i \in I,$$

$$(\lambda_{i1}, \ldots, \lambda_{in}) \text{ SOS2} \qquad \forall i \in I.$$

The above modeling aspects were explored in [309, 310] for tackling a chance-constrained single-sourcing facility location problems of relevance when planning for sheltering in the context of the preparedness phase in Humanitarian Logistics.

The approximation just described relies on assumptions that, in practice, can be strong: (i) a large number of demand points is assigned to each operating facility; (ii) the demands are independent. If this is not the case, it is desirable to seek a solution that is "distributionally" robust and thus that also reflects some covariance structure. This topic will be explored in Chap. 8.

The chance constraints considered so far were imposed independently on each facility. In some cases, a global reliability level is desired. This calls for considering joint chance constraints. One such setting was explored in [63] in the context of locating emergency facilities (e.g., ambulances). The problem is to decide where to locate such equipment and the number of vehicles to house in each location. We keep using location variables y_i, $i \in I$. Nevertheless, we redefine the allocation variables: x_{ij} denoting the number of vehicles located at $i \in I$ that will be used to serve demand point $j \in J$. The demand at node $j \in J$ is represented by a random variable ξ_j described by a discrete probability distribution. In emergency systems, the demand nodes can be served only by facilities within a certain distance or travel time. For this reason, we define I_j as the set of locations that cover demand node $j \in J$ and J_i set of demand nodes covered by location $i \in I$. The problem can be formulated as follows:

$$\text{minimize} \quad \sum_{i \in I} f_i y_i + \sum_{i \in I} \sum_{j \in J} c_{ij} x_{ij}, \tag{7.12}$$

$$\text{subject to} \quad \sum_{j \in J_i} x_{ij} \leq q_i y_i \qquad \forall j \in I, \tag{7.13}$$

$$\mathbb{P}\left[\sum_{i \in I_j} x_{ij} \geq \xi_j,\ j \in J\right] \geq \alpha, \tag{7.14}$$

$$x_{ij} \geq 0 \text{ and integer} \qquad \forall i \in I,\ j \in I_j, \tag{7.15}$$

$$y_i \in \{0, 1\} \qquad \forall i \in I. \tag{7.16}$$

Finding a deterministic equivalent to the above problem may be challenging because of the need to invert the joint CDF at a point that depends on the decision to make. Nevertheless, we can do it if we assume independent demands. In such a case, the probabilistic constraint (7.14) can be written as

$$\prod_{j \in J} F_j \left(\sum_{i \in I_j} x_{ij} \right) \geq \alpha, \tag{7.17}$$

where $F_j(.)$ stands for the (marginal) CDF of ξ_j, $j \in J$. Let $w_j = \sum_{i \in I_j} x_{ij}$, $j \in J$. Replacing in (7.17), we get

$$\prod_{j \in J} F_j \left(w_j\right) \geq \alpha, \tag{7.18}$$

and applying the logarithm function, we have

$$\sum_{j \in J} \ln(F_j(w_j)) \geq \ln(\alpha). \tag{7.19}$$

Given that the values w_j $(j \in J)$ must satisfy (7.18), we must have $w_j \geq \zeta_j(\alpha)$, where $\zeta_j(\alpha)$ is the α-quantile of the marginal distribution $F_j(.)$. Handling the above inequality in the general case may be challenging. If demands are discrete random variables, then there is a simple way to write a tractable model. We proceed with that case due to its relevance in many applications.

Assuming that demands are discrete, let κ_j be an upper bound for w_j that exists in our case due to the limited capacity of each potential location. We can resort to a discretization of w_j to reformulate the model (see [245]). Hence, introduce a set of binary variables, $z_{j\ell}$, such that for $j \in J$, $\ell \in \{\zeta_j, \ldots, \kappa_j\}$, $z_{j\ell}$ is equal to one if and only if w_j is equal to ℓ. In this case, (7.19), and thus (7.14), can be replaced by

$$\sum_{j \in J} \sum_{\ell=\zeta_j}^{\kappa_j} \ln(F_j(\ell))\, z_{j\ell} \geq \alpha, \tag{7.20}$$

$$\sum_{\ell=\zeta_j}^{\kappa_j} z_{j\ell} \leq 1 \quad \forall j \in J. \tag{7.21}$$

Beraldi et al. [63] explore these aspects. The authors obtain a reformulation that they tackle using a general-purpose solver. The authors also assume they are working with log-concave distribution functions, which allow fine-tuning the reformulation (see [436] for additional details on handling log-concave distribution functions).

7.3 Probabilistic Set Covering Location

In a covering location problem, a demand node is said to be covered when a facility is open within a given radius, distance, or travel time. We have already discussed in this volume that congestion is an aspect that has been much dealt with in the context of such problems. In emergency facility location, this is of great relevance.

Like in Sect. 6.6.1, let us denote by π the probability that the equipment or server is busy (not operational). For the moment, we assume this probability to be the same for all servers, irrespective of their location. We also assume that servers operate independently.

A natural chance constraint to consider is one ensuring that the probability of having equipment available for serving a demand node $j \in J$ when that node calls for being served is above a given threshold, α:

$$\mathbb{P}\left[\text{At least one server is available for serving customer } j\right] \geq \alpha \quad \forall j \in J. \tag{7.22}$$

Next, we analyze different settings. We follow the seminal work [451].

Suppose we seek to cover all customers with a set of locations of minimum cost. Chance constraints (7.22) can be used instead of the "usual" hard covering constraints

$$\sum_{i \in I_j} y_i \geq 1 \quad \forall j \in J, \tag{7.23}$$

where I_j stands for the potential locations $i \in I$ that cover demand node $j \in J$. A (numerical) deterministic equivalent for (7.22) is given by

$$\sum_{i \in I_j} y_i \geq \beta \quad \forall j \in J, \tag{7.24}$$

with $\beta = \lceil \ln(1-\alpha)/\ln \pi \rceil$. The probability that no location among those covering customer $j \in J$ is available to serve a customer is given by $\pi^{\sum_{i \in I_j} y_i}$. Therefore, the probability that at least one location among those covering customer $j \in J$ can serve it immediately is equal to $1 - \pi^{\sum_{i \in I_j} y_i}$, which, together with (7.22), leads to the deterministic equivalent just presented. These developments and variants have been proposed in [22, 111].

Unlike the hard covering constraints (7.23), Inequalities (7.24) often requires multiple coverage at the nodes. Consequently, multiple facilities or servers must be allowed at each location. Otherwise, Constraints (7.24) may become infeasible. For these reasons, in the probabilistic set covering problem, the binary domain for the location variables must be replaced by

$$y_i \ \text{integer} \quad \forall i \in I. \tag{7.25}$$

The extension of the above setting to the case in which we have customer- or location-dependent probabilities has been dealt with in the literature as it is the case in [451]. For each $j \in J$, let π_j denote the probability that all servers that can serve demand node j (i.e., lie in the corresponding covering area) are busy. Those authors show that for each $j \in J$ the corresponding chance constraints (7.22) are equivalent to

$$\sum_{i \in I_j} y_i \geq \beta_j \quad \forall j \in J, \tag{7.26}$$

with $\beta_j = \lceil \ln(1-\alpha)/\ln \pi_j \rceil$. Note that we can also make the desirable reliability level α dependent on j, which is straightforward and thus omitted.

Different ways for estimating or computing the probabilities π_j have been proposed in the literature. ReVelle and Hogan [451] leveraging [153] are led to $\hat{\pi}_j = (\tau \times \phi_j)/(24 \times \nu_j)$, where τ is the average service time, ϕ_j is the number of daily service calls originated at j or in other demand nodes within the coverage radius of j, and ν_j is the number of servers operating that can serve j. The above computation is for a daily busy probability (one day = 24 hours).

The number of servers covering j, ν_j, is given by $\sum_{i \in I_j} y_i$ and thus depends on the decision to make. Accordingly, in [451] the authors devise an iterative procedure that allows fine-tuning the estimates starting from any given estimate (such as a uniform busy probability across the system). Those estimates can be embedded in Constraints (7.26) and the problem solved, which leads to a new solution $\{\hat{y}\}$ that can be used to calculate new values for the estimates. This process can continue and hopefully provide converging values of the busy probabilities in the different demand areas $j \in J$.

Marianov and ReVelle [366] propose an alternative for determining β_j, i.e., the minimum number of servers to locate within the coverage radius of each demand node $j \in J$. The analysis is motivated by the likely dependence between the busy probabilities for servers belonging to the same "neighborhood"—servers covering the same node j. By modeling the probabilistic behavior of a neighborhood as an $M/M/s/s$ queuing system (Poisson arrivals, exponential service times, s servers, and up to s customers waiting), the authors explicitly capture congestion and go further in terms of the busy probabilities. Let λ_j be the expected number of calls generated by area j per time unit and μ_j, the expected number of calls served (also per time unit). Analyzing each demand node separately, the probability that all s servers in area j are busy is given by

$$\pi_j(s) = \frac{\frac{1}{s!}\frac{\lambda_j}{\mu_j}}{1 + \frac{\lambda_j}{\mu_j} + \frac{1}{2!}\left(\frac{\lambda_j}{\mu_j}\right)^2 + \cdots + \frac{1}{s!}\left(\frac{\lambda_j}{\mu_j}\right)^s}.$$

Hence, the number of servers required to guarantee the reliability α (for a fixed demand node $j \in J$) should be larger than or equal to the value β_j, which is the smallest integer satisfying

$$\pi_j(s) = \frac{\frac{1}{\beta_j!}\frac{\lambda_j}{\mu_j}}{1 + \frac{\lambda_j}{\mu_j} + \frac{1}{2!}\left(\frac{\lambda_j}{\mu_j}\right)^2 + \cdots + \frac{1}{\beta_j!}\left(\frac{\lambda_j}{\mu_j}\right)^{\beta_j}}.$$

Also in [366] the authors show that the probabilities $\pi_j(s)$ decrease when s increases. Hence, the integers β_j can be pre-computed and then embedded into the optimization model: We first consider $s = 1$. If $\pi_j(1) > 1 - \alpha$, we consider $s = 2$. If $\pi_j(2) > 1 - \alpha$, we proceed for $s = 3$. The process continues until the probability is smaller than or equal to $1 - \alpha$. The estimate proposed in [451] is at the core of the estimation of the values λ_j and μ_j, $j \in J$. In particular, λ_j is set to the number of daily calls; μ_j is set to 24 (hours) divided by the average service time (also in hours). If historical data are available, the above computations can be easily accomplished.

If we consider again a set covering location problem but facing location-dependent busy fractions, we must deal with probabilities π_i, $i \in I$. The chance constraints would be written as

$$1 - \prod_{i \in I_j} \pi_i^{y_i} \geq \alpha, \quad j \in J, \tag{7.27}$$

which, nonetheless, still have a deterministic equivalent:

$$\sum_{i \in I_j} y_i \log \pi_i \leq \log(1 - \alpha), \quad j \in J. \tag{7.28}$$

Again, estimating the probabilities π_i is of relevance. In [85] the authors estimate such a probability using $\hat{\pi}_i = (\tau \times \phi_i)/(24 \times \nu_i)$ with τ representing the average service time, ϕ_i is the average number of calls originated at the demand nodes covered by location i, and ν_i is the total number of servers installed at location i. As in the case of busy fractions dependent on the demand point, an iterative procedure can be applied to find the above estimates.

The chance-constrained set covering location problem can be used to solve other problems such as the α-reliable p-center problem and the maximum reliability location problem. In the first case, we seek to locate p facilities minimizing the maximum distance or time within which service is available with probability (reliability) α. The second consists of locating p facilities to maximize the capability of providing a service within a given coverage radius. The interested reader can refer to [453] to see how these problems easily leverage the chance-constrained set covering location problem discussed above.

Ball and Lin [42] also assume location-dependent busy fractions. The authors consider two-indexed location variables, $y_{i\ell}$ for $i \in I$ and $\ell \in \{1, \ldots, L_i\}$ and L_i denoting the maximum number of servers that can be located at i. There is a cost, say $f_{i\ell}$ for installing ℓ servers at location i. The problem can be formulated as follows:

$$\text{minimize} \quad \sum_{i \in I} \sum_{\ell=1}^{L_i} f_{i\ell} y_{i\ell}, \tag{7.29}$$

$$\text{subject to} \quad \prod_{i \in I_j} \prod_{\ell=1}^{L_i} (\mathbb{P}[D(i) \geq \ell])^{y_{i\ell}} \leq 1 - \alpha \quad \forall j \in J, \tag{7.30}$$

$$\sum_{\ell=1}^{L_i} y_{i\ell} \leq 1 \quad \forall i \in I, \tag{7.31}$$

$$y_{i\ell} \in \{0, 1\} \quad \forall i \in i,\ \ell \in \{1, \ldots, L_i\}. \tag{7.32}$$

In the above model, $D(i)$ is the number of calls generated by all demand nodes covered by location $i \in I$. Thus, $\prod_{\ell=1}^{L_i} (\mathbb{P}[D(i) \geq \ell])^{y_{i\ell}}$ stands for an upper bound on the busy probability of location $i \in I$. Constraints (7.30) can be easily linearized using a logarithm function. For tackling the model, in [42] it is assumed that service calls arriving at a location follow a Poisson distribution. The authors also enhance the model by applying several preprocessing tools that include coefficients round-down and valid inequalities.

The set covering location problem under stochastic demand has been further explored by considering unit-demand nodes. This means that we do not know the nodes that will call for being served. Nevertheless, a set of facilities must be set up here and now. In this setting, we cannot ensure that all demand is covered unless we assume that all demand nodes will call for being served, thus resorting to a classical set covering location problem. However, this may lead to overly conservative solutions.

This can be avoided by imposing a service level, in the sense that the facilities to be opened should ensure coverage with at least some given probability α.

We assume a stochastic random vector $\boldsymbol{\xi} = (\xi_1, \ldots, \xi_{|J|})^\top$ such that each x_j is described by a Bernoulli distribution. Leveraging the classical set covering location model, we formulate this stochastic extension as follows:

$$\text{minimize} \quad \sum_{i \in I} f_i y_i, \tag{7.33}$$

$$\text{subject to} \quad \mathbb{P}\left[\sum_{i \in I_j} y_i \geq \xi_j,\ j \in J\right] \geq \alpha, \tag{7.34}$$

$$y_i \in \{0, 1\} \qquad \forall i \in I. \tag{7.35}$$

This problem was investigated in [64, 476]. First, we note that (7.34) is actually (7.14), and thus, the analysis already presented for the latter in case of independent demands can again be applied here. Accordingly, the challenge is to tackle the general case when the components of $\boldsymbol{\xi} = (\xi_1, \ldots, \xi_{|J|})^\top$ are dependent.

In [64] the authors introduce an algorithm that consists of enumerating the set of α-efficient points.[1] The reasoning underlying their procedure is that if all those

[1] A point $\boldsymbol{\nu} \in \{0, 1\}^m$ is an α-efficient point of a cumulative distribution function $F(.)$ if $F(\boldsymbol{\nu}) \geq \alpha$ and there is no other binary point $\boldsymbol{\chi}$, $\boldsymbol{\chi} \neq \boldsymbol{\nu}$, such that $F(\boldsymbol{\chi}) \geq \alpha$ (see [163, 435]).

points, say $\nu^1, \ldots, \nu^N$, are known, then the probabilistic set covering problem resorts to solving the problems

$$\begin{aligned} &\text{minimize} \quad (7.33), \\ &\text{subject to} \quad \sum_{i \in I_j} y_i \geq \nu_j^\ell, \qquad j \in J, \\ &\qquad\qquad\quad (7.35), \end{aligned} \tag{7.36}$$

for $\ell = 1, \ldots, N$, and $\nu^\ell = (\nu_1^\ell, \ldots, \nu_{|J|}^\ell)$. The optimal solution to the original problem is the best among those N optimal solutions.

The set covering location problem is NP-hard. Thus, solving it many times renders the above solution problem inefficient. In [64] the authors propose a so-called bounding–pruning scheme. They start with a sequence of α-efficient points. Additionally, they assume that an upper bound, say $\hat{V}$, has been determined (e.g., obtained using a heuristic). At each iteration, they select an α-efficient point, say ν^ℓ, and solve the relaxation of the corresponding problem. The optimal value is rounded up, yielding a value we denote by V^ℓ. If the solution to the relaxed problem is not integer and $V^\ell \geq \hat{V}$, then the problem is pruned. If the solution to the relaxed problem is not integer but $V^\ell < \hat{V}$, then the problem is added to the list of candidate problems to solve as integer.

Unfortunately, some discrete distributions can have many α-efficient points (even for a high value of α), which makes the above enumeration phase hardly affordable from a computational perspective. For this reason, some hybrid techniques are also proposed in [64] to improve their base algorithm. Nevertheless, the enumeration of α-efficient points is still a drawback.

Saxena et al. [476] overcome that drawback by integrating both phases of [64] in one MIP. The authors investigate a generalized version of the probabilistic set covering problem stemming from assuming that the stochastic vector $\boldsymbol{\xi} = (\xi_1, \ldots, \xi_{|J|})^\top$ can be decomposed into L blocks, $\{\boldsymbol{\xi}^1, \ldots, \boldsymbol{\xi}^L\}$, $(1 \leq L \leq |J|)$, such that $\boldsymbol{\xi}^{l'}$ and $\boldsymbol{\xi}^{l''}$ are independent random vectors for $l' \neq l''$. Each random vector $\boldsymbol{\xi}^l$ refers to a subset J^l of customers in J. Moreover, $J^1, \ldots, J^L$ define a partition of J. This partition induces a set of distribution functions, $F^1(.), \ldots, F^L(.)$ such that F^l is the restriction of $F(.)$ (CDF of $\boldsymbol{\xi}$) to the customers in J^l.

For each $l \in \{1, \ldots, L\}$, let $\overline{S}^l$ denote the set of $|J^l|$-dimensional binary vectors that are either α-efficient or dominate an α-efficient point of $F^l(.)$. Additionally, let $\underline{S}^l$ denote the set of $|J^l|$-dimensional binary vectors that are α-inefficient points of $F^l(.)$.[2]

[2] A point $\boldsymbol{\nu} \in \{0, 1\}^m$ is called α-inefficient point of a cumulative distribution function $F(.)$ if $F(\boldsymbol{\nu}) < \alpha$ and there is no other binary point $\boldsymbol{\chi} \geq \boldsymbol{\nu}$, $\boldsymbol{\chi} \neq \boldsymbol{\nu}$, such that $F(\boldsymbol{\chi}) < \alpha$. The concept was introduced in [476]. Note that $\boldsymbol{\chi} \geq \boldsymbol{\nu}$ indicates that each component of the first vector is larger than or equal to the corresponding component of the second.

Theorem 7.1 *The probabilistic set covering problem can be formulated as follows:*

$$\text{minimize} \quad (7.33), \tag{7.37}$$

$$\textit{subject to} \quad \sum_{i \in I_j} y_i \geq z_j \qquad \forall j \in J, \tag{7.38}$$

$$\sum_{l=1}^{L} w^l \geq \ln \alpha, \tag{7.39}$$

$$w^l \leq \ln(F^l(\boldsymbol{v})) \left(1 - \sum_{j \in J^l,\, v_j = 0} z_j\right) \qquad \forall \boldsymbol{v} \in \overline{S}^l,\ l \in \{1, \ldots, L\}, \tag{7.40}$$

$$\sum_{j \in J^l,\, v_j = 0} z_j \geq 1 \qquad \forall \boldsymbol{v} \in \underline{S}^l,\ l \in \{1, \ldots, L\}, \tag{7.41}$$

$$z_j \in \{0, 1\} \qquad \forall j \in J, \tag{7.42}$$

$$(7.35).$$

Proof See [476]. □

Theorem 7.2 *The probabilistic set covering problem can be reformulated as follows:*

$$\text{minimize} \quad (7.33),$$

$$\text{subject to} \quad (7.38),\ (7.39), \tag{7.43}$$

$$(\boldsymbol{z}^l, w^l) \in \mathcal{P}^l \qquad \forall l \in \{1, \ldots, L\}, \tag{7.44}$$

$$0 \leq z_j \leq 1 \qquad \forall j \in J, \tag{7.45}$$

$$(7.35),$$

where $\boldsymbol{z}^l$ is a vector containing the components of $\boldsymbol{z} = (z_1, \ldots, z_{|J|})^\top$ that are in J^l and $\mathcal{P}^l$ is the closed convex hull of the set of points in

$$\{(\boldsymbol{v}, w) \mid \boldsymbol{v} \in \overline{S}^l,\ w \leq \ln(F^l(\boldsymbol{v}))\}.$$

Proof See [476]. □

Still in [476] the authors point out that the linear relaxation of the above reformulation dominates the linear relaxation of the formulation stated in Theorem 7.1. The authors also note that (7.44) can be replaced by the system of inequalities defining each $\mathcal{P}^l$. The latter can be efficiently separated, as stated in the following result.

Theorem 7.3 *Let $l \in \{1, \ldots, L\}$ and let $(\hat{z}^l, \hat{w}^l) \in \mathbb{R}^{|J^l|} \times \mathbb{R}$ be such that $\mathbf{0} \leq \hat{z}^l \leq \mathbf{1}$ and $\sum_{j \in J^l, v_j=0} \hat{z}^l_j \geq 1, \forall \boldsymbol{v} \in \underline{S}^l$. Define*

$$\hat{J}^l = \{j \in J^l \mid \hat{z}_j = 1 \, \forall \boldsymbol{z}^l \in \{0,1\}^{|J^l|} \text{ such that } F^l(\boldsymbol{z}^l) \geq \alpha\}.$$

$(\hat{z}^l, \hat{w}^l) \in \mathcal{P}^l$ if and only if the optimal value of the following linear problem is non-negative:

$$\text{minimize} \quad \boldsymbol{a}^\top \hat{z}^l - b\,\hat{w}^l - c, \tag{7.46}$$

$$\text{subject to} \quad \boldsymbol{a}^\top \boldsymbol{z}^l - b \ln(F^l(\boldsymbol{z}^l)) - c \geq 0 \quad \forall \boldsymbol{z}^l \in \overline{S}^l, \tag{7.47}$$

$$\sum_{j \in J \setminus \hat{J}} a_j + b = 1, \tag{7.48}$$

$$a_j \geq 0 \qquad \forall j \in J^l \setminus \hat{J}^l, \tag{7.49}$$

$$a_i = 0 \qquad \forall j \in \hat{J}^l, \tag{7.50}$$

$$b \geq 0. \tag{7.51}$$

Moreover, if $\{\boldsymbol{a}, b, c\}$ is a feasible solution to the above problem satisfying $\boldsymbol{a}^\top \hat{z}^l - b\hat{w}^l - c < 0$, then $\boldsymbol{a}^\top \boldsymbol{z}^l - bw^l \geq c$ is a valid inequality for $\mathcal{P}^l$ that cuts off $(\hat{z}^l, \hat{w}^l)$.

Proof See [476]. □

Finally, another reformulation for the probabilistic set covering problem (7.33)–(7.35) can be derived if the underlying cumulative distribution function is stationary, i.e., if $F(\boldsymbol{v}') = F(\boldsymbol{v}''), \forall \boldsymbol{v}', \boldsymbol{v}'' \in \{0,1\}^{|J|}$ such that $\sum_{j \in J} v'_j = \sum_{j \in J} v''_j$ (in each point its value only depends on the number of "1" involved in the point).

Proposition 7.1 *Suppose that $F^l(.)$ is a stationary distribution $\forall l \in \{1, \ldots, L\}$. Assume also that $F^l(.)$ is defined by the vector of probabilities (each component indicating the value of $F^l(.)$ at a point with the corresponding number of ones) $(\pi^l_0, \pi^l_1, \ldots, \pi^l_{|J^l|})$. Let $\kappa^l = \min\{\kappa \mid 0 \leq \kappa \leq |J^l|, \pi^l_\kappa \geq \alpha\}$. The probabilistic set covering problem can be formulated as follows:*

$$\text{minimize} \quad (7.33),$$

$$\textit{subject to} \quad (7.38), (7.39), (7.42), (7.35),$$

$$u^l = \sum_{j \in J^l} z_j \quad \forall l \in \{1, \ldots, L\}, \tag{7.52}$$

$$u^l \geq \kappa^l \qquad \forall l \in \{1, \ldots, L\}, \tag{7.53}$$

$$u^l \in \mathbb{Z} \qquad \forall l \in \{1, \ldots, L\}, \tag{7.54}$$

$$u^l + v_\ell^l(|I| - \ell) \leq |I| \quad \forall \ell \in \{1, \ldots, |J^l|\},\ l \in \{1, \ldots, L\}, \tag{7.55}$$

$$u^l + v_\ell^l(\ell + 1) \geq \ell + 1 \quad \forall \ell \in \{1, \ldots, |J^l|\},\ l \in \{1, \ldots, L\}, \tag{7.56}$$

$$w^l \leq (\ln(\pi_\ell^l)) v_\ell^l \qquad \forall \ell \in \{1, \ldots, |J^l|\},\ l \in \{1, \ldots, L\}, \tag{7.57}$$

$$v_\ell^l \in \{0, 1\} \qquad \forall \ell \in \{1, \ldots, |J^l|\},\ l \in \{1, \ldots, L\}. \tag{7.58}$$

Proof See [476]. □

We call (M) the model stated in the above proposition. Algorithm 7.1 solves the probabilistic set covering problem.

Algorithm 7.1: Solving the probabilistic set covering problem

Compute $\overline{S}^l$ and $\underline{S}^l$, $\forall l \in \{1, \ldots, L\}$.
Generate model (M).
Solve the linear relaxation of (M).
repeat
  Solve (7.46)–(7.51) to find violated cuts from $\mathcal{P}^l$, $l \in \{1, \ldots, L\}$.
  Add violated cuts to (M).
until *no violated cuts found*;
Obtain a strengthened model by retaining only the binding cuts.
Solve the resulting model using a general-purpose solver.

7.4 Probabilistic Maximal Covering Location

As we have seen for set covering location problems, reliability maximal covering location problems have also been handled using chance constraints. In this case, we seek to maximize the total coverage, satisfying some budget constraints. To simplify, we assume that at most p facilities or servers can be installed. In Sect. 6.6.3, one such problem emerged when stochasticity was assumed for travel times and demands. The so-called (γ, ϵ)-stochastic maximal covering location problem introduced in [567] was presented:

$$\text{maximize} \quad d, \tag{7.59}$$

$$\text{subject to} \quad \mathbb{P}\left[\sum_{j \in J} x_j \xi_j \geq d\right] \geq \epsilon, \tag{7.60}$$

$$\sum_{i \in I_j(\gamma)} y_i \geq x_j \qquad \forall j \in J, \tag{7.61}$$

$$\sum_{i \in I} y_i \leq p, \tag{7.62}$$

$$x_j \in \{0, 1\} \qquad \forall j \in J, \tag{7.63}$$

$$y_i \in \{0, 1\} \qquad \forall i \in I. \tag{7.64}$$

As mentioned in Sect. 6.6.3, the tractability of the above model by standard optimization tools relies on the possibility of finding a (tractable) deterministic equivalent. One such case is that in which the demand vector $\boldsymbol{\xi} = (\xi_1, \ldots, \xi_{|J|})^\top$ has a multivariate Gaussian distribution $N(\boldsymbol{\mu}, \Sigma)$ with $\boldsymbol{\mu} = (\mu_1 \ldots, \mu_{|J|})^\top$, $\mu_j = \mathbb{E}[\xi_j]$, $j \in J$, and $\Sigma = \mathbb{E}\left[(\boldsymbol{\xi} - \boldsymbol{\mu})(\boldsymbol{\xi} - \boldsymbol{\mu})^\top\right]$.

Let $\boldsymbol{x} = (x_1, \ldots, x_j)^\top$. As pointed out in [436], the problem can be converted into a convex problem if $\epsilon \geq 0.5$. This is a reasonable assumption in our case. Given that $\mathbb{E}[\boldsymbol{\xi}^\top \boldsymbol{x}] = \boldsymbol{\mu}^\top \boldsymbol{x}$ and $\mathbb{V}[\boldsymbol{\xi}^\top \boldsymbol{x}] = \boldsymbol{x}^\top \Sigma \boldsymbol{x}$, if for an $\boldsymbol{x}$ we have $\boldsymbol{x}^\top \Sigma \boldsymbol{x} > 0$, then

$$\mathbb{P}[\boldsymbol{\xi}^\top \boldsymbol{x} \geq d] = \mathbb{P}\left[\frac{(\boldsymbol{\xi} - \boldsymbol{\mu})^\top \boldsymbol{x}}{\sqrt{\boldsymbol{x}^\top \Sigma \boldsymbol{x}}} \geq \frac{d - \boldsymbol{\mu}^\top \boldsymbol{x}}{\sqrt{\boldsymbol{x}^\top \Sigma \boldsymbol{x}}}\right] = 1 - \Phi\left(\frac{d - \boldsymbol{\mu}^\top \boldsymbol{x}}{\sqrt{\boldsymbol{x}^\top \Sigma \boldsymbol{x}}}\right),$$

and thus, Constraint (7.60) is equivalent to

$$\boldsymbol{\mu}^\top \boldsymbol{x} + \Phi^{-1}(1 - \epsilon)\sqrt{\boldsymbol{x}^\top \Sigma \boldsymbol{x}} \geq d, \tag{7.65}$$

with $\Phi^{-1}(1 - \epsilon)$ standing for the $(1 - \epsilon)$-quantile of a standardized Gaussian distribution. If $\boldsymbol{x}^\top \Sigma \boldsymbol{x} = 0$, then the above constraint is still equivalent to the chance constraint. Thus, the problem that consists of minimizing (7.59) subject to (7.60)–(7.64) is equivalent to

$$\begin{aligned} \text{maximize} \quad & \boldsymbol{\mu}^\top \boldsymbol{x} + \Phi^{-1}(1 - \epsilon)\sqrt{\boldsymbol{x}^\top \Sigma \boldsymbol{x}}, \\ \text{subject to} \quad & (7.61) - (7.64). \end{aligned} \tag{7.66}$$

Given that Σ is a positive semidefinite matrix, the function $\sqrt{\boldsymbol{x}^\top \Sigma \boldsymbol{x}}$ is convex. Furthermore, we are assuming $\epsilon \geq 0.5$. This ensures $\Phi^{-1}(1-\epsilon) \leq 0$. Consequently, the above problem is a convex programming problem.

In general, when the technological constraint matrix is uncertain, we cannot easily derive a deterministic equivalent. This may call to resort to approximate procedures involving simulation.

We proceed with our analysis considering congestion, i.e., that a server may be busy when a demand node calls for service. Using the terminology already presented in this chapter, we can formulate the problem as

$$\text{maximize} \quad \sum_{j \in J} d_j x_j, \tag{7.67}$$

$$\text{subject to} \quad \sum_{i \in I_j} y_i \geq x_j \qquad \forall j \in J, \tag{7.68}$$

$$\sum_{i \in I} y_i \leq p, \tag{7.69}$$

$$(7.22),\ (7.25),\ (7.63).$$

In this case, d_j stands for the number of calls (per time unit) originating at node j; the binary variable x_j is equal to one if and only if demand node j is covered by at least one server ($j \in J$). We keep denoting by β_j the minimum number of facilities or servers that should cover demand node $j \in J$ to ensure service availability with probability at least equal to α. The discussion above presented on how to find this number and estimate the busy probabilities can be directly applied here. Thus, again, (7.22) can be replaced by (7.26).

ReVelle and Hogan [452] propose a reformulation considering two-indexed binary x variables with $x_{j\ell}$ equal to one if and only if demand node $j \in J$ is covered by at least ℓ servers, $\ell \in \{1, \ldots, \beta_j\}$. This allows re-writing the model as

$$\text{maximize} \quad \sum_{j \in J} d_j x_{j\beta_j}, \tag{7.70}$$

$$\text{subject to} \quad \sum_{\ell=1}^{\beta_j} x_{j\ell} \leq \sum_{i \in I_j} y_i \quad \forall j \in J, \tag{7.71}$$

$$x_{j\ell} \leq x_{j\ell-1} \qquad \forall j \in J,\ \ell \in \{2, \ldots, \beta_j\}, \tag{7.72}$$

$$(7.25),\ (7.69),$$

$$x_{j\ell} \in \{0, 1\} \qquad \forall j \in J,\ \ell \in \{1, \ldots, \beta_j\}. \tag{7.73}$$

In the objective function, a variable $x_{j\beta_j}$ equal to one indicates that demand node j is covered by β_j servers with reliability α.

ReVelle and Marianov [450] combine the above ideas with the Facility Location Equipment Emplacement Technique (FLEET) model proposed in [482] to investigate a reliability maximal covering location problem with two types of facilities. They consider equipment-specific busy fractions and chance constraints. In the proposed model, the coverage satisfying the reliability level is maximized. Liu et al. [346] prefer to adopt a double standard model (DSM). Two coverage radii are considered: a tighter one for coverage reaching a guaranteed (local) reliability level and a looser one for full coverage. This extension seeks to maximize coverage at a guaranteed service level while guaranteeing full coverage according to the loose radius.

7.5 Uncertainty Captured by a Finite Set of Scenarios

Consider the generic problem formulated in (7.12)–(7.16). Let us assume now that demand can be captured by a finite set of scenarios, Ω, such that each scenario fully determines the number of calls originated at each demand node.

For each scenario $\omega \in \Omega$, we have a demand vector $(d_{1\omega}, \ldots, d_{|J|,\omega})^\top$, with J again denoting the set of demand nodes. For each scenario $\omega \in \Omega$, we denote by π_ω its probability ($\pi_\omega \geq 0 \; \forall \omega \in \Omega$ and $\sum_{\omega \in \Omega} \pi_\omega = 1$).

In this case, we can easily find a deterministic equivalent formulation. To accomplish this, consider one additional set of binary variables, δ_ω, $\omega \in \Omega$. The deterministic equivalent writes as follows:

$$\text{minimize} \quad (7.12),$$

$$\text{subject to} \quad (7.13),\ (7.15),\ (7.16),$$

$$\sum_{i \in I_j} x_{ij} \geq d_{j\omega} - \delta_\omega M \qquad \forall j \in J, \tag{7.74}$$

$$\sum_{\omega \in \Omega} \pi_\omega \delta_\omega \leq 1 - \alpha, \tag{7.75}$$

$$\delta_\omega \in \{0, 1\} \qquad \forall \omega \in \Omega. \tag{7.76}$$

This model ensures that the total probability of satisfying the demand is larger than or equal to α. Scenarios ω in which demand may not be satisfied are those such that $\delta_\omega = 1$ whose total probability cannot exceed $1 - \alpha$.

Beraldi and Bruni [62] introduce a facility location problem under demand uncertainty represented by a finite set of scenarios. The authors intertwine stochastic programming with recourse and chance-constrained programming. Like in [63], the problem emerges in the context of emergency medical systems planning. It consists of deciding where to locate some bases for providing a service and the number of units to house in each selected location. The decision about the locations to select and the number of vehicles to house in each must be made here and now, i.e., before knowing the exact demand. Once the occurring scenario is revealed, a recourse decision is implemented, which regards the assignment of the demand nodes to the selected bases. A fixed cost is incurred for every assignment. A demand node may be assigned to more than one facility. It is also assumed that not all locations can cover all demand points. Finally, there is a given joint probability for having all locations capable of serving all demand nodes assigned to them and having all demand nodes assigned to at least one location. The goal is to minimize the cost of setting up the facilities and the in-house vehicles, plus the expected cost of allocating demand nodes to the open facilities.

For this problem, a branch-and-bound procedure is devised in [62]. It is expected to solve up to proven optimality small- to medium-sized instances. For large-scale instances, the authors devise three heuristics. Two of them are constructive. The third one makes use of the above-mentioned branch-and-bound procedure. Nevertheless, a warm start is considered using a feasible solution. Moreover, the algorithm is allowed to terminate before finding a proven optimal solution. The principles underlying these three heuristics can be easily adapted to other problems.

We consider again the problem that consists of minimizing (7.12) subject to (7.13), (7.15), (7.16), and (7.74)–(7.76).

Let us define $\Omega^*_\alpha = \{\omega \in \Omega \mid \delta_\omega = 0\}$ to be the set of scenarios included in the optimal solution of the problem. If that set was known beforehand, the optimal solution to the problem could be obtained by solving the problem restricted to the scenarios in Ω^*_α. In practice, we may try to find an approximate set of scenarios.

One possibility for finding a feasible solution to the problem is then to look for the subset of scenarios of minimum cardinality ensuring a total probability of at least α, i.e., we consider an optimal solution to the problem

$$\begin{aligned} \text{minimize} \quad & |\Omega_\alpha|, \\ \text{subject to} \quad & \sum_{\omega \in \Omega_\alpha} \pi_\omega \geq \alpha. \end{aligned}$$

If $\hat{\Omega}_\alpha$ is an optimal solution to the above problem, an approximate solution to our problem is obtained by solving it restricted to the scenarios in $\hat{\Omega}_\alpha$.

Beraldi and Bruni [62] report good results applying this heuristic to their problem. However, the same authors propose a more sophisticated method that can be easily adapted to the problem we have been considering. They suggest considering a variable splitting reformulation and thus state the non-anticipativity constraints explicitly. This calls for scenario-indexing the variables representing the here-and-now decision:

$$\text{minimize} \quad \sum_{\omega \in \Omega} \pi_\omega \left(\sum_{i \in I} f_i y_{i\omega} + \sum_{i \in I} \sum_{j \in J} c_{ij} x_{ij\omega} \right), \tag{7.77}$$

$$\text{subject to} \quad \sum_{j \in J_i} x_{ij\omega} \leq q_i y_{i\omega} \qquad \forall j \in I,\ \omega \in \Omega, \tag{7.78}$$

$$\sum_{i \in I_j} x_{ij\omega} \geq d_{j\omega} - \delta_\omega M \qquad \forall j \in J,\ \omega \in \Omega \tag{7.79}$$

$$\sum_{\omega\in\Omega} \pi_\omega(1-\delta_\omega) \geq \alpha, \tag{7.80}$$

$$x_{ij\omega'} = x_{ij\omega''} \qquad \forall i \in I,\ j \in J_i,\ \omega', \omega'' \in \Omega, \tag{7.81}$$

$$y_{i\omega'} = y_{i\omega''} \qquad \forall i \in I,\ \omega', \omega'' \in \Omega, \tag{7.82}$$

$$x_{ij\omega} \geq 0 \text{ and integer} \qquad \forall i \in I,\ j \in J_i,\ \omega \in \Omega, \tag{7.83}$$

$$y_{i\omega} \in \{0,1\} \qquad \forall i \in I,\ \omega \in \Omega, \tag{7.84}$$

(7.76).

If we relax Constraints (7.80)–(7.82), the problem can be decomposed in ω defining a subproblem P_ω for each $\omega \in \Omega$.

We can now solve the linear relaxation of P_ω for every $\omega \in \Omega$. Let us denote the optimal values by $V(\bar{\mathrm{P}}_\omega)$. By solving the knapsack problem

$$\begin{aligned} \text{minimize} \quad & \sum_{\omega\in\Omega} V(\bar{\mathrm{P}}_\omega)\chi_\omega, \\ \text{subject to} \quad & \sum_{\omega\in\Omega} \chi_\omega \pi_\omega \geq \alpha, \\ & \chi_\omega \in \{0,1\} \quad \forall \omega \in \Omega. \end{aligned}$$

We obtain a subset of scenarios (those for which the corresponding χ-variable was selected to be equal to one) that again we denote by $\hat{\Omega}_\alpha$. Hopefully, this set is a good approximation for the optimal set Ω^*_α. By solving the original problem restricted to the scenarios in $\hat{\Omega}$, we obtain a feasible solution to the whole problem.

The above process corresponds to considering the model that consists of minimizing (7.12) subject to (7.13), (7.15), (7.16), (7.74)–(7.76), and solving it for each scenario individually. In this case, the single scenario to consider is the only one selected, i.e., the corresponding δ-variable is set equal to 0. Nevertheless, the variable splitting representation has much more potential than we have described here, namely when it is set at the core of a dual decomposition-based algorithm (see, e.g., [106]).

7.6 Conclusions and Further Reading

Chance-constraint programming was introduced in [112] and has been attracting the attention of many scholars leading to significant breakthroughs (see, e.g., [126, 358, 421]). Nevertheless, from an optimization perspective, many problems are challenging. These include many discrete facility location problems.

Despite the advances observed, chance-constrained facility location is a field where much research still needs to be done. In other chapters of this volume, namely in Chap. 13, we will have the opportunity to discuss some recent developments in the context of specific application areas.

Chapter 8
Distributionally Robust Facility Location

Abstract This chapter discusses distributional robust optimization (DRO) models and techniques for facility location problems. The classical capacitated facility location problem is extended. Different possibilities are considered for capturing uncertainty. In particular, a data-driven DRO model based on a Wasserstein ambiguity set, a decision-dependent DRO model, and a DRO model using a state-wise ambiguity set are presented. A nested Benders decomposition algorithm for solving the model exactly is discussed, which leverages the subgradients of the worst-case expected second-stage cost at the location decisions formed insightfully by the associated worst-case distributions. The nested Benders decomposition procedure ensures a finite-step convergence, which can also be regarded as an extension of the classic L-shaped algorithm for two-stage stochastic programming to the state-wise robust stochastic facility location problem with a conic representable ambiguity set. Finally, some suggestions for further reading are presented.

Keywords Data-driven decision-making · Decision-dependent ambiguity sets · State-wise ambiguity sets · Moment-based ambiguity sets · Wasserstein-metric-based ambiguity sets

8.1 Introduction

A central assumption in this chapter is that the uncertainty underlying a facility location problem is associated with a probability distribution that is not perfectly known. Instead, it can be described by a so-called ambiguity set of distributions. In other words, we focus on distributionally robust facility location problems. These are problems that can be looked at as an integration of the robust facility location as discussed in Chap. 5 and the stochastic facility location covered in Chap. 6. In this chapter, we present the modeling paradigm in the context of facility location and discuss an exact solution scheme.

We focus the analysis on a slightly more general version of the classical capacitated facility location. In many logistics applications, a service level below

© The Author(s), under exclusive license to Springer Nature Switzerland AG 2024
F. Saldanha-da-Gama, S. Wang, *Facility Location Under Uncertainty*,
International Series in Operations Research & Management Science 356,
https://doi.org/10.1007/978-3-031-55927-3_8

100% is acceptable (or even desirable), although a cost may be incurred (opportunity loss cost, outsourcing, extraordinary ordering costs, etc.). Additionally, constraints may be imposed restricting the facilities that we can select (e.g., a budget constraint).

As customary in this volume, we let I denote the set of potential locations for the facilities and J the set of customers. Assume that the capacity of a facility operating at $i \in I$ is q_i and the demand of customer $j \in I$ is d_j. Regarding the costs, f_i is the fixed setup cost for a facility installed at $i \in I$ and c_{ij} the unit transportation (and operational) cost for demand satisfaction from facility $i \in I$ to customer $j \in J$. The unit penalty cost for the unsupplied demand is denoted by r_j, $j \in J$. The problem can be formulated using two sets of decision variables: y_i is a binary variable equal to 1 if and only if a facility is installed at $i \in I$; x_{ij} is a non-negative real-valued quantity representing the amount of demand shipped from facility i to demand node $j, i \in I, j \in J$. The mathematical model is the following:

$$\text{minimize} \quad \sum_{i\in I} f_i y_i + \sum_{i\in I}\sum_{j\in J} c_{ij}x_{ij} + \sum_{j\in J} r_j\left(d_j - \sum_{i\in I} x_{ij}\right), \tag{8.1}$$

$$\text{subject to} \quad \sum_{i\in I} x_{ij} \le d_j \qquad \forall j \in J, \tag{8.2}$$

$$\sum_{j\in J} x_{ij} \le q_i y_i \qquad \forall i \in I, \tag{8.3}$$

$$x_{ij} \ge 0 \qquad \forall i \in I,\ j \in J, \tag{8.4}$$

$$\boldsymbol{y} \in \mathcal{Y}. \tag{8.5}$$

In this model, $\boldsymbol{y} = (y_1, \ldots, y_{|I|})^\top$ and the set $\mathcal{Y}$ defines the feasible choices for the y variables as explained above. More formally, we assume $\mathcal{Y} = \{\boldsymbol{y} \in \{0,1\}^{|I|} \mid \mathbf{G}\boldsymbol{y} \ge \boldsymbol{h}\}$. The objective function (8.1) accounts for the total cost. It includes the setup cost for the facilities, the transportation cost from facilities to customers, and the cost for unsupplied demand. Constraints (8.2) ensure that the total amount shipped to each customer does not exceed its demand. Inequalities (8.3) are the capacity constraints. Finally, we have the domain constraints for the decision variables.

It is worth noticing that the above problem can be reformulated as a full-demand satisfaction problem by considering a dummy facility operating at a cost zero "to supply" the unsupplied demand, i.e., with the shipment costs to the customers equal to the corresponding penalties. Nevertheless, the above form is convenient for some particular insights, so we keep it.

The remainder of the chapter is organized as follows. Section 8.2 presents a data-driven DRO facility location problem using a Wasserstein ambiguity set. Section 8.3 discusses a decision-dependent DRO facility location model. A state-wise distributional robust optimization problem is formalized and analyzed in Sect. 8.4. Some related algorithmic challenges and an exact procedure are detailed in Sect. 8.5. The chapter ends with some conclusions and directions for further reading.

8.2 Data-Driven Distributionally Robust Optimization Model

A main ingredient of a DRO model is the so-called ambiguity set. In Sect. 4.5, we discussed moment-based ambiguity sets as well as a Wasserstein-metric-based one. In this chapter, we leverage the latter by building a data-driven DRO facility location model.

In a DRO problem, we aim to find a solution that optimizes the expected value of an objective function for the worst-case probability distribution among a specific set of distributions. One possibility for the latter is the following:

$$\mathcal{F}_{\varepsilon}(\widehat{\mathbb{P}}_N, \Xi) = \left\{\mathbb{P} \in \mathcal{M}(\Xi) : d_{\mathrm{W}}\left(\widehat{\mathbb{P}}_N, \mathbb{P}\right) \leq \varepsilon\right\}. \tag{8.6}$$

In this set, $d_{\mathrm{W}}(\widehat{\mathbb{P}}_N, \mathbb{P})$ is the Wasserstein metric (see Definition 4.6) and $\widehat{\mathbb{P}}_N$ is an empirical distribution such that each discrete point in the sample $\widehat{\Xi} = \{\widehat{\boldsymbol{d}}_1, \ldots, \widehat{\boldsymbol{d}}_N\}$ is associated with a probability of $\frac{1}{N}$. In other words, $\widehat{\mathbb{P}}_N = \frac{1}{N}\sum_{n=1}^{N} \delta_{\widehat{\boldsymbol{d}}_n}$, where $\delta_{\widehat{\boldsymbol{d}}_n}$ is a Dirac measure concentrating unit mass at $\widehat{\boldsymbol{d}}_n$. The empirical distribution can be estimated based on historical data.

The Wasserstein ambiguity set $\mathcal{F}_{\varepsilon}(\widehat{\mathbb{P}}_N, \Xi)$ includes all probability distributions supported on $\Xi \subset \mathbb{R}^{|J|}$ that are within a distance $\varepsilon \geq 0$ of the reference/empirical distribution $\widehat{\mathbb{P}}^N$. Intuitively, when $\varepsilon = 0$, the ambiguity set contains only the empirical distribution—the DRO problem reduces to a stochastic programming problem. With a very large ε, one can place the entire probability mass on a vertex that maximizes the cost—the problem becomes a robust optimization problem.

When considering the Wasserstein distributional ambiguity set $\mathcal{F}_{\varepsilon}(\widehat{\mathbb{P}}_N, \Xi)$, we assume that (i) the support is a bounded polyhedron defined as $\Xi = \{\boldsymbol{d} \in \mathbb{R}^J \mid A\boldsymbol{d} \leq \boldsymbol{b}\}$, for some $A \in \mathbb{R}^{L\times J}$ and $\boldsymbol{b} \in \mathbb{R}^L$, and (ii) the ℓ_1-norm is used in the Wasserstein metric.

Thinking about a risk-averse decision-maker who aims to avoid future disappointments and thus desires a probabilistic guarantee on the out-of-sample performance, we can formulate a two-stage data-driven DRO facility location problem as follows:

$$\min_{\boldsymbol{y}\in\mathcal{Y}} \left\{\sum_{i\in I} f_i y_i + \sup_{\mathbb{P}\in\mathcal{F}_{\varepsilon}(\widehat{\mathbb{P}}_N, \Xi)} \mathbb{E}_{\mathbb{P}}[g(\boldsymbol{y}, \boldsymbol{d})]\right\}, \tag{8.7}$$

where $g(\boldsymbol{y}, \boldsymbol{d})$ is the recourse function, evaluated after uncertainty is revealed by solving the second-stage problem

$$g(\boldsymbol{y}, \boldsymbol{d}) = \min \quad \sum_{i\in I}\sum_{j\in J} c_{ij}x_{ij} + \sum_{j\in J} r_j\left(d_j - \sum_{i\in I} x_{ij}\right), \tag{8.8}$$

$$\text{s. t.} \quad \sum_{i\in I} x_{ij} \leq d_j \qquad \forall j \in J, \tag{8.9}$$

$$\sum_{j\in J} x_{ij} \leq q_i y_i \qquad \forall i \in I, \tag{8.10}$$

$$x_{ij} \geq 0 \qquad \forall i \in I, \ j \in J. \tag{8.11}$$

We also need to ensure that the recourse problem remains feasible for all feasible first-stage decisions, i.e., has relatively complete recourse. We enforce this property by adding the following valid inequality to the first-stage problem:

$$\sup_{d\in\Xi} e^\top \boldsymbol{d} \leq \boldsymbol{q}^\top \boldsymbol{y},$$

with $\boldsymbol{q} = (q_1, \ldots, q_{|I|})^\top$, which, under the assumptions above mentioned for Wasserstein distributional ambiguity sets, can be reformulated through LP duality using multipliers $\boldsymbol{w}$ as

$$\begin{aligned} \inf_{\boldsymbol{w}\in\mathbb{R}^L_+} \ & \boldsymbol{b}^\top \boldsymbol{w} \leq \boldsymbol{q}^\top \boldsymbol{x}, \\ \text{s.t.} \ & A^\top \boldsymbol{w} = e. \end{aligned} \tag{8.12}$$

The adaptive two-stage DRO problem (8.7) has been proven to be NP-hard. Nevertheless, in [468] develop two exact solution approaches are developed for it. The first starts by dualizing the recourse problem. Then, a column-and-constraint generation algorithm is considered to implicitly enumerate the vertices of the polyhedral feasible set of the dual recourse problem. The second approach uses a lifting of the support set and then implements a column generation algorithm to enumerate the vertices of the lifted support set.

Moreover, the authors in [468] also propose conservative approximations of the adaptive two-stage DRO problem (8.7) based on support-set relaxation. This is another approximation. It uses affine decision rules and a relaxation stemming from a support-set restriction. The interested reader can refer to that work for further details.

8.3 Decision-Dependent Distributionally Robust Optimization Model

While the traditional facility location problem considers exogenous demand, in some applications, the location decisions can affect the willingness of customers to use certain services or facilities, e.g., carsharing. In this case, the location decisions affect the realizations of the underlying random demand. Moreover, a decision-maker may not know the exact distribution of such endogenous demand and how it is affected by location choices. In this section, we consider DRO facility location problems, in which the moments of stochastic demand are discussed as functions of facility location decisions.

We first introduce an ambiguity set describing the distributional information of demand. Specifically, the actual demand distribution comes from a set of possible distributions, where the random demand at each customer site $j \in J$ can take values in finite support set $\Xi = \{\boldsymbol{d}_1, \ldots, \boldsymbol{d}_K\}$ with probabilities $\pi_{j1}, \ldots, \pi_{jK}$. Therefore, the ambiguity set $U(\boldsymbol{y})$ of the unknown probabilities $\pi_{j1}, \ldots, \pi_{jK}$ is given by

$$\begin{aligned} U(\boldsymbol{y}) = \Bigg\{ \{\pi_j\}_{j \in J} : & \pi_j \in \mathbb{R}_+^{|K|}, \sum_{k \in K} \pi_{jk} = 1 \quad \forall j \in J, \\ & \left| \sum_{k \in K} \pi_{jk} \boldsymbol{d}_k - \mu_j(\boldsymbol{y}) \right| \leq \epsilon_j^{\mu} \quad \forall j \in J, \\ & \left(\sigma_j^2(\boldsymbol{y}) + (\mu_j(\boldsymbol{y}))^2 \right) \underline{\epsilon}_j^{\sigma} \leq \sum_{k \in K} \pi_{jk} \\ & \boldsymbol{d}_k^2 \leq \left(\sigma_j^2(\boldsymbol{y}) + (\mu_j(\boldsymbol{y}))^2 \right) \bar{\epsilon}_j^{\sigma} \quad \forall j \in J \Bigg\}, \end{aligned} \tag{8.13}$$

where $\mu_j(\boldsymbol{y})$ and $\sigma_j^2(\boldsymbol{y})$ are the mean and variance of the demand of node j. They can be any function of the location decision vector $\boldsymbol{y}$.

The constraints in set (8.13) guarantee that (i) the probabilities at all customer sites over the support set sum up to 1, (ii) the true mean of $d_j(\boldsymbol{y})$ is within an ℓ_1-based distance ϵ_j^{μ} to the mean $\mu_j(\boldsymbol{y})$, and (iii) the actual second moment of $d_j(\boldsymbol{y})$ is bounded by the sum of $(\mu_j(\boldsymbol{y}))^2$ and $\sigma_j^2(\boldsymbol{y})$ with upper- and lower bound parameters satisfying $0 \leq \underline{\epsilon}_j^{\sigma} \leq 1 \leq \bar{\epsilon}_j^{\sigma}$, for each $j \in J$. Parameters $\epsilon_j^{\mu}, \underline{\epsilon}_j^{\sigma}, \bar{\epsilon}_j^{\sigma}$ determine the ambiguity set robustness, for each customer site $j \in J$. In case we have the perfect knowledge regarding the first and second moments of random demand at site j, then $\epsilon_j^{\mu} = 0$, and $\epsilon_j^{\sigma} = \bar{\epsilon}_j^{\sigma} = 1$; otherwise we can adjust these parameters to consider distributions within a certain proximity to the desired moment information, consequently impacting the conservatism of the facility location decisions.

To broaden the discussion about modeling possibilities, instead of assuming a total capacity at each location, we consider a relaxed capacity constraint at each facility pre-divided for individual customer sites. For example, to prepare for shipments, different sizes of vehicle fleets are pre-booked and scheduled to serve each specific customer region. We denote the total capacity in each location i as $\sum_{j \in J} q_{ij}$, where q_{ij} is the capacity at location i dedicated to customer site j. For notation convenience, w.l.o.g., we assume the same amount of capacity pre-allocated to serve each customer (i.e., q_{ij} is equal to q_i for all customers $j \in J$).

The overall decision-dependent distributionally robust facility location problem is formulated as

$$\min_{\boldsymbol{y} \in \mathcal{Y}} \left\{ \sum_{i \in I} f_i y_i + \sup_{\pi \in U(\boldsymbol{y})} \mathbb{E}_{\pi}[h(\boldsymbol{y}, \boldsymbol{d}(\boldsymbol{y}))] \right\}, \tag{8.14}$$

where

$$h(\boldsymbol{y}, \boldsymbol{d}(\boldsymbol{y})) = \min \quad \sum_{i\in I}\sum_{j\in J} c_{ij}x_{ij} + \sum_{j\in J}\left(p_j s_j - r_j d_j(\boldsymbol{y})\right), \tag{8.15}$$

$$\text{s. t.} \quad \sum_{i\in I} x_{ij} + s_j = d_j(\boldsymbol{y}) \qquad \forall j \in J, \tag{8.16}$$

$$x_{ij} \le q_i y_i \qquad \forall i \in I, j \in J, \tag{8.17}$$

$$s_i, x_{ij} \ge 0 \qquad \forall i \in I, j \in J. \tag{8.18}$$

The objective function accounts for the total cost of locating facilities plus the maximum (expected) cost for any demand distribution $\pi \in U(y)$, of transportation and unmet demand minus revenue. The first set of constraints in the subproblem ensures that demand at each customer site is either satisfied by other locations or penalized. The second set of constraints enforces capacity restriction for each open facility $i \in I$.

Basciftci et al. [47] derive a monolithic formulation for solving the DRO model with decision-dependent demand whose moments can be any function of facility location decisions. Furthermore, when demand mean and variance are piecewise linear functions of location solutions, the authors obtain an exact mixed-integer linear programming (MILP) reformulation of the DRO model through duality and convex envelopes. Readers can refer to that work for more details.

8.4 A State-Dependent Distributionally Robust Optimization Model

Facility location decisions are usually associated with a strategic and long-term planning problem. In such a case, the future demand may be significantly affected by seasonal or socio-economic factors in different states (or periods). To manage the demand uncertainty more effectively with a more accurate description, we can incorporate such state dependency into the modeling of distributional uncertainty of demand. This yields a so-called event-wise ambiguity set (see [121]), i.e., an ambiguity set conditional on the state variables. In the context of our problem, we define it as follows:

$$\mathcal{F} = \left\{ \mathbb{P} \in \mathcal{P}(\mathbb{R}^{|J|} \times K) \;\middle|\; \begin{array}{ll} (\tilde{\boldsymbol{d}}, \tilde{k}) \sim \mathbb{P} & \\ \mathbb{E}_{\mathbb{P}}\left[\tilde{\boldsymbol{d}} \,\middle|\, \tilde{k} = k\right] \in [\boldsymbol{d}_k^-, \boldsymbol{d}_k^+], & \forall k \in K \\ \mathbb{E}_{\mathbb{P}}\left[\boldsymbol{g}_k(\tilde{\boldsymbol{d}}) \,\middle|\, \tilde{k} = k\right] \le \boldsymbol{\sigma}_k, & \forall k \in K \\ \mathbb{P}\left[\tilde{\boldsymbol{d}} \in \mathcal{D}_k \,\middle|\, \tilde{k} = k\right] = 1, & \forall k \in K \\ \mathbb{P}\left[\tilde{k} = k\right] = p_k, & \forall k \in K \end{array} \right\}. \tag{8.19}$$

In this ambiguity set, $\mathcal{P}$ represents the collection of probability distributions of $(\tilde{\boldsymbol{d}}, \tilde{k})$ supported on $\mathbb{R}^{|J|} \times K$, with each $k \in K$ representing a state. The second set of conditions captures the range (with lower bound $\boldsymbol{d}_k^-$ and upper bound $\boldsymbol{d}_k^+$) of the conditional expected demand in each state k. The third set of constraints, for each k, uses a general proper convex mapping $\boldsymbol{g}_k(\cdot) : \mathbb{R}^{|J|} \mapsto \mathbb{R}^{n_k}$, together with $\boldsymbol{\sigma}_k \in \mathbb{R}^{n_k}$ to describe the possible conditional dispersion or correlation in each state k. We illustrate this aspect below. The fourth set of constraints assumes a general bounded polyhedral support for $\tilde{\boldsymbol{d}}$ conditional on each state $\tilde{k} = k \in K$, which is specified as $\mathcal{D}_k = \left\{\boldsymbol{d} \in \mathbb{R}_+^{|J|} \mid \boldsymbol{A}_k\boldsymbol{d} \le \boldsymbol{b}_k\right\}$, $\boldsymbol{A}_k = [\boldsymbol{a}_{1k}, \boldsymbol{a}_{2k}, \cdots, \boldsymbol{a}_{|J|k}] \in \mathbb{R}^{m_k \times |J|}$, and also satisfies $[\boldsymbol{d}_k^-, \boldsymbol{d}_k^+] \subseteq \mathcal{D}_k$. The final set of constraints specifies the probabilities estimated for each state.

For technical convenience, we assume the Slater condition holds.[1] We also note that, as discussed in [121], the above state-wise ambiguity set $\mathcal{F}$ enjoys a state-wise decomposable structure such that $\mathcal{F}$ can also be represented as an ambiguity set of multimodal distributions (or mixture distributions) proposed in [258]. We exploit that structure later in this chapter.

Noticeably, though in a simple format, the third set of expectation constraints with dispersion mappings $\boldsymbol{g}_k(\cdot)$ provides valuable modeling flexibility for incorporating a variety of useful distributional information, such as the possible correlation among several random demands and data-driven scheme using demand forecasts directly. To provide more constructive insights on set $\mathcal{F}$, we next present a few examples of the mapping $\boldsymbol{g}_k(\cdot)$.

Example 8.1 (Absolute Dispersion) Let us assume that the mapping $\boldsymbol{g}_k(\boldsymbol{d}) : \mathbb{R}^{|J|} \mapsto \mathbb{R}^{n_k}$ in the ambiguity set $\mathcal{F}$ is defined as

$$\boldsymbol{g}_k(\boldsymbol{d}) = \left(\left\|\boldsymbol{U}_{lk}\boldsymbol{d} - \boldsymbol{\theta}_{lk}\right\|_1\right)_{l \in N_k}, \quad \forall k \in K, \tag{8.20}$$

where $N_k = \{1, \ldots, n_k\}$, $\|\cdot\|_1$ is the ℓ_1-norm, $\boldsymbol{U}_{lk} \in \mathbb{R}^{L_l \times |J|}$, $k \in K, l \in N_k$, and each $\boldsymbol{\theta}_{lk} \in \mathbb{R}^{L_l}$ is an estimated input. The third set of expectation constraints in set $\mathcal{F}$ becomes

$$\mathbb{E}_{\mathbb{P}}\left[\left\|\boldsymbol{U}_{lk}\tilde{\boldsymbol{d}} - \boldsymbol{\theta}_{lk}\right\|_1 \middle| \tilde{k} = k\right] \le \sigma_{lk}, \quad \forall l \in N_k,\ k \in K.$$

In particular, if $\boldsymbol{U}_{lk} = \frac{1}{|J|}\boldsymbol{e}^\top \in \mathbb{R}^{1 \times |J|}$ and $\boldsymbol{\theta}_{lk} = \bar{d}_{lk} \in \mathbb{R}$, for some l and k with $\bar{d}_{lk}$ denoting a nominal value for the demand, then we have

$$\mathbb{E}_{\mathbb{P}}\left[\left\|\boldsymbol{U}_{lk}\tilde{\boldsymbol{d}} - \boldsymbol{\theta}_{lk}\right\|_1 \middle| \tilde{k} = k\right] = \mathbb{E}_{\mathbb{P}}\left[\left|\frac{1}{|J|}\sum_{j \in J}\tilde{d}_j - \bar{d}_{lk}\right| \mid \tilde{k} = k\right] \le \sigma_{lk}.$$

[1] This condition establishes that there exists $\boldsymbol{d}_k^\dagger \in (\boldsymbol{d}_k^-, \boldsymbol{d}_k^+)$, $k \in K$ such that $\boldsymbol{g}_k(\boldsymbol{d}_k^\dagger) < \boldsymbol{\sigma}_k$, $k \in K$. It ensures strong duality. In the case of $\boldsymbol{d}_k^- = \boldsymbol{d}_k^+ = \bar{\boldsymbol{d}}_k$, the Slater condition reduces to $\boldsymbol{g}_k(\bar{\boldsymbol{d}}_k) < \boldsymbol{\sigma}_k$.

This reduces to the first-order cross-central moment metric. Therefore, for each state $k \in K$, these constraints capture (at least partially) the conditional correlation among the demands induced by the different scenarios in that state.

Example 8.2 (Mahalanobis Distance) The Mahalanobis distance (see [157]) can be used to measure the disturbance of demand $\tilde{\boldsymbol{d}}$ concerning the estimated mean $\bar{\boldsymbol{d}}_k$ over the estimated covariance $\boldsymbol{\Sigma}_k$ conditional on state $\tilde{k} = k$. In particular, for simplicity, we take $n_k = 1, \forall k \in K$ and define

$$g_k(\boldsymbol{d}) = \sqrt{\left(\boldsymbol{d} - \bar{\boldsymbol{d}}_k\right)^\top \boldsymbol{\Sigma}_k^{-1} \left(\boldsymbol{d} - \bar{\boldsymbol{d}}_k\right)}, \tag{8.21}$$

where we assume $\boldsymbol{\Sigma}_k \succ \mathbf{0}$ for each $k \in K$. Then we have

$$\mathbb{E}_{\mathbb{P}}\left[\sqrt{\left(\tilde{\boldsymbol{d}} - \bar{\boldsymbol{d}}_k\right)^\top \boldsymbol{\Sigma}_k^{-1} \left(\tilde{\boldsymbol{d}} - \bar{\boldsymbol{d}}_k\right)} \;\middle|\; \tilde{k} = k\right] \leq \sigma_k, \quad \forall k \in K,$$

as the third set of expectation constraints in the ambiguity set (8.19).

The mappings $\boldsymbol{g}_k(\cdot)$ in the above two examples are defined using dispersion statistics on the demand $\tilde{\boldsymbol{d}}$ to construct set $\mathcal{F}$. The following example illustrates that $\mathcal{F}$ can also be constructed using directly the demand forecasts $\{\hat{\boldsymbol{d}}_k, k \in K\}$, where each forecast is treated as a "state" and the mappings $\boldsymbol{g}_k(\cdot)$ are used to describe a statistical distance between the actual distribution of demand and an empirical distribution in terms of the demand forecasts.

Example 8.3 (∞-Wasserstein Ball) Consider a random vector $\tilde{\boldsymbol{d}} \sim \mathbb{P}$ and an empirical distribution $\tilde{\boldsymbol{\xi}} \sim \widehat{\mathbb{P}}_K$ with $\widehat{\mathbb{P}}_K = \sum_{k \in K} p_k \delta_{\hat{\boldsymbol{d}}_k}$, where $\delta_{\hat{\boldsymbol{d}}_k}$ is the Dirac distribution taking on $\hat{\boldsymbol{d}}_k$ with mass 1, both supported in a set $\mathcal{D}$. The ∞-Wasserstein distance [74, 235, 545] between the two distributions $\mathbb{P}$ and $\widehat{\mathbb{P}}_K$ is defined as follows:

$$W^\infty(\mathbb{P}, \widehat{\mathbb{P}}_K) = \inf_{\mathbb{Q} \in \mathcal{P}(\mathcal{D} \times \mathcal{D})} \left\{ \mathbb{Q}\text{-ess.sup } \|\tilde{\boldsymbol{d}} - \tilde{\boldsymbol{\xi}}\|_q \;\middle|\; \begin{array}{l} \mathbb{Q} \text{ is the joint distribution} \\ \text{of } (\tilde{\boldsymbol{d}}, \hat{\boldsymbol{\xi}}) \text{ with marginals} \\ \mathbb{P} \text{ and } \widehat{\mathbb{P}}_K, \text{ respectively} \end{array} \right\},$$

where $\|\cdot\|_q$ is the q-norm for $q \in [1, \infty]$ and $\mathbb{Q}$-ess.sup is the essential supremum of $\|\cdot\|_q$ with respect to probability measure $\mathbb{Q}$, i.e., $\inf\{U : \mathbb{Q}[\|\tilde{\boldsymbol{d}} - \tilde{\boldsymbol{\xi}}\|_q > U] = 0\}$. Given a radius $\theta \geq 0$, the associated ∞-Wasserstein ball can be defined as $\mathcal{F}_{\mathrm{W}}^\infty(\theta) = \{\mathbb{P} \in \mathcal{P}(\mathcal{D}) | W^\infty(\mathbb{P}, \widehat{\mathbb{P}}_K) \leq \theta\}$. It has been showed in [121] that $\mathcal{F}_{\mathrm{W}}^\infty(\theta)$ has an equivalent representation using the event-wise ambiguity set

$$\mathcal{F}_{\mathrm{W}}^\infty(\theta) = \left\{ \mathbb{P} \in \mathcal{P}(\mathbb{R}^{|J|} \times K) \;\middle|\; \begin{array}{ll} (\tilde{\boldsymbol{d}}, \tilde{k}) \sim \mathbb{P}, \mathbb{P}\left[\tilde{k} = k\right] = p_k, & \forall k \in K \\ \mathbb{P}\left[\tilde{\boldsymbol{d}} \in \mathcal{D}, \|\tilde{\boldsymbol{d}} - \hat{\boldsymbol{d}}_k\|_q \leq \theta | \tilde{k} = k\right] = 1, & \forall k \in K \end{array} \right\}. \tag{8.22}$$

Note that this ambiguity set does not contain any expectation constraints. Now in our ambiguity set in (8.19), if we let $\sigma_k = \theta, \mathcal{D}_k = \mathcal{D}, \forall k \in K, [\boldsymbol{d}_k^-, \boldsymbol{d}_k^+]$ be large enough, and

$$g_k(\boldsymbol{d}) = \left\| \boldsymbol{d} - \hat{\boldsymbol{d}}_k \right\|_q, \forall k \in K,$$

then the resulting ambiguity set in (8.19), denoted by $\mathcal{F}(\theta)$, actually recovers a *lifted* version of the above ∞-Wasserstein ball $\mathcal{F}_{\mathrm{W}}^{\infty}(\theta)$ as follows:

$$\mathcal{F}(\theta) = \left\{ \mathbb{P} \in \mathcal{P}(\mathbb{R}^{|J|} \times \mathbb{R} \times K) \left| \begin{array}{ll} (\tilde{\boldsymbol{d}}, \tilde{w}, \tilde{k}) \sim \mathbb{P}, \mathbb{P}\left[\tilde{k} = k\right] = p_k, & \forall k \in K \\ \mathbb{E}_{\mathbb{P}}\left[\tilde{w} | \tilde{k} = k\right] \leq \theta, & \forall k \in K \\ \mathbb{P}\left[\tilde{\boldsymbol{d}} \in \mathcal{D}, \|\tilde{\boldsymbol{d}} - \hat{\boldsymbol{d}}_k\|_q \leq \tilde{w} | \tilde{k} = k\right] = 1, & \forall k \in K \end{array} \right. \right\} \tag{8.23}$$

It is not difficult to see that $\mathcal{F}(\theta)$ and $\mathcal{F}_{\mathrm{W}}^{\infty}(\theta)$ can be treated equivalently in the sense that the projection of $\mathcal{F}(\theta)$ on the space of $(\tilde{\boldsymbol{d}}, \tilde{k})$ is equivalent to $\mathcal{F}_{\mathrm{W}}^{\infty}(\theta)$.

As above exemplified, different dispersion mappings choices $\boldsymbol{g}_k(\cdot)$ induce different conic representable sets. This leads to different tractable reformulations of the worst-case expected second-stage cost. In turn, this affects the computational effort of an exact solution algorithm, to be further elaborated in Sect. 8.5.2.

To complement the above discussion, we present different possibilities for building the state-wise ambiguity set based on demand historical data and the possible covariates:

1. *State-wise ambiguity set with labeled states.* In many practical situations, the states are explicitly labeled, e.g., when the state refers to the season, working condition (e.g., temperature), or the weather, to mention a few examples. In such cases, it is natural to group the samples directly for each state (e.g., one of four seasons) and evaluate the sample statistics of the (conditional) mean, dispersion, and support for each state. Here, dispersion statistics could have multiple choices. For instance, we can consider the *Absolute Dispersion* illustrated in Example 8.1 and the *Mahalanobis Distance* illustrated in Example 8.2. Furthermore, for each state, the lower and upper bounds for the mean and the upper bound for the dispersion can be estimated by scaling the corresponding sample mean and sample dispersion, respectively.
2. *State-wise ambiguity set without labeled states.* In some situations, the state variables cannot be observed. In this case, one can employ an unsupervised machine learning technique to *learn* the different states first. For instance, we can use the K-means clustering approach to classify the samples into different states (see, e.g., [121, 221, 260, 427]). Once the states are determined, we can follow the mechanism above mentioned for the "labeled state" case to form the ambiguity set. A practical issue in this case is the trade-off between the number of

samples and the number K of states determined. If there are too many states after the clustering, the sample sizes in some states could be insufficient to estimate the state-wise ambiguity parameters. Therefore, if the number of states determined is relatively large, some statistical methods such as *bootstrapping* (see, e.g., [471]) or some machine learning method may be necessary to deal with the imbalance dataset (see, e.g., [252]).

3. *State-wise ambiguity set with states being forecasts.* As mentioned in Example 8.3, when the decision-maker intends to use each of the forecasts directly as the state, s/he/it can build the state-wise ambiguity set as a type-∞ Wasserstein ambiguity set. In this case, the Wasserstein ball radius θ can be justified by using the cross-validation approach that is popular for hyper-parameter learning in machine learning (see, e.g., [385]).

Given set $\mathcal{F}$, we formulate the adaptive distributionally robust optimization model for the capacitated facility location under uncertainty $\tilde{\boldsymbol{d}}$ associated with the ambiguous distributions $\mathbb{P} \in \mathcal{F}$, which we term *Robust Stochastic Facility Location Problem* (RS-FLP), as

$$\Pi^* = \min_{\boldsymbol{y} \in \mathcal{Y}} \left\{ \sum_{i \in I} f_i y_i + \sup_{\mathbb{P} \in \mathcal{F}} \mathbb{E}_{\mathbb{P}} \left[\mathrm{Q}(\boldsymbol{y}, \tilde{\boldsymbol{d}}) \right] \right\}. \tag{8.24}$$

For every location decision $\boldsymbol{y}$ and for every realization $\boldsymbol{d}$ of the demand, the second-stage optimization problem is defined as

$$\mathrm{Q}(\boldsymbol{y}, \boldsymbol{d}) = \min_{\boldsymbol{x} \in \mathcal{X}(\boldsymbol{y}, \boldsymbol{d})} \sum_{i \in I} \sum_{j \in J} c_{ij} x_{ij} + \sum_{j \in J} r_j \left(d_j - \sum_{i \in I} x_{ij} \right), \tag{8.25}$$

where $\mathcal{X}(\boldsymbol{y}, \boldsymbol{d})$ is the second-stage feasibility set formally given by

$$\mathcal{X}(\boldsymbol{y}, \boldsymbol{d}) = \left\{ \boldsymbol{x} \in \mathbb{R}^{|I| \times |J|} \;\middle|\; \begin{array}{ll} \sum_{i \in I} x_{ij} \le d_j, & \forall j \in J, \\ \sum_{j \in J} x_{ij} \le q_i y_i, & \forall i \in I, \\ x_{ij} \ge 0, & \forall i \in I,\ j \in J \end{array} \right\}. \tag{8.26}$$

Problem (8.25) seeks to determine the shipping decisions that minimize the transportation cost plus the penalty cost for unsupplied demand given a set of open facilities and an observation of the demand. The overall problem RS-FLP (8.24) aims at finding a location decision that minimizes the sum of setup cost for the facilities and the worst-case expected cost for shipping the commodity from the facilities to the customers and for the penalty incurred due to unsupplied demand.

As the name implies, the robust stochastic facility location model (8.24), leveraging the structure of the state-wise ambiguity set $\mathcal{F}$, reduces naturally to the adaptive distributionally robust model when $K = \{1\}$ while reduces to the classical two-stage stochastic programming model under an empirical distribution $\widehat{\mathbb{P}}_K = \sum_{k \in K} p_k \delta_{\hat{d}_k}$ when the conditional support $\mathcal{D}_k$ shrinks to a single scenario, i.e., $\mathcal{D}_k = \{\hat{d}_k\}$. Therefore, problem RS-FLP (8.24) can be seen as a trade-off between adaptive distributionally robust and stochastic facility location.

Next, we derive the dual reformulation of the worst-case expected second-stage cost function, which will be used in the following sections. Applying a standard duality argument for the moment problem (see [490]), the following Lemma holds.

Lemma 8.1 *Given a location decision* $\boldsymbol{y}$*, the worst-case expected second-stage cost satisfies*

$$\sup_{\mathbb{P} \in \mathcal{F}} \mathbb{E}_{\mathbb{P}} \left[\mathrm{Q}(\boldsymbol{y}, \tilde{\boldsymbol{d}}) \right] = \sum_{k \in K} p_k \Pi_k(\boldsymbol{d}_k^+, \boldsymbol{d}_k^-, \boldsymbol{\sigma}_k | \boldsymbol{y}) \tag{8.27}$$

with

$$\Pi_k(\boldsymbol{d}_k^+, \boldsymbol{d}_k^-, \boldsymbol{\sigma}_k | \boldsymbol{y})$$

$$= \begin{bmatrix} \min & \alpha_k + \overline{\boldsymbol{\beta}}_k^\top \boldsymbol{d}_k^+ - \underline{\boldsymbol{\beta}}_k^\top \boldsymbol{d}_k^- + \boldsymbol{\rho}_k^\top \boldsymbol{\sigma}_k, \\ s.\,t. & \alpha_k + (\overline{\boldsymbol{\beta}}_k - \underline{\boldsymbol{\beta}}_k)^\top \boldsymbol{d} + \boldsymbol{\rho}_k^\top \boldsymbol{w} \geq \mathrm{Q}(\boldsymbol{y}, \boldsymbol{d}), \forall (\boldsymbol{d}, \boldsymbol{w}) \in \mathcal{W}_k \\ & \alpha_k \in \mathbb{R},\ \overline{\boldsymbol{\beta}}_k, \underline{\boldsymbol{\beta}}_k \in \mathbb{R}_+^{|J|},\ \boldsymbol{\rho}_k \in \mathbb{R}_+^{n_k} \end{bmatrix}, \ k \in K,$$

where $\mathcal{W}_k$*,* $k \in K$*, is a lifted uncertainty set consisting of the support set* $\mathcal{D}_k$ *and the epigraph of* $\boldsymbol{g}_k$*, i.e.,*

$$\mathcal{W}_k = \left\{ (\boldsymbol{d}, \boldsymbol{w}) \in \mathbb{R}^{|J|} \times \mathbb{R}^{n_k} \,\middle|\, \boldsymbol{d} \in \mathcal{D}_k,\ \boldsymbol{g}_k(\boldsymbol{d}) \leq \boldsymbol{w} \right\}. \tag{8.28}$$

Additionally, $\overline{\boldsymbol{\beta}}_k, \underline{\boldsymbol{\beta}}_k, \boldsymbol{\rho}_k, \alpha_k$ $(k \in K)$ *are the dual variables associated with the ambiguity constraints in set* (8.19)*.* $\mathrm{Q}(\boldsymbol{y}, \boldsymbol{d})$ *is defined according to* (8.25)*.*

8.5 An Exact Algorithm: Nested Benders Decomposition

In this section, we discuss an exact solution algorithm for our robust stochastic capacitated facility location problem (8.24), which consists of a nested Benders decomposition in the sense that a Benders decomposition is derived for solving the Benders subproblems.

We consider the capacitated RS-FLP (8.24) embedding the dual reformulation (8.27) presented in Lemma 8.1. This leads to

$$\text{minimize} \quad \sum_{i \in I} f_i y_i + \sum_{k \in K} p_k \left(\alpha_k + \overline{\boldsymbol{\beta}}_k^\top \boldsymbol{d}_k^+ - \underline{\boldsymbol{\beta}}_k^\top \boldsymbol{d}_k^- + \boldsymbol{\rho}_k^\top \boldsymbol{\sigma}_k \right), \tag{8.29}$$

$$\text{subject to} \quad \alpha_k + (\overline{\boldsymbol{\beta}}_k - \underline{\boldsymbol{\beta}}_k)^\top \boldsymbol{d} + \boldsymbol{\rho}_k^\top \boldsymbol{w} \geq \mathrm{Q}(\boldsymbol{y}, \boldsymbol{d}) \; \forall k \in K, \; (\boldsymbol{d}, \boldsymbol{w}) \in \mathcal{W}_k, \tag{8.30}$$

$$\alpha_k \in \mathbb{R}, \; \overline{\boldsymbol{\beta}}_k, \underline{\boldsymbol{\beta}}_k \in \mathbb{R}_+^{|J|}, \; \boldsymbol{\rho}_k \in \mathbb{R}_+^{n_k} \; \forall k \in K, \; \boldsymbol{y} \in \mathcal{Y}. \tag{8.31}$$

In particular, if each lifted uncertainty set $\mathcal{W}_k$ is a polyhedron (e.g., the situation stemming from Example 8.1, where the ℓ_1-norm-based mappings $\boldsymbol{g}_k(\cdot)$ are assumed) that we denote by $\mathcal{W}_k^{\mathrm{P}}$ $(k \in K)$, we can devise different schemes to construct the master problem with a finite set of cuts.

One such scheme exploits the polyhedral structure of each $\mathcal{W}_k^{\mathrm{P}}$ to formulate a master problem as a mixed-integer linear program (MILP) as follows:

$$\text{minimize} \quad \sum_{i \in I} f_i y_i + \sum_{k \in K} p_k \left(\alpha_k + \overline{\boldsymbol{\beta}}_k^\top \boldsymbol{d}_k^+ - \underline{\boldsymbol{\beta}}_k^\top \boldsymbol{d}^- + \boldsymbol{\rho}_k^\top \boldsymbol{\sigma}_k \right), \tag{8.32}$$

$$\text{subject to} \quad \alpha_k \geq \sum_{i \in I} \sum_{j \in J} (c_{ij} - r_j) x_{ij}^{e_k} - \boldsymbol{\rho}_k^\top \hat{\boldsymbol{w}}^{e_k} + (\boldsymbol{r} - \overline{\boldsymbol{\beta}}_k + \underline{\boldsymbol{\beta}}_k)^\top \hat{\boldsymbol{d}}^{e_k}, \quad \forall k \in K, \; e_k \in E_k, \tag{8.33}$$

$$\boldsymbol{x}^{e_k} \in \mathcal{X}(\boldsymbol{y}, \hat{\boldsymbol{d}}^{e_k}) \quad \forall k \in K, \; e_k \in E_k, \tag{8.34}$$

$$\alpha_k \in \mathbb{R}, \; \overline{\boldsymbol{\beta}}_k, \underline{\boldsymbol{\beta}}_k \in \mathbb{R}_+^{|J|}, \; \boldsymbol{\rho}_k \in \mathbb{R}_+^{n_k} \quad \forall k \in K, \tag{8.35}$$

$$\boldsymbol{y} \in \mathcal{Y}, \tag{8.36}$$

where $\mathcal{X}(\boldsymbol{y}, \boldsymbol{d})$ is the second-stage feasible set (8.26) given $(\boldsymbol{y}, \boldsymbol{d})$, and $(\hat{\boldsymbol{d}}^1, \hat{\boldsymbol{w}}^1)$, $\ldots$, $(\hat{\boldsymbol{d}}^{|E_k|}, \hat{\boldsymbol{w}}^{|E_k|})$ are the extreme points of the polyhedral-representable sets $\mathcal{W}_k^{\mathrm{P}}$ with index set E_k, for each $k \in K$. This is similar to the scheme employed in the exact solution approaches developed in [73, 564] that are tailored for the resulting two-stage robust linear program with a polyhedral uncertainty set.

Alternatively, we can also exploit the polyhedral structure of the second-stage dual linear program $\mathrm{Q}(\boldsymbol{y}, \boldsymbol{d})$ to derive the Benders cuts, which also leads to an MILP master problem. This possibility has been exploited in [468] considering the ℓ_1-norm operator in the type-1 Wasserstein ambiguity set in a facility location problem, thus ensuring the MILP structure of the master problem.

The RS-FLP (8.24) assumes a general conic representation for each $\mathcal{W}_k, k \in K$. This is not particularly appealing for using the first scheme. First, without the polyhedral structure of $\mathcal{W}_k$, the resulting dual reformulation (8.29)–(8.31) would contain an infinite number of constraints (cuts) with $(\boldsymbol{d}, \boldsymbol{w}) \in \mathcal{W}_k$. Second, even for the case of polyhedral lifted uncertainty sets $\mathcal{W}_k^{\mathrm{P}}$, our problem considers different states $k \in K$, which would lead to $|K|$ sets of extreme points to consider. This becomes clear by observing the MILP reformulation (8.32)–(8.36). This can seriously limit the stability of the resulting algorithm. Therefore, the second scheme seems more promising as a starting point for developing an exact solution algorithm for RS-FLP (8.24). Next, we discuss a subgradient-based nested decomposition algorithm. We closely follow [345].

The nested Benders decomposition approach leverages (i) the subgradient cuts formed by extreme conditional distributions in different states used to decompose the relaxed master problem and (ii) the complementary approach introduced in [564] to cope with the separation problem in the conic distributionally robust setting we are facing. In [73] the authors also extend the approach in [564] within the same polyhedral setting by focusing on the feasibility issues. As we show later, this is unnecessary in our location problem context.

8.5.1 The Relaxed Master Problem and Its Decomposition

For a clearer exposition, we re-write the worst-case expected second-stage cost given $\boldsymbol{y} \in \mathcal{Y}$ as follows:

$$\Pi(\boldsymbol{y}) = \sum_{k \in K} p_k \sup_{\mathbb{P} \in \mathcal{F}_k} \mathbb{E}_{\mathbb{P}}\left[\mathrm{Q}(\boldsymbol{y}, \tilde{\boldsymbol{d}})\right] = \sup_{\mathbb{P} \in \mathcal{F}} \mathbb{E}_{\mathbb{P}}\left[\mathrm{Q}(\boldsymbol{y}, \tilde{\boldsymbol{d}})\right], \tag{8.37}$$

where $\mathcal{F}_k$ is the demand distribution ambiguity set under the state $k \in K$. It consists of the distributional information for state k in the ambiguity set $\mathcal{F}$ given by (8.19):

$$\mathcal{F}_k = \left\{ \mathbb{P} \in \mathcal{P}(\mathbb{R}^{|J|}) \;\middle|\; \begin{array}{l} \tilde{\boldsymbol{d}} \sim \mathbb{P} \\ \mathbb{E}_{\mathbb{P}}\left[\tilde{\boldsymbol{d}}\right] \in [\boldsymbol{d}_k^-, \boldsymbol{d}_k^+], \mathbb{E}_{\mathbb{P}}\left[\boldsymbol{g}_k(\tilde{\boldsymbol{d}})\right] \le \boldsymbol{\sigma}_k \\ \mathbb{P}\left[\tilde{\boldsymbol{d}} \in \mathcal{D}_k\right] = 1 \end{array} \right\}. \tag{8.38}$$

The reformulation (8.37) can be easily achieved by dualizing back the formulation $\Pi_k(\boldsymbol{d}_k^+, \boldsymbol{d}_k^-, \boldsymbol{\sigma}_k | \boldsymbol{y})$ in (8.27) separately for each $k \in K$ given $\boldsymbol{y} \in \mathcal{Y}$. In particular, we dualize the second-stage transportation problem given $(\boldsymbol{y}, \boldsymbol{d})$ as

$$\begin{aligned} \mathrm{Q}(\boldsymbol{y}, \boldsymbol{d}) &= \boldsymbol{r}^\top \boldsymbol{d} + \min_{\boldsymbol{x} \in \mathcal{X}(\boldsymbol{y}, \boldsymbol{d})} \sum_{i \in I} \sum_{j \in J} (c_{ij} - r_j) x_{ij} \\ &= \boldsymbol{r}^\top \boldsymbol{d} + \max_{(\boldsymbol{v}, \boldsymbol{\lambda}) \in \mathcal{H}} \left\{ -\boldsymbol{v}^\top \boldsymbol{d} - \sum_{i \in I} q_i y_i \lambda_i \right\}, \end{aligned} \tag{8.39}$$

where $\boldsymbol{r} = (r_1, \ldots, r_{|J|})^\top$ and $\mathcal{H}$ is the dual feasibility set for the second-stage transportation problem:

$$\mathcal{H} = \left\{(\boldsymbol{v}, \boldsymbol{\lambda}) \in \mathbb{R}_+^{|J|} \times \mathbb{R}_+^{|I|} \mid -v_j - \lambda_i \leq c_{ij} - r_j \quad \forall i \in I,\ j \in J\right\}. \tag{8.40}$$

Due to the complete-recourse structure of our location problem, the right-hand side maximization dual problem in (8.39) is always bounded. Since the variables $\boldsymbol{v}$ and $\boldsymbol{\lambda}$ are bounded from below and given the nonnegativity and boundedness of $\boldsymbol{d}$ and $q_i y_i$, $\forall i \in I$, the maximization problem is bounded. Hence, an optimal solution can be obtained at an extreme point.

We now consider the following reformulation of problem (8.24):

$$\Pi^* = \min_{\boldsymbol{y} \in \mathcal{Y}} \left\{ \sum_{i \in I} f_i y_i + \sum_{k \in K} p_k \sup_{\mathbb{P} \in \mathcal{F}_k} \mathbb{E}_{\mathbb{P}} \left[\mathrm{Q}(\boldsymbol{y}, \tilde{\boldsymbol{d}}) \right] \right\}, \tag{8.41}$$

where each $\mathcal{F}_k$ stands for the ambiguity subset (8.38) corresponding to the demand distribution under the state $k \in K$.

To construct the relaxed master problem, we consider a lower bound for each worst-case expected second-stage cost $\sup_{\mathbb{P} \in \mathcal{F}_k} \mathbb{E}_{\mathbb{P}} \left[\mathrm{Q}(\boldsymbol{y}, \tilde{\boldsymbol{d}}) \right]$ in state $k \in K$. In particular, let E be the index set of extreme points of the polyhedral set $\mathcal{H}$ given by (8.40), for any $k \in K$ and $(\boldsymbol{y}, \boldsymbol{d}) \in \mathcal{Y} \times \mathcal{D}_k$. We have

$$\mathrm{Q}(\boldsymbol{y}, \boldsymbol{d}) = \mathrm{Q}^E(\boldsymbol{y}, \boldsymbol{d}) = \boldsymbol{r}^\top \boldsymbol{d} + \max_{e \in E} \left\{ -\boldsymbol{d}^\top \boldsymbol{v}^e - \sum_{i \in I} q_i y_i \lambda_i^e \right\}. \tag{8.42}$$

Accordingly, by considering a subset $L \subset E$ of extreme points of $\mathcal{H}$, we obtain a lower bound for $\mathrm{Q}(\boldsymbol{y}, \boldsymbol{d})$:

$$\mathrm{Q}(\boldsymbol{y}, \boldsymbol{d}) \geq \mathrm{Q}^L(\boldsymbol{y}, \boldsymbol{d}) = \boldsymbol{r}^\top \boldsymbol{d} + \max_{e \in L} \left\{ -\boldsymbol{d}^\top \boldsymbol{v}^e - \sum_{i \in I} q_i y_i \lambda_i^e \right\}, \tag{8.43}$$

which provides a *relaxed master problem* as follows:

$$\Pi_L^* = \min_{\boldsymbol{y} \in \mathcal{Y}} \left\{ \sum_{i \in I} f_i y_i + \sum_{k \in K} p_k \sup_{\mathbb{P} \in \mathcal{F}_k} \mathbb{E}_{\mathbb{P}} \left[\mathrm{Q}^L(\boldsymbol{y}, \tilde{\boldsymbol{d}}) \right] \right\}. \tag{8.44}$$

In particular, when $L = E$, problem (8.44) becomes the *master problem*, i.e., the overall problem (8.41). Considering the above discussion, we can harness jointly the polyhedral structure of the second-stage dual feasible set $\mathcal{H}$ and the conic representable structure of each $\mathcal{W}_k$ shown as

$$\mathcal{W}_k = \left\{(\boldsymbol{d}, \boldsymbol{w}) \in \mathbb{R}_+^{|J|} \times \mathbb{R}^{n_k} \middle| \boldsymbol{A}_k \boldsymbol{d} \leq \boldsymbol{b}_k,\ (\boldsymbol{d}, \boldsymbol{w}, t) \in \mathcal{K}_k, t = 1\right\},\ \forall k \in K, \tag{8.45}$$

where $\mathcal{K}_k$ is a proper cone[2] defined as

$$\mathcal{K}_k = \text{cl}\left\{(\boldsymbol{d}, \boldsymbol{w}, t) \in \mathbb{R}_+^{|J|} \times \mathbb{R}^{n_k} \times \mathbb{R} \mid \boldsymbol{g}_k(\boldsymbol{d}/t) \leq \boldsymbol{w}/t,\ t > 0\right\} \quad \forall k \in K, \tag{8.46}$$

where cl(A) stands for the closure of A.

This allows reformulating problem (8.44) as a mixed-integer conic linear program. The following result states this.

Proposition 8.1 *Given a subset L of extreme points $\{(\hat{\boldsymbol{v}}^{e_k}, \hat{\boldsymbol{\lambda}}^{e_k})_{e_k \in L}\}$ of $\mathcal{H}$, the relaxed master problem* (8.44) *is equivalent to the following mixed-integer conic linear program:*

$$\Pi_L^* = \min \quad \sum_{i\in I} f_i y_i + \sum_{k\in K} p_k \left(\alpha_k + \overline{\boldsymbol{\beta}}_k^\top \boldsymbol{d}_k^+ - \underline{\boldsymbol{\beta}}_k^\top \boldsymbol{d}_k^- + \boldsymbol{\rho}_k^\top \boldsymbol{\sigma}_k\right), \tag{8.47}$$

$$\text{s. t.} \quad \alpha_k - \boldsymbol{b}_k^\top \boldsymbol{\xi}^{e_k} - \gamma^{e_k} + \sum_{i\in I} q_i y_i \hat{\lambda}_i^{e_k} \geq 0 \qquad \forall k \in K,\ e_k \in L, \tag{8.48}$$

$$\boldsymbol{\psi}^{e_k} - \boldsymbol{A}_k^\top \boldsymbol{\xi}^{e_k} - \overline{\boldsymbol{\beta}}_k + \underline{\boldsymbol{\beta}}_k + \boldsymbol{r} - \hat{\boldsymbol{v}}^{e_k} = \boldsymbol{0} \qquad \forall k \in K,\ e_k \in L, \tag{8.49}$$

$$(\boldsymbol{\psi}^{e_k}, \boldsymbol{\rho}_k, \gamma^{e_k}) \in \mathcal{K}_k^* \qquad \forall k \in K,\ e_k \in L, \tag{8.50}$$

$$\boldsymbol{\xi}^{e_k} \in \mathbb{R}_+^{m_k}, \boldsymbol{\psi}^{e_k} \in \mathbb{R}^{|J|}, \gamma^{e_k} \in \mathbb{R} \qquad \forall k \in K,\ e_k \in L, \tag{8.51}$$

$$\alpha_k \in \mathbb{R},\ \overline{\boldsymbol{\beta}}_k, \underline{\boldsymbol{\beta}}_k \in \mathbb{R}_+^{|J|},\ \boldsymbol{\rho}_k \in \mathbb{R}_+^{n_k} \qquad \forall k \in K, \tag{8.52}$$

$$\boldsymbol{y} \in \mathcal{Y}, \tag{8.53}$$

where, for every $k \in K$, $\mathcal{K}_k^*$ denotes the dual cone of $\mathcal{K}_k$.

Proof See [345]. □

In most cases, it might not be practical to deal directly with the relaxed master problem as presented in the above result because its size will grow as more cuts are added. This motivates the development of a decomposition scheme to solve that problem iteratively. Leveraging such a decomposition scheme, it is possible to show that in each iteration of the exact algorithm being discussed for the overall problem, it suffices to solve separately K small conic linear programs and an MILP problem.

---

[2] A cone is proper if it is convex, closed, solid (i.e., nonempty interior), and pointed (i.e., $x, -x \in \mathcal{K}_k \Rightarrow x = 0$). See [86] for additional insights.

Given (8.43), it is not difficult to conclude that the lower bound of the worst-case expected second-stage cost for each state k in (8.44), i.e.,

$$\Pi^L(\boldsymbol{y}) = \sup_{\mathbb{P}\in\mathcal{F}_k} \mathbb{E}_{\mathbb{P}}\left[\mathrm{Q}^L(\boldsymbol{y}, \tilde{\boldsymbol{d}})\right] = \mathbb{E}_{\mathbb{P}_k^\star}\left[\mathrm{Q}^L(\boldsymbol{y}, \tilde{\boldsymbol{d}})\right], \tag{8.54}$$

is a convex function in $\boldsymbol{y}$. Furthermore, the decomposition scheme being discussed builds on its subgradient at the given $\boldsymbol{y} \in \mathcal{Y}$, which is formed by the associated worst-case distribution $\mathbb{P}_k^\star$. This is formally presented in the following result.

Proposition 8.2 *Let L be a subset of extreme points* $\{(\hat{\boldsymbol{v}}^{e_k}, \hat{\boldsymbol{\lambda}}^{e_k})_{e_k\in L}\}$ *of* $\mathcal{H}$*. Consider a facility location decision* $\boldsymbol{y} \in \mathcal{Y}$*. For each* $k \in K$*, the worst-case distribution* $\mathbb{P}_k^\star$ *invoked in* (8.54) *admits an* $|L|$*-point discrete distribution*

$$\mathbb{P}_k^\star\left[\tilde{\boldsymbol{d}} = \frac{\boldsymbol{\eta}_\star^{e_k}}{\zeta_\star^{e_k}}\right] = \zeta_\star^{e_k}, \forall e_k \in L, \tag{8.55}$$

where $(\boldsymbol{\eta}_\star^{e_k}, \zeta_\star^{e_k})_{e_k\in L}$ *solve a related conic linear program*

$$\mathbb{E}_{\mathbb{P}_k^\star}\left[\mathrm{Q}^L(\boldsymbol{y}, \tilde{\boldsymbol{d}})\right] = \max_{\zeta,\eta} \quad \sum_{e_k\in E}(\boldsymbol{r} - \hat{\boldsymbol{v}}^{e_k})^\top \boldsymbol{\eta}^{e_k} - \sum_{e_k\in E}\sum_{i\in I} q_i y_i \hat{\lambda}_i^{e_k}\zeta^{e_k}, \tag{8.56}$$

$$\text{s. t.} \quad \sum_{e_k\in E} \zeta^{e_k} = 1, \tag{8.57}$$

$$\sum_{e_k\in E} \boldsymbol{\eta}^{e_k} \le \boldsymbol{d}_k^+, \tag{8.58}$$

$$\sum_{e_k\in E} \boldsymbol{\eta}^{e_k} \ge \boldsymbol{d}_k^-, \tag{8.59}$$

$$\sum_{e_k\in E} \boldsymbol{\tau}^{e_k} \le \boldsymbol{\sigma}_k, \tag{8.60}$$

$$\boldsymbol{A}_k\boldsymbol{\eta}^{e_k} - \zeta^{e_k}\boldsymbol{b}_k \le \boldsymbol{0} \qquad \forall e_k \in L, \tag{8.61}$$

$$(\boldsymbol{\eta}^{e_k}, \boldsymbol{\tau}^{e_k}, \zeta^{e_k}) \in \mathcal{K}_k \qquad \forall e_k \in L, \tag{8.62}$$

$$\zeta^{e_k} \in \mathbb{R}_+, \boldsymbol{\eta}^{e_k} \in \mathbb{R}^{|J|}, \boldsymbol{\tau}^{e_k} \in \mathbb{R}^{|n_k|} \quad \forall e_k \in L, \tag{8.63}$$

where each $\mathcal{K}_k$ *is the proper cone given by* (8.46)*. Furthermore, we have*

$$\left(-q_i \sum_{e^k\in L} \hat{\lambda}_i^e \zeta_\star^{e_k}\right)_{i\in I} \in \partial\Pi^L(\boldsymbol{y}). \tag{8.64}$$

Proof See [345]. □

This result provides important insights for the development of the decomposition procedure:

(i) Each subproblem $\sup_{\mathbb{P}\in\mathcal{F}_k} \mathbb{E}_{\mathbb{P}}[\mathrm{Q}^L(\boldsymbol{y}, \tilde{\boldsymbol{d}})]$ given $\boldsymbol{y}$ is a small-scale conic linear program (8.56)–(8.63), which can be solved separately.
(ii) Solving (8.56)–(8.63) provides us with a subgradient for $\sup_{\mathbb{P}\in\mathcal{F}_k} \mathbb{E}_{\mathbb{P}}[\mathrm{Q}^L(\boldsymbol{y}, \tilde{\boldsymbol{d}})]$ at $\boldsymbol{y}$. Harnessing the binary structure of $\boldsymbol{y} \in \mathcal{Y}$, the subgradient allows us to form a finite piecewise affine reformulation of $\sup_{\mathbb{P}\in\mathcal{F}_k} \mathbb{E}_{\mathbb{P}}[\mathrm{Q}^L(\boldsymbol{y}, \tilde{\boldsymbol{d}})]$, which becomes the finite linear cuts that fully represent and decompose our relaxed master problem (8.44).
(iii) In addition, by (8.56)–(8.63), in [345] the authors show that $\sup_{\mathbb{P}\in\mathcal{F}_k} \mathbb{E}_{\mathbb{P}}[\mathrm{Q}^L(\boldsymbol{y}, \tilde{\boldsymbol{d}})]$ is always bounded. In particular, for each $\boldsymbol{y} \in \mathcal{Y}$ and $L \subseteq E$, we have

$$\left| \sup_{\mathbb{P}\in\mathcal{F}_k} \mathbb{E}_{\mathbb{P}}[\mathrm{Q}^L(\boldsymbol{y}, \tilde{\boldsymbol{d}})] \right| < \infty, \ \forall k \in K. \tag{8.65}$$

This boundedness result makes it clear that it is not necessary to consider any feasibility cuts in the forthcoming decomposition procedure.

That procedure stems from the following result, proved also in [345]:

Proposition 8.3 *The relaxed master problem* (8.44) *with the subset L of vertices of $\mathcal{H}$ is equivalent to the following mixed-integer linear program:*

$$\text{minimize} \quad \sum_{i\in I} f_i y_i + \sum_{k\in K} p_k \theta_k, \tag{8.66}$$

$$\textit{subject to} \quad \theta_k \geq \sum_{e_k\in L} (\boldsymbol{r} - \hat{\boldsymbol{v}}^{e_k})^\top \boldsymbol{\eta}_\star^{e_k}(\boldsymbol{y}^\diamond) - \sum_{e_k\in L} \sum_{i\in I} q_i y_i \hat{\lambda}_i^{e_k} \zeta_\star^{e_k}(\boldsymbol{y}^\diamond)$$
$$\forall \boldsymbol{y}^\diamond \in \mathcal{Y}, k \in K, \tag{8.67}$$

$$y \in \mathcal{Y}, \boldsymbol{\theta} \in \mathbb{R}^{|K|}, \tag{8.68}$$

where $(\boldsymbol{\eta}_\star^{e_k}(\boldsymbol{y}^\diamond), \zeta_\star^{e_k}(\boldsymbol{y}^\diamond))_{e_k\in L}$ *result from the resolution of* (8.56)–(8.63) *under the location decision* $\boldsymbol{y}^\diamond$.

The MILP (8.66)–(8.68) together with the conic linear program (8.56)–(8.63) for each $k \in K$, can be set at the core of a decomposition scheme for the relaxed master problem (8.44).

Given a (non-optimal) location decision $\boldsymbol{y}^n \in \mathcal{Y}$, we can solve the conic subproblem (8.56)–(8.63) separately for each $k \in K$, to obtain $\sup_{\mathbb{P}\in\mathcal{F}_k} \mathbb{E}_{\mathbb{P}}[\mathrm{Q}^L(\boldsymbol{y}^n, \tilde{\boldsymbol{d}})]$. This induces an upper bound on the optimal value of the relaxed master problem (8.44) or (8.66)–(8.68):

$$\mathrm{UB}_L = \sum_{i\in I} f_i y_i + \sum_{k\in K} p_k \sup_{\mathbb{P}\in\mathcal{F}_k} \mathbb{E}_{\mathbb{P}}[\mathrm{Q}^L(\boldsymbol{y}^n, \tilde{\boldsymbol{d}})]. \tag{8.69}$$

Algorithm 8.1: Decomposition scheme for the relaxed master problem (8.44)

Data: Model parameters, a subset L of extreme points $\{(\boldsymbol{v}^{e_k}, \boldsymbol{\lambda}^{e_k})_{e_k \in L}\}$ of $\mathcal{H}$
Result: Optimal solution $\boldsymbol{y}^{\star}$.
Consider an initial location decision $\boldsymbol{y}^{\diamond} \in \mathcal{Y}$;
$n \leftarrow 0$; $\mathrm{LB}_L \leftarrow -\infty$; $\mathrm{UB}_L \leftarrow +\infty$;
repeat
  $n \leftarrow n+1$;
  $\boldsymbol{y}^n \leftarrow \boldsymbol{y}^{\diamond}$;
  For each $k \in K$, solve the conic linear program (8.56)–(8.63) and obtain $\left(-q_i \sum_{e_k \in L} \hat{\lambda}_i^{e_k} \zeta_{\star}^{e_k}(\boldsymbol{y}^n)\right)_{i \in I}$;
  $\mathrm{UB}_L \leftarrow \min\left\{\mathrm{UB}, \sum_{i \in I} f_i y_i^n + \sum_{k \in K} p_k \sup_{\mathbb{P} \in \mathcal{F}_k} \mathbb{E}_{\mathbb{P}}[\mathrm{Q}^L(\boldsymbol{y}^n, \tilde{\boldsymbol{d}})]\right\}$;
  Add the new cut with $\left(-q_i \sum_{e_k \in L} \hat{\lambda}_i^{e_k} \zeta_{\star}^{e_k}(\boldsymbol{y}^n)\right)_{i \in I}$ into the MILP (8.70)–(8.72);
  Solve it obtaining a new solution $(\boldsymbol{y}^{\diamond}, \boldsymbol{\theta}^{\diamond})$;
  $\mathrm{LB}_L \leftarrow \sum_{i \in I} f_i y_i^{\diamond} + \sum_{k \in K} p_k \theta_k^{\diamond}$; //update lower bound:
until UB $\leq$ LB;
$\boldsymbol{y}^{\star} \leftarrow \boldsymbol{y}^{\diamond}$;

The solution of (8.56)–(8.63) separately for $k \in K$ also leads to the subgradient

$$\left(-q_i \sum_{e_k \in L} \hat{\lambda}_i^{e_k} \zeta_{\star}^{e_k}(\boldsymbol{y}^n)\right)_{i \in I},$$

which, in turn, provides a cut for the Lower Bound of MILP problem (8.66)–(8.68):

$$\mathrm{LB}_L = \min \quad \sum_{i \in I} f_i y_i + \sum_{k \in K} p_k \theta_k, \tag{8.70}$$

$$\text{s. t.} \quad \theta_k \geq \sum_{e_k \in L} (\boldsymbol{r} - \hat{\boldsymbol{v}}^{e_k})^{\top} \boldsymbol{\eta}_{\star}^{e_k}(\boldsymbol{y}^n) - \sum_{e_k \in L} \sum_{i \in I} q_i y_i \hat{\lambda}_i^{e_k} \zeta_{\star}^{e_k}(\boldsymbol{y}^n) \quad \forall n \in N, k \in K, \tag{8.71}$$

$$\boldsymbol{y} \in \mathcal{Y}, \boldsymbol{\theta} \in \mathbb{R}^{|K|}. \tag{8.72}$$

Solving this problem, we obtain a new location decision, $\boldsymbol{y}^{n+1}$, thus triggering a new iteration. We solve the conic subproblem (8.56)–(8.63) using that solution and generate new cuts with subgradient and update the upper bound.

The described decomposition procedure for the problem (8.44) is summarized in Algorithm 8.1.

8.5.2 The Separation Problem

Given a location decision $\boldsymbol{y}$, by the equivalent form (8.29)–(8.31) of the RS-FLP, the worst-case expected second-stage cost can be written as follows:

$$\sum_{k\in K} p_k \min_{\overline{\boldsymbol{\beta}}_k,\underline{\boldsymbol{\beta}}_k,\boldsymbol{\rho}_k} \left\{ \max_{(\boldsymbol{d},\boldsymbol{w})\in\mathcal{W}_k} \left(\mathrm{Q}(\boldsymbol{y},\boldsymbol{d}) - (\overline{\boldsymbol{\beta}}_k - \underline{\boldsymbol{\beta}}_k)^\top \boldsymbol{d} - \boldsymbol{w}^\top \boldsymbol{\rho}_k \right) + \overline{\boldsymbol{\beta}}_k^\top \boldsymbol{d}_k^+ - \underline{\boldsymbol{\beta}}_k^\top \boldsymbol{d}_k^- + \boldsymbol{\rho}_k^\top \boldsymbol{\sigma}_k \right\},$$

which reduces to solving $|K|$ "min-maxmin" problems. This differs from the structure of the adaptive robust linear optimization problem, where the worst-case second-stage cost reduces to a "maximin" problem. Therefore, to apply a complementary approach like that proposed in [564]—thus achieving a linear (conic) separation problem—we do the following: For each $\boldsymbol{y}^\star$ obtained from solving the relaxed master problem (8.44) using Algorithm 8.1, we consider the equivalent relaxed master problem formulation (8.47)–(8.53) by fixing $\boldsymbol{y} = \boldsymbol{y}^\star$. Then we solve the $|K|$ conic linear problems separately to obtain the optimal values of the distributional ambiguity dual variables $(\alpha_k, \overline{\boldsymbol{\beta}}_k, \underline{\boldsymbol{\beta}}_k, \boldsymbol{\rho}_k)_{k\in K}$ for the fixed $\boldsymbol{y}^\star$. We can now formulate a *separation problem* that checks whether the solution just found, i.e., $((\alpha_k^\star, \overline{\boldsymbol{\beta}}_k^\star, \underline{\boldsymbol{\beta}}_k^\star, \boldsymbol{\rho}_k^\star)_{k\in K}, \boldsymbol{y}^\star)$ is feasible (therefore optimal) to the master problem (8.29)–(8.31). That is to check whether that solution satisfies all constraints in (8.29)–(8.31):

$$\alpha_k^\star \geq \mathrm{Q}(\boldsymbol{y}^\star, \boldsymbol{d}) - (\overline{\boldsymbol{\beta}}_k^\star - \underline{\boldsymbol{\beta}}_k^\star)^\top \boldsymbol{d} - \boldsymbol{w}^\top \boldsymbol{\rho}_k^\star, \ \forall (\boldsymbol{d},\boldsymbol{w}) \in \mathcal{W}_k, \ k \in K. \tag{8.73}$$

For every $k \in K$, this inequality can be re-written as $\alpha_k^\star \geq \Omega_k, \ k \in K$ with

$$\Omega_k = \max_{(\boldsymbol{d},\boldsymbol{w})\in\mathcal{W}_k} \left\{ \min_{\boldsymbol{x}\in\mathcal{X}(\boldsymbol{y}^\star,\boldsymbol{d})} \left\{ \sum_{i\in I}\sum_{j\in J} (c_{ij} - r_j) x_{ij} \right\} + (\boldsymbol{r} - \overline{\boldsymbol{\beta}}_k^\star + \underline{\boldsymbol{\beta}}_k^\star)^\top \boldsymbol{d} - \boldsymbol{w}^\top \boldsymbol{\rho}_k^\star \right\}. \tag{8.74}$$

We can consider the linear programming complementary slackness conditions to re-write the inner (linear programming) problem. Moreover, we also enhance the practical value of the complementary approach by deriving valid inequalities (to avoid using Big-M terms), which is useful when performing the separation problems along the iterations. This leads to the following result, also proved in [345]:

Proposition 8.4 *For every $k \in K$, and given the solution $(\overline{\boldsymbol{\beta}}_k^\star, \underline{\boldsymbol{\beta}}_k^\star, \boldsymbol{\rho}_k^\star, \boldsymbol{y}^\star)$ from the relaxed master problem* (8.47)–(8.53)*, the maxmin optimization problem* (8.74) *is equivalent to the following mixed-integer conic linear program:*

$$\Omega_k = \max_{\boldsymbol{d},\boldsymbol{w},\boldsymbol{x},\boldsymbol{v},\boldsymbol{\lambda},\boldsymbol{\pi}} \sum_{i\in I}\sum_{j\in J} (c_{ij} - r_j) x_{ij} + (\boldsymbol{r} - \overline{\boldsymbol{\beta}}_k^\star + \underline{\boldsymbol{\beta}}_k^\star)^\top \boldsymbol{d} - \boldsymbol{w}^\top \boldsymbol{\rho}_k^\star, \tag{8.75}$$

$$\text{s. t.} \quad \sum_{i\in I} x_{ij} \leq d_j \qquad \forall j \in J, \tag{8.76}$$

$$\sum_{j \in J} x_{ij} \leq q_i y_i^\star \qquad \forall i \in I, \tag{8.77}$$

$$\nu_j + \lambda_i \geq r_j - c_{ij} \qquad \forall i \in I,\ j \in J, \tag{8.78}$$

$$\nu_j \leq \overline{r}\pi_j^1 \qquad \forall j \in J, \tag{8.79}$$

$$\lambda_i \leq \overline{r}\pi_i^2 \qquad \forall i \in I, \tag{8.80}$$

$$x_{ij} \leq \overline{q}(1 - \pi_{ij}^3) \qquad \forall i \in I,\ j \in J, \tag{8.81}$$

$$d_j - \sum_{i \in I} x_{ij} \leq \overline{d}(1 - \pi_j^1) \qquad \forall j \in J, \tag{8.82}$$

$$\sum_{j \in J} x_{ij} \geq q_i y_i^\star - \overline{q}(1 - \pi_i^2) \qquad \forall i \in I, \tag{8.83}$$

$$\nu_j + \lambda_i \leq (2\overline{r} - \underline{r})\pi_{ij}^3 + r_j - c_{ij} \qquad \forall i \in I,\ j \in J, \tag{8.84}$$

$$\boldsymbol{A}_k \boldsymbol{d} \leq \boldsymbol{b}_k,\ (\boldsymbol{d}, \boldsymbol{w}, 1) \in \mathcal{K}_k, \tag{8.85}$$

$$\boldsymbol{\nu} \in \mathbb{R}_+^{|J|}, \boldsymbol{\lambda} \in \mathbb{R}_+^{|I|}, \boldsymbol{x} \in \mathbb{R}_+^{|I| \times |J|}, \tag{8.86}$$

$$\boldsymbol{\pi}^1 \in \{0,1\}^{|J|}, \boldsymbol{\pi}^2 \in \{0,1\}^{|I|}, \boldsymbol{\pi}^3 \in \{0,1\}^{|I| \times |J|}, \tag{8.87}$$

where $\overline{d}$, $\overline{q}$, $\overline{r}$, and $\underline{r}$ are constant inputs given by:

$$\overline{q} = \max_{i \in I}\{q_i y_i^\star\}$$
$$\overline{d} = \max_{j \in J, k \in K}\{\overline{d}_{jk}\}$$
$$\overline{r} = \max_{i \in I, j \in J}\{0, r_j - c_{ij}\}$$
$$\underline{r} = \min_{i \in I, j \in J}\{0, r_j - c_{ij}\}$$
$$\overline{d}_{jk} = \max\{\boldsymbol{e}_j^\top \boldsymbol{d} \mid \boldsymbol{A}_k \boldsymbol{d} \leq \boldsymbol{b}_k,\ \boldsymbol{d} \geq \boldsymbol{0}\}\ \ \forall k \in K, j \in J$$

Proof See [345]. □

When $\boldsymbol{g}_k(\cdot)$ is given as in Examples 8.1 and 8.3 with $\|\cdot\|_1$ or $\|\cdot\|_\infty$ operators being used in the ∞-Wasserstein ball, the conic constraint of problem (8.75)–(8.87) defines a polyhedron.

When $\boldsymbol{g}_k(\cdot)$ is given as in Examples 8.2 and 8.3 with $\|\cdot\|_2$ operator being used in the ∞-Wasserstein ball, the conic constraint of problem (8.75)–(8.87) defines a second-order cone. Hence, the model becomes an MILP and a mixed-integer SOCP, respectively.

Let $((\alpha_k^\star, \overrightarrow{\boldsymbol{\beta}}_k^\star, \underline{\boldsymbol{\beta}}_k^\star, \boldsymbol{\rho}_k^\star)_{k \in K}, \boldsymbol{y}^\star)$ be an optimal solution of the relaxed master problem (8.47)–(8.53) with a total cost of Π^L. Let Ω_k be obtained by solving (8.75)–

(8.87) fixing $((\overline{\boldsymbol{\beta}}^{\star}_k, \underline{\boldsymbol{\beta}}^{\star}_k, \rho^{\star}_k)_{k\in K}, \boldsymbol{y}^{\star})$. Liu et al. [345] show that in this case, we have $\Omega_k \geq \alpha^{\star}_k, \forall k \in K$ and also that the optimality gap for our RS-FLP is bounded by

$$\Pi^* - \Pi^L \leq \sum_{k\in K} p_k(\Omega_k - \alpha^{\star}_k). \tag{8.88}$$

By solving the problem (8.75)–(8.87) for each $k \in K$, not only can we check the optimality of the solution ($\alpha^{\star}_k \geq \Omega_k$, $k \in K$) but we also obtain an optimal solution $(\boldsymbol{\nu}^{\star}, \boldsymbol{\lambda}^{\star})$ of the inner dual subproblem (8.39) given $(\boldsymbol{y}^{\star}, \boldsymbol{d}^{\star})$. This allows us to readily recover an extreme point $(\boldsymbol{\nu}^{\star\star}, \boldsymbol{\lambda}^{\star\star})$ of the polyhedron $\mathcal{H}$ that can then be used for generating an optimality cut if the solution is not optimal. In other words, this extreme point $(\boldsymbol{\nu}^{\star\star}, \boldsymbol{\lambda}^{\star\star})$ will be included in the set of extreme points L to define an enhanced relaxed master problem (8.47)–(8.53). Formally, if $(\boldsymbol{x}^{\star}, \boldsymbol{\nu}^{\star}, \boldsymbol{\lambda}^{\star}, \boldsymbol{d}^{\star}, \boldsymbol{w}^{\star}, \boldsymbol{\pi}^{\star})$ is an optimal solution of problem (8.75)–(8.87) given a solution $((\alpha^{\star}_k, \overline{\boldsymbol{\beta}}^{\star}_k, \underline{\boldsymbol{\beta}}^{\star}_k, \rho^{\star}_k)_{k\in K}, \boldsymbol{y}^{\star})$, then $(\boldsymbol{\nu}^{\star}, \boldsymbol{\lambda}^{\star})$ is an optimal solution of the dual subproblem (8.39) given $(\boldsymbol{y}^{\star}, \boldsymbol{d}^{\star})$. Furthermore, any basic feasible solution (BFS) of the following linear program:

$$\min_{\boldsymbol{\nu},\boldsymbol{\lambda}} \left\{ \boldsymbol{\nu}^\top \boldsymbol{d}^{\star} + \sum_{i\in I} q_i y^{\star}_i \lambda_i \;\middle|\; \begin{array}{l} (\boldsymbol{\nu}, \boldsymbol{\lambda}) \in \mathcal{H} \\ [\boldsymbol{d}^{\star}]^\top \boldsymbol{\nu} + \sum_{i\in I} q_i y^{\star}_i \lambda_i = [\boldsymbol{d}^{\star}]^\top \boldsymbol{\nu}^{\star} + \sum_{i\in I} q_i y^{\star}_i \lambda^{\star}_i \end{array} \right\} \tag{8.89}$$

is an extreme point of the polyhedral set $\mathcal{H}$.

From this discussion, we conclude that with the solution $(\boldsymbol{\nu}^{\star}, \boldsymbol{\lambda}^{\star})$ obtained from solving each separation subproblem (8.75)–(8.87), instead of solving the dual subproblem (8.39) given $(\boldsymbol{y}^{\star}, \boldsymbol{d}^{\star})$, we only need to find an initial basic feasible solution (BFS) $(\boldsymbol{\nu}^{\star\star}, \boldsymbol{\lambda}^{\star\star})$ of the LP (8.89). This can be easily accomplished using the two-phase Simplex method or the Big-M method—well-known initializations for the Simplex Algorithm.

8.5.3 The Overall Algorithm Procedure

We summarize the overall algorithm procedure for solving our RS-FLP (8.41):

(i) We start with an initial subset L of extreme points of $\mathcal{H}$. It can be easily obtained by solving the (simple) second-stage dual linear program:

$$\max_{(\boldsymbol{\nu},\boldsymbol{\lambda})\in\mathcal{H}} \left\{ -\boldsymbol{\nu}^\top \boldsymbol{d} - \sum_{i\in I} q_i y_i \lambda_i \right\}$$

with any feasible $\boldsymbol{y}$ and demand realization $\boldsymbol{d}$, to obtain an extreme point $(\widehat{\boldsymbol{\lambda}}^0, \widehat{\boldsymbol{\nu}}^0)$ of $\mathcal{H}$.

(ii) With a set L, we then solve the relaxed master problem (8.47)–(8.53) using the decomposition scheme (Algorithm 8.1). A solution $(\boldsymbol{y}^\star, \{\alpha_k^\star, \overline{\boldsymbol{\beta}}_k^\star, \underline{\boldsymbol{\beta}}_k^\star, \boldsymbol{\rho}_k^\star\}_{k\in K})$ is obtained.

(iii) We then check the optimality (gap) of $\boldsymbol{y}^\star$ using inequalities (8.73) by solving $|K|$ separation subproblems (8.75)–(8.87) (or using the gap bound provided by (8.88)).

(iv) With the solutions of the $|K'|$ violated separation subproblems, we can form $|K'|$ initial BFSs of (8.89) as the extreme points $(\widehat{\boldsymbol{\lambda}}^k, \widehat{\boldsymbol{\nu}}^k)$, $k \in K'$ of $\mathcal{H}$, which can be added into L to induce new cuts that will enhance the relaxed master problem.

(v) The algorithm terminates when no violation occurs in inequality (8.73) or when a solution with an acceptable gap tolerance is found.

The overall procedure is formalized in the following Algorithm 8.2. In particular, the algorithm terminates in a finite number of steps (see [345]).

There are several aspects worth noticing regarding the nested decomposition procedure just discussed. The first regards the connection to the L-shaped algorithm for two-stage stochastic programming. In fact, when the support $\mathcal{D}_k$ reduces to a singleton $\hat{\boldsymbol{d}}_k$, the state-wise ambiguity set becomes

$$\mathcal{F} = \left\{ \widehat{\mathbb{P}}_K = \sum_{k\in K} p_k \delta_{\hat{\boldsymbol{d}}_k} \right\}.$$

As we have already noticed in this chapter, in this case, the RS-FLP reduces to a stochastic (capacitated) facility location problem under empirical distribution $\widehat{\mathbb{P}}_K$.

Algorithm 8.2: Nested Benders decomposition for the RS-FLP (8.41)

Data: Model parameters, gap tolerance ϵ.
Result: Facility location decision $\boldsymbol{y}^{\star\star}$.

1 Find an initial extreme point of $\mathcal{H}$, say $e_0 = (\hat{\boldsymbol{\nu}}^0, \hat{\boldsymbol{\lambda}}^0)$
2 $L \leftarrow \{e_0\}$.
3 Execute Algorithm 8.1 to solve the relaxed master problem (8.44) to obtain a location solution $\boldsymbol{y}^\star$.
4 Solve $|K|$ conic linear programs separately using (8.47)–(8.53) by setting $\boldsymbol{y} = \boldsymbol{y}^\star$ to obtain $(\alpha^\star, \overline{\boldsymbol{\beta}}_k^\star, \underline{\boldsymbol{\beta}}_k^\star, \boldsymbol{\rho}_k^\star)_{k\in K}$;
5 Solve the separation subproblem (8.75)–(8.87) to obtain Ω_k and extreme point $(\widehat{\boldsymbol{\nu}}^k, \widehat{\boldsymbol{\lambda}}^k)$ for each $k \in K$;
6 **if** $\sum_{k\in K} p_k \left(\Omega_k - \alpha_k^\star\right) \leq \epsilon$ **then**
7     $\boldsymbol{y}^{\star\star} \leftarrow \boldsymbol{y}^\star$;
8   STOP.
9 $L \leftarrow L \cup \{(\widehat{\boldsymbol{\nu}}_k, \widehat{\boldsymbol{\lambda}}_k) \mid \alpha_k^\star < \Omega_k, k \in K\}$;
10 Enhance the relaxed master problem (8.47)–(8.53) with L and go to 3;

Interestingly, in this case, the above solution approach reduces to the classical L-shaped algorithm for two-stage stochastic programming [78]. The resulting separation problem solves $|K|$ second-stage linear programs given $\boldsymbol{y}$.

Furthermore, we note that the set K may also correspond to the samples in the robust stochastic case (e.g., in Example 8.3 of using the ∞-Wasserstein ball for defining $\mathcal{F}$). Therefore, the proposed solution procedure inherits the powerful decomposability of the L-shaped algorithm to deal with our broad class of robust stochastic capacitated location problems.

On the subgradient cuts with $\boldsymbol{y} \in \mathcal{Y}$, it is noted that the number of iterations of the algorithm depends on the number $|E|$ of extreme points of the second-stage dual feasible set $\mathcal{H}$ and the number of the binaries $\boldsymbol{y}$. The former is implicit; the latter is explicit. Therefore, when solving the relaxed master problem, we may "manually" add some "good" location decisions $\boldsymbol{y}$ (as the cuts) to improve the convergence. Additionally, the subgradient decomposition approach can also be extended to the other two-stage optimization problems with continuous first-stage decisions, e.g., $\mathcal{Y}$ is a polyhedron, which leads to a continuous conic linear relaxed master problem (8.47)–(8.53).

Finally, in [345] the authors perform a series of computational experiments that justify using state-wise distributional information incorporated in the capacitated facility location problem considered throughout this chapter. The same authors also assess the performance of the described nested Benders decomposition and its applicability in realistic instances.

8.6 Conclusions and Further Reading

In this chapter, we have introduced base concepts related to distributionally robust optimization in the context of facility location problems. An extension of the classical capacitated facility location problem was used to discuss and illustrate many challenges in the field. Different modeling aspects related to how uncertainty is captured were analyzed. These included a data-driven DRO model based on the Wasserstein ambiguity set, a decision-dependent DRO model, and a DRO model using the state-wise ambiguity set. An exact algorithm was detailed with different implications. In particular, it extends to the well-known L-shaped algorithm for two-stage stochastic programming.

The topic addressed in this chapter is still recent. Some notable cases should be highlighted.

Additional insights on data-driven distributionally robust facility location can be found in [468, 499, 576]. Further details on decision-dependent distributionally robust facility location can be found in [47], and a deeper discussion on state-dependent distributionally robust facility location can be found in [125] and [345].

Distributionally robust chance-constrained facility location intertwines the contents of this and the previous chapter. Some work in that direction includes [342, 573].

Other works that complement this chapter include [244, 249, 342, 344, 496, 497]. In Chap. 15, we also go beyond the content of the current chapter by introducing a distributionally robust multi-period facility location model based on the time series Wasserstein ambiguity set mentioned.

Part III
Extended Models

Chapter 9
Location-Inventory Problems

Abstract In this chapter, inventory and facility location decisions are intertwined. Different problems, models, and algorithms are discussed. The impact of risk pooling in optimal decisions is also considered. After introducing several base concepts related to the location-inventory problems, continuous review policies are analyzed separately for uncapacitated and capacitated facilities. A periodic-review model with extensions is also presented. Demand is the usual source of uncertainty, assumed to be represented by a random variable with some given distribution.

Keywords Continuous review · Periodic review · Risk pooling

9.1 Introduction

Facility location problems are prone to integrate other decisions, thus seeking solutions better than those obtained if the decisions were made separately. One relevant stream regards location-inventory problems, in which inventory decisions are intertwined with facility location. This is an area in logistics and supply chain management where several decisions are holistically tackled.

Work on location-inventory problems has been developed since the 1990s. Some such work seeks to find the inventory costs in a preprocessing stage before considering a classical facility location problem reflecting those costs (e.g., in the fixed setup costs for the facilities). This is the case in [410, 411], where a linear approximation for the safety stock costs is considered. This is embedded in the fixed costs for the facilities.

In other cases, the parameters are deterministic. In [45] the authors consider a multi-product setting and a fixed cost for storing a product at a distribution center. They develop a column-generation-based algorithm. Teo and Shu [521] investigate the problem of finding the number and location of warehouses to set up, how to serve the retailers using these warehouses, and to determine the optimal inventory policies for the warehouses and retailers. The objective is to minimize the total multi-echelon inventory, transportation, and facility location costs. This seems to be the first work

© The Author(s), under exclusive license to Springer Nature Switzerland AG 2024

F. Saldanha-da-Gama, S. Wang, *Facility Location Under Uncertainty*,
International Series in Operations Research & Management Science 356,
https://doi.org/10.1007/978-3-031-55927-3_9

considering a two-echelon inventory cost function arising from the coordination of replenishment activities between the warehouses and the retailers. The authors develop a column generation procedure.

In this chapter, we are interested in problems integrating facility location and inventory decisions, with uncertainty playing a central role. An important aspect motivating this integration regards the need to seek a trade-off between the costs of locating the facilities and the inventory costs. The latter typically favor the selection of a few facilities, thus taking advantage of economies of scale in the cycle stock costs leading to the so-called risk polling effect (see, e.g., [189]). However, opening several facilities may call for a trade-off to be sought.

In the upcoming sections, we explore this topic by discussing models, solution techniques, and applications. The sources of uncertainty usually at stake are the demand and the lead time. We assume that the reader has some fundamental background in inventory theory. Nevertheless, an effort is made to ensure that the chapter is self-contained. Still, the interested reader can refer to [461, 511, 579].

The remainder of this chapter is organized as follows. In Sect. 9.2, several preliminary aspects and assumptions related to inventory theory are reviewed to lay the ground for the discussion to follow. Section 9.3 covers risk pooling as an unavoidable aspect when intertwining facility location decisions and inventory management. Section 9.4 introduces capacitated facilities in location inventory. In all the sections, continuous review models are discussed. Section 9.5 considers periodic-review location-inventory problems. The chapter ends with some conclusions and directions for further reading.

9.2 Preliminaries and Assumptions

In this chapter, we are assuming that inventory management is being conducted at a set of facilities (e.g., warehouses, or distribution centers) to satisfy the demand of a set of customers, J. One single stable item is considered.

Demand is assumed to be stochastic and stationary. Thus, it can be represented by a random variable with some known cumulative distribution function. The demand observed at a facility results from the demands of customers that such facility must supply, i.e., single allocation is assumed for the customers to the facilities.

Apart from Sect. 9.5, a continuous review policy is adopted at the facilities. In particular, a so-called (r, Q) policy is sought at each facility, indicating that when the stock level drops to r, a batch/lot of size Q is ordered or produced. In the former case (ordering), we assume that facilities are supplied from a single supplier. The inventory costs to be accounted for at each facility are: (i) acquisition or production costs, (ii) the holding costs (for the average inventory level over some time), and (iii) the safety stock kept to hedge against shortages due to uncertain demand. A service level type 1 is assumed: A maximum threshold is imposed for the probability of observing a stockout during the lead time.[1]

For each facility $i \in I$, the following parameters are considered:

b_i cost for each unit ordered by the facility;
K_i fixed ordering cost at facility i;
h_i per unit holding cost per unit time;
l_i lead time for orders issued by facility i;
p_i shortage cost per unit missing irrespective of the shortage time;
α service level.

A central quantity for our analysis is the demand observed at each facility during the lead time. This is a random quantity since the customers' demand also is. For each $i \in I$, let φ_i denote that quantity and $F_i(.)$ its CDF that we assume to be continuous. Let $\nu_i = \mathbb{E}[\varphi_i]$. Note that $\nu_i = l_i \mathbb{E}[D_i]$, with D_i standing for the demand per unit time observed at facility i. Finally, we assume the safety stock at facility $i \in I$ is given by $r_i - \nu_i$. This form is often assumed in the literature, with r_i denoting the reorder point at facility $i \in I$ (to be determined). Note, however, that the specific way to compute the safety stock often varies. The interested reader can refer to [241] for further details and additional references.

For $i \in I$, denote by Q_i the batch size at facility i. We seek to find the values of Q_i and r_i, $i \in I$ (and possibly others) minimizing the total cost per time unit.

Let $\eta_i(r_i)$ be the expected outstanding amount per cycle at facility $i \in I$. Given that the (expected) length of a cycle is given by $\frac{Q}{\mathbb{E}[D_i]}$, the total cost per cycle results from summing three components:

Acquisition cost: $K_i + b_i Q_i$
Holding cost: $h\left(\frac{Q}{2} + r_i - \nu_i\right)\frac{Q}{\mathbb{E}[D_i]}$
Shortage cost: $p_i \times \eta_i(r_i)$

Hence, the total cost per cycle is given by

$$K_i + b_i Q_i + h\left(\frac{Q_i}{2} + r_i - \nu_i\right)\frac{Q_i}{\mathbb{E}[D_i]} + p_i \times \eta_i(r_i).$$

Under our assumptions, we can write $\eta_i(r_i)$ explicitly as follows:

$$\eta_i(r_i) = \int_{r_i}^{\infty} (x - r_i) f_i(x) dx, \tag{9.1}$$

with $f_i(.)$ standing for the density function of the demand during the lead time at facility $i \in I$. The above expression can be computed exactly in some cases (e.g.,

[1] A service level type 2 would involve a fill rate proportion—demand satisfied from the on-hand stock.

demand during the lead time is described by a Gaussian distribution). In practice, the upper limit of the integral is finite due to a limited ordering capacity.

Finally, the cost per unit time is obtained by dividing the cost per cycle by the length of a cycle ($Q/\mathbb{E}[D_i]$):

$$\Psi(Q_i, r_i) = \frac{K_i \mathbb{E}[D_i]}{Q_i} + b_i \mathbb{E}[D_i] + h\left(\frac{Q_i}{2} + r_i - \nu_i\right) + \frac{p_i \eta_i(r_i)\mathbb{E}[D_i]}{Q_i}.$$

Given that we are imposing a service level type 1, α, i.e., we are setting $\alpha = \mathbb{P}[\varphi_i \leq r_i]$, we conclude that $r_i^* = F_i^{-1}(\alpha)$. Consequently, we can now easily derive the optimal batch size at facility $i \in I$ is

$$Q_i^* = \sqrt{\frac{2(K_i + p_i \eta_i(r_i^*))\mathbb{E}[D_i]}{h_i}}$$

and the optimal cost per time unit is (recall that $\nu_i = l_i \mathbb{E}[D_i]$)

$$\begin{aligned}\Psi(Q_i^*, r_i^*) = &\sqrt{\frac{h_i K_i^2 \mathbb{E}[D_i]}{2(K_i + p_i \eta_i(r_i^*))}} + b_i \mathbb{E}[D_i] \\ &+ h_i \left(\sqrt{\frac{(K_i + p_i \eta_i(r_i^*))\mathbb{E}[D_i]}{2h_i}} + r_i^* - l_i \mathbb{E}[D_i]\right) \\ &+ \sqrt{\frac{h_i p_i^2 \eta_i^2(r_i^*)\mathbb{E}[D_i]}{2[K_i + p_i \eta_i(r_i^*)]}}.\end{aligned} \tag{9.2}$$

Although the above expression is involved, some particular cases of interest emerge. The first regards shortages leading to lost demand without penalty. In this case, we have $p_i = 0$, $i \in I$, and the above expression reduces to

$$\Psi(Q_i^*, r_i^*) = \sqrt{2h_i K_i \mathbb{E}[D_i]} + b_i \mathbb{E}[D_i] + h_i \left(r_i^* - l_i \mathbb{E}[D_i]\right). \tag{9.3}$$

Furthermore, if the demand observed at facility i per unit time, D_i, is distributed according to a Gaussian distribution with parameters $\mathbb{E}[D_i]$ and $\mathbb{V}[D_i]$ then the demand during the lead time is described by a Gaussian distribution with parameters $\nu_i = \mathbb{E}[D_i] \times l_i$ and $\varsigma_i^2 = \mathbb{V}[D_i] * l_i$. Accordingly,

$$r_i^* = \nu_i + \varsigma_i z_\alpha$$

with z_α denoting the α-quantile of a standardized Gaussian distribution. With this additional assumption, we obtain

$$\Psi(Q_i^*, r_i^*) = \sqrt{2h_i K_i \mathbb{E}[D_i]} + b_i \mathbb{E}[D_i] + h_i z_\alpha \sqrt{l_i \mathbb{V}[D_i]}, \quad i \in I. \tag{9.4}$$

We have just gathered the main elements from inventory theory that allow us to proceed. From the perspective of facility location, our starting point is the uncapacitated facility location problem whose formulation we recall from Sect. 2.1:

$$\text{minimize} \quad \sum_{i \in I} f_i y_i + \sum_{i \in I} \sum_{j \in J} c_{ij} d_j x_{ij}, \tag{9.5}$$

$$\text{subject to} \quad \sum_{i \in I} x_{ij} = 1 \qquad \forall j \in J, \tag{9.6}$$

$$x_{ij} \leq y_i \qquad \forall i \in I, \tag{9.7}$$

$$x_{ij} \in \{0, 1\} \qquad \forall i \in I,\ j \in J, \tag{9.8}$$

$$y_i \in \{0, 1\} \qquad \forall i \in I. \tag{9.9}$$

As customary in this volume, for $i \in I$, y_i is a binary variable equal to one if and only if a facility is installed at i. The variable x_{ij} is also binary, equal to one if and only if demand node $j \in J$ is allocated to facility i. The cost f_i is the fixed costs (per time unit) for having a facility operating at $i \in I$, and c_{ij} is the unit shipping cost from facility $i \in I$ to customer $j \in J$. The demand of customer $j \in J$ is represented by d_j. The interpretation of the objective function and the constraints should be straightforward, given the contents of this volume so far. Note that given the single sourcing assumed, the second term of the objective function can be written more concisely. Nevertheless, we keep the above form due to its convenience regarding the inventory decisions to embed in the model.

Conceptually, embedding inventory costs can be accomplished by considering the following formulation:

$$\text{minimize} \quad \sum_{i \in I} f_i y_i + \sum_{i \in I} \sum_{j \in J} c_{ij} d_j x_{ij} + \Gamma(\boldsymbol{x}, \boldsymbol{y}), \tag{9.10}$$

$$\text{subject to} \quad (9.6)\text{–}(9.9),$$

with $\Gamma(\boldsymbol{x}, \boldsymbol{y})$ denoting the total costs for managing the inventory (purchasing, holding, and safety stock) in the entire system given that decision $\{\boldsymbol{x}, \boldsymbol{y}\}$ is adopted. The parameter d_j synthesizes in some way the (stochastic) demand. Typically, the expected value is adopted. We do the same in this chapter.

The challenge is to find a tractable expression for $\Gamma(\boldsymbol{x}, \boldsymbol{y})$. This depends on the specific setting considered, namely the underlying probability distributions.

From the preliminary analysis presented in the previous section, it becomes clear that the inventory costs depend on the actual demand assigned to each facility, e.g., to its expected value and variance. However, in our case, that results from a decision to make: facilities to open and assignment of the customers to those facilities. Therefore, the best we can do at this stage is to write the expected value and variance of the demand at each facility in each time unit as a function of our decision variables $\boldsymbol{x}$ and $\boldsymbol{y}$.

We denote by ξ_j the random variable representing the demand originated at customer $j \in J$, $\mu_j = \mathbb{E}[\xi_j]$, and $\sigma_j^2 = \mathbb{V}[\xi_j]$. Accordingly, the expected demand observed at each facility can be written as

$$\mathbb{E}[D_i] = \sum_{j \in J} \mu_j x_{ij} \quad \forall i \in I.$$

Ignoring the penalty costs for stockouts, from function (9.3), we obtain

$$\Gamma(\boldsymbol{x}, \boldsymbol{y}) = \sum_{i \in I} \left(\sqrt{2h_i K_i \sum_{j \in J} \mu_j x_{ij}} + b_i \sum_{j \in J} \mu_j x_{ij} \right.$$
$$\left. + h_i \left(F_i^{-1}(\alpha) - l_i \sum_{j \in J} \mu_j x_{ij} \right) \right). \tag{9.11}$$

Unfortunately, the above expression is worrisome. First, it is nonlinear. This is not necessarily a drawback if the function has some features, making it friendly from an optimization perspective. The challenge is the cumulative distribution function $F_i(.)$. Even if known, the characteristics of the family it belongs to (e.g., expectation and variance) depend on the decision to make (facilities to open and assignments to make). For this reason, the literature has focused on some specific cases. In fact, from this stage onward, the developments we can make depend on the exact setting being investigated.

We conclude this section by noting that if customer demands are independent, then given an assignment decision $\boldsymbol{x}$, we can also write the variance of the demand observed at each facility as

$$\mathbb{V}[D_i] = \sum_{j \in J} \sigma_j^2 x_{ij} \quad \forall i \in I.$$

9.3 Risk Pooling

A motivation for embedding inventory decisions in a facility location problem is to capture the risk-pooling effect. This is usually sought in terms of safety stocks.

We start by considering the case in which the facilities can be supplied from several capacitated suppliers. Nevertheless, we assume that the fixed ordering cost at a facility is constant irrespective of the number and location of the plants supplying it. In addition to the notation already introduced, we consider:

K set of existing plants that are to supply the DCs;
b_{ik} unit cost for supplying DC $i \in I$ from plant $k \in K$;
q_k supplying capacity at plan $k \in K$.

All the quantities are per unit time (e.g., month or year). We also consider transportation variables from plants to facilities:

w_{ik} = amount to be shipped from plan $k \in K$ to facility $i \in I$.

The problem can be formulated as follows:

$$\text{minimize} \quad (9.10),$$
$$\text{subject to} \quad (9.6)\text{–}(9.9),$$

$$\sum_{i \in I} w_{ik} \le q_k \qquad \forall k \in K, \tag{9.12}$$

$$\sum_{k \in K} w_{ik} = \sum_{j \in J} \mu_j x_{ij} \quad \forall i \in I, \tag{9.13}$$

$$w_{ik} \ge 0 \qquad \forall i \in I,\ k \in K. \tag{9.14}$$

What remains to define is the function $\Gamma(\boldsymbol{x}, \boldsymbol{y})$ that now is slightly different from (9.11) due to the explicit supplying decisions we are considering. We keep assuming that the demands are independent. Moreover, the demand observed at each facility during the lead time is assumed to be described by a Gaussian distribution. In this case, we can explicitly write the inventory costs (cycle stock, acquisition, and safety stock):

$$\Gamma(\boldsymbol{x}, \boldsymbol{y}) = \sum_{i \in I} \left(\sqrt{2 h_i K_i \sum_{j \in J} \mu_j x_{ij}} + \sum_{k \in K} b_{ik} w_{ik} + h_i \left(F_i^{-1}(\alpha) - l_i \sum_{j \in J} \mu_j x_{ij} \right) \right). \tag{9.15}$$

The above model is defined in a discrete context, i.e., the potential locations for the facilities are finite and given in advance. Nevertheless, it can be easily adapted to other settings, such as the one introduced in [190]. In that work, the authors assume a setting similar to the above one. However, each demand node lies at the center of a unit-length square. The set of squares (each of which specifies one customer) defines a grid. The distribution centers can be located anywhere in this grid. Such locations are to be determined.

Also in [190] the authors assume all the setup costs f_i to be the same and thus equal to a value f. Second, they consider per unit commodity and per unit distance transportation costs. We denote such unit costs by b and c for shipments between plants and facilities and between facilities and customers, respectively. Therefore, we take

$$b_{ij} = b \times w_{ik} \times \ell_{ik} \quad \forall k \in K,\ i \in I,$$

and

$$c_{ij} = c \times \mu_j \times \ell_{ij} \quad \forall i \in I,\ j \in J.$$

In these expressions, ℓ denotes the adequate distance between the two nodes indicated. Rectilinear distances are considered in [190]. The above model can also be fine-tuned to other settings of interest. Note that in [190] problem, the coordinates of the facilities are to be determined. This fact makes the formulation highly nonlinear. We note that the component of the safety stock costs presented in that paper seems different from what we are discussing. However, that is not the case. Those authors use the variance of the entire demand in the system during the lead time and take as the variance observed at facility i the proportion of demand that goes to that facility. They are proposing an estimate for the variance of the demand observed at a facility. In the expression we have presented, it is assumed that such an estimate was already obtained. Thus, we work directly with $\mathbb{V}[D_i]$ for facility $i \in I$.

Erlebacher and Meller [190] develop an approximation scheme. By simplifying some assumptions, they derive a model seeking a good indication of the number of facilities to open. Once this is accomplished, they look for upper bounds on that number and thus end up with a reasonable range of values for the number of facilities to open that they consider further in their analysis. Finally, they propose heuristics to assign customers to the facilities. The interested reader can resort to that work for all details.

Teo et al. [520] investigate a location-inventory problem like the one that we generically defined as consisting of minimizing (9.10) subject to (9.6)–(9.9), but they ignore the transportation costs. Their objective function becomes $\sum_{i \in I} f_i y_i + \Gamma(\boldsymbol{x}, \boldsymbol{y})$. In this case, they take the long-run average inventory costs (see, e.g., [220]) and thus assume

$$\Gamma(\boldsymbol{x}, \boldsymbol{y}) = \sum_{i \in I} \Gamma_i(\boldsymbol{x}, \boldsymbol{y})$$

with

$$\Gamma_i(\boldsymbol{x}, \boldsymbol{y}) = \min_{r_i, Q_i} \frac{g_i(Q_i) \sum_{j \in J} \mu_j x_{ij} + \int_{r_i}^{r_i + Q_i} \mathbb{E}[h_i(s - \varphi_i(\boldsymbol{x}))^+ + p_i(\varphi_i(\boldsymbol{x}) - s)^+] ds}{Q_i},$$

$i \in I$ and $(a)^+ = \max\{0, a\}$. The function $g_i(Q_i)$ gives the unit acquisition cost at facility $i \in I$ if a quantity Q_i is ordered, and $\varphi_i(\boldsymbol{x})$ stands for the demand observed at facility $i \in I$ during the lead time given an assignment decision $\boldsymbol{x}$. Note that this model now includes shortage costs. Furthermore, it is assumed that the inventory level is uniformly distributed between r_i and $r_i + Q_i$ and that such a level is independent of the demand during the lead time.

In [520] the authors prove that if the demands $\xi_1, \ldots, \xi_{|J|}$ are independent and identically distributed, then there is an optimal solution to the problem in which all demand nodes are consolidated at a single facility. This is interesting even though, in practice, the demand distributions are not identical. Nevertheless, as pointed out by the same authors, the result still applies if the demands are generated according to Poisson processes, not necessarily with the same mean rate. This is a relevant detail because, in a general setting, demands are generated from many identical individual points. Thus, a Poisson process provides a good approximation for demand generation. Hence, the above theorem may be helpful.

In the general case, the expression defining the optimal inventory cost is much involved. Also in [520] the authors derive a bound on that cost considering tractable functions. Considering such bounds, the authors show that consolidation at a single facility can be infinitely worse than the optimal decentralized solution. Furthermore, they derive an algorithm for the general case that finds a distribution strategy with an inventory cost within $\sqrt{2}$ of the optimal one. Despite the intricate proofs (the interested reader can refer to that paper for all the details), the results are easy to follow. First, the authors show that

$$\sum_{i\in I}\left(\min_{Q_i}\left(\frac{g_i(Q_i)\sum_{j\in J}\mu_j x_{ij}}{Q_i}+\frac{h_i p_i}{h_i+p_i}\frac{Q_i}{2}\right)+\Upsilon_i(\boldsymbol{x})\right) \tag{9.16}$$

with $\Upsilon_i(\boldsymbol{x}) = \min_\zeta \mathbb{E}[h_i(\zeta - \varphi_i(\boldsymbol{x})) + p_i(\varphi_i(\boldsymbol{x}) - \zeta)]$ is an upper bound on $\Gamma(\boldsymbol{x}, \boldsymbol{y})$. Then, they prove that an optimal solution to the problem that consists of minimizing (9.16) subject to (9.6)–(9.9) is feasible to the original problem, and its value provides a $\sqrt{2}$ approximation to the optimal value of the problem.

Although finding a closed form for $\Upsilon_i(\boldsymbol{x})$ may not be easy in general, it can be derived in some cases. This holds when the demands ξ_j are independent and distributed according to a Gaussian distribution. Then,

$$\Upsilon_i(\boldsymbol{x}) = \sigma\sqrt{l_i}\min_\zeta \mathbb{E}\left[h_i\left(\zeta-\frac{\varphi(\boldsymbol{x})-\mu l_i}{\sigma\sqrt{l_i}}\right)^+ + p_i\left(\frac{\varphi(\boldsymbol{x})-\mu l_i}{\sigma\sqrt{l_i}}-\zeta\right)^+\right],$$

with $\mu = \sum_{j\in J, x_{ij}=1}\mu_j$ and $\sigma = \sqrt{\sum_{j\in J, x_{ij}=1}\sigma_j^2}$. In this case, the upper bound problem can be reformulated as a concave partitioning problem: Find a partition $J_1, \ldots, J_{|I|}$ of J that minimizes the function

$$\sum_{i\in I}\gamma_i\Big(\sum_{j\in J_i}\mu_j, \sum_{j\in J_i}\sigma_j^2\Big),$$

with

$$\gamma_i(a,b) = f_i\delta(a) + \min_{Q_i}\left(\frac{ag_i(Q_i)}{Q_i} + \frac{h_i p_i}{h_i+p_i}\frac{Q_i}{2}\right) + \kappa_i\sqrt{b},$$

$$\kappa_i = \sqrt{l_i}\min_{\zeta}\mathbb{E}\left[h_i\left(\zeta - \frac{\varphi(\boldsymbol{x}) - \mu l_i}{\sigma\sqrt{l_i}}\right)^+ + p_i\left(\frac{\varphi(\boldsymbol{x}) - \mu l_i}{\sigma\sqrt{l_i}} - \zeta\right)^+\right],$$

$i \in I$, and $\delta(a)$ equal to one if $a > 0$ and zero otherwise.

This partitioning problem and extensions have been studied in the literature (e.g., [30, 109, 218]). Chakravarty et al. [109] show that if $\gamma_i(a,b)$ is concave in a and b for all $i \in I$, and if $\mu_1/\sigma_1^2 \leq \mu_2/\sigma_2^2 \leq \cdots \leq \mu_{|J|}/\sigma_{|J|}^2$, then there is an optimal partition of J such that the sets J_i are consecutive, i.e., if $j', j'' \in S_i$, $j' < j''$, and if $j' < \hat{j} < j''$, then $\hat{j} \in S_i$.

Theorem 9.1 *Let* $\theta(a,b) = \min_{i\in I}\gamma_i(a,b)$. *Then*

$$\min_{J_1,\ldots,J_n}\sum_{i\in I}\gamma_i\left(\sum_{j\in J_i}\mu_j, \sum_{j\in J_i}\sigma_j^2\right) = \min_{J_1,\ldots,J_n}\sum_{i\in I}\theta\left(\sum_{j\in J_i}\mu_j, \sum_{j\in J_i}\sigma_j^2\right),$$

and there is a consecutive optimal partition (in the sense above described) to the right-hand side minimization problem.

Proof See [520]. □

Using the above result, the optimal partition of customers can be obtained by determining the shortest path between two nodes in a specific graph. Indeed, an oriented graph can be considered with $|J| + 1$ nodes as depicted in Fig. 9.1. An arc is considered between every pair of nodes j' and j'' such that $j' < j''$. The corresponding cost is set equal to

$$\text{cost}_{(j',j'')} = \theta\left(\sum_{\ell=j'+1}^{j''}\mu_\ell, \sum_{\ell=j'+1}^{j''}\sigma_\ell^2\right) = \min_{i\in I}\gamma_i\left(\sum_{\ell=j'+1}^{j''}\mu_\ell, \sum_{\ell=j'+1}^{j''}\sigma_\ell^2\right).$$

The above results show that although the location-inventory problem we are discussing can be formulated in a tractable form under some assumptions, it remains a challenging problem. Two complicating aspects are: (i) the risk-pooling effect calling for a consolidation of the customer demands at the facilities and (ii) the inclusion of the shortage costs. As we observed in Sect. 9.2, the inventory cost structure simplifies significantly when the penalty for shortages is equal to zero, i.e., the demand is lost with no additional cost incurred.

Daskin et al. [155] and Shen et al. [501] introduce a so-called location model with risk pooling. They consider a location-inventory problem, ignoring the shortage

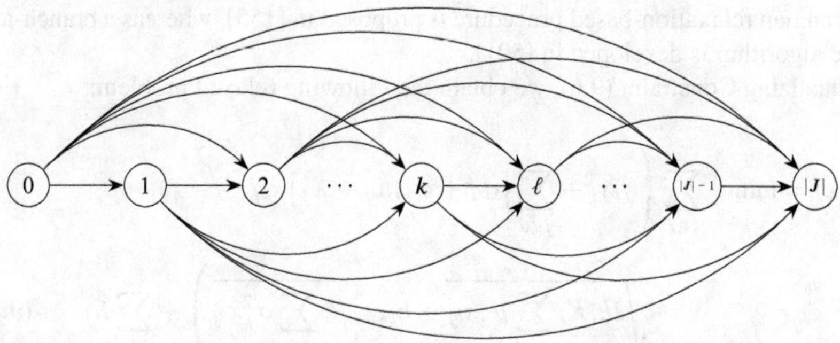

Fig. 9.1 Graph for determining the optimal consecutive partition of customers

costs but considering transportation costs between facilities and customers. They assume that each facility has a single supplier. Thus, the information regarding the latter can be omitted.

The investigated problem consists of minimizing (9.10) subject to (9.6)–(9.9) with $\Gamma(\boldsymbol{x}, \boldsymbol{y})$ given by (9.11). Assuming that demands $\xi_1, \ldots, \xi_{|J|}$ are independent and Gaussian distributed, we can take advantage from (9.4). The problem can be formulated as follows:

$$\text{minimize} \quad \sum_{i\in I} f_i y_i + \sum_{i\in I}\sum_{j\in J} c_{ij}\mu_j x_{ij} + \sum_{i\in I}\left(\sqrt{2h_i K_i \sum_{j\in J}\mu_j x_{ij}} + b_i \sum_{j\in J}\mu_j x_{ij} + h_i z_\alpha \sqrt{l_i \sum_{j\in J}\sigma_j^2 x_{ij}}\right) \tag{9.17}$$

subject to (9.6)–(9.9).

The objective function can be re-written as follows:

$$\sum_{i\in I}\left(f_i y_i + \sum_{j\in J}(b_i + c_{ij})\mu_j x_{ij} + \sqrt{2h_i K_i \sum_{j\in J}\mu_j x_{ij}} + h_i z_\alpha \sqrt{l_i \sum_{j\in J}\sigma_j^2 x_{ij}}\right). \tag{9.18}$$

Also in [155, 501] the authors further specify the objective function because they consider that the variable cost associated with an order at facility $i \in I$ is a function $a_i + b_i Q_i$ instead of $b_i Q$ as we are considering (see Sect. 9.2). Additionally, those authors assume exogenous weights for the transportation and inventory costs. This is a way to fine-tune such costs, particularly the fixed ordering costs. Nevertheless, the model is essentially the same as the above one. Hence, we proceed with the analysis without changing it. For solving the model (or a similar one), a

Lagrangian relaxation-based procedure is proposed in [155], whereas a branch-and-price algorithm is developed in [501].

Dualizing Constraint (9.6), we obtain the following relaxed problem:

$$\mathcal{L}(\boldsymbol{\lambda}) = \min \quad \sum_{i \in I} \left(f_i y_i + \sum_{j \in J} \left[(b_i + c_{ij})\mu_j - \lambda_j \right] x_{ij} + \sqrt{2h_i K_i \sum_{j \in J} \mu_j x_{ij}} + h_i z_\alpha \sqrt{l_i \sum_{j \in J} \sigma_j^2 x_{ij}} \right) + \sum_{j \in J} \lambda_j, \tag{9.19}$$

$$\text{s. t.} \quad (9.7)–(9.9).$$

This problem can be decomposed into $i \in I$ as for the classical uncapacitated facility location problem (see [209]), and we only need to focus on the problems:

$$\mathcal{L}_i(\boldsymbol{\lambda}) = \min \quad \sum_{j \in J} \left[(b_i + c_{ij})\mu_j - \lambda_j \right] x_{ij} + \sqrt{2h_i K_i \sum_{j \in J} \mu_j x_{ij}} + h_i z_\alpha \sqrt{l_i \sum_{j \in J} \sigma_j^2 x_{ij}}, \tag{9.20}$$

$$\text{s. t.} \quad x_{ij} \in \{0, 1\} \qquad \forall j \in J, \tag{9.8}$$

for $i \in I$. For each $i \in I$, we face a concave integer minimization problem. This problem appears as a pricing problem in a branch-and-price algorithm proposed in [505] that we discuss later. For a fixed i the problem can be solved as follows.

For each $S \subseteq J$, let

$$\mathcal{G}_i(S) = \sum_{j \in S} \left[(b_i + c_{ij})\mu_j - \lambda_j \right] + \sqrt{2h_i K_i \sum_{j \in S} \mu_j} + h_i z_\alpha \sqrt{l_i \sum_{j \in S} \sigma_j^2}.$$

Then, the above concave integer minimization problem can be reformulated as

$$\min_{S \subseteq J} \mathcal{G}_i(S). \tag{9.21}$$

Shu et al. [505] show that given an $i \in I$, there exists an optimal solution $S^* \subseteq J$ to (9.21) such that $(b_i + c_{ij})\mu_j - \lambda_j < 0 \quad \forall j \in S^*$. The proof stems from the fact that given an optimal solution to the problem that does not satisfy the inequality, it is possible to modify it to verify the condition without deteriorating the objective function value. This property restricts the search for the best subset $S \subseteq I$ to the customers in $J^- = \{j \in J \mid (b_i + c_{ij})\mu_j - \lambda_j < 0\}$.

Hereafter, we simplify the notation defining

$$\varpi_S = \sum_{j \in S} \left[(b_i + c_{ij})\mu_j - \lambda_j \right],$$

$$\rho_S = 2h_i K_i \sum_{j \in S} \mu_j,$$

$$\varrho_S = h_i^2 z_\alpha^2 l_i \left(\sum_{j \in S} \sigma_j^2 \right)^2,$$

$$H(\varpi_S, \rho_S, \varrho_S) = \varpi_S + \sqrt{\rho_S} + \sqrt{\varrho_S}.$$

The set $\{(\varpi_S, \rho_S, \varrho_S) \mid S \subseteq J^-\}$ is finite. Hence, its convex hull is a convex polyhedron that we denote by $\mathcal{H}$. Moreover, since $H(\varpi_S, \rho_S, \varrho_S)$ is a concave function in $\varpi_S, \rho_S, \varrho_S$, the problem (9.21) has a solution that is an extreme point of $\mathcal{H}$. Several procedures have been proposed in the literature to solve such a concave minimization problem (see, e.g., [513]). Shu et al. [505] develop a polynomial time algorithm (with complexity $O(|J|^2 \log |J|)$) for finding the optimal subset S^*. They define

$$\chi_j = -\frac{\mu_j}{(b_i + c_{ij})\mu_j - \lambda_j}, \quad \phi_j = -\frac{\sigma_j^2}{(b_i + c_{ij})\mu_j - \lambda_j}, \quad \forall j \in J^-,$$

and show that Algorithm 9.1 finds the optimal S^*.

The value of $\mathcal{L}_i(\boldsymbol{\lambda})$ for all $i \in I$ allows solving the Lagrangian problem for a fixed vector of multipliers as customarily done for the UFLP: We add the fixed-charge costs f_i to the optimal value of the Lagrangian subproblem associated with facility $i \in I$ thus obtaining the value of using candidate location i in this Lagrangian solution. Then, we select the locations with a negative value and fix them open. If all values are non-negative, we select a single facility, namely the one yielding the smallest value. Our goal is to solve the Lagrangian dual: $\max_{\boldsymbol{\lambda}} \mathcal{L}(\boldsymbol{\lambda})$. A standard subgradient algorithm can be used. It requires an upper bound at each iteration. Such bound can be constructed in a similar way as proposed in [155, 505]:

- Set open the facilities selected in the Lagrangian solution.
- Reassign each demand node assigned to at least one open facility in the Lagrangian solution to the facility that ensures a minimum increase in the cost given the allocation made so far.
- Assign each demand node not assigned to any open facility in the current Lagrangian solution to the facility that ensures a minimum increase in the cost given the allocation made so far.

Note that the final upper bound obtained can still be improved (e.g., using local search techniques).

Algorithm 9.1: Solving problem (9.21)

$V^* \leftarrow \infty$;
$S' \leftarrow \emptyset$;
for $\kappa_1 \in J^-$ **do**
$\hat{S} \leftarrow \emptyset$;
for $\kappa \in J^- \setminus \{\kappa_1\}$ **do**
if $\frac{\chi_\kappa - \chi_{\kappa_1}}{\phi_\kappa - \phi_{\kappa_1}} > 0 \wedge \chi_\kappa < \chi_{\kappa_1} \wedge \phi_\kappa < \phi_{\kappa_1}$ **then**
$S' \leftarrow S' \cup \{\kappa\}$;
else
if $\phi_\kappa = \phi_{\kappa_1} \wedge \chi_\kappa < \chi_{\kappa_1}$ **then**
$S' \leftarrow S' \cup \{\kappa\}$;
else
if $\frac{\chi_\kappa - \chi_{\kappa_1}}{\phi_\kappa - \phi_{\kappa_1}} \leq 0$ **then**
$\hat{S} \leftarrow \hat{S} \cup \{\kappa\}$;
if $\hat{S} \neq \emptyset$ **then**
Sort the elements of $\hat{S}$ and relabel the indices so that $\frac{\chi_1 - \chi_{\kappa_1}}{\phi_1 - \phi_{\kappa_1}} \leq \frac{\chi_2 - \chi_{\kappa_1}}{\phi_2 - \phi_{\kappa_1}} \leq \cdots \leq \frac{\chi_{|\hat{S}|} - \chi_{\kappa_1}}{\phi_{|\hat{S}|} - \phi_{\kappa_1}}$;
for $\kappa_2 \in \{1, 2, \ldots, |\hat{S}|\}$ **do**
$S'' \leftarrow \emptyset$; $S''' \leftarrow \{\kappa_2\}$;
for $\kappa \in \{1, 2, \ldots, |\hat{S}|\}$ **do**
if $\kappa \leq \kappa_2 \wedge \phi_\kappa - \phi_{\kappa_1} > 0$ **then**
$S'' \leftarrow S'' \cup \{\kappa\}$;
if $\kappa > \kappa_2 \wedge \phi_{\kappa_3} - \phi_{\kappa_1} < 0$ **then**
$S'' \leftarrow S'' \cup \{\kappa\}$;
if $\frac{\chi_\kappa - \chi_{\kappa_1}}{\phi_\kappa - \phi_{\kappa_1}} = \frac{\chi_{\kappa_2} - \chi_{\kappa_1}}{\phi_{\kappa_2} - \phi_{\kappa_1}}$ **then**
$S''' \leftarrow S''' \cup \{\kappa\}$;
if $\kappa < \kappa_2 \wedge \frac{\chi_\kappa - \chi_{\kappa_1}}{\phi_\kappa - \phi_{\kappa_1}} < \frac{\chi_{\kappa_2} - \chi_{\kappa_1}}{\phi_{\kappa_2} - \phi_{\kappa_1}} \wedge \phi_\kappa - \phi_{\kappa_1} > 0$ **then**
$S''' \leftarrow S''' \cup \{\kappa\}$;
if $\kappa > \kappa_2 \wedge \frac{\chi_\kappa - \chi_{\kappa_1}}{\phi_\kappa - \phi_{\kappa_1}} > \frac{\chi_{\kappa_2} - \chi_{\kappa_1}}{\phi_{\kappa_2} - \phi_{\kappa_1}} \wedge \phi_\kappa - \phi_{\kappa_1} < 0$ **then**
$S''' \leftarrow S''' \cup \{\kappa\}$;
$S \leftarrow S' \cup S''$;
if $\mathcal{G}_i(S) < V^*$ **then**
$V^* \leftarrow \mathcal{G}_i(S)$; $S^* \leftarrow S$;
$S \leftarrow S' \cup S'''$;
if $\mathcal{G}_i(S) < V^*$ **then**
$V^* \leftarrow \mathcal{G}_i(S)$; $S^* \leftarrow S$;
return S^*.

Shu et al. [505] formulate as a set covering problem the location-inventory problem we are discussing. Let S be a subset of customers. The cost of serving all customers in S by a facility open at $i \in I$ is immediate from (9.18):

$$C_{S,i} = f_i y_i + \sum_{j \in S} (b_i + c_{ij}) \mu_j + \sqrt{2 h_i K_i \sum_{j \in S} \mu_j} + h_i z_\alpha \sqrt{l_i \sum_{j \in S} \sigma_j^2}.$$

Hence, the cheapest cost for serving the customers in S from a single facility is given by $C_S^* = \min_{i\in I} C_{S,i}$.

Let $\mathcal{S}$ be the set of subsets of J. For each $S \in \mathcal{S}$, let Z_S be a binary variable equal to one if and only if subset S is in the solution. The location-inventory problem can now be formulated as follows:

$$\text{minimize} \quad \sum_{S\in\mathcal{S}} C_S^* Z_S, \tag{9.22}$$

$$\text{subject to} \quad \sum_{S\in\mathcal{S}:j\in S} Z_S \geq 1 \quad \forall j \in J, \tag{9.23}$$

$$Z_S \in \{0, 1\} \quad \forall S \in \mathcal{S}. \tag{9.24}$$

A restricted master problem is obtained by considering the linear relaxation of the model restricted to a subset of $\mathcal{S}$. Interestingly, the pricing problem (consisting of finding a subset S with a positive reduced cost) is the same as the Lagrangian subproblem (9.20), (9.8) (with the fixed cost added to the objective function). Thus, the pricing problem can be solved efficiently (see the above discussion).

A third approach for solving the location-inventory problem (9.18), (9.6)–(9.9) takes advantage from the recent advances in Conic Programming namely the fact that many optimization solvers can now handle second-order conic programming problems very efficiently.

A challenge in our problem concerns the nonlinear terms of (9.18). Let us define two sets of auxiliary variables as follows:

$$W_{1i} = \sqrt{\sum_{j\in J} \mu_j x_{ij}} \quad \forall i \in I;$$

$$W_{2i} = \sqrt{\sum_{j\in J} \sigma_j^2 x_{ij}} \quad \forall i \in I.$$

In this case, the problem can be reformulated as follows:

$$\text{minimize} \quad \sum_{i\in I} \left(f_i y_i + \sum_{j\in J} (b_i + c_{ij}) \mu_j x_{ij} + \sqrt{2h_i K_i} W_{1i} + h_i z_\alpha \sqrt{l_i} W_{2i} \right), \tag{9.25}$$

subject to (9.6)–(9.9),

$$\sum_{j\in J} \mu_j x_{ij}^2 \leq W_{1i}^2 \quad \forall i \in I, \tag{9.26}$$

$$\sum_{j \in J} \sigma_j^2 x_{ij}^2 \leq W_{2i}^2 \quad \forall i \in I, \tag{9.27}$$

$$W_{1i}, W_{2i} \geq 0 \qquad \forall i \in I. \tag{9.28}$$

Note that x-variables are binary and thus equal to their squared values. The above problem can be tackled directly by several general-purpose solvers. This fact was explored in [36] for solving this problem and an extension discussed later in this chapter.

9.3.1 Constant Variance-to-Mean Ratio

When considering the location-inventory problem that consists of minimizing (9.18) subject to (9.6)–(9.9), some authors explore the particular case in which the variance-to-mean ratio is similar for all demand nodes (see, e.g., [155, 501]). In fact, assuming that $\sigma_j^2/\mu_j = \eta \; \forall j \in J$ leads to simplifying the objective function (9.18), which writes as follows:

$$\sum_{i \in I} \left(f_i y_i + \sum_{j \in J} (b_i + c_{ij}) \mu_j x_{ij} + \left(\sqrt{2h_i K_i} + h_i z_\alpha \sqrt{l_i \eta} \right) \sqrt{\sum_{j \in J} \mu_j x_{ij}} \right). \tag{9.29}$$

The above simplification positively impacts the resolution of the problem. It simplifies Algorithm 9.1 considerably. This is relevant, either using a Lagrangian relaxation-based approach or a branch-and-price algorithm. In fact, under a variance-to-mean constant ratio, the Lagrangian subproblem for facility $i \in I$ becomes

$$\mathcal{L}_i(\lambda) = \min \quad \sum_{j \in J} \left[(b_i + c_{ij}) \mu_j - \lambda_j \right] x_{ij} + \sqrt{\sum_{j \in J} \left(2h_i K_i + h_i^2 (z_\alpha)^2 l_i \eta \right) \mu_j x_{ij}} \tag{9.30}$$

$$\text{s. t.} \quad x_{ij} \in \{0, 1\} \qquad \forall j \in J. \tag{9.8}$$

For each $j \in J$, define

$$A_j = (b_i + c_{ij}) \mu_j - \lambda_j, \quad B_j = \left(2h_i K_i + h_i^2 (z_\alpha)^2 l_i \eta \right) \mu_j.$$

The Lagrangian subproblem can be solved optimally by Algorithm 9.2 (see [155]).

Algorithm 9.2: Lagrangian subproblem resolution under a constant variance-to-mean ratio

---

Partition the set J into three subsets:
$J^+ = \{j \in J \mid A_j \geq 0\}$;
$J^0 = \{j \in J \mid A_j < 0 \wedge B_j = 0\}$;
$J^- = \{j \in J \mid A_j < 0 \wedge B_j > 0\}$;
Sort the elements of J^- so that $\frac{A_{[1]}}{B_{[1]}} \leq \frac{A_{[2]}}{B_{[2]}} \leq \cdots \leq \frac{A_{[|J^-|]}}{B_{[|J^-|]}}$;
$V^* \leftarrow \infty$;
for $\ell = 1$ **to** $|J^-|$ **do**
    $S^\ell \leftarrow J^0 \cup \{[1], \ldots, [m]\}$
    Compute the cost $C(S^\ell)$ is only the demand notes in S^ℓ are served from i
    if $C(S^\ell) < V^*$ **then**
        $V^* \leftarrow C(S^\ell)$; $\ell^* \leftarrow \ell$;
$S^* \leftarrow J^0 \cup \{[1], \ldots, [\ell^*]\}$;
return S^*. // Optimal set of demand nodes to be allocated to i.

9.4 Embedding Capacities

In the problems considered so far, facilities are uncapacitated. This is unrealistic in many applications.

For each facility $i \in I$, assume a storage capacity q_i (maximum amount that can be at the facility at any time). In what follows, we keep ignoring the shortage costs (i.e., assuming them equal to zero). A challenge when capacities are included in the problem is that the economic order quantities may render a solution infeasible. For this reason, the lot sizes Q_i must now be assumed as decision variables. We recall the general expression we obtained for the costs with the inventory at a facility $i \in I$:

$$\frac{K_i \mathbb{E}[D_i]}{Q_i} + b_i \mathbb{E}[D_i] + h\left(\frac{Q_i}{2} + r_i - \nu_i\right).$$

This includes acquisition, holding, and safety stock costs.

We keep imposing a service level type 1, α, i.e., setting $\alpha = \mathbb{P}[\varphi_i \leq r]$. Thus $r^* = F_i^{-1}(\alpha)$. We keep assuming the demand observed at facility i per time unit, D_i, is distributed according to a Gaussian distribution. Hence, the demand during the lead time is described by a Gaussian distribution with parameters $\mathbb{E}[D_i] \times l_i$ and $\mathbb{V}[D_i] * l_i$. Accordingly,

$$r_i^* = l_i \mathbb{E}[D_i] + z_\alpha \sqrt{l_i \mathbb{V}[D_i]},$$

with z_α denoting the α-quantile of a standardized Gaussian distribution. Therefore, the above cost becomes

$$\frac{K_i \mathbb{E}[D_i]}{Q_i} + b_i \mathbb{E}[D_i] + h\left(\frac{Q_i}{2} + z_\alpha \sqrt{l_i \mathbb{V}[D_i]}\right).$$

Given that the maximum amount observable at a facility occurs when the stock is replenished if the demand during the lead time is zero. Thus, the maximum observable amount at a facility is equal to the order quantity plus the safety stock:

$$Q_i + z_\alpha \sqrt{l_i \mathbb{V}[D_i]} + l_i \mathbb{E}[D_i].$$

This quantity cannot exceed q_i.

In our problem, the expected demand per time unit at a facility depends on how the demand nodes are allocated. As done before, we take $\mathbb{E}[D_i] = \sum_{j\in J} \mu_j x_{ij}$ and $\mathbb{V}[D_i] = \sum_{j\in J} \sigma_j^2 x_{ij}$. Finally, replacing the above expressions for costs and maximum inventory level, we can formulate the problem as follows:

$$\begin{aligned} \text{minimize} \quad & \sum_{i\in I} f_i y_i + \sum_{i\in I}\sum_{j\in J} c_{ij} x_{ij} \\ & + \sum_{i\in I} \left(\frac{K_i}{Q_i} \sum_{j\in J} \mu_j x_{ij} + b_i \sum_{j\in J} \mu_j x_{ij} + \frac{hQ_i}{2} + z_\alpha h \sqrt{l_i \sum_{j\in J} \sigma_j^2 x_{ij}} \right), \end{aligned} \tag{9.31}$$

$$\text{subject to} \quad Q_i + z_\alpha \sqrt{l_i \sum_{j\in J} \sigma_j^2 x_{ij}} + l_i \sum_{j\in J} \mu_j x_{ij} \le q_i y_i \quad \forall i \in I, \tag{9.32}$$

(9.6)–(9.9),

$$Q_i \ge 0 \qquad \forall i \in I. \tag{9.33}$$

The objective function can be re-written as follows:

$$\sum_{i\in I} \left(f_i y_i + \sum_{j\in J} \left(c_{ij} + b_i \mu_j\right) x_{ij} + \frac{K_i}{Q_i} \sum_{j\in J} \mu_j x_{ij} + \frac{hQ_i}{2} + z_\alpha h \sqrt{l_i \sum_{j\in J} \sigma_j^2 x_{ij}} \right). \tag{9.34}$$

A similar problem (apart from some weights given to the transportation and inventory costs) was investigated in [420], where a Lagrangian relaxation-based procedure for finding lower and upper bounds is proposed. Nevertheless, a relevant assumption in that work is that the demands occur according to a Poisson process, which allows setting their variance equal to the mean value. This allows a significant simplification in the above model.

Atamtürk et al. [36] reformulate the above problem as a second-order conic programming problem. The authors note that

$$\frac{K_i}{Q_i} \sum_{j\in J} \mu_j x_{ij} + \frac{hQ_i}{2} = \frac{h}{2} \left(\frac{2K_i}{h} \frac{\sum_{j\in J} \mu_j x_{ij}}{Q_i} + Q_i \right).$$

Thus, they replace the term inside parenthesis by a variable W_{1i} in the objective function. Furthermore, they consider the following auxiliary variables:

$$W_{2i} = \sqrt{\sum_{j \in J} \sigma_j^2 x_{ij}} \quad i \in I.$$

The new model becomes

$$\text{minimize} \quad \sum_{i \in I} \left(f_i y_i + \sum_{j \in J} (c_{ij} + b_i \mu_j) x_{ij} + z_\alpha h \sqrt{l_i} W_{2i} + \frac{h}{2} W_{1i} \right), \tag{9.35}$$

$$\text{subject to} \quad (9.6)–(9.9),\ (9.32),\ (9.33),$$

$$\sum_{j \in J} \sigma_j^2 x_{ij}^2 \leq W_{2i} \qquad \forall i \in I, \tag{9.36}$$

$$\frac{2K_i}{h} \frac{\sum_{j \in J} \mu_j x_{ij}}{Q_i} + Q_i \leq W_{1i} \quad \forall i \in I, \tag{9.37}$$

$$W_{1i}, W_{2i} \geq 0 \qquad \forall i \in I. \tag{9.38}$$

Constraint (9.37) can be re-written as

$$\frac{2K_i}{h} \sum_{j \in J} \mu_j x_{ij} + \left(Q_i - \frac{W_{1i}}{2} \right)^2 \leq \frac{W_{1i}^2}{4} \quad \forall i \in I. \tag{9.39}$$

This allows solving the problem using several general-purpose solvers. Nevertheless, the model can be enhanced by adding several valid inequalities derived from the capacity constraints. Let $i \in I$. Following some analysis already presented, for $S \subset J$, we define

$$\mathcal{F}_i(S) = z_\alpha \sqrt{l_i \sum_{j \in S} \sigma_j^2} + l_i \sum_{j \in S} \mu_j,$$

and

$$\mathcal{H}_i = \left\{ (x_{i1}, \ldots, x_{i|J|}) \in \{0, 1\}^{|J|} \mid z_\alpha \sqrt{l_i \sum_{j \in J} \sigma_j^2 x_{ij}} + l_i \sum_{j \in J} \mu_j x_{ij} \leq q_i \right\}.$$

A set $S \subset J$ is called a cover for $\mathcal{H}_i$ if $\mathcal{F}_i(S) > q_i$. In this case, the so-called cover inequality

$$\sum_{j \in J} x_{ij} \leq |S| - 1$$

is valid for $\mathcal{H}_i$. Atamtürk et al. [36] prove these results and suggest separating the cover inequalities heuristically following [37].

The above second-order conic programming model is quite powerful—it can be easily adapted to more complex settings. In the same paper [36] the authors show how to accommodate in the above model aspects such as correlated demands (the demand vector is assumed to be represented by a multidimensional Gaussian distribution), stochastic lead times, multiple commodities, and multiple commodities with correlated demands. The reader can refer to that paper for the changes to perform in the conic programming model. In all cases, the authors present computational tests using the branch-and-bound solver for second-order conic programming provided by IBM ILOG CPLEX.

Also related to the inclusion of capacities in a location-inventory problem, we refer to [381], where a probabilistic capacity constraint instead of planning for the absolute maximum inventory level is considered. The authors assume a similar setting as we have been discussing. However, they replace the capacity constraint (9.32) by probabilistic constraints, namely

$$\mathbb{P}[r_i - l_i D_i + Q_i \leq q_i] = \beta \quad \forall i \in I. \tag{9.40}$$

To ensure that the probabilistic constraint is satisfied, the authors replace the above constraint with the following deterministic one:

$$Q_i + (z_\alpha + z_\beta)\sqrt{l_i \mathbb{V}[D_i]} \leq q_i \quad \forall i \in I. \tag{9.41}$$

For the resulting problem, a Lagrangian relaxation-based procedure is developed in line with what we have already discussed.

9.5 Periodic-Review Location-Inventory Problems

So far, we have discussed location-inventory problems under a continuous review policy. Thinking of many applications in logistics and supply chain management that rely on discrete-time forecasting systems (e.g., weekly, monthly), we focus now on periodic-review policies. Thus, there are specific moments in time for replenishing the stock. In particular, there is an underlying planning horizon divided into several periods. A single commodity is considered. In each period, up to a given number of new facilities (e.g., distribution centers) can be installed. The operating facilities in each period can be supplied from a set of suppliers subject to a single-allocation pattern. In each period, the customers are supplied from a single facility. The supplier-facility and facility-customer assignments may change from one period to another. Backlogging is allowed—upon the occurrence of a stockout, the missing demand can be supplied in a subsequent period (by the facility running out of stock). The demand throughout the planning horizon is uncertain. We assume it to be captured by a set of given scenarios, each holding with a given probability.

The problem seeks to find a location plan for the facilities and the best policy for managing the inventory throughout the planning horizon to minimize the total expected cost. A risk-neutral decision-maker is assumed. The costs involved in the problem include: (i) fixed operating costs for the facilities, (ii) cost for allocating a supplier to a facility (e.g., vehicle acquisition), (iii) cost for allocating a customer to a facility (e.g., vehicle acquisition), (iv) fixed costs for issuing an order at the facilities, (v) cost per unit ordered at a facility, (vi) unit holding cost at the facilities (per unit remaining at the end of each period), and (vii) unit revenues for supplying a customer (that may depend on whether demand is satisfied when it occurs or with a delay). Holding costs regard the inventory level at a facility at the end of each period.

The decisions to make are time-dependent and include: (i) timing and location for the facilities, (ii) allocation of the open facilities to the suppliers, (iii) allocation of the demand nodes to the operating facilities, (iv) when to supply the facilities, (v) batch size when an order is issued, (vi) inventory level at a facility at the end of a period, (vii) amount to be shipped to the customers, and (viii) backlogging. Decisions (iv)–(vii) are customary in inventory planning and management.

Finally, while some decisions are non-adaptive (facility location and facility-supplier/customer-facility allocation), the others depend on how demand materializes, adapting to it. Accordingly, we adopt a two-stage stochastic programming model.

In what follows, we assume that a facility starts operating at the beginning of the period it is set up. Initially, we assume negligible lead times—instantaneous replenishment. Later, we adapt the base model to positive lead times. The following notation is used hereafter:

Sets:

M, a set of suppliers;
I, a set of potential locations for the facilities;
J, a set of demand nodes;
T, a set of periods in the planning horizon;
Ω, a set of scenarios capturing the demand throughout the planning horizon;

Parameters (Deterministic):

f_{it}, fixed cost for setting up facility $i \in I$ at the beginning of period $t \in T$ (may include maintenance and operating costs from period t until the end of the planning horizon);
g^1_{imt}, fixed cost for allocating facility $i \in I$ to supplier $m \in M$ in period $t \in T$;

g^2_{ijt}, fixed cost for allocating demand node $j \in J$ to facility $i \in I$ in period $t \in T$;
K_{imt}, fixed cost for issuing an order by facility $i \in I$ to supplier $m \in M$ in period $t \in T$;
c_{imt}, unit cost for the items ordered by facility $i \in I$ to supplier $m \in M$ in period $t \in T$;
h_{it}, unit holding cost per unit held in stock at facility $i \in I$ at period $t \in T$;
$s_{ijt\tau}$, unit selling price or revenue for each item sent from facility $i \in I$ to customer $j \in J$ in period $t \in T$ to satisfy the demand originated in period $\tau \in T_t$, with $T_t = \{1, \ldots, t\}$;
n_t, a maximum number of facilities we can install in period $t \in T$.

Parameters (Stochastic):

$d_{jt\omega}$, demand of node $j \in J$ in period $t \in T$ under scenario $\omega \in \Omega$;
π_ω, probability that scenario $\omega \in \Omega$ occurs.

First-stage decision variables:

$$y_{it} = \begin{cases} 1, & \text{if facility } i \in I \text{ is operating in period } t \in T, \\ 0, & \text{otherwise.} \end{cases}$$

$$w_{imt} = \begin{cases} 1, & \text{if facility } i \in I \text{ is allocated to supplier } m \in M \text{ in period } t \in T, \\ 0, & \text{otherwise.} \end{cases}$$

$$x_{ijt} = \begin{cases} 1, & \text{if demand node } j \in J \text{ is allocated to facility } i \in I \text{ in period} \\ & t \in T, \\ 0, & \text{otherwise.} \end{cases}$$

Second-stage decision variables:

$$r_{imt\omega} = \begin{cases} 1, & \text{if facility } i \in I \text{ issues an order at supplier} m \in M \text{ at the} \\ & \text{beginning of period } t \in T \text{ under scenario } \omega \in \Omega, \\ 0, & \text{otherwise.} \end{cases}$$

$z_{imt\omega}$ = amount shipped to facility $i \in I$ from supplier $m \in M$ at the beginning of period $t \in T$ under scenario $\omega \in \Omega$.
$u_{it\omega}$ = level of the stock at facility $i \in I$ at the end of period $t \in T$ under scenario $\omega \in \Omega$.
$v_{ijt\tau\omega}$ = amount sent from facility $i \in I$ to customer $j \in J$ in period $t \in T$ to supply demand originated at period $\tau \in T_t$, with $T_t = \{1, \ldots, t\}$, under scenario $\omega \in \Omega$.

The extensive form of the deterministic equivalent can be formulated as follows:

$$\text{minimize} \quad \sum_{t\in T}\left[\sum_{i\in I} f_{it}(y_{it}-y_{i,t-1}) + \sum_{i\in I}\left(\sum_{m\in M} g^1_{imt} w_{imt} + \sum_{j\in J} g^2_{ijt} x_{ijt}\right)\right]$$

$$+\sum_{\omega\in\Omega}\pi_\omega\left(\sum_{t\in T}\sum_{m\in M}\sum_{i\in I}(K_{imt} r_{imt\omega} + c_{imt} z_{imt\omega})\right.$$

$$\left.+\sum_{t\in T}\sum_{i\in I} h_{it} u_{it\omega} - \sum_{t\in T}\sum_{\tau\in T_t}\sum_{i\in I}\sum_{j\in J} s_{ijt\tau} v_{ijt\tau\omega}\right), \tag{9.42}$$

$$\text{subject to} \quad \sum_{i\in I}(y_{it}-y_{i,t-1}) \le n_t \quad \forall t\in T, \tag{9.43}$$

$$y_{it} \ge y_{i,t-1} \quad \forall i\in I,\ t\in T, \tag{9.44}$$

$$\sum_{i\in I} x_{ijt} = 1 \quad \forall j\in J,\ t\in T, \tag{9.45}$$

$$x_{ijt} \le y_{it} \quad \forall i\in I,\ j\in J,\ t\in T, \tag{9.46}$$

$$\sum_{m\in M} w_{imt} = y_{it} \quad \forall i\in I,\ t\in T, \tag{9.47}$$

$$x_{ijt} \in \{0,1\} \quad \forall i\in I,\ j\in J,\ t\in T, \tag{9.48}$$

$$w_{imt} \in \{0,1\} \quad \forall i\in I,\ m\in M,\ t\in T, \tag{9.49}$$

$$y_{it} \in \{0,1\} \quad \forall i\in I,\ t\in T, \tag{9.50}$$

$$\sum_{m\in M} z_{imt\omega} + u_{i,t-1,\omega} = u_{it\omega} + \sum_{j\in J}\sum_{\tau\in T_t} v_{ij\tau\omega}$$

$$\forall i\in I,\ t\in T,\ \omega\in\Omega \tag{9.51}$$

$$z_{imt\omega} \le \Theta r_{imt\omega} \quad \forall i\in I,\ m\in M,\ t\in T,\ \omega\in\Omega, \tag{9.52}$$

$$r_{imt\omega} \le w_{imt} \quad \forall i\in I,\ m\in M,\ t\in T,\ \omega\in\Omega, \tag{9.53}$$

$$u_{it\omega} \le \Theta y_{it} \quad \forall i\in I,\ t\in T,\ \omega\in\Omega, \tag{9.54}$$

$$v_{ijtt\omega} + \sum_{\tau\in T\setminus T_t} v_{ij\tau t\omega} \le \Theta x_{ijt} \quad \forall i\in I,\ j\in J,\ t\in T,\ \omega\in\Omega, \tag{9.55}$$

$$\sum_{i \in I} \left(v_{ijtt\omega} + \sum_{\tau \in T \setminus T_t} v_{ijt\tau\omega} \right) = d_{jt\omega} \qquad \forall j \in J,\ t \in T,\ \omega \in \Omega, \tag{9.56}$$

$$r_{imt\omega} \in \{0, 1\} \qquad \forall i \in I,\ m \in M,\ t \in T,\ \omega \in \Omega, \tag{9.57}$$

$$z_{imt\omega} \geq 0 \qquad \forall i \in I,\ m \in M,\ \omega \in \Omega, \tag{9.58}$$

$$u_{it\omega} \geq 0 \qquad \forall i \in I,\ t \in T,\ \omega \in \Omega, \tag{9.59}$$

$$v_{ijt\tau\omega} \geq 0 \qquad \forall i \in I,\ j \in J,\ t \in T,\ \tau \in T_t. \tag{9.60}$$

The objective function accounts for the total cost. The first row includes the cost of opening facilities and the cost to the facility-supplier and customer-facility allocations. The second row stands for the acquisition costs at the facilities (adaptive to the occurring scenario). Finally, the third row accounts for the holding costs and the revenues. Constraints (9.43) ensure that the number of new facilities opened respects the given limit in each period. Inequalities (9.44) guarantee that the system is purely phased-in. Constraints (9.45) and (9.46) ensure that all customers are allocated to one and only one open facility in each period. These constraints are quite relevant. If a facility cannot supply a demand node in some period, it should do it in some later period (which will be eventually ensured by a set of second-stage constraints). Equalities (9.47) state that in each period, an open facility must be allocated to some supplier. The domain constraints (9.48)–(9.50) end the set of first-stage constraints, thus ensuring that a here-and-now feasible solution is obtained.

Regarding the second-stage constraints, we have a set for each scenario. Constraints (9.51) ensure flow conservation at each facility in each period. Inequalities (9.52) guarantee that a facility receives the commodity only from the supplier it is allocated to. Constraints (9.53) state that in each period and irrespective of the scenario, a facility can only order from the supplier it is allocated to. Inequalities (9.54) ensure the consistency in the choice of the u and y variables. (9.55) ensure that the demand of some node in some period must be supplied by the facility the node is allocated to in that period, even if with delay. Equalities (9.56) ensure demand satisfaction in all periods and scenarios. Finally, we have (9.57)–(9.60), the domain constraints for the second-stage decision variables.

In the above model, Θ stands for a sufficiently large number, which can be easily tuned. Also, in the above model for $i \in I$ and $\omega \in \Omega$, $u_{i0\omega} = 0$ is the initial stock at each facility (which should be equal in all scenarios). Furthermore, y_{i0} is the status of location $i \in I$ at the beginning of the planning horizon. If some facilities are already operating before the beginning of the planning horizon, then such variables should be set equal to one.

Coelho [132] reports on instances of the above problem with up to 10 suppliers, 50 potential locations for the facilities, 250 demand nodes, and six scenarios solved up to proven optimality by a general solver.

The inclusion of lead times in the above problem makes it more interesting from a practical perspective. Let Λ_{im} be the lead time for an order issued by facility $i \in I$ to supplier $m \in M$ at the beginning of a time period. To simplify the exposition, assume that the lead times are a multiple of the period length. Nonetheless, the content we present next can be easily extended to other settings.

Given the lead times, we can define for each pair (i, m) with $i \in I$ and $m \in M$ a set $T_{im} = \{1 + \Lambda_{im}, 2 + \Lambda_{im}, \ldots, |T|\}$ that contains the periods in which facility i can receive commodity from supplier m.

From a modeling perspective, most of what we have presented so far remains the same. The changes involve the variables w, r, and z. Regarding the acquisition costs (second row of the objective function (9.42)), we consider (for each scenario $\omega \in \Omega$)

$$\sum_{t \in T} \sum_{m \in M} \sum_{i \in I} \left(K_{im,t-\Lambda_{im}} r_{im,t-\Lambda_{im},\omega} + c_{im,t-\Lambda_{im}} z_{imt\omega} \right). \tag{9.61}$$

This change reflects on some constraints:

Constraint (9.52) now becomes

$$z_{imt\omega} \leq \Theta r_{im,t-\Lambda_{im},\omega} \quad \forall i \in I,\ m \in M,\ t \in T_{im},\ \omega \in \Omega, \tag{9.62}$$

and Constraint (9.53) should be replaced by

$$r_{im,t-\Lambda_{im}\omega} \leq w_{im,t-\Lambda_{im}\omega} \quad \forall i \in I,\ m \in M,\ t \in T_{im},\ \omega \in \Omega. \tag{9.63}$$

These new constraints ensure consistency between the time an order is issued, the facility-supplier allocations, and the quantities received.

9.6 Conclusions and Further Reading

This chapter aimed to discuss fundamental concepts, models, and solution procedures in location inventory. The contents of the previous sections can be complemented with further reading.

Yao et al. [556] investigate a location-inventory problem with a setting similar to that analyzed in Section (9.3) but assuming multiple commodities and multiple allocation for the facilities to the suppliers. Continuous review is assumed. The fact that a facility may be supplied from different sources raises some challenges when quantifying the safety stock. In that work, an approximation is proposed for it. Furthermore, the authors consider a constraint on the weighted average of the lead time at the customers. A heuristic is proposed for finding feasible solutions to the problem.

Diabat et al. [168] consider an inventory policy at facilities and the demand nodes. This calls for coordination between orders at the different levels. The well-known power-of-two rule (see [464]) can be adopted. The authors do so and prove that for a fixed assignment of retailers to warehouses, such a policy is 98% effective (like for the one-warehouse multi-retailer problem). The authors develop a Lagrangian relaxation-based algorithm for finding lower and upper bounds. This work would be extended in [167] by considering capacities at the facilities. This results in a nonlinear optimization problem handled by linearizing exactly or approximately (using a piecewise approximation) the different nonlinear terms.

Berman et al. [67] investigate a location-inventory model where facilities follow a periodic-review inventory policy. System coordination is achieved by choosing review intervals at the facilities from a set of choices. Two schemes are studied: partial and complete coordination. In the former, each facility can choose its review interval from a set of possibilities; in the latter, all the facilities have an identical review interval. While full coordination increases the location and inventory costs, it likely reduces the overall costs of running the system, namely when operational costs such as those related to delivery schedules are considered. The problem consists of determining the location of the facilities, the assignment of demand nodes to open facilities, and the inventory policy at the facilities so that the total cost is minimized. A nonlinear optimization model is proposed, which is tackled by a Lagrangian relaxation-based procedure.

Despite all the progress observed in the location-inventory problems, this chapter also reveals some aspects that have not been dealt with in the literature or have scarcely been considered. One such aspect regards the uncertainty that is often observed in lead times. Another aspect of relevance in applications stemming from logistics and supply chain management regards the need to consider multiple commodities and possible interactions among them (e.g., correlation). We also conclude that in terms of multi-layered facility location, embedding inventory decisions is still an open research area. Finally, we note that most continuous review location-inventory problems still rely on the Gaussian distribution of the demands. Although this may be a reasonable assumption in many applications, it may not be the case in others. Nevertheless, this poses many challenges when finding tractable optimization models. Finally, we note that uncertainty keeps being captured by random variables described by a given CDF. Robust optimization models in the context of location inventory remain to be explored.

Chapter 10
Facility Location with Routing Decisions

Abstract This chapter focuses on integrating location and routing decisions. These are the problems that have been extensively studied in deterministic settings. In practice, several sources of uncertainty emerge, which include demands, travel times, and infrastructure operationality. These can be accounted for in different ways. Several families of problems are considered. As far as location routing is concerned, cost- and time-oriented objectives and reliability problems are discussed. Uncertainty in location-arc routing is also mentioned, although this is still much unexplored in the literature. Finally, some recent trends and developments in the field are reviewed.

Keywords Location routing · Cost-oriented objectives · Time-oriented objectives · Time-dependent decisions · Reliability · Arc routing

10.1 Introduction

In most contents presented in this volume, we have considered facility location problems, assuming that the demand nodes are supplied from the facilities according to a direct-allocation pattern. However, in many applications (e.g., in logistics and transportation), demand nodes are served as part of routes originated at and destined to a facility that acts as a depot. This calls for making routing decisions while deciding the location of the depots (facilities, as we keep calling them). We start working in the domain of location-routing problems. Furthermore, we may even consider the case in which the demands occur throughout the edges of a network. In this case, we face a location-arc routing problem.

Facility location problems embedding routing decisions have received significant attention in the literature. The interested reader can refer to [12, 364] as well as to the references therein. In this chapter, our focus is on hedging against uncertainty in such problems. Uncertainty typically stems from demands, costs, and travel times. Nevertheless, it may also be associated with the infrastructure, such as road disruptions due to unexpected events, e.g., a flood.

© The Author(s), under exclusive license to Springer Nature Switzerland AG 2024
F. Saldanha-da-Gama, S. Wang, *Facility Location Under Uncertainty*,
International Series in Operations Research & Management Science 356,
https://doi.org/10.1007/978-3-031-55927-3_10

From the perspective of vehicle routing and arc routing (with the depots given), uncertainty is an aspect that has been considered in the literature. It is not our goal to review that work. The interested reader can refer to [158, 226, 418, 419] for recent related overviews.

In a vehicle routing problem, when a demand node (or a link) is visited by a vehicle, a delivery may occur, a collection, or both. In the latter case, we have a pickup and delivery problem. The first two cases can be cast in a similar setting. In this chapter, we assume that demand at a demand point is related to some commodity to be delivered. The models and algorithms remain unchanged if something is to be collected instead.

Throughout this chapter, unless stated otherwise, we assume that the traveling costs or distances are symmetric and satisfy the triangle inequality.

The remainder of this chapter is organized as follows. Section 10.2 discusses how to hedge against uncertainty in cost-oriented location-routing problems. Section 10.3 turns the attention to time-oriented objectives. Section 10.4 considers uncertainty in location-routing problems with time-dependent decisions. We proceed with the analysis of reliability issues. This is done in Sect. 10.5, where disruption is considered at a depot or a vehicle. Section 10.6 focuses on location-arc routing under uncertainty. The chapter ends with an overview and some references for further reading.

10.2 Cost-Oriented Location Routing

The location of facilities is often considered a strategic decision not only for the time the facilities should operate but also because of the costs involved. On the other hand, routing decisions are often looked at an operational level because, in many applications, they can easily adjust to short-term information. When uncertainty stems from the demand, a two-stage decision-making process can be adopted. We decide about the location of the facilities and the allocation of the customers to them and implement such a solution here and now. When demands are eventually observed, the routing decisions adapt to the actual "future" faced. Naturally, we seek integrated modeling frameworks to hedge against uncertainty. Therefore, the possible future scenarios should not be ignored when decisions are made here and now.

We denote by I the locations for the facilities (or depots in this chapter) and by J the set of demand nodes. We start by working with uncapacitated facilities. Let us consider, for the time being, that for each demand node $j \in J$, we know its demand that we denote by d_j. Let f_i be the fixed setup cost for facility $i \in I$. As presented in Sect. 2.2, a directed connected graph $G = (V, E)$ with $V = I \cup J$ is assumed to represent the underlying transportation network. A unit transportation cost is associated with each arc of the network. We assume we have a set M of homogeneous vehicles available, each with capacity H. Thus, we can design up to $|M|$ routes for supplying the customers. Additionally, we incur a fixed cost of F for every vehicle used. We assume that a single visit (of a vehicle) can supply a demand node—no demand is higher than H.

We keep using binary location variables y_i $(i \in I)$. The binary allocation variables, x_{ij}, are also considered to indicate whether demand node $j \in J$ is supplied as part of a route originated at depot $i \in I$. Additionally, for $m \in M$ and $\ell, \ell' \in V, \ell \neq \ell'$, $v_{\ell\ell'm}$ is equal to one if route m visits node ℓ' immediately after visiting node ℓ and zero otherwise. Finally we consider parameters $c_{\ell\ell'}$ representing the shortest unit transportation cost in the network G between nodes ℓ and ℓ' $(\ell, \ell' \in V, \ell \neq \ell')$. Assuming that each route starts and ends at the same depot, we consider the following (deterministic) location-routing problem (resulting directly from the model presented in Sect. 2.2 when the capacity of the facilities is ignored):

$$\text{minimize} \quad \sum_{i \in I} f_i y_i + \sum_{m \in M} \sum_{\substack{\ell, \ell' \in V \\ \ell \neq \ell'}} c_{\ell\ell'} v_{\ell\ell'm} + F \sum_{m \in M} \sum_{\ell \in I} \sum_{\ell' \in J} v_{\ell\ell'm}, \tag{10.1}$$

$$\text{subject to} \quad x_{ij} \leq y_i \qquad \forall i \in I,\ j \in J, \tag{10.2}$$

$$\sum_{m \in M} \sum_{\ell \in V : \ell \neq j} v_{\ell jm} = 1 \qquad \forall j \in J, \tag{10.3}$$

$$\sum_{j \in J} \left(d_j \sum_{\ell \in V : \ell \neq j} v_{\ell jm} \right) \leq H \qquad \forall m \in M, \tag{10.4}$$

$$\sum_{\ell' \in V : \ell' \neq \ell} v_{\ell\ell'm} - \sum_{\ell' \in V : \ell' \neq \ell} v_{\ell'\ell m} = 0 \qquad \forall m \in M,\ \ell \in V, \tag{10.5}$$

$$\sum_{\ell \in I} \sum_{\ell' \in J} v_{\ell\ell'm} \leq 1 \qquad \forall m \in M, \tag{10.6}$$

$$\sum_{m \in M} \sum_{\ell \in S} \sum_{\ell' \in V \setminus S} v_{\ell\ell'm} \geq 1 \qquad \forall I \subseteq S \subset V, \tag{10.7}$$

$$\sum_{\ell' \in J} v_{i\ell'm} + \sum_{\ell' \in V : \ell' \neq j} v_{j\ell'm} \leq 1 + x_{ij} \qquad \forall i \in I,\ j \in J,\ m \in M, \tag{10.8}$$

$$v_{\ell\ell'm} \in \{0, 1\} \qquad \forall \ell, \ell' \in V,\ \ell \neq \ell',\ m \in M, \tag{10.9}$$

$$x_{ij} \in \{0, 1\} \qquad \forall i \in I,\ j \in J, \tag{10.10}$$

$$y_i \in \{0, 1\} \qquad \forall i \in I. \tag{10.11}$$

The objective function (10.1) represents the total cost incurred, which includes the facility setup costs (the first term), transportation (the second term), and vehicle usage (the third term). Inequalities (10.2) ensure that nodes can only be supplied from operating facilities. Equalities (10.3) ensure that every demand node is included in one and only one route. Inequalities (10.4) guarantee that no route exceeds the vehicle capacity. Equalities (10.5) state that a set of routes is indeed created. Inequalities (10.6) ensure that each route uses (at most) a single facility. Constraints (10.7) ensure that every route involves a facility. Constraints (10.8) impose that if a route includes the depot i and the demand node j, then x_{ij} is equal to one. In this case, j is served by a route originated at (and destined to) depot i. Conversely, if x_{ij} is equal to zero, no route can involve simultaneously the depot i and the demand node j. Finally, we specify the domain of the decision variables.

We consider now that the demands are stochastic. In particular, we assume that ξ_j is a random variable representing the demand of node $j \in J$. Furthermore we assume that the random vector $\boldsymbol{\xi} = (\xi_1, \ldots, \xi_{|J|})$ has some known joint CDF.

Assuming that routing is a process that repeats many times, we can think of optimizing the total routing cost on average, leveraging the Law of the Large Numbers and thus adopting a risk-neutral attitude of the decision-maker toward risk. The problem can be reformulated as follows:

$$\text{minimize} \quad \sum_{i \in I} f_i y_i + \mathcal{Q}(\boldsymbol{x}, \boldsymbol{y}), \tag{10.12}$$

$$\text{subject to} \quad (10.2), (10.10), (10.11),$$

$$\sum_{i \in I} x_{ij} = 1 \qquad \forall j \in J. \tag{10.13}$$

The first-stage problem consists of an uncapacitated facility location problem with single sourcing. The recourse function $\mathcal{Q}(\boldsymbol{x}, \boldsymbol{y})$ accounts for the expected routing cost by making the decision $\{\boldsymbol{x}, \boldsymbol{y}\}$. Hence, we can write $\mathcal{Q}(\boldsymbol{x}, \boldsymbol{y}) = \mathbb{E}_{\boldsymbol{\xi}}[Q(\boldsymbol{x}, \boldsymbol{y})]$ with $Q(\boldsymbol{x}, \boldsymbol{y})$ standing for the optimal value of the stochastic problem

$$\text{minimize} \quad \sum_{m \in M} \sum_{\substack{\ell, \ell' \in V \\ \ell \neq \ell'}} c_{\ell\ell'} v_{\ell\ell' m} + F \sum_{m \in M} \sum_{\ell \in I} \sum_{\ell' \in J} v_{\ell\ell' m}, \tag{10.14}$$

$$\text{subject to} \quad (10.3), (10.5)\text{–}(10.9),$$

$$\sum_{j \in J} \left(\xi_j \sum_{\ell \in V : \ell \neq j} v_{\ell j m} \right) \leq H \qquad \forall m \in M. \tag{10.15}$$

In this second-stage problem, the values of variables $\boldsymbol{x}$ are fixed in (10.8).

This model is relevant if we can capture demand uncertainty by a finite set of scenarios, say Ω, with each scenario $\omega \in \Omega$ occurring with a given probability π_ω. As customary, we can index the second-stage parameters and variables in the scenarios to obtain the extensive form of the deterministic equivalent model. Keeping the costs deterministic, we denote by $d_{j\omega}$ the demand of node $j \in J$ under scenario $\omega \in \Omega$ and by $v_{\ell\ell' m\omega}$ the binary variable equal to one if and only if route

$m \in M$ serves node ℓ' immediately after serving node ℓ under scenario $\omega \in \Omega$, $\ell, \ell' \in V, \ell \neq \ell'$. The extensive form of the deterministic equivalent becomes

$$\text{minimize} \quad \sum_{i \in I} f_i y_i + \sum_{\omega \in \Omega} \pi_\omega \left(\sum_{m \in M} \sum_{\substack{\ell,\ell' \in V \\ \ell \neq \ell'}} c_{\ell\ell'} v_{\ell\ell' m\omega} + F \sum_{m \in M} \sum_{\ell \in I} \sum_{\ell' \in J} v_{\ell\ell' m\omega} \right), \tag{10.16}$$

subject to (10.2), (10.10), (10.11),

$$\sum_{m \in M} \sum_{\ell \in V : \ell \neq j} v_{\ell j m\omega} = 1 \qquad \forall j \in J,\ \omega \in \Omega, \tag{10.17}$$

$$\sum_{j \in J} \left(d_{j\omega} \sum_{\ell \in V : \ell \neq j} v_{\ell j m\omega} \right) \leq H \qquad \forall m \in M,\ \omega \in \Omega, \tag{10.18}$$

$$\sum_{\ell' \in V : \ell' \neq \ell} v_{\ell\ell' m\omega} - \sum_{\ell' \in V : \ell' \neq \ell} v_{\ell'\ell m\omega} = 0 \qquad \forall m \in M,\ \ell \in V,\ \omega \in \Omega, \tag{10.19}$$

$$\sum_{\ell \in I} \sum_{\ell' \in J} v_{\ell\ell' m\omega} \leq 1 \qquad \forall m \in M,\ \omega \in \Omega, \tag{10.20}$$

$$\sum_{m \in M} \sum_{\ell \in S} \sum_{\ell' \in V \setminus S} v_{\ell\ell' m\omega} \geq 1 \qquad \forall I \subseteq S \subset V,\ \omega \in \Omega, \tag{10.21}$$

$$\sum_{\ell' \in J} v_{i\ell' m\omega} + \sum_{\ell' \in V : \ell' \neq j} v_{j\ell' m\omega} \leq 1 + x_{ij} \qquad \forall i \in I,\ j \in J,\ m \in M,\ \omega \in \Omega, \tag{10.22}$$

$$v_{\ell\ell' m\omega} \in \{0, 1\} \qquad \forall \ell, \ell' \in V,\ \ell \neq \ell',\ m \in M,\ \omega \in \Omega. \tag{10.23}$$

The role played by the constraints is straightforward from the other contents above.

This model is a convenient one. In addition to being compact, it represents a problem with complete recourse, i.e., every feasible first-stage solution has a second-stage (feasible) completion for each scenario that may occur (assuming that the number of routes to design is large enough so that irrespective of the scenario there is always a partition of the customers ensuring that the total demand in each group does not exceed H). The above model can be easily adapted to the case where we face scenario-dependent costs. Unfortunately, this type of model grows fast in size, preventing it from being handled using the existing knowledge and technology. Nevertheless, if we can efficiently solve the model for a few scenarios, we can explore it in more comprehensive scenario settings through a sample-based procedure.

In addition to the fast growth in size, the above model has some potential drawbacks. First, it does not consider an a priori definition of routes. This is sought in some applications even if corrections are performed after demands become known. Second, it neglects the capacity of the facilities, which, again, may be of practical relevance. Finally, a scenario set may be a too restrictive way of capturing uncertainty.

Let us assume now that facilities are capacitated. In particular, let q_i represent the capacity of a facility installed at $i \in I$. Additionally, suppose that for each $j \in J$, ξ_j is a random variable described by a non-negative upper-bounded probability distribution but whose specific values are only revealed when the vehicles are en route. Furthermore, consider the setting in which the routes are *ex ante* decisions.

Due to the uncertain nature of the demand, an a priori planned route may become infeasible. This is the case if a vehicle does not carry enough quantity to serve a node. This may be the case due to the limited capacity of the vehicle. Accordingly, a corrective course of action must be implemented. We assume it to call for the vehicle to complete a round-trip to its depot for additional loading/delivery and resume the route afterward.

Given that the demands are only known while the routes are in progress, we can adopt either a reactive or a preventive strategy. In the first case, the round-trip to the depot occurs only when the vehicle arrives at a demand node and observes it cannot be served. In the second case, after serving a customer, we realize that the expected demand at the following node in the route is above the amount available at the vehicle. The round-trip to the depot occurs at that point. In any case, the cost of a route becomes uncertain due to the cost associated with the corrective actions.

A question remains: If we seek to locate the depots and define a priori routes, what information about the demands should be used? A simple answer is the expected demand of the customers. If we consider it, the problem can be formulated as follows:

$$\text{minimize} \quad \sum_{i \in I} f_i y_i + F \sum_{m \in M} \sum_{\ell \in I} \sum_{\ell' \in J} v_{\ell\ell'm} + \mathcal{Q}(\boldsymbol{v}, \boldsymbol{x}, \boldsymbol{y}), \tag{10.24}$$

$$\text{subject to} \quad \sum_{j \in J} \mathbb{E}[\xi_j] x_{ij} \le q_i y_i \qquad \forall i \in I, \tag{10.25}$$

$$\sum_{j \in J} \mathbb{E}[\xi_j] \sum_{\ell \in V} v_{\ell jm} \le H \qquad \forall m \in M, \tag{10.26}$$

$$(10.3), (10.5)\text{–}(10.11).$$

The recourse function can be defined as

$$\mathcal{Q}(\boldsymbol{v}, \boldsymbol{x}, \boldsymbol{y}) = \sum_{m \in M} \mathbb{E}[\mathcal{R}_m],$$

with $\mathbb{E}[\mathcal{R}_m]$ standing for the expected cost of route $m \in M$. In fact, $\mathcal{R}_m$ is a random variable that we can define as follows:

$$\mathcal{R}_m = \begin{cases} \sum_{\ell,\ell' \in V:\ell \neq \ell'} c_{\ell\ell'} v_{\ell\ell' m} & \text{if} \sum_{j \in J} \sum_{\ell \in V:\ell \neq j} \xi_j v_{\ell jm} \leq H, \\ \sum_{\ell,\ell' \in V:\ell \neq \ell'} c_{\ell\ell'} v_{\ell\ell' m} + \rho & \text{otherwise.} \end{cases} \tag{10.27}$$

Quintero-Araujo et al. [441] consider a safety stock at the vehicles defined as a proportion of the capacity H. When the remaining supply in the vehicle drops below that proportion, it returns to the depot to fully replenish (or unload if this is the case) before the route is resumed. Denoting such proportion by γ, the idea is to replace H with $(1 - \gamma)H$ in (10.26) and (10.27). This is a conservative way to hedge against uncertainty. Parameter γ is adjustable by the decision-maker.

In the same work, the authors overcome the difficulty in finding a tractable expression for the cost of a route by simulating it. Additionally, those authors simulate the so-called reliability of a route that they define as the estimated percentage of non-route failures. The reliability of a feasible solution (location of depots, allocation of demand nodes to depots, and a priori routes) is then the product of the reliabilities of the underlying routes.

The above elements trigger the development of a simheuristic (see, e.g., [292]). The proposed procedure has three main phases. In the first one, many subsets of facilities are selected and assessed using fast allocation and routing heuristics. In the second phase, the most promising sets of facilities are considered. The allocation and routing components are further improved using more intensive search algorithms. A final set of solutions is then assessed through simulation to find the best ones and estimate the routing costs and solution reliability. The authors assume log-normal demands in the computational tests whose results they report on.

For a setting similar to that considered in [441], a particle swarm optimization algorithm is developed in [369]. In this case, the author assumes Poisson demands.

A closer look at expression (10.27) leads to writing the objective function (10.24) in a different way:

$$\sum_{i \in I} f_i y_i + F \sum_{m \in M} \sum_{\ell \in I} \sum_{\ell' \in J} v_{\ell\ell' m} + \sum_{\ell,\ell' \in V:\ell \neq \ell'} c_{\ell\ell'} v_{\ell\ell' m} + \mathcal{Q}(\boldsymbol{v}, \boldsymbol{x}, \boldsymbol{y}), \tag{10.28}$$

but now $Q(\boldsymbol{v}, \boldsymbol{x}, \boldsymbol{y})$ contains only the total expected extra cost paid for round-trips to the depot for reloading. Hence, in this case,

$$Q(\boldsymbol{v}, \boldsymbol{x}, \boldsymbol{y}) = \sum_{m \in M} \mathbb{E}[\mathcal{R}_m^p]$$

with $\mathcal{R}_m^p$ denoting the penalty paid at route m. The third term of the objective function is the total cost for the a priori routes and thus the total routing cost if nothing fails—no round-trip to the depots is needed.

Suppose we are given a route $\mathcal{R}_m = (j_0, j_1, \ldots, j_T, j_0)$ defined by a sequence of nodes starting and ending at a depot (j_0). We say that a failure occurs at a node j_t when we conclude that the vehicle must perform a round-trip to the depot before serving that node. Such can occur (i) when the vehicle arrives at a customer and observes that the demand of the customer is beyond the amount available at the vehicle or (ii) when the vehicle is leaving a node and receives information about the demand of the following node in the route concluding that a round-trip to the depot is needed. In the first case, we have a penalty $p(t) = 2c_{j_t j_0}$. In the second, we get $p(t) = c_{j_t j_0} + c_{j_0 j_{t+1}} - c_{j_t j_{t+1}}$. Laporte et al. [323] show how to compute the expected penalty of a route.

Proposition 10.1 *Denoting by E_t the event "a failure occurs at node j_t," we have*

$$\mathbb{E}[\mathcal{R}_m^p] = \sum_{t=1}^{T} \mathbb{P}[E_t]p(t),$$

with

$$\mathbb{P}[E_t] = (1 - \Psi_{t-1,t}(H))\mathbb{P}[E_{t-1}] + \sum_{\ell=2}^{t-1} \left(\Psi_{t-\ell,t-1}(H) - \Psi_{t-\ell,t}(H)\right)\mathbb{P}[E_{t-1}] + \left(\Psi_{1,t-1}(H) - \Psi_{1,t}(H)\right),$$

$\mathbb{P}[E_1] = \mathbb{P}[\xi_{j_1} > H] = 0$, *and* $\Psi_{t't''}(q) = \mathbb{P}[\sum_{\ell=t'}^{t''} \xi_\ell \leq q]$.

Proof See [323]. □

The above result establishes a recursive equation for computing the probability of facing a route failure in any route node. In the same work the authors also note that although a route $\mathcal{R}_m$ and its reverse $\mathcal{R}_m^{-1} = (j_0, j_T, \ldots, j_1, j_0)$ have equal costs due to the assumed symmetry of the cost matrix, the expected penalties are not necessarily the same. For this reason, it can be assumed that a route is decided according to the orientation of the least expected penalty. However, once starting according to that orientation, the route follows it.

This discussion shows that we can hardly expect to find tractable optimization models for such a general problem. Nevertheless, some cases have been treated successfully, as we show next.

Laporte et al. [323] investigate a stochastic location-routing problem where a single uncapacitated depot is to be opened and a priori routes are to be defined. This is a pioneering work on the topic. With all routes starting and ending in the same depot, we can simplify the notation we have been considering. First, assume that the nodes in $V = I \cup J$ are numbered from 1 to $|V|$. Then, for every $\ell, \ell' \in V$ with $\ell < \ell'$, we define $v_{\ell\ell'}$ as the number of times that edge $\{\ell, \ell'\}$ is used in the optimal solution (irrespective of the direction). Finally, we define a variable w representing the number of vehicles (routes) used.

The uncertainty underlying the problem can be handled in different ways. One possibility is to impose that the expected penalty of a route cannot exceed a given proportion, say γ, of its a priori cost. The problem can be formulated as follows:

$$\text{minimize} \quad \sum_{i \in I} f_i y_i + F\, w + \sum_{\substack{\ell,\ell' \in V \\ \ell < \ell'}} c_{\ell\ell'} v_{\ell\ell'}, \tag{10.29}$$

$$\text{subject to} \quad \sum_{i \in I} y_i = 1, \tag{10.30}$$

$$\sum_{\ell < i} v_{\ell i} + \sum_{\ell > i} v_{i\ell} \leq 2|M| y_i \qquad \forall i \in I, \tag{10.31}$$

$$\sum_{\ell < i} v_{\ell i} + \sum_{\ell > i} v_{i\ell} \leq 2w \qquad \forall i \in I, \tag{10.32}$$

$$\sum_{\ell < j} v_{\ell j} + \sum_{\ell > j} v_{j\ell} = 2w \qquad \forall j \in J, \tag{10.33}$$

$$\sum_{\substack{\ell,\ell' \in S \\ \ell < \ell'}} v_{\ell\ell'} \leq |S| - 1 \qquad \forall S \subseteq J,\ |S| \geq 2, \tag{10.34}$$

$$\sum_{t=1}^{T} v_{j_t j_{t+1}} \leq T \quad \forall \mathcal{R} = (j_0, j_1, \ldots, j_T, j_{T+1} = j_0),\ j_0 \in I,$$

$$j_1, \ldots, j_T \in J,\ \min\{\mathbb{E}[\mathcal{R}^p], \mathbb{E}[\mathcal{R}^{-1^p}]\} > \gamma \sum_{t=0}^{T} c_{j_t j_{t+1}}, \tag{10.35}$$

$$v_{\ell\ell'} \in \{0, 1\} \qquad \forall \ell, \ell' \in J, \tag{10.36}$$

$$v_{\ell\ell'} \in \{0, 1, 2\} \qquad \forall \ell \in I \text{ or } \ell' \in \mathrm{I}, \tag{10.37}$$

$$1 \leq w \leq |M| \text{ and integer}, \tag{10.38}$$

$$y_i \in \{0, 1\} \qquad \forall i \in I. \tag{10.39}$$

In this formulation, the objective function accounts for the costs associated with the depot, the vehicles, and the a priori routes. Constraint (10.30) ensures that a single depot is installed. Constraints (10.31)–(10.33) guarantee that the a priori routes are actually created. Inequalities (10.34) eliminate subtours disconnected from the depot. Constraints (10.35) ensure that the routes satisfy the expected penalty constraint. Finally, we have the domain constraints.

An alternative for hedging against uncertainty in the location-routing problem we are considering is to impose a maximum probability, α, of having a planned route failing at least once. This can be accomplished by replacing (10.35) with

$$\mathbb{P}[\text{A planned route fails at least once}] \leq \alpha. \tag{10.40}$$

The tractability of the new formulation depends on the possibility of finding a "tractable" deterministic equivalent. This can be accomplished by replacing (10.34) and (10.40) with

$$\sum_{\substack{\ell \in S, \ell' \in V \setminus S \\ \text{or} \\ \ell \in V \setminus S, \ell' \in S}} v_{\ell\ell'} \geq 2V_\alpha(S) \qquad \forall S \subseteq J,\ |S| \geq 2. \tag{10.41}$$

In the above constraints, $V_\alpha(S)$ stands for the minimum number of vehicles required to serve S such that the probability of failure in S does not exceed α. $V_\alpha(S)$ can be found as the smallest integer satisfying

$$\mathbb{P}\left[\sum_{j \in S} \xi_j > HV_\alpha(S)\right] \leq \alpha.$$

Hence, Constraints (10.41) ensure that there are at least $2V_\alpha(S)$ edges between S and $V \setminus S$ if at least $V_\alpha(S)$ vehicles visit the demand nodes in S.

If the distribution of $\sum_{j \in S} \xi_j$ is known or can be approximated, then computing $V_\alpha(S)$ poses no problem. For instance, if demands are independent and the cardinality of A is fairly large, we can invoke the central limit theorem and set $V_\alpha(S)$ equal to the smallest integer greater than or equal to

$$\frac{\sum_{j \in S} \mu_j + z_{1-\alpha}\sqrt{\sum_{j \in S} \sigma_j^2}}{H},$$

with $\mu_j = \mathbb{E}[\xi_j]$ and $\sigma_j^2 = \mathbb{V}[\xi_j]$, $j \in J$.

A challenging feature of the above models is the exponential number of constraints (10.34) and (10.35) in the first variant and (10.41) in the second. However, we can incorporate these constraints on the fly within a branch-and-cut algorithm. Algorithm 10.1 does so and solves the problem optimally.

Algorithm 10.1: Solving the single-depot stochastic location-routing problem

```
V* ← ∞;
Define a problem with the objective function (10.29) and constraints (10.30)–(10.33),
  (10.36)–(10.39);
Insert the problem in a stack;
while Stack is not empty do
    Select the top problem in the stack, say P;
    repeat
        integerFeasible ← false;
        violatedConstraintsFound ← false;
        Solve the linear programming relaxation of P;
        if problem is feasible then
            Let V' be its optimal value.
            if V' < V* then
                if optimal solution is fractional then
                    Branch on a fractional variable;
                    For each subproblem, compute a lower bound and discard the
                      subproblem if the lower bound is larger than V*;
                    Add the non-discarded subproblems to the stack;
                else
                    if solution violates (10.34) and (10.35) (first variant) or (10.41)
                      (second variant) then
                        Add the violated constraints to P;
                        violatedConstraintsFound ← true;
                    else
                        integerFeasible ← true;
    until violatedConstraintsFound = false;
    if integerFeasible = true then
        V* ← V';
        Store the best incumbent solution:
Return Best incumbent;
```

So far in this section, we have considered a recourse action to hedge against uncertainty in location-routing problems. That action calls for a round-trip to the depot.

An alternative that avoids such round-trips is to skip some demand nodes. However, again, a penalty is incurred. It may correspond, e.g., to a loss opportunity cost for not serving a customer. Such a setting was investigated in [8]. Multiple (capacitated) depots can be installed, each with a single vehicle. The latter are uncapacitated although they have an implicit capacity given but the capacity of the depot that serves as their base. The authors assume that demand regards a service and not a quantity of some commodity. This leads to unit demand customers. In our context, this indicates that the demands follow a Bernoulli distribution. Furthermore, a single route is to be designed for each installed depot.

When the customers who ask to be served are known, the a priori planned routes are implemented but possibly skipping some customers. The objective is to minimize the sum of the fixed costs for opening the depots plus the total expected transportation costs and penalties. The fact that some customers can be skipped makes the routing cost adaptative to the demand scenario observed.

The problem can be formulated as follows:

$$\text{minimize} \quad \sum_{i\in I} f_i y_i + Q(\boldsymbol{v}, \boldsymbol{x}, \boldsymbol{y}), \tag{10.42}$$

$$\text{subject to} \quad \sum_{i\in I} x_{ij} = 1 \qquad \forall j \in J, \tag{10.43}$$

$$(10.2), (10.3), (10.5)\text{–}(10.11).$$

A first-stage solution may allocate to one depot more demand nodes than the service capacity of the depot. This is the way to hedge against uncertainty. Given an a priori solution, for each $i \in I$ such that $y_i = 1$, the actual demand allocated to the depot is itself a random variable, $\eta_i = \sum_{j\in J|x_{ij}=1} \xi_j$. Let n_i be the number of demand nodes allocated to i. Then η_i is a Binomial random variable with parameters n_i and π. The latter is the probability that a demand node calls for being served. We are assuming the same probability for all demand nodes.

Let $J(i)$ denote the set of demand nodes out of those allocated to depot i that are visited in the *ex post* routes. Denoting by q_i the capacity of depot i (in terms of the number of demand nodes that it can serve), two cases are distinguished in [8]:

If $\eta_i \le q_i$, then $J(i) = \{j \in J : x_{ij} = 1 \wedge \xi_j = 1\}$;

If $\eta_i > q_i$, then $J(i)$ is a subset of $\{j \in J : x_{ij} = 1 \wedge \xi_j = 1\}$ with cardinality q_i.

If the penalty for skipping a demand node that calls for being served is equal to g, then the recourse function can be written as follows:

$$\begin{aligned} Q(\boldsymbol{v}, \boldsymbol{x}, \boldsymbol{y}) &= \mathbb{E}_{\xi}\left[\sum_{i\in I:y_i=1} g(\eta_i - q_i)^+ + \mathbb{E}_{J(i)}\left[\sum_{i\in I|y_i=1} \boldsymbol{c}'\boldsymbol{v}_i\right]\right] \\ &= g\sum_{i\in I|y_i=1} \mathbb{E}_{\xi}[(\eta_i - q_i)^+] + \sum_{i\in I|y_i=1} \mathbb{E}_{\xi}\left[\mathbb{E}_{J(i)}[\boldsymbol{c}'\boldsymbol{v}_i]\right], \end{aligned} \tag{10.44}$$

with $(a)^+ = \max\{0, a\}$ and $\boldsymbol{v}_i$ standing for the incidence vector of the arcs from the a priori route associated with depot i that will be part of the *a posteriori* route by skipping the demand nodes not in $J(i)$.

Looking into the above function, we conclude that the involved part is the expected cost of the *a posteriori* route associated with each open depot.

Proposition 10.2 *For each $i \in I$ such that $y_i = 1$,*

$$\mathbb{E}_{\xi}\left[\mathbb{E}_{J(i)}\left[\boldsymbol{c}'\boldsymbol{v}_i\right]\right] = \sum_{h=1}^{n_i} R_{h-1} c_{ij_h} + \sum_{h=1}^{n_i-1}\sum_{t=h+1}^{n_i} P_{t-h-1} c_{j_h j_t} + \sum_{h=1}^{n_i} R_{n_i-h} c_{j_h i}$$

with

$$R_\ell = \pi \left[\sum_{k=1}^{\kappa_1} \binom{n_i - \ell - 1}{k-1} \pi^{k-1} (1-\pi)^{n_i - k} \right.$$
$$\left. + \sum_{k=q_i+1}^{n_i} \pi^{k-1} (1-\pi)^{n_i-k} \frac{\sum_{t=t_0}^{t_1} \binom{\ell}{t} \binom{n_i-\ell-1}{k-t-1} \binom{k-t-1}{q_i-1}}{\binom{k}{q_i}} \right],$$

$$P_\ell = \pi^2 \left[\sum_{k=2}^{\kappa_1} \binom{n_i - \ell - 2}{k-2} \pi^{k-2} (1-\pi)^{n_i - k} \right.$$
$$\left. + \sum_{k=q_i+1}^{n_i} \pi^{k-2} (1-\pi)^{n_i-k} \frac{\sum_{t=t_0}^{t_1} \binom{\ell}{t} \binom{n_i-\ell-2}{k-t-2} \binom{k-t-2}{q_i-2}}{\binom{k}{q_i}} \right],$$

$$\kappa_1 = \min\{q_i, n_i - \ell\},\ t_0 = \min\{k - n_i + \ell, 0\},\ \textit{and}\ t_1 = \min\{k - q_i, \ell\}.$$

Proof See [8]. □

In the above result, R_ℓ is the probability of using an arc incident to the depot directly skipping ℓ intermediate demand nodes that were part of the a priori route. Furthermore, P_ℓ is the probability of using an arc that connects two demand nodes having ℓ intermediate nodes in the a priori route.

Although finding a closed-form expression for the recourse function is interesting, the above expressions become computationally unaffordable in practice. For this reason, the same authors propose approximating $\mathbb{E}_\xi \left[\mathbb{E}_{J(i)} \left[c' v_i \right] \right]$ with

$$\sum_{k=1}^{n_i} \bar{\pi} (1-\bar{\pi})^{k-1} c_{ij_k} + \sum_{k=1}^{n_i-1} \sum_{t=k+1}^{n_i} \bar{\pi}^2 (1-\bar{\pi})^{t-k-1} c_{j_k j_t} + \sum_{k=1}^{n_i} \bar{\pi} (1-\bar{\pi})^{n_i-k} c_{j_k i},$$

where

$$\bar{\pi} = \pi \left(\sum_{k=1}^{q_i} \binom{n_i - 1}{k-1} \pi^{k-1} (1-\pi)^{n-k} + \frac{\sum_{k=q_1+1}^{n_i} \binom{n_i-1}{k-1} \pi^{k-1} (1-\pi)^{n_i-k} q_i}{k} \right).$$

In these expressions, $\bar{\pi}$ approximates the probability of visiting a customer.

For the above problem, a heuristic algorithm that consists of two phases: i) a constructive phase in which the location, the allocation, and the routes are decided using different exact and heuristic procedures, and ii) an improvement phase based on local search. Lower bounds are also proposed in that work. The interested reader should refer to that paper for the algorithmic details.

10.3 Time-Oriented Location Routing

In the previous section, we focused on cost-minimization objectives in location-routing problems under uncertainty. In this section, we turn our attention to the travel times. There are many applications in which travel time is an aspect to account for and a related measure is to be optimized.

A recent research trend in location-routing problems involves latency, i.e., the sum of the arrival times at the demand nodes. The latency location-routing problem extends the cumulative capacitated vehicle routing problem [401], which, in turn, generalizes the traveling repairman problem [210]. The cumulative capacitated vehicle routing problem involves a single depot and a fleet of capacitated vehicles dispatched to serve a set of demand nodes. The traveling repairman problem considers a single depot and a single uncapacitated vehicle.

Two formulations for the (deterministic) latency location-routing problem were reviewed in Sect. 2.2. The first results from adapting to the problem the model considered in [437, 438] for capacitated location-routing:

$$\text{minimize} \quad \sum_{m \in M} \sum_{j \in J} t_{jm}, \tag{10.45}$$

$$\text{subject to} \quad \sum_{j \in J} d_j x_{ij} \leq q_i y_i \qquad \forall i \in I, \tag{10.46}$$

$$(10.3)\text{–}(10.6),\ (10.8)\text{–}(10.11),$$

$$t_{jm} \geq t_{\ell m} + b_{\ell j} - \Theta(1 - v_{\ell jm}) \qquad \forall j \in J,\ \ell \in V,\ \ell \neq j,\ m \in M, \tag{10.47}$$

$$t_{jm} \geq 0 \qquad \forall j \in J,\ m \in M. \tag{10.48}$$

In this model, $b_{\ell\ell'}$ is the travel time between nodes ℓ and ℓ' with $\ell, \ell' \in V, \ell \neq \ell'$. Furthermore, t_{jm} is a decision variable representing the arrival time at demand node $j \in J$ if visited by vehicle $m \in M$. Θ is a large-enough value.

A notable attempt to handle the latency location-routing problem under uncertainty presented in [574], who consider uncertainty stemming from the demand. In addition to a latency objective, the authors also consider a cost-oriented one. The latter accounts for installing the facilities, using the vehicles, transportation, and shortage/surplus at the demand nodes. The authors assume that a decision must be made about the delivery quantities before the demand is known. Such quantities must lie within a given range. The authors also assume a vehicle-dependent fixed cost and capacity.

We can adapt the above model to the problem investigated in [574]. Consider H_m and F_m as the capacity and fixed cost of vehicle $m \in M$, respectively. Let ϕ_1 (ϕ_2) be the unit penalty cost for shortage (surplus) at the demand nodes. Regarding the decision variables, let s_j be the amount shipped to demand node $j \in J$. Finally, we denote by $\underline{d}_j$ and $\overline{d}_j$ the lower and upper thresholds for the amount shipped to demand node $j \in J$. If the demands are given, we can consider the following bi-objective nonlinear model:

$$\text{minimize} \quad \sum_{j\in J}\sum_{m\in M} t_{jm}, \tag{10.49}$$

$$\begin{aligned}\text{minimize} \quad & \sum_{i\in I} f_i y_i + \sum_{\ell,\ell'\in V:\ell\neq\ell'} c_{\ell\ell'} v_{\ell\ell' m} + \sum_{m\in M}\left(F_m \sum_{i\in I}\sum_{j\in J} v_{ijm}\right)\\ & + \phi_1 \sum_{j\in J}(d_j - s_j)^+ + \phi_2 \sum_{j\in J}(s_j - d_j)^+,\end{aligned} \tag{10.50}$$

$$\text{subject to} \quad \sum_{j\in J} s_j x_{ij} \leq q_i y_i, \qquad \forall i \in I, \tag{10.51}$$

$$\sum_{j\in J}\left(s_j \sum_{\ell\in V:\ell\neq j} v_{\ell jm}\right) \leq H_m \quad \forall m \in M, \tag{10.52}$$

$$\underline{d}_j \leq s_j \leq \overline{d}_j \qquad \forall j \in J, \tag{10.53}$$

(10.3), (10.5), (10.6), (10.8)–(10.11), (10.47), (10.48).

The objective function (10.49) accounts for the total system latency, while (10.50) stands for the total cost. Constraints (10.51) regard the capacity of the facilities, whereas (10.52) ensure that the capacity of the vehicles is satisfied as well. Constraints (10.53) guarantees that the amounts shipped to the demand nodes lie in the given bounds.

The nonlinearity of Constraints (10.51) and (10.52) can be easily overcome. The former can be replaced by

$$\sum_{j\in J} \varrho_{ij} \leq q_i y_i \qquad \forall i \in I, \tag{10.54}$$

$$\varrho_{ij} \geq s_j - \Theta(1 - x_{ij}) \qquad \forall i \in I,\ j \in J, \tag{10.55}$$

$$\varrho_{ij} \leq s_j + \Theta(1 - x_{ij}) \qquad \forall i \in I,\ j \in J, \tag{10.56}$$

$$\varrho_{ij} \geq 0 \qquad \forall i \in I,\ j \in J. \tag{10.57}$$

Constraint (10.52) can be replaced by

$$\sum_{j\in J}\sum_{\ell\in V:\ell\neq j} \vartheta_{\ell jm} \leq H_m \qquad \forall m \in M, \tag{10.58}$$

$$\vartheta_{\ell jm} \geq s_j - \Theta(1 - v_{\ell jm}) \qquad j \in J,\ \ell \in V : \ell \neq j,\ m \in M, \tag{10.59}$$

$$\vartheta_{\ell jm} \leq s_j + \Theta(1 - v_{\ell jm}) \qquad j \in J,\ \ell \in V : \ell \neq j,\ m \in M, \tag{10.60}$$

$$\vartheta_{\ell jm} \geq 0 \qquad j \in J,\ \ell \in V : \ell \neq j,\ m \in M. \tag{10.61}$$

In the above expressions, Θ represents a large enough number. Let us assume that a finite set of scenarios, say Ω, has been identified for the demand vector, and let π_ω be the probability associated with scenario ω ($\pi_\omega \geq 0$, $\forall \omega \in \Omega$ and $\sum_{\omega \in \Omega} \pi_\omega = 1$). Under this setting, a bi-objective model to minimize the conditional value-at-risk of the regrets concerning time and cost is proposed in [574]. In fact, for each scenario $\omega \in \Omega$, we can compute the optimal latency and cost by solving the above model considering the corresponding single-objective function and the demands according to the scenario given. For each scenario ω, let us denote by V_ω^{T*} and V_ω^{T*} the optimal values obtained for the single-objective problems (cost and time) considering the demands determined by the scenario.

Now, following the terminology used in Sect. 5.2, for a feasible solution $\hat{\boldsymbol{\Upsilon}} = \{\hat{\boldsymbol{s}}, \hat{\boldsymbol{t}}, \hat{\boldsymbol{v}}, \hat{\boldsymbol{x}}, \hat{\boldsymbol{y}}\}$, we can define its regret in each scenario w.r.t. time and w.r.t. cost:

$$\Gamma_\omega^T(\hat{\boldsymbol{\Upsilon}}) = V_\omega^T(\hat{\boldsymbol{\Upsilon}}) - V_\omega^{T*}, \tag{10.62}$$

$$\Gamma_\omega^C(\hat{\boldsymbol{\Upsilon}}) = V_\omega^C(\hat{\boldsymbol{\Upsilon}}) - V_\omega^{C*}, \tag{10.63}$$

where $V_\omega^T(\hat{\boldsymbol{\Upsilon}})$ ($V_\omega^C(\hat{\boldsymbol{\Upsilon}})$) is the value of the objective function (10.49) ((10.50)) computed using solution $\hat{\boldsymbol{\Upsilon}}$ and the demands in scenario ω. From here, we can, for instance, define a bi-objective minmax regret model and thus seek to find Pareto solutions considering as objective functions the maximum regret w.r.t. time and cost.

Zhong et al. [574] generalize this idea: They consider the CVaR$_\alpha$ for the regret concerning time cost. Using all the elements gathered so far (and also recalling the contents of Sect. 3.5), we can formulate the problem as follows:

$$\text{minimize} \quad \eta^T + \frac{1}{1-\alpha} \sum_{\omega \in \Omega} \pi_\omega \zeta_\omega^T, \tag{10.64}$$

$$\text{minimize} \quad \eta^C + \frac{1}{1-\alpha} \sum_{\omega \in \Omega} \pi_\omega \zeta_\omega^C, \tag{10.65}$$

subject to (10.3), (10.5), (10.6), (10.8)–(10.11), (10.47), (10.48), (10.53)–(10.61),

$$\zeta_\omega^C \geq \Gamma_\omega^C(\boldsymbol{\Upsilon}) - \eta^C \qquad \forall \omega \in \Omega, \tag{10.66}$$

$$\zeta_\omega^T \geq \Gamma_\omega^T(\boldsymbol{\Upsilon}) - \eta^T \qquad \forall \omega \in \Omega, \tag{10.67}$$

$$\zeta_\omega^T \geq 0 \qquad \forall \omega \in \Omega, \tag{10.68}$$

$$\zeta_\omega^C \geq 0 \qquad \forall \omega \in \Omega. \tag{10.69}$$

In this model, $V_\omega^C(\boldsymbol{\Upsilon})$ (relevant for computing $\Gamma_\omega^C(\boldsymbol{\Upsilon})$ is given by

$$\sum_{i\in I} f_i y_i + \sum_{\ell,\ell'\in V:\ell\neq\ell'} c_{\ell\ell'} v_{\ell\ell' m} + \sum_{m\in M}\left(F_m \sum_{i\in I}\sum_{j\in J} v_{ijm}\right) + \phi_1 \sum_{j\in J}(d_j - s_j)^+ + \phi_2 \sum_{j\in J}(s_j - d_j)^+.$$

Zhong et al. [574] develop a metaheuristic for finding approximate Pareto solutions to the above problem. The interested reader should refer to that work for the algorithmic details.

Despite considering a time-oriented objective function, the above contents still neglect the travel times as a source of uncertainty. Nevertheless, the latter may also be of practical relevance. In this context, we recall another formulation discussed in Chap. 2 for the (deterministic) latency minimization location-routing problem:

$$\text{minimize} \quad \sum_{i\in I}\sum_{j\in J}\sum_{r\in R} r b_{ij} v_{ijr} + \sum_{\substack{j,j'\in J\\ j'\neq j}}\sum_{r\in R} r b_{jj'} v_{jj'r}, \tag{10.70}$$

$$\text{subject to} \quad \sum_{r\in R} z_{jr} = 1 \qquad \forall j \in J, \tag{10.71}$$

$$\sum_{r\in R}\sum_{i\in I}\sum_{j\in J} v_{ijr} \leq |M|, \tag{10.72}$$

$$\sum_{j\in J} z_{j1} = \sum_{r\in R}\sum_{i\in I}\sum_{j\in J} v_{ijr}, \tag{10.73}$$

$$\sum_{j'\in J:j'\neq j} v_{jj'r} = z_{j,r+1} \qquad \forall j \in J,\ r \in \{1,\ldots,|R|-1\}, \tag{10.74}$$

$$\sum_{\ell\in V:\ell\neq j} v_{\ell jr} = z_{jr} \qquad \forall j \in J,\ r \in \{1,\ldots,|R|-1\}, \tag{10.75}$$

$$v_{ij,|R|} = z_{j,|R|} \qquad \forall i \in I,\ j \in J, \tag{10.76}$$

$$v_{ijr} \leq y_i \qquad \forall i \in I,\ j \in J,\ r \in R, \tag{10.77}$$

$$\sum_{i\in I} w_{ij} + \sum_{j'\in J:j\neq j} u_{j'j} - \sum_{j'\in J:j'\neq j} u_{jj'} = d_j \qquad \forall j \in J, \tag{10.78}$$

$$d_j \sum_{r\in R} v_{ijr} \leq w_{ij} \qquad \forall i \in I,\ j \in J, \tag{10.79}$$

$$d_j \sum_{r \in R} v_{j'jr} \leq u_{j'j} \qquad \forall j, j' \in J,\ j \neq j', \tag{10.80}$$

$$w_{ij} \leq H \sum_{r \in R} v_{ijr} \qquad \forall i \in I,\ j \in J, \tag{10.81}$$

$$u_{jj'} \leq (H - d_j) \sum_{r \in R} v_{jj'r} \qquad \forall j, j' \in J,\ j \neq j', \tag{10.82}$$

$$u_{jj'} \geq 0 \qquad \forall j, j' \in J,\ j \neq j', \tag{10.83}$$

$$v_{\ell jr} \in \{0, 1\} \qquad \forall j \in J,\ \ell \in V,\ \ell \neq j,\ r \in R, \tag{10.84}$$

$$w_{ij} \geq 0 \qquad \forall i \in I,\ j \in J, \tag{10.85}$$

$$y_i \in \{0, 1\} \qquad \forall i \in I, \tag{10.86}$$

$$z_{jr} \in \{0, 1\} \qquad \forall j \in J,\ r \in R. \tag{10.87}$$

This model stems from considering a multi-layer network representing the problem. Each layer represents the backward position of a node in a route. The set of layers is denoted by $R = \{1, 2, \ldots\}$ with its cardinality established by an upper bound on the number of demand nodes we may observe in a route. Each layer from 2 to $|R|$ contains $|I| + |J|$ nodes, with the first $|I|$ corresponding to the potential locations for the depots and the remaining $|J|$ corresponding to the demand nodes. The first layer only contains a copy of the demand nodes.

In the above formulation, for $r \in R$, $j \in J$, $\ell \in V$, $\ell \neq j$, $v_{\ell jr}$ is a binary variable that is equal to one if and only if demand node $j \in J$ is visited at layer r immediately after the route has visited node ℓ. For $i \in I$, y_i keeps indicating whether location i is selected for installing a depot. z_{jr} is a binary variable equal to one if and only if demand node $j \in J$ is visited at layer $r \in R$. For $i \in I$ and $j \in J$, w_{ij} denotes the load of a vehicle leaving depot i to visit demand node j next. Finally, $u_{jj'}$ is the load of a vehicle when it is departing from demand node $j \in J$ to immediately visit demand node $j' \in J$, $j' \neq j$. The reader should refer to Sect. 2.2 for further details.

Embedding uncertainty in this problem raises some concerns. First, we should ensure at least relatively complete recourse. One possibility is to install the depots here and now, leaving the design of the routes to a second-stage decision level.

In a latency minimization location-routing problem, the travel times represent a source of uncertainty particularly relevant that we have not considered so far in this chapter. We could keep considering uncertainty in the demands. Nevertheless, for the moment, we assume these to be deterministic to better focus the discussion on the travel times.

In search of tractable optimization models, we assume uncertainty captured by a finite set of scenarios, say Ω. As customary in this volume, when such an assumption is made, we note that even if the number of scenarios is infinite (e.g., there is a continuous distribution describing the underlying random vector) or if it is finite but too large to be tractable directly, then the discussion that follows may still be of interest if embedded in a sampling scheme.

We can now consider scenario-indexed uncertain parameters and second-stage decision variables. Accordingly, $b_{\ell\ell'\omega}$ is the traveling time between nodes ℓ and ℓ' under scenario $\omega \in \Omega$, $\ell, \ell' \in V, \ell \neq \ell'$. Regarding the decision variables, for each scenario $\omega \in \Omega$, we consider $v_{\ell jr\omega}$, a binary variable that is equal to one if and only if demand node $j \in J$ is visited at layer $r \in R$ immediately after the route has visited node $\ell \in V, \ell \neq j$ under scenario ω, $z_{jr\omega}$ is a binary variable equal to one if and only if demand node $j \in J$ is visited at layer $r \in R$ under scenario ω, $w_{ij\omega}$ is the load of a vehicle that is leaving depot $i \in I$ to immediately visit demand node $j \in J$ under scenario ω, and $u_{jj'\omega}$ is the load of a vehicle when it is departing from demand node j to immediately visit demand node j' under scenario ω, $j, j' \in J$, $j' \neq j$.

Assuming a risk-neutral attitude toward risk, the extensive form of the deterministic equivalent writes as follows:

$$\text{minimize} \quad \sum_{\omega\in\Omega} \pi_\omega \left(\sum_{i\in I}\sum_{j\in J}\sum_{r\in R} r b_{ij\omega} v_{ijr\omega} + \sum_{\substack{j,j'\in J \\ j'\neq j}} \sum_{r\in R} r b_{jj'\omega} v_{jj'r\omega} \right), \tag{10.88}$$

$$\text{subject to} \quad \sum_{r\in R} z_{jr\omega} = 1 \qquad \forall j \in J,\ \omega \in \Omega, \tag{10.89}$$

$$\sum_{r\in R}\sum_{i\in I}\sum_{j\in J} v_{ijr\omega} \leq |M| \qquad \forall \omega \in \Omega, \tag{10.90}$$

$$\sum_{j\in J} z_{j1\omega} = \sum_{r\in R}\sum_{i\in I}\sum_{j\in J} v_{ijr\omega} \qquad \forall \omega \in \Omega, \tag{10.91}$$

$$\sum_{j'\in J:j'\neq j} v_{jj'r\omega} = z_{j,r+1,\omega} \qquad \forall j \in J,\ r \in R,\ \omega \in \Omega, \tag{10.92}$$

$$\sum_{\ell\in V:\ell\neq j} v_{\ell jr\omega} = z_{jr\omega} \qquad \forall j \in J,\ r \in R,\ \omega \in \Omega, \tag{10.93}$$

$$v_{ij,|R|,\omega} = z_{j,|R|,\omega} \qquad \forall i \in I,\ j \in J,\ \omega \in \Omega, \tag{10.94}$$

$$v_{ijr\omega} \leq y_i \qquad \forall i \in I,\ j \in J,\ r \in R,\ \omega \in \Omega, \tag{10.95}$$

$$\sum_{i\in I} w_{ij\omega} + \sum_{j'\in J:j\neq j} u_{j'j\omega} - \sum_{j'\in J:j'\neq j} u_{jj'\omega} = d_j \qquad \forall j \in J,\ \omega \in \Omega, \tag{10.96}$$

$$d_j \sum_{r\in R} v_{ijr\omega} \leq w_{ij\omega} \qquad \forall i \in I,\ j \in J,\ \omega \in \Omega, \tag{10.97}$$

$$d_j \sum_{r \in R} v_{j'jr\omega} \leq u_{j'j\omega} \qquad \forall j, j' \in J,\ j \neq j',\ \omega \in \Omega, \tag{10.98}$$

$$w_{ij\omega} \leq H \sum_{r \in R} v_{ijr\omega} \qquad \forall i \in I,\ j \in J,\ \omega \in \Omega, \tag{10.99}$$

$$u_{jj'} \leq (H - d_j) \sum_{r \in R} v_{jj'r\omega} \qquad \forall j, j' \in J,\ j \neq j', \omega \in \Omega, \tag{10.100}$$

$$u_{jj'\omega} \geq 0 \qquad \forall j, j' \in J,\ j \neq j',\ \omega \in \Omega, \tag{10.101}$$

$$v_{\ell jr\omega} \in \{0, 1\} \qquad \forall j \in J,\ \ell \in V,\ \ell \neq j,\ r \in R,\ \omega \in \Omega, \tag{10.102}$$

$$w_{ij\omega} \geq 0 \qquad \forall i \in I,\ j \in J,\ \omega \in \Omega, \tag{10.103}$$

$$y_i \in \{0, 1\} \qquad \forall i \in I, \tag{10.86}$$

$$z_{jr\omega} \in \{0, 1\} \qquad \forall j \in J,\ r \in R,\ \omega \in \Omega. \tag{10.104}$$

Note that the above model becomes ill-defined if there is uncertainty associated with the demands since the upper bound on the number of layers in the multi-layered network representing the problem depends on that. In such a case, it may be better to resort to the model initially presented in Sect. 2.2 for the latency routing problem. The scarcity of work in the above domains motivates more research to explore these features.

10.4 Multi-period Location Routing

Location-routing problems involve two types of decisions often associated with different time scales. The location of the facilities and the allocation of demand nodes are often a strategic or tactical decision that should last for a considerable time, e.g., a month or a year. On the other hand, the routing decisions can often be made on an operational basis, e.g., daily. Nevertheless, under uncertainty, it is desirable to explicitly include the routing decisions in the model seeking an integrated solution to hedge against the uncertainty.

Next, we adopt most of the notation in use in this volume. We consider a finite planning horizon divided into several periods. There is a set of customers originating demand in each such period. Under uncertainty, such demands define a compound stochastic process that we assume to be stationary. Furthermore, we assume that a set of relevant realizations of that process was identified beforehand that constitute a finite set of scenarios, say Ω.

To keep the notation simple, we work with time-invariant transportation costs. Nevertheless, it is straightforward to relax this assumption. Let T be a finite set representing the multi-period planning horizon. We denote by $d_{jt\omega}$ the demand on node $j \in J$ in period $t \in T$ under scenario $\omega \in \Omega$ and let $v_{\ell\ell' mt\omega}$ be a binary variable equal to one if and only if route $m \in M$ serves node ℓ' immediately after serving node ℓ in period $t \in T$ under scenario $\omega \in \Omega$, $\ell, \ell' \in V$, $\ell \neq \ell'$. Model (10.16)–(10.23) can be easily adapted to this setting leading to

$$\text{minimize} \quad \sum_{i\in I} f_i y_i + \sum_{\omega\in\Omega} \pi_\omega \left(\sum_{t\in T}\sum_{m\in M}\sum_{\substack{\ell,\ell'\in V\\ \ell\neq\ell'}} c_{\ell\ell'} v_{\ell\ell' mt\omega} + F \sum_{t\in T}\sum_{m\in M}\sum_{\ell\in I}\sum_{\ell'\in J} v_{\ell\ell' mt\omega} \right), \tag{10.105}$$

subject to (10.2), (10.10), (10.11),

$$\sum_{m\in M}\sum_{\ell\in V:\ell\neq j} v_{\ell jmt\omega} = 1 \qquad \forall j \in J,\ t \in T,\ \omega \in \Omega, \tag{10.106}$$

$$\sum_{j\in J}\left(d_{jt\omega} \sum_{\ell\in V:\ell\neq j} v_{\ell jmt\omega} \right) \leq H \quad \forall m \in M,\ t \in T,\ \omega \in \Omega, \tag{10.107}$$

$$\sum_{\ell'\in V:\ell'\neq\ell} v_{\ell\ell' mt\omega} - \sum_{\ell'\in V:\ell'\neq\ell} v_{\ell'\ell mt\omega} = 0 \qquad \forall m \in M,\ \ell \in V,\ t \in T,\ \omega \in \Omega, \tag{10.108}$$

$$\sum_{\ell\in I}\sum_{\ell'\in J} v_{\ell\ell' mt\omega} \leq 1 \qquad \forall m \in M,\ t \in T,\ \omega \in \Omega, \tag{10.109}$$

$$\sum_{m\in M}\sum_{\ell\in S}\sum_{\ell'\in V\setminus S} v_{\ell\ell' mt\omega} \geq 1 \qquad \forall I \subseteq S \subset V,\ t \in T,\ \omega \in \Omega, \tag{10.110}$$

$$\sum_{\ell' \in J} v_{i\ell' mt\omega} + \sum_{\ell' \in V:\ell' \neq j} v_{j\ell' mt\omega} \leq 1 + x_{ij}$$

$$\forall i \in I,\ j \in J,\ m \in M,\ t \in T,\ \omega \in \Omega, \qquad (10.111)$$

$$v_{\ell\ell' mt\omega} \in \{0, 1\} \qquad \forall \ell, \ell' \in V,\ \ell \neq \ell',\ m \in M,\ t \in T,\ \omega \in \Omega. \qquad (10.112)$$

This model is quite challenging due to the five-index binary variables involved.

As pointed out in [313, 314] in many logistics applications, the commodities produced or transformed at a company are to be shipped to the customers using external transportation companies. In that case, the design of the available routes is assumed to be given—they are exogenous—and the decision-maker seeks to select those to be used out of a set of possibilities offered by the external partners. Furthermore, the offer may change over time: One route may be available in some periods and not in others. This is the case with less-than-truck-load transportation often provided by maritime transportation carriers or freight transportation companies.

The same authors introduce what they call a stochastic multi-period transportation problem. The problem consists of locating a set of depots and assigning a set of demand nodes to them to satisfy a set of time-varying demands. The total cost is to be minimized. A single commodity is considered. The depots and their configuration (demand nodes assigned to them) are a decision to be implemented now and should hold for the entire planning horizon. Single allocation of demand nodes is assumed. Then, depending on the orders received in each period, the distribution plan is adapted throughout time, possibly changing from one period to another.

Once an order is received, a shipping mode is selected. Orders that are larger than a full-truck-load (FTL) are shipped as much as possible directly to the demand nodes. The remaining amount is then included in one of the routes available at the depot, other than direct full-truck-load. The transportation cost corresponds to the freights selected. There is a unit profit for the commodities shipped from a depot to a demand node. This profit reflects the production and acquisition cost at the origin and revenue at the destination.

We assume that uncertainty stems from the demands with a finite set of scenarios capturing it. Each scenario determines the demand for the entire planning horizon.

In addition to the notation already introduced, we consider the following:

O, set of routes offered by the external partners (direct full-truck-loads and other routes)

O^F_{ijt}, ($\subset O$) set of full-truck-load transportation offers between depot $i \in I$ and demand node $j \in J$ in period $t \in T$;

O^L_{it}, ($\subset O$) set of routes different from direct full-truck-load offered to depot $i \in I$ in period $t \in T$;

g_o fixed cost for using transportation offer $o \in O$;

H_o capacity of one vehicle performing route $o \in O$;

p_{ij} unit profit made for the commodity shipped from depot $i \in I$ to demand node $j \in J$.

Regarding the decision variables, in addition to the customary location and allocation variables (y_i and x_{ij}), we define $u_{oit\omega}$ representing the number of full-truck-load shipments using offer $o \in O_{it}^F$ originated at depot $i \in I$ in period $t \in T$ under scenario $\omega \in \Omega$; a binary variable $v_{oit\omega}$ is also adopted that is equal to one if and only if route $o \in O_{it}^L$ is used by the depot $i \in I$ in period $t \in T$ under scenario $\omega \in \Omega$.

Additionally, we consider an auxiliary variable $d'_{jt\omega}$ denoting the amount to be shipped to demand node $j \in J$ in period $t \in T$ under scenario $\omega \in \Omega$ by routes other than direct full-truck-load. Finally, we use an indicator δ_{oj} equal to one if demand node $j \in J$ is part of route $o \in O$ and zero otherwise.

The problem can be formulated as follows:

$$\text{minimize} \quad \sum_{i \in I} f_i y_i + \sum_{\omega \in \Omega} \pi_\omega \Bigg(\sum_{t \in T} \sum_{i \in I} \sum_{o \in O_{ijt}^F} g_o u_{oit\omega} + \sum_{o \in O_{it}^L} g_o v_{oit\omega} - \sum_{j \in J} p_{ij} d_{j\omega} x_{ij} \Bigg), \tag{10.113}$$

$$\text{subject to} \quad \sum_{i \in I} x_{ij} = 1 \qquad \forall j \in J, \tag{10.114}$$

$$x_{ij} \le y_i \qquad \forall i \in I,\ j \in J, \tag{10.115}$$

$$d'_{jtw} = d_{jt\omega} x_{ij} - \sum_{o \in O_{ijt}^F} H_o u_{oit\omega} \qquad \forall j \in J,\ t \in T,\ \omega \in \Omega, \tag{10.116}$$

$$\sum_{o \in O_{it}^L} \delta_{oj} v_{oit\omega} = x_{ij} \qquad \forall t \in T,\ i \in I,\ j \in J : d_{jt\omega} > 0, \tag{10.117}$$

$$\sum_{j \in o} \vartheta_{ijtw} \le H_o v_{oit\omega} \qquad \forall o \in O_{it}^L,\ i \in I,\ t \in T, \omega \in \Omega, \tag{10.118}$$

$$\vartheta_{ijtw} \ge d'_{jt\omega} - \Theta(1 - x_{ij}) \qquad \forall i \in I,\ j \in J,\ t \in T,\ \omega \in \Omega, \tag{10.119}$$

$$\vartheta_{ijtw} \le d'_{jt\omega} + \Theta(1 - x_{ij}) \qquad \forall i \in I,\ j \in J,\ t \in T,\ \omega \in \Omega, \tag{10.120}$$

$$\vartheta_{ijtw} \ge 0 \qquad \forall i \in I,\ j \in J,\ t \in T,\ \omega \in \Omega, \tag{10.121}$$

$$d'_{jt\omega} \ge 0 \qquad \forall j \in J,\ t \in T,\ \omega \in \Omega, \tag{10.122}$$

$$x_{ij} \in \{0, 1\} \qquad \forall i \in I\ j \in J, \tag{10.123}$$

$$y_i \in \{0, 1\} \qquad \forall i \in I. \tag{10.124}$$

The objective function (10.113) accounts for the fixed cost for the depots plus the total expected cost with transportation minus profit. Constraints (10.114) ensure that every demand node is assigned to a depot (for the entire planning horizon). Inequalities (10.115) guarantee that demand is allocated to open depots. Equalities (10.116) account for the demand of a customer in some period and scenario that will be shipped through routes other than direct FTL. With that purpose, Constraints (10.117) impose that every demand node with demand in some period and scenario should be allocated to one route other than direct FTL. Inequalities (10.118)–(10.121) guarantee that the capacity of the routes is respected. Finally, we have the domain constraints.

Klibi et al. [313, 314] use a formulation slightly different from the above one. Moreover, they consider the possibility that not all depots can serve all demand nodes. In the first work, a sample average approximation scheme is proposed for finding feasible solutions to the problem. In fact, given that each scenario determines the demand for all periods of the planning horizon, one may end up with a large number of possibilities. Still, some limitations may emerge from the need to solve many mixed-integer programs. For this reason, in [313] the authors develop a heuristic algorithm capable of tackling problems of realistic sizes. In [314] they analyze different possibilities for approximating the original model. These include (i) scenario reduction (e.g., via sample average approximation), (ii) demand aggregation into a single planning period, and (iii) customer aggregation in a small number of customer zones. The interested reader should refer to that work for all details.

10.5 Reliable Location Routing

An LRP seeking to hedge against disruptions in its infrastructure is called a reliable location-routing problem (RLRP). Such disruptions may occur both at a depot level and at a vehicle level. In both cases, a decrease in the operational capacity is observed.

Let us start by considering model (10.1)–(10.11) and the underlying notation. We now assume the capacities of the vehicles to be vehicle-specific and thus denoted by H_m, $m \in M$.

In this section, we assume that demands and costs are deterministic, i.e., uncertainty stems only from the actual capacity of the depots and vehicles. In particular, for $i \in I$, let ϱ_i be the random variable representing the actual capacity of depot $i \in I$ (smaller than or equal to q_i); for $m \in M$, let γ_m be the random variable representing the actual capacity of vehicle (smaller than or equal to H_m).

Due to the possibility of facing decreased capacities in the depots or vehicles, we may end up with a demand shortage, i.e., an amount shipped to a customer below its actual demand. If this is the case, a penalty is incurred (opportunity loss cost, delay cost, outsourcing, *et cetera*). For a customer $j \in J$, let b_j be the unit shortage cost. Additionally, in case of a vehicle disruption, an extra cost, say β, is incurred for each unitary decrease in its capacity.

Assuming a risk-neutral decision-maker, we can formulate a reliable location-routing problem as follows:

$$\text{minimize} \quad \sum_{i \in I} f_i y_i + \sum_{m \in M} \sum_{\substack{\ell,\ell' \in V \\ \ell \neq \ell'}} c_{\ell\ell' m} v_{\ell\ell' m} + F \sum_{m \in M} \sum_{i \in I} \sum_{j \in J} v_{ijm} + Q(\boldsymbol{v}, \boldsymbol{x}, \boldsymbol{y}), \tag{10.125}$$

$$\text{subject to} \quad (10.1)\text{–}(10.11),$$

with

$$Q(\boldsymbol{v}, \boldsymbol{x}, \boldsymbol{y}) = \mathbb{E}\left[\beta \sum_{m \in M} \left((H_m - \gamma_m) \sum_{i \in I} \sum_{j \in J} v_{ijm} \right) + \sum_{i \in I} b_i \max\left\{ \sum_{j \in J} d_j x_{ij} - \varrho_i, 0 \right\} \right].$$

The recourse function computes the total expected cost related to disruptions—vehicles and depots. Note that the first-stage decision is a feasible solution to the location-routing problem induced by the original (no disruption) capacities. If the underlying random variables have a finite support, we can go beyond and write the extensive form of the deterministic equivalent, as shown next.

Let Ω be the set of scenarios, each determining a vector of capacities at the depots and capacities for the vehicles. In particular, for $\omega \in \Omega$, let $\varrho_{i\omega}$ be the actual capacity at depot $i \in I$ and $H_{m\omega}$ the actual capacity of vehicle $m \in M$ under that scenario. If π_ω is the probability that scenario $\omega \in \Omega$ occurs, the problem can be reformulated. We omit that simple exercise here. Instead, we proceed to a risk-averse measure, namely the conditional value-at-risk for a given confidence level α, which generalizes the expectation considered above:

$$\text{minimize} \quad \sum_{i \in I} f_i y_i + \sum_{m \in M} \sum_{\substack{\ell,\ell' \in V \\ \ell \neq \ell'}} c_{\ell\ell' m} v_{\ell\ell' m} + F \sum_{m \in M} \sum_{i \in I} \sum_{j \in J} v_{ijm} + \eta + \frac{1}{1-\alpha} \sum_{\omega \in \Omega} \pi_\omega \zeta_\omega, \tag{10.126}$$

$$\text{subject to} \quad (10.1)\text{–}(10.11),$$

$$\zeta_\omega \geq \left(\beta \sum_{m \in M} \left((H_{m\omega} - \gamma_{m\omega}) \sum_{i \in I} \sum_{j \in J} v_{ijm} \right) + \sum_{i \in I} b_i \max \left\{ \sum_{j \in J} d_j x_{ij} - \varrho_{i\omega}, 0 \right\} \right) - \eta \qquad \forall \omega \in \Omega, \tag{10.127}$$

$$\zeta_\omega \geq 0 \qquad \forall \omega \in \Omega. \tag{10.128}$$

Ahmadi-Javid and Seddighi [3] investigate a reliable location-routing problem in line with what we have just described, i.e., considering a first-stage decision given by a feasible solution to the location-routing problem induced by the original capacities. The authors assume that the problem has been devised for some time, during which the vehicles should make multiple visits to the demand nodes. The capacity of the vehicles is given by the number of visits they can do (irrespective of the customers). Furthermore, demand is represented by the number of visits to each customer. There are no fixed costs for using the vehicles ($F_m = 0$, $m \in M$). With these changes, the objective function fitting a risk-neutral decision-maker becomes

$$f_i y_i + \sum_{m \in M} \left(H_m \sum_{\substack{\ell, \ell' \in V \\ \ell \neq \ell'}} c_{\ell\ell' m} v_{\ell\ell' m} \right) + \mathcal{Q}(\boldsymbol{v}, \boldsymbol{x}, \boldsymbol{y}), \tag{10.129}$$

with

$$\mathcal{Q}(\boldsymbol{v}, \boldsymbol{x}, \boldsymbol{y}) = \mathbb{E} \left[\beta \sum_{m \in M} \left((H_m - \gamma_m) \sum_{\substack{\ell, \ell' \in V \\ \ell \neq \ell'}} c_{\ell\ell' m} v_{\ell\ell' m} \right) + \sum_{i \in I} b_i \max \left\{ \sum_{j \in J} d_j x_{ij} - \varrho_i, 0 \right\} \right].$$

In addition to the above risk-neutral oriented objective, the authors in [3] also consider the CVaR$_\alpha$ as well as the extreme risk-aversion case seeking the minimization of the maximum penalty across the scenarios. Small- to medium-sized instances of the models are solved using a general-purpose solver. Large-scale instances are tackled by a heuristic algorithm based on simulated annealing. For the computational tests, the authors assume that the disruptions at the depots and those related to the vehicles are described by a uniform and binomial probability distribution, respectively. The interested reader can refer to that work for all the algorithmic details.

Under potential disruptions, a first-stage decision induced by the non-disruptive setting may be unsuitable for some scenarios. A more comprehensive second-stage decision may be desirable to devise better adaptive schemes and thus to better hedge against uncertainty. This is done in [572]. The authors consider possible disruptions at the depots. A disruption renders a depot non-operational. The depots to install define the first decision level. Then, we find the second-state decisions that comprise the allocation of customers to depots and the design of the routes. These decisions adapt to the occurring scenario—operational depots. Due to the limited capacity of the depots, some demand nodes may not be served. In this case, a penalty is incurred.

We can adopt the notation we have been considering, namely that underlying the base model (10.1)–(10.11). Nevertheless, the new setting calls for some changes.

A scenario is now a subset of depots that will become disrupted. If Ω is the complete set of scenarios, for each $\omega \in \Omega$, let δ_{is} be an indicator that is equal to one if facility $i \in I$ is disrupted in scenario ω and zero otherwise. If ς_i is the probability that a depot fails, then the probability associated with a specific scenario is given by

$$\pi_\omega = \prod_{i \in I: \delta_{i\omega}=1} \varsigma_i \cdot \prod_{i \in I: \delta_{i\omega}=0} (1 - \varsigma_i).$$

Given that both the allocation of the demand nodes to the depots and the routes adapt to the scenarios, we consider the corresponding decision variables indexed accordingly. For each scenario $\omega \in \Omega$, let $I_\omega = \{i \in I \mid \delta_{i\omega} = 0\}$. In this case, we consider $x_{ij\omega}$ as a binary variable equal to one if and only if demand node $j \in J$ is allocated to depot $i \in I_\omega$ under scenario $\omega \in \Omega$. Additionally, $v_{\ell\ell' m\omega}$ is a binary variable equal to one if route $m \in M$ visits node ℓ' immediately after visiting node ℓ under scenario $\omega \in \Omega$, $\ell, \ell' \in V$, $\ell \neq \ell'$.

To accommodate the case in which a demand node cannot be served due to the limited capacity of the operating depots, we consider an artificial depot, ν, to which the non-served customers in a scenario are allocated. In this case, the allocation cost is given by the penalty for not supplying a demand node, denoted by b_j for node $j \in J$.

Assuming a homogeneous fleet of vehicles, each with capacity H, the reliable location-routing problem we are considering can now be formulated as follows:

$$\text{minimize} \quad \sum_{i \in I} f_i y_i + \sum_{\omega \in \Omega} \pi_\omega \left(\sum_{m \in M} \sum_{\substack{\ell, \ell' \in V \\ \ell \neq \ell'}} c_{\ell\ell'} v_{\ell\ell' m\omega} \right.$$

$$\left. + F \sum_{m \in M} \sum_{i \in I_\omega} \sum_{j \in J} v_{ijm} + \sum_{j \in J} b_j x_{\nu j\omega} \right), \tag{10.130}$$

$$\text{subject to} \quad \sum_{j \in J} d_j x_{ij\omega} \leq (1 - \delta_{i\omega}) q_i y_i \qquad \forall i \in I_\omega,\ \omega \in \Omega, \tag{10.131}$$

$$\sum_{j \in J} \left(d_j \sum_{\ell \in V : \ell \neq j} v_{\ell j m \omega} \right) \leq H \qquad \forall m \in M,\ \omega \in \Omega, \tag{10.132}$$

$$\sum_{\ell' \in V : \ell' \neq \ell} v_{\ell \ell' m \omega} - \sum_{\ell' \in V : \ell' \neq \ell} v_{\ell' \ell m \omega} = 0 \qquad \forall m \in M,\ \ell \in V,\ \omega \in \Omega, \tag{10.133}$$

$$\sum_{\ell \in I} \sum_{\ell' \in J} v_{\ell \ell' m \omega} \leq 1 \qquad \forall m \in M,\ \omega \in \Omega, \tag{10.134}$$

$$\sum_{m \in M} \sum_{\ell \in S} \sum_{\ell' \in V \setminus S} v_{\ell \ell' m \omega} \geq 1 \qquad \forall I_\omega \subseteq S \subset V,\ \omega \in \Omega, \tag{10.135}$$

$$\sum_{\ell' \in J} v_{i \ell' m \omega} + \sum_{\ell' \in V : \ell' \neq j} v_{j \ell' m \omega} \leq 1 + x_{ij\omega} \qquad \forall i \in I_\omega,\ j \in J,\ m \in M,\ \omega \in \Omega, \tag{10.136}$$

$$v_{\ell \ell' m \omega} \in \{0, 1\} \qquad \forall \ell, \ell' \in V,\ \ell \neq \ell',\ m \in M,\ \omega \in \Omega, \tag{10.137}$$

$$x_{ij\omega} \in \{0, 1\} \qquad \forall i \in I_\omega,\ j \in J,\ \omega \in \Omega, \tag{10.138}$$

$$y_i \in \{0, 1\} \qquad \forall i \in I. \tag{10.139}$$

Zhang et al. [572] consider a similar problem, ignoring the fixed costs of using the vehicles. They consider a different set of routing variables. While ours are indexed in the vehicles ($m \in M$), theirs are indexed in the depots ($i \in I$).

The complexity of the problem calls for using an approximate procedure. A sampling technique is the choice made. It can be seen as a particular case of the sample average approximation method (see [312]). In particular, the authors use a so-called maximum-likelihood sampling and thus reduce the set of scenarios to a certain number that includes the most probable ones.

A heuristic is then proposed for tackling the reduced problem. First, an initial solution is generated. Here, two main decisions are made: the location of the depots and the allocation of the demand notes to them. This is accomplished by solving a capacitated facility location problem for a reference scenario (the one involving the highest number of non-disrupted depots) and then solving an allocation problem for each scenario considering the direct allocation costs to the depots.

When the demand nodes are clustered around the depots, the routing decisions are made (again, for each scenario in the reduced set). This process is initialized using the modified insertion criterion 1 (I1) algorithm proposed in [512] and then improved by a granular tabu search heuristic. Given an improved solution, a route-allocation problem is solved to optimize the number of depots and the allocation of the routes for the depots. This is done using a Lagrangian relaxation-based procedure. An attempt to further improve the solution is accomplished using simulated annealing that incorporates depot- and customer-based moves.

A different model is introduced in [548] for a reliable location-routing problem assuming possible depot disruptions (rendering them non-usable). The authors assume that depots are uncapacitated. This allows re-assigning to the operating depots the routes initially assigned to depots that turned out to be disrupted. The vehicles are assumed to be homogeneous.

A subset of customers put into a sequence is called a string if their total demand is below H. It corresponds to a set sequence that can be handled by one vehicle.

Let O denote the set of all strings. For each string $o \in O$, let δ_{oj} be a binary indicator equal to one if and only if demand node $j \in J$ is part of string o. The total demand of a string $o \in O$ is denoted by d_o; the cost of allocating that string to a depot $i \in I$ is represented by c_{oi}.

Once a depot fails, its allocated customers are served from other depots. Alternatively, the service is lost with a penalty cost incurred. For each string, define its rth choice as its level-r depot. Accordingly, each string is assigned up to $R \geq 1$ depots. A string ends up being served by its level-r depot ($r \leq R - 1$) if all its level-0, level-1, ..., level-$r - 1$ depots fail. If all R depots fail, a unit penalty cost of b is incurred for the unsupplied demand.

Let us denote by ς the probability that a depot fails (equal for all depots). The probability that a string is served by its level-r depot ($0 \leq r \leq R - 1$) is equal to $\varsigma^r(1 - \varsigma)$; the probability of incurring a penalty is ς^R.

To formulate the problem, we keep using the location variables y_i, $i \in I$. Additionally, we consider a binary variable z_o equal to one if and only if string $o \in O$ is used. Finally, we consider a binary variable w_{oir} equal to one if and only if depot $i \in I$ is the level-r choice for string $o \in O$, $0 \leq r \leq R - 1$. The reliable location-routing problem can now be formulated as follows:

$$\text{minimize} \quad \sum_{i \in I} f_i y_i + \sum_{o \in O} \left(\sum_{i \in I} \sum_{r=0}^{R-1} \varsigma^r (1 - \varsigma) c_{oi} w_{oir} + b \varsigma^R d_o \right), \tag{10.140}$$

$$\text{subject to} \quad \sum_{o \in O} \sum_{i \in I} \delta_{oj} w_{oir} = 1 \qquad \forall r = 0, \ldots, R-1, \ j \in J, \tag{10.141}$$

$$\sum_{o \in O} \sum_{r=0}^{R-1} \delta_{oj} w_{oir} \leq y_i \qquad \forall i \in I, \ j \in J, \tag{10.142}$$

$$\sum_{o \in O} \delta_{oj} z_o = 1 \qquad \forall j \in J, \tag{10.143}$$

$$\sum_{i \in I} w_{oir} \leq z_o \qquad \forall o \in O, \ r = 0, \ldots, R-1, \tag{10.144}$$

$$\sum_{o \in O} z_o \leq |M|, \tag{10.145}$$

$$w_{oir} \in \{0, 1\} \qquad \forall o \in O,\ i \in I,\ r = 0, \ldots, R-1, \tag{10.146}$$

$$y_i \in \{0, 1\} \qquad \forall i \in I, \tag{10.147}$$

$$z_o \in \{0, 1\} \qquad \forall o \in O. \tag{10.148}$$

The objective function (10.140) accounts for the total cost. It includes the setup cost for the depots plus the expected service cost. Constraints (10.141) ensure that every demand node is in a selected string that, in turn, is assigned to a depot for every possible level. Inequalities (10.142) guarantee that all customers are served within a string allocated only to open depots. Constraints (10.143) state that each customer should belong to one and only one string. Inequalities (10.144) state that an assignment of a string, irrespective of the level, can occur only if the string is used. Constraints (10.145) limit the number of activated strings to the number of available vehicles. Finally, (10.146)–(10.148) are the domain constraints.

Regarding the above model, note that $\sum_{o \in O} b\varsigma^R d_o$ is a constant. This can be dropped from the objective function. Furthermore, the following (intuitive) result holds:

Proposition 10.3 *There is always an optimal solution to the above problem such that* $c_{oi_1} \leq c_{oi_2}$, $\forall o \in O$, $\forall i_1, i_2 \in I$, $i_1 \neq i_2$ *if* $w_{oi_1r_1} = w_{oi_2r_2}$, $0 \leq r_1 < r_2 < R$.

Proof If an optimal solution does not satisfy the above property, we can reconfigure it so that it does without deteriorating the total cost. □

Xie et al. [548] relax Constraints (10.141) and (10.144) in a Lagrangian fashion. Considering a vector of non-negative multipliers for each one and denoting them by $\boldsymbol{\mu}$ and $\boldsymbol{\lambda}$, respectively, we obtain

$$\mathcal{L}(\boldsymbol{\mu}, \boldsymbol{\lambda}) = \min \quad \sum_{i \in I} f_i y_i + \sum_{o \in O} \sum_{i \in I} \sum_{r=0}^{R-1} \left(\varsigma^r (1 - \varsigma) c_{oi} - \sum_{j \in J} \mu_{rj} \delta_{oj} + \lambda_{or} \right) w_{oir}$$
$$- \sum_{o \in O} \sum_{r=1}^{R-1} \lambda_{or} z_o + \sum_{r=0}^{R-1} \sum_{j \in J} \mu_{rj}, \tag{10.149}$$

s. t. (10.142), (10.143), (10.145)–(10.148).

In turn, this problem can be decomposed into two subproblems:

$$\mathcal{L}_1(\boldsymbol{\lambda}) = \min \quad -\sum_{o\in O}\sum_{r=1}^{R-1} \lambda_{or} z_o + \sum_{r=0}^{R-1}\sum_{j\in J} \mu_{rj}, \tag{10.150}$$

$$\text{s. t.} \quad (10.143), (10.144), (10.148),$$

and

$$\mathcal{L}_2(\boldsymbol{\mu}, \boldsymbol{\lambda}) = \min \sum_{i\in I} f_i y_i + \sum_{o\in O}\sum_{i\in I}\sum_{r=0}^{R-1}\left(\varsigma^r(1-\varsigma)c_{oi} - \sum_{j\in J}\mu_{rj}\delta_{oj} + \lambda_{or}\right) w_{oir}, \tag{10.151}$$

$$\text{s. t.} \quad (10.142), (10.146), (10.147).$$

The second subproblem can be decomposed in the facilities. For each $i \in I$, we get

$$\mathcal{L}_{2i}(\boldsymbol{\mu}, \boldsymbol{\lambda}) = \min \quad f_i y_i + \sum_{o\in O}\sum_{r=0}^{R-1}\left(\varsigma^r(1-\varsigma)c_{oi} - \sum_{j\in J}\mu_{rj}\delta_{oj} + \lambda_{or}\right) w_{oir}, \tag{10.152}$$

$$\text{s. t.} \quad \sum_{o\in O}\sum_{r=0}^{R-1} \delta_{oj} w_{oir} \le y_i \qquad \forall j \in J, \tag{10.153}$$

$$w_{oir} \in \{0, 1\} \qquad \forall o \in O,\ r = 0, \ldots, R-1, \tag{10.154}$$

$$y_i \in \{0, 1\}. \tag{10.155}$$

Following [548], a column generation procedure can now be devised for the linear relaxation of subproblem 1 to obtain a tight bound on the optimal value of the subproblem. To see the way to accomplish this, we formulate the dual of the linear relaxation of subproblem 1:

$$\text{maximize} \quad \sum_{j\in J} \tau_j - M\theta, \tag{10.156}$$

$$\text{subject to} \quad \sum_{j\in J} \delta_{oj}\tau_j - \theta \le -\sum_{r=0}^{R-1} \lambda_{or} \qquad \forall o \in O, \tag{10.157}$$

$$\theta \ge 0. \tag{10.158}$$

From the above, we conclude that given a dual solution $(\hat{\boldsymbol{\tau}}, \hat{\theta})$, the reduced cost of string (column) $o \in O$ is

$$\hat{\gamma}_o = -\sum_{r=0}^{R-1} \lambda_{or} - \sum_{j \in J} \delta_{oj} \hat{\tau}_j + \hat{\theta}.$$

If $\hat{\gamma}_o \geq 0$, $\forall o \in O$, then the current dual solution is optimal; otherwise, we should find the string with the smallest reduced cost, which can be done by solving the following knapsack problem:

$$\text{minimize} \quad -\sum_{r=0}^{R-1} \lambda_{or} - \sum_{j \in J} \hat{\tau}_j \varphi_j + \hat{\theta}, \tag{10.159}$$

$$\text{subject to} \quad \sum_{j \in J} d_j \varphi_j \leq H, \tag{10.160}$$

$$\varphi_j \in \{0, 1\} \qquad \forall j \in J, \tag{10.161}$$

where φ_j is a binary variable equal to one if and only if demand node $j \in J$ is part of the selected string.

Consider now the second Lagrangian subproblem associated with a location $i \in I$. Naturally, its optimal objective function is zero if $y_i = 0$. Thus, we only need to focus on the case $y_i = 1$. Setting y_i accordingly in $\mathcal{L}_{2i}(\boldsymbol{\mu}, \boldsymbol{\lambda})$, the dual of the linear relaxation is given by

$$\text{maximize} \quad -\sum_{j \in J} \kappa_j, \tag{10.162}$$

$$\text{subject to} \quad -\sum_{j \in J} \delta_{oj} \kappa_j \leq \varsigma^r (1 - \varsigma) c_{oi} - \sum_{j \in J} \mu_{rj} \delta_{oj} + \lambda_{or}$$
$$\forall o \in O,\ r = 0, \ldots, R-1, \tag{10.163}$$

$$\kappa_j \geq 0 \quad \forall j \in J. \tag{10.164}$$

As above, a dual solution $\hat{\boldsymbol{\kappa}}$ to this problem induces a reduced cost for every string $o \in O$ and level $r = 0, \ldots, R-1$. The following problem can now be solved to find string-level pairs with negative reduced costs and identifying the one with the most negative value (if some exists):

$$\text{minimize} \quad \varsigma^r (1 - \varsigma) \tilde{c}_\varphi + \sum_{j \in J} (\hat{\kappa}_k - \mu_{rj}) \varphi_j + \lambda_{or}, \tag{10.165}$$

$$\text{subject to} \quad \sum_{j \in J} d_j \varphi_j \leq H, \tag{10.166}$$

$$\varphi_j \in \{0, 1\} \qquad \forall j \in J. \tag{10.167}$$

Again, φ_j is a binary variable equal to one if the demand node $j \in J$ is part of the "optimal" string and zero otherwise. $\tilde{c}_\varphi$ stands for the optimal tour length of the demand nodes $j \in J$ such that $\varphi_j = 1$. As pointed out in [548], the above problem can be seen as a prize-collecting TSP. The authors suggest finding approximate bounds for the problem given that several options are available (see, e.g., [422] and also the references therein).

To obtain feasible solutions to the whole problem (and thus collect all the ingredients for optimizing the Lagrangian dual), also in [548], the authors suggest taking the strings obtained by solving the subproblem 1 partitioning the customers using such strings. Then, a feasible solution to the problem can be found in terms of the $\boldsymbol{w}$- and $\boldsymbol{y}$-variables. As soon as a feasible solution is obtained, a local search procedure is used to improve it.

10.6 Arc-Routing Problems

In the previous sections, we focused on location problems with routing decisions seeking to satisfy the demand at a set of demand nodes. Nevertheless, as we reviewed in Chap. 2, in some cases, demand is found along the arcs of the underlying network. When location decisions are to be made for the depots supporting the service, we face a location-arc routing.

We recall the optimization model presented for the problem:

$$\text{minimize} \quad \sum_{i \in I} f_i y_i + \sum_{m \in M} \sum_{(\ell,\ell') \in \tilde{E}} \tilde{c}_{\ell\ell'} v_{\ell\ell'm} + F \sum_{m \in M} \sum_{\ell \in I} \sum_{\substack{\ell' \in J: \\ (\ell,\ell') \in \tilde{E}}} v_{\ell\ell'm}, \tag{10.168}$$

$$\text{subject to} \quad \sum_{(\ell,\ell') \in E_d} d_{(\ell,\ell')} x_{i(\ell,\ell')} \le q_i y_i \qquad \forall i \in I, \tag{10.169}$$

$$\sum_{m \in M} v_{\ell\ell'm} = 1 \qquad \forall (\ell,\ell') \in E_d, \tag{10.170}$$

$$\sum_{(\ell,\ell') \in E_d} d_{(\ell,\ell')} v_{\ell\ell'm} \le H \qquad \forall m \in M, \tag{10.171}$$

$$\sum_{\ell' \in \tilde{V}} v_{\ell\ell'm} - \sum_{\ell' \in \tilde{V}} v_{\ell'\ell m} = 0 \qquad \forall m \in M,\ \ell \in \tilde{V}, \tag{10.172}$$

$$\sum_{\ell \in I} \sum_{\ell' \in \tilde{V}} v_{\ell\ell'm} \le 1 \qquad \forall m \in M, \tag{10.173}$$

$$\sum_{m \in M} \sum_{\ell \in S} \sum_{\ell' \in \tilde{V} \setminus S} v_{\ell\ell'm} \ge 1 \qquad \forall I \subseteq S \subset \tilde{V}, \tag{10.174}$$

$$\sum_{\ell' \in \tilde{V} \setminus I} v_{i\ell' m} + v_{\ell\ell' m} \leq 1 + x_{i(\ell,\ell')} \quad \forall i \in I,\ (\ell,\ell') \in E_d,\ m \in M, \tag{10.175}$$

$$y_i \in \{0,1\} \qquad \forall i \in I, \tag{10.176}$$

$$x_{i(\ell,\ell')} \in \{0,1\} \qquad \forall i \in I,\ (\ell,\ell') \in E_d, \tag{10.177}$$

$$v_{\ell\ell' m} \in \{0,1\} \qquad \forall (\ell,\ell') \in E_d,\ m \in M. \tag{10.178}$$

There is an underlying transportation network $G = (V, E)$ with V containing I—the given potential locations for the depots—and possibly other nodes. E contains the available arcs. M is again the set of (homogeneous) vehicles available, each with capacity H. In the above model, $d_{(\ell,\ell')}$ stands for the demand in arc $(\ell,\ell') \in E$. Furthermore, E_d is the set of arcs in E that have demand; the other arcs can still be used as part of the routes to design, but no service will be provided. A complete graph, $\tilde{G} = (\tilde{V}, \tilde{E})$, is defined such that $\tilde{V}$ contains all nodes in I as well as all nodes that correspond to extremes of an arc in E_d. The unit cost associated with an arc $(\ell,\ell') \in \tilde{E}$ that we denote by $\tilde{c}_{\ell\ell'}$ is equal to $c_{\ell\ell'}$ if $(\ell,\ell') \in E_d$ and it is equal to the smallest transportation cost between ℓ and ℓ' in $G = (V, E)$ otherwise. If both $\ell, \ell' \in I$, then we set $\tilde{c}_{\ell\ell'} = \infty$.

To formulate the problem, the allocation variables were redefined so that they refer to arcs: $x_{i(\ell,\ell')}$ is equal to 1 if and only if the demand in arc $(\ell,\ell') \in E_d$ is supplied from facility $i \in I$. Regarding the routing variables v, they are now defined only in $\tilde{E}$ as follows: $v_{\ell\ell' m}$ is equal to one if route m uses arc (ℓ,ℓ') and zero otherwise, $\ell, \ell' \in \tilde{E}$.

The objective function accounts for the total cost, which includes facility location, vehicle usage, and transportation. Constraints (10.169) regard the capacity of the selected depots; Equalities (10.170) guarantee that all demand arcs are visited. The following constraints, (10.171), refer to the capacity of the vehicles. Constraints (10.172) ensure that routes are defined, and Constraints (10.173) impose that each route is assigned to at most a single depot, whereas (10.174) guarantee that every route is associated with a depot. Constraints (10.175) ensure that if a depot and a demand arc are in the same route, the arc is allocated to the depot. Finally, (10.176)–(10.178) are the domain constraints.

As customary, we assume we have enough vehicles to ensure a feasible partition of the demand arcs into routes.

A natural source of uncertainty in this problem is the demand. This poses a challenge since the graph $\tilde{G} = (\tilde{V}, \tilde{E})$ becomes itself uncertain and thus dependent on the actual demand observed, even the set E_d may change.

Let us denote by $\boldsymbol{\xi} = (\xi_{(\ell,\ell')})_{(\ell,\ell') \in E}$ the random vector associated with the demand (in the original graph). Moreover, assume that the probability distribution of this vector has a finite support. In this case, we can index the demand in a finite set of scenarios, Ω. For each scenario, we can define a set $E_{d\omega}$ containing the arcs with demand in scenario $\omega \in \Omega$. Furthermore we can define a scenario-dependent complete graph $\tilde{G}_\omega = (\tilde{V}_\omega, \tilde{E}_\omega)$. In this case, a natural way to hedge against uncertainty is to anticipate it when deciding on the location of the depots and then adjust the routing to the actual scenario observed. Let us assume deterministic costs associated with the vehicles and transportation, although what we present next can be easily extended. We keep denoting by π_ω the probability associated with scenario $\omega \in \Omega$.

Indexing the demand in the scenario set as well as the routing variables $\boldsymbol{v}$ and the allocation variables $\boldsymbol{x}$ and assuming a risk-neutral decision-maker, we obtain the following model:

$$\text{minimize} \quad \sum_{i \in I} f_i y_i + \sum_{\omega \in \Omega} \pi_\omega \Bigg[\sum_{m \in M} \sum_{(\ell,\ell') \in \tilde{E}_\omega} \tilde{c}_{\ell\ell'} v_{\ell\ell' m\omega} + F \sum_{m \in M} \sum_{\ell \in I} \sum_{\substack{\ell' \in \tilde{V}_\omega \setminus I \\ (\ell,\ell') \in \tilde{E}_\omega}} v_{\ell\ell' m\omega} \Bigg], \tag{10.179}$$

$$\text{subject to} \quad \sum_{(\ell,\ell') \in E_{d\omega}} d_{(\ell,\ell')\omega} x_{i(\ell,\ell')\omega} \leq q_i y_i \qquad \forall i \in I,\ \omega \in \Omega, \tag{10.180}$$

$$\sum_{m \in M} v_{\ell\ell' m\omega} = 1 \qquad \forall (\ell,\ell') \in E_{d\omega},\ \omega \in \Omega, \tag{10.181}$$

$$\sum_{(\ell,\ell') \in E_{d\omega}} d_{(\ell,\ell')\omega} v_{\ell\ell' m\omega} \leq H \qquad \forall m \in M,\ \omega \in \Omega, \tag{10.182}$$

$$\sum_{\ell' \in \tilde{V}_\omega} v_{\ell\ell' m\omega} - \sum_{\ell' \in \tilde{V}_\omega} v_{\ell'\ell m\omega} = 0 \qquad \forall m \in M,\ \ell \in \tilde{V}_\omega,\ \omega \in \Omega, \tag{10.183}$$

$$\sum_{\ell \in I} \sum_{\ell' \in \tilde{V}_\omega} v_{\ell\ell' m\omega} \leq 1 \qquad \forall m \in M,\ \omega \in \Omega, \tag{10.184}$$

$$\sum_{m \in M} \sum_{\ell \in S} \sum_{\ell' \in \tilde{V}_\omega \setminus S} v_{\ell\ell' m\omega} \geq 1 \qquad \forall I \subseteq S \subset \tilde{V}_\omega,\ \omega \in \Omega, \tag{10.185}$$

$$\sum_{\ell' \in \tilde{V}_\omega \setminus I} v_{i\ell' m\omega} + v_{\ell\ell' m\omega} \leq 1 + x_{i(\ell,\ell')\omega} \qquad \forall i \in I,\ (\ell,\ell') \in E_{d\omega},\ m \in M,\ \omega \in \Omega, \tag{10.186}$$

$$y_i \in \{0,1\} \qquad \forall i \in I, \tag{10.176}$$

$$x_{i(\ell,\ell')\omega} \in \{0,1\} \qquad \forall i \in I,\ (\ell,\ell') \in E_{d\omega},\ \omega \in \Omega, \tag{10.187}$$

$$v_{\ell\ell' m\omega} \in \{0,1\} \qquad \forall (\ell,\ell') \in E_{d\omega},\ m \in M,\ \omega \in \Omega. \tag{10.188}$$

To the best of the authors' knowledge, a stochastic programming framework such as the above one for location-arc routing was never considered in the literature.

Some attempts have been made to tackle the problem from a robust optimization perspective. Mirzaei-khafri et al. [384] investigate the problem assuming polyhedral demand uncertainty.

Amini et al. [23] consider budget uncertainty for both the demands and the travel costs. The authors propose lower and upper bounds as well as a metaheuristic.

10.7 Conclusion and Further Reading

This chapter was dedicated to facility location problems involving routing decisions. Despite the progress we have observed in recent decades, the contents of this chapter highlight much that remains to be done. In particular, we can easily highlight several potential research avenues:

- Cost-oriented problems with uncertain travel times;
- Time-oriented problems with demand uncertainty;
- Combinations of different sources of uncertainty;
- Models and solution techniques for risk-averse decision-making;
- The usage of probability distributions other than the Gaussian distribution;
- Robust optimization models and solution techniques;
- Distributionally robust optimization location routing and location-arc routing.

Overall, location routing and location-arc routing under uncertainty benefit directly from the advances observed in vehicle routing and arc routing under uncertainty. These still face many challenges, such as the surge of electric vehicles. In particular, the involved technology has called for new aspects to be considered, such as consumption rates and the need for recharging the vehicles. Some developments can be summarized in problems that also involve location decisions.

Schiffer and Walther [481] investigate a problem that consists of deciding the location of a set of recharging stations and the routes of the vehicles, considering linear recharging and consumption rates as well as uncertainty in the demands. Time windows and service times are also captured. The problem is cast as a robust optimization problem. The robustness method used by the authors was the so-called adversarial approach. A parallelized adaptive large neighborhood search (ALNS) algorithm is developed to find feasible solutions to the problem.

More recently, a location-routing problem for electric vehicles with stochastic demands is investigated in [569]. The facilities to locate correspond to battery-swap stations. Two recourse policies are considered for hedging against uncertainty. An algorithm hybridizing VNS and PSO is proposed. In the computational tests whose results are reported, the demands are uniformly distributed in a three-value discrete set.

As observed in the previous sections, capturing uncertainty in facility location problems embedding routing decisions remains challenging.

In addition to the developments presented in this chapter, several attempts have been made to consider the uncertainty captured by fuzzy sets. Mehrjerdi and Nadizadeh [376] seek to minimize the total cost, which includes the fixed opening cost for depots and vehicles and variable routing costs. The authors consider an FCCP model.

Nevertheless, due to the complexity of the problem and the difficulty in tackling it using a general-purpose solver, the authors considered a relaxation for providing a lower bound on the optimal value. For the whole problem, a heuristic algorithm is devised to find feasible solutions.

Nadizadeh and Nasab [393] study the dynamic version of the above problem. Location decisions must be made in the first period and remain fixed throughout the planning horizon. The routing decision can change according to the demands observed during the different periods. A penalty is assumed for not serving customers, which is included in the objective function. An FCCP model is derived and tackled using an off-the-shelf solver. However, this is inefficient for large-scale instances. Thus, a heuristic procedure similar to the one proposed in [376] is devised.

A CLRP with time windows (CLRP-TW) under uncertainty in demands and travel times is investigated in [563]. The problem involves a fleet of homogeneous capacitated vehicles and heterogeneous capacitated depots. Uncertainty is captured using fuzzy numbers. The objective function consists of minimizing the total cost. It includes the location of the depots, the routing, and the additional travel distance due to route failures. The authors introduce an FCCP model and a simulated annealing (SA) algorithm for tackling the problem. A similar problem is analyzed in [426]. Nevertheless, now only the demands are assumed to be fuzzy. The objective function seeks to minimize the total cost that results from adding the transportation costs, the fixed costs for using depots and vehicles, and the penalties for violating the time windows set for the customers. The problem is tackled using a variable neighborhood search (VNS) algorithm hybridized with an evolutionary algorithm.

A stochastic CLRP with pickup and deliveries for multiple products is investigated in [280]. Both the depots and the vehicles have specific capacities for each product. The authors consider stochastic demands following a Poisson distribution. The problem is modeled as a two-stage stochastic programming problem. In the first stage, the location, the number of vehicles, and the routes are decided, and in the second stage, a recourse action considers a penalty for exceeding the capacities of vehicles and depots. A tabu search (TS) heuristic is developed for approximating the optimal solution.

A different type of LRP under uncertainty is investigated in [110]. The problem captures features such as a maximum length acceptable for the routes and split deliveries (i.e., the customers can be visited more than once by a vehicle). The stochastic demands are captured using a queuing network for each customer. The problem is tackled using a heuristic that combines the three-dimensional space-filling curve algorithm and an extended Clarke Wright procedure.

Chapter 11
Hub Location

Abstract This chapter discusses models and solution techniques for hedging against uncertainty in hub location problems. This area grew significantly in a little more than a decade. Thus, a rich literature exists. It requires segmenting the chapter starting from some base settings and progressively discussing more comprehensive models. Accordingly, the chapter starts by considering uncapacitated problems with complete hub networks. A distinction is made between problems featuring fixed costs, hub median problems, and hub center and covering problems. The chapter proceeds, discussing the inclusion of capacity constraints and hub network design decisions. Reliability issues are then discussed, namely those related to unexpected disruptions at the hubs or edges. Distributionally robust hub location is a recent trend that is also presented. The chapter concludes with hints for further reading and promising research avenues.

Keywords Complete hub networks · Hub network design · Single allocation · Multiple allocation · Fixed-charge problems · Median objectives · Covering objectives · Endogenous capacities · Reliability

11.1 Introduction

A hub location problem stems from the need to ship some commodity from many to many nodes in a network. Instead of shipping the commodity directly between the nodes, a subset of them is selected to become hubs, thus consolidating and redistributing the flow. This induces a structure to the network that typically renders transportation more efficient and less costly. The fundamental decisions to make in such a problem include where to locate the hubs, the allocation of the non-hub nodes to the hubs, the connection between the hubs, and the flow routing via the hubs.

Hub location is a field with applications in telecommunications, logistics, and transportation, to mention a few. The topic has been extensively studied in the literature, as attested by different overviews, which include [18, 100, 134, 137, 470]. Those papers also highlight different perspectives that have been studied: single

© The Author(s), under exclusive license to Springer Nature Switzerland AG 2024
F. Saldanha-da-Gama, S. Wang, *Facility Location Under Uncertainty*,
International Series in Operations Research & Management Science 356,
https://doi.org/10.1007/978-3-031-55927-3_11

versus multiple allocation, complete hub networks versus network design decisions, capacitated versus uncapacitated hubs, direct shipment allowed (up to some extent), deterministic versus uncertain parameters/settings, minsum versus minmax objectives, multicriteria hub location, *et cetera.*

In this chapter, we assume that the reader is familiar with the basic concepts and models in the field. Given the nature of this volume, we narrow the discussion by directly focusing on hub location under uncertainty. An overview of this specific area is presented in [514]. In this volume, we go beyond that book chapter both in terms of the comprehensiveness of the analysis and the modeling paradigms adopted.

Several sources of uncertainty may emerge in a hub location problem that we can categorize into three groups: demands, costs, and network structures. We discuss models and algorithms for hedging against uncertainty in such different settings.

The remainder of this chapter is organized as follows. In Sect. 11.2, we consider uncapacitated hubs with complete hub-level networks. We distinguish between multiple- and single-allocation models. We also discuss hub median and hub center problems. In Sect. 11.3, hub capacities are included. A distinction is made between endogenous and exogenous capacities. Network design decisions are presented in Sect. 11.4. Section 11.5 analyzes reliability aspects related to disruptions at the hubs or edges and mechanisms to hedge against those. Section 11.6 introduces distributional robust optimization in the context of hub location. Both independent and correlated demands and costs are considered. Capacities at the hubs are also covered. The chapter ends with an overview and suggestions for further reading.

11.2 Uncapacitated Problems with Complete Hub-Level Networks

In this section, we focus on some fundamental hub location problems and discuss how they can be extended to hedge against uncertainty.

11.2.1 Multiple-Allocation Hub Location with Fixed Costs

Our starting point is the uncapacitated multiple-allocation hub location problem. As discussed in Sect. 2.3, a model that represents a good trade-off between tractability and tightness was proposed in [193]:

$$\begin{aligned}\text{minimize} \quad & \sum_{k \in K} f_k y_k + \sum_{i \in V} \sum_{k \in K} c_{ik} x_{ik} + \alpha \sum_{i \in V} \sum_{k,\ell \in K} c_{k\ell} u_{ik\ell} \\ & + \sum_{i,j \in V} \sum_{\ell \in K} c_{\ell j} u'_{i\ell j}, \end{aligned} \tag{11.1}$$

$$\text{subject to} \quad \sum_{k \in K} x_{ik} = O_i \qquad \forall i \in V, \tag{11.2}$$

$$\sum_{\ell \in K} u'_{i\ell j} = d_{ij} \qquad \forall i, j \in V, \tag{11.3}$$

$$\sum_{\substack{\ell \in K \\ \ell \neq k}} u_{ik\ell} - \sum_{\substack{\ell \in K \\ \ell \neq k}} u_{i\ell k} = x_{ik} - \sum_{j \in V} u'_{ikj} \qquad \forall i \in V,\ k \in K, \tag{11.4}$$

$$\sum_{\ell \in K} u_{ik\ell} \leq x_{ik} \qquad \forall i \in V,\ k \in K, \tag{11.5}$$

$$x_{ik} \leq O_i y_k \qquad \forall i, k \in K, \tag{11.6}$$

$$u'_{i\ell j} \leq d_{ij} y_\ell \qquad \forall i, j \in V,\ \ell \in K, \tag{11.7}$$

$$u_{ik\ell} \geq 0 \qquad \forall i \in V,\ k, \ell \in K, \tag{11.8}$$

$$u'_{i\ell j} \geq 0 \qquad \forall i, j \in V,\ \ell \in K. \tag{11.9}$$

$$x_{ik} \geq 0 \qquad \forall i \in V,\ k \in K, \tag{11.10}$$

$$y_k \in \{0, 1\} \qquad \forall k \in K. \tag{11.11}$$

In the above model, V is the set of nodes, and $K \subseteq V$ contains those that are potential hubs. The location variables are denoted by y_k: For each $k \in K$, it indicates whether node k is selected as hub; x_{ik} represents the amount of flow sent from node $i \in V$ to hub $k \in K$; $u_{ik\ell}$ is the amount of flow originated at $i \in V$ that is routed via hubs k and ℓ in this sequence, $k, \ell \in K$; $u'_{i\ell j}$ is the flow originated at $i \in V$ and destined to node $j \in V$ that is distributed via hub $\ell \in K$. The fixed cost for installing a hub at node $k \in K$ is f_k. For $\ell, \ell' \in V$, $c_{\ell\ell'}$ is the unit transportation cost for shipping directly from ℓ to ℓ', and α is the discount factor applied to the inter-hub flow. It is assumed that the above costs satisfy the triangle inequality. The flow to be shipped from i to j ($i, j \in V$) is denoted by d_{ij} and keeps being called "demand." The total flow originated at node $i \in V$, i.e., $\sum_{j \in J} d_{ij}$, is denoted by O_i, $i \in V$.

The objective function accounts for the total cost (setup of hubs and transportation); Constraints (11.2) (Constraints (11.3)) ensures that all flow is collected at (distributed from) the hubs. Constraints (11.4) regard flow divergence. Inequalities (11.5) impose a consistent choice for the values of the u and x variables. Constraints (11.6) and (11.7) guarantee a consistent choice in terms of the values of the involved variables. Finally, we have the domain constraints.

An alternative formulation for the problem that we also discussed in Sect. 2.3 is the multi-commodity formulation introduced in [256]:

$$\text{minimize} \quad \sum_{k \in K} f_k y_k + \sum_{i, j \in V} \sum_{k, \ell \in K} d_{ij} C_{ijk\ell} w_{ijk\ell}, \tag{11.12}$$

$$\text{subject to} \quad \sum_{k,\ell \in K} w_{ijk\ell} = 1 \qquad \forall i, j \in V, \tag{11.13}$$

$$\sum_{\ell \in K} w_{ijk\ell} + \sum_{\ell \in K \setminus \{k\}} w_{ij\ell k} \leq y_k \qquad \forall i, j \in V,\ k \in K, \tag{11.14}$$

$$w_{ijk\ell} \geq 0 \qquad \forall i, j \in V,\ k, \ell \in K, \tag{11.15}$$

$$y_k \in \{0, 1\} \qquad \forall k \in K. \tag{11.16}$$

A commodity corresponds to the flow from some node $i \in V$ to another node $j \in V$. Variable $w_{ik\ell j}$ represents the proportion of the flow from i to j that is shipped via hubs k and ℓ in this sequence $(i, j \in V, k, \ell \in K)$. Parameter $C_{ik\ell j}$ represents the unit cost for shipping commodity from i to j through hubs k (as first hub) and ℓ (as the second) $(i, j \in V, k, \ell \in K)$. This cost is computed easily: $C_{ijk\ell} = c_{ik} + \alpha c_{k\ell} + c_{\ell j}$. We recall that Constraints (11.13) ensure that all flow is routed through the network, whereas Constraints (11.14) guarantee that if a node is used as a hub, the fixed cost is paid. Finally, we have the domain constraints.

Note that the hubs are uncapacitated. Thus, there is always a solution with all w variables equal to zero or one. Note also that by removing the fixed costs in the above two formulations and inserting a constraint on the number of hubs to open, we obtain a formulation for the p-hub median problem.

Looking into the parameters, we realize that uncertainty can stem from demands, costs, or both. In the particular case of the multi-commodity formulation, we realize that such sources of uncertainty appear only in the objective function.

11.2.1.1 Robust Optimization Models

We start by assuming interval uncertainty for the demands. In particular, for $i, j \in V$, we assume $d_{ij} \in [\overline{d}_{ij}, \overline{d}_{ij} + \Delta_{ij}]$ with $\overline{d}_{ij}$ standing for the nominal value and Δ_{ij} representing the scaling factor (refer to Sect. 5.3). Zetina et al. [565] assume budget uncertainty by imposing a maximum number D of demands that can differ from their nominal values. Considering (11.12)–(11.16) as the base model, a robust hub location problem can be formulated accordingly:

$$\begin{aligned}\text{minimize} \quad & \sum_{k \in K} f_k y_k + \sum_{i,j \in V} \sum_{k,\ell \in K} d_{ij} C_{ijk\ell} w_{ijk\ell} \\ & + \max_{\substack{S \subseteq V \times V \\ |S| \leq D}} \left\{ \sum_{i,j \in V} \sum_{k,\ell \in K} \Delta_{i,j} C_{ijk\ell} w_{ijk\ell} \right\},\end{aligned} \tag{11.17}$$

subject to (11.13)–(11.16).

This problem can be reformulated by considering auxiliary binary variables, say δ_{ij}, equal to one if and only if the scaling factor for the flow to be shipped from i to j is accounted for in the objective function. The new model becomes

$$\text{minimize} \quad \sum_{k \in K} f_k y_k + \sum_{i,j \in V} \sum_{k,\ell \in K} d_{ij} C_{ijk\ell} w_{ijk\ell}$$

$$+ \max \left\{ \sum_{(i,j) \in S} \sum_{k,\ell \in K} \Delta_{i,j} C_{ijk\ell} w_{ijk\ell} \delta_{ij} \right\}, \tag{11.18}$$

subject to (11.13)–(11.16),

$$\sum_{i,j \in V} \delta_{ij} \leq D, \tag{11.19}$$

$$0 \leq \delta_{ij} \leq 1 \qquad \forall i, j \in V. \tag{11.20}$$

Note that the integrality constraints in the δ-variables have been relaxed because the constraint matrix underlying (11.19) is totally unimodular (see [565]). For a given solution $\{\boldsymbol{w}, \boldsymbol{y}\}$, the inner maximization problem can be linearized, which allows re-writing the whole problem as follows:

$$\text{minimize} \quad \sum_{k \in K} f_k y_k + \sum_{i,j \in V} \sum_{k,\ell \in K} d_{ij} C_{ijk\ell} w_{ijk\ell} + \sum_{i,j \in V} \lambda_{ij} + \mathrm{d}\mu, \tag{11.21}$$

subject to (11.13)–(11.16),

$$\lambda_{ij} + \mu \geq \sum_{k,\ell \in K} C_{ijk\ell} \Delta_{ij} w_{ijk\ell} \qquad \forall i, j \in V, \tag{11.22}$$

$$\lambda_{ij} \geq 0 \qquad \forall i, j \in V, \tag{11.23}$$

$$\mu \geq 0. \tag{11.24}$$

A similar technique is also explored in [565] to account for a similar type of uncertainty in transportation costs. Unfortunately, when uncertainty stems simultaneously from the demand and the transportation costs, that technique cannot be applied. Nevertheless, an MILP model with an exponential number of constraints can be derived for which a standard branch-and-cut algorithm is proposed. We refer the reader to that paper for the algorithmic details.

We consider now a more complex setting assuming uncertain demands, represented by ξ_{ij}, $i, j \in V$ and given parametrically by

$$\xi_{ij} = d_{ij} + \sum_{s=1}^{S} a_{ijs} \theta_s \quad \forall i, j, \in V.$$

$\theta_1, \ldots, \theta_S$ represent independent random variables with a symmetric and bounded non-specified distribution. This representation allows incorporating different sources of uncertainty in the demands, each with a certain weight. Given that the random variables θ_s are symmetric and bounded, we can assume w.l.o.g. that $\mathbb{E}[\theta_s] = 0$, and $|\theta_s| < 1$, $s = 1, \ldots, S$ (see [118]). This structure is adopted in [488], who investigates the robust optimization problem formulated mathematically as

$$\text{minimize} \quad \sum_{k \in K} f_k y_k + \max_{\|\boldsymbol{\theta}\| \le L} \left\{ \sum_{i,j \in V} \sum_{k,\ell \in K} \left(d_{ij} + \sum_{s=1}^{S} a_{ijs}\theta_s \right) C_{ijk\ell} w_{ijk\ell} \right\}, \tag{11.25}$$

$$\text{subject to} \quad (11.13)\text{–}(11.16).$$

The authors show that under the stated assumptions on the random variables θ_s $(s = 1, \ldots, S)$, the objective function (11.25) is equivalent to

$$\sum_{k \in K} f_k y_k + \sum_{i,j \in V} \sum_{k,\ell \in K} d_{ij} C_{ijk\ell} w_{ijk\ell} + L \sqrt{\sum_{s=1}^{S} \left(\sum_{i,j \in V} \sum_{k,\ell \in K} a_{ijs} C_{ijk\ell} w_{ijk\ell} \right)^2}. \tag{11.26}$$

This can be concluded by explicitly finding

$$\max_{\|\boldsymbol{\theta}\| \le L} \left\{ \sum_{i,j \in V} \sum_{k,\ell \in K} \sum_{s=1}^{S} a_{ijs}\theta_s C_{ijk\ell} w_{ijk\ell} \right\},$$

which can be straightforwardly accomplished using Lagrangian multipliers.

Shahabi and Unnikrishnan [488] explore the structure of the robust optimization multiple-allocation hub location problem by reformulating it as a mixed-integer conic quadratic problem. To do so, they introduce non-negative auxiliary variables φ_s and ϕ, $s = 1, \ldots, S$ thus writing the problem as follows:

$$\text{minimize} \quad \sum_{k \in K} f_k y_k + \sum_{i,j \in V} \sum_{k,\ell \in K} d_{ij} C_{ijk\ell} w_{ijk\ell} + L\phi, \tag{11.27}$$

$$\text{subject to} \quad (11.13)\text{–}(11.15),$$

$$\sum_{i,j \in V} \sum_{k\ell \in K} a_{ijs} C_{ijk\ell} w_{ijk\ell} \le \varphi_s \qquad \forall s = 1, \ldots, S, \tag{11.28}$$

$$\sum_{s=1}^{S} \varphi_s^2 \le \phi^2, \tag{11.29}$$

$$\varphi_s \geq 0 \qquad \forall s = 1, \ldots, S, \tag{11.30}$$

$$\phi \geq 0. \tag{11.31}$$

This reformulation enables tackling the problem with existing general-purpose solvers such as those found in IBM ILOG CPLEX Optimization Studio (https://www.ibm.com/products/ilog-cplex-optimization-studio). Nevertheless, the same authors also propose relaxations for obtaining improved lower bounds, which they accomplish in three steps: (i) replace φ_s^2 with ψ_{ss} for $s = 1, \ldots, S$ and replace ϕ^2 with ϱ, adding constraints $\varphi_s^2 = \psi_{ss}$, $s = 1, \ldots, S$, and $\phi^2 = \varrho$; (ii) Replace the added equality constraints with $\varphi_s^2 \leq \psi_{ss}$, $s = 1, \ldots, S$, and $\phi^2 \leq \varrho$; (iii) Assume $|\varphi_s| \geq 1$, $s = 1, \ldots, S$ and thus replacing the inequalities of (ii) with $\varphi_s \leq \psi_{ss}$, $s = 1, \ldots, S$, and $\phi \leq \varrho$. In synthesis, we obtain the following relaxed problem:

$$\text{minimize} \quad (11.27), \tag{11.32}$$

$$\text{subject to} \quad (11.13)–(11.15), (11.28), (11.31)$$

$$\sum_{s=1}^{S} \psi_{ss} \leq \varrho, \tag{11.33}$$

$$|\varphi_s| \leq \psi_{ss} \qquad \forall s = 1, \ldots, S, \tag{11.34}$$

$$\phi \leq \varrho, \tag{11.35}$$

$$\psi_{ss} \geq 0 \qquad \forall s = 1, \ldots, S, \tag{11.36}$$

$$\varrho \geq 0. \tag{11.37}$$

A key assumption for considering this relaxation is that $|\varphi_s| \geq 1$, $s = 1, \ldots, S$. This holds if $\sum_{i,j \in V} \sum_{k,\ell \in K} a_{ijs} C_{ijk\ell} w_{ijk\ell}$. In case the cost component $C_{ijk\ell}$ is significantly larger than one, the assumption is reasonable.

11.2.1.2 Stochastic Programming Models

We assume now that there is a random event that we represent by a random variable ξ, which determines some parameters (e.g., the demands or the costs). In particular, one observation of ξ fully specifies the uncertainty underlying the problem. Let us assume that we know the probability distribution of ξ and let Ξ be its support.

We start by assuming that demands are stochastic, i.e., that we are working with random variables $d_{ij}(\xi)$, each representing the flow to be sent from i to j, $\forall i, j \in V$.

Consider the model (11.12)–(11.16). Assuming a neutral attitude of the decision-maker toward risk and also that a decision about the hubs to select should be implemented here and now, we can formulate a stochastic multiple-allocation hub location problem as follows:

$$\text{minimize} \quad \sum_{k \in K} f_k y_k + \mathbb{E}_{\boldsymbol{\xi}} \left[\sum_{i,j \in V} \sum_{k,\ell \in K} d_{ij}(\boldsymbol{\xi}) C_{ijk\ell} w_{ijk\ell}(\boldsymbol{\xi}) \right], \tag{11.38}$$

$$\text{subject to} \quad \sum_{k,\ell \in K} w_{ijk\ell}(\xi) = 1 \qquad \forall i, j \in V,\ \xi \in \Xi, \tag{11.39}$$

$$\sum_{\ell \in K} w_{ijk\ell}(\xi) + \sum_{\ell \in K \setminus \{k\}} w_{ij\ell k}(\xi) \leq y_k \qquad \forall i, j \in V,\ k \in K,\ \xi \in \Xi, \tag{11.40}$$

$$w_{ijk\ell}(\xi) \geq 0 \qquad \forall i, j \in V,\ k, \ell \in K,\ \xi \in \Xi, \tag{11.41}$$

$$y_k \in \{0, 1\} \qquad \forall k \in K. \tag{11.42}$$

In this model, $w_{ijk\ell}(\xi)$ is the path to be followed by the flow between i and j $(i, j \in V)$, which is considered for every possible observation ξ of $\boldsymbol{\xi}$.

Even if relevant to capture uncertainty in the above uncapacitated multiple-allocation hub location problem (e.g., a high value is found for the EVPI), still, it is not worth solving the model because it has an exact simplification obtained by replacing the stochastic demands with their expectations. This is stated in the following result:

Proposition 11.1 *The value of the stochastic solution of the above uncapacitated multiple-allocation hub location problem is zero.*

Proof The proof can be found in [136]. It stems from the fact that we have a multiple-allocation pattern with uncapacitated hubs. Therefore, given a set of open hubs, the optimal path for the flow does not depend on the demand—it is simply the shortest path from the origin to the destination. In other words, if $\hat{\mathbf{y}}$ is a first-stage solution (i.e., a set of open hubs), then there is a second-stage solution verifying $w_{ijk\ell}(\xi) = \hat{w}_{ijk\ell}$, $\forall \xi \in \Xi$, with $\hat{w}_{ijk\ell}$ indicating a routing solution strictly following the shortest path between every pair of nodes (given $\hat{\mathbf{y}}$). Hence,

$$\mathbb{E}_{\boldsymbol{\xi}} \left[\sum_{i,j \in V} \sum_{k,\ell \in K} d_{ij}(\boldsymbol{\xi}) C_{ijk\ell} w_{ijk\ell}(\boldsymbol{\xi}) \right]$$

$$= \mathbb{E}_{\boldsymbol{\xi}} \left[\sum_{i,j \in V} \sum_{k,\ell \in K} d_{ij}(\boldsymbol{\xi}) C_{ijk\ell} \hat{w}_{ijk\ell} \right] = \sum_{i,j \in V} \sum_{k,\ell \in K} \mathbb{E}_{\boldsymbol{\xi}} \left[d_{ij}(\boldsymbol{\xi}) \right] C_{ijk\ell} \hat{w}_{ijk\ell}.$$

Thus, problem (11.38)–(11.42) is equivalent to

$$\text{minimize} \quad \sum_{i,j \in V} \sum_{k,\ell \in K} \mathbb{E}_{\xi} \left[d_{ij}(\boldsymbol{\xi}) \right] C_{ijk\ell} w_{ijk\ell}, \tag{11.43}$$

$$\text{subject to} \quad (11.13)\text{–}(11.16). \tag{11.44}$$

□

Contreras et al. [136] prove a similar result for the case in which the unit transportation costs are the source of uncertainty. The authors assume dependent costs ruled by a common stochastic factor: $C_{ijk\ell}(\xi) = \gamma(\xi)(c_{ij} + \alpha c_{k\ell} + c_{\ell j})$, where ξ stands for an observation of the random variable $\boldsymbol{\xi}$. A multiplication factor is also assumed for the unit collection and distribution costs. We are omitting that trivial extension. Again, in this setting, the value of the stochastic solution is equal to zero: the expected value problem provides an optimal solution to the stochastic problem.

From a stochastic programming perspective, a more challenging setting emerges if uncertainty stems from independent unit transportation costs. In this case, for every $\ell, \ell' \in V$, the unit cost $c_{\ell\ell'}$ is replaced by the random variable $c_{\ell\ell'}(\boldsymbol{\xi})$. In particular, for $\xi \in \Xi$, $c_{\ell\ell'}$ is the unit transportation cost between nodes ℓ and ℓ' if the event ξ occurs, and thus we need to consider costs[1] $C_{ijk\ell}(\xi) = c_{ij}(\xi) + \alpha c_{k\ell}(\xi) + c_{\ell j}(\xi)$. The cost-independence structure makes the triangle inequality no longer guaranteed—it may be optimal to route flow using more than two hubs. We keep following [136] and impose a maximum of two nodes for routing the flow between any two nodes, which is a reasonable assumption in practice. In this case, the stochastic problem can be formulated as

$$\text{minimize} \quad \sum_{k \in K} f_k y_k + \mathbb{E}_{\boldsymbol{\xi}} \left[\sum_{i,j \in V} \sum_{k,\ell \in K} d_{ij} C_{ijk\ell}(\boldsymbol{\xi}) w_{ijk\ell}(\boldsymbol{\xi}) \right], \tag{11.45}$$

$$\text{subject to} \quad (11.39)\text{–}(11.42). \tag{11.46}$$

The value of the stochastic solution in this problem may be positive because the paths selected for the flows depend on the occurring event, ξ.

For this problem, a reformulation with fewer variables is proposed in [136] together with a sample average approximation algorithm. The reformulation is as follows:

$$\text{minimize} \quad \sum_{k \in K} f_k y_k + \mathbb{E}_{\boldsymbol{\xi}} \left[\sum_{i,j \in V} \sum_{e \in E_{ij}(\boldsymbol{\xi})} d_{ij} C_{ije}(\boldsymbol{\xi}) w_{ije}(\boldsymbol{\xi}) \right], \tag{11.47}$$

[1] We keep using the notation $C_{ijk\ell}(\xi)$ as above to avoid additional one.

$$\text{subject to} \quad \sum_{e \in E_{ij}(\xi)} w_{ije}(\xi) = 1 \qquad \forall i, j \in V, \ \xi \in \Xi, \tag{11.48}$$

$$\sum_{e \in E_{ij}(\xi) | k \in e} w_{ije}(\xi) \leq y_k \qquad \forall i, j \in V, \ k \in K, \ \xi \in \Xi, \tag{11.49}$$

$$w_{ije}(\xi) \geq 0 \qquad \forall i, j \in V, \ e \in E_{ij}(\xi), \ \xi \in \Xi, \tag{11.50}$$

$$y_k \in \{0, 1\} \qquad \forall k \in K. \tag{11.51}$$

In the above model, "e" stands for a hub edge. For $i, j \in V$, $E_{ij}(\xi)$ contains the set of hub edges that are candidates to route the flow from i to j. For each $\xi \in \Xi$, this set is obtained as follows:

(i) Compute the set E that contains all subsets of K of cardinality one or two.
(ii) For each $e \in E$, define $C_{ije}(\xi) = \min\{C_{ijk\ell}(\xi), C_{ij\ell k}(\xi\}$ if $e = \{k, \ell\}$ (2-node edge) and $C_{ije}(\xi) = C_{ijkk}(\xi)$ if $e = \{k\}$ (1-node edge).
(iii) If $i = j$, include in $E_{ij}(\xi)$ all $e \in E$ such that $|e| = 1$; if $i \neq j$, include in $E_{ij}(\xi)$ all $e \in E$ such that $|e| = 1$ as well as all 2-node hub edges $e = \{k, \ell\}$ such that $C_{ije}(\xi) < \min\{C_{ijk}(\xi), C_{ij\ell}(\xi)\}$.

A sample average approximation procedure is detailed in Algorithm 11.1. In this procedure, μ_R^S provides a lower bound on the optimal solution to the stochastic problem; $\sigma^2_{\mu_R^S}$ estimates the variance. The sample size S' used in step 4 can be much larger than the original samples because the hubs are fixed. Thus, we have a linear programming problem. This way, $V^{N'}(\hat{y})$ is an estimate of the upper bound induced by the chosen solution $\hat{y}$ and $\sigma^2_{N'}(\hat{y})$ is an estimate of the variance of the estimator. Finally, $\text{gap}_{S,Ri,S'}(\hat{y})$ is an estimate of the absolute optimality gap for the original problem and σ^2_{gap} is an estimate of the variance of that estimator. The last two quantities allow the construction of a confidence interval for the optimality gap using the SAA algorithm.

Algorithm 11.1: An SAA for the multiple-allocation hub location problem with independent stochastic travel costs.

---

1 Generate a set $\{N_1, \ldots, N_R\}$ of independent samples, each of size S. In particular, the r-th sample is given by $N_r = (\xi_r^1, \ldots, \xi_r^S)$

2 For each $r = 1, \ldots, R$, solve the SAA problem corresponding to sample N_r denoting by V^{N_r} and $\hat{\mathbf{y}}^{N_r}$ its optimal value and optimal hub solution, respectively:

$$\begin{aligned}
\text{minimize} \quad & \sum_{k \in K} f_k y_k + \frac{1}{S} \sum_{\xi \in N_r} \sum_{i,j \in V} \sum_{e \in E_{ij}(\xi)} C_{ije}(\xi) w_{ije}(\xi), \\
\text{subject to} \quad & \sum_{e \in E_{ij}(\xi)} w_{ije}(\xi) = 1 && \forall i, j \in V,\ \xi \in N_r, \\
& \sum_{e \in E_{ij}(\xi) | k \in e} w_{ije}(\xi) \le y_k && \forall i, j \in V,\ k \in K,\ \xi \in N_r, \\
& w_{ije}(\xi) \ge 0 && \forall i, j \in V,\ e \in E_{ij}(\xi),\ \xi \in N_r, \\
& y_k \in \{0, 1\} && \forall k \in K.
\end{aligned}$$

3 Compute

$$\mu_R^S = \frac{1}{R} \sum_{r=1}^{R} V^{N_r}$$

and

$$\sigma^2_{\mu_R^S} = \frac{1}{R(R-1)} \sum_{r=1}^{R} (V^{N_r} - \mu_R^S)^2.$$

4 Select a feasible solution to the problem, say $\hat{\mathbf{y}}$ (*e.g.*, one of the solutions $\hat{\mathbf{y}}^{N_r}, r = 1, \ldots, R$). Generate a new sample N' randomly of size $S' >>> S$ and solve the problem, denoting by $V^{N'}(\hat{\mathbf{y}})$ its optimal value:

$$\begin{aligned}
\text{minimize} \quad & \sum_{k \in K} f_k \hat{y}_k + \frac{1}{S'} \sum_{\xi \in N'} \sum_{i,j \in V} \sum_{e \in E_{ij}(\xi)} C_{ije}(\xi) w_{ije}(\xi), \\
\text{subject to} \quad & \sum_{e \in E_{ij}(\xi)} w_{ije}(\xi) = 1 && \forall i, j \in V,\ \xi \in N', \\
& \sum_{e \in E_{ij}(\xi) | k \in e} w_{ije}(\xi) \le \hat{y}_k && \forall i, j \in V,\ k \in K,\ \xi \in N', \\
& w_{ije}(\xi) \ge 0 && \forall i, j \in V,\ e \in E_{ij}(\xi),\ \xi \in N'.
\end{aligned}$$

(Continued)

Algorithm 11.1: Continued

5 Compute

$$\sigma^2_{N'}(\hat{y}) = \frac{1}{S'(S'-1)} \sum_{\xi \in N'} \left(\sum_{k \in K} f_k \hat{y}_k + \sum_{i,j \in V} \sum_{e \in E_{ij}(\xi)} C_{ije}(\xi) w_{ije}(\xi) - V^{N'}(\hat{y}) \right)^2$$

6 Compute

$$\text{gap}_{S,R,S'}(\hat{y}) = V^{N'}(\hat{y}) - \mu_R^S$$

and

$$\sigma^2_{\text{gap}} = \sigma^2_{\mu_R^S} + \sigma^2_{N'}(\hat{y}).$$

Contreras et al. [136] point out that the SAA problem in step 2 can still be challenging to be tackled by a general-purpose solver. Thus, they suggest using a Benders decomposition procedure. The distinguishing feature of that procedure is the efficient resolution of the dual subproblem, which is inspired by the procedure proposed in [379] that we detail in Sect. 11.2.3 in the context of hub median problems. Contreras et al. [136] use the methodology for tackling instances based upon the AP data set (see [192]) assuming Gaussian and Gamma distributed transportation costs.

So far, we have assumed uncertainty stemming either from the demands or from the transportation costs. Nevertheless, as well noted in [19], we may also face uncertainty from the side of the hub setup costs. This may stem from land acquisition or construction and thus, in practice, hardly will it be described by a probability distribution. Therefore, robust optimization appears as the most natural way to hedge against this source of uncertainty. Let us assume that a finite set of scenarios is identified for the setup costs of the hubs. Denote it by Ω_f. Each such scenario fully specifies the setup costs. In particular, we denote by f_k^ω the cost for setting up a hub at $k \in K$ under scenario $\omega \in \Omega_f$.

Let us consider the model (11.1)–(11.11). If we knew the exact occurring scenario, we could solve this model considering the appropriate setup costs and obtain the best possible solution under that scenario. Let us denote by V^ω the optimal value of the single-scenario problem induced by scenario $\omega \in \Omega_f$.

Due to uncertainty, we can consider determining a solution that minimizes the maximum regret across all scenarios. In other words, we seek to solve the following problem:

$$\min_{u,u',x,y} \max_{\omega\in\Omega} \left\{ \left(\sum_{k\in K} f_k^{\omega} y_k + \sum_{i\in V}\sum_{k\in K} c_{ik}x_{ik} + \alpha \sum_{i\in V}\sum_{k\in K}\sum_{\ell\in K} c_{k\ell}u_{ik\ell} + \sum_{i\in V}\sum_{\ell\in K}\sum_{j\in V} c_{\ell j}u'_{i\ell j} \right) - V^{\omega} \right\}, \tag{11.52}$$

s. t. (11.2)–(11.11).

In the objective function, we find the value of the objective function under each scenario for the solution selected and the best that could be achieved in terms of cost under that scenario.

We can now easily combine the above setting with some other already discussed. Consider a random event, induced by a random variable $\boldsymbol{\xi}$, which determines the demands and the transportation costs. We can even assume a known probability distribution for $\boldsymbol{\xi}$ with support Ξ. In this case, apart from the total hub setup cost, the objective function (11.52) becomes ill-defined. We assume a risk-neutral decision-maker. If we know in advance the exact scenario, say ω, we are facing for the setup costs, then we should solve the following problem:

$$\text{minimize} \quad \sum_{k\in K} f_k^{\omega} y_k + \mathbb{E}_{\boldsymbol{\xi}} \left[\sum_{i\in V}\sum_{k\in K} c_{ik}(\boldsymbol{\xi})x_{ik}(\boldsymbol{\xi}) + \alpha \sum_{i\in V}\sum_{k\in K}\sum_{\ell\in K} c_{k\ell}(\boldsymbol{\xi})u_{ik\ell}(\boldsymbol{\xi}) + \sum_{i\in V}\sum_{\ell\in K}\sum_{j\in V} c_{\ell j}(\boldsymbol{\xi})u'_{i\ell j}(\boldsymbol{\xi}) \right], \tag{11.53}$$

$$\text{subject to} \quad \sum_{k\in K} x_{ik}(\xi) = O_i(\xi) \qquad \forall i \in V,\ \xi \in \Xi, \tag{11.54}$$

$$\sum_{\ell\in K} u'_{i\ell j}(\xi) = d_{ij}(\xi) \qquad \forall i, j \in V,\ \xi \in \Xi, \tag{11.55}$$

$$\sum_{\substack{\ell\in K\\ \ell\neq k}} u_{ik\ell}(\xi) - \sum_{\substack{\ell\in K\\ \ell\neq k}} u_{i\ell k}(\xi) = x_{ik}(\xi) - \sum_{j\in V} u'_{ikj}(\xi) \qquad \forall i \in V,\ k \in K,\ \xi \in \Xi, \tag{11.56}$$

$$\sum_{\ell\in K} u_{ik\ell} \le x_{ik}(\xi) \qquad \forall i \in V,\ k \in K,\ \xi \in \Xi, \tag{11.57}$$

$$x_{ik}(\xi) \le O_i(\xi) y_k \qquad \forall i, k \in K,\ \xi \in \Xi, \tag{11.58}$$

$$u'_{i\ell j}(\xi) \le d_{ij}(\xi) y_\ell \qquad \forall i, j \in V,\ \ell \in K,\ \xi \in \Xi, \tag{11.59}$$

$$u_{ik\ell}(\xi) \geq 0 \qquad \forall i \in V,\ k, \ell \in K,\ \xi \in \Xi, \tag{11.60}$$

$$u'_{i\ell j}(\xi) \geq 0 \qquad \forall i, j \in V,\ \ell \in K,\ \xi \in \Xi, \tag{11.61}$$

$$x_{ik}(\xi) \geq 0 \qquad \forall i \in V,\ k \in K,\ \xi \in \Xi, \tag{11.62}$$

$$y_k \in \{0, 1\} \qquad \forall k \in K. \tag{11.11}$$

Let us denote by V_{ξ}^{ω} the optimal value of the above problem.

Given that we do not know in advance the exact scenario occurring in terms of the setup costs, we can consider the following problem that seeks to minimize the minmax regret across all setup-cost scenarios:

$$\min_{u,u',x,y} \quad \max_{\omega\in\Omega} \left\{ \left(\sum_{k\in K} f_k^{\omega} y_k + \mathbb{E}_{\boldsymbol{\xi}} \left[\sum_{i\in V} \sum_{k\in K} c_{ik}(\boldsymbol{\xi}) x_{ik}(\boldsymbol{\xi}) \right.\right.\right.$$

$$\left.\left.\left. +\alpha \sum_{i\in V} \sum_{k\in K} \sum_{\ell\in K} c_{k\ell}(\boldsymbol{\xi}) u_{ik\ell}(\boldsymbol{\xi}) + \sum_{i\in V} \sum_{\ell\in K} \sum_{j\in V} c_{\ell j}(\boldsymbol{\xi}) u'_{i\ell j}(\boldsymbol{\xi}) \right] \right) - V_{\boldsymbol{\xi}}^{\omega} \right\}, \tag{11.63}$$

s. t. (11.11), (11.54)–(11.62).

As discussed above, in case $\boldsymbol{\xi}$ determines only the demands, then the problem reduces to minimizing (11.52) subject to (11.2)–(11.11) with the demands d_{ij} in that problem replaced by $\mathbb{E}_{\boldsymbol{\xi}}[d_{ij}(\boldsymbol{\xi})]$ (and O_i replaced with $\mathbb{E}_{\boldsymbol{\xi}}[O_i(\boldsymbol{\xi})]$). In that case, the single-scenario problems used to find $V_{\boldsymbol{\xi}}^{\omega}$ can be simplified likewise. Alumur et al. [19] consider a particular case of the above model with stochasticity associated only with the demands and captured by a finite number of scenarios. The authors report on the resolution of instances based on the CAB data set (see [416]) with five scenarios for the setup costs and five scenarios for the demands.

11.2.2 Single-Allocation Hub Location with Fixed Costs

We recall the formulation for the uncapacitated single-allocation hub location problem introduced in [194] (ignoring the capacity constraints) and incorporating the additional constraints pointed out in [143]:

$$\text{minimize} \quad \sum_{k\in K} f_k x_{kk} + \sum_{i\in V} \sum_{k\in K} c_{ik}(O_i + D_i) x_{ik} + \alpha \sum_{i\in V} \sum_{k,\ell\in K} c_{k\ell} u_{ik\ell}, \tag{11.64}$$

$$\text{subject to} \quad \sum_{k \in K} x_{ik} = 1 \qquad \forall i \in V, \tag{11.65}$$

$$x_{ik} \le x_{kk} \qquad \forall i \in V,\ k \in K, \tag{11.66}$$

$$\sum_{\substack{\ell \in K \\ \ell \neq k}} u_{ik\ell} - \sum_{\substack{\ell \in K \\ \ell \neq k}} u_{i\ell k} = O_i x_{ik} - \sum_{j \in V} d_{ij} x_{jk} \quad \forall i \in V,\ k \in K, \tag{11.67}$$

$$\sum_{\ell \in K, \ell \neq k} u_{ik\ell} \le O_i x_{ik} \qquad \forall i \in V,\ k \in K, \tag{11.68}$$

$$u_{ik\ell} \ge 0 \qquad \forall i \in V,\ k, \ell \in K, \tag{11.69}$$

$$x_{ik} \in \{0, 1\} \qquad \forall i \in V,\ k \in K. \tag{11.70}$$

In this model, the elements should all be clear. Given the single-allocation nature of the problem, instead of location variables $\boldsymbol{y}$, we use allocation variables x_{ik}, with x_{kk} as a proxy for y_k $(i \in V, k \in K)$.

Furthermore, defining $C_{ijk\ell} = c_{ik} + \alpha c_{k\ell} + c_{\ell j}$ and considering four-indexed variables $w_{ijk\ell}$ to represent the proportion of the flow from i to j going through hubs k and ℓ $(i, j \in V, k, \ell \in K)$, we recall an alternative formulation (see Sect. 2.3):

$$\text{minimize} \quad \sum_{k \in K} f_k x_{kk} + \sum_{i,j \in V} \sum_{k,\ell \in K} d_{ij} C_{ijk\ell} w_{ijk\ell}, \tag{11.71}$$

$$\text{subject to} \quad \sum_{k \in K} x_{ik} = 1 \qquad \forall i \in V, \tag{11.72}$$

$$x_{ik} \le x_{kk} \qquad \forall i \in V,\ k \in K, \tag{11.73}$$

$$\sum_{\ell \in K} w_{ijk\ell} = x_{ik} \qquad \forall i, j \in V,\ k \in K, \tag{11.74}$$

$$\sum_{k \in K} w_{ijk\ell} = x_{j\ell} \qquad \forall i, j \in V,\ \ell \in K, \tag{11.75}$$

$$x_{ik} \in \{0, 1\} \qquad \forall i \in V,\ k \in K, \tag{11.76}$$

$$w_{ijk\ell} \ge 0 \qquad \forall i, j \in V,\ \forall k, \ell \in K. \tag{11.77}$$

Like in the multiple-allocation case, we start by assuming uncertain demands, say ξ_{ij}, $i, j \in V$, and the parametric representation

$$\xi_{ij} = d_{ij} + \sum_{s=1}^{S} a_{ijs} \theta_s \quad \forall i, j, \in V.$$

The parameters $\theta_1, \ldots, \theta_S$ represent independent random variables with a symmetric and bounded non-specified distribution. This leads to the assumptions already discussed. Like for multiple allocation, a mixed-integer conic quadratic formulation is derived in [488] as follows:

$$\text{minimize} \quad \sum_{k \in K} f_k x_{kk} + \sum_{i,j \in V} \sum_{k,\ell \in K} d_{ij} C_{ijk\ell} w_{ijk\ell} + L\phi, \tag{11.78}$$

$$\text{subject to} \quad (11.72)\text{–}(11.77),$$
$$(11.28)\text{–}(11.31).$$

The same authors also propose a relaxation of this model seeking sharp lower bounds. This is accomplished exactly in the same way as described for the multiple-allocation case. Thus, we omit the details.

Suppose now that there is some random event induced by a random variable $\boldsymbol{\xi}$ that determines some parameters (e.g., the demands or the costs). One observation of that random variable fully specifies the uncertain parameters underlying the problem. Assume we know the probability distribution underlying $\boldsymbol{\xi}$ and denote its support by Ξ.

We start by considering the case in which only demands are stochastic, i.e., we are working with random variables $d_{ij}(\boldsymbol{\xi})$, each representing the flow to be sent from i to j, $\forall i, j \in V$.

Let us take model (11.64)–(11.70). Assuming a neutral attitude of the decision-maker toward risk and also that a decision about the hubs to select and the allocation of the nodes to the hubs should be implemented here and now, we can formulate a stochastic multiple-allocation hub location problem as follows:

$$\text{minimize} \quad \sum_{k \in K} f_k x_{kk} + \mathbb{E}_{\boldsymbol{\xi}} \left[\sum_{i \in V} \sum_{k \in K} c_{ik} \left(O_i(\boldsymbol{\xi}) + D_i(\boldsymbol{\xi}) \right) x_{ik} \right.$$
$$\left. + \alpha \sum_{i \in V} \sum_{k,\ell \in K} c_{k\ell} u_{ik\ell} \right], \tag{11.79}$$

$$\text{subject to} \quad (11.65), (11.66), (11.70),$$

$$\sum_{\substack{\ell \in K \\ \ell \neq k}} u_{ik\ell}(\xi) - \sum_{\substack{\ell \in K \\ \ell \neq k}} u_{i\ell k}(\xi) = O_i(\xi) x_{ik} - \sum_{j \in V} d_{ij}(\xi) x_{jk}$$
$$\forall i \in V,\ k \in K,\ \xi \in \Xi, \tag{11.80}$$

$$\sum_{\ell \in K, \ell \neq k} u_{ik\ell}(\xi) \leq O_i(\xi) x_{ik} \qquad \forall i \in V,\ k \in K,\ \xi \in \Xi, \tag{11.81}$$

$$u_{ik\ell}(\xi) \geq 0 \qquad \forall i \in V,\ k, \ell \in K,\ \xi \in \Xi. \tag{11.82}$$

In this model, $u_{ik\ell}(\xi)$ is the amount of flow originated at $i \in V$ that is routed via hubs k and ℓ ($k, \ell \in K$), which is considered for every possible observation ξ of the random variable $\boldsymbol{\xi}$.

As for the multiple-allocation case, there is a simplified model that is exact.

Proposition 11.2 *The value of the stochastic solution of the above single-allocation hub location problem is zero.*

Proof A proof can be found in [463], which reduces to verifying that irrespective of the support of $\boldsymbol{\xi}$, there is an optimal solution to the stochastic problem in which

$$u_{ik\ell}(\xi) = x_{ik} \sum_{j \in V} \mathbb{E}_{\boldsymbol{\xi}}[d_{ij}(\boldsymbol{\xi})] x_{j\ell} \quad \forall i \in V,\ k, \ell \in K, \xi \in \Xi.$$

□

Given the above result, a more interesting problem emerges when the node allocation to hubs is an adaptive decision. To analyze this case, we detach the location and allocation by considering location variables y_k that are binary and indicate whether node $k \in K$ is selected as a hub. The problem can be formulated as follows:

$$\begin{aligned}
\text{minimize} \quad & \sum_{k \in K} f_k y_k + \mathbb{E}_{\boldsymbol{\xi}} \Bigg[\sum_{i \in V} \sum_{k \in K} c_{ik} (O_i(\boldsymbol{\xi}) + D_i(\boldsymbol{\xi})) x_{ik}(\boldsymbol{\xi}) \\
& + \alpha \sum_{i \in V} \sum_{k, \ell \in K} c_{k\ell} u_{ik\ell}(\boldsymbol{\xi}) \Bigg], && (11.83) \\
\text{subject to} \quad & \sum_{k \in K} x_{ik}(\xi) = 1 && \forall i \in V,\ \xi \in \Xi, \quad (11.84) \\
& x_{ik}(\xi) \le y_k && \forall i \in V,\ k \in K,\ \xi \in \Xi, \quad (11.85) \\
& \sum_{\substack{\ell \in K \\ \ell \neq k}} u_{ik\ell}(\xi) - \sum_{\substack{\ell \in K \\ \ell \neq k}} u_{i\ell k}(\xi) = O_i(\xi) x_{ik}(\xi) - \sum_{j \in V} d_{ij}(\xi) x_{jk}(\xi) \\
& && \forall i \in V,\ k \in K,\ \xi \in \Xi, \quad (11.86) \\
& \sum_{\ell \in K, \ell \neq k} u_{ik\ell}(\xi) \le O_i(\xi) x_{ik}(\xi) && \forall i \in V,\ k \in K,\ \xi \in \Xi, \quad (11.87) \\
& u_{ik\ell}(\xi) \ge 0 && \forall i \in V,\ k, \ell \in K,\ \xi \in \Xi, \quad (11.88)
\end{aligned}$$

$$x_{ik}(\xi) \in \{0, 1\} \qquad \forall i \in V,\ k \in K,\ \xi \in \Xi, \tag{11.89}$$

$$y_k \in \{0, 1\} \qquad \forall k \in K. \tag{11.90}$$

If the support of the random vector is finite, we can index the possible observations of $\boldsymbol{\xi}$ in a finite set Ω as we have customarily done in this volume. In particular, we index the demands and the second-stage decision variables in Ω, thus considering $d_{ij\omega}$, $O_{i\omega}$, $D_{i\omega}$, $x_{ik\omega}$, $u_{ik\ell\omega}$, $i, j \in V$, $k, \ell \in K$, $\omega \in \Omega$. If π_ω is the probability associated with scenario $\omega \in \Omega$, the extensive form of the deterministic equivalent becomes

$$\text{minimize} \quad \sum_{k \in K} f_k y_k + \sum_{\omega \in \Omega} \pi_\omega \left[\sum_{i \in V} \sum_{k \in K} c_{ik}(O_{i\omega} + D_{i\omega}) x_{ik\omega} \right.$$

$$\left. + \alpha \sum_{i \in V} \sum_{k, \ell \in K} c_{k\ell} u_{ik\ell\omega} \right], \tag{11.91}$$

$$\text{subject to} \quad \sum_{k \in K} x_{ik\omega} = 1 \qquad \forall i \in V,\ \omega \in \Omega, \tag{11.92}$$

$$x_{ik\omega} \le y_k \qquad \forall i \in V,\ k \in K,\ \omega \in \Omega, \tag{11.93}$$

$$\sum_{\substack{\ell \in K \\ \ell \ne k}} u_{ik\ell\omega} - \sum_{\substack{\ell \in K \\ \ell \ne k}} u_{i\ell k\omega} = O_{i\omega} x_{ik\omega} - \sum_{j \in V} d_{ij\omega} x_{jk\omega}$$

$$\forall i \in V,\ k \in K,\ \omega \in \Omega, \tag{11.94}$$

$$\sum_{\ell \in K, \ell \ne k} u_{ik\ell\omega} \le O_{i\omega} x_{ik\omega} \qquad \forall i \in V,\ k \in K,\ \omega \in \Omega, \tag{11.95}$$

$$u_{ik\ell\omega} \ge 0 \qquad \forall i \in V,\ k, \ell \in K,\ \omega \in \Omega, \tag{11.96}$$

$$x_{ik\omega} \in \{0, 1\} \qquad \forall i \in V,\ k \in K,\ \omega \in \Omega, \tag{11.97}$$

$$y_k \in \{0, 1\} \qquad \forall k \in K. \tag{11.90}$$

Rostami et al. [463] consider a (nonlinear) projection of the above model into the $\boldsymbol{x}$–$\boldsymbol{y}$ space, which is an extension to the finite-scenario stochastic setting of the binary quadratic formulation introduced in [416] that was also explored in [462]. Note that in the latter work, the authors consider a multiplication factor for collecting and distributing the flow, which is straightforward and thus omitted here. Defining $\delta(i) = y_i \, \forall i \in K$ and $\delta(i) = 0 \, \forall i \in V \setminus \{K\}$, we have

$$\text{minimize} \quad \sum_{k\in K} f_k y_k + \sum_{\omega\in\Omega} \pi_\omega \Bigg[\sum_{i\in V} \sum_{\substack{k\in K\\k\neq i}} c_{ik}(O_{i\omega} + D_{i\omega}) x_{ik\omega}$$

$$+ \alpha\Bigg(\sum_{k,\ell\in K} d_{k\ell\omega} y_k y_\ell + \sum_{j\in V} \sum_{k,\ell\in K} d_{kj\omega} c_{k\ell} y_k x_{j\ell\omega}$$

$$+ \sum_{i\in V} \sum_{k,\ell\in K} d_{i\ell\omega} c_{k\ell} x_{ik\omega} y_\ell + \sum_{i,j\in V} \sum_{\substack{k,\ell\in K\\k\neq i,\ell\neq j}} c_{k\ell} x_{ik\omega} x_{j\ell\omega} \Bigg) \Bigg], \tag{11.98}$$

$$\text{subject to} \quad \sum_{\substack{k\in K\\k\neq i}} x_{ik\omega} = 1 - \delta(i) \qquad \forall i \in V,\ \omega \in \Omega, \tag{11.99}$$

$$x_{ik\omega} \leq y_k \qquad \forall i \in V,\ k \in K,\ k \neq i,\ \omega \in \Omega, \tag{11.100}$$

$$x_{ik\omega} \qquad \forall i \in V,\ k \in K,\ \omega \in \Omega, \tag{11.101}$$

$$y_k \in \{0, 1\} \qquad \forall k \in K. \tag{11.102}$$

Rostami et al. [463] also reformulate the problem by partitioning the inter-hub transportation cost between the two hubs in a hub edge. Furthermore, they moved the quadratic term to the constraints. The reformulation is the following (we keep assuming that the set of potential locations for the hubs is $K \subseteq V$ and thus adapt the literature accordingly):

$$\text{minimize} \quad \sum_{k\in K} f_k y_k + \sum_{\omega\in\Omega} \sum_{i\in V} \sum_{\substack{k\in K\\k\neq i}} \pi_\omega c_{ik}(O_{i\omega} + D_{i\omega}) x_{ik\omega}$$

$$+ \alpha \sum_{\omega\in\Omega} \sum_{i\in V} \pi_\omega \Upsilon_{i\omega}, \tag{11.103}$$

$$\text{subject to} \quad \Upsilon_{i\omega} \geq \sum_{j\in V} d_{ij\omega} \Bigg(u_{ii\omega}\delta(i) + \sum_{\substack{k\in K\\k\neq i}} u_{ik\omega} x_{ik\omega}$$

$$+ v_{ij\omega}\delta(j) + \sum_{\substack{\ell\in K\\\ell\neq j}} v_{i\ell\omega} x_{j\ell\omega} \Bigg) \quad \forall i \in V,\ \omega \in \Omega, \tag{11.104}$$

$$u_{ik\omega} + v_{i\ell\omega} \geq c_{k\ell} \qquad \forall i \in V,\ k,\ell \in K,\ \omega \in \Omega, \tag{11.105}$$

$$(11.99)\text{–}(11.102). \tag{11.106}$$

In the above model, the $\boldsymbol{u}$ and $\boldsymbol{v}$ variables help compute the inter-hub transportation cost for the flow originated at each node $i \in V$. The problem is an MINLP that can be solved efficiently within a branch-and-cut framework by separating inequalities (11.104) and deriving subgradient cuts.

Proposition 11.3 *Let $\hat{\boldsymbol{x}}_\omega \in [0,1]^{|V|\times|K|}$ be a feasible allocation decision for scenario $\omega \in \Omega$. Then, $\forall i \in V$,*

$$\Upsilon_{i\omega} \geq \sum_{j\in V} d_{ij\omega}\Big(\overline{u}_{ii\omega}\delta(i) + \sum_{\substack{k\in K\\ k\neq i}} \overline{u}_{ik\omega}x_{ik\omega} + \overline{v}_{ij\omega}\delta(j) + \sum_{\substack{\ell\in K\\ \ell\neq j}} \overline{v}_{i\ell\omega}x_{j\ell\omega}\Big)$$

with

$$\overline{v}_{i\ell\omega} = \sum_{k\in K} c_{k\ell}\hat{x}_{ik\omega} \quad \forall \ell \in K$$

and

$$\overline{u}_{ik\omega} = \min_{\ell\in K}\{c_{k\ell} - \overline{v}_{i\ell\omega}\} \quad \forall k \in K,$$

is a valid (subgradient) cut for the MINLP.

Proof See [463]. □

Based upon the above result, a branch-and-cut algorithm can be devised with Constraints (11.104) being initially ignored (the inter-hub transportation costs are initially ignored). In turn, this information is progressively considered by adding on the fly the cuts presented in the above proposition, which are violated.

Like done for the multiple-allocation problem, a single-allocation pattern with uncertainty stemming from the setup costs for the hubs and from the demands is considered in [19]. Regarding the former, for the reasons already mentioned, it makes sense to cast them within a robust optimization modeling framework. Again, we denote by Ω_f a finite set of scenarios identified for those costs.

Let us consider model (11.64)–(11.70). If we know the exact setup-cost occurring scenario, we can solve this model considering the appropriate setup costs. We would obtain the best possible solution under that scenario. Let us denote by V^ω the optimal value of the single-scenario problem induced by scenario $\omega \in \Omega_f$.

Due to the uncertainty, we can consider determining a solution that minimizes the maximum regret across all scenarios. In other words, we seek to solve the following problem:

$$\min_{\boldsymbol{u},\boldsymbol{x}} \quad \max_{\omega\in\Omega}\Bigg\{\Bigg(\sum_{k\in K} f_k^\omega x_{kk} + \sum_{i\in V}\sum_{k\in K} c_{ik}(O_i + D_i)x_{ik} +$$

$$+\alpha \sum_{i\in V}\sum_{k,\ell\in K} c_{k\ell}u_{ik\ell}\Bigg) - V^{\omega}\Bigg\}, \tag{11.107}$$

$$\text{s.t.} \quad (11.65)\text{–}(11.70).$$

In the objective function, we find the value of the objective function under each scenario for the solution selected and the best that could be achieved in terms of cost under that scenario.

As suggested for the multiple-allocation case, we can combine the above setting with another where uncertainty stems from demands and transportation costs. Suppose we face a random event, represented by a random variable $\boldsymbol{\xi}$, that determines the demands and the transportation costs. We assume a given probability distribution for $\boldsymbol{\xi}$ with support Ξ. Assuming a risk-neutral decision-maker when it comes to assessing the total transportation cost, in case we know in advance the exact occurring scenario in terms of the setup costs, we can solve the following problem:

$$\text{minimize} \quad \sum_{k\in K} f_k^{\omega} x_{kk} + \mathbb{E}_{\boldsymbol{\xi}}\Bigg[\sum_{i\in V}\sum_{k\in K} c_{ik}(\boldsymbol{\xi})\left(O_i(\boldsymbol{\xi}) + D_i(\boldsymbol{\xi})\right) x_{ik}(\boldsymbol{\xi})$$

$$+\alpha \sum_{i\in V}\sum_{k,\ell\in K} c_{k\ell}(\boldsymbol{\xi}) u_{ik\ell}(\boldsymbol{\xi})\Bigg], \tag{11.108}$$

$$\text{subject to} \quad \sum_{k\in K} x_{ik}(\xi) = 1 \qquad \forall i \in V,\ \xi \in \Xi, \tag{11.109}$$

$$x_{ik}(\xi) \le x_{kk}(\xi) \qquad \forall i \in V,\ k \in K,\ \xi \in \Xi, \tag{11.110}$$

$$\sum_{\substack{\ell\in K\\ \ell\neq k}} u_{ik\ell}(\xi) - \sum_{\substack{\ell\in K\\ \ell\neq k}} u_{i\ell k}(\xi) = O_i(\xi)x_{ik}(\xi) - \sum_{j\in V} d_{ij}(\xi)x_{jk}(\xi)$$

$$\forall i \in V,\ k \in K,\ \xi \in \Xi, \tag{11.111}$$

$$\sum_{\ell\in K,\ell\neq k} u_{ik\ell}(\xi) \le O_i(\xi)x_{ik}(\xi) \qquad \forall i \in V,\ k \in K,\ \xi \in \Xi, \tag{11.112}$$

$$u_{ik\ell}(\xi) \ge 0 \qquad \forall i \in V,\ k,\ell \in K,\ \xi \in \Xi, \tag{11.113}$$

$$x_{ik}(\xi) \in \{0, 1\} \qquad \forall i \in V,\ k \in K, \xi \in \Xi. \tag{11.70}$$

Let us denote by $V_{\boldsymbol{\xi}}^{\omega}$ the optimal value of the above problem.

Considering that we do not know in advance the exact scenario occurring in terms of the setup costs, we can adopt the following model, to minimize the minmax regret across all setup-cost scenarios:

$$\min_{\boldsymbol{u},\boldsymbol{x}} \quad \max_{\omega\in\Omega} \left\{ \left(\sum_{k\in K} f_k^{\omega} x_{kk} + \mathbb{E}_{\xi} \left[\sum_{i\in V} \sum_{k\in K} c_{ik}(\boldsymbol{\xi}) \left(O_i(\boldsymbol{\xi}) + D_i(\boldsymbol{\xi})\right) x_{ik}(\boldsymbol{\xi}) \right.\right.\right.$$

$$\left.\left.\left. + \alpha \sum_{i\in V} \sum_{k\in K} \sum_{\ell\in K} c_{k\ell}(\boldsymbol{\xi}) u_{ik\ell}(\boldsymbol{\xi}) \right] \right) - V_{\xi}^{\omega} \right\}, \qquad (11.114)$$

$$\text{s.t.} \quad (11.70), (11.109)–(11.113).$$

Alumur et al. [19] consider a particular case of the above model with stochasticity stemming only from the demands and captured by a finite number of scenarios. The authors report on results for instances based upon the CAB data set (see [416] with five scenarios for the setup costs and five scenarios for the demands).

All the problems analyzed so far assume that the decision-maker is neutral toward risk. In practice, this is often not true. Nevertheless, the literature considering risk aversion in the context of hub location is scarce.

Assume that uncertainty stems from the demand and is captured by a finite set of scenarios, Ω. Thinking of a risk-averse decision-maker, we focus on the CVaR minimization for a given confidence level α. Consider the base model (11.71)–(11.77). If the allocation decisions are adaptive, we must detach location and allocation and thus re-use binary variables y_k indicating whether a hub is installed at $k \in K$. The problem that consists of selecting the hubs and routing the flow to minimize the CVaR of the transportation cost can be formulated as follows:

$$\text{minimize} \quad \sum_{k\in K} f_k y_{kk} + \eta + \frac{1}{1-\alpha} \sum_{\omega\in\Omega} \pi_\omega \zeta_\omega, \qquad (11.115)$$

$$\text{subject to} \quad \zeta_\omega \geq \sum_{i,j\in V} \sum_{k,\ell\in K} d_{ij\omega} C_{ijk\ell} w_{ijk\ell\omega} - \eta \qquad \forall \omega \in \Omega, \qquad (11.116)$$

$$\sum_{k\in K} x_{ik\omega} = 1 \qquad \forall i \in V,\ \omega \in \Omega, \qquad (11.117)$$

$$x_{ik\omega} \leq y_k \qquad \forall i \in V,\ k \in K,\ \omega \in \Omega, \qquad (11.118)$$

$$\sum_{\ell\in K} w_{ijk\ell\omega} = x_{ik\omega} \qquad \forall i, j \in V,\ k \in K,\ \omega \in \Omega, \qquad (11.119)$$

$$\sum_{k\in K} w_{ijk\ell\omega} = x_{j\ell\omega} \qquad \forall i, j \in V,\ \ell \in K,\ \omega \in \Omega, \qquad (11.120)$$

$$y_k \in \{0, 1\} \qquad \forall k \in K, \qquad (11.121)$$

$$x_{ik\omega} \in \{0, 1\} \qquad \forall i \in V,\ k \in K,\ \omega \in \Omega, \tag{11.122}$$

$$w_{ijk\ell\omega} \geq 0 \qquad \forall i, j \in V,\ \forall k, \ell \in K,\ \omega \in \Omega. \tag{11.123}$$

Ghaffarinasab et al. [233] extend this problem in different ways: (i) In addition to CVaR, the authors also consider a mean transportation cost in the objective function that is weighted with CVaR (the sum of weights being equal to 1); (ii) Economies of scale are assumed for the inter-hub traffic captured via piecewise linear concave functions. For the resulting problem, the authors develop a Benders decomposition whose distinctive element is the efficient algorithm proposed for solving the dual subproblem as done for other hub location problems already discussed.

11.2.3 *Hub Median Problems*

In the previous section, we discussed fixed-charge hub location problems. Now, we consider that there are no fixed costs for the hubs, although there is a specific number of them to be installed. This leads to the family of p-hub median problems.

Meraklı and Yaman [379] investigate a p-hub median problem under uncertainty. The authors assume a multiple-allocation pattern and cast the problem in the context of robust optimization. The deterministic version can be formulated but performing a few adjustments in the model (11.12)–(11.16):

$$\text{minimize} \quad \sum_{i,j \in V} \sum_{k,\ell \in K} d_{ij} C_{ijk\ell} w_{ijk\ell}, \tag{11.124}$$

$$\text{subject to} \quad \sum_{k \in K} y_k = p, \tag{11.125}$$

$$(11.13)\text{–}(11.16).$$

In [379] the authors assume $d_{ii} = 0\ \forall i \in V$, thus excluding from the demand vector and from the above model all pairs (i, i), $i \in V$. Furthermore, the authors assume the so-called hose uncertainty for the demands:

$$\mathcal{U}^{\text{hose}} = \left\{ \boldsymbol{d} \in \mathbb{R}_+^{|V|(|V|-1)} \mid \sum_{j \in V \setminus \{i\}} d_{ij} + \sum_{j \in V \setminus \{i\}} d_{ji} \leq L_i \ \ \forall i \in V \right\}.$$

In practice, this set may be combined with additional demand specifications such as box uncertainty, imposing acceptable limits for the individual demands (see, e.g., [17]):

$$\mathcal{U}^{B} = \left\{ \boldsymbol{d} \in \mathbb{R}_+^{|V|(|V|-1)} \mid \underline{d}_{ij} \leq d_{ij} \leq \overline{d}_{ij} \ \ \forall i, j \in V,\ i \neq j \right\}.$$

A hybridized uncertainty set is thus

$$\mathcal{U} = \mathcal{U}^{\text{hose}} \cap \mathcal{U}^{B}.$$

The multiple-allocation p-hub median problem under the above uncertainty set is formulated as

$$\text{minimize} \quad \left(\max_{d \in \mathcal{U}} \sum_{\substack{i,j \in V, \\ i \neq j}} \sum_{k\ell \in K} d_{ij} C_{ijk\ell} w_{ijk\ell} \right), \tag{11.126}$$

$$\text{subject to} \quad (11.13)\text{–}(11.16),\ (11.125),$$

with all pairs (i, i), $i \in V$ being removed from the constraints. For a given vector $(\boldsymbol{w}, \boldsymbol{y})$, the inner maximization problem is a linear programming problem that can be dualized, thus allowing reformulating the problem to be as follows:

$$\text{minimize} \quad \sum_{i \in V} \lambda_i L_i + \sum_{\substack{i,j \in V, \\ i \neq j}} (\overline{d}_{ij} \beta_{ij} - \underline{d}_{ij} \gamma_{ij}), \tag{11.127}$$

$$\text{subject to} \quad (11.13)\text{–}(11.16),\ (11.125),$$

$$\lambda_i + \lambda_j + \beta_{ij} - \gamma_{ij} \geq \sum_{k,\ell \in K} C_{ijk\ell} w_{ijk\ell} \quad \forall i, j \in V,\ i \neq j, \tag{11.128}$$

$$\lambda_i \geq 0 \qquad \forall i \in V, \tag{11.129}$$

$$\beta_{ij}, \gamma_{ij} \geq 0 \qquad \forall i, j \in v,\ i \neq j. \tag{11.130}$$

In this model, variables λ_i are the dual variables associated with the constraints defining the hose uncertainty set, and β_{ij} (γ_{ij}) are the dual variables associated with the upper (lower) bounds on the demands according to the box uncertainty set.

For a given location solution, say $\hat{\boldsymbol{y}}$, the above problem becomes

$$\text{minimize} \quad \sum_{i \in V} \lambda_i L_i + \sum_{\substack{i,j \in V, \\ i \neq j}} (\overline{d}_{ij} \beta_{ij} - \underline{d}_{ij} \gamma_{ij}), \tag{11.127}$$

$$\text{subject to} \quad \sum_{k,\ell \in K} w_{ijk\ell} = 1 \qquad \forall i, j \in V,\ i \neq J, \tag{11.131}$$

$$\sum_{\ell \in K} w_{ijk\ell} + \sum_{\ell \in K \setminus \{k\}} w_{ij\ell k} \leq \hat{y}_k \quad \forall i, j \in V,\ i \neq j,\ k \in K, \tag{11.132}$$

$$\lambda_i + \lambda_j + \beta_{ij} - \gamma_{ij} \geq \sum_{k,\ell \in K} C_{ijk\ell} w_{ijk\ell} \quad \forall i, j \in V,\ i \neq j, \tag{11.128}$$

$$w_{ijk\ell} \geq 0 \quad \forall i, j \in V,\ i \neq j,\ k, \ell \in K, \tag{11.133}$$

$$\lambda_i \geq 0 \quad \forall i \in V, \tag{11.129}$$

$$\beta_{ij}, \gamma_{ij} \geq 0 \quad \forall i, j \in v,\ i \neq j. \tag{11.130}$$

We can change the type of Constraints (11.131) to "$\geq$" due to the nature of the problem (there is always an optimal solution satisfying those inequalities as equalities). Note also that the above problem is a linear programming problem that is feasible and bounded if $\sum_{k \in K} \hat{y}_k \geq 1$, $\overline{d}_{ij} \geq \underline{d}_{ij}$ $\forall i, j \in V$, $i \neq j$, and $L_i \geq \sum_{j \in V \setminus \{i\}} (\underline{d}_{ij} + \underline{d}_{ji})$.

Associating dual variables $\boldsymbol{\rho}$ to Constraints (11.131) (written as "$\geq$"), $\boldsymbol{\nu}$ to (11.132), and $\boldsymbol{\varpi}$ to (11.128), we can formulate the dual problem:

$$\text{maximize} \quad \sum_{\substack{i,j \in V, \\ i \neq j}} \left(\rho_{ij} - \sum_{k \in K} \hat{y}_k \nu_{ijk} \right), \tag{11.134}$$

$$\text{subject to} \quad \sum_{j \in V \setminus \{i\}} \varpi_{ij} - \sum_{j \in V \setminus \{i\}} \varpi_{ji} \leq L_i \quad \forall i \in V, \tag{11.135}$$

$$\underline{d}_{ij} \leq \varpi_{ij} \leq \overline{d}_{ij} \quad \forall i, j \in V,\ i \neq j, \tag{11.136}$$

$$\rho_{ij} - \nu_{ijk} - \nu_{ij\ell} \leq C_{ijk\ell} \varpi_{ij} \quad \forall i, j \in V,\ i \neq j,\ k, \ell \in K,\ k \neq \ell, \tag{11.137}$$

$$\rho_{ij} - \nu_{ijk} \leq C_{ijkk} \varpi_{ij} \quad \forall i, j \in V,\ i \neq j,\ k \in K, \tag{11.138}$$

$$\rho_{ij} \geq 0 \quad \forall i, j \in V,\ i \neq j, \tag{11.139}$$

$$\nu_{ijk} \geq 0 \quad \forall i, j \in V,\ i \neq j,\ k \in K. \tag{11.140}$$

Consequently, we can reformulate the robust uncapacitated p-hub median problem we are considering as

$$\text{minimize} \quad \upsilon, \tag{11.141}$$

$$\text{subject to} \quad \upsilon \geq \sum_{\substack{i,j \in V, \\ i \neq j}} \left(\rho_{ij}^q - \sum_{k \in K} y_k \nu_{ijk}^q \right) \quad \forall q = 1, \ldots, Q, \tag{11.142}$$

$$\sum_{k \in K} y_k = p, \tag{11.125}$$

$$y_k \in \{0, 1\} \qquad \forall k \in K, \tag{11.16}$$

with $(\boldsymbol{\rho}^q, \boldsymbol{\nu}^q, \boldsymbol{\varpi}^q)$ denoting the q-th extreme point of the polyhedron defined by (11.135)–(11.140). This problem can be solved within a branch-and-cut algorithmic framework embedding the cuts (11.142) on the fly when needed. In fact, for a given $(\hat{\upsilon}, \hat{\boldsymbol{y}})$, we can check whether there is an inequality (11.142) violated by solving the above dual problem (11.134)–(11.140). Meraklı and Yaman [379] suggest reformulating the dual problem as

$$\text{maximize} \quad \sum_{\substack{i,j \in V,\\ i \neq j}} \varpi_{ij} \left(\bar{\rho}_{ij} - \sum_{k \in K} \hat{y}_k \bar{\nu}_{ijk} \right), \tag{11.143}$$

$$\text{subject to} \quad (11.135), (11.136),$$

$$\bar{\rho}_{ij} - \bar{\nu}_{ijk} - \bar{\nu}_{ij\ell} \leq C_{ijk\ell} \quad \forall i, j \in V,\ i \neq j,\ k, \ell \in K,\ k \neq \ell, \tag{11.144}$$

$$\bar{\rho}_{ij} - \bar{\nu}_{ijk} \leq C_{ijkk} \qquad \forall i, j \in V,\ i \neq j,\ k \in K, \tag{11.145}$$

$$\bar{\rho}_{ij} \geq 0 \qquad \forall i, j \in V,\ i \neq j, \tag{11.146}$$

$$\bar{\nu}_{ijk} \geq 0 \qquad \forall i, j \in V,\ i \neq j,\ k \in K, \tag{11.147}$$

with $\bar{\rho} = \rho_{ij}/\varpi_{ij}$, $\bar{\nu}_{ijk} = \nu_{ijk}/\varpi_{ij}$ $\forall i, j \in I,\ i \neq j,\ k \in K$. Note also that Constraints (11.135) and (11.136) simply state that $\boldsymbol{\varpi} \in \mathcal{U}$. Finally, we can rewrite the dual as

$$\max_{\boldsymbol{\varpi} \in \mathcal{U}} \sum_{\substack{i,j \in V,\\ i \neq j}} \varpi_{ij} \Psi_{ij},$$

with

$$\Psi_{ij} = \max \quad \bar{\rho}_{ij} - \sum_{k \in K} \hat{y}_k \bar{\nu}_{ijk}, \tag{11.148}$$

$$\text{s.t.} \quad \bar{\rho}_{ij} - \bar{\nu}_{ijk} - \bar{\nu}_{ij\ell} \leq C_{ijk\ell} \quad \forall k, \ell \in K,\ k \neq \ell, \tag{11.149}$$

$$\bar{\rho}_{ij} - \bar{\nu}_{ijk} \leq C_{ijkk} \qquad \forall k \in K, \tag{11.150}$$

$$\bar{\rho}_{ij} \geq 0, \tag{11.151}$$

$$\bar{\nu}_{ijk} \geq 0 \qquad \forall k \in K. \tag{11.152}$$

In turn, the latter is the dual of

$$\text{minimize} \quad \sum_{k\ell \in K} C_{ijk\ell} w_{ijk\ell}, \tag{11.153}$$

$$\text{subject to} \quad \sum_{k\ell} w_{ijk\ell} \geq 1, \tag{11.154}$$

$$\sum_{\ell \in K} w_{ijk\ell} + \sum_{\ell \in K \setminus \{k\}} w_{ij\ell k} \leq \hat{y}_k \qquad \forall k \in K, \tag{11.155}$$

$$w_{ijk\ell} \geq 0 \qquad \forall k, \ell \in K. \tag{11.156}$$

As pointed out in [135], given that the hubs are uncapacitated, if we are given a set of open hubs, the optimal flow routing does not depend on the actual demand observed. This is clear from this last subproblem: The flow follows the unit-cost-wise shortest path between origin and destination.

Therefore, the dual problem can be efficiently solved in two steps: (i) We compute the shortest path (in terms of unit costs) between each origin and destination. (ii) Given that the constraints defining $\mathcal{U}$ are linear, define a linear programming problem seeking the demand vector in $\mathcal{U}$ that maximizes the total flow costs using the paths obtained in (i).

Meraklı and Yaman [379] show empirically that the branch-and-cut procedure described is efficient for tackling large instances (up to 150 nodes and five hubs).

For the problem just described and thinking of large-scale instances, a tabu search procedure is proposed in [232] for finding feasible solutions. The interested reader can refer to that work for all details.

We proceed this chapter with the so-called r-allocation p-hub median problem, first introduced in [553]. It unifies in a single modeling framework different allocation strategies. An advantage of considering this generalization is that it contains the single- and multiple-allocation settings as particular cases. Thus, the discussion to be presented next is valid for both. The problem can be formulated using two sets of decision variables. Adopting the customary notation, we keep using x_{ik} as a binary variable equal to one if and only if node $i \in V$ is allocated to hub $k \in K$. A variable x_{kk} equal to one indicates that $k \in K$ is selected as a hub. Regarding the flow, we consider $w_{ijk\ell}$ representing the proportion of the flow to be sent from i to j that is routed via hubs k and ℓ, $i, j \in V$, $k, \ell \in K$. A mathematical formulation is the following (we adopt the notation underlying the single-allocation hub location problem already discussed):

$$\text{minimize} \quad \sum_{i,j \in V} \sum_{k,\ell \in K} d_{ij} C_{ijk\ell} w_{ijk\ell}, \tag{11.157}$$

$$\text{subject to} \quad \sum_{k \in K} x_{ik} \leq r \qquad \forall i \in V, \tag{11.158}$$

$$x_{ik} \leq x_{kk} \qquad \forall i \in V,\ k \in K, \tag{11.159}$$

$$\sum_{k\in K} x_{kk} = p, \tag{11.160}$$

$$\sum_{k,\ell\in K} w_{ijk\ell} = 1 \qquad \forall i, j \in V, \tag{11.161}$$

$$\sum_{\ell\in K} w_{ijk\ell} \le x_{ik} \qquad \forall i, j \in V,\ k \in K, \tag{11.162}$$

$$\sum_{k\in K} w_{ijk\ell} \le x_{j\ell} \qquad \forall i, j \in V,\ \ell \in K, \tag{11.163}$$

$$x_{ik} \in \{0, 1\} \qquad \forall i \in V,\ k \in K, \tag{11.164}$$

$$w_{ijk\ell} \ge 0 \qquad \forall i, j \in V,\ \forall k, \ell \in K. \tag{11.165}$$

In this model, Constraints (11.158) impose the allocation of each node to at most r-hubs. If r is one, we obtain the single-allocation problem; in case $r = p$, we fall into the multiple-allocation case.

Peiró et al. [425] enrich the above model in different ways: (i) by assuming the existence of a fixed cost for connecting the nodes to the hubs (e.g., a vehicle); (ii) by assuming a fixed cost for the inter-hub edges (e.g., a vehicle); (iii) by allowing direct shipments (that the authors call non-stop service) between non-hub nodes although incurring a fixed and a variable cost.

To capture those extensions, we introduce some additional notation. Let a_{ik} denote the cost for allocating node $i \in V$ to hub $k \in K$. Regarding the hub edges, $g_{k\ell}$ is the fixed cost for having an edge operating between hubs k and ℓ $(k, \ell \in K)$. Finally, sending flow directly from node i to j incurs a fixed cost b_{ij} and a unit flow cost e_{ij}. Two different sets of decision variables need to be included. A binary variable z_{ij} is equal to one if and only if a direct service is established between nodes i and j, $\forall i, j \in V$. Furthermore, $y_{k\ell}$ is a variable that is equal to one if both k and ℓ are hubs and zero otherwise, $k, \ell \in K, k < \ell$.

The r-allocation p-hub median problem can be extended to the new setting:

$$\text{minimize} \quad \sum_{i\in V}\sum_{\substack{k\in K\\ k\neq i}} a_{ik}x_{ik} + \sum_{\substack{k,\ell\in K\\ k<\ell}} g_{k\ell}y_{k\ell} + \sum_{i,j\in V}\sum_{k,\ell\in K} d_{ij}C_{ijk\ell}w_{ijk\ell}$$
$$+ \sum_{i,j\in V} (b_{ij} + e_{ij}d_{ij})z_{ij}, \tag{11.166}$$

$$\text{subject to} \quad (11.158)\text{–}(11.165), \tag{11.167}$$

$$\sum_{k,\ell\in K} w_{ijk\ell} + z_{ij} = 1 \qquad \forall i, j \in V, \tag{11.168}$$

$$\sum_{j\in V} z_{ij} + \sum_{j\in V} z_{ji} \le \Theta(1 - x_{ii}) \qquad \forall i \in K, \tag{11.169}$$

$$y_{k\ell} \leq x_{kk} \qquad \forall k, \ell \in K,\ k < \ell, \tag{11.170}$$

$$y_{k\ell} \leq x_{\ell\ell} \qquad \forall k, \ell \in K,\ k < \ell, \tag{11.171}$$

$$x_{kk} + x_{\ell\ell} \leq y_{k\ell} + 1 \qquad \forall k, \ell \in K,\ k < \ell, \tag{11.172}$$

$$y_{k\ell} \in \{0, 1\} \qquad \forall k, \ell \in K,\ k < \ell, \tag{11.173}$$

$$z_{ij} \in \{0, 1\} \qquad \forall i, j \in V. \tag{11.174}$$

The meaning of the additional components of the objective function is immediate. Constraints (11.168) state that the flow between a pair of nodes is routed either through the hub network or using a direct service but not both. Constraints (11.169) prevent the hubs from being involved in direct services (this is a possibility only between non-hub nodes). Inequalities (11.170)–(11.172) ensure that a hub edge connects two operating hubs. Finally, we have the domain constraints for the additional decision variables.

In this problem, the location and allocation decisions (when some investment is involved) have a long-lasting effect. They are typically made before having perfect information about the future. On the other hand, transportation decisions are typically made at an operational (or at most tactical) level and thus can adapt to the actual observed information. This motivates a two-stage decision-making process: The *ex ante* decisions concern network design; the *ex post* ones regard transportation.

Let us assume uncertainty stemming from the demands and transportation costs. In particular, we assume that d_{ij}, $C_{ijk\ell}$, b_{ij}, e_{ij} are random variables ($i, j \in V$, $k, \ell \in K$) that define a random vector $\boldsymbol{\xi} = (\boldsymbol{d}, \boldsymbol{C}, \boldsymbol{b}, \boldsymbol{e})$ that has a known CDF.

Given the assumptions made, we can cast the problem as a two-stage stochastic optimization problem. Let us assume that $\boldsymbol{\xi}$ has a finite support. In this case, we face a finite set of scenarios that we denote by Ω to be consistent with the notation adopted. Furthermore, we can index in Ω both the uncertain parameters and the second-stage decisions. Accordingly, we define:

- $d_{ij\omega}$, flow to be shipped from i to j ($i, j \in V$) under scenario $\omega \in \Omega$;
- $C_{ijk\ell\omega}$, unit cost for the traffic routed along the path $i \to k \to \ell \to j$ ($i, j \in V$, $k, \ell \in K$) under scenario $\omega \in \Omega$;
- $b_{ij\omega}$, fixed cost for using a direct service between nodes i and j ($i, j \in V$) under scenario $\omega \in \Omega$;
- $e_{ij\omega}$, unit cost for the traffic shipped between nodes i and j ($i, j \in V$) using a direct service under scenario $\omega \in \Omega$.

Regarding the flow variables, we define $w_{ijk\ell\omega}$ as the proportion of the flow from i to j ($i, j \in V$) that is routed via hubs k and ℓ ($k, \ell \in K$) under scenario $\omega \in \Omega$. Finally, we redefine the direct-service variables: $z_{ij\omega}$ is a binary variable equal to one if a direct service is adopted between nodes i and j ($i, j \in V$) under scenario $\omega \in \Omega$ and zero otherwise. Denoting by π_ω the probability associated with scenario $\omega \in \Omega$ ($\pi_\omega > 0\ \forall \omega \in \Omega$, and $\sum_{\omega \in \Omega} \pi_\omega = 1$), the extensive form of the deterministic equivalent can be formulated mathematically as follows:

$$\text{minimize} \quad \sum_{i\in V}\sum_{\substack{k\in K\\ k\neq i}} a_{ik}x_{ik} \sum_{\substack{k,\ell\in K\\ k<\ell}} g_{k\ell}y_{k\ell} + \sum_{\omega\in\Omega}\pi_\omega \left(\sum_{i,j\in V}\sum_{k,\ell\in K} d_{ij\omega}C_{ijk\ell\omega}w_{ijk\ell\omega} + \sum_{i,j\in V}(b_{ij\omega}+e_{ij\omega}d_{ij\omega})z_{ij\omega}\right) \tag{11.175}$$

subject to (11.158)–(11.160), (11.170)–(11.172), (11.173), (11.174),

$$\sum_{\ell\in K} w_{ijk\ell\omega} \leq x_{ik} \qquad \forall i,j\in V,\ k\in K,\ \omega\in\Omega \tag{11.176}$$

$$\sum_{k\in K} w_{ijk\ell\omega} \leq x_{j\ell} \qquad \forall i,j\in V,\ \ell\in K,\ \omega\in\Omega, \tag{11.177}$$

$$\sum_{k,\ell\in K} w_{ijk\ell} + z_{ij\omega} = 1 \qquad \forall i,j\in V,\ \omega\in\Omega, \tag{11.178}$$

$$\sum_{j\in V} z_{ij\omega} + \sum_{j\in V} z_{ji\omega} \leq \Theta(1-x_{ii}) \qquad \forall i\in K,\ \omega\in\Omega, \tag{11.179}$$

$$w_{ijk\ell\omega}\in\{0,1\} \qquad \forall i,j\in V,\ k\ell\in K,\ \omega\in\Omega, \tag{11.180}$$

$$z_{ij\omega}\in\{0,1\} \qquad \forall i,j\in V,\ \omega\in\Omega. \tag{11.181}$$

Given the above model, we seek to select a set of hubs and define the assignment of the terminals to these hubs (first-stage decisions) so that the resulting network design performs well in every possible scenario.

Given a network G, with $|V| = 15$ and nine scenarios, out-of-memory errors were reported in [425] when tackling the model using an off-the-shelf solver. However, the same authors notice that deterministic instances with more than 400 nodes can be solved within a reasonable computing time. This is approximately the number of commercial airports in the USA. A heuristic procedure for the stochastic problem based on the single-scenario problems is proposed accordingly. Next, we follow that work.

Given that the whole problem (let us denote it by $\mathcal{HLP}$) seeks a solution that hedges against uncertainty, it is reasonable to think that single-scenario problems, say $\mathcal{HLP}_\omega$ ($\omega\in\Omega$), can provide information about attributes of a solution to $\mathcal{HLP}$. This motivates the construction of a solution to $\mathcal{HLP}$ using solutions to $\mathcal{HLP}_\omega$. Nevertheless, solutions for different problems $\mathcal{HLP}_\omega$ may lead to different network designs regarding the selected hubs or the allocation of terminals to them. Even the number of hubs allocated to a terminal may change from one scenario to another.

The above challenges motivated the authors in [425] to devise a new type of heuristic procedure that they call a greedy attributive scenario-based constructive method for finding solutions to the stochastic problem, which is sketched in

Algorithm 11.2. The scheme is general enough to be extended to other stochastic programming problems with binary first-stage decisions.

Algorithm 11.2 is a randomized procedure that can be executed repeatedly. The final solution is the best among those found in the different runs.

In this algorithm, β^* denotes the best upper bound known for $\mathcal{HLP}$, whereas β_ω denotes the best upper bound for problem $\mathcal{HLP}_\omega$ ($\omega \in \Omega$). Furthermore, $(\boldsymbol{w}, \boldsymbol{x}, \boldsymbol{y}, \boldsymbol{z})$ represents the best solution known for problem $\mathcal{HLP}_\omega$ and $(\boldsymbol{w}, \boldsymbol{x}, \boldsymbol{y}, \boldsymbol{z})^{\mathcal{HLP}}$ a feasible solution for $\mathcal{HLP}$. Finally, $(\boldsymbol{w}, \boldsymbol{x}, \boldsymbol{y}, \boldsymbol{z})^*$ is the best solution for $\mathcal{HLP}$ found by the algorithm.

Lines 3–11 seek a feasible solution for every scenario $\omega \in \Omega$. In line 12, the greedy attributive approach is called to build a feasible solution to the stochastic problem. In lines 13–15, the incumbent solution for the stochastic problem is updated when a better solution is found.

Since the procedure is randomized, we repeat the entire process $iter_1^{\max}$ times, keeping the best solution found. The same holds when determining a feasible solution to the single-scenario problem (lines 5–11)—the construction procedure is randomized.

The above algorithm leaves two main aspects for discussion. The first is related to the single-scenario problems. Since these are deterministic hub location problems, we refer the reader to existing heuristics to r-allocation p-hub median problems and their variants, which can be extended to the setting we are discussing. Moreover, the reader can refer to [425] and the constructive and improvement procedures presented there.

Algorithm 11.2: Main loop

Input: Problem data, t_{max}, $itermax$

1 $\beta^* \leftarrow +\infty$; $\beta_\omega \leftarrow \infty \; \forall \omega \in \Omega$
2 **while** t_{max} is not reached **do**
3   **foreach** $\omega \in \Omega$ **do**
4         $\beta_\omega \equiv (\boldsymbol{w}, \boldsymbol{x}, \boldsymbol{y}, \boldsymbol{z})^\omega \leftarrow +\infty$
5   **for** $iter \leftarrow 1$ **to** $itermax$ **do**
6   Construct a solution $(\boldsymbol{w}, \boldsymbol{x}, \boldsymbol{y}, \boldsymbol{z})^\omega_{iter}$ // heuristic
7   Improve the solution $(\boldsymbol{w}, \boldsymbol{x}, \boldsymbol{y}, \boldsymbol{z})^\omega_{iter}$ // local search
8   // Update incumbent if an improved solution is found:
9   **if** value of $(\boldsymbol{w}, \boldsymbol{x}, \boldsymbol{y}, \boldsymbol{z})^\omega_{iter} <$ value of $(\boldsymbol{w}, \boldsymbol{x}, \boldsymbol{y}, \boldsymbol{z})^\omega$ **then**
10                 $(\boldsymbol{w}, \boldsymbol{x}, \boldsymbol{y}, \boldsymbol{z})^\omega \leftarrow (\boldsymbol{w}, \boldsymbol{x}, \boldsymbol{y}, \boldsymbol{z})^\omega_{iter}$
11   update β_ω
12   Obtain $(\boldsymbol{w}, \boldsymbol{x}, \boldsymbol{y}, \boldsymbol{z})^{\mathcal{HLP}}$ from $(\boldsymbol{w}, \boldsymbol{x}, \boldsymbol{y}, \boldsymbol{z})^\omega$, $\omega \in \Omega$ // core part
13   **if** value of $(\boldsymbol{w}, \boldsymbol{x}, \boldsymbol{y}, \boldsymbol{z})^{\mathcal{HLP}} < \beta^*$ **then**
14         $\beta^* \leftarrow$ value of $\mathcal{HLP}$
15         $(\boldsymbol{w}, \boldsymbol{x}, \boldsymbol{y}, \boldsymbol{z})^* \leftarrow (\boldsymbol{w}, \boldsymbol{x}, \boldsymbol{y}, \boldsymbol{z})^{\mathcal{HLP}}$

Output: $(\boldsymbol{w}, \boldsymbol{x}, \boldsymbol{y}, \boldsymbol{z})^*$

Given the perspective of this volume, we concentrate on the second open aspect for discussion: how to obtain a solution to the stochastic problem from the solutions to the single-scenario problems.

Let $\hat{\boldsymbol{x}}^{\omega}$ be the values of the allocation variables $\boldsymbol{x}$ in the solution found for $\mathcal{HLP}_{\omega}$, $\omega \in \Omega$. Consider the $|V| \times |K|$ matrix

$$U = \begin{pmatrix} u_{11} & u_{12} & \cdots & u_{1|K|} \\ u_{21} & u_{22} & \cdots & a_{2|K|} \\ \vdots & \vdots & u_{ij} & \vdots \\ u_{|V|1} & u_{|V|2} & \cdots & u_{|V||K|} \end{pmatrix}$$

with generic entry given by

$$u_{ij} = \sum_{\omega \in \Omega} \pi_{\omega} x_{ij}^{\omega} = \pi_1 x_{ij}^1 + \pi_2 x_{ij}^2 + \ldots + \pi_{|\Omega|} x_{ij}^{|\Omega|}.$$

We can look at each row of this matrix representing a node and each column representing a potential hub. The information in U can be explored to assess the "attractiveness" of a node to become hub in $\mathcal{HLP}$—we can compute the marginal vector, $\overline{u}$, resulting from summing all rows of U:

$$\overline{u} = \left(\overline{u}_1, \overline{u}_2, \ldots, \overline{u}_j, \ldots, \overline{u}_{|K|}\right).$$

Generically, we have $\overline{u}_j = \sum_{i \in V} u_{ij} \; \forall j \in K$.

The starting point for building a feasible solution to $\mathcal{HLP}$ is to randomly select a scenario ω^* according to the probabilities $\pi_1, \ldots, \pi_{|\Omega|}$. Then, a process inspired by the path relinking methodology (see [236]) that gradually transforms an initiating solution by incorporating attributes of some guiding solutions is devised. In particular, we consider attributes of the best solutions obtained in each scenario averaged according to their probabilities. This information is contained in matrix U. The steps of the process can be summarized as follows:

(i) Consider the set of p-hubs, H_{ω^*}, in the feasible solution known for $\mathcal{HLP}_{\omega^*}$.
(ii) Find the p indices associated with the largest values of $\overline{u}_j$ for $j \in V \setminus H_{\omega^*}$.
(iii) Denote by $J^p = (j_1, \ldots, j_p)$ a sequence of the indices found in (ii) resulting from sorting the corresponding values of $\overline{u}_j$ non-increasingly, and set $F = \emptyset$.
Repeat steps (iv)–(vii) $p - 1$ times:
(iv) For each $j \in J^p \setminus F$.
(v) For each $h \in H_{\omega^*} \setminus F$:

  (a) Denote by $H_{\omega^*}(h, j)$ the set of hubs resulting from replacing h by j in H_{ω^*}.
  (b) For each $k \in H_{\omega^*}(h, j)$ and for each $i \in V$, if $u_{ik} > 0$, set $z_{ik} = 1$, i.e., allocate node i to hub k.

(c) For each node $i \in V$, denote by $K(i)$ the set of hubs the node is allocated to (if some) and check whether this allocation set is feasible. If not, repair it. Three cases can be distinguished:

c1. Terminal i has been assigned to more than r-hubs (i.e., $|K(i)| > r$).
In this case, the $|K(i)| - r$ assignments of i to the hubs in $H_{\omega^*}(h, j)$ of the smallest values u_{ij} are discarded by setting the corresponding z variables to 0.

c2. Terminal i has not been assigned to any hub (i.e., $|K(i)| = 0$).
This may be the case of a node i assigned only to hub h. The assignment of i to an existing hub say k is evaluated for every scenario using the following measure:

$$a_{ik\omega} + c_{ik\omega} \sum_{j \in V} d_{ij\omega} + \alpha c_{ki\omega} \sum_{j \in V} d_{ji\omega}$$
$$+ \sum_{j \in V} (d_{ij\omega} + d_{ji\omega}) \alpha \left(\frac{\sum_{\ell \in H_{\omega^*}(h,j) \setminus \{k\}} c_{k\ell\omega}}{|H_{\omega^*}(h, j)| - 1} \right).$$

The hub that appears the most across all scenarios is the one selected for allocating i.

c3. Node i is a hub and was allocated to other hubs.
In this case, all the assignments of i to other hubs are removed.

(d) Consider $H_{\omega^*}(h, j)$ as the set of operating hubs together with the allocations z_{ik} resulting from (b) and (c). For each scenario $\omega \in \Omega$, do the following:

d1. Compute the cost of the solution if we implement such a feasible network design for that particular scenario. Denote that cost by $c^{\omega}(h, j)$.

d2. Compute the value $r^{\omega}(h, j) = \frac{c^{\omega}(h,j) - \beta_{\omega}}{\beta_{\omega}}$. This value is in some sense a relative "regret" if the network design implemented in scenario ω is the one induced by $H_{\omega^*}(h, j)$ instead of the best network design known so far for that scenario. (Recall that β_{ω} is the incumbent upper bound on the optimal value for scenario ω.)

d3. Compute $r^{\max}(h, j) = \max_{\omega \in \Omega}\{r^{\omega}(h, j)\}$ as the maximum relative "regret" across all scenarios if we take the network design induced by $H_{\omega^*}(h, j)$ with the allocations resulting from (b) and (c). Update β_{ω} if the value of the solution found for problem $\mathcal{HLP}_{\omega}$ has improved.

(e) Update the best incumbent solution for $\mathcal{HLP}$ if the cost found is smaller than the current cost.

(vi) Compare the above sets of hubs and decide on one of them. To do so, we use the values $r^{\max}(h, j)$ computed in (v)(d)(3.). In particular, we select the hub set $H_{\omega^*}(h', j')$ associated with the minimum value of $r^{\max}(h, j)$.

(vii) Update F and H_{ω^*} as follows: $F = F \cup \{j'\}$ and $H_{\omega^*} = H_{\omega^*} \setminus \{h'\} \cup \{j'\}$.

The mechanism defined by steps (i)–(vii) results in a feasible solution to the stochastic $\mathcal{HLP}$. The complete heuristic repeats this process a certain number of iterations defined beforehand.

It is worth noticing that the above-presented scheme is quite flexible. It is modular, i.e., some parts can be replaced without changing the global structure. For instance, if a different approach is considered for tackling the single-scenario problems, the above structure can be adopted exactly as presented.

The whole procedure is summarized in Algorithm 11.3. Peiró et al. [425] report computational results using this heuristic for tackling instances generated from the AP dataset (see [192]) with up to 50 nodes and up to ten scenarios.

Algorithm 11.3: Procedure to obtain feasible solutions to $\mathcal{HLP}$

```
Input: (w, x, y, z)^ω, ∀ω ∈ Ω; matrix U = [u_ij]_{i∈V, j∈K} = (∑_ω π_ω x_ij^(ω))_{i∈V, j∈K}
for repetition=1 to repetition = ⌈|Ω|/2⌉ do
    Randomly select a scenario ω*, according to the probabilities π_1, ..., π_|Ω|, as starting
      point of the process. Let H_ω* be the set of hubs in the solution obtained for scenario
      ω*.
    Compute (ū_j)_{j∈K\H_ω*} = (∑_{i∈V} u_ij)_{j∈K\H_ω*} and let J^p = (j_1, ..., j_p) be the p
      indices with the largest values. Let F = ∅.
    for times=1 to times = p - 1 do
        foreach j ∈ J^p \ F do
            foreach h ∈ H_ω* \ F do
                Let H_ω*(h, j) = H_ω* \ {h} ∪ {j}.
                Allocate nodes i ∈ V to hubs in H_ω*(h, j): node i is allocated to hub k
                  if u_ik > 0.
                If i has been assigned to more than r-hubs, or it has not been assigned to
                  any hub, or it is a hub allocated to other hubs, apply a repair process
                  (see above).
                foreach ω ∈ Ω do
                    Solve the flow routing subproblem and denote the cost of the
                      solution by c^ω(h, j).
                    Compute the maximum relative "regret" across all scenarios
                      r^max(h, j) = max_{ω∈Ω}{(c^ω(h, j) − β_ω)/β_ω}, where β_ω is the incumbent
                      upper bound on the optimal value for scenario ω.
                    Update β_ω if the value of the solution found for the single-scenario
                      problem HLP_ω has improved.
                The best solution for HLP is updated if a new solution with a lower
                  cost is found.
        Among all the explored sets of p-hubs H_ω*(h, j), the set H_ω*(h', j') associated
          with the minimum value of r^max(h, j) is selected.
        Update F = F ∪ {j'} and H_ω* = H_ω* \ {h'} ∪ {j'}.
Output: (w, x, y, z)^HLP
```

Ghaffarinasab and Kara [234] introduce a modeling framework for the stochastic multiple-allocation p-hub median problem assuming a risk-averse decision-maker. Removing the fixed costs from model (2.116)–(2.119) and adding a constraint on the number of hubs to open, the authors consider the following base model:

$$\text{minimize} \quad \sum_{i,j\in V}\sum_{k,\ell\in K} d_{ij}C_{ijk\ell}w_{ijk\ell}, \tag{11.182}$$

$$\text{subject to} \quad \sum_{k\in K} y_k = p, \tag{11.183}$$

$$\sum_{k,\ell\in K} w_{ijk\ell} = 1 \qquad \forall i,j\in V, \tag{11.184}$$

$$\sum_{\ell\in K} w_{ijk\ell} + \sum_{\ell\in K\setminus\{k\}} w_{ij\ell k} \le y_k \qquad \forall k\in K,\ i,j\in V, \tag{11.185}$$

$$w_{ijk\ell} \ge 0 \qquad \forall i,j\in V,\ k,\ell\in K, \tag{11.186}$$

$$y_k \in \{0,1\} \qquad \forall k\in K. \tag{11.187}$$

Based upon the above model, those authors introduce a so-called conditional β-mean criterion. They consider demand as the source of uncertainty and assume that the latter can be captured by a finite set of scenarios indexed in a set Ω. Given a tolerance $0 < \beta \le 1$, the goal is to find the hubs to locate and the flow routing to optimize the average performance across the worst $\lceil\beta|\Omega|\rceil$ scenarios. Defining a set of additional binary variables, r_ω equal to one if and only if scenario $\omega\in\Omega$ is within the worst $\lceil\beta|\Omega|\rceil$ scenarios, let $\mathcal{R} = \{\boldsymbol{r}\in\{0,1\}^{|\Omega|} \mid \sum_{\omega\in\Omega} r_\omega = \lceil\beta|\Omega|\rceil\}$. The problem can be formulated as follows:

$$\text{minimize} \quad \max_{r\in\mathcal{R}} \frac{1}{\lceil\beta|\Omega|\rceil}\sum_{\omega\in\Omega}\sum_{i,j\in V}\sum_{k,\ell\in K}\left(d_{ij\omega}C_{ijk\ell}w_{ijk\ell\omega}\right) r_\omega, \tag{11.188}$$

subject to (11.183), (11.187),

$$\sum_{k,\ell\in K} w_{ijk\ell\omega} = 1 \qquad \forall i,j\in V,\ \omega\in\Omega, \tag{11.189}$$

$$\sum_{\ell\in K} w_{ijk\ell\omega} + \sum_{\ell\in K\setminus\{k\}} w_{ij\ell k\omega} \le y_k \qquad \forall k\in K,\ i,j\in V,\ \omega\in\Omega, \tag{11.190}$$

$$w_{ijk\ell\omega} \ge 0 \qquad \forall i,j\in V,\ k,\ell\in K,\ \omega\in\Omega. \tag{11.191}$$

The inner maximization problem resulting from fixing the $\boldsymbol{w}$ and $\boldsymbol{y}$ variables is a binary knapsack problem whose matrix is totally unimodular—the integrality constraints in the $\boldsymbol{r}$ variables can be relaxed, thus allowing dualizing the inner problem. Defining a dual multiplier μ for constraint $\sum_{\omega \in \Omega} r_\omega = 1$ and multipliers λ_ω for $r_\omega \leq 1$, $\omega \in \Omega$, the resulting MILP formulation for the whole problem becomes

$$\text{minimize} \quad \lceil \beta |\Omega| \rceil \mu + \sum_{\omega \in \Omega} \lambda_\omega, \tag{11.192}$$

$$\text{subject to} \quad (11.183), (11.187), (11.189)–(11.191),$$

$$\mu + \lambda_\omega \geq \frac{1}{\lceil \beta |\Omega| \rceil} \sum_{i,j \in V} \sum_{k,\ell \in K} d_{ij\omega} C_{ijk\ell} w_{ijk\ell\omega} \quad \forall \omega \in \Omega, \tag{11.193}$$

$$\lambda_\omega \geq 0 \quad \forall \omega \in \Omega. \tag{11.194}$$

Ghaffarinasab and Kara [234] propose a Benders decomposition algorithm for this problem. The structure of the dual subproblems allows solving them efficiently: A greedy heuristic can be used for determining the optimal values of one set of dual variables and the resulting problem solved by inspection using a mechanism similar to that proposed in [135].

11.2.4 Hub Center and Covering Problems

The p-hub center problem consists of selecting p nodes for becoming hubs so that the maximum of some observable measure is minimized. The problem is particularly relevant when travel times are accounted for, and we seek to design the network to minimize the maximum traveling time among all origin–destination pairs. We recall a model presented in Sect. 2.3 for the single-allocation problem:

$$\text{minimize} \quad B, \tag{11.195}$$

$$\text{subject to} \quad B \geq C_{ijk\ell} w_{ijk\ell} \quad \forall i, j \in V,\ k\ell \in K, \tag{11.196}$$

$$\sum_{k \in K} x_{ik} = 1 \quad \forall i \in V, \tag{11.197}$$

$$\sum_{k \in K} x_{kk} = p, \tag{11.198}$$

$$x_{ik} \leq x_{kk} \quad \forall i \in v,\ k \in K, \tag{11.199}$$

$$w_{ijk\ell} \geq x_{ik} + x_{j\ell} - 1 \quad \forall i, j \in V,\ k\ell \in K, \tag{11.200}$$

$$w_{ijk\ell} \in \{0, 1\} \quad \forall i, j \in V,\ k\ell \in K, \tag{11.201}$$

$$x_{ik} \in \{0, 1\} \qquad \forall i \in V,\ k \in K. \tag{11.202}$$

In this model, the costs $C_{ijk\ell}$ may represent, for instance, the travel time between nodes i and j if the route between them uses hubs located at k and ℓ ($i, j \in V$, $k, \ell \in K$). The variables x_{ik} indicate whether a node $i \in V$ is allocated to hub $k \in K$, whereas variables $w_{ijk\ell}$ indicate whether the flow between i and j is routed via hubs k and ℓ ($i, j \in V$, $k, \ell \in K$). The meaning of the objective function and the constraints is aligned with the discussion we have been presenting.

Looking into the above model, we realize that uncertainty can stem only from the costs $C_{ijk\ell}$. Such a setting is considered in [507]. The authors assume the travel times in between any two nodes ℓ and ℓ' ($\ell, \ell' \in V$) independent and represented by a Gaussian random variable, say $\boldsymbol{\zeta}_{\ell\ell'}$ with expected value $\mu_{\ell\ell'}$ and standard deviation $\sigma_{\ell\ell'}$. Thus, assuming as customary in this volume $C_{ijk\ell} = c_{ik} + \alpha c_{k\ell} + c_{\ell j}$, we conclude that $C_{ijk\ell}$ is also distributed according to a Gaussian random variable, say $\boldsymbol{\xi}_{ijk\ell}$, with parameters $\mu_{ik} + \alpha\mu_{k\ell} + \mu_{\ell j}$ and $\sqrt{\sigma_{ij}^2 + \alpha^2\sigma_{k\ell} + \sigma_{\ell j}^2}$.

A major issue under uncertain costs $\boldsymbol{\xi}_{ijk\ell}$ is that Constraints (11.196) becomes ill-defined. From a practical perspective, we can seek a network design that minimizes the maximum service cost that can be guaranteed with some given probability. This calls for replacing (11.196) with

$$\mathbb{P}[\boldsymbol{\xi}_{ijk\ell} w_{ijk\ell} \leq B] \geq \gamma \qquad \forall i, j \in V,\ k, \ell \in K, \tag{11.203}$$

keeping the other components of the model unchanged including the objective function. In other words, we are hedging against uncertainty using chance constraints.

Given the Gaussian distribution of the costs, we can easily derive a deterministic equivalent for the probabilistic constraints (11.203):

$$B \geq \left(\mu_{ik} + \alpha\mu_{k\ell} + \mu_{\ell j} + z_\gamma \sqrt{\sigma_{ij}^2 + \alpha^2\sigma_{k\ell} + \sigma_{\ell j}^2}\right) w_{ijk\ell} \quad \forall i, j \in V,\ k\ell \in K, \tag{11.204}$$

with z_γ denoting the γ-quantile of a standardized Gaussian distribution.

Despite being possible to find a deterministic equivalent for the chance-constrained problem, the resulting optimization model is still computationally quite demanding to the four-indexed binary variables involved. Sim et al. [507] prove that for a given selection of hubs, say H, the problem that consists of finding the optimal allocation of the nodes to the hubs can be formulated as follows:

$$\text{minimize} \quad B, \tag{11.205}$$

$$\begin{aligned} \text{subject to} \quad B \geq & \sum_{k \in H} \left(\mu_{ik} + \alpha\mu_{k\ell} + z_\gamma \sqrt{\sigma_{ik}^2 + \alpha^2\sigma_{k\ell}^2 + \sigma_{\ell j}^2}\right) x_{il} \\ & + (\mu_{\ell j} + z_\gamma \sigma_{\ell j}) x_{j\ell} - z_\gamma \sigma_{\ell j} \qquad \forall i, j \in V,\ \ell \in H, \end{aligned} \tag{11.206}$$

$$\sum_{k \in H} x_{ik} = 1 \qquad \forall i \in V,\ k \in H, \tag{11.207}$$

$$x_{ik} \in \{0, 1\} \qquad \forall i \in V,\ k \in H. \tag{11.208}$$

Constraints (11.206) are inspired by the formulation proposed in [301] for the p-hub center problem.

Sim et al. [507] propose a heuristic based upon the well-known exchange Teitz and Bart procedure (see [519]). The interested reader can refer to those papers for all the details.

In the context of the stochastic p-hub center problem, it is worth revisiting a linearization also considered in [507]. In addition to the distribution assumptions above presented, those authors also assume a stochastic equivalent to the triangle inequality:

$$\zeta_{\ell\ell'} \prec\prec \zeta_{\ell\ell''} + \zeta_{\ell''\ell'} \quad \forall \ell, \ell', \ell'' \in V,$$

which determines stochastic dominance. In the case of a Gaussian assumption, it is equivalent to

$$\mu_{ik} + z_\gamma \sigma_{ik} \leq \mu_{ij} + \mu_{jk} + z_\gamma \sqrt{\sigma_{ij}^2 + \sigma_{jk}^2}.$$

Under this stochastic dominance, in addition to the other distributional assumptions, the authors in [507] also show that (11.203) is equivalent to

$$\begin{aligned} B \geq & \left(\mu_{ik} + \alpha\mu_{k\ell} + z_\gamma \sqrt{\sigma_{ik}^2 + \alpha^2\sigma_{k\ell}^2 + \sigma_{\ell j}^2}\right) x_{ik} \\ & + \left(\mu_{\ell j} + z_\gamma \sqrt{\sigma_{ik}^2 + \alpha^2\sigma_{k\ell}^2 + \sigma_{\ell j}^2}\right) x_{j\ell} \\ & - z_\gamma \sqrt{\sigma_{ik}^2 + \alpha^2\sigma_{k\ell}^2 + \sigma_{\ell j}^2}. \end{aligned}$$

As discussed in Sect. 2.3, hub covering location problems are somehow related to center problems, although now the coverage radius is exogenous. Ghaffarinasab and Kara [234] investigate such a problem assuming a risk-averse decision-maker. Those authors consider a so-called β-mean objective for the p-hub covering problem. The deterministic problem stems directly from (11.182)–(11.187) by replacing the objective function with

$$\text{maximize} \quad \sum_{i,j \in V} \sum_{k,\ell \in K} d_{ij} a_{ijk\ell} w_{ijkl}, \tag{11.209}$$

where $a_{ijk\ell}$ is a binary parameter that for each $i, j \in V$ and $k, \ell \in K$ is equal to one if and only if the cost for shipping demand d_{ij} is smaller than or equal to a given acceptable threshold.

Ghaffarinasab and Kara [234] consider a similar mechanism for tackling the above problem as we explained in the case of the p-hub median problem. Assuming demand as the only source of uncertainty and captured by a finite set of scenarios indexed in a finite set Ω, the problem can be formulated as follows:

$$\text{maximize} \quad \min_{r \in \mathcal{R}} \frac{1}{\lceil \beta |\Omega| \rceil} \sum_{\omega \in \Omega} \sum_{i,j \in V} \sum_{k,\ell \in K} \left(d_{ij\omega} a_{ijk\ell} w_{ijk\ell\omega} \right) r_\omega, \tag{11.210}$$

$$\text{subject to} \quad (11.183), (11.187), (11.189)–(11.191).$$

In this problem, for each $\omega \in \Omega$, r_ω is a binary variable equal to one if and only if scenario $\omega \in \Omega$ is within the worst $\lceil \beta |\Omega| \rceil$ scenarios. Also, $\mathcal{R} = \{\boldsymbol{r} \in \{0,1\}^{|\Omega|} \mid \sum_{\omega \in \Omega} r_\omega = \lceil \beta |\Omega| \rceil\}$. As for the p-hub median problem, the same authors note that the inner maximization problem resulting from fixing the $\boldsymbol{w}$ and $\boldsymbol{y}$ variables is a binary knapsack problem whose matrix is totally unimodular, and thus the integrality constraints in the $\boldsymbol{r}$ variables can be relaxed. In turn, this allows dualizing the inner problem. Defining a multiplier μ for constraint $\sum_{\omega \in \Omega} r_\omega = 1$ and multipliers λ_ω for $r_\omega \leq 1$, $\omega \in \Omega$, the resulting MILP formulation for the problem becomes

$$\text{maximize} \quad \lceil \beta |\Omega| \rceil \mu - \sum_{\omega \in \Omega} \lambda_\omega, \tag{11.211}$$

$$\text{subject to} \quad (11.183), (11.187), (11.189)–(11.191),$$

$$\mu - \lambda_\omega \geq \frac{1}{\lceil \beta |\Omega| \rceil} \sum_{i,j \in V} \sum_{k,\ell \in K} d_{ij\omega} a_{ijk\ell} w_{ijk\ell\omega} \qquad \forall \omega \in \Omega, \tag{11.212}$$

$$\lambda_\omega \geq 0 \qquad \forall \omega \in \Omega. \tag{11.213}$$

For tackling this problem, an algorithm similar to that already summarized above for the p-hub median problem is proposed in [234]. Thus, we omit the details here.

11.3 Embedding Capacities

An aspect of practical relevance in hub location is the possibility of having capacitated hubs. Unfortunately, this extension poses some challenges, as others discussed in the following pages. This explains that the work capturing uncertainty in the context of capacitated hub location problems is still scarce.

In a capacitated hub location problem, the types of resources or operations involved stipulate the appropriate way to model capacity limitations. For instance, in airline applications, runaway availability limits the entire traffic going through a hub.

However, in transportation systems, the bottleneck capacity at the hubs typically concerns incoming traffic. The reason is that passengers must leave a train or a plane and proceed to the corresponding connection. In this case, the outbound is often not a problem since people go in different directions and over some time. Another situation occurs when the hubs perform sorting operations as in postal services. In this case, the capacity of a hub node refers only to the incoming non-processed flow, i.e., flow sent from the non-hub nodes to that hub or flow originating at the hub (that has to be processed before being sent to other hubs or non-hub nodes). The incoming flow from other hubs and the outgoing flow do not pose capacity challenges because it does not require any processing. This last case has mostly been considered in the hub location literature in general. It has also been the case where uncertainty has been investigated the most.

11.3.1 Exogenous Capacities

We first discuss aspects related to hedging against uncertainty in hub location problems with exogenous capacity limits. We consider both robust optimization and stochastic programming paradigms since those are the ones that have been considered.

We keep using the notation introduced so far unless stated otherwise. If q_k is the capacity of a hub installed at $k \in K$, the problem can be formulated by adding capacity constraints to the models already discussed for hub location. Let us assume the existence of fixed costs for installing the hubs. By embedding capacity constraints in model (2.116)–(2.120), we obtain

$$\text{minimize} \quad \sum_{k \in K} f_k y_k + \sum_{i,j \in V} \sum_{k,\ell \in K} d_{ij} C_{ijk\ell} w_{ijk\ell}, \tag{11.214}$$

$$\text{subject to} \quad \sum_{k,\ell \in K} w_{ijk\ell} = 1 \qquad \forall i, j \in V, \tag{11.215}$$

$$\sum_{\ell \in K} w_{ijk\ell} + \sum_{\ell \in K \setminus \{k\}} w_{ij\ell k} \leq y_k \qquad \forall i, j \in V,\ k \in K, \tag{11.216}$$

$$\sum_{i,j \in V} \sum_{\ell \in K} d_{ij} w_{ijk\ell} \leq q_k y_k \qquad \forall k \in K, \tag{11.217}$$

$$w_{ijk\ell} \geq 0 \qquad \forall i, j \in V,\ k, \ell \in K, \tag{11.218}$$

$$y_k \in \{0, 1\} \qquad \forall k \in K. \tag{11.219}$$

In the above model, capacity refers to the incoming flow into the hubs. Meraklı and Yaman [380] consider this to extend their work in [379]. Again, they assume

$d_{ij} = 0, \forall i, j \in V, i \neq j$, and thus adjust the above model accordingly. They assume hose uncertain demand like that discussed in Sect. 11.2.3:

$$\mathcal{U}^{\text{hose}} = \left\{ d \in \mathbb{R}_+^{|V|(|V|-1)} \mid \sum_{j \in V \setminus \{i\}} d_{ij} + \sum_{j \in V \setminus \{i\}} d_{ji} \leq L_i \ \forall i \in V \right\}.$$

This leads to the following robust optimization model:

$$\text{minimize} \quad \sum_{k \in K} f_k y_k + \max_{d \in \mathcal{U}^{\text{hose}}} \left\{ \sum_{\substack{i,j \in V \\ i \neq j}} \sum_{k,\ell \in K} d_{ij} C_{ijk\ell} w_{ijk\ell} \right\}, \tag{11.220}$$

$$\text{subject to} \quad (11.215), (11.216), (11.218), (11.219),$$

$$\max_{d \in \mathcal{U}^{\text{hose}}} \left\{ \sum_{i,j \in V} \sum_{\ell \in K} d_{ij} w_{ijk\ell} \right\} \leq q_k y_k \qquad \forall k \in K. \tag{11.221}$$

By dualizing both maximization problems (note that the hose constraints are linear and thus strongly duality holds), we obtain the following MILP robust counterpart (the reasoning is similar to that discussed in Sect. 11.2.3):

$$\text{minimize} \quad \sum_{k \in K} f_k y_k + \sum_{i \in V} \lambda_i L_i, \tag{11.222}$$

$$\text{subject to} \quad (11.215), (11.216), (11.218), (11.219),$$

$$\lambda_i + \lambda_j \geq \sum_{k,\ell \in K} C_{ijk\ell} w_{ijk\ell} \qquad \forall i, j \in V, \ i \neq j, \tag{11.223}$$

$$\sum_{i \in V} \beta_{ik} L_i \leq q_k y_k \qquad \forall k \in K, \tag{11.224}$$

$$\beta_{ik} + \beta_{jk} \geq \sum_{\ell \in K} w_{ijk\ell} \qquad \forall i, j \in V, \ i \neq j, \ k \in K, \tag{11.225}$$

$$\lambda_i \geq 0 \qquad \forall i \in V, \tag{11.226}$$

$$\beta_{ik} \geq 0 \qquad \forall i \in V, \ k \in K. \tag{11.227}$$

Meraklı and Yaman [380] prove that

$$\sum_{k \in K} q_k y_k \geq \min \left\{ \left(\sum_{i \in V} L_i - \max_{i \in V} L_i \right), \frac{1}{2} \sum_{i \in V} L_i \right\}$$

is a valid inequality for the above model. The relevance of this result is that it adapts to the robust optimization model the well-known cut in discrete facility location (and hub location), stating that the operational capacity must be equal to or greater than the total demand. In the above inequality, the worst-case demand observation is used.

Meraklı and Yaman [380] also develop two Benders decomposition algorithms. In the first one, they decompose by fixing variables $\boldsymbol{y}$ and $\boldsymbol{\beta}$. In the second one, they project out the flow variables. We omit the details here since, from the perspective of optimization under uncertainty, those authors leverage the discussion above presented in the context of the robust p-hub median problem with the necessary changes and additional algorithmic aspects required by including fixed costs and the capacity constraints.

When it comes to stochastic programming approaches for capacitated hub location, we start by discussing the extension of the capacitated facility location problem with Bernoulli demands discussed in Sect. 6.3.4. Accordingly, hubs are uncapacitated, and we face unit origin–destination demands (a service is to be provided). We keep considering complete hub-level networks. When the actual demand allocated to a hub exceeds its capacity, a recourse action is conducted that we generically call outsourcing. Following the terminology used in Sect. 6.3.4 (that in turn follows [9]), we may have facility outsourcing, indicating that some extra capacity is acquired and the origin–destination pairs are served as usual from the hub. In the case of customer outsourcing, upon exceeding the demand, some origin–destination pairs will be served by a third party. Outsourcing costs incur for both policies.

Considering the single allocation of non-hubs to hubs and also the usual allocation variables x_{ik} considered in hub location, the problem can be conceptually formulated as follows:

$$\text{minimize} \quad \sum_{k \in K} f_k y_k + Q(\boldsymbol{x}) \tag{11.228}$$

$$\text{subject to} \quad \sum_{k \in K} x_{ik} = 1 \qquad \forall i \in V, \tag{11.229}$$

$$x_{ik} \leq x_{kk} \qquad \forall i \in V,\ k \in K, \tag{11.230}$$

$$l_k x_{kk} \leq \sum_{i \in V} x_{ik} \qquad \forall k \in K, \tag{11.231}$$

$$x_{ik} \in \{0, 1\} \qquad \forall i \in V,\ k \in K. \tag{11.232}$$

The objective function includes the setup costs for the hubs and the recourse function. Constraints (11.229) impose the allocation of all nodes to a hub irrespective of having demand or not. In turn, the following Inequalities (11.230) guarantee that allocation is performed only to installed hubs. Constraints (11.231) impose a minimum number of allocations to a hub to justify opening it. Finally, we have the domain constraints.

Let us assume that the flow from node $i \in V$ to $j \in V$ is represented by a Bernoulli random variable, ξ_{ij} with parameter π_{ij}. Let OD_k denote the set of origin–destination pairs that are assigned to hub $k \in K$ for being served if calling for it, i.e., $OD_k = \{(i, j) \in V \times V \mid x_{ik} = 1\}$. Furthermore, for each $k \in K$, let q_k be its service capacity and η_k be the random variable representing the demand allocated to hub k, i.e., $\eta_k = \sum_{(i,j)\in OD_k} \xi_{ij}$. We have

$$\begin{aligned}\mathbb{P}[\eta_k = s] &= \mathbb{P}\left[\sum_{i\in V}\left(\sum_{j\in V} \xi_{ij}\right) x_{ik} = s\right]\\ &= \sum_{S\subset OD_k:|S|=s} \prod_{(i,j)\in S} \pi_{ij} \prod_{(i,j)\in OD_k\setminus S} (1-\pi_{ij}).\end{aligned}$$

Under facility (customer) outsourcing, let g'_k (g''_k) be the cost incurred at hub $k \in K$ per origin–destination pair outsourced.

Leveraging the contents of Sect. 6.3.4 and assuming a neutral attitude of the decision-maker toward risk, it is easy to conclude that the expected outsourcing cost is equal to

$$\sum_{k\in K} g'_k \left(\sum_{s=q_k+1}^{|OD_k|} \mathbb{P}[\eta_k = s](s - q_k)\right) x_{kk},$$

in the case of facility outsourcing and the same but with g''_k replacing g'_k in the case of customer outsourcing.

Regarding the expected transportation cost, again following the same work is straightforward to conclude that it is given by

$$\sum_{i,j\in V} \pi_{ij} \left(\sum_{k\in K} c_{ik}x_{ik} + \sum_{\ell\in K} c_{\ell j}x_{j\ell} + \alpha \sum_{k,\ell\in K} c_{k\ell}x_{ik}x_{j\ell}\right),$$

in the case of facility outsourcing. Under customer outsourcing, we obtain

$$\begin{aligned}\sum_{k\in K}\sum_{s=l_k}^{|OD_k|} \mathbb{P}[\eta_k = s] &\left(\sum_{i,j\in V} \mathbb{P}[\xi_{ij} = 1 \mid \eta_k = s] \times \mathbb{P}_k[s]\right.\\ &\left.\times \left[c_{ik}x_{ik} + \sum_{\ell\in K} c_{\ell j}x_{j\ell} + \alpha \sum_{k\in K} c_{k\ell}x_{ik}x_{j\ell}\right]\right).\end{aligned}$$

Unfortunately, the deterministic equivalent problems resulting from the above expressions are intractable in the general case. This is an open research avenue. If the demands are homogeneous $\pi_{ij} = \pi\ \forall i, j \in V$, then it is possible to derive compact formulations to the deterministic equivalent problems. This is done in [231].

Proposition 11.4 *Let Y_k^ν be a binary variable equal to one if and only if a hub is installed at $k \in K$ and $\nu \in \{l_k, \ldots, |V|^2\}$ origin–destination services are assigned to it. Consider again variables $u_{ik\ell}$ representing the amount of flow originated at node i that is processed at hub $k \in K$ and distributed via hub $\ell \in K$. For each $k \in K$ and $\nu \in \{l_k, \ldots, |V|^2\}$, let $f_k^{\prime\nu} = f_k + g_k^{\prime} \sum_{s=q_k+1}^{\nu} B_{\nu s} \cdot (s - q_k)$, and $B_{\nu s} = \binom{\nu}{s}\pi^s(1-\pi)^{\nu-s}\ \forall s = 0, \ldots, \nu$. Under homogeneous Bernoulli flows and facility outsourcing, the following model is a deterministic equivalent for the stochastic single-allocation hub location problem:*

$$\text{minimize} \quad \sum_{k \in K} \sum_{\nu=l_k}^{|V|^2} f_k^{\prime\nu} Y_k^\nu + \sum_{i \in V} \sum_{k \in K} |V|\pi(c_{ik} + c_{ki})x_{ik}$$

$$+ \alpha \sum_{i \in V} \sum_{k,\ell \in K} \pi d_{k\ell} u_{ik\ell}, \tag{11.233}$$

$$\text{subject to} \quad (11.229),\ (11.230),\ (11.232),$$

$$\sum_{\ell \in K} u_{ik\ell} - \sum_{\ell \in K} u_{i\ell k} = |V|x_{ik} - \sum_{j \in V} x_{jk} \qquad \forall i \in V,\ k \in K, \tag{11.234}$$

$$\sum_{\nu=l_k}^{|V|^2} \nu Y_k^\nu = |V| \sum_{i \in V} x_{ik} \qquad \forall k \in K, \tag{11.235}$$

$$\sum_{\nu=l_k}^{|V|^2} Y_k^\nu \leq 1 \qquad \forall k \in K,\ \nu \in \{l_k, \ldots, |V|^2\}, \tag{11.236}$$

$$Y_k^\nu \in \{0, 1\} \qquad \forall k \in K,\ t \in \{l_k, \ldots, |V|^2\}, \tag{11.237}$$

$$u_{ik\ell} \geq 0 \qquad \forall i \in V,\ k, \ell \in K. \tag{11.238}$$

Proof The proof is very much aligned with that for a similar result presented in [9] and can be found in [231]. Overall, it consists of discretizing the (hub) location variables (see [245] for the basics on the application of that technique in the context of discrete facility location). In other words, not only do the location variables indicate whether some location is chosen but also whether some specific number of origin–destination pairs are assigned to it. □

Proposition 11.5 *Let X^{ν}_{ik} be a binary variable equal to one if and only if node $i \in V$ is allocated to hub $k \in K$ and $\nu \in \{l_k, \ldots, |V|^2\}$ origin–destination flows are assigned to that same hub. Define $U^{\nu}_{ik\ell}$ as the amount of flow originated at node i that is processed at hub $k \in K$ and distributed via hub $\ell \in K$ if ν origin–destination pairs are assigned to k. Let $U'^{\nu}_{ik\ell}$ be the traffic received at node $j \in V$ from hubs $k \in K$ and $\ell \in K$ ($\ell \neq k$) and given that ν origin–destination pairs are assigned to hub k for processing (if they call for it).*

For each $k \in K$ and $\nu \in \{l_k, \ldots, |V|^2\}$, let $f''^{\nu}_k = f_k + g''_k \sum_{s=l_k}^{\nu} B_{\nu s} \cdot (s - q_k)$ and $B_{\nu s} = \binom{\nu}{s} \pi^s (1-\pi)^{\nu-s}$ $\forall s = 0, \ldots, \nu$. Finally, let $c^{\nu}_{ijk} = c_{ij} \left(\sum_{s=l_k}^{\nu} B_{\nu s} \frac{\min\{q_k, s\}}{\nu} \right)$ $\forall i, j \in V,\ k \in K,\ \nu \in \{l_k, \ldots, |V|^2\}$. Under homogeneous Bernoulli flows and customer outsourcing, the following model is a deterministic equivalent for the stochastic single-allocation hub location problem:

$$
\begin{aligned}
\text{minimize} \quad & \sum_{k\in K} \sum_{\nu=l_k}^{|V|^2} f''^{\nu}_k X^{\nu}_{kk} + \sum_{i\in V} \sum_{k\in K} \sum_{\nu=l_k}^{|V|^2} |V| c^{\nu}_{ijk} X^{\nu}_{ik} + \alpha \sum_{i\in V} \sum_{k,\ell\in K} \sum_{\nu=l_k}^{|V|^2} c^{\nu}_{k\ell k} U^{\nu}_{ik\ell} \\
& + \sum_{k,\ell\in K} \sum_{j\in V} \sum_{\nu=l_k}^{|V|^2} c^{\nu}_{j\ell k} U'^{\nu}_{k\ell j} + \sum_{j\in V} \sum_{\ell\in K} \sum_{\nu=l_\ell}^{|V|^2} \frac{\nu}{|V|} c^{\nu}_{j\ell\ell} X^{\nu}_{j\ell},
\end{aligned} \tag{11.239}
$$

$$
\text{subject to} \quad \sum_{k\in K} \sum_{\nu=l_k}^{|V|^2} X^{\nu}_{ik} = 1 \qquad \forall i \in V, \tag{11.240}
$$

$$
X^{\nu}_{ik} \leq X^{\nu}_{kk} \qquad \forall i \in V,\ k \in K,\ \nu \in \{l_k, \ldots, |V|^2\}, \tag{11.241}
$$

$$
\sum_{\ell\in K} \sum_{\nu=l_k}^{|V|^2} U^{\nu}_{ik\ell} - \sum_{\ell\in K} \sum_{\nu=l_k}^{|V|^2} U^{\nu}_{i\ell k} = |V| \sum_{\nu=l_k}^{|V|^2} X^{\nu}_{ik} - \sum_{j\in V} \sum_{\nu=l_k}^{|V|^2} X^{\nu}_{jk} \qquad \forall i \in V,\ k \in K, \tag{11.242}
$$

$$
\sum_{k\in K} \sum_{\nu=l_k}^{|V|^2} U'^{\nu}_{k\ell j} - \sum_{k\in K} \sum_{\nu=l_k}^{|V|^2} U'^{\nu}_{\ell kj} = |V| \sum_{\nu=l_\ell}^{|V|^2} X^{\nu}_{j\ell} - \sum_{i\in V} \sum_{\nu=l_\ell}^{|V|^2} X^{\nu}_{i\ell} \qquad \forall j \in V,\ \ell \in K, \tag{11.243}
$$

$$
\sum_{i\in V} \sum_{\ell\in K} U^{\nu}_{ik\ell} \leq |V|^2 X^{\nu}_{kk} \qquad \forall k \in K,\ \nu \in \{l_k, \ldots, |V|^2\}, \tag{11.244}
$$

$$\sum_{j\in V}\sum_{\ell\in K} U'^{\nu}_{k\ell j} \le |V|^2 X^{\nu}_{kk} \qquad \forall k \in K,\ \nu \in \{l_k,\dots,|V|^2\}, \tag{11.245}$$

$$\nu X^{\nu}_{kk} = \sum_{i\in V} |V| X^{\nu}_{ik} \qquad \forall k \in K,\ \nu \in \{l_k,\dots,|V|^2\}, \tag{11.246}$$

$$\sum_{\nu=l_k}^{|V|^2} X^{\nu}_{kk} \le 1 \qquad \forall k \in K, \tag{11.247}$$

$$X^{\nu}_{ik} \in \{0,1\} \qquad \forall i \in V,\ k \in K,\ \nu \in \{l_k,\dots,|V|^2\}, \tag{11.248}$$

$$U^{\nu}_{ik\ell} \ge 0 \qquad \forall i \in V,\ k,\ell \in K,\ \nu \in \{l_k,\dots,|V|^2\}, \tag{11.249}$$

$$U'^{\nu}_{k\ell j} \ge 0 \qquad \forall j \in V,\ k,\ell \in K,\ \nu \in \{l_k,\dots,|V|^2\}. \tag{11.250}$$

Proof See [231] and the hints provided in the proof of Proposition 11.4. □

The relevance of capturing uncertainty can be assessed by the additional cost incurred for exceeding the service capacities. The expected fixed cost under facility outsourcing is given by

$$\overline{f}'^{\nu}_{k} = f_k + g'_k \sum_{s=q_k}^{\nu} \max\{0, \pi\, s - q_k\} \quad \forall k \in K,\ \nu \in \{l_k,\dots,|V|^2\},$$

whereas under customer outsourcing we obtain

$$\overline{f}''^{\nu}_{k} = f_k + g''_k \sum_{s=l_k}^{\nu} \max\{0, \pi\, s - q_k\} \quad \forall k \in K,\ \nu \in \{l_k,\dots,|V|^2\}.$$

When evaluating the expected transportation cost, we must consider

$$c^{\nu}_{ijk} = c_{ij} \sum_{s=l_k}^{\nu} \frac{\max\{0, \pi\, s - q_k\}}{\nu} \quad \forall i,j \in N,\ k \in K,\ \nu \in \{l_k,\dots,|V|^2\}.$$

Using these modified elements in the objective functions, we obtain the so-called expected value problem, which, in turn, allows us to assess the relevance of considering a stochastic programming modeling framework.

Ghaffarinasab [231] develops a Benders decomposition to the problem. Andaryan et al. [26] propose a tabu search heuristic for tackling large-scale instances. We refer the interested reader to those papers for all algorithmic details.

A challenge when facing non-unit-demand capacitated problems is that the problem easily misses complete or even relatively complete recourse, i.e., unless an overly conservative solution is chosen, a decision implemented before uncertainty is revealed may easily do not allow a feasible completion for some scenarios. A natural way for capturing uncertainty in this case is via chance constraints. We keep considering single allocation of non-hubs to hubs and stochastic demands.

We also keep considering unit costs satisfying the triangle inequality, which, as mentioned in Sect. 2.3, allows halving the number of flow variables by considering $w_{ijk\ell}$, $i \le j$ $(i, j \in V)$. This is possible because the hub edge used by the flow between two nodes i and j is the same for the flow between j and i $(i, j \in V)$. Let us consider again model (11.71)–(11.77) with variables $w_{ijk\ell}$, $i \le j$ as above stated. Additionally, let $\boldsymbol{\xi}_{ij}$ be the random variable representing the flow to be sent from i to j $(i, j \in V)$. We represent by $\boldsymbol{\xi}_{i\cdot}$ $(\boldsymbol{\xi}_{\cdot i})$ the random variable representing the total flow originated at (destined to) $i \in V$.

A chance-constraint single-allocation hub location problem can be formulated as follows:

$$\text{minimize} \quad \sum_{k \in K} f_k x_{kk} + Q(\boldsymbol{w}, \boldsymbol{x}), \tag{11.251}$$

$$\text{subject to} \quad \sum_{k \in K} x_{ik} = 1 \qquad \forall i \in V, \tag{11.252}$$

$$x_{ik} \le x_{kk} \qquad \forall i \in V,\ k \in K, \tag{11.253}$$

$$\sum_{\ell \in K} w_{ijk\ell} = x_{ik} \qquad \forall i, j \in V,\ i \le j,\ k \in K, \tag{11.254}$$

$$\sum_{k \in K} w_{ijk\ell} = x_{j\ell} \qquad \forall i, j \in V,\ i \le j,\ \ell \in K, \tag{11.255}$$

$$\mathbb{P}\left[\sum_{i \in V} \boldsymbol{\xi}_{i\cdot} x_{ik} \le q_k x_{kk}\right] \ge \beta, \qquad \forall k \in K, \tag{11.256}$$

$$x_{ik} \in \{0, 1\} \qquad \forall i \in V,\ k \in K, \tag{11.257}$$

$$w_{ijk\ell} \ge 0 \qquad \forall i, j \in V,\ \forall k, \ell \in K. \tag{11.258}$$

In the above model, q_k is the capacity of a hub installed at $k \in K$. We are facing a single-stage problem, given that all is decided here and now. Nevertheless, this raises an issue concerning the objective function. The transportation cost can only be computed after the demands are observed. Assuming a risk-neutral decision-maker, a possible choice for the second term of the objective function is

$$Q(\boldsymbol{w}, \boldsymbol{x}) = \sum_{i \in V} \sum_{k \in K} (\mathbb{E}[\boldsymbol{\xi}_{i\cdot}] + \mathbb{E}[\boldsymbol{\xi}_{\cdot i}])\, c_{ik} x_{ik}$$
$$+ \sum_{\substack{i,j \in V \\ i \leq j}} \sum_{k,\ell \in K} \alpha \left(\mathbb{E}[\boldsymbol{\xi}_{ij}] c_{k\ell} + \mathbb{E}[\boldsymbol{\xi}_{ji}] c_{\ell k}\right) w_{ijk\ell}.$$

As discussed in Chap. 7, in some cases, a deterministic equivalent can be found for the chance constraints. For instance, in case of independent demands distributed according to a Gaussian distribution, $\boldsymbol{\xi}_{ij} \sim N(\mu_{ij}, \sigma_{ij})$, $i, j \in V$, then the same holds with $\boldsymbol{\xi}_{i\cdot}$, $i \in V$, which is also distributed according to a Gaussian random variable with parameters $\mu_{i\cdot} = \sum_{j \in V} \mu_{ij}$ and $\sigma_{i\cdot} = \sqrt{\sum_{j \in J} \sigma_{ij}^2}$. Accordingly, Constraints (11.256) can be written as follows:

$$\sum_{i \in V} \mu_{i\cdot} x_{ik} + \zeta_\beta \sqrt{\sum_{i \in V} \sigma_{i\cdot}^2 x_{ik}^2} \leq q_k x_{kk} \quad \forall k \in K, \tag{11.259}$$

where ζ_β is the β-quantile of a standardized Gaussian distribution. Furthermore, given that in practice β is fairly large (clearly larger than 0.5), then we can re-write the above constraints as second-order conic constraints

$$\|\Sigma^{\frac{1}{2}} \boldsymbol{x}_{\cdot k}\| \leq \frac{q_k x_{kk} - \sum_{i \in V} \mu_{i\cdot} x_{ik}}{\zeta_\beta} \qquad \forall k \in K, \tag{11.260}$$

with $\boldsymbol{x}_{\cdot k} = (x_{1k}, \ldots, x_{|V|k})^\top$ and $\Sigma^{\frac{1}{2}} = \text{Diag}(\sigma_{1\cdot}, \ldots, \sigma_{|V|\cdot})$.

Note that also as explained in Chap. 7, an MILP formulation can be derived for the deterministic equivalent of our stochastic hub location problem by considering an auxiliary variable $v_k = \sum_{i \in V} \sigma_{i\cdot}^2 x_{ik}^2$ and using a piecewise linear approximation for the resulting quadratic functions (see [309, 310, 338]) as well as the contents of that chapter.

Hu et al. [279] present a work that extends these models by imposing a maximum and minimum fraction of the capacities to be satisfied by all hubs with some joint probability.

11.3.2 Endogenous Capacities

We focus now on the case in which the capacity of the hubs is a decision to make. Moreover, we directly discuss a broader multi-period setting. In Part IV of the current volume, this setting is relevant for one application discussed.

Let us consider the uncapacitated multiple-allocation hub location problem that started this chapter, given by the model (11.1)–(11.11). Assuming that a hub must process its outbound flow and distribute its inbound flow, then, as discussed in Sect. 2.3, we need to include the following inequalities:

$$x_{ik} \leq O_i(1 - y_i) \qquad \forall i, k \in K,\ i \neq k, \tag{11.261}$$

$$u'_{i\ell j} \leq D_j(1 - y_j) \qquad \forall j, \ell \in K,\ j \neq \ell. \tag{11.262}$$

Assuming the capacity of a hub specified by the highest flow it can process, denote by q_k that quantity for a hub operating at $k \in K$. The capacity constraints to be added to the model are

$$\sum_{i \in V} x_{ik} \leq q_k y_k \qquad \forall k \in K. \tag{11.263}$$

As done in Sect. 11.2.2, we consider generically a random event induced by a random variable $\boldsymbol{\xi}$ with each observation fully determining the uncertain parameters underlying our problem. Let Ξ denote the support of $\boldsymbol{\xi}$. Assuming that we know the probability distribution of $\boldsymbol{\xi}$ and considering uncertainty stemming from the demands, a natural wait to embed uncertainty in a model is to consider a two-stage stochastic programming with the network design decisions being implemented here and now and the flow routing being looked at as adaptive. Unfortunately, this setting raises the usual concern: A first-stage decision is no longer guaranteed to have a second-stage feasible completion. This should be avoided.

Correia et al. [144] propose a generic two-stage stochastic modeling framework overcoming that drawback and being aligned to some practical needs: They study a stochastic multi-period hub location problem with modular capacities. The capacities can adapt to the demand faced throughout time and thus represent adaptive decisions. At the beginning of each period, we can open new hubs that remain operating until the end of the planning horizon.

When a hub is installed, we must decide the number of modules that define the initial capacity of the hub. At the beginning of the following periods, we can install additional modules. A limit exists for the number of modules installed at each location. At any point, the number of modules installed in a hub determines its capacity. During the planning horizon, neither can the hubs be closed nor have their capacities reduced—i.e., we are dealing with a pure phase-in problem.

The capacity of each hub refers to the maximum flow it can process. Each non-hub node can be allocated to more than one hub in each period, although the allocations can change over the planning horizon. The transportation costs are functions of the distances between nodes. Those distances are assumed to satisfy the triangle inequality. As a result, all the flow shipped through the network traverses at least one and at most two hubs. The decisions to be made are: (i) when and where to install new hubs, (ii) the initial capacity for new hubs, (iii) when and where to expand capacity (for hubs that have already been installed), and (iv) how to route the flow between O–D pairs in each period. The objective is to minimize the total cost, which has several components: (i) setup cost for new hubs and their initial capacity, (ii) capacity expansion at the existing hubs, (iii) operation of the hubs, and (iv) flow routing.

Let T denote the set of periods in the planning horizon. Additionally, we define several deterministic parameters:

Γ, capacity of a module;
N_k, maximum number of modules that can be installed in a hub at location $k \in K$;
α_t, transfer cost discount factor in period $t \in T$;
f_{kt}, fixed cost for installing a hub in location $k \in K$ in period $t \in T$;
f'_{knt}, fixed cost for installing n modules in a hub at location $k \in K$ in period $t \in T, n = 1, \ldots, N_k$;
m'_{knt}, operating cost incurred when n modules are installed in a hub at location $k \in K$ in period $t \in T, n = 1, \ldots, N_k$;
f''_{knt}, fixed cost for installing n additional modules in a hub at location $k \in K$ in period $t \in T \setminus \{1\}, n = 1, \ldots, N_k - 1$;
m''_{knt}, operating cost incurred when n additional modules are installed in a hub at location $k \in K$ in period $t \in T \setminus \{1\}, n = 1, \ldots, N_k - 1$;

Regarding the stochastic parameters, we define:

$d_{ijt}(\boldsymbol{\xi})$, random variable representing the flow to be sent from node $i \in V$ to node $j \in V$ in period $t \in T$;
$O_{it}(\boldsymbol{\xi})$, random variable representing the total flow originated at node $i \in V$ in period $t \in T$ $(O_{it}(\xi) = \sum_{j\in V} d_{ijt}(\xi)\ \forall \xi \in \Xi)$;
$D_{it}(\boldsymbol{\xi})$, random variable representing the total flow destined to node $i \in I$ in period $t \in T$ $(D_{it}(\xi) = \sum_{j\in N} d_{jit}(\xi)\ \forall \xi \in \Xi)$.

By combining the fixed costs for the modules and the corresponding operating costs over an appropriate number of periods, the following aggregated costs are obtained:

$$g_{knt} = f'_{knt} + \sum_{\tau=t}^{|T|} m'_{kn\tau} \quad \forall k \in K,\ n = 1, \ldots, N_k,\ t \in T,$$

$$h_{knt} = f''_{knt} + \sum_{\tau=t}^{|T|} m''_{kn\tau} \quad \forall k \in K,\ n = 1, \ldots, N_k - 1,\ t \in T \setminus \{1\}.$$

For each $k \in K$, g_{knt} denotes the total cost resulting from setting up the initial capacity of a hub with n modules to be located at k in period t. This cost includes the related operational cost until the end of the planning horizon. Analogously, fixed and operating costs are also part of h_{knt} that stands for the cost to be paid from period t until the end of the planning horizon when the capacity of (an existing) hub k is expanded by installing n additional modules.

We note that the above cost structure is general. It can capture several suitable cases (e.g., nonlinear capacity costs).

Regarding the variables representing non-adaptive decisions, we consider the following:

$$y_{kt} = \begin{cases} 1, \text{ if a hub is installed at } k \in K \text{at the beginning of period } t \in T, \\ 0, \text{ otherwise.} \end{cases}$$

$$z_{knt} = \begin{cases} 1, \text{ if hub } k \in K \text{receives } n \in \{1, \ldots, N_k\} \text{ modules in period } t \in T, \\ 0, \text{ otherwise.} \end{cases}$$

We also consider the adaptive decision variables:

$$r_{knt}(\boldsymbol{\xi}) = \begin{cases} 1, \text{ if hub } k \in K \text{receives } n \text{ additional modules in period } t \in T, \\ 0, \text{ otherwise.} \end{cases}$$

$$r_{knt}(\boldsymbol{\xi}) = \begin{cases} 1, \text{ if hub } k \in K \text{receives } n \text{ additional modules in period } t \in T, \\ 0, \text{ otherwise.} \end{cases}$$

$x_{ikt}(\boldsymbol{\xi})$ = amount of flow originated at $i \in V$ that is collected at hub $k \in K$ in period $t \in T$.

$u_{ik\ell t}(\boldsymbol{\xi})$ = amount of flow originated at $i \in V$ that is collected at hub $k \in K$ and distributed by hub $\ell \in K$ in period $t \in T$.

$u'_{ij\ell t}(\boldsymbol{\xi})$ = amount of flow from $i \in V$ to $j \in V$ that is distributed by hub $\ell \in K$ in period $t \in T$.

Finally, we can formulate mathematically the problem we are discussing:

$$\text{minimize} \quad \sum_{k\in K}\sum_{t\in T}\left(f_{kt}y_{kt} + \sum_{n=1}^{N_k} g_{knt}z_{knt} \right) + \mathcal{Q}(\boldsymbol{y}, \boldsymbol{z}), \tag{11.264}$$

$$\text{subject to} \quad \sum_{n=1}^{N_k} z_{knt} = y_{kt} \qquad \forall k \in K,\ t \in T, \tag{11.265}$$

$$\sum_{t\in T} y_{kt} \le 1 \qquad \forall k \in K, \tag{11.266}$$

$$y_{kt} \in \{0, 1\} \qquad \forall k \in K,\ t \in T, \tag{11.267}$$

$$z_{knt} \in \{0, 1\} \qquad \forall k \in K,\ t \in T,\ n \in \{1, \ldots, N_k\}. \tag{11.268}$$

The objective function includes the cost of setting up the hubs and initially equipping them. It also includes the recourse function. Constraints (11.265) ensure that when a hub is installed, it is directly equipped with some operational capacity. Inequalities (11.265) guarantee that no hub is installed more than once during the planning horizon. Finally, (11.267) and (11.267) define the domain of the first-stage decision variables.

Assuming a neutral attitude of the decision-maker toward risk, we can set $Q(\mathbf{y}, \mathbf{z}) = E_{\boldsymbol{\xi}}[Q(\mathbf{y}, \mathbf{z}, \boldsymbol{\xi})]$ such that for each $\xi \in \Xi$:

$$Q(\mathbf{y}, \mathbf{z}, \xi) = \min \quad \sum_{t \in T} \sum_{i \in V} \sum_{k \in K} c_{ik} x_{ikt}(\xi) + \sum_{t \in T} \alpha_t \sum_{i \in N} \sum_{k,\ell \in K} c_{k\ell} u_{ik\ell t}(\xi)$$

$$+ \sum_{t \in T} \sum_{i,j \in V} \sum_{\ell \in K} c_{\ell j} u'_{ij\ell t}(\xi) + \sum_{k \in K} \sum_{t \in T \setminus \{1\}} \sum_{n=1}^{N_k - 1} h_{knt} r_{knt}(\xi), \tag{11.269}$$

$$\text{s.t.} \quad \sum_{k \in K} x_{ikt}(\xi) = O_{it}(\xi) \qquad \forall i \in N,\ t \in T, \tag{11.270}$$

$$\sum_{\ell \in K} u'_{ij\ell t}(\xi) = d_{ijt}(\xi) \qquad \forall i, j \in N,\ t \in T, \tag{11.271}$$

$$\sum_{n=1}^{N_k - 1} r_{knt}(\xi) \leq \sum_{\tau=1}^{t-1} y_{k\tau} \qquad \forall k \in K,\ t \in T \setminus \{1\}, \tag{11.272}$$

$$\sum_{i \in N} x_{ikt}(\xi) \leq \Gamma \sum_{n=1}^{N_k} \sum_{\tau=1}^{t} n \cdot z_{kn\tau} + \Gamma \sum_{n=1}^{N_k - 1} \sum_{\tau=2}^{t} n \cdot r_{kn\tau}(\xi)$$

$$\forall k \in N,\ t \in T, \tag{11.273}$$

$$\sum_{n=1}^{N_k} \sum_{t \in T} n \cdot z_{knt} + \sum_{n=1}^{N_k - 1} \sum_{t \in T \setminus \{1\}} n \cdot r_{knt}(\xi) \leq N_k$$

$$\forall k \in K, \tag{11.274}$$

$$\sum_{\substack{\ell \in K \\ \ell \neq k}} u_{ik\ell t}(\xi) - \sum_{\substack{\ell \in K \\ \ell \neq k}} u_{i\ell kt}(\xi) = x_{ikt}(\xi) - \sum_{j \in V} u'_{ijkt}(\xi)$$

$$\forall i \in V,\ k \in K,\ t \in T, \tag{11.275}$$

$$\sum_{\substack{\ell \in K \\ \ell \neq k}} u_{ik\ell t}(\xi) \leq x_{ikt}(\xi) \qquad \forall i \in V,\ k \in K,\ t \in T, \tag{11.276}$$

$$x_{ikt}(\xi) \leq O_{it}(\xi) \left(1 - \sum_{\tau=1}^{t} y_{i\tau} \right)$$

$$\forall i, k \in K : i \neq k,\ t \in T, \tag{11.277}$$

$$\sum_{i \in V} u'_{ij\ell t}(\xi) \leq D_{jt}(\xi) \sum_{\tau=1}^{t} y_{\ell\tau} \qquad \forall j \in V, \ \ell \in K, \ t \in T, \tag{11.278}$$

$$u'_{ij\ell t}(\xi) \leq D_{jt}(\xi) \left(1 - \sum_{\tau=1}^{t} y_{j\tau}\right) \qquad \forall i \in V, \ j, \ell \in K : \ell \neq j, \ t \in T, \tag{11.279}$$

$$r_{knt}(\xi) \in \{0, 1\} \qquad \forall k \in K, \ n = 1, \ldots, N_k - 1, \ t \in T \setminus \{1\}, \tag{11.280}$$

$$x_{ikt}(\xi) \geq 0 \qquad \forall i \in V, \ k \in K, \ t \in T, \tag{11.281}$$

$$u_{ik\ell t}(\xi) \geq 0 \qquad \forall i \in V, \ k, \ell \in K, \ t \in T, \tag{11.282}$$

$$u'_{ij\ell t}(\xi) \geq 0 \qquad \forall i, j \in V, \ \ell \in K, \ t \in T. \tag{11.283}$$

In the second-stage problem, variables y_{kt} and z_{knt} are fixed. The second-stage objective function accounts for the total transportation and capacity expansion costs. Most of the constraints have a meaning already discussed in this chapter. Regarding Constraints (11.272), they guarantee that capacity expansion occurs only at previously installed hubs. Inequalities (11.273) state that in each period, the incoming flow at a hub and the flow it originates cannot exceed the operating capacity installed. Inequalities (11.274) regard the maximum number of capacity modules that can be installed at each hub.

If the random vector underlying our stochastic problem has a finite support, we can index the second-stage decision variables in the scenario set. This is now a straightforward exercise given the similar examples in this volume. Thus, we omit it here.

Correia et al. [144] use the above model to solve small- to medium-sized instances. The problem can be seen as an extension of the single-period problem in which the hubs are installed and equipped with some initial capacity that can be adjusted when the demand is revealed.

11.4 Embedding Network Design Decisions

We assume now that the hub-level network is not necessarily a complete graph. Particularly, hub network design decisions are also part of the decision-making process. This is assumed in [284] who, nonetheless, consider uncapacitated hubs.

The authors analyze both the single- and multiple-allocation cases. For the former, the base model considered is that given by (2.67), (2.68), (2.70)–(2.76); for the latter, the base model used is (2.77)–(2.79), (2.81)–(2.91). In both cases, those authors propose a simheuristic for tackling the problem (see [292]). It starts with a network configuration found by solving a deterministic approximation to the problem. The actual cost of the resulting network design is estimated by simulation. In turn, the simulation phase provides information to guide a change in the network design. Such a change is imposed by adding a set of additional constraints to the deterministic problem. Again, simulation is used for estimating the actual cost. The algorithm proceeds until no further improvement is suggested as an outcome of the simulation phase. The authors test their methodology assuming continuous uniform distributions for different parameters such as the demands, the transportation costs, and the fixed costs for hubs and hub edges.

Combining capacities at the hubs with network design decisions raises many challenges, which may explain why little work can be found in the literature. Nevertheless, a notable exception is [51]. The authors investigate a capacitated single-allocation hub network design problem under stochastic demands. The authors model the problem as a network of M/M/1 queues where each hub is looked at as a single server and an exponential service rate is assumed (resulting from a capacity decision to be made). A first-come-first-served arrival process with a rate is also adopted that depends on how the allocation decisions to the hubs. The model introduced in [51] is comprehensive and realistic. Next, we revisit it.

We adopt most of the notation presented so far, changing it or introducing new only when necessary. A capacity level can be decided for each installed hub. The available levels are listed in a set L with q^l denoting the capacity corresponding to level $l \in L$ that, in turn, determines the service rate at a hub. Given that a steady state is achieved only if the arrival rate is lower than the service rate, we do not allow assigning to a hub any set of nodes inducing an arrival rate higher than the installed service capability at the hub.

Let P_{ij} be the set of feasible paths that can be used to route the flow from i to j $(i, j \in V)$. A feasible path is required to visit at least one and at most a certain given number κ of hubs. For each origin–destination pair, the set of feasible paths that can be used depends on the selected hubs. If A_p is the set of arcs visited by a path p, we define the unit transportation cost through this path as C_p. Note that a discount factor α is applied to the cost for traversing hub edges. We consider uncertainty captured by a finite set of scenarios Ω, each specifying a flow rate for every origin–destination pair (i, j) $\forall i, j \in V$.

To represent the decisions to make, the authors in [51] consider a variable $v_{p\omega}$ that for each $p \in P_{ij}$ and $i, j \in V$ represent the proportion of the flow from i to j that goes through path p. Another relevant quantity is represented by $u_{k\omega}$ that stands for the arrival rate at hub $k \in K$ under scenario $\omega \in \Omega$. These quantities are a function of the flows and the routes chosen for them. Specifically, we have $u_{k\omega} = \sum_{i,j \in V} \sum_{p \in P_{ij} | k \in p} d_{ij\omega} v_{p\omega}$. Note also that the mean transit time through a hub under steady-state conditions is given by $1/(\sum_{l \in L} q^l y_k^l - u_{k\omega})$, $k \in K, \omega \in \Omega$.

Regarding the location variables, we now index them in the capacity levels: y_k^l is a binary equal to one if and only if a hub is installed at $k \in K$ and equipped with capacity level $l \in L$.

The problem can be formulated mathematically as follows:

$$\text{minimize} \quad \sum_{k \in K} \sum_{l \in L} f_k^l y_k^l + \sum_{\omega \in \Omega} \pi_\omega \left(\sum_{k \in K} b_k \frac{u_{k\omega}}{\sum_{l \in L} q^l y_k^l - u_{k\omega} + \varepsilon} + \sum_{i,j \in V} \sum_{p \in P_{ij}} c_p v_{p\omega} \right), \quad (11.284)$$

$$\text{subject to} \quad \sum_{p \in P_{ij}} v_{p\omega} = 1 \qquad \forall i, j \in V,\ \omega \in \Omega, \quad (11.285)$$

$$\sum_{i,j \in V} \sum_{p \in P_{ij} | k \in p} d_{ij\omega} v_{p\omega} = u_{k\omega} \qquad \forall k \in K,\ \omega \in \Omega, \quad (11.286)$$

$$u_{k\omega} \leq \sum_{l \in L} q^l y_k^l \qquad \forall k \in K,\ \omega \in \Omega, \quad (11.287)$$

$$\sum_{l \in L} y_k^l \leq 1 \qquad \forall k \in K, \quad (11.288)$$

$$y_k^l \in \{0, 1\} \qquad \forall k \in K,\ l \in L, \quad (11.289)$$

$$v_{p\omega} \geq 0 \qquad \forall i, j \in V,\ p \in P_{ij},\ \omega \in \Omega, \quad (11.290)$$

$$u_{k\omega} \geq 0 \qquad \forall k \in K,\ \omega \in \Omega. \quad (11.291)$$

In the above model, the congestion effect is captured using the cost function used in [187], which, in turn, is based upon the expected demand in the system, assuming M/M/1 queues. In that function, b_k is a scaling factor used to compute the congestion cost at a hub. The objective function seeks to minimize the cost of opening the hubs (with some prescribed capacity) and expected congestion and transportation costs. Constraints (11.285) state that all the flow is routed through the network. Constraints (11.286) compute the total flow entering each hub under each scenario. Constraints (11.287) limit the flow shipped through a hub in each scenario according to the capacity decided for it. Constraints (11.288) impose at most one capacity level selected for an installed hub. Finally, we have the domain constraints.

Variables $\boldsymbol{u}$ can be discretized by indexing them in the capacity levels. This leads to a more tractable model. Define $u_{k\omega}^l$ as a total flow shipped through hub $k \in K$ under scenario $\omega \in \Omega$ having the hub equipped with capacity level $l \in L$. The resulting model is the following:

$$\text{minimize} \quad \sum_{k\in K}\sum_{l\in L} f_k^l y_k^l + \sum_{\omega\in\Omega}\pi_\omega\left(\sum_{k\in K}\sum_{l\in L} b_k \frac{u_{k\omega}^l}{q^l - u_{k\omega}^l} + \sum_{i,j\in V}\sum_{p\in P_{ij}} c_p v_{p\omega}\right), \quad (11.292)$$

subject to (11.285), (11.288)–(11.290),

$$\sum_{i,j\in V}\sum_{p\in P_{ij}|k\in p} d_{ij\omega} v_{p\omega} = \sum_{l\in L} u_{k\omega}^l \quad \forall k \in K,\ \omega\in\Omega, \quad (11.293)$$

$$u_{k\omega}^l \leq q^l y_k^l \quad \forall k \in K,\ l\in L,\ \omega\in\Omega, \quad (11.294)$$

$$u_{k\omega}^l \geq 0 \quad \forall k \in K,\ l\in L,\ \omega\in\Omega. \quad (11.295)$$

Bayram et al. [51] show how to go even further in terms of modeling the problem. They consider auxiliary variables $r_{k\omega}^l$:

$$r_{k\omega}^l \geq 0 \quad \forall k \in K,\ l\in L,\ \omega\in\Omega, \quad (11.296)$$

$$r_{k\omega}^l \geq \frac{u_{k\omega}^l}{q^l - u_{k\omega}^l} \quad \forall k \in k,\ l\in L,\ \omega\in\Omega. \quad (11.297)$$

The authors also prove that Inequalities (11.297) can be reformulated as the following second-order cone constraints:

$$\|2u_{k\omega}^l, q^l r_{k\omega}^l - q^l\| \leq q^l r_{k\omega}^l + q^l - 2u_{k\omega}^l \quad \forall k \in K,\ l\in L,\ \omega\in\Omega. \quad (11.298)$$

This triggers the following reformulation:

$$\text{minimize} \quad \sum_{k\in K}\sum_{l\in L} f_k^l y_k^l + \sum_{\omega\in\Omega}\pi_\omega\left(\sum_{k\in K}\sum_{l\in L} b_k r_{k\omega}^l + \sum_{i,j\in V}\sum_{p\in P_{ij}} c_p v_{p\omega}\right), \quad (11.299)$$

subject to (11.285), (11.289), (11.290), (11.293)–(11.296), (11.298).

Note also that Constraints (11.298) can be represented as

$$2u_{k\omega}^l - s_{k\omega}^l = 0 \quad \forall k \in K,\ l\in L, \quad (11.300)$$

$$q^l r_{k\omega}^l - s_{k\omega}^{l-} = q^l \quad \forall k \in K,\ l\in L, \quad (11.301)$$

$$q^l r_{k\omega}^l - 2u_{k\omega}^l - s_{k\omega}^{l+} = -q^l \quad \forall k \in K,\ l\in L, \quad (11.302)$$

$$(s^{l}_{k\omega})^2 + (s^{l-}_{k\omega})^2 \leq (s^{l+}_{k\omega})^2 \qquad \forall k \in K,\ l \in L, \tag{11.303}$$

$$s^{l+}_{k\omega} \geq 0 \qquad \forall k \in K,\ l \in L. \tag{11.304}$$

Bayram et al. [51] develop a Benders decomposition algorithm for this problem. The subproblem is an SOCP problem. The same holds with its dual. The Benders subproblem is solved by column generation. Given that both the primal and dual problems are feasible and bounded and that the Slater condition holds for the former, strong duality applies. Thus, the Benders decomposition algorithm is guaranteed to terminate after a finite number of steps. The reader should refer to that work for all details, including those related to a column generation procedure proposed for tackling the Benders subproblem (in practice, we cannot consider all the feasible paths for each origin–destination pair).

11.5 Reliability Problems

So far in this chapter, we have assumed that once installed, the hubs and the edges are reliable structures that operate as we expect them to do. However, this is not necessarily true. In practice, disruptions may occur, affecting such structures and thus decreasing their operationality. In this section, we discuss this aspect. In particular, we consider disruptions at the network edges and the hubs.

To better emphasize the mechanisms that allow accounting for possible structural disruptions in a hub location problem, we consider a simple problem as a basis for the analysis, namely the single-allocation p-hub median problem. Let us consider its (nonlinear) formulation using the allocation variables only:

$$\text{minimize} \quad \sum_{i,j\in V}\sum_{k,\ell\in K} d_{ij} C_{ijk\ell} x_{ik} x_{j\ell}, \tag{11.305}$$

$$\text{subject to} \quad \sum_{k\in K} x_{ik} = 1 \qquad \forall i \in V, \tag{11.306}$$

$$\sum_{k\in K} x_{kk} = p, \tag{11.307}$$

$$x_{ik} \leq x_{kk} \qquad \forall i \in V,\ k \in K, \tag{11.308}$$

$$x_{ik} \in \{0, 1\} \qquad \forall i \in V,\ k \in K. \tag{11.309}$$

The terminology has been used many times in this chapter. Thus, we omit the details. We also know how to linearize the above model (assuming at most two hubs for connecting an origin–destination pair). In what follows, we assume deterministic demands. Again, this allows emphasizing the discussion on the reliability issues. Hence, for $i \in V$ and $j \in V$, d_{ij} is the flow to be shipped from i to j.

11.5.1 Edge Disruptions

Often, hub location problems emerge as part of transportation systems design. In such a context, the edges used (spoke–hub and hub–hub) have a capacity that may be affected (fully or partially) by some unexpected event or just by a natural, although uncertain, level of degradation. For $\ell \in V$ and $\ell' \in K$ with $\ell' > \ell$ if $\ell \in K$, let $q_{\ell\ell'}$ be the capacity of the (bidirectional) edge connecting ℓ and ℓ'. Furthermore, assume that $q_{\ell\ell'}$ is a random variable with a support included in $[0, \tilde{q}_{\ell\ell'}]$ where $\tilde{q}_{\ell\ell'}$ is the disruption-free capacity of the edge. Let $\gamma_{\ell\ell'}$ be a given edge-specific maximum probability for exceeding the actual edge capacity.

If $w_{\ell\ell'}$ is a decision variable representing the total flow traversing the edge connecting ℓ and ℓ ($\ell \in V$, $\ell' \in K$, and $\ell' > \ell$ if $\ell \in K$), a reliability-aware single-allocation p-hub median problem can be formulated as follows:

$$\text{minimize} \quad (11.305),$$

$$\text{subject to} \quad (11.306)\text{–}(11.309),$$

$$\mathbb{P}[w_{\ell\ell'} \geq q_{\ell\ell'}] \leq \gamma_{\ell\ell'} \qquad \forall \ell \in V,\ \ell' \in K,\ \text{and } \ell' > \ell \text{ if } \ell \in K, \tag{11.310}$$

$$w_{ik} = \sum_{\ell \in K} \sum_{j \in V} d_{ij} x_{ik} x_{\ell j} + \sum_{\ell \in K} \sum_{j \in V} d_{ji} x_{j\ell} x_{ik} \qquad \forall i \in V,\ k \in K, \tag{11.311}$$

$$w_{k\ell} = \sum_{i,j \in V} d_{ij} x_{ik} x_{\ell j} + \sum_{i,j \in V} d_{ji} x_{j\ell} x_{ik} \qquad \forall k, \ell \in K, k < \ell, \tag{11.312}$$

$$w_{ik} \geq 0 \qquad \forall i \in V,\ k \in K, \tag{11.313}$$

$$w_{k\ell} \geq 0 \qquad \forall k, \ell \in K,\ k < \ell. \tag{11.314}$$

In the above model, the chance constraints (11.310) ensure that in each edge, the probability that the total assigned flow (in both directions) exceeds the edge capacity is smaller than or equal to the given threshold. Constraints (11.311) and (11.312) assess the total flow traversing each edge. Finally, we have the domain constraints for the w variables.

Sadeghi et al. [466] investigate a problem like the above one, assuming that the capacities of the edges are represented by an Erlang random variable truncated at the value of the disruption-free capacities. In particular, they assume

$$q_{\ell\ell'} \sim \text{TruncErl}(\beta, \alpha, \tilde{q}_{\ell\ell'}) \quad \forall \ell \in V,\ \ell' \in K \text{ and } \ell' > \ell \text{ if } \ell \in K.$$

Proposition 11.6 *Under the above assumptions and setting, a deterministic equivalent for Constraints* (11.310) *is given by*

$$y_{\ell\ell'} \leq \frac{\Theta}{\beta}\left[(1 - \gamma_{\ell\ell'}\mathbb{P}[0 < q_{\ell\ell'} < \tilde{q}_{\ell\ell'}])\,(1-\alpha) - \kappa - 1\right]$$

$$\forall \ell \in V,\ \ell' \in K, \quad \text{and } \ell < \ell' \text{ if } \ell \in K,$$

where Θ *is a value greater than* $\tilde{q}_{\ell\ell'}$ *and* κ *is a small constant.*

Proof The proof stems from using the density function of a truncated Erlang distribution:

$$\phi(\chi \mid \beta, \alpha, \tilde{q}) = \frac{\beta^{\alpha}}{(\alpha-1)!}\frac{\chi^{\alpha-1}e^{-\beta\chi}}{\mathbb{P}[0 < \chi < \tilde{q}]} \qquad 0 < \chi < \tilde{q}.$$

The interested reader can refer to [466] for all details. □

The above problem assumes that the number of hubs to install is given. Moreover, the hub-level network is complete.

Blanco et al. [81] investigate a reliability multiple-allocation hub location problem with fixed costs and hub network design decisions. The hubs are uncapacitated. Thus, there is always an optimal solution in which all the flow between an origin–destination pair is routed via a single path. The authors assume that the flow between an origin–destination pair is routed by at least one and at most two hubs (recall the discussion in Sect. 11.2.1.2 inspired in [136]). They assume the inter-hub edges to install also include loops, which, as explained by the authors, makes sense in many applications. Thus, each origin–destination flow is assumed to make use of one and only one hub edge (a loop indicates that the flow is processed and routed via one hub only). The hub edges may fail. Thus, in addition to an "original" routing path, a "backup" path must also be decided for each origin–destination. In particular, for $k, \ell \in K$, $k \leq \ell$, the edge between the hubs fails with probability $\pi_{k\ell}$. We denote $\overline{\pi}_{k\ell} = 1 - \pi_{k\ell}$.

We adopt the terminology we have been using, making the necessary adjustments when needed. For $k \in K$, we use y_k as a binary location variable indicating whether the hub is installed, and f_k is the corresponding fixed cost. For $k, \ell \in K$, $k \leq \ell$, $z_{k\ell}$ is a binary variable equal to one if and only if an edge is set between hubs k and ℓ. In this case, $g_{k\ell}$ is the incurred fixed cost. Regarding the flows, we use the variables $w_{ijk\ell}$ that are binary and equal to one if and only if the flow from i to j is routed via hubs k and ℓ in this sequence ($i, j \in V$, $k, \ell \in K$). Regarding the backup paths, for $i, j \in V$ and $k, \ell \in K$, we consider $\overline{w}_{ijk\ell}$ that as a binary variable that is equal to one if and only if the alternative flow from i to j is routed via hubs k and ℓ in this order. We keep using the notation $C_{ijk\ell}$ to indicate the cost for shipping the flow from i to j via hubs k and ℓ in this order, $i, j \in V$, $k, \ell \in K$.

The problem can now be formulated as follows:

$$\text{minimize} \quad \sum_{k\in K} f_k y_k + \sum_{\substack{k,\ell\in K\\ k\le\ell}} g_{k\ell} z_{k\ell}$$

$$+ \sum_{i,j\in V}\left[\sum_{\substack{k,\ell\in K\\ k<\ell}} \overline{\pi}_{k\ell}(C_{ijk\ell}w_{ijk\ell} + C_{ij\ell k}w_{ij\ell k}) + \sum_{k\in K}\overline{\pi}_{kk}C_{ijkk}u_{ijkl}\right]$$

$$+ \sum_{i,j\in V}\left[\sum_{\substack{k,\ell\in K\\ k<\ell}}\left(\pi_{k\ell}(w_{ijk\ell}+w_{ij\ell k}) \sum_{\substack{k',\ell'\in K\\ k'\neq k\vee \ell'\neq\ell}} C_{ijk'\ell'}\overline{w}_{ijk'\ell'}\right)\right.$$

$$\left. + \sum_{k\in K}\left(\pi_{kk}w_{ijkk} \sum_{\substack{k',\ell'\in K\\ k'\neq k\vee \ell'\neq k}} C_{ijk'\ell'}\overline{w}_{ijk'\ell'}\right)\right], \tag{11.315}$$

$$\text{subject to} \quad \sum_{k,\ell\in K} w_{ijk\ell} = 1 \qquad \forall i,j\in V, \tag{11.316}$$

$$\sum_{k,\ell\in K} \overline{w}_{ijk\ell} = 1 \qquad \forall i,j\in V, \tag{11.317}$$

$$w_{ijk\ell} + w_{ij\ell k} + \overline{w}_{ijk\ell} + \overline{w}_{ij\ell k} \le z_{k\ell}$$
$$\forall i,j\in V,\ k,\ell\in K,\ k<\ell, \tag{11.318}$$

$$w_{ijkk} + \overline{w}_{ijkk} \le z_{kk} \qquad \forall i,j\in V,\ k\in K, \tag{11.319}$$

$$z_{k\ell} \le y_k \qquad \forall k,\ell\in K,\ k\le\ell, \tag{11.320}$$

$$z_{k\ell} \le y_\ell \qquad \forall k,\ell\in K,\ k<\ell, \tag{11.321}$$

$$w_{ijk\ell}, \overline{w}_{ijk\ell} \in \{0,1\} \qquad \forall i,j\in V,\ k,\ell\in K, \tag{11.322}$$

$$y_k \in \{0,1\} \qquad \forall k\in K, \tag{11.323}$$

$$z_{k\ell} \in \{0,1\} \qquad \forall k,\ell\in K,\ k\le\ell. \tag{11.324}$$

In the above model, the first line of the objective function accounts for the setup cost for hubs and the edges connecting them. The remainder of the objective function computes the total expected routing cost (using the original path or the alternative one). The other elements should be clear, given the contents of the chapter so far.

Blanco et al. [81] go much further in terms of the above problem: (i) They linearize the objective function by considering an additional set of decision variables to represent the probability that an inter-hub arc is used as a backup path for some flow; (ii) they consider a so-called λ-connectivity backbone network ensuring that the hub-level network contains at least λ routing paths connecting any pair of activated hubs (relaxing the assumption of the single hub-edge usage for routing the flow is relaxed); (iii) for the latter case, a branch-and-cut algorithm is proposed. The interested reader should refer to that work for all details.

11.5.2 Hub Disruptions

Let us now consider the case in which the unexpected disruption may affect the hubs. In particular, assume that a hub installed at $k \in K$ can fail with a given probability π_k independent from the other hubs. We define $\overline{\pi}_k = 1 - \pi_k \; \forall k \in K$. Assume additionally that if a hub fails, it cannot be used. In this case, some alternative hub is needed to route the flow initially planned to be routed through the missing hub.

We consider again the single-allocation p-hub median problem as a starting point for the discussion. Given that hubs can fail, the allocation of the nodes to the hubs should follow a sequence with a customer being allocated to its level-s-hub only if the first $s - 1$ assigned hubs fail. At each assignment level, a node is allocated to a single hub. If all hubs fail, a global penalty is incurred. The objective is to minimize the expected transportation cost plus that penalty.

To formulate the problem, we follow [526]. We consider binary location variables, again denoted by y_k, indicating whether a hub is installed at $k \in K$. Additionally, we redefine the allocation variables and also introduce additional ones. For $s \in \{1, \ldots, p\}$, let x_{ik}^s be a binary variable equal to one if and only if a hub installed at $k \in K$ is the level-s-hub for node $i \in V$. Finally, we also consider the following decision variables:

$$V_{ijk}^{sr} = \begin{cases} 1, & \text{if location } k \text{ is the level-}r \text{ hub for node } j \text{ but not the level-}s \\ & \text{hub for node } i, \\ 0, & \text{otherwise,} \qquad k \in K, i, j \in V, i < j, r, s \in \{1, \ldots, p\} \end{cases}$$

$$W_{ijk}^{sr} = \begin{cases} 1, & \text{if location } k \text{ is the level-}r \text{ hub for node } j \text{ and the level-}s \\ & \text{hub for node } i, \\ 0, & \text{otherwise,} \qquad k \in K, i, j \in V, i < j, r, s \in \{1, \ldots, p\} \end{cases}$$

$\theta_{ijk\ell}^{sr}$ = probability that node i can allocate to hub k as its level-s-hub and node j can allocate to hub ℓ as its level-r-hub.

Given the single-allocation nature of the problem, we have $\theta_{ijk\ell}^{sr} = \theta_{ji\ell k}^{rs} \; \forall i, j \in V, \; k, \ell \in K, \; s, r \in \{1, \ldots, p\}$. This allows for halving those decision variables. Accordingly, we denote $C'_{ijk\ell} = d_{ij} C_{ijk\ell} + d_{ji} C_{ji\ell k}$ and $d'_{ij} = d_{ij} + d_{ji} \; \forall i, j \in V, \; i < j, \; k, \ell \in K$.

The reliable single-allocation p-hub median problem can be formulated as follows:

$$\text{minimize} \quad \sum_{\substack{i,j\in V\\ i<j}} \sum_{k,\ell\in K} \sum_{r,s=1}^{p} C'_{ijk\ell}\theta^{sr}_{ijk\ell} + \sum_{\substack{i,j\in V\\ i<j}} \phi d'_{ij} \prod_{k\in K} \overline{\pi}_k\, y_k, \tag{11.325}$$

$$\text{subject to} \quad \sum_{k\in K} y_k = p, \tag{11.326}$$

$$\sum_{k\in K} x^s_{ik} = 1 \qquad \forall i \in V,\ s \in \{1,\ldots,p\}, \tag{11.327}$$

$$\sum_{s=1}^{p} x^s_{ik} \le y_k \qquad \forall i \in V,\ k \in K, \tag{11.328}$$

$$\theta^{sr}_{ijk\ell} = \begin{cases} \displaystyle\prod_{\rho=1}^{s-1}\prod_{m\in K}(1-\overline{\pi}_m x^{\rho}_{im})\overline{\pi}_k x^s_{ik} \\ \quad \displaystyle\times \prod_{\rho=1}^{r-1}\prod_{m\in K}(1-\overline{\pi}_m V^{s\rho}_{ijm})(1-W^{sr}_{ijm})\overline{\pi}_\ell V^{sr}_{ij\ell} & \ell \ne k \\ \displaystyle\prod_{\rho=1}^{s-1}\prod_{m\in K}(1-\overline{\pi}_m x^{\rho}_{im})\overline{\pi}_k x^s_{ik} \\ \quad \displaystyle\times \prod_{\rho=1}^{r-1}\prod_{m\in K}(1-\overline{\pi}_m V^{s\rho}_{ijm})(1-W^{sr}_{ijm})W^{sr}_{ij\ell} & \ell = k \end{cases}$$

$$\forall i, j \in V,\ i < j,\ k, \ell \in K,\ s, r \in \{1,\ldots,p\}, \tag{11.329}$$

$$1 + V^{sr}_{ij\ell} \le x^r_{j\ell} + \sum_{\rho=s+1}^{p} x^{\rho}_{ik}$$

$$\forall i, j \in V,\ i < j,\ k \in K,\ s, r \in \{1,\ldots,p\}, \tag{11.330}$$

$$V^{sr}_{ijk} \le x^r_{jk}$$

$$\forall i, j \in V,\ i < j,\ k \in K,\ s, r \in \{1,\ldots,p\}, \tag{11.331}$$

$$V^{sr}_{ijk} \le \sum_{\rho=s+1}^{p} x^{\rho}_{ik}$$

$$\forall i, j \in V,\ i < j,\ k \in K,\ s, r \in \{1,\ldots,p\}, \tag{11.332}$$

$$1 + W^{sr}_{ijk} \geq x^{r}_{jk} + x^{s}_{ik} \qquad \forall i, j \in V,\ i < j,\ k \in K,\ s, r \in \{1, \ldots, p\}, \tag{11.333}$$

$$W^{sr}_{ijk} \leq x^{r}_{jk} \qquad \forall i, j \in V,\ i < j,\ k \in K,\ s, r \in \{1, \ldots, p\}, \tag{11.334}$$

$$W^{sr}_{ijk} \leq x^{s}_{ik} \qquad \forall i, j \in V,\ i < j,\ k \in K,\ s, r \in \{1, \ldots, p\}, \tag{11.335}$$

$$y_k \in \{0, 1\} \qquad \forall k \in K, \tag{11.336}$$

$$x^{s}_{ik} \in \{0, 1\} \qquad \forall i \in V,\ k \in K,\ s \in \{1, \ldots, p\}, \tag{11.337}$$

$$V^{sr}_{ijk} \geq 0 \qquad \forall i, j \in V,\ i < j,\ k \in K,\ s, r \in \{1, \ldots, p\}, \tag{11.338}$$

$$W^{sr}_{ijk} \geq 0 \qquad \forall i, j \in V,\ i < j,\ k \in K,\ s, r \in \{1, \ldots, p\}. \tag{11.339}$$

In this model, ϕ is the unit penalty incurred for non-routed flow. The objective function (11.325) seeks to minimize the total expected transportation cost plus the penalty if all hubs fail. Constraints (11.326)–(11.328) have a straightforward meaning. For each $i, j \in V$ and $s, r \in \{1, \ldots, p\}$, Equalities (11.329) compute the probability that node i is allocated to hub k as its level-s-hub and node j is allocated to hub ℓ as its level-r-hub. Constraints (11.330)–(11.335) ensure a consistent choice for the values of V and W variables that are part of (11.329). Finally, we have the domain constraints.

The above model is studied in [526]. The authors linearize it, reduce its size, and propose a tabu search algorithm to find approximate solutions. The interested reader can refer to that work for all details.

Finally, aligned with reliability, we cite [40]. The authors consider a problem in which the hubs can fail with some given probability. In that case, the demand nodes should be allocated to some alternative hub. The authors assume a complete hub-level network but enrich the problem by considering endogenous discrete capacities at the hubs. In particular, when a hub is installed, we need to decide the capacity level to equip it. A flow-dependent inter-hub cost discount factor is considered, making the setting more realistic. A piecewise linear function is adopted in this case. The complexity of the problem calls for an approximate procedure to be developed. A variable neighborhood search is proposed accordingly.

11.6 Distributionally Robust Hub Location

As discussed in Chap. 8, the central assumption of a given probability distribution in stochastic programming is arguable in many applications. In practice, it is often difficult, if not impossible, to acquire precise knowledge of the underlying distribution because of the insufficiency of historical data and the long planning horizon a hub location is usually planned for. Next, we discuss distributionally robust optimization models for both uncapacitated and capacitated hub location problems. The models allow the joint distribution of the uncertain demand and cost to be ambiguous, which can be characterized by distributional information, such as the support, mean, and dispersion.

We start the discussion considering the multiple-allocation hub location problem given by (11.12)–(11.16) although making a few changes to help make the notation lighter throughout the section. In particular, we consider the multi-commodity formulation for the problem. Each commodity m is characterized by an origin o_m, a destination ς_m, and a demand d_m that is the value d_{o_m,ς_m} using the notation of the previous section, $o_m, \varsigma_m \in V$. We represent by M the set of commodities. The unit cost for routing a commodity $m \in M$ through hubs k and ℓ in this sequence is denoted by $C_{k\ell m}$. This value is $C_{o_m \varsigma_m k\ell}$ considered in the previous sections. Model (11.12)–(11.16) can be reformulated as follows:

$$\text{minimize} \quad \sum_{k\in K} f_k y_k + \sum_{m\in M}\sum_{k,\ell\in K} d_m C_{k\ell m} w_{k\ell m}, \tag{11.340}$$

$$\text{subject to} \quad \sum_{k,\ell\in K} w_{k\ell m} = 1 \qquad \forall m \in M, \tag{11.341}$$

$$\sum_{\ell\in K} w_{k\ell m} + \sum_{\ell\in K\setminus\{k\}} w_{\ell k m} \le y_k \qquad \forall m \in M,\ k \in K, \tag{11.342}$$

$$w_{k\ell m} \ge 0 \qquad \forall m \in M,\ k,\ell \in K, \tag{11.343}$$

$$y_k \in \{0,1\} \qquad \forall k \in K. \tag{11.344}$$

In what follows, we denote by $\mathbb{R}_+^N$ the set of N-dimensional vectors of non-negative real numbers. Demand uncertainty is now captured using an ambiguity set, say $\mathcal{F}_0$, that is characterized by the partial distributional information about the demand random vector that we denote by $\tilde{\boldsymbol{d}}$:

$$\mathcal{F}_0 = \left\{ \mathbb{P} \in \mathcal{P}\left(\mathbb{R}^{|M|}\right) \left| \begin{array}{l} \tilde{\boldsymbol{d}} \sim \mathbb{P} \\ \mathbb{E}_{\mathbb{P}}[\tilde{\boldsymbol{d}}] \in [\boldsymbol{d}^-, \boldsymbol{d}^+] \\ \mathbb{E}_{\mathbb{P}}\left[\|\boldsymbol{D}_{0l}(\tilde{\boldsymbol{d}} - \overline{\boldsymbol{d}})\|_1\right] \le \sigma_{0l} \quad \forall l \in L_0 \\ \mathbb{P}[\tilde{\boldsymbol{d}} \in \mathcal{D}_0] = 1 \end{array} \right. \right\}, \tag{11.345}$$

where $L_0 = \{1, \ldots, n_0\}$. The vectors $\boldsymbol{d}^-$, $\boldsymbol{d}^+$, $\overline{\boldsymbol{d}}$, $(\sigma_{0l})_{l \in L_0}$ are estimated inputs and $\|.\|_1$ stands for the one-norm. The first condition in set (11.345) captures the range of the expected demand; the last constraint specifies the bounded support of $\tilde{\boldsymbol{d}}$, which we assume to be a general polyhedron $\mathcal{D}_0 = \{\boldsymbol{d} \in \mathbb{R}_+^{|M|} \mid B_0 \boldsymbol{d} \leq b_0\}$ with $B_0 \in \mathbb{R}^{N_0 \times |M|}$, $b_0 \in \mathbb{R}^{N_0}$ and satisfying $[\boldsymbol{d}^-, \boldsymbol{d}^+] \subseteq \mathcal{D}_0$.

By varying the matrix $\boldsymbol{D}_{0l} \in \mathbb{R}^{N'_{0l} \times |M|}$, the second set of expectation constraints (which are labeled by the collection of indices L_0) though in a simple format provides us with powerful modeling flexibility in specifying useful distributional information, such as the marginal dispersion of random demand and even possible correlation across demands.

Likewise, we can consider uncertainty in the transportation costs. We recall that for each commodity $m \in M$ we have the cost $C_{k\ell m}$ given by $C_{k\ell m} = c_{o_m k} + \alpha c_{k\ell} + c_{\ell \varsigma_m}$. This is already a simplification of a structure in which the unit transportation costs are multiplied by the distances, with the latter satisfying the triangle inequality. We recover that original structure by writing $C_{k\ell m} = \delta_{o_m k} c_{m1} + \delta_{k\ell} c_{m2} + \delta_{\ell \varsigma_m} c_{m3} = \boldsymbol{\delta}_{k\ell m}^\top \boldsymbol{c}_m$. Now, c_{m1} represents the unit collection cost for commodity m, whereas c_{m2} is the unit transfer cost, and c_{m3} is the unit distribution cost. The vector $\boldsymbol{\delta}_{k\ell m}$ contains the three distances involved in the path for routing commodity m.

For a commodity $m \in M$, we assume the ambiguous distribution of its uncertainty transportation cost vector, $\tilde{\boldsymbol{c}}_m$, in the following ambiguity set:

$$\mathcal{F}_m = \left\{ \mathbb{P} \in \mathcal{P}\left(\mathbb{R}^3\right) \left| \begin{array}{l} \tilde{\boldsymbol{c}}_m \sim \mathbb{P} \\ \mathbb{E}_{\mathbb{P}}[\tilde{\boldsymbol{c}}_m] \in [\boldsymbol{c}_m^-, \boldsymbol{c}_m^+] \\ \mathbb{E}_{\mathbb{P}}\left[\|\boldsymbol{D}_{ml}(\tilde{\boldsymbol{c}}_m - \overline{\boldsymbol{c}}_m)\|_1\right] \leq \sigma_{ml} \quad \forall l \in L_m \\ \mathbb{P}[\tilde{\boldsymbol{c}}_m \in \mathcal{D}_m] = 1 \end{array} \right. \right\}, \tag{11.346}$$

where $L_m = \{1, \ldots, n_m\}$. The vectors $\boldsymbol{c}_m^-$, $\boldsymbol{c}_m^+$, $\overline{\boldsymbol{c}}_m$, $(\sigma_{ml})_{l \in L_m}$ are estimated inputs, and $\|.\|_1$ stands for the one-norm. The first condition involving an expectation captures the range of the expected cost vector; the last condition specifies the bounded support of $\tilde{\boldsymbol{c}}_m$ which we assume to be a general polyhedron $\mathcal{D}_m = \{\boldsymbol{c} \in \mathbb{R}_+^3 \mid B_m \boldsymbol{c} \leq b_m\}$ with $B_m \in \mathbb{R}^{N_m \times 3}$, $b_m \in \mathbb{R}^{N_m}$ and satisfying $[\boldsymbol{c}_m^-, \boldsymbol{c}_m^+] \subseteq \mathcal{D}_m$.

By varying the matrix $\boldsymbol{D}_{ml} \in \mathbb{R}^{N'_{ml} \times |M|}$, the second set of expectation constraints (which are labeled by the collection of indices L_m) though in a simple format indeed provides us with powerful modeling flexibility in specifying useful distributional information, such as the marginal dispersion of random demand and even possible correlation across demands.

11.6.1 Independent Demands and Costs

We assume the uncertain demand $\tilde{\boldsymbol{d}}$ and the uncertain costs $\tilde{\boldsymbol{c}} = (\tilde{\boldsymbol{c}}_m)_{m \in M}$ are independent and $\tilde{\boldsymbol{c}}_1, \ldots, \tilde{\boldsymbol{c}}_{|M|}$ are also mutually independent. Hence, the uncertain demand and cost, $(\tilde{\boldsymbol{d}}, \tilde{\boldsymbol{c}})$, follow a joint distribution belonging to the ambiguity set

$$\mathcal{F} = \left\{ \mathbb{P} \in \mathcal{P}\left(\mathbb{R}^{|M|} \times \mathbb{R}^{3|M|}\right) \middle| \mathbb{P} = \otimes_{m \in M \cup \{0\}} \mathbb{P}_m,\ \mathbb{P}_0 \in \mathcal{F}_0,\ \mathbb{P}_m \in \mathcal{F}_m,\ m \in M \right\}. \tag{11.347}$$

In this ambiguity set, "$\otimes$" denotes the direct product operator for distributions.

Once having defined the underlying ambiguity set, we can formulate the adaptive distributionally robust hub location problem as follows:

$$\text{minimize} \quad \sum_{k \in K} f_k y_k + \sup_{\mathbb{P} \in \mathcal{F}} \mathbb{E}_{\mathbb{P}} \left[Q(\boldsymbol{y}, \tilde{\boldsymbol{d}}, \tilde{\boldsymbol{c}}) \right], \tag{11.348}$$

$$\text{subject to} \quad y_k \in \{0, 1\} \qquad \forall k \in K, \tag{11.349}$$

such that for each fixed $\boldsymbol{y}$ and uncertainty realization $(\boldsymbol{d}, \boldsymbol{c})$,

$$Q(\boldsymbol{y}, \boldsymbol{d}, \boldsymbol{c}) = \sum_{m \in M} d_m \left(\min_{\boldsymbol{w}_m \in \mathcal{W}(\boldsymbol{y})} \sum_{k, \ell \in K} \boldsymbol{\delta}_{k\ell m}^{\top} \boldsymbol{c}_m w_{k\ell m} \right), \tag{11.350}$$

with

$$\mathcal{W}(\boldsymbol{y}) = \left\{ \boldsymbol{w} \in \mathbb{R}_+^{|K|^2} \middle| \begin{array}{l} \sum_{k, \ell \in K} w_{k\ell} = 1 \\ \sum_{\ell \in K} w_{k\ell} + \sum_{\ell \in K \setminus \{k\}} w_{\ell k} \le y_k \ \ \forall k \in K \end{array} \right\}.$$

It is well-known that when the ambiguity set specifies solely the support information about the uncertainty, distributionally robust optimization reduces to robust optimization (see, e.g., [122, 543]). Interestingly, under the independence assumption, the adaptive distributionally robust UHLP is equivalent to a robust UHLP, even when the ambiguity set contains other distributional information (such as mean and dispersion specified through the expectation constraints) beyond the support.

Theorem 11.1 *The adaptive distributionally robust optimization model* (11.348), (11.349) *is equivalent to an adaptive classical robust model*

$$\min_{\boldsymbol{y} \in \{0,1\}^{|K|}} \left\{ \sum_{k \in K} f_k y_k + \max_{(\boldsymbol{d}, \boldsymbol{c}) \in \mathcal{U}} \mathrm{Q}(\boldsymbol{y}, \boldsymbol{d}, \boldsymbol{c}) \right\}, \tag{11.351}$$

where given the ambiguity set $\mathcal{F}$ *in* (11.347), *we define accordingly uncertainty sets*

$$\bar{\mathcal{U}}^d = \left\{ \boldsymbol{d} \in \mathbb{R}_+^{|M|} \mid \boldsymbol{B}_0 \boldsymbol{d} \le \boldsymbol{b}_0,\ \boldsymbol{d} \in [\boldsymbol{d}^-, \boldsymbol{d}^+],\ \|\boldsymbol{D}_{0l}(\boldsymbol{d} - \bar{\boldsymbol{d}})\|_1 \le \sigma_{0l}\ \forall l \in L_0 \right\},$$

$$\bar{\mathcal{U}}_m^c = \left\{ \boldsymbol{c}_m \in \mathbb{R}_+^3 \,\middle|\, \begin{array}{l} \boldsymbol{B}_m \boldsymbol{c}_m \leq \boldsymbol{b}_m,\ \boldsymbol{c}_m \in [\boldsymbol{c}_m^-, \boldsymbol{c}_m^+] \\ \|\boldsymbol{D}_{ml}(\boldsymbol{c}_m - \bar{\boldsymbol{c}}_m)\|_1 \leq \sigma_{ml}\ \forall l \in L_m \end{array} \right\}, \forall m \in M,$$

and an uncertainty set for $(\boldsymbol{d}, \boldsymbol{c})$*,*

$$\mathcal{U} \triangleq \{(\boldsymbol{d}, \boldsymbol{c}) \in \mathbb{R}^{|M|} \times \mathbb{R}^{3|M|} \mid \boldsymbol{d} \in \bar{\mathcal{U}}^d,\ \boldsymbol{c} = (\boldsymbol{c}_1, \ldots, \boldsymbol{c}_{|M|}) : \boldsymbol{c}_m \in \bar{\mathcal{U}}_m^c\ \forall m \in M\}. \tag{11.352}$$

Proof See [535]. □

The above theorem reveals an interesting structure of the adaptive distributionally robust model of the UHLP under the independence condition: (i) the optimal value of its distributionally robust counterpart (11.348), (11.349) over the ambiguity set $\mathcal{F}$ is equivalent to that of its adaptive robust counterpart (11.351) over an uncertainty set $\mathcal{U}$; and (ii) the uncertainty set $\mathcal{U}$ integrates the distributional information of mean and dispersion from the ambiguity set $\mathcal{F}$ by using constraints that consistently correspond to expectation constraints in the ambiguity set $\mathcal{F}$.

The above theorem can also be interpreted from the stochastic programming perspective. Suppose that the mean of the uncertainty $(\tilde{\boldsymbol{d}}, \tilde{\boldsymbol{c}})$ is treated as a Dirac distribution with mass 1 assigned to its value, and this Dirac distribution is feasible in $\mathcal{F}$. Then, the adaptive distributionally robust UHLP can be viewed as the adaptive robust counterpart of the mean problem of stochastic UHLP, where the mean is now treated as the uncertainty that varies within the uncertainty set $\mathcal{U}$. This observation generalizes the result in stochastic UHLP with a known demand distribution (see Proposition 11.1) for an ambiguous demand–cost distribution.

The following result evidences that the adaptive robust model (11.351) has a hidden *non-adaptive* structure.

Theorem 11.2 *Given a hub location design* $\boldsymbol{y}$*, each component* $\boldsymbol{w}_m^\star$ *of the optimal routing decision* $\boldsymbol{w}^\star$ *is non-adaptive and solves the following robust problem:*

$$\min_{\boldsymbol{w}_m \in \mathcal{W}(\boldsymbol{y})} \max_{\boldsymbol{c}_m \in \bar{\mathcal{U}}_m^c} \sum_{k,\ell \in K} \boldsymbol{\delta}_{k\ell m}^\top \boldsymbol{c}_m w_{k\ell m}. \tag{11.353}$$

Moreover, the adaptive robust model (11.351) *is equivalent to the non-adaptive robust problem:*

$$\text{minimize} \quad \sum_{k \in K} f_k y_k + \max_{\boldsymbol{d} \in \bar{\mathcal{U}}^d} \sum_{m \in M} d_m \lambda_m, \tag{11.354}$$

$$\text{subject to} \quad \lambda_m \geq \sum_{k,\ell \in K} \boldsymbol{\delta}_{k\ell m}^\top \boldsymbol{c}_m w_{k\ell m} \quad \forall \boldsymbol{c}_m \in \bar{\mathcal{U}}_m^c,\ m \in M, \tag{11.355}$$

$$\boldsymbol{w}_m \in \mathcal{W}(\boldsymbol{y}) \quad \forall m \in M, \tag{11.356}$$

$$\lambda_m \in \mathbb{R} \quad \forall m \in M, \tag{11.357}$$

$$y_k \in \{0, 1\} \qquad \forall k \in K, \tag{11.358}$$

where λ_m are auxiliary variables ($m \in M$).

Proof See [535]. □

This theorem and Theorem 11.1 have some implications. *(i)* The adaptive distributionally robust UHLP (11.348), (11.349), under the independence assumption, is equivalent to a classical robust optimization problem (11.354)–(11.358). *(ii)* There exists a static policy such that $\boldsymbol{w}^\star(\boldsymbol{d}, \boldsymbol{c}) = \boldsymbol{w}^\star$ for all $(\boldsymbol{d}, \boldsymbol{c}) \in \mathcal{U}$. This policy can be efficiently obtained by solving a set of subproblems in the form (11.353). *(iii)* In the spirit of the adaptive distributionally robust model, there is a worst-case distribution $\mathbb{P}^\star \in \mathcal{F}$ of $(\tilde{\boldsymbol{d}}, \tilde{\boldsymbol{c}})$ to which the optimal routing policy $\boldsymbol{w}^\star(\cdot)$ "statically" adapts to. The established equivalence between distributionally robust and robust models implies the existence of a worst-case Dirac distribution $\mathbb{P}^\star = \delta_{(\boldsymbol{d}^\star, \boldsymbol{c}^\star)}$ with $(\boldsymbol{d}^\star, \boldsymbol{c}^\star) \in \mathcal{U}$. Moreover, using Theorem 11.2, we can efficiently extract this worst-case distribution.

Wang et al. [535] end this analysis by providing an MILP reformulation for the distributionally robust hub location problem we are discussing. Particularly, given the ambiguity set $\mathcal{F}$ (11.347), those authors prove that the distributionally robust optimization problems given by (11.348), (11.349) can be reformulated as follows:

$$\text{minimize} \quad \sum_{k \in K} f_k y_k$$

$$+ \overline{\boldsymbol{\phi}}_0^\top \boldsymbol{d}^+ - \underline{\boldsymbol{\phi}}_0^\top \boldsymbol{d}^- + \boldsymbol{\gamma}_0^\top \boldsymbol{b}_0 + \sum_{l \in L_0} (\eta_l \sigma_{0l} - \boldsymbol{\psi}_{0l}^\top \boldsymbol{D}_{0l} \bar{\boldsymbol{d}}), \tag{11.359}$$

$$\text{subject to} \quad \overline{\boldsymbol{\phi}}_0 - \underline{\boldsymbol{\phi}}_0 + \boldsymbol{B}_0^\top \boldsymbol{\gamma}_0 - \sum_{l \in L_0} \boldsymbol{D}_{0l}^\top \boldsymbol{\psi}_{0l} \geq \boldsymbol{\lambda}, \tag{11.360}$$

$$\lambda_m \geq \overline{\boldsymbol{\phi}}_m^\top \boldsymbol{c}_m^+ - \underline{\boldsymbol{\phi}}_m^\top \boldsymbol{c}_m^- + \boldsymbol{\gamma}_m^\top \boldsymbol{b}_m + \sum_{l \in L_m} (\eta_{ml} \varepsilon_{ml} - \boldsymbol{\psi}_{ml}^\top \boldsymbol{D}_{ml} \bar{\boldsymbol{c}}_m)$$

$$\forall m \in M, \tag{11.361}$$

$$\overline{\boldsymbol{\phi}}_m - \underline{\boldsymbol{\phi}}_m + \boldsymbol{B}_m^\top \boldsymbol{\gamma}_m - \sum_{l \in L_m} \boldsymbol{D}_{ml}^\top \boldsymbol{\psi}_{ml} - \sum_{k, \ell \in K} \boldsymbol{\delta}_{k\ell m} w_{k\ell m} \geq \boldsymbol{0}$$

$$\forall m \in M, \tag{11.362}$$

$$\eta_l \in \mathbb{R}, \ \boldsymbol{\psi}_{0l} \in \mathbb{R}_+^{N'_{0l}}, \ \eta_l \geq \|\boldsymbol{\psi}_{0l}\|_\infty \qquad \forall l \in L_0, \tag{11.363}$$

$$\overline{\boldsymbol{\phi}}_0, \underline{\boldsymbol{\phi}}_0 \in \mathbb{R}_+^{|M|}, \ \boldsymbol{\gamma}_0 \in \mathbb{R}_+^{N_0}, \ \boldsymbol{\lambda} \in \mathbb{R}^{|M|}, \tag{11.364}$$

$$\eta_{ml} \in \mathbb{R}, \ \boldsymbol{\psi}_{ml} \in \mathbb{R}_+^{N'_{ml}}, \ \eta_{ml} \geq \|\boldsymbol{\psi}_{ml}\|_\infty \quad \forall m \in M, \ l \in L_k, \tag{11.365}$$

$$\overline{\boldsymbol{\phi}}_m, \underline{\boldsymbol{\phi}}_m \in \mathbb{R}^3_+,\ \boldsymbol{\gamma}_m \in \mathbb{R}^{N_m}_+,\ \boldsymbol{w}_m \in \mathcal{W}(\boldsymbol{y}) \quad \forall m \in M, \tag{11.366}$$

$$y_k \in \{0, 1\} \quad \forall k \in K, \tag{11.367}$$

where $\boldsymbol{\lambda}, \overline{\boldsymbol{\phi}}_0, \underline{\boldsymbol{\phi}}_0, \boldsymbol{\gamma}_0, \eta_l, \boldsymbol{\psi}_{0l}, \overline{\boldsymbol{\phi}}_m, \underline{\boldsymbol{\phi}}_m, \boldsymbol{\gamma}_m, \eta_{ml}, \boldsymbol{\psi}_{ml}$ are auxiliary decision variables.

For a given hub location decision, $\boldsymbol{y}$, Algorithm 11.4 allows finding the worst-case distribution.

Algorithm 11.4: Obtaining the worst-case distribution for a given hub location decision

---

Input: A hub location decision $\boldsymbol{y}$

for $m \in M$ **do**

  solve the robust problem

$$\lambda^\star_m = \min_{\boldsymbol{w}_m \in \mathcal{W}(\boldsymbol{y})} \max_{\boldsymbol{c}_m \in \bar{\mathcal{U}}^c_m} \sum_{k\ell \in K} \boldsymbol{\delta}^\top_{k\ell m} \boldsymbol{c}_m w_{k\ell m},$$

  to obtain the optimal solution $\boldsymbol{w}^\star_m$.

for $m \in M$ **do**

  solve the linear program

$$\boldsymbol{c}^\star_m \in \arg\max_{\boldsymbol{c}_m \in \bar{\mathcal{U}}^c_m} \sum_{k\ell \in K} \boldsymbol{\delta}^\top_{k\ell m} \boldsymbol{c}_m w^\star_{k\ell m}$$

  to obtain the worst-case scenario $\boldsymbol{c}^\star_m$ for the cost associated with commodity m.

solve the linear program

$$\boldsymbol{d}^\star \in \arg\max_{\boldsymbol{d} \in \bar{\mathcal{U}}^d} \sum_{m \in M} d_m \lambda^\star_m$$

to obtain the worst-case scenario for the demand

Output: Worst-case distribution $\mathbb{P}^\star$ given by the Dirac delta function at $(\boldsymbol{d}^\star, \boldsymbol{c}^\star)$.

11.6.2 Relaxing the Independent Assumption for Demands and Costs

We consider the situation where the routing policy $\boldsymbol{w}(\cdot)$ is not static and becomes adaptive to the realizations of the demand–cost uncertainty. Such a situation emerges when the independence assumption is violated. Fortunately, under some mild assumptions on the support of the uncertain demand or uncertain cost, we can still obtain a tractable reformulation for the adaptive distributionally robust UHLP.

We consider now a joint ambiguity set as follows:

$$\mathcal{F}^{\dagger}=\left\{\mathbb{P}\in\mathcal{P}(\mathbb{R}^{|M|}\times\mathbb{R}^{3|M|})\;\middle|\;\begin{array}{ll}(\tilde{\boldsymbol{d}},\tilde{\boldsymbol{c}})\sim\mathbb{P} & \\ \mathbb{E}_{\mathbb{P}}\left[\tilde{\boldsymbol{d}}\right]\in[\boldsymbol{d}^{-},\boldsymbol{d}^{+}] & \\ \mathbb{E}_{\mathbb{P}}\left[\tilde{\boldsymbol{c}}_m\right]\in[\boldsymbol{c}_m^{-},\boldsymbol{c}_m^{+}] & \forall m\in M\\ \mathbb{E}_{\mathbb{P}}\left[\|\boldsymbol{D}_l(\tilde{\boldsymbol{c}}-\bar{\boldsymbol{c}})\|_1\right]\leq\sigma_l & \forall l\in L,\\ \mathbb{P}\left[\tilde{\boldsymbol{d}}\in\mathcal{U}_{\mathrm{CH}},\tilde{\boldsymbol{c}}\in\mathcal{D}\right]=1 & \end{array}\right\},\tag{11.368}$$

where $\mathcal{U}_{\mathrm{CH}}=\mathrm{CH}(\{\boldsymbol{d}_\omega\}_{\omega\in\Omega})$ is the convex hull of a set of given scenarios $\{\boldsymbol{d}_\omega\}_{\omega\in\Omega}$ and $\mathcal{D}=\{\boldsymbol{c}\in\mathbb{R}_+^{3|M|}\mid\boldsymbol{Bc}\leq\boldsymbol{b}\}$.

For ease of exposition, we require the matrix $\boldsymbol{B}\in\mathbb{R}^{|M|\times 3|M|}$ in the description of the support set $\mathcal{D}$ to satisfy $\boldsymbol{B}=(\boldsymbol{B}_1\ \boldsymbol{B}_2\ \cdots\ \boldsymbol{B}_{|M|})$ such that $\boldsymbol{B}_m\in\mathbb{R}^{|M|\times 3}$ for every $m\in M$ and $\boldsymbol{Bc}=\sum_{m\in M}\boldsymbol{B}_m\boldsymbol{c}_m$. Similarly, for every $l\in L$, the matrix $\boldsymbol{D}_l\in\mathbb{R}^{N_l\times 3|M|}$ satisfies $\boldsymbol{D}_l=(\boldsymbol{D}_{l1}\ \boldsymbol{D}_{l2}\ \cdots\ \boldsymbol{D}_{l|M|})$ such that $\boldsymbol{D}_{lm}\in\mathbb{R}^{|M|\times 3}$ for every $m\in M$ and $\boldsymbol{D}_l\boldsymbol{c}=\sum_{m\in M}\boldsymbol{D}_{lm}\boldsymbol{c}_m$.

There are differences between the above ambiguity set $\mathcal{F}^{\dagger}$ and that stated in (11.347). First, the ambiguity set $\mathcal{F}^{\dagger}$ does not impose the independence between demands and costs. Hence, $\tilde{\boldsymbol{d}}$ and $\tilde{\boldsymbol{c}}$ can be correlated in the worst-case distribution. Second, the shipping costs $\tilde{\boldsymbol{c}}_m,\ m\in M$ of different commodities are not necessarily mutually independent because in $\mathcal{F}^{\dagger}$, they are jointly supported in a single support set $\mathcal{D}$, while in $\mathcal{F}$ they are separately supported in polyhedra $\mathcal{D}_m,\ m\in M$. Note that we only incorporate in $\mathcal{F}^{\dagger}$ the dispersion constraints of $\tilde{\boldsymbol{c}}$ for tractability issues. Finally, instead of residing in a general polyhedral support set, the random demand in $\mathcal{F}^{\dagger}$ is supported on the convex hull $\mathcal{U}_{\mathrm{CH}}$. Such an approach is not uncommon and has been used in robust optimization models for many practical applications. The reader can refer to [535] for a deep discussion.

The following result identifies the tractable reformulation for the adaptive distributionally robust UHLP we are considering.

Theorem 11.3 *Given the ambiguity set $\mathcal{F}^{\dagger}$, the distributionally robust UHLP is equivalent to the following mixed-integer linear program:*

$$\text{minimize}\quad\sum_{k\in K}f_k y_k+\alpha+\overline{\boldsymbol{\beta}}^{\top}\boldsymbol{d}^{+}-\underline{\boldsymbol{\beta}}^{\top}\boldsymbol{d}^{-}+\sum_{m\in M}(\overline{\boldsymbol{\gamma}}_m^{\top}\boldsymbol{c}_m^{+}-\underline{\boldsymbol{\gamma}}_m^{\top}\boldsymbol{c}_m^{-})+\sum_{l\in L}\rho_l\sigma_l,\tag{11.369}$$

$$\text{subject to}\quad\alpha+(\overline{\boldsymbol{\beta}}-\underline{\boldsymbol{\beta}})^{\top}\boldsymbol{d}_\omega-\boldsymbol{b}^{\top}\boldsymbol{\zeta}_\omega+\bar{\boldsymbol{c}}^{\top}\sum_{l\in L}\boldsymbol{D}_l^{\top}\boldsymbol{\xi}_{l\omega}\geq 0\quad\forall\omega\in\Omega,\tag{11.370}$$

$$\boldsymbol{B}_m^\top \boldsymbol{\zeta}_\omega - \sum_{l\in L} \boldsymbol{D}_{lm}^\top \boldsymbol{\xi}_{l\omega} + \overline{\boldsymbol{\gamma}}_m - \underline{\boldsymbol{\gamma}}_m \geq \sum_{k,\ell\in K} \boldsymbol{\delta}_{k\ell m} d_{m\omega} w_{k\ell m\omega}$$

$$\forall \omega \in \Omega,\ m \in M, \tag{11.371}$$

$$\rho_l \geq \|\boldsymbol{\xi}_{l\omega}\|_\infty,\ \boldsymbol{\xi}_{l\omega} \in \mathbb{R}^{N_l} \quad \forall \omega \in \Omega,\ l \in L, \tag{11.372}$$

$$\boldsymbol{\zeta}_\omega \in \mathbb{R}_+^N,\ \overline{\boldsymbol{\gamma}}_m, \underline{\boldsymbol{\gamma}}_m \in \mathbb{R}_+^3,\ \boldsymbol{w}_{m\omega} \in \mathcal{W}(\boldsymbol{y}) \quad \forall \omega \in \Omega,\ m \in M, \tag{11.373}$$

$$\alpha \in \mathbb{R},\ \overline{\boldsymbol{\beta}}, \underline{\boldsymbol{\beta}} \in \mathbb{R}_+^{|M|},\ \boldsymbol{\rho} \in \mathbb{R}_+^L,\ \boldsymbol{y} \in \{0,1\}^{|K|}, \tag{11.374}$$

where $\alpha, \overline{\boldsymbol{\beta}}, \underline{\boldsymbol{\beta}}, \overline{\boldsymbol{\gamma}}_m, \underline{\boldsymbol{\gamma}}_m, \boldsymbol{\rho}, \boldsymbol{\zeta}_\omega, \boldsymbol{\xi}_{l\omega}$ are auxiliary decision variables.

Proof The proof relies on using the extreme points of $\mathcal{U}_{\mathrm{CH}}$. This allows simplifying the analysis. The details can be found in [535]. □

This theorem implies that when the independence assumption is relaxed, the optimal routing decision $\boldsymbol{w}^\star(\cdot)$ effectively adapts to the realization of the uncertain demands and costs. In particular, with the support set being $\mathcal{U}_{\mathrm{CH}}$, the optimal routing decision $\boldsymbol{w}^\star(\cdot)$ follows a scenario-wise policy such that $\boldsymbol{w}^\star(\boldsymbol{d}_\omega, \boldsymbol{c}_\omega) = \boldsymbol{w}_\omega^\star$ for every $\omega \in \Omega$, where $\boldsymbol{w}_\omega^\star = (\boldsymbol{w}_{m\omega}^\star)_{m\in M}$ is part of the optimal solution of the MILP problem stated in the theorem.

The scenarios $(\boldsymbol{d}_\omega, \boldsymbol{c}_\omega)$, $\omega \in \Omega$ are the possible realizations from the support set of the worst-case distribution $\mathbb{P}^\dagger \in \mathcal{F}^\dagger$, which can be obtained via the following proposition.

Proposition 11.7 *Let $(\boldsymbol{w}_\omega^\star)_{\omega\in\Omega}, (\overline{\boldsymbol{\gamma}}_m^\star, \underline{\boldsymbol{\gamma}}_m^\star)_{m\in M}, (\rho_l^\star)_{l\in L}$ be an optimal solution of the problem* (11.369)–(11.374)*. The worst-case distribution $\mathbb{P}^\dagger \in \mathcal{F}^\dagger$ for the optimal hub location design $\boldsymbol{y}^\star$ takes*

$$\mathbb{P}^\dagger[(\tilde{\boldsymbol{d}}, \tilde{\boldsymbol{c}}) = (\boldsymbol{d}_\omega, \boldsymbol{c}_\omega)] = \pi_\omega \quad \forall \omega \in \Omega,$$

where for every $\omega \in \Omega$, the worst-case scenario $\boldsymbol{c}_\omega$ solves the linear program

$$\max_{\boldsymbol{c}\in\mathcal{D}} \left\{ \sum_{k,\ell\in K} \sum_{m\in M} \boldsymbol{d}_{k\ell m}^\top \boldsymbol{c}_m d_{m\omega} w_{k\ell m\omega}^\star \right.$$

$$\left. - \sum_{m\in M} (\overline{\boldsymbol{\gamma}}_m^\star - \underline{\boldsymbol{\gamma}}_m^\star)^\top \boldsymbol{c}_m - \sum_{l\in L} \rho_l^\star \|\boldsymbol{D}_l(\boldsymbol{c} - \bar{\boldsymbol{c}})\|_1 \right\}, \tag{11.375}$$

and the probabilities $(\pi_\omega)_{\omega\in\Omega} = \boldsymbol{\pi}^\star$ solve the following linear problem using the obtained values $(\boldsymbol{c}_\omega)_{\omega\in\Omega}$:

$$\text{maximize} \quad \sum_{\omega\in\Omega} \pi_\omega \sum_{k,\ell\in K} \sum_{m\in M} \boldsymbol{\delta}_{k\ell m}^\top \boldsymbol{c}_{m\omega} d_{m\omega} w_{k\ell m\omega}^\star, \tag{11.376}$$

$$\text{subject to} \quad \sum_{\omega \in \Omega} \pi_\omega = 1, \tag{11.377}$$

$$\sum_{\omega \in \Omega} \pi_\omega \boldsymbol{d}_\omega \in [\boldsymbol{d}^-, \boldsymbol{d}^+], \tag{11.378}$$

$$\sum_{\omega \in \Omega} \pi_\omega \boldsymbol{d}_{m\omega} \in [\boldsymbol{c}_m^-, \boldsymbol{c}_m^+] \qquad \forall m \in M, \tag{11.379}$$

$$\sum_{\omega \in \Omega} \pi_\omega \|\boldsymbol{D}_l(\boldsymbol{c}_\omega - \bar{\boldsymbol{c}})\|_1 \leq \sigma_l \qquad \forall l \in L, \tag{11.380}$$

$$\pi_\omega \geq 0 \qquad \forall \omega \in \Omega. \tag{11.381}$$

Proof See [535]. □

Remark 11.1 In the description of the ambiguity set $\mathcal{F}^\dagger$, we can alternatively consider a probability constraint $\mathbb{P}\left[\tilde{\boldsymbol{d}} \in \mathcal{D}, \tilde{\boldsymbol{c}} \in \mathcal{U}_{\text{CH}}\right] = 1$, where $\mathcal{D}$ is a general polyhedron and $\mathcal{U}_{\text{CH}} = \text{CH}(\{\boldsymbol{c}_\omega\}_{\omega\in\Omega})$ is the convex hull of $\{\boldsymbol{c}_\omega\}_{\omega\in\Omega}$, and replace the dispersion constraint of $\tilde{\boldsymbol{c}}$, $\mathbb{E}_{\mathbb{P}}\left[\|\boldsymbol{D}_l(\tilde{\boldsymbol{c}} - \bar{\boldsymbol{c}})\|_1\right] \leq \sigma_l$, with that of $\tilde{\boldsymbol{d}}$, i.e., $\mathbb{E}_{\mathbb{P}}\left[\|\boldsymbol{D}_l(\tilde{\boldsymbol{d}} - \bar{\boldsymbol{d}})\|_1\right] \leq \sigma_l$. The results presented naturally extend to this case. These two alternatives are relevant because they allow the decision-maker to flexibly choose the demand or the cost to be supported on the scenario-based convex hull, depending on the available data.

11.6.3 Embedding Capacities at the Hubs

We assume that each potential hub $k \in K$ has a capacity q_k that limits the incoming flow into the hub. We assume a multiple-allocation pattern, i.e., commodities can be split and follow multiple routes.

Under some ambiguity sets, the second-stage routing decision still follows an optimal scenario-wise policy. We redefine the $\boldsymbol{w}$ variables to represent the amount of commodity $m \in M$ routed via hubs k and ℓ in this sequence, $k, \ell \in K$. The deterministic capacitated hub location problem (CHLP) can be formulated as follows:

$$\text{minimize} \quad \sum_{k \in K} f_k y_k + \sum_{m \in M} \sum_{k,\ell \in K} \boldsymbol{\delta}_{k\ell m}^\top \boldsymbol{c}_m w_{k\ell m} + \sum_{m \in M} h_m \left(d_m - \sum_{k,\ell \in K} w_{k\ell m} \right), \tag{11.382}$$

$$\text{subject to} \quad \sum_{k,\ell \in K} w_{k\ell m} = d_k \qquad \forall m \in M, \tag{11.383}$$

$$\sum_{m\in M}\left(\sum_{\ell\in K} w_{k\ell m} + \sum_{\ell\in K\setminus\{k\}} w_{\ell km}\right) \leq y_k q_k \qquad \forall k\in K, \tag{11.384}$$

$$w_{ijk} \geq 0 \qquad \forall k,\ell\in K,\ m\in M, \tag{11.385}$$

$$y_k \in \{0,1\} \qquad \forall k\in K. \tag{11.386}$$

In this model, for $m \in M$, h_m is the unit penalty cost for not supplying commodity m. The other elements in the model should now be familiar to the reader.

As we saw in the previous section, it is possible to formulate this problem using variables representing the proportion of a commodity routed via any two hubs (fractional model). However, the capacity constraints prevent eliminating the demand parameters from the constraints. Hence, the symmetric structure for the uncertain demand and cost in the fraction model no longer exists in the CHLP. In particular, we cannot exploit the independence condition as done for UHLP. Nevertheless, there are still practically interesting cases for which we can obtain a tractable reformulation for the adaptive distributionally robust CHLP.

We directly consider the ambiguity set (11.368) and investigate the following adaptive distributionally robust CHLP:

$$\text{minimize} \quad \sum_{k\in K} f_k y_k + \sup_{\mathbb{P}\in\mathcal{F}^\dagger} \mathbb{E}_{\mathbb{P}}\left[\mathrm{Q}(\boldsymbol{y},\tilde{\boldsymbol{d}},\tilde{\boldsymbol{c}})\right], \tag{11.387}$$

$$\text{subject to} \quad y_k\in\{0,1\} \qquad \forall k\in K, \tag{11.388}$$

where, for a given hub location design $\boldsymbol{y}$ and realization $(\boldsymbol{d},\boldsymbol{c})$ of the demands and costs, the second-stage flow routing problem is the following:

$$\mathrm{Q}(\boldsymbol{y},\boldsymbol{d},\boldsymbol{c}) = \min_{\boldsymbol{w}\in\mathcal{W}(\boldsymbol{y},\boldsymbol{d})} \left\{\sum_{m\in M}\sum_{k,\ell\in K}\left(\boldsymbol{\delta}_{k\ell m}^\top \boldsymbol{c}_m - h_m\right) w_{k\ell m} + \boldsymbol{h}^\top\tilde{\boldsymbol{d}}\right\}, \tag{11.389}$$

with

$$\mathcal{W}(\boldsymbol{y},\boldsymbol{d}) = \left\{\boldsymbol{w}\in\mathbb{R}_+^{|K|^2\times|M|} \;\middle|\; \begin{array}{ll} \displaystyle\sum_{k,\ell\in K} w_{k\ell m} = d_k & \forall m\in M, \\ \displaystyle\sum_{m\in M}\left(\sum_{\ell\in K} w_{k\ell m} + \sum_{\ell\in K\setminus\{k\}} w_{k\ell m}\right) \leq y_k q_k & \forall k\in K \end{array}\right\}.$$

The following result states that this adaptive distributionally robust optimization problem admits a computationally tractable reformulation.

Theorem 11.4 *Given the ambiguity set* (11.368)*, the distributionally robust CHLP can be formulated as the following mixed-integer linear program:*

$$\text{minimize} \quad \sum_{k \in K} f_k y_k + \alpha + \overline{\boldsymbol{\beta}}^\top \boldsymbol{d}^+ - \underline{\boldsymbol{\beta}}^\top \boldsymbol{d}^- + \sum_{m \in M} (\overline{\boldsymbol{\gamma}}_m^\top \boldsymbol{c}_m^+ - \underline{\boldsymbol{\gamma}}_m^\top \boldsymbol{c}_m^-) + \sum_{l \in L} \rho_l \sigma_l, \tag{11.390}$$

$$\text{subject to} \quad \alpha + (\overline{\boldsymbol{\beta}} - \underline{\boldsymbol{\beta}} - \boldsymbol{h})^\top \boldsymbol{d}_\omega + \sum_{m \in M} \sum_{k,\ell \in K} h_m w_{k\ell m\omega} - \boldsymbol{b}^\top \boldsymbol{\zeta}_\omega + \bar{\boldsymbol{c}}^\top \sum_{l \in L} \boldsymbol{D}_l^\top \boldsymbol{\xi}_{l\omega} \geq 0 \qquad \forall \omega \in \Omega, \tag{11.391}$$

$$\boldsymbol{B}_\omega^\top \boldsymbol{\zeta}_\omega - \sum_{l \in L} \boldsymbol{D}_{lm}^\top \boldsymbol{\xi}_{l\omega} + \overline{\boldsymbol{\gamma}}_m - \underline{\boldsymbol{\gamma}}_m \geq \sum_{k,\ell \in K} \delta_{k\ell m} w_{k\ell m\omega} \qquad \forall \omega \in \Omega,\ m \in M, \tag{11.392}$$

$$\rho_l \geq \|\boldsymbol{\xi}_{l\omega}\|_\infty,\ \boldsymbol{\xi}_{l\omega} \in \mathbb{R}^{N_l} \qquad \forall \omega \in \Omega,\ l \in L, \tag{11.393}$$

$$\boldsymbol{\zeta}_\omega \in \mathbb{R}_+^N,\ \overline{\boldsymbol{\gamma}}_m, \underline{\boldsymbol{\gamma}}_m \in \mathbb{R}_+^3,\ z_\omega \in \mathcal{W}(\boldsymbol{y}, \boldsymbol{d}_\omega) \quad \forall \omega \in \Omega,\ m \in M, \tag{11.394}$$

$$\alpha \in \mathbb{R},\ \overline{\boldsymbol{\beta}}, \underline{\boldsymbol{\beta}} \in \mathbb{R}_+^{|M|},\ \boldsymbol{\rho} \in \mathbb{R}_+^{|L|},\quad \boldsymbol{y} \in \{0,1\}^{|K|}, \tag{11.395}$$

where $\alpha, \overline{\boldsymbol{\beta}}, \underline{\boldsymbol{\beta}}, \overline{\boldsymbol{\gamma}}_m, \underline{\boldsymbol{\gamma}}_m, \boldsymbol{\rho}, \boldsymbol{\zeta}_\omega, \boldsymbol{\xi}_{l\omega}$ *are auxiliary decision variables.*

Proof See [535]. □

This tractability result stems from the assumption that the support set of the uncertain demand $\tilde{\boldsymbol{d}}$ is the convex hull of given scenarios. It does not hold when we assume the support set of the uncertain cost $\tilde{\boldsymbol{c}}$ to be a convex hull of the available scenarios. This differs from the UHLP case. Due to the absence of the symmetric structure for the uncertain demand and cost in the fraction model of CHLP, we cannot shift the demand $\boldsymbol{d}$ to the objective function. Moreover, the feasible set $\mathcal{W}(\boldsymbol{y}, \boldsymbol{d})$ for the second-stage routing problem depends on both the hub location design $\boldsymbol{y}$ and the demand realization $\boldsymbol{d}$.

It is still interesting to note that given the ambiguity set $\mathcal{F}^\dagger$ and the optimal hub location design $\boldsymbol{y}^\star$, the resulting optimal routing decision $\boldsymbol{w}^\star(\cdot)$ of the distributionally robust CHLP is adaptive to the scenarios $\omega \in \Omega$, namely, those that originate from the worst-case distribution of $(\tilde{\boldsymbol{d}}, \tilde{\boldsymbol{c}})$. This is consistent with the optimal routing decision $\boldsymbol{y}^\star(\cdot)$ in the distributionally robust UHLP. Besides, by a similar treatment as that in Proposition 11.7, we can obtain the worst-case distribution $\mathbb{P}_v^\dagger \in \mathcal{F}^\dagger$ of $(\tilde{\boldsymbol{d}}, \tilde{\boldsymbol{c}})$ in an efficient manner. This is stated in the following result.

Proposition 11.8 *Let $(\boldsymbol{w}^\star_\omega)_{\omega\in\Omega}$, $(\overline{\boldsymbol{\gamma}}^\star_m, \underline{\boldsymbol{\gamma}}^\star_m)_{m\in M}$, $(\rho^\star_l)_{l\in L}$ be the optimal solution to problem* (11.390)–(11.395). *The worst-case distribution $\mathbb{P}^\dagger \in \mathcal{F}^\dagger$ for the optimal hub location design $\boldsymbol{y}^\star$ takes*

$$\mathbb{P}^\dagger[(\tilde{\boldsymbol{d}}, \tilde{\boldsymbol{c}}) = (\boldsymbol{d}_\omega, \boldsymbol{c}_\omega)] = \pi_\omega \quad \forall \omega \in \Omega.$$

Here for every $\omega \in \Omega$, the worst-case scenario $\boldsymbol{c}_\omega$ solves the linear program

$$\max_{\boldsymbol{c}\in\mathcal{D}} \left\{ \sum_{k,\ell\in K} \sum_{m\in M} \boldsymbol{\delta}^\top_{k\ell m} \boldsymbol{c}_m w^\star_{k\ell m\omega} - \sum_{m\in M} (\overline{\boldsymbol{\gamma}}^\star_m - \underline{\boldsymbol{\gamma}}^\star_m)^\top \boldsymbol{c}_m - \sum_{l\in L} \rho^\star_l \| \boldsymbol{D}_l(\boldsymbol{c} - \bar{\boldsymbol{c}})\|_1 \right\},$$

and the probabilities $(\pi_\omega)_{\omega\in\Omega} = \boldsymbol{\pi}^\star$ solve the linear problem using $(\boldsymbol{c}_\omega)_{\omega\in\Omega}$:

$$\text{maximize} \quad \sum_{\omega\in\Omega} \pi_\omega \sum_{k,\ell\in K} \sum_{m\in M} \boldsymbol{\delta}^\top_{k\ell m} \boldsymbol{c}_{m\omega} w^\star_{k\ell m\omega}, \tag{11.396}$$

$$\text{subject to} \quad \sum_{\omega\in\Omega} \pi_\omega = 1, \tag{11.397}$$

$$\sum_{\omega\in\Omega} \pi_\omega \boldsymbol{d}_\omega \in [\boldsymbol{d}^-, \boldsymbol{d}^+], \tag{11.398}$$

$$\sum_{\omega\in\Omega} \pi_\omega \boldsymbol{c}_{m\omega} \in [\boldsymbol{c}^-_m, \boldsymbol{c}^+_m] \qquad \forall m \in M, \tag{11.399}$$

$$\sum_{\omega\in\Omega} \pi_\omega \| \boldsymbol{D}_l(\boldsymbol{c}_\omega - \bar{\boldsymbol{c}})\|_1 \le \varepsilon_l \qquad \forall l \in L. \tag{11.400}$$

Proof See [535]. □

11.7 Conclusions and Further Reading

In this chapter, we discussed mechanisms for hedging against uncertainty in hub location problems. The discussion is far from exhaustive. More could be written on the topic. Nevertheless, we have provided the reader with some hints for further reading. In Chap. 15, we analyze in detail a specific application that intertwines hub location and time series in a multi-period setting.

With the development of new models for hedging against uncertainty in hub location, we have also observed some novelty regarding the objective functions analyzed. Motivated by sustainability issues, two new environmental-based cost functions are introduced in [387], accounting for air and noise pollution of. Multiple sources of uncertainty are considered. Regarding the demands, they are assumed to be represented by random variables. The other uncertain parameters are modeled using fuzzy numbers or random fuzzy data (fuzzy numbers with randomized boundaries). Approximate algorithms are devised for tackling the resulting problem.

Another trend we have observed in the hub location literature regards the inclusion of profit-oriented objective functions. In this context, we also find work seeking to hedge against uncertainty. This is the case in [516], where the authors introduce robust stochastic models for profit-maximizing capacitated hub location problems. Uncertainty stems from the demands and revenues. First, a two-stage stochastic program is developed in which uncertainty is assumed to be represented by a random vector with some given joint probability distribution. The authors also propose hybrid robust stochastic models to capture revenue uncertainty, assuming stochastic demand. Exact algorithms based on Benders decomposition and coupled with sample average approximation procedure are presented.

Li et al. [336] investigate a hub location problem to integrate urban and rural logistics networks considering demand uncertainty. The latter is assumed to be captured by a finite set of scenarios. A hierarchical hub location model is proposed. It seeks to minimize the expected total system cost by optimizing the locations, number, and capacities of hubs seeking to support an "urban-town-village" logistics system. The interactions between the logistics hubs and among the hub-and-spoke connections are explicitly considered in the presence of logistics demand uncertainty. Hub capacity constraints are also considered. A branch-and-Benders-cut algorithm is developed to solve the proposed model.

Habibi et al. [253] focus on another trend in distribution logistics: collaboration. In particular, the authors focus on a hub location problem in a collaborative context of two distribution networks from different supply chains. Uncertainty stems from the supplementary cost to include in the setup cost for establishing a shared hub. Different types of collaboration are analyzed. Several cost-sharing strategies are also investigated.

Kargar and İrfan Mahmutoğulları [302] investigate risk-neutral and risk-averse two-stage stochastic formulations for the uncapacitated multiple-allocation p-hub median problem and discuss the impact of risk aversion on the optimal solution. Although stochastic models are relevant for tackling uncertainty in problem parameters, the solution of these models requires higher computational effort than their deterministic counterparts. Therefore, the authors present a scenario-decomposition algorithm for the stochastic formulations. To evaluate the performance of the proposed solution algorithm, a set of computational experiments is conducted on real datasets. The results show that the proposed algorithm is very effective in finding optimal or near-optimal solutions in significantly shorter computation time than that of deterministic equivalent problems.

Finally, we mention [573], in which the authors investigate a distributionally robust chance-constrained p-hub center problem. The use of chance constraints has not been much explored in the context of hub location problems, although that makes much sense in the center and covering problems, as discussed in Chap. 7. In that work, it becomes clear that these problems are challenging. For some particular cases, tractable models can be obtained. This is the case when a Wasserstein ambiguity set is considered.

Hub location under uncertainty has been one of the most prolific fields within facility location under uncertainty. The contents of this chapter show that the existing knowledge has evolved immensely in just one decade. Nevertheless, there is still much growth potential because some research directions are unexplored. This is the case with (i) center and covering problems, (ii) the use of risk-averse objective functions, (iii) problems embedding hub network design decisions, (iv) resilience-oriented problems (e.g., unexpected disruptions at the hubs or network connections), and (v) distributionally robust optimization.

Chapter 12
Logistics and Supply Chain Management

Abstract This chapter focuses on the role of facility location under uncertainty in logistics and supply chain management. Several prototype models are discussed, gathering relevant elements of practical relevance, including multiple facility echelons, multiple products or families of products, and time-dependent decisions. This comes along with two relevant aspects: resilience and sustainability. The first requires careful anticipation of potential disruptions in the underlying network and the development of mechanisms for efficiently and effectively overcoming them. The second is aligned with the vision of a progressively more circular economy, ensuring that reverse logistics operations complement the forward ones, thus contributing to a better utilization of the available resources while preserving the environment. Financial risk and supply chain risk analytics are concepts also discussed in this chapter. Humanitarian supply chains call for specific types of decisions, objective functions, and constraints. These are also discussed. The chapter also covers particular families of problems, such as those resulting from gathering the elements discussed in the previous chapters.

Keywords Supply chain network design · Production · Inventory · Distribution · Supply chain risk analytics · Resilience · Sustainability · Humanitarian supply chains · Location-inventory routing

12.1 Introduction

Logistics and Supply Chain Management are two fields very much intertwined. The former regards planning, implementation, and control of storage and flow of goods (forward and reverse) in a supply chain network. The related processes also include services and related information between the point of origin and the point of consumption to meet the customer's requirements (https://cscmp.org/). Accordingly, logistics encompasses different activities such as purchasing, warehousing, inventory management, packaging, distribution fleet management,

© The Author(s), under exclusive license to Springer Nature Switzerland AG 2024

F. Saldanha-da-Gama, S. Wang, *Facility Location Under Uncertainty*,
International Series in Operations Research & Management Science 356,
https://doi.org/10.1007/978-3-031-55927-3_12

and transportation, to mention a few. Furthermore, logistics activities typically seek customer satisfaction.

On the other hand, Supply Chain Management refers to the activities that allow the coordination of suppliers, manufacturers, customers, and possibly other stakeholders. The goal is to achieve a competitive advantage. The activities usually considered include sourcing raw materials, procurement of consumer requirements, converting raw materials into final products, and timely delivery to customers.

Overall, logistics is crucial for synchronizing a supply chain. All the activities rely on a network—the supply chain network—of distinct but interrelated facilities (suppliers, factories, central distribution centers, warehouses, *et cetera*) and also on moving the commodities across different partners in the network. The design of a supply chain network is typically a strategic problem. This is where Facility Location, Logistics, and Transportation strongly complement one another (see [470]).

The role of facility location in the design of production and distribution networks is not a new topic. We observe this in [315] where the authors identify several related elements of relevance: multiple products, capacity decisions, inventory management, multiple layers of facilities, multiple objectives whose trade-off must be sought, *et cetera*.

Melo et al. [377] develop a unified optimization model for supply chain network design, comprising many aspects of practical relevance, including those above mentioned. The concept of "selectable" and "non-selectable" facilities is introduced in that article. For instance, a customer or a supplier corresponds to a non-selectable facility; a distribution center or a manufacturing plant refers to facilities whose location can be decided upon. Thus, they represent selectable facilities. More recently, in [264] an integrated model was discussed again featuring several aspects of relevance in logistics network design.

In this chapter, we leverage the above references and discuss the role of uncertainty when making facility location decisions in the context of strategic supply chain planning. The remainder of the chapter is organized as follows. In Sect. 12.2, we generically discuss the role of facility location in the strategic supply chain network design. A prototype problem is initially discussed that is progressively extended, eventually leading to a multi-stage stochastic decision-making process. In Sect. 12.3, the focus is on building resilient supply chains. The concepts of financial risk in supply chain management and supply chain risk analytics are explored. This analysis is complemented by Sect. 12.4, where sustainability is the focus. This is accomplished by designing adequate networks for supporting reverse logistics operations and eventually integrating them with the forward ones. Section 12.5 presents specific aspects and challenges from humanitarian supply chain and disaster operations management. The chapter proceeds with a review of the work done on some extensions and applications that include location-inventory-routing, multi-modal transportation, and hazardous materials transportation. The chapter closes with an overview and some hints for further research.

12.2 Strategic Supply Chain Network Design

In this section, we follow [470] to derive a model capturing a minimal set of elements of relevance in strategic supply chain planning under uncertainty. To make this section self-contained, we start from the classical capacitated facility location problem and then progressively add building blocks until getting the final comprehensive model. We recall the classical capacitated facility location problem (CFLP) described in Chap. 2 and formulated as follows:

$$\text{minimize} \quad \sum_{i\in I} f_i y_i + \sum_{i\in I}\sum_{j\in J} c_{ij}x_{ij}, \tag{12.1}$$

$$\text{subject to} \quad \sum_{i\in I} x_{ij} = d_j \qquad \forall j \in J, \tag{12.2}$$

$$\sum_{j\in J} x_{ij} \le q_i y_i \qquad \forall i \in I, \tag{12.3}$$

$$y_i \in \{0, 1\} \qquad \forall i \in I, \tag{12.4}$$

$$x_{ij} \ge 0 \qquad \forall i \in I,\ j \in J. \tag{12.5}$$

In this model, I is a finite set of potential locations for the facilities, and J is a set of demand nodes. Furthermore, y_i is a binary variable equal to one if and only if a facility is established at $i \in I$, and x_{ij} is the amount shipped from location $i \in I$ to demand node $j \in J$. Regarding the parameters, f_i is the fixed cost for setting up a facility at $i \in I$, c_{ij} is the unit cost for supplying demand node $j \in J$ from facility $i \in J$, d_j is the demand of node $j \in J$, and q_i is the capacity of location $i \in I$. The objective function accounts for the total cost; the constraints regard demand satisfaction and capacity at the facilities.

As emphasized in [378], time is a feature of relevance in supply chain network design due to decisions such as those related to capacity adjustments (e.g., expansion and reduction), inventory management, facilities opening or closing, *et cetera*. Moreover, for different reasons, time is usually discretized (see [181]). Hence, we seek to plan for a time horizon, T, divided into several periods. In addition to time, we also directly consider the existence of multiple products or families of products (see [142]).

In many settings, we need to plan for an already operating system. Hence, in addition to installing new facilities over time, we can close the existing ones. This is a way to model facility or capacity relocation as explained in [377]. Accordingly, we partition the set of locations I into two subsets: I^c is the set of locations where facilities are currently operating—which can be removed during the planning horizon—and I^o is the set of potential locations for facilities to be installed during the planning horizon. We consider time-dependent facility costs: f_{it} denotes the cost for having facility $i \in I$ operating in period $t \in T$. This cost may include fixed setup costs for the facilities paid only once when the facility is established or removed (see [403] for further details).

Another aspect to capture in logistics and supply chain management regards the existence of multiple products or product families. This calls for considering time- and product-dependent demands and costs: d_{jpt} represents the demand of customer $j \in J$ for product $p \in P$ in period $t \in T$, and c_{ijpt} is the unit transportation cost for product p from location $i \in I$ to demand node $j \in J$ in period $t \in T$.

Often, products differ in terms of capacity consumption or production factors. Hence, for each $p \in P$, we can consider a value ρ_{ip} denoting the unit capacity/production consumption factor for product p at facility $i \in I$. This facility-dependent factor allows modeling features such as technology choice or variable production rates.

Concerning the decision variables, we adapt those considered above for the CFLP. We consider a binary variable y_{it} equal to one if and only if a facility is operating at $i \in I$ in period $t \in T$, and x_{ijpt} stands for the amount of product $p \in P$ to send from location $i \in I$ to demand node $j \in J$ in period $t \in T$.

We assume that all facilities can ship/produce all products. Moreover, the capacity available at the facilities is shared by all such products. The CFLP extends as follows:

$$\text{minimize} \quad \sum_{t \in T} \sum_{i \in I} f_{it} y_{it} + \sum_{t \in T} \sum_{i \in I} \sum_{j \in J} \sum_{p \in P} c_{ijpt} x_{ijpt}, \tag{12.6}$$

$$\text{subject to} \quad \sum_{j \in J} \sum_{p \in P} \rho_{ip} x_{ijpt} \leq q_i y_{it} \quad \forall i \in I,\ t \in T, \tag{12.7}$$

$$\sum_{i \in I} x_{ijpt} = d_{jpt} \quad \forall j \in J,\ p \in P,\ t \in T, \tag{12.8}$$

$$y_{it} \leq y_{it-1} \quad \forall i \in I^c,\ t \in T \setminus \{1\}, \tag{12.9}$$

$$y_{it} \geq y_{it-1} \quad \forall i \in I^o,\ t \in T \setminus \{1\}, \tag{12.10}$$

$$y_{it} \in \{0, 1\} \quad \forall i \in I,\ t \in T, \tag{12.11}$$

$$x_{ijpt} \geq 0 \quad \forall i \in I,\ j \in J,\ p \in P,\ t \in T. \tag{12.12}$$

The meaning of most elements in this model is clear from other models presented and the related explanations provided. Regarding constraints (12.9) and (12.10), they guarantee that once an existing facility is closed, it is no longer established (until the end of the planning horizon) and vice versa: once installed, a new facility remains operating until the end of the planning horizon.

In addition to time dependency and multiple-product flow planning, we consider a third element of relevance in strategic supply chain planning: the multi-layered facility structure of the network, with each layer having a specialized function. We can unify all facility locations using a set L, irrespective of their layer and function. This set includes (i) existing facilities that can be removed, (ii) locations when new facilities can be installed, (iii) existing facilities that cannot be removed, (iv) customers, (v) suppliers, *et cetera.* Following [377], we distinguish among selectable locations, say S, and non-selectable locations, $L \setminus S$. Location decisions are made only for the former. Also in that case, we distinguish among locations where facilities are currently operating (which can be removed during the planning horizon) S^c and those where new facilities can be established, S^o.

We keep using the facility costs f_{it} but now assuming $i \in L$. The transportation costs are denoted by c_{ijpt} again, with $i, j \in L$. The same index set is assumed for the capacities q_i, for the unit production or capacity consumption factors, ρ_{ip}, and for the demands d_{ipt}. Apart from i, the other indices belong to the same sets.

The integrated model we are building assumes the "products" to include final commodities, raw materials, and in-process materials/semi-products. For this reason, we need to consider the so-called Bill-of-Materials (BoM): a_{qp} stands for the number of units of product $q \in P$ required to produce one unit of product $p \in P$, $p \neq q$.

Regarding the costs and in addition to those already introduced, we consider (i) b^{PC}_{ipt} representing the unit cost paid at facility $i \in L$ for buying/procuring product $p \in P$ from an external supplier in period $t \in T$, (ii) b^{PD}_{ipt} standing for the unit cost for producing one unit of product $p \in P$ at facility $i \in L$ in period $t \in T$, and (iii) b^{I}_{ipt} accounting for the unit cost for storing one unit of product $p \in P$ at facility $i \in L$ at the end of period $t \in T$.

We keep using variables y_{it} but now assuming $i \in S$. We also keep the x-variables, but taking $i, j \in L$, $i \neq j$. Additionally, we need variables for the logistics-related activities: u_{ipt} represents the quantity of product $p \in P$ procured by facility $i \in L$ from an external supplier in period $t \in T$, r_{ipt} is the amount of product $p \in P$ produced at facility $i \in L$ in period $t \in T$, and s_{ipt} is the amount of product $p \in P$ stored at location $i \in L$ at the end of period $t \in T$. The extended problem can be formulated mathematically as follows:

$$\text{minimize} \quad \sum_{t \in T} \sum_{i \in L} f_{it} y_{it} + \sum_{t \in T} \sum_{i \in L} \sum_{j \in L, j \neq i} \sum_{p \in P} c_{ijpt} x_{ijpt} + \sum_{t \in T} \sum_{i \in L} \sum_{p \in P} (b^{\text{PC}}_{ipt} u_{ipt} + b^{\text{PD}}_{ipt} r_{ipt} + b^{\text{I}}_{ipt} s_{ipt}), \tag{12.13}$$

$$\text{subject to} \quad u_{ipt} + \sum_{j \in L, j \neq i} x_{jipt} + r_{ipt} + s_{ipt-1} = \sum_{j \in L, j \neq i} x_{ijpt} + \sum_{q \in P} a_{pq} r_{iqt} + s_{ipt} + d_{ipt} \quad \forall i \in L,\ p \in P,\ t \in T, \tag{12.14}$$

$$\sum_{p \in P} \rho_{ip} r_{ipt} \leq q_i y_{it} \qquad \forall i \in L,\ t \in T, \tag{12.15}$$

$$y_{it} \geq y_{it-1} \qquad \forall i \in S^c,\ t \in T \setminus \{1\}, \tag{12.16}$$

$$y_{it} \leq y_{it-1} \qquad \forall i \in S^o,\ t \in T \setminus \{1\}, \tag{12.17}$$

$$\sum_{p \in P} \left(u_{ipt} + s_{ipt} + \sum_{j \in L, j \neq i} \left[x_{ijpt} + x_{jipt} \right] \right) \leq M y_{it}$$

$$\forall i \in L,\ t \in T, \tag{12.18}$$

$$y_{it} \in \{0, 1\} \qquad \forall i \in L,\ t \in T, \tag{12.19}$$

$$x_{ijpt} \geq 0 \qquad \forall i, j \in L,\ j \neq i,\ p \in P,\ t \in T, \tag{12.20}$$

$$u_{ipt}, r_{ipt}, s_{ipt} \geq 0 \qquad \forall i \in L,\ p \in P,\ t \in T. \tag{12.21}$$

The objective function has been extended to include the costs related to logistics activities: production, storage, and procurement. Constraints (12.14) ensure flow conservation constraints, which include production, storage, procurement, and demands. Inequalities (12.15) refer to the production capacity at the facilities. Constraints (12.16) and (12.17) ensure that the selectable facilities can have their status changed (establishment or removal depending on the case) at most once. Inequalities (12.18) guarantee that any activity at a facility in some period can only exist if the facility is operating in that period (and the corresponding costs are accounted for). Finally, (12.19)–(12.21) are the domain constraints.

The model just presented allows a dynamic management of the customers. We can directly consider in L demand nodes that will emerge later in the planning horizon and thus have null activity (e.g., demand) before. Furthermore, the above model allows for the existence of stocks before the beginning of the planning horizon through variables s_{ip0}. If this is not the case, these variables are assumed to value zero.

The above model represents a minimal setting for integrating location, transportation, and logistics decisions. Much work published in the context of Logistics and Supply Chain Management considers additional decisions to make. The interested reader can refer to [139, 181, 264], among other references.

12.2.1 *Hedging Against Uncertainty*

We proceed with our discussion by showing how uncertainty can be accounted for in the described problem.

We keep adding building blocks to the model. We now assume that both the demands and the costs are sources of uncertainty. They define a random vector with a given joint distribution function.

For each $t \in T$, let $\boldsymbol{\xi}_t = (\boldsymbol{d}_t, \boldsymbol{f}_t, \boldsymbol{c}_t, \boldsymbol{b}_t^{\text{PC}}, \boldsymbol{b}_t^{\text{PD}}, \boldsymbol{b}_t^{\text{I}})$ with

$$\boldsymbol{d}_t = (d_{ipt})_{i \in L, p \in P}, \quad \boldsymbol{f}_t = (f_{it})_{i \in L}, \quad \boldsymbol{c}_t = (c_{ijpt})_{i,j \in L, j \neq i, p \in P},$$

$$\boldsymbol{b}_t^{\text{PC}} = (b_{ipt}^{\text{PC}})_{i \in L, p \in P}, \quad \boldsymbol{b}_t^{\text{PD}} = (b_{ipt}^{\text{PD}})_{i \in L, p \in P}, \quad \boldsymbol{b}_t^{\text{I}} = (b_{ipt}^{\text{I}})_{i \in L, p \in P}.$$

We assume that $(\boldsymbol{\xi}_1, \boldsymbol{\xi}_2, \ldots)$ is a discrete-time stochastic process. A popular approach for tackling the resulting optimization problem consists of discretizing the distribution of the this stochastic process (e.g., by sampling), thus constructing a so-called scenario tree, which, in turn, allows the formulation of a deterministic optimization problem. In what follows, we assume that a scenario tree has been identified or sampled beforehand, capturing a meaningful set of future scenarios and thus the possible evolution of the above stochastic process [266, 277, 306]. We refer the reader to the discussion presented in Sect. 6.4. Nevertheless, we recall the illustration of a scenario tree in Fig. 12.1 for a two-period planning horizon and thus with three decision stages. The first stage corresponds to the beginning of period 1. The second refers to the end of period 1 ($\boldsymbol{\xi}_1$ has been observed) and to the beginning of period 2. Finally, stage 3 refers to the end of period 2 ($\boldsymbol{\xi}_2$ has been observed).

In our setting, one scenario is a sequence of events from the beginning of the first period until the end of the last one, i.e., one scenario fully determines all the information for the entire planning horizon and thus induces a deterministic multi-period multi-commodity problem that can be formulated according to (12.13)–(12.21).

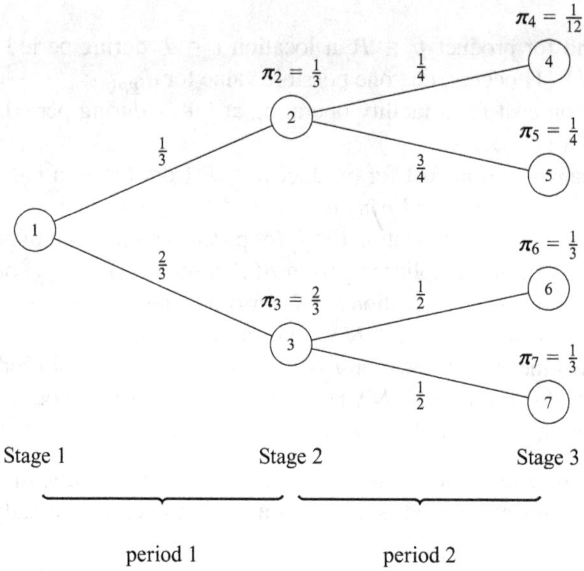

Fig. 12.1 A three-stage scenario tree—two periods in the planning horizon

Based upon this representation of uncertainty, we can formulate the extensive form of the deterministic equivalent of our problem. We consider the following notation (some already introduced in Sect. 6.4 and some specific to our problem):

$\mathcal{N}$, set of nodes in the scenario tree;
M, number of stages ($M = |T| + 1$);
Ω, set of scenarios—leaf nodes;
$\gamma(n)$, immediate predecessor of node $n \in \mathcal{N} \setminus \{1\}$;
$\sigma(n)$, stage to which node n belongs, $n \in \mathcal{N}$;
$\rho(n)$, period ending at node $n \in \mathcal{N} \setminus \{1\}$. Due to the relation between stages and periods, we have $\rho(n) = \sigma(n) - 1$.
π^n, probability that node $n \in \mathcal{N}$ occurs. Given that in each stage $m \in \{1, \ldots, M\}$ some node will certainly occur, then we have $\sum_{n \in \mathcal{N} | \sigma(n)=m} \pi^n = 1$, $m \in M$. In particular, the probability associated with node one is 1: the *status quo*.

A node $n \in \mathcal{N} \setminus \{1\}$ in the scenario tree represents a possible sequence of events from the beginning of period 1 to the end of period $\rho(n)$. Assuming independence across different periods, the probability associated with each node n is the product of all the probabilities of the events whose sequence leads to node n. This is again emphasized in Fig. 12.1, where these values are illustrated. In particular, we associate with each edge the probability associated with the events. Next to the nodes, we indicate the probability of the sequence of events leading to the node.

We can now formally introduce the stochastic parameters associated with the problem.

d_{ip}^n, demand for product $p \in P$ at location $i \in L$ during period $\rho(n)$ if node $n \in \mathcal{N} \setminus \{1\}$ occurs, i.e., one possible value for $d_{ip\rho(n)}$;
f_i^n, operation cost for a facility operating at $i \in I$ during period $\rho(n)$ if node $n \in \mathcal{N} \setminus \{1\}$ occurs;
c_{ijp}^n, unit transportation cost for product $p \in P$ from location $i \in L$ to location $j \in L$ ($j \neq i$) in period $\rho(n)$ if node $n \in \mathcal{N} \setminus \{1\}$ occurs;
$b_{ip}^{\mathrm{PC}^n}$, unit cost incurred at location $i \in L$ for purchasing one unit of product $p \in P$ from an external supplier in period $\rho(n)$ if node $n \in \mathcal{N} \setminus \{1\}$ occurs;
$b_{ip}^{\mathrm{PD}^n}$, unit cost incurred at location $i \in L$ for producing one unit of product $p \in P$ in period $\rho(n)$ if node $n \in \mathcal{N} \setminus \{1\}$ occurs;
$b_{ip}^{\mathrm{I}^n}$, unit cost incurred at location $i \in L$ for storing one unit of product $p \in P$ in period $\rho(n)$ if node $n \in \mathcal{N} \setminus \{1\}$ in the scenario tree occurs;
α^n, discount factor to apply to the costs in stage $\sigma(n)$ (period $\rho(n)$).

The decision variables to consider must reflect the uncertainty. In particular, the decisions made in some period (stage) depend on the decisions already made in the previous periods and the expected evolution of the stochastic process until the end of the planning horizon. Formally, we define

$$y_i^n = \begin{cases} 1, & \text{if facility } i \in S \text{ is operating in period } \sigma(n) \text{ and node } n \in \mathcal{N} \setminus \{\Omega\} \\ & \text{occurs;} \\ 0, & \text{otherwise.} \end{cases}$$

$x_{ijp}^n =$ amount of product $p \in P$ transported from location $i \in L$ to location $j \in L, (j \neq i)$ in period $\rho(n)$ if node $n \in \mathcal{N} \setminus \{1\}$ occurs.

$u_{ip}^n =$ amount of product $p \in P$ procured by location $i \in L$ from an external supplier in period $\rho(n)$ if node $n \in \mathcal{N} \setminus \{1\}$ occurs.

$r_{ip}^n =$ amount of product $p \in P$ produced at location $i \in L$ in period $\rho(n)$ if node $n \in \mathcal{N} \setminus \{1\}$ occurs.

$s_{ip}^n =$ amount of product $p \in P$ stored at location $i \in L$ at the end of period $\rho(n)$ if node $n \in \mathcal{N} \setminus \{1\}$ occurs.

Note that the decision about whether a selectable facility operates in some period must be implemented before knowing the exact operational costs that will be incurred that information is disclosed during the period and the other costs and parameters that, due to uncertainty, will be known also during the period. Using the above notation and terminology, we derive the following multi-stage stochastic programming model for the location-logistics-transportation problem we are considering:

$$\text{minimize} \quad \sum_{n \in \mathcal{N} \setminus \{1\}} \pi^n \alpha^n \Bigg(\sum_{i \in L} f_i^n y_i^{\gamma(n)} + \sum_{i \in L} \sum_{j \in L, j \neq i} \sum_{p \in P} c_{ijp}^n x_{ijp}^n ,$$

$$+ \sum_{i \in L} \sum_{p \in P} \left[b_{ip}^{\mathrm{PC}^n} u_{ip}^n + b_{ip}^{\mathrm{PD}^n} r_{ip}^n + b_{ip}^{\mathrm{I}^n} s_{ip}^n \right] \Bigg) , \tag{12.22}$$

$$\text{subject to} \quad u_{ip}^n + \sum_{j \in L, j \neq i} x_{jip}^n + r_{ip}^n + s_{ip}^{\gamma(n)} = \sum_{j \in L, j \neq i} x_{ijp}^n$$

$$+ \sum_{q \in P} a_{pq} r_{iq}^n + s_{ip}^n + d_{ipn} \qquad \forall i \in L,\ p \in P,\ n \in \mathcal{N} \setminus \{1\}, \tag{12.23}$$

$$\sum_{p \in P} \rho_{ip} r_{ip}^n \leq q_i y_i^{\gamma(n)} \qquad \forall i \in L, n \in \mathcal{N} \setminus \{1\}, \tag{12.24}$$

$$y_i^n \leq y_i^{\gamma(n)} \qquad \forall i \in S^c,\ n \in \mathcal{N} \setminus \{\Omega \cup \{1\}\}, \tag{12.25}$$

$$y_i^n \geq y_i^{\gamma(n)} \qquad \forall i \in I^o,\ n \in \mathcal{N} \setminus \{\Omega \cup \{1\}\}, \tag{12.26}$$

$$\sum_{p\in P}\left(u_{ip}^{n}+s_{ip}^{n}+\sum_{j\in L, j\neq i}\left[x_{ijp}^{n}+x_{jip}^{n}\right]\right)\leq My_{i}^{\gamma(n)}$$

$$\forall i\in L,\ n\in\mathcal{N}\setminus\{1\}, \tag{12.27}$$

$$y_i^n\in\{0,1\} \qquad \forall i\in S,\ n\in\mathcal{N}\setminus\{\Omega\}, \tag{12.28}$$

$$x_{ijp}^{n}\geq 0 \qquad \forall i\in L,\ j\in L,\ j\neq i,\ p\in P,\ n\in\mathcal{N}\setminus\{1\}, \tag{12.29}$$

$$u_{ip}^{n}, r_{ip}^{n}, s_{ip}^{n}\geq 0 \qquad \forall i\in L,\ p\in P,\ n\in\mathcal{N}\setminus\{1\}. \tag{12.30}$$

Recall the discussion presented in Sect. 6.4 that allows re-writing the objective function, making it more intuitive:

$$\sum_{n\in\mathcal{N}\setminus\{1\}}\pi^n\alpha^n\sum_{i\in I}f_i^n y_i^{\gamma(n)}=\sum_{n\in\mathcal{N}\setminus\Omega}\left(\sum_{i\in I}y_i^n\sum_{\nu\in\mathcal{N}|\gamma(\nu)=n}\pi^\nu\alpha^\nu f_i^\nu\right)$$
$$=\sum_{n\in\mathcal{N}\setminus\Omega}\left(\pi^n\sum_{i\in I}y_i^n\sum_{\nu\in\mathcal{N}|\gamma(\nu)=n}\frac{\pi^\nu}{\pi^n}\alpha^\nu f_i^\nu\right)=\sum_{n\in\mathcal{N}\setminus\Omega}\left(\pi^n\sum_{i\in I}y_i^n\,\mathbb{E}[\boldsymbol{F}_i^n]\right). \tag{12.31}$$

$\boldsymbol{F}_i^n$ represents the (discounted) operational cost for facility $i\in I$ when we decide to operate it in the period starting at node n.

In the above model, the objective function (12.22) accounts for the total expected cost for the entire planning horizon, which has several components: (i) the operation cost for the facilities, (ii) the transportation costs, and (iii) procurement, production, and storage costs. Note that when a decision is made for setting a facility operating in some period, we still do not know the corresponding operational cost, which explains the expected value observed in (12.31). On the other hand, demand satisfaction is accomplished at the end of the period. Thus, we already know the corresponding costs and values for the demand. This explains the second term of the objective function (12.22). Constraints (12.23) ensure flow conservation in all periods and scenarios. Inequalities (12.24) concern the capacity of the facilities in every period and scenario. Constraints (12.25) ensure the phasing out for initially existing facilities (once closed, they cannot be open again). Constraints (12.26) guarantees that once a new facility is open, it should remain so until the end of the planning horizon. Inequalities (12.27) guarantee that in every period and scenario, we can only observe activities (procurement, production, transportation, or storage) in facilities that are operating. Finally, (12.28)–(12.30) define the domain of the decision variables.

Note that we allow a previously existing facility to never operate during the planning horizon. Additionally, no facility removal is considered at the end of the last period since it would not impact the optimal value of the objective function.

The above model raises a relevant issue in practice: The number of scenarios can be large, which is quite likely given that we are considering a multi-period stochastic problem—the scenario tree quickly grows beyond tractability. As pointed out in [491], even building the scenario tree may be problematic. Nevertheless, the above model can always be set at the core of a sampling approach, thus contributing to finding approximate solutions to the problem.

Another remark about the model concerns the capacity of the facilities. These parameters can also be a source of uncertainty. This fact is taken into account in the next section.

Particular cases or extensions of the modeling frameworks discussed so far have been considered in the literature for hedging against uncertainty in the strategic supply chain network design.

Kiya and Davoudpour [311] investigate a warehouse network re-design problem that includes integrating or eliminating existing warehouses and establishing new sites. Uncertainty is assumed for demand levels and operational costs. A two-stage stochastic programming problem is obtained. A Sample Average Approximation (SAA) algorithm is developed that is intertwined with Benders decomposition. The reported results highlight the value of the stochastic solutions in the investigated setting.

Escudero and Monge [199] investigate a two-echelon multi-stage stochastic supply chain network design problem. The facilities include manufacturing plants and distribution centers (DCs). Multiple products are considered that can be produced at the plants. The uncertainty stems from several costs (i) building the plants and the DCs, (ii) initial and additional capacity modules at the plants over time, (iii) the residual value of the plants and DCs, (iv) raw material supply, (v) transportation to and manufacturing at the plants, (vi) processing coefficients of the raw material, (vii) the cost of plant maintenance, and (viii) transportation from the plans to the DCs. Uncertainty is also considered for the demand at the DCs. The above uncertainty is captured using a so-called strategic multi-stage scenario tree with operational two-stage scenario trees. There is an underlying multi-stage scenario tree capturing the uncertainty related to strategic parameters (i, ii, and iii above). Then at each node of that tree, a two-stage scenario tree is considered for capturing uncertainty related to operational parameters (iv–viii above). The motivation is realistic: The strategic decisions should not depend on how individual operational uncertainty was revealed in the previous stages. On the other hand, strategic decisions should look ahead to the realizations of the strategic uncertain parameters in the stages and to the possible realizations of the operational parameters in the following stages. Hence, when working with a multi-stage scenario tree, one should consider that the "strategic" nodes in the tree should not be successors of the individual operational scenarios.

The strategic decision variables in this problem concern the location of plants and DCs and the modular capacity expansion at the plants over the planning horizon. The operational decision variables regard raw material supply, transportation to and manufacturing at the operational plants, the product volume and its transportation to the DCs, and the product demand shortfall at the DCs.

The problem is tackled heuristically using a procedure based on Fix-and-Relax (see [201] and the content of Sect. 6.4).

12.3 Supply Chain Risk Analytics: Toward Resilience

It is a common understanding that uncertainty leads to risk. Nevertheless, usually, risk is perceived as financial risk. This is also the case in supply chain management, given that large financial amounts are involved in building and managing the supply chain. As pointed out in [493], financial decisions may strongly interact with supply chain planning. Structuring and managing a supply chain is often just part of a whole set of activities associated with a company. Accordingly, the investments in the supply chain must be integrated with other profitable investments. The assessment of the investments made in a supply chain is usually based on their return rate. This suggests including revenues in optimization models used for the supply chain network design. In turn, this gives rise to the possibility of setting a target on the Return On Investment (ROI). Aligned with the above views, an optimization model for embedding financial decisions in strategic supply chain planning is presented in [404, 405]. Next, we review this model, although reformulating it consistently with the contents of this volume.

As in the previous section, a multi-period supply chain network design problem is considered. Two echelons are adopted, comprising facilities (e.g., distribution centers) and customers. In each period, customers demand multiple products, which should be supplied from the operating facilities respecting their capacities. The capacities are assumed to be time-dependent (e.g., a technology change during the planning horizon). Moreover, there is a minimum throughput for using the facility. This is also time-dependent. It is possible to make direct investments in the supply chain (e.g., set facilities operating) and other alternative investments.

Uncertainty is associated with demands and return rates. It is assumed to be captured by a scenario tree, like in the previous section. To capture potentially interesting investments, loans are also possible. Note that a loan can be seen as an investment (with a negative return).

A measure is considered for the service level. In fact, by including revenues related to the supply chain activities and investments in alternative activities, it may no longer be "profitable" to fully satisfy the demand. The actual service level is itself a consequence of the decisions to make. A target is set for the ROI. The risk of falling below it is measured and embedded in the decision-making process.

The decisions to make are the following: (i) location of the facilities, (ii) flow of commodities, and (iii) investments to make in alternative activities to those directly related to the supply chain design.

We adopt the notation of the previous section. Nevertheless, additional notation is required; other can be simplified. Given the two-echelon network we are considering, we use again I to denote the set of potential locations for the facilities and J the set of customers. The notation related to the scenario tree is the same as in the previous section. Nevertheless, now we also consider

$\Gamma(n)$, set of all nodes in the unique path connecting node 1 to node $n \in \Omega$ (including 1 and n)

Regarding the parameters, we consider the fixed costs f_{it} deterministic and thus known beforehand. Additionally, we consider the following deterministic parameters:

- q_{it}, capacity of a facility operating at $i \in I$ in period $t \in T$;
- $\underline{q}_{it}$, minimum throughput of a facility operating at $i \in I$ in period $t \in T$;
- a_p, unit capacity consumption for product $p \in P$;
- β_j, weight/importance of customer $j \in J$;
- ROI, target for the return on investment (e.g., 0.03=3%);
- g_{ijpt}, unit profit for the product $p \in P$ shipped from facility $i \in I$ to customer $j \in J$ in period $t \in T$. This profit results, for instance, from subtracting the unit transportation costs from the unit revenues. Furthermore, to help simplify the model, we denote $G_{ijpt} = g_{ijpt}(1+ROI)^{|T|-t}$, i.e., the profits are reported to the end of the planning horizon using the ROI as a target rate.
- B_t, exogenous budget available at the beginning of period $t \in T$. To help when writing the model, we denote $B = \sum_{t \in T} B_t(1 + ROI)^{|T|-t+1}$. B represents the total budget throughout the planning horizon with the values reported to the end of the planning horizon using the target ROI.
- K_t, set of all investments available at the beginning of period $t \in T$. Loans are included in this set.
- ϱ, weight for the downside risk

The stochastic parameters involved in the problem are:

- d^n_{ijp}, demand of customer $j \in J$ for product $p \in P$ in period $\rho(n)$ given that node $n \in \mathcal{N} \setminus \{1\}$ occurs;
- ζ^n_κ, interest rate of an investment $\kappa \in K_\tau$, $\tau \leq \rho(n)$ that is achieved (or has to be paid in case of a loan) at the end of period $\rho(n)$ given than node n occurs. Note that due to the definition, ζ^n_κ may be positive or negative (the latter for loans).

We keep using decision variables y^n_i, $n \in \mathcal{N} \setminus \Omega$ as well as variables x^n_{ijp}, $i \in I$, $j \in J$, $p \in P$, $n \in \mathcal{N} \setminus \{1\}$. Additionally, we consider w^n_κ standing for the amount of money to spend (or obtain in case of a loan) in investment $\kappa \in K_{\sigma(n)}$ at the beginning of period $\sigma(n)$ if node $n \in \mathcal{N} \setminus \{\Omega\}$ occurs. Finally, we consider two sets of auxiliary variables: For each $n \in \Omega$, δ^n stands for the downside risk if node n occurs (each scenario induces a downside risk). For each $n \in \Omega$, α^n_j denotes the service level achieved for customer $j \in J$ if scenario n occurs. In Fig. 12.2, we illustrate the decisions made in each node of the scenario tree. We do not explicitly

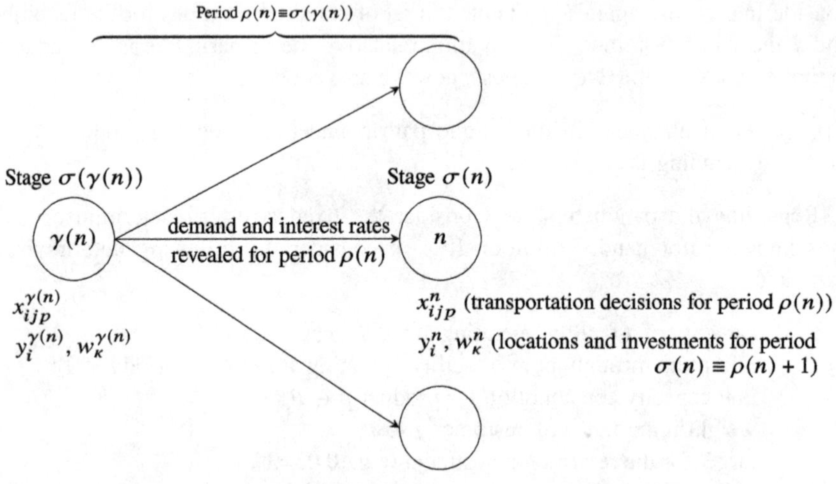

Fig. 12.2 Illustration of the decisions related to a node n of the scenario tree

represent the auxiliary variables because they are computed only at the leaves of the scenario tree.

The problem can be formulated as follows:

$$\text{maximize} \sum_{n\in\mathcal{N}\setminus\{1\}} \pi^n \left(\sum_{i\in I}\sum_{j\in J}\sum_{p\in P} G_{ijp,\rho(n)} x_{ijp}^n \right) + \sum_{n\in\Omega} \pi^n \left(\sum_{j\in J} \beta_j \alpha_j^n + \gamma\delta^n + \sum_{n'\in\Gamma(n)\setminus\{n\}} \sum_{\kappa\in K_{\sigma(n')}} \zeta_\kappa^n w_\kappa^{n'} \right), \tag{12.32}$$

$$\text{subject to} \sum_{i\in I} f_{i1} y_i^1 + \sum_{\kappa\in K_1} w_\kappa^1 \le B_1, \tag{12.33}$$

$$\sum_{i\in I} f_{i,\sigma(n)} y_i^n + \sum_{\kappa\in K_{\sigma(n)}} w_\kappa^n \le B_{\sigma(n)} + \sum_{n'\in\Gamma(n)\setminus\{n\}} \sum_{\kappa\in K_{\sigma(n')}} \zeta_\kappa^n w_\kappa^{n'} \quad \forall n \in \mathcal{N}\setminus\{1,\Omega\}, \tag{12.34}$$

$$\delta^n \ge 1 - \frac{\displaystyle\sum_{n'\in\Gamma(n)\setminus\{n\}} \sum_{\kappa\in K_{\sigma(n')}} \zeta_\kappa^n w_\kappa^{n'} + \sum_{n'\in\Gamma(n)\setminus\{1\}} \sum_{i\in I}\sum_{j\in J}\sum_{p\in P} G_{ijp,\rho(n')} x_{ijp}^{n'}}{B} \quad \forall n \in \Omega, \tag{12.35}$$

$$\sum_{n' \in \Gamma(n) \setminus \{1\}} x_{ijp}^{n'} \geq \alpha_j^n \sum_{n' \in \Gamma(n) \setminus \{1\}} d_{jp}^{n'} \qquad \forall n \in \Omega,\ j \in J, \tag{12.36}$$

$$\sum_{p \in P} a_p \sum_{j \in J} x_{ijp}^n \leq q_{i,\rho(n)} y_i^{\gamma(n)} \qquad \forall n \in \mathcal{N} \setminus \{1\}, \tag{12.37}$$

$$\sum_{p \in P} a_p \sum_{j \in J} x_{ijp}^n \geq \underline{q}_{i,\rho(n)} y_i^{\gamma(n)} \qquad \forall n \in \mathcal{N} \setminus \{1\}, \tag{12.38}$$

$$\sum_{i \in I} x_{ijp}^n \leq d_{jp}^n \qquad \forall n \in \mathcal{N} \setminus \{1\},\ j \in J,\ p \in P, \tag{12.39}$$

$$y_i^n \in \{0, 1\} \qquad \forall i \in I,\ n \in \mathcal{N} \setminus \{\Omega\}, \tag{12.40}$$

$$x_{ijp}^n \geq 0 \qquad \forall i \in I,\ j \in J,\ p \in P,\ n \in \mathcal{N} \setminus \{1\}, \tag{12.41}$$

$$w_\kappa^n \geq 0 \qquad \forall n \in \mathcal{N} \setminus \{\Omega\},\ \kappa \in K_{\sigma(n)}, \tag{12.42}$$

$$0 \leq \alpha_j^n \leq 1 \qquad \forall n \in \Omega,\ j \in J, \tag{12.43}$$

$$\delta^n \geq 0 \qquad \forall n \in \Omega. \tag{12.44}$$

Nickel et al. [405] show that a general-purpose solver can tackle the above model for small- to medium-sized instances. The authors consider up to three periods, five potential locations, 20 customer zones, eight investments, three loans, and 729 scenarios (leaves in the scenario tree). The value of the multi-stage stochastic solution for the above problem can be computed following the contents of Sect. 6.4.

The above model seeks a comprehensive quantification of financial risk in supply chain management. Nevertheless, in recent years, much debate can be found about the exact meaning of risk in SCM and its quantification in optimization models. Heckmann et al. [265] present an overview of this analysis. In the same work, the authors propose an innovative definition of risk in SCM:

Definition 12.1 Supply chain risk is the time-dependent potential loss for a supply chain in terms of its target values of profitability and functionality evaluated by the decision-maker's nature and evoked by uncertain changes of the affected supply chain and its processes whose changes were caused by occurring triggering events.

This definition highlights four core elements: time dependency, risk objectives, the nature of the decision-maker, and risk exposition. An outcome of the analysis conducted in [265] is a new concept: supply chain risk analytics.

Definition 12.2 Supply chain risk analytics is a bundle of mathematical methods and measurement techniques tailored for determining risk-aware solutions for supply chain design planning and execution.

Next, we discuss a model that materializes these concepts. The starting point is, again, the capacitated facility location problem that we directly extend to a multi-period setting and demand uncertainty. The latter is assumed to be captured by a finite set of scenarios. The problem can be formulated as follows:

$$\text{minimize} \quad \sum_{i\in I} f_i y_i + \sum_{\omega\in\Omega} \pi_\omega \left(\sum_{t\in T} \sum_{i\in I} \sum_{j\in J} (c_{ij} - r_j)\, d_{jt\omega} x_{ijt\omega} \right), \tag{12.45}$$

$$\text{subject to} \quad \sum_{i\in I} x_{ijt\omega} \le 1 \qquad \forall j \in J,\ t \in T,\ \omega \in \Omega, \tag{12.46}$$

$$\sum_{j\in J} d_{jt\omega} x_{ijt\omega} \le q_i y_i \quad \forall i \in I,\ t \in T,\ \omega \in \Omega, \tag{12.47}$$

$$x_{ijt\omega} \ge 0 \qquad \forall i \in I,\ j \in J,\ t \in T,\ \omega \in \Omega, \tag{12.48}$$

$$y_i \in \{0, 1\} \qquad \forall i \in I. \tag{12.49}$$

In the above model, most elements should be clear, given the contents already discussed in this chapter. We are facing a two-stage stochastic programming problem. The first stage seeks facility location. The second stage involves adaptive transportation decisions. A single product is considered. In the objective function, c_{ij} is the unit transportation cost between facility $i \in I$ and customer $j \in J$, whereas r_j is the unit revenue obtained from selling the product to customer $j \in J$. Given that revenues are at stake, a complete demand supply may not be desirable, which explains the inequality stated in Constraints (12.46). Then, we have capacity constraints at the facilities. Finally, we find the domain constraints.

This model captures supply chain efficiency because the objective function represents the total expected cost. Furthermore, the model embeds time dependency and uncertainty. However, it misses an important element: effectiveness. The latter is related to the ability of the supply chain to fulfill its function—to supply the customers according to their demand.

A way to account for effectiveness is to consider service level as a decision to make and then to use the objective function to penalize any shortage w.r.t. a minimum threshold defined by the decision-maker—the target service level. To extend the model, we introduce some additional notation, namely $\alpha^0 \in [0, 1]$ denoting the target service level, and h standing for the unit cost for shortage service level shortage (w.r.t. target). The additional decision variables include α_ω representing the service level achieved under scenario $\omega \in \Omega$, and Δ_ω is the service level shortage (w.r.t. the target value) under scenario $\omega \in \Omega$ and $u_{jt\omega}$ the proportion of unsupplied demand for customer $j \in J$, in period $t \in T$ under scenario $\omega \in \Omega$.

Using the above notation, for each $\omega \in \Omega$ we have

$$\alpha_\omega = 1 - \frac{\sum_{j\in J}\sum_{t\in T} d_{jt\omega}u_{jt\omega}}{\sum_{j\in J}\sum_{t\in T} d_{jt\omega}} \quad \text{and} \quad \Delta_\omega = \max\left\{0, \alpha^0 - \alpha_\omega\right\}.$$

For $\omega \in \Omega$, $\alpha_\omega \in [0, 1]$, with 0 indicating that no demand is supplied and 1 indicating that all demand is supplied. A more comprehensive model can now be considered:

$$\text{minimize} \quad \sum_{i\in I} f_i y_i + \sum_{\omega\in\Omega} \pi_\omega \left(h\,\Delta_\omega + \sum_{t\in T}\sum_{i\in I}\sum_{j\in J} (c_{ij} - r_j)\, d_{jt\omega} x_{ijt\omega} \right), \tag{12.50}$$

$$\text{subject to} \quad \sum_{i\in I} x_{ijt\omega} + u_{jt\omega} = 1 \qquad \forall j \in J,\ t \in T,\ \omega \in \Omega, \tag{12.51}$$

$$\Delta_\omega \geq \alpha^0 - \left(1 - \frac{\sum_{j\in J}\sum_{t\in T} d_{jt\omega}u_{jt\omega}}{\sum_{j\in J}\sum_{t\in T} d_{jt\omega}}\right) \qquad \forall \omega \in \Omega, \tag{12.52}$$

(12.47)–(12.49),

$$u_{jt\omega} \geq 0 \qquad \forall j \in J,\ t \in T,\ \omega \in \Omega, \tag{12.53}$$

$$\Delta_\omega \geq 0 \qquad \forall \omega \in \Omega. \tag{12.54}$$

When capacity becomes tight, we must decide about the customers to serve. In the model, the unit revenues steer that decision. Note that a customer may be served by multiple facilities, which is of particular interest, when, for instance, disruptions prevent the supply from a major facility (see [27]).

At first glance, the extended model presented seems to indicate that we are considering a risk-neutral decision-maker. The objective function measures the expected value of the future assets. However, a closer look shows that we can do more than that. A risk-aware optimization model for the supply chain network design must seek a trade-off between customer satisfaction and cost. This suggests the use of a multicriteria optimization model. However, looking closely into the model, we realize this is being accomplished through scalarization. The trade-off is assured and "regulated" by parameter h. By changing it, we can capture the extent to which the decision-maker is willing to invest to get more operational capacity to increase the value of customer satisfaction. A decision-maker with a higher level of risk aversion concerning investments in the supply chain chooses smaller values for h. Accordingly, the objective function allows the definition of different risk profiles for decision-makers regarding financial risk. This analysis is still valid (and even better highlighted) when other costs are included in the objective function, which we consider next.

As stated in Definition 12.1, risk exposition depends on the impact of disruptive triggers in the supply chain structures. The most common form of disruption in a supply chain concerns its capacity for fulfilling demand. Concerning the facility location problem, this corresponds to a capacity reduction, thus affecting constraints (12.47).

Let us assume that we can identify a set of scenarios in terms of one or several disruptive triggers and their impact on the operating capacity of the facilities. If we combine each of these with each scenario already in Ω, we get an extended set of scenarios, each determining all the uncertainty parameters. For this reason, w.l.o.g., we keep using the notation Ω for the enlarged set of scenarios. In particular, each scenario now specifies a capacity reduction (in $[0, 1]$) for each facility in each period. We need one extra set of parameters: $\gamma_{it\omega}$ stands for the proportion of the operational capacity of facility $i \in I$ in period $t \in T$ that is available under scenario $\omega \in \Omega$. Thus, $\gamma_{it\omega} \times q_i$ becomes the actual capacity of facility $i \in I$ in period $t \in T$ under scenario $\omega \in \Omega$.

A disruptive trigger can also affect the demand (either by increasing or by decreasing it). Moreover, we may face a situation in which a combination of disruptive triggers leads to simultaneous changes in the operating capacity and the demand. The notation adopted before for the demand can already capture this situation since we use scenario-indexed parameters.

Also related to the risk exposition is the affected supply chain and its resilience. In this case, additional decisions may be required to ensure resilience, i.e., bringing the system back to a status of efficiency and effectiveness. This can be incorporated in our modeling framework by considering temporary capacity expansions associated with extra capacity options (see, e.g., [551] for supplier selection) and thus corresponding to a here-and-now decision. Preparedness measures adequate to hedge against a disruption call for a corporation to decide in advance about the possibilities that may need to be activated to mitigate sudden changes in the underlying setting. Such decisions are typically part of a contingency plan and thus are identified in advance. On the other hand, the specific decisions concerning when and at which level to activate an option are recourse decisions calling for a more detailed course of action that often can be made only after a disruption occurs, i.e., the actual future scenario becomes known.

We define one extra set L, containing the capacity expansion levels available; each level determines a temporary increase in the operational capacity of a facility. Additional parameters are also required: g_i is the fixed cost associated with an option contract for facility $i \in I$ to ensure a temporary capacity increase if necessary, b_ℓ stands for the unit capacity cost associated with capacity expansion level $\ell \in L$, and k_ℓ represents the amount of extra capacity associated with expansion level $\ell \in L$.

The extended model makes use of the following two additional sets of binary variables: z_i is a binary variable equal to 1 if and only if a capacity expansion option is contracted for facility $i \in I$, and $w_{it\ell\omega}$ is also binary and equal to 1 if and only if in scenario $\omega \in \Omega$ expansion level $\ell \in L$ is installed at facility $i \in I$ in period $t \in$

T. We can now formulate a risk-aware multi-period stochastic capacitated facility location problem:

$$\text{minimize} \quad \sum_{i\in I}(f_i y_i + g_i z_i) + \sum_{\omega\in\Omega}\pi_\omega \left[(h\,\Delta_\omega + \sum_{i\in I}\sum_{\ell\in L}\left(b_\ell k_\ell \sum_{t\in T} w_{it\ell\omega}\right) \right.$$

$$\left. + \sum_{t\in T}\sum_{i\in I}\sum_{j\in J}\left(c_{ij} - r_j\right) d_{jt\omega} x_{ijt\omega} \right], \tag{12.55}$$

$$\text{subject to} \quad \sum_{i\in I} x_{ijt\omega} + u_{jt\omega} = 1 \qquad \forall j \in J,\ t\in T,\ \omega\in\Omega, \tag{12.51}$$

$$\sum_{j\in J} d_{jt\omega} x_{ijt\omega} \le \gamma_{it\omega}\, q_i\, y_i + \sum_{\ell\in L} k_\ell w_{it\ell\omega} \qquad \forall i\in I,\ t\in T,\ \omega\in\Omega, \tag{12.56}$$

$$z_i \le y_i \qquad \forall i\in I, \tag{12.57}$$

$$\sum_{\ell\in L} w_{it\ell\omega} \le z_i \qquad \forall i\in I,\ t\in T,\ \omega\in\Omega, \tag{12.58}$$

$$\Delta_\omega \ge \alpha^0 - \left(1 - \frac{\sum_{j\in J}\sum_{t\in T} d_{jt\omega} u_{jt\omega}}{\sum_{j\in J}\sum_{t\in T} d_{jt\omega}}\right) \qquad \forall\omega\in\Omega, \tag{12.52}$$

$$x_{ijt\omega} \ge 0 \qquad \forall i\in I,\ j\in J,\ t\in T,\ \omega\in\Omega, \tag{12.48}$$

$$y_i \in \{0,1\} \qquad \forall i\in I, \tag{12.49}$$

$$u_{jt\omega} \ge 0 \qquad \forall j\in J,\ t\in T,\ \omega\in\Omega, \tag{12.53}$$

$$\Delta_\omega \ge 0 \qquad \forall\omega\in\Omega, \tag{12.54}$$

$$z_i \in \{0,1\} \qquad \forall i\in I, \tag{12.59}$$

$$w_{it\ell\omega} \in \{0,1\} \qquad \forall i\in I,\ t\in T,\ \ell\in L,\ \omega\in\Omega. \tag{12.60}$$

In the above model, the first-stage decision includes expansion options to contract and the facilities that should operate. The recourse decisions include unsupplied demand, capacity expansion, and demand satisfaction. We seek to minimize the total expected cost. In addition to the constraints introduced before, we consider now also constraints (12.56) ensuring that even with the temporary capacity expansions, we cannot satisfy more demand than the actual operating capacity (in every facility, period, and scenario). Constraints (12.57) ensure that options can only be contracted for operating facilities, whereas Constraints (12.58) guarantee that operating capacity can only be expanded if a contract option was previously established. Finally, constraints (12.59) and (12.60) are the domain of the decision variables associated with capacity expansion.

Although uncertainty is an "ingredient" for risk, it is not the only one. For this reason, the Value of the Stochastic Solution (VSS) and the Expected Value of Perfect Information (EVPI) are not appropriate for assessing the relevance of a modeling framework such as the one presented above.

The need for a risk-aware supply chain design if of relevance if hedging against risk has some expected "positive value" quantifying it. Therefore, we must be able to measure such a value. Suppose we know in advance the exact scenario, say ω, we will be facing. In that case, the single-scenario model induced by the above formulation is inappropriate. The single-scenario problems induced by the above model do not represent a non-risk-aware counterpart. What is more, the solution produced by a single scenario (e.g., stemming from replacing the random variables by their expectation) may turn out to be infeasible for some actual scenarios (disruption profiles). Additionally, under a single-scenario, risk-free setting, it makes no sense to contract options for expanding the capacity since that feature stems from the uncertainty underlying the problem. Moreover, in a deterministic setting, the service level is no longer uncertain; it directly results from the decision to make. Therefore, setting a target value for it does not make sense either.

Heckmann et al. [265] suggest that an adequate deterministic (non-risk-aware) counterpart should be an extension of the single-scenario version of model (12.45)–(12.49) that accounts for unsupplied demand. This way, it becomes possible to evaluate a solution induced by some scenario rather than simply concluding that for some other scenarios, it is infeasible. The authors propose using the following model for each scenario $\omega \in \Omega$ as a first step in finding a measure of the relevance for a risk-aware solution:

$$\text{minimize} \quad \sum_{i \in I} f_i y_{i\omega} + \sum_{t \in T} \sum_{j \in J} \hat{h} d_{jt\omega} u_{jt\omega} + \sum_{t \in T} \sum_{i \in I} \sum_{j \in J} (c_{ij} - r_j) d_{jt\omega} x_{ijt\omega}, \tag{12.61}$$

$$\text{subject to} \quad \sum_{i \in I} x_{ijt\omega} + u_{jt\omega} = 1 \qquad \forall j \in J,\ t \in T, \tag{12.62}$$

$$\sum_{j \in J} d_{jt\omega} x_{ijt\omega} \leq \gamma_{it\omega} q_i y_i \qquad \forall i \in I,\ t \in T, \tag{12.63}$$

$$x_{ijt\omega} \geq 0 \qquad \forall i \in I,\ j \in J,\ t \in T, \tag{12.64}$$

$$y_{is} \in \{0, 1\} \qquad \forall i \in I. \tag{12.65}$$

For each $\omega \in \Omega$, we denote by V_ω the optimal value of the corresponding problem and by V the optimal value of the full risk-aware model presented.

Definition 12.3 The value of a risk-aware supply chain solution is the difference between the optimal value of the risk-aware model and the weighted value of the models induced by every scenario, i.e.,

$$V - \sum_{\omega \in \Omega} \pi_\omega V_\omega.$$

This definition is in some sense similar to the EVPI. However, the differences are clear. First, the way this value is computed is not the same as for the EVPI since the single-scenario problems do not result from considering one scenario in the original stochastic model. Second, this measure can, in principle, be computed for any risk-aware supply chain network design model.

The model we have been discussing materializes a concept in modern supply chain planning: resilience. This is the capability of a system to hedge against disruptions by anticipating them or, if this is not possible, then to recover from them in the most cost- and time-efficient way [265, 282, 296, 431, 580]. The complexity of the modern supply chains has increased due to many aspects. Uncertainty is one of them. In particular, due to the globalization observed in the past decades, supply chains are more vulnerable to disruptions, namely those triggered by natural disasters such as earthquakes or floods. In turn, a disruption may easily propagate, which is a concept relatively new [48, 176].

Although a prototype, the above model shows that it is possible to prevent and control disruptions and efficiently return to the *status quo*. In general, the strategies for designing resilient supply chains include (i) multiple sourcing [283], (ii) contracting with backup facilities [477] (2021), (iii) fortification/protection of facilities [265], (iv) capacity expansion [265], (v) development of contingency plans [207], (vi) fortification [571], *et cetera*.

Regarding disruption propagation quantification, we have observed an increasing trend in the related literature. It is not the purpose of this volume to discuss that context in detail. Nevertheless, the interest reader can refer to [229, 275, 276, 332, 334, 500].

12.4 Closing the Loop: Toward Sustainability

Reverse logistics comprehends the activities associated with collecting, recovering, or treating product returns or disposals. This type of activity is often motivated by three aspects: (i) economic revenue by recapturing value from used products, (ii) environmental concerns (e.g., the well-known European Union WEEE Directive) encouraging sustainability, and (iii) consumer pressure (e.g., return of defective products).

Reverse logistics activities are supported by specific facilities such as collection, inspection, treatment, and recovery centers. The resulting networks often involve multiple facility layers (see, e.g., [20, 264]). One aspect of relevance in many cases is the semi-obnoxious nature of the reverse facilities. On the one hand, no population wants such facilities nearby; on the other hand, these facilities should be located close to the populations they serve to lower the transportation costs. Therefore, a compromise must be reached.

Uncertainty is inherent to a reverse logistics planning problem, namely the amount of returned products (see, e.g., [212, 530]). Next, we discuss a prototype model from a facility location perspective, seeking strategic reverse logistics planning and featuring various facility echelons, multiple commodities, technology selection, stochasticity, and several objectives. This prototype can be easily adjusted for more complex situations. Moreover, it can also be combined with forward logistics activities, as discussed in the previous sections.

We consider a general framework with a set of origins generating different types of disposal. We focus on waste and end-of-life products (disposals). Such type of commodity can either be collected in a collection center and then sent to a recovery center or can be sent directly to a recovery center. Collection centers may be convenient for several reasons. For example, some products may be disassembled so that part of them is sent to the recovery center and the rest adequately treated. We assume in this discussion that collection centers are facilities where disposal is compacted. The resulting disposal volume decrease leads to lower transportation costs, which may offset the costs associated with collection centers.

Once treated and separated in a recovery center, each type of disposal can lead to what we call *valuable commodities* or to a residual disposal that will be sent to a landfill. Valuable commodities can be sold in the market, e.g., glass obtained from a recovery center can be sold to a related manufacturing factory. A particular case of a valuable commodity is material to be incinerated, thus being converted into energy.

Often, in practice, there are both minimum and maximum conversion rates. In particular, depending on the context, there may be a maximum amount for each valuable commodity beyond which we cannot sell it to the market, profitable though it may be. Likewise, maximum values for incineration may be imposed by law. Besides, the current legislation often imposes minimum conversion rates from every waste product into every valuable commodity.

The prototype problem we discuss next emerges in the context of a so-called *management integrated system*, i.e., a company or an organization is in charge of the whole system. Nevertheless, the discussion is general enough to allow immediate application to other settings.

We assume a given set of locations where collection and recovery centers can be installed or selected (if already installed). A collection center can be geographically close to a recovery center (e.g., a storage area in the latter) and can receive any disposal. A recovery center can deal with any disposal as far as the appropriate technology is installed. We assume that different types of technology are available for each type of disposal (e.g., manual, semi-automatic, and automatic). Thus, for each selected recovery center, we must decide the types of disposal to be treated and the technology that will be used. Different technologies typically offer different conversion rates and costs.

Finally, we consider a set of potential landfills. The residual disposal that cannot be further processed and the valuable commodities that cannot be sold should be sent to the selected landfills, which we assume uncapacitated.

We consider a fixed cost for installing/selecting a collection center irrespective of the disposals to be treated there. This may include land acquisition, labor costs, *et cetera*. Besides, for each type of disposal that will flow through a collection center, we consider a setup cost for preparing the collection center since we may need specific equipment (mainly presses) for each waste product.

Regarding the recovery centers, we consider a setup cost irrespective of the disposals to be treated there, such as land acquisition or meeting law requirements to operate. For each type of disposal, a fixed cost is assumed for each technology available to process it, regardless of the capacity of that technology. The capacity to treat each product with each technology is modular: It is measured by the number of modules installed (for example, that may refer to having one line or two lines to separate products). A cost is considered for each module of each technology. In a recovery center, we assume that only one technology can be chosen for each type of disposal.

In addition to the costs associated with the facilities, we consider transportation costs (i) from the origins to the facilities (collection and recovery centers), (ii) from the collection centers to the recovery centers, and (iii) from the recovery centers to the landfills.

Unitary transportation costs depend on the types of disposal since this influences some aspects such as volume, risk management procedures, and weight. Transportation costs are lower if products are compacted. Finally, we assume that the transportation costs between two points in the underlying network result from the shortest road distance and the unit transportation cost. Figure 12.3 depicts typical flows involved in a reverse network like the one we are considering. One aspect to point out in this figure is its convergent nature. Two objectives emerge in the

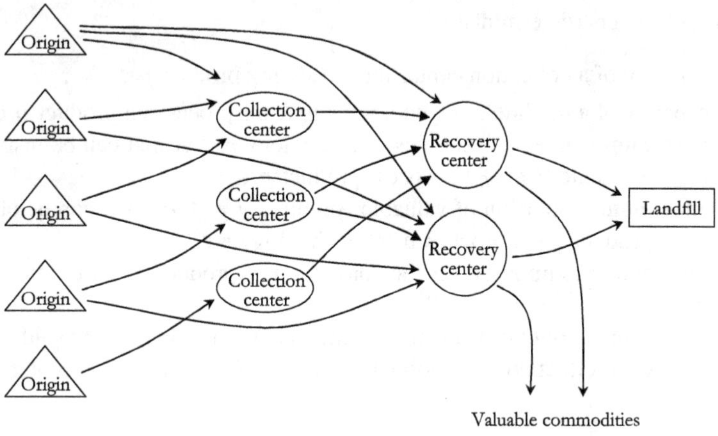

Fig. 12.3 Reverse logistics flows

problem: the total (expected) cost and the obnoxious effect of the installed/selected facilities.

Obnoxious effects are typically a function of the Euclidean distance because undesirable effects like odors and noise *do not travel by road*—the closer in terms of the Euclidean distance a population is to a facility, the more the effect is felt. Different functions of the Euclidean distance can be considered, which usually depend on the specific type of products flowing in the reverse network. The interested reader can refer to [130] for many details on the location of this type of facility.

The obnoxious effect can be mitigated by compensating the affected populations. For instance, a population may be more willing to have some facility in their geographical area if this entails a tax reduction or impacts (positively) on employment. This aspect can also be included in the model. We also assume that the obnoxious effect is weighted by correcting factors that reflect specific aspects associated, such as the type of disposal treated and the weight of the population affected. Finally, we assume that landfills have a higher obnoxious effect than recovery centers. The obnoxious effects in a recovery center are assumed to be additive. The same is considered for the incentives mentioned. Uncertainty is assumed both for the amounts at the origins and for the transportation costs.

In synthesis, we face a multi-echelon, multi-product, bi-objective stochastic location problem with technology choice. We consider the following notation:
Sets:

J, origins for the returning products;
I_C, potential locations for collection centers;
I_R, potential locations for recovery centers;
I_L, potential locations for landfills;
P_W, disposal types;
P_F, valuable commodities;
R, set of different technologies available for the recovery centers.

General parameters (deterministic):

q_p^C, capacity of a collection center for processing product $p \in P_W$;
q_{pr}^R, capacity of a module of technology $r \in R$ for processing product $p \in P_W$;
n_{ipr}, maximum number of modules of technology $r \in R$ that can be installed at recovery center $i \in I_R$ to process product $p \in P_W$;
a_{pqr}, maximum proportion of valuable commodity $p \in P_F$ that can be obtained from product $q \in P_W$ when using technology $r \in R$;
μ_{pq}, minimum rate imposed for the conversion of product $q \in P_W$ into product $p \in P_F$;
s_p, maximum amount of valuable commodity $p \in P_F$ that can be sold;
o_{ip}^C, obnoxious effect due to sending product $p \in P_W$ to collection center $i \in I_C$;

o_{ip}^R, obnoxious effect due to sending product $p \in P_W$ to recovery center $i \in I_R$;
o_i^L, obnoxious effect due to sending the final residual to landfill $i \in I_L$;
$l_{kk'}$, road distance between nodes k and k', with $k, k' \in J \cup I_C \cup I_R \cup I_L$ (according to the transportation channels depicted in Fig. 12.3).

In addition to the Euclidean distances mentioned above, other factors contribute to the obnoxious effects, namely the populations involved and the type of disposal considered. We can define the following additional notation:

α_p^C, impact of processing product $p \in P_W$ in a collection center;
α_p^R, impact of processing product $p \in P_W$ in a recovery center;
α^L, impact caused by a landfill;
β_{ij}^C, reduction factor $\in (0, 1]$ on the obnoxious effect due to the possible benefits for population $j \in J$ for locating/selecting collection center in $i \in I_C$;
β_{ij}^R, reduction factor $\in (0, 1]$ on the obnoxious effect due to the possible benefits for population $j \in J$ for locating/selecting recovery center in $i \in I_R$;
β_{ij}^L, reduction factor $\in (0, 1]$ on the obnoxious effect due to the possible benefits for population $j \in J$ for locating/selecting landfill in $i \in I_L$;
γ_{jp}, joint weight/importance associated with origin $j \in J$ and disposal $p \in P_W$;
γ_j^L, joint weight/importance associated with origin $j \in J$ and the residual product (sent to a landfill);

Parameters α_p^C, α_p^R, and α^L convey the idea that the obnoxious effect depends on the type of facility. Moreover, the obnoxious effect produced by recovery and collection centers depends on the type of waste they process.

Regarding the reduction of the obnoxious effect due to possible benefits from having facilities nearby, we assume $\beta_{ij} = 1$ if location i is not in the geographical area (e.g., province or municipality) of population j and $\beta_{ij} < 1$ otherwise. Moreover, for each product and a facility location in the geographical area of two different populations, we assume $\gamma_{jp} < \gamma_{j'p} \Rightarrow \beta_{ij} < \beta_{ij'}$. The reasoning underlying this is twofold:

(i) The impact on a population can only be reduced if a facility is installed in its municipality area.
(ii) Two incentives are considered for reducing the obnoxious effect: tax reduction and a decrease in the unemployment rate. Tax reduction leads to a reduction in the obnoxious effect independent of the relative importance of the population. The decrease in the unemployment rate is more significant in smaller populations than in larger ones. Therefore, β_{ij} can depend to some extent on the weight of the population.

Using the information presented, the obnoxious effect can be defined as follows:

$$O_{ip}^{C} = \alpha_{p}^{C} \sum_{j \in J} \beta_{ij}^{C} \cdot \gamma_{jp} \cdot \varphi_{p}(\delta_{ij}) \quad \forall i \in I_C, \ p \in P_W,$$

$$O_{ip}^{R} = \alpha_{p}^{R} \sum_{j \in J} \beta_{ij}^{R} \cdot \gamma_{jp} \cdot \varphi_{p}(\delta_{ij}) \quad \forall i \in I_R, \ p \in P_W,$$

$$O_{i}^{L} = \alpha^{L} \sum_{j \in J} \beta_{ij}^{L} \cdot \gamma_{j}^{L} \cdot \varphi^{L}(\delta_{ij}) \quad \forall i \in I_L.$$

Functions $\varphi_p(.)$ and $\varphi^L(.)$ are typically inverse functions of the square or cube of the Euclidean distances. Nevertheless, other possibilities exist depending on the specific setting considered.

We proceed by introducing the notation related to the deterministic costs and profits.

f_i^C, fixed cost for selecting/installing collection center $i \in I_C$;
f_i^R, fixed cost for selecting/installing recovery center $i \in I_R$;
f_i^L, fixed cost for selecting/installing landfill $i \in I_L$;
g_p^C, fixed cost for preparing a collection center for receiving disposal $p \in P_W$;
g_{ipr}^R, fixed cost for installing a module of technology $r \in R$ in recovery center $i \in I_R$ for processing disposal $p \in P_W$;
b_p, unit profit for valuable commodity $p \in P_F$;
e, unit residual disposal tax.

As far as uncertainty is concerned, we assume it to be captured by a finite set of scenarios Ω and also that we know the probability π_ω of each scenario $\omega \in \Omega$ ($\pi_\omega > 0$, $\omega \in \Omega$; $\sum_{\omega \in \Omega} \pi_\omega = 1$). This allows casting the problem as a two-stage stochastic programming problem. Accordingly, we consider:
Stochastic parameters:

$d_{jp\omega}$, amount of product $p \in P_W$ originated at $j \in J$ under scenario $\omega \in \Omega$;
$c_{p\omega}^1$, cost (per product unit and km) for shipping product $p \in P_W$ from an origin to a collection or a recovery center under scenario $\omega \in \Omega$;
$c_{p\omega}^2$, cost (per product unit and km) for shipping product $p \in P_W$ from a collection center to a recovery center under scenario $\omega \in \Omega$;
c_ω^3, cost (per product unit and km) for shipping product from a recovery center to a landfill under scenario $\omega \in \Omega$;

Finally, we introduce the decision variables that allow us to formulate the problem mathematically.
First-stage decision variables:

$$y_i^C = \begin{cases} 1, & \text{if collection center } i \in I_C \text{ is selected/installed,} \\ 0, & \text{otherwise.} \end{cases}$$

$$y_i^R = \begin{cases} 1, & \text{if recovery center } i \in I_R \text{ is selected/installed,} \\ 0, & \text{otherwise.} \end{cases}$$

$$y_i^L = \begin{cases} 1, & \text{if landfill } i \in I_L \text{ is selected/installed,} \\ 0, & \text{otherwise.} \end{cases}$$

$$w_{ip}^C = \begin{cases} 1, & \text{if collection center } i \in I_C \text{ processes disposal } p \in P_W, \\ 0, & \text{otherwise.} \end{cases}$$

$$z_{ipr}^R = \begin{cases} 1, & \text{if recovery center } i \in I_R \text{ processes disposal } p \in P_W \text{ using} \\ & \text{technology } r \in R, \\ 0, & \text{otherwise.} \end{cases}$$

u_{ipr} = number of modules of technology $r \in R$ installed in recovery center $i \in I_R$ to treat disposal $p \in P_W$.

Second-stage decision variables:

$x_{ijp\omega}^{OC}$ = amount of disposal $p \in P_W$ sent from origin $j \in J$ to collection center $i \in I_C$ under scenario $\omega \in \Omega$.

$x_{ijp\omega}^{OR}$ = amount of disposal $p \in P_W$ sent from origin $j \in J$ to recovery center $i \in I_R$ under scenario $\omega \in \Omega$.

$x_{ii'p\omega}^{CR}$ = amount of disposal $p \in P_W$ sent from collection center $i \in I_C$ to recovery center $i' \in I_R$ under scenario $\omega \in \Omega$.

$x_{ii'\omega}^{RL}$ = amount of residual disposal obtained in recovery center $i \in I_R$ sent to landfill $i' \in I_L$ under scenario $\omega \in \Omega$.

$x_{ipqr\omega}^{RV}$ = amount of valuable commodity $p \in P_F$ obtained in recovery center $i \in I_R$ from product $q \in P_W$ using technology $r \in R$ under scenario $\omega \in \Omega$.

The amount of disposal is measured in different ways (e.g., number of units or tons) depending on its specificity. Figure 12.4 depicts the variables representing

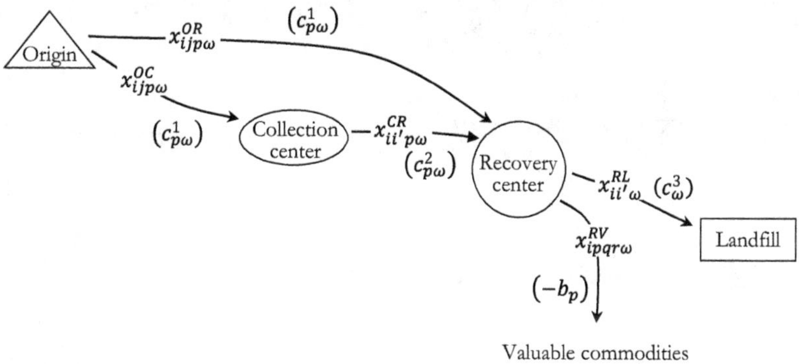

Fig. 12.4 Flow variables and the corresponding unitary costs

the flows in the reverse network. Inside parenthesis we find the corresponding unit costs. Note that the unit costs associated with the valuable commodities are negative because they represent revenue. We can finally present the extensive form of the deterministic equivalent of the reverse logistics network design problem under uncertainty that we have described:

$$
\begin{aligned}
\text{minimize} \quad & \sum_{i \in I_C} f_i^C y_i^C + \sum_{i \in I_R} f_i^R y_i^R + \sum_{i \in I_L} f_i^L y_i^L \\
& + \sum_{p \in P_W} \left(g_p^C \sum_{i \in I_C} w_{ip}^C \right) + \sum_{i \in I_R} \sum_{p \in P_W} \sum_{r \in R} g_{ipr}^R n_{ipr}^R \\
& + \sum_{\omega \in \Omega} \pi_\omega \Bigg[\sum_{j \in J} \sum_{i \in I_C} \sum_{p \in P_W} c_{p\omega}^1 l_{ij} x_{ijp\omega}^{OC} + \sum_{j \in J} \sum_{i \in I_R} \sum_{p \in P_W} c_{p\omega}^1 l_{ij} x_{ijp\omega}^{OR} \\
& + \sum_{i \in I_C} \sum_{i' \in I_R} \sum_{p \in P_W} c_{p\omega}^2 l_{ii'} x_{ii'p\omega}^{CR} + \sum_{i \in I_R} \sum_{i' \in I_L} c_\omega^3 l_{ii'} x_{ii'\omega}^{RL} \\
& + e \sum_{i \in I_R} \sum_{i' \in I_L} x_{ii'\omega}^{RL} - \sum_{i \in I_R} \sum_{p \in P_F} \sum_{q \in P_W} b_p \sum_{r \in R} x_{ipqr\omega}^{RV} \Bigg],
\end{aligned} \tag{12.66}
$$

$$
\text{minimize} \quad \sum_{i \in I_C} \sum_{p \in P_W} o_{ip}^C w_{ip}^C + \sum_{i \in I_R} \sum_{p \in P_W} o_{ip}^R \sum_{r \in R} z_{ipr}^R + \sum_{i \in I_L} o_i^L y_i^L, \tag{12.67}
$$

$$
\text{subject to} \quad w_{ip}^C \le y_i^C \qquad \forall i \in I_C,\ p \in P_W, \tag{12.68}
$$

$$
\sum_{r \in R} z_{ipr}^R \le y_i^R \qquad \forall i \in I_R,\ p \in P_W, \tag{12.69}
$$

$$
u_{ipr}^R \le n_{ipr} z_{ipr}^R \qquad \forall i \in I_R,\ p \in P_W,\ r \in R, \tag{12.70}
$$

$$
\sum_{i \in I_C} x_{ijp\omega}^{OC} + \sum_{i \in I_R} x_{ijp\omega}^{OR} = d_{jp\omega} \qquad \forall p \in P_W,\ j \in J,\ \omega \in \Omega, \tag{12.71}
$$

$$
\sum_{j \in J} x_{ijp\omega}^{OC} = \sum_{i' \in I_R} x_{ii'p\omega}^{CR} \qquad \forall i \in I_C,\ p \in P_W,\ \omega \in \Omega, \tag{12.72}
$$

$$
a_{pqr} \left(\sum_{j \in J} x_{ijq\omega}^{OR} + \sum_{i' \in I_C} x_{i'iq\omega}^{CR} \right) \ge x_{ipqg\omega}^{RV} \qquad \forall i \in I_R,\ q \in P_W,\ p \in P_F,\ r \in R,\ \omega \in \Omega, \tag{12.73}
$$

$$x^{RV}_{ipqr\omega} \le \left(\sum_{j\in J} d_{jq\omega}\right) z^{R}_{iqr} \qquad \forall i \in I_R,\ q \in P_W,\ p \in P_F,\ r \in R,\ \omega \in \Omega, \tag{12.74}$$

$$\sum_{i'\in I_L} x^{RL}_{ii'\omega} = \sum_{q\in P_W}\left(\sum_{j\in J} x^{OR}_{ijq\omega} + \sum_{i'\in I_C} x^{CR}_{i'iq\omega}\right) - \sum_{q\in P_W}\sum_{p\in P_F}\sum_{r\in R} x^{RV}_{ipqr\omega} \qquad \forall i \in I_R,\ \omega \in \Omega, \tag{12.75}$$

$$\sum_{j\in J} x^{OR}_{ijp\omega} + \sum_{i'\in I_C} x^{CR}_{i'ip\omega} \le \sum_{r\in R} q^{R}_{pr} u_{ipr} \qquad \forall i \in I_R,\ p \in P_W,\ \omega \in \Omega, \tag{12.76}$$

$$\sum_{i\in I_R}\sum_{r\in R} x^{RV}_{ipqr\omega} \ge \mu_{pq}\left(\sum_{j\in J} d_{jq\omega}\right) \qquad \forall q \in P_W,\ p \in P_F,\ \omega \in \Omega, \tag{12.77}$$

$$\sum_{i\in I_R}\sum_{q\in P_W}\sum_{r\in R} x^{RV}_{ipqr\omega} \le s_p \qquad \forall p \in P_F,\ \omega \in \Omega, \tag{12.78}$$

$$\sum_{j\in J} x^{OC}_{ijp\omega} \le q^{C}_{p} w^{C}_{ip} \qquad \forall i \in I_C,\ p \in P_W,\ \omega \in \Omega, \tag{12.79}$$

$$x^{RL}_{ii'\omega} \le \left(\sum_{j\in J}\sum_{p\in P_W} d_{jp\omega}\right) y^{L}_{i'} \qquad \forall i \in I_R,\ i' \in I_L,\ \omega \in \Omega, \tag{12.80}$$

$$y^{C}_{i} \in \{0,1\} \qquad \forall i \in I_C, \tag{12.81}$$

$$y^{R}_{i} \in \{0,1\} \qquad \forall i \in I_R, \tag{12.82}$$

$$y^{L}_{i} \in \{0,1\} \qquad \forall i \in I_L, \tag{12.83}$$

$$w^{C}_{ip} \in \{0,1\} \qquad \forall i \in I_C,\ p \in P_W, \tag{12.84}$$

$$z^{R}_{ipr} \in \{0,1\} \qquad \forall i \in I_R,\ p \in P_W,\ r \in R, \tag{12.85}$$

$$u_{ipr}^{R} \in \mathbb{Z}_0^+ \qquad \forall i \in I_R,\ p \in P_W,\ r \in R, \tag{12.86}$$

$$x_{ijp\omega}^{OC} \geq 0 \qquad \forall i \in I_C,\ j \in J,\ p \in P_W,\ \omega \in \Omega, \tag{12.87}$$

$$x_{ijp\omega}^{OR} \geq 0 \qquad \forall i \in I_R,\ j \in J,\ p \in P_W,\ \omega \in \Omega, \tag{12.88}$$

$$x_{ii'p\omega}^{CR} \geq 0 \qquad \forall i \in I_C,\ i' \in I_R,\ p \in P_W,\ \omega \in \Omega, \tag{12.89}$$

$$x_{ii'\omega}^{RL} \geq 0 \qquad \forall i \in I_R,\ i' \in I_L,\ \omega \in \Omega, \tag{12.90}$$

$$x_{ipqr\omega}^{RV} \geq 0 \qquad \forall i \in I_R,\ q \in P_W,\ p \in P_F,\ r \in R,\ \omega \in \Omega. \tag{12.91}$$

In the above formulation, the first objective function (12.66) represents the total cost associated with the first-stage decisions plus the expected cost of the second-stage decisions. The objective function (12.67) assesses the total obnoxious effect. Constraints (12.68) guarantee that disposals are collected only in installed collection centers. Constraints (12.69) ensure that disposals are transformed only in operating recovery centers and impose that at most one technology can be chosen in each recovery center for each disposal. Constraints (12.70) impose a consistent choice for the values of the u- and the z-variables. They also state that the number of modules of a particular technology installed in a recovery center cannot exceed a determined maximum value. For each type of disposal and each origin, Constraints (12.71) guarantee that the amount generated in the origin is sent to a collection or recovery center. Constraints (12.72) stand for the flow conservation at each collection center and for each type of disposal. Constraints (12.73) restrict the amount of every valuable commodity obtained from every waste product in a recovery center for every technology. It depends on the amount of received waste product and the conversion rates allowed by the technology. Constraints (12.74) guarantee that the amount of some valuable commodity obtained from some disposal in a recovery center can only be greater than zero if that recovery center is treating that disposal. Constraints (12.75) state that for each recovery center, the amount of disposal not transformed into valuable commodities becomes a residual product to send to a landfill. Constraints (12.76) limit the amount of each waste product delivered to a recovery center, given its capacity. Constraints (12.77) and (12.78) impose a minimum proportion of the total amount of every disposal to be transformed into valuable commodities although it can only be sold in the market up to a given limit. These constraints prevent only the most valuable commodities from being recycled and avoid sending large amounts of disposal to landfills without recycling (which is cheaper in most cases and thus more attractive). Constraints (12.79) guarantee that the maximum capacity of each collection center for each type of disposal is not exceeded. Constraints (12.80) ensure that the residual disposal is sent (from the recovery centers) only to installed landfills. The amount $\sum_{j\in J}\sum_{p\in P_W} d_{jp\omega}$ is an upper bound for the total amount sent to the landfills. Constraints (12.81)–(12.91) define the domain of the decision variables.

Fonseca et al. [212] use data from the province of Cordoba in Spain involving 74 municipalities. Eight out of those are identified for locating facilities. The commodities include urban solid waste, plastics or similar products, glass, and paper. The valuable commodities are sellable glass, compost, sellable paper, sellable plastic, and incineration material. Two different technologies were considered at the recovery centers: manual and automatic. Nine scenarios are identified for the underlying random vector. Using a general-purpose solver, the authors successfully identify Pareto solutions.

12.4.1 Closed-Loop Supply Chain Planning

A first formulation of a facility location problem involving forward and reverse flows in a distribution system is found in [368]. A single primary commodity is considered that is to be shipped from a set of facilities to a set of customers. Part of the shipped amount returns to the facilities. A reverse flow from a customer does not necessarily use the forward channels used to supply it. Nevertheless, the facilities receiving the returning flows are those open and thus those supplying the customers. The problem can be formulated as follows:

$$\text{minimize} \quad \sum_{i\in I} f_i y_i + \sum_{i\in I}\sum_{j\in J}\left(c^1_{ij}x^1_{ij} + c^2_{ij}x^2_{ij}\right), \tag{12.92}$$

$$\text{subject to} \quad \sum_{i\in I} x^1_{ij} = d_j \qquad \forall j \in J, \tag{12.93}$$

$$\sum_{i\in I} x^2_{ij} = \gamma d_j \qquad \forall j \in J, \tag{12.94}$$

$$x^1_{ij} \le d_j y_i \qquad \forall i \in I,\ j \in J, \tag{12.95}$$

$$x^2_{ij} \le \gamma d_j y_i \qquad \forall i \in I,\ j \in J, \tag{12.96}$$

$$\sum_{j\in J} x^2_{ij} = \gamma \sum_{j\in J} x^1_{ij} \qquad \forall i \in I, \tag{12.97}$$

$$x^1_{ij} \ge 0 \qquad \forall i \in I,\ j \in J, \tag{12.98}$$

$$x^2_{ij} \ge 0 \qquad \forall i \in I,\ j \in J. \tag{12.99}$$

$$y_i \in \{0,1\} \qquad \forall i \in I. \tag{12.100}$$

In this model, I is the set of potential locations for the facilities, and J is the set of populations. The costs include fixed quantities for installing facilities (f_i, $i \in I$) and transportation costs for forward and reverse flows (x^1_{ij} and x^2_{ij}, respectively, $i \in I$, $j \in J$). The objective function accounts for the total cost involved. Constraints (12.93) ensure demand satisfaction. Equalities (12.94) guaranteee that a fixed proportion, γ of the total direct flow, returns to the facilities. Constraints (12.95)–(12.97) impose consistent values chosen for the decision variables. Finally, we have the domain constraints.

As we have discussed, a natural source of uncertainty in reverse logistics planning is related to the returned amounts. In the above problem, this means that the return rate γ becomes uncertain. Furthermore, we can assume that the return rates are population-dependent.

For each $j \in J$, let us assume that γ_j is a random variable with some given probability distribution. Let $\boldsymbol{\xi} = (\gamma_1, \ldots, \gamma_{|J|})$. Assuming a risk-neutral decision-maker, we can formulate a first stochastic programming problem seeking to hedge against uncertainty:

$$\text{minimize} \quad \sum_{i \in I} f_i y_i + \sum_{i \in I} \sum_{j \in J} \left(c_{ij}^1 x_{ij}^1 \right) + \mathcal{Q}(\boldsymbol{y}, \boldsymbol{x}^1), \tag{12.101}$$

$$\text{subject to} \quad (12.93),\ (12.95),\ (12.98),\ (12.100),$$

with $\mathcal{Q}(\boldsymbol{y}, \boldsymbol{x}^1) = \mathbb{E}_{\boldsymbol{\xi}}[Q(\boldsymbol{y}, \boldsymbol{x}^1)]$ and

$$Q(\boldsymbol{y}, \boldsymbol{x}^1, \hat{\boldsymbol{\xi}}) = \min \quad \sum_{i \in I} \sum_{j \in J} c_{ij}^2 x_{ij}^2(\hat{\boldsymbol{\xi}}), \tag{12.102}$$

$$\text{s. t.} \quad \sum_{i \in I} x_{ij}^2(\hat{\boldsymbol{\xi}}) = \gamma_j(\hat{\boldsymbol{\xi}}) d_j, \qquad \forall j \in J, \tag{12.103}$$

$$x_{ij}^2(\hat{\boldsymbol{\xi}}) \leq \gamma_j(\hat{\boldsymbol{\xi}}) d_j y_i, \qquad \forall i \in I,\ j \in J, \tag{12.104}$$

$$\sum_{j \in J} x_{ij}^2(\hat{\boldsymbol{\xi}}) = \gamma_j(\hat{\boldsymbol{\xi}}) \sum_{j \in J} x_{ij}^1, \qquad \forall i \in I, \tag{12.105}$$

$$x_{ij}^2(\hat{\boldsymbol{\xi}}) \geq 0, \qquad \forall i \in I,\ j \in J. \tag{12.106}$$

In the second-stage model, $\hat{\boldsymbol{\xi}}$ is an observed value of the random vector $\boldsymbol{\xi}$. The notation adopted for the second-stage variables, $x_{ij}^2(\hat{\boldsymbol{\xi}})$, emphasizes that they react/adapt to the actual occurring return rates.

We do not proceed further with the analysis of this model since several similar discussions are presented in this volume. Nevertheless, for the above setting, it is a simple exercise to formulate the extensive form of the deterministic equivalent in case the support of the random vector $\boldsymbol{\xi}$ is finite. Furthermore, in that case, the Lagrangian relaxation procedure proposed [368] for the deterministic case can be easily embedded in a dual-decomposition procedure resulting from relaxing the non-anticipativity constraints (see Sect. 4.3.2).

The above facility location problem is rather simplistic in the context of closed-loop supply chain planning as the latter often calls for multiple products, multiple echelons, capacity decisions, selection of transportation mode, *et cetera*. Nowadays, the literature focusing on closed-loop supply chain network design is abundant (see, e.g., [362]). Nevertheless, from the perspective of facility location under uncertainty, closed-loop supply chain planning leverages the contents of this and the previous sections. The resulting models are often quite involved and thus tackled using approximate algorithms. We refer the reader to the specialized literature on supply chain management.

12.5 Humanitarian Supply Chain Planning

One particular area within logistics and supply chain management that has attracted much attention is humanitarian logistics. It refers to the efficient and cost-effective planning, implementation, and control of the flow and storage of goods/materials and the related information from the point of origin to that of destination to alleviate the suffering of vulnerable people [524].

The development of humanitarian logistics has led more recently to the concept of the humanitarian supply chain [300] and to that of humanitarian operations. In this context, facility location under uncertainty plays a central role. In [177] six types of facilities relevant in that context are identified: suppliers, distribution centers, points of distribution, shelters, field hospitals, and blood centers. Furthermore, those authors also identify and discuss in detail the three sources of uncertainty that often emerge: that related to demand (e.g., population affected by a catastrophic event), supply (e.g., due to unknown donations), and network connectivity (e.g., roads blocked).

In humanitarian logistics, two major planning phases are distinguished within disaster operations management: preparedness, which includes preventive measures taken before a disaster occurs, and response, which seeks to react to the consequences of the disaster. In the first phase, we find location and inventory decisions that stem from the need to pre-position relief items. In the second phase, we often find distribution decisions. They may emerge as allocation decisions (demand nodes to facilities) or routing. In the former, we end up with a location-allocation problem possible with other decisions. In the latter, we end up with a location-routing problem.

Finally, a distinguishing feature of facility location problems in humanitarian logistics regards the criteria to optimize. In addition to cost (seeking efficiency), equity and reliability objectives are often of primordial relevance. Often, these different objectives need to be considered simultaneously, thus calling for using multicriteria models and methods.

Dönmez et al. [177] discuss several modeling aspects of relevance when considering a facility location problem in the context of humanitarian logistics. Several sets are often considered:

I, set of facilities (existing or potential);
J, set of beneficiaries (e.g., populations that need or may need help);
P, set of items (e.g., drinkable water of first-aid kits);
M, fleet of vehicles (of relevance when the beneficiaries are to be supplied as part of routes).

Related problems usually involve a network such as the road network in some geographical area. Thus, we can define the set of arcs in such a case:

$$A = \{(\ell, \ell') \mid \ell, \ell' \in I \cup J \text{ and } \ell' \text{ is reachable from } \ell\}.$$

Depending on the specific setting, several decision variables may be of interest.

- Location:

$$y_i = \begin{cases} 1, & \text{if facility } i \in I \text{ is installed/used;} \\ 0, & \text{otherwise.} \end{cases}$$

- Pre-positioning:

$u_{ip} =$ amount of product $p \in P$ to store at location $i \in I$.

- Allocation/Transportation:

$$w_{ij} = \begin{cases} 1, & \text{if population } j \in J \text{ is assigned to facility } i \in I; \\ 0, & \text{otherwise.} \end{cases}$$

$x_{ijp} =$ amount of product $p \in P$ shipped from location $i \in I$ to population $j \in J$.

$$v_{\ell\ell'm} = \begin{cases} 1, & \text{if population } j \in J \text{ is assigned to facility } i \in I; \\ 0, & \text{otherwise.} \end{cases}$$

As far as the parameters are concerned, the relevant ones depend on a specific context. Nevertheless, we can identify a few possibilities.

- Related to the demand:

d_{jp}, demand of beneficiary $j \in J$ in terms of product type $p \in P$.
This quantity is usually uncertain because it depends on the magnitude and intensity of the disaster.

h^-_{jp}, unit shortage cost at beneficiary $j \in J$ w.r.t. product $p \in P$.
A related quantity is the so-called deprivation cost (see, e.g., [270]), which is a measure of the human suffering due to being deprived of some product(s).

- Related to the supply:

β_i, risk or vulnerability factor associated with facility $i \in I$.
Depending on the value of this parameter, some or all pre-positioned items may become unusable.
s_{ip}, supply of product $p \in P$ in facility $i \in I$;
The type, amount, and availability time may be uncertain;
f_i, fixed cost for installing/using facility $i \in I$;
g_i, operating cost of facility $i \in I$;
h^+_{ip}, unit inventory holding cost for item type $p \in P$ at facility $i \in I$.

- Related to the transportation:

$\zeta_{\ell,\ell'}$ risk or vulnerability factor associated with link (ℓ, ℓ');
$c_{\ell,\ell'}$, traveling cost between ℓ and ℓ', with $(\ell, \ell') \in A$;
$t_{\ell,\ell'}$, traveling time between ℓ and ℓ', with $(\ell, \ell') \in A$;
θ, link reliability threshold;
δ, accessibility threshold in terms of distance or time;
α, service level (e.g., a ratio or a probability).

When formulating an optimization model, the constraints depend on the problem being investigated. Nevertheless, several types of constraints are pervasive.

- Related to the direct allocation or routing:
Direct allocation of a population can be made to a facility if and only if the facility is used, i.e.,

$$w_{ij} \leq y_i \quad \forall i \in I, \ j \in J.$$

All populations must be included in a route, i.e.,

$$\sum_{m \in M} \sum_{\ell | (j,\ell) \in A} v_{j\ell m} = 1 \quad \forall j \in J.$$

Each route is associated with a single facility, i.e.,

$$\sum_{(i,j) \in A} v_{ijm} \leq 1 \quad \forall m \in M.$$

A population and a facility are in the same route if and only if the population is allocated to the facility, i.e.,

$$\sum_{\ell \in I \cup J} (v_{i\ell m} + v_{\ell jm}) \leq 1 + w_{ij} \quad \forall i \in I \ : j \in J.$$

- Related with pre-positioning and supply:
  The amount of each product shipped to a population from a facility cannot exceed the amount available at the facility, i.e.,

$$\sum_{j \in J} x_{ijp} \leq (1 - \beta_i) u_{ip} + s_{ip} \quad \forall p \in P.$$

- Related to the demand:
  The amount of each product shipped to each population should not exceed the corresponding demand, i.e.,

$$x_{ijp} \leq d_{jp} w_{ij} \quad \forall i \in I,\ j \in J,\ p \in P.$$

Different performance measures may be of interest depending on the specific problem. We can generically represent some possibilities:

- Cost-oriented. Some examples are

$$\varphi^C(\boldsymbol{u}, \boldsymbol{w}, \boldsymbol{y}) = \sum_{i \in I} (f_i + g_i) y_i + \sum_{i \in I} \sum_{p \in P} h^+_{ip} u_{ip} + \sum_{i \in I} \sum_{j \in J} c_{ij} w_{ij};$$

$$\varphi^C(\boldsymbol{x}) = \sum_{j \in J} \sum_{p \in P} h^-_{jp} \left(d_{jp} - \sum_{i \in I} x_{ijp} \right);$$

$$\varphi^C(\boldsymbol{u}, \boldsymbol{v}, \boldsymbol{y}) = \sum_{i \in I} (f_i + g_i) y_i + \sum_{i \in I} \sum_{p \in P} h^+_{ip} u_{ip} + \sum_{m \in M} \sum_{\ell, \ell' \in C} c_{\ell\ell'} v_{\ell\ell' m}.$$

- Equity-oriented. The emphasis can be on demand satisfaction or accessibility. Some examples are

$$\varphi^S(\boldsymbol{x}) = \max_{j \in J} \left\{ d_{jp} - \sum_{i \in I} x_{ijp} \right\};$$

$$\varphi^A(\boldsymbol{x}) = \max_{i \in i,\ j \in J} \{ t_{ij} w_{ij} \}.$$

- Reliability-oriented: $\varphi^R(\boldsymbol{u}, \boldsymbol{v}, \boldsymbol{w}, \boldsymbol{x}, \boldsymbol{y})$.

The above performance measures can be used as objective functions in an optimization model or may be part of the constraint set with some threshold imposed. For instance, a budget constraint on facilities to use would impose

$$\sum_{i \in I} f_i y_i \leq B,$$

with B standing for the available budget. Regarding demand satisfaction, we may need to impose a certain service level, e.g.,

$$\frac{1}{d_{jp}} \sum_{i \in I} x_{ijp} \geq \alpha \quad \forall j \in J,\ p \in P.$$

Finally, regarding accessibility, we can impose that each population has a facility operating within a certain travel time, i.e.,

$$\sum_{i \in I | t_{ij} \leq \delta} w_{ij} \geq 1 \quad \forall j \in J.$$

Uncertainty is a central aspect of the current volume. Regarding the above-introduced parameters, let us assume the most generic case in which uncertainty stems from demands, supply, and vulnerability of the facilities and links. Accordingly, we can define $\boldsymbol{\xi} = (\boldsymbol{d}, \boldsymbol{s}, \boldsymbol{\beta}, \boldsymbol{\zeta})$. Assuming that $\boldsymbol{\xi}$ is a random vector with a given distribution, we may cast a problem as a stochastic programming problem.

In this case, we can think of several possibilities.

- Two-stage stochastic programming
  A cost-oriented objective can be generically written as

$$\text{(min) } \varphi^C(\boldsymbol{u}, \boldsymbol{v}, \boldsymbol{w}, \boldsymbol{y}) + \mathbb{E}_{\boldsymbol{\xi}}[\phi^C(\boldsymbol{u}, \boldsymbol{v}, \boldsymbol{w}, \boldsymbol{y}, \boldsymbol{\xi})].$$

  An equity objective with emphasis on demand satisfaction could be

$$\text{(min) } \mathbb{E}_{\boldsymbol{\xi}}[\varphi^S(\boldsymbol{x})].$$

  An equity objective with emphasis to accessibility is

$$\text{(min) } \mathbb{E}_{\boldsymbol{\xi}}[\varphi^A(\boldsymbol{w})].$$

  Under uncertainty we may now also think of reliability-oriented objectives, which generically can be written as

$$\varphi^R(\boldsymbol{u}, \boldsymbol{v}, \boldsymbol{w}, \boldsymbol{y}) + \mathbb{E}_{\boldsymbol{\xi}}[\phi^R(\boldsymbol{u}, \boldsymbol{v}, \boldsymbol{w}, \boldsymbol{y}, \boldsymbol{\xi})].$$

- Chance-constrained programming
  A cost-oriented chance constraint could be

$$\mathbb{P}\left[\varphi^C(.) \geq B\right] \leq \gamma.$$

An equity-oriented chance constraint with emphasis to demand satisfaction can be written, for instance, as

$$\mathbb{P}\left[\sum_{i\in I} x_{ijp} \geq d_{jp}\right] \geq \alpha \quad \forall j \in j\ p \in P,$$

or

$$\mathbb{P}\left[\sum_{j\in J} x_{ijp} \leq \beta_i u_{ip} + s_{ip}\right] \geq \alpha \quad \forall i \in I,\ p \in P.$$

An equity-oriented chance constraint with emphasis to accessibility can be

$$\mathbb{P}\left[\sum_{i\in I|t_{ij}\leq\delta} (1-\zeta_{ij})w_{ij} \geq 1\right] \geq \theta \quad \forall j \in J.$$

A reliability chance constraint could be

$$\mathbb{P}\left[\sum_{(\ell,\ell')\in A} (1-\zeta_{\ell,\ell'}) \sum_{m\in M} v_{\ell\ell' m} \geq 1\right] \geq \theta \quad \forall \ell' \in J,$$

or

$$\mathbb{P}\left[\sum_{i\in I}(1-\beta_i)y_i \geq 1\right] \geq \theta.$$

The interested reader can refer to [177] for a synthesis of the literature capturing all the above elements.

We conclude this section by citing some recent work in this area. Paul and Wang [424] develop robust optimization models for earthquake preparedness. The decisions include the number, location, and capacity of distribution centers. The goal is to minimize the total social cost, which includes setup, initial supplies, and deprivation. Earthquake magnitude-specific uncertainty is captured. The latter may regard facility damage, casualty by severity, and travel time. Robustness is captured in two ways: (i) a finite set of scenarios specifying all the uncertain parameters—a minmax regret model is adopted, and (ii) parameters captured by box uncertainty—a robust counterpart is derived.

Demange et al. [162] investigate a robust p-center problem in the context of humanitarian supply chains. The goal is to decide about the location of shelters in case of a wildfire. Uncertainty is related to network connectivity. An MILP model is proposed. For large instances, a decomposition procedure is proposed for solving the problem exactly.

Mohammadi et al. [388] investigate a multi-objective humanitarian logistics network design problem under uncertainty. Three different types of facilities must be located. Routes for the delivery of relief items must also be decided. The goal is to minimize the total cost of the system, as well as the time used in the relief operations and a workload balance objective. Several sources of uncertainty are considered: capacities, demands, and amount of injured people. Uncertainty sets are assumed for those unknowns. The costs and the transportation times are captured using fuzzy numbers. A so-called robust neutrosophic model is formulated and solved using a commercial solver.

The combination of location routing with the pre-positioning of relief items often found in humanitarian logistics applications leads to a subproblem that belongs to the family of the so-called location-inventory-routing problems. This area emerged independently from humanitarian logistics due to its importance in logistics and supply chain management. For this reason, we discuss it in a separate section.

12.6 Location-Inventory-Routing and Beyond

The location-inventory-routing problem (LIRP) is an extension of location routing in which inventory decisions are also to be made. Thus, we find contents from Chaps. 9 and 10 gathered in a problem.

As we observed in Chap. 9, inventory decisions often lead to nonlinear cost objective functions. Furthermore, many LIRPs emerge in multi-echelon network design problems with different types of facilities to be located (e.g., suppliers and distribution centers). A pioneering work in this area is presented in [502]. The authors consider demands distributed according to a Gaussian distribution and account for economies of scale for the transportation and inventory costs. To simplify the problem, approximate costs are derived for the routing problem. An MINLP formulation is derived that is tackled using a Lagrangian relaxation-based algorithm.

An LIRP with stochastic demands seeking to minimize the total cost of the system is presented in [285]. The authors also assume Gaussian distributed demands. In addition to the location, routing, and inventory decisions, the capacity of the open depots must also be decided. The problem is formulated using a mixed-integer-convex programming model and solved with a commercial solver. A heuristic combining tabu search and simulated annealing is also proposed.

A similar problem was considered in [396]. A bi-objective multi-product multi-period LIRP with stochastic demands is studied. Again, Gaussian distributed demands are assumed. The objective functions (to minimize) are the total cost and the maximum time for delivering commodities to customers. A bi-objective mixed-integer nonlinear programming model is proposed, which is solved (for small instances) using the ϵ-constrained method. For dealing with large-scale instances, four heuristics are proposed and compared.

Caunhye et al. [108] investigate an LIRP in the context of a humanitarian logistics application. A two-stage decision-making process is assumed. In the first stage, the location and the inventory associated with the depots must be decided. The second-stage decision regards the routes and transshipment of flow between the depots. The objective is to minimize the weighted setup cost of the depots plus the worst-case response time. The second-stage "cost" also accounts for a weighted penalty due to not serving the customers. The authors propose a linearization for their model and tackle it using an off-the-shelf solver. Computational experiments are presented for small instances.

A multi-objective closed-loop supply chain network design problem under uncertainty is studied in [566]. This problem can be seen as a particular case of the multi-period LIRP in which additional decisions must be made. The distribution centers to open must be decided together with the ones that will be used as remanufacturing facilities. The capacities of these facilities are also a decision to make. The routing is only performed for the delivery from the distribution centers to the retailers (customers); the products to be remanufactured are sent directly from the retailers to the adequate centers. Uncertainty is associated with the demands and the amounts to remanufacture. Such uncertainty is assumed to be described by a Gaussian distribution. The authors seek to optimize economic, environmental, and social objectives. To solve the problem, a hybrid two-stage approach is proposed that solves only small- and medium-sized instances. For larger ones, a self-adaptive genetic algorithm metaheuristic is proposed.

A multi-objective stochastic LIRP for hazardous waste management is presented in [442]. The objectives to minimize are the total cost and the environmental risks associated with opening facilities and hazmat transportation. The authors consider uncertainty both in the demands and in the risk parameters. The problem is tackled using a multi-objective simheuristic.

Biuki et al. [79] investigate a multi-period multi-depot four-echelon supply chain network design problem for perishable products assuming fuzzy demands. In addition to location, routing, and inventory, the capacity of the facilities is a decision to make. An MILP model is formulated. Small instances are tackled using a general-purpose solver. Thinking of large instances, a hybrid heuristic combining a genetic algorithm and PSO is proposed.

Chavez et al. [113] study a multi-objective location-inventory-routing problem with stochastic demands in the agricultural biofuel supply chain. The problem is modeled as a multi-echelon network design problem featuring multiple periods and products. In addition to the location, routing, and inventory decisions, capacity expansion and opening/closing facilities in different periods are decisions to make. A two-phase heuristic algorithm is proposed based on simulated annealing.

A robust multi-period LIRP is studied in [489] motivated by healthcare logistics. The problem consists of deciding the location of warehouses, the inventory level at those facilities, and the routes from warehouses to the hospitals. Demand for multiple products is assumed to be uncertain. Direct shipment from the supplier (thus not part of a route) is allowed when the inventory level cannot satisfy the demand. Nevertheless, in such a case, a high cost is incurred. A robust MILP model is formulated whose objective function accounts for the underlying total cost. A general solver is used for tackling the model. The robust optimization setting adopted in this work follows the methodology proposed in [75].

Kaya and Ozkok [304] focus on the design of a blood supply chain network to supply a set of health facilities. Some facilities are selected as local blood banks. Every other facility are to be assigned to one such bank. A mixed-integer nonlinear programming (MINLP) model is used to determine the optimal selection of blood banks, assignment of facilities to them, optimal inventory levels at each bank, and routing decisions among the facilities. The goal is to minimize the total cost. Uncertainty is associated with the demand for blood at health facilities, which calls for using stochastic inventory planning. A piecewise linear approximation procedure is also proposed with a simulated annealing heuristic to find approximate solutions to the problem.

Last but not least, we cite [562] that presents a review on location-inventory models in the context of supply chain planning.

12.7 Conclusion and Further Reading

In this chapter, we discussed the role of facility location under uncertainty in logistics and supply chain management. The contents show that this area has been deeply investigated. Above all, it is clear that facility location plays a central role in strategic supply chain planning, which, in turn, is relevant for ensuring efficient and effective logistics operations.

Despite all the work done, we still observe some areas where further research needs to be done. Chance-constrained programming and distributionally robust optimization have been considered to a much smaller extent when compared to robust optimization and stochastic programming with recourse. Nevertheless, the two first paradigms are becoming progressively more relevant for hedging against uncertainty when planning for complex production and distribution systems. Often, some constraints are not as "hard" as it may be indicated by much existing work (e.g., capacity constraints). Finding solutions that guarantee their satisfaction with a given probability is a possibility to solve the related problems. This has been explored in the context of humanitarian logistics but still to a limited extent and often with strong assumptions. The use of distributionally robust optimization allows for considering a set of distributions—those in the adopted ambiguity set. The latter can capture aspects of practical relevance, such as the correlation between the underlying random variables.

The problems often faced in supply chain planning have a dimension that makes them computationally challenging. We still observe much work done on modeling for capturing uncertainty but not so much concerning efficient solution procedures.

Despite the above aspects calling for additional research, some notable progress has been attained recently, which aims at making models more comprehensive and realistic. Next, we briefly summarize that work.

The existence of several facility layers is pervasive in supply chain network design. In [261] the authors consider a bi-objective multi-echelon location-routing with time windows. Uncertainty stems from the demands and service times. The problem is inspired by the SARSCoV-2 (COVID-19) pandemic, where hazardous materials had to be managed. The setting calls for two types of facilities to be located: treatment centers (that can be partially or permanently open) and temporary transfer stations. The authors consider that large customers (e.g., hospitals) can be directly allocated to the treatment centers; small customers (e.g., laboratories) must be served as part of a route collecting the hazardous materials at the transfer stations before sending them to the treatment centers. The objectives to minimize are the total cost and the risk of infection to which the populations close to open facilities are exposed. Three scenarios are considered. A variant of the ϵ-constrained method with a branch-and-price algorithm is proposed for determining the Pareto front.

Mohamed et al. [386] combine a multi-layered facility system with time-dependent decisions. The authors investigate a two-echelon multi-period capacitated location-routing problem with stochastic demands. The problem corresponds to an extension of the classical capacitated location-routing problem with an upper-level facility to be located (in addition to the base setting). A two-stage stochastic programming model is derived considering the location and capacity decisions at the first stage and the routing decisions at the second stage. The demands are assumed to follow a Gaussian distribution. A Benders decomposition algorithm is proposed.

Multimodal transportation is another feature of relevance in many logistics and supply chain management applications. In [204] the authors deal with a multimodal network design problem under uncertainty. Again, a two-stage decision-making process is adopted. The first phase seeks to determine how to ship the products from a supplier to a set of distribution centers (depots for the second phase) with the possibility of combining several transportation modes. Nevertheless, that requires the location of intermodal changing points. The second phase corresponds to a location-routing problem. Time windows are considered. Uncertainty in demand is captured using fuzzy numbers. The goal is to minimize the total cost. A fuzzy mixed-integer programming model is derived. It is solved (for small instances) using a commercial solver. The complexity of the problem also motivates the development of a genetic algorithm seeking feasible solutions.

The location transportation problem (LTP) is a generalization of the location-routing problem when several transportation modes are available. A multi-period LTP with stochastic demands is presented in [313]. The problem is cast as a two-stage stochastic problem with recourse in which the objective is to maximize the expected profit. The authors consider the location of the depots at the first stage. The transportation planning is the recourse decision. The uncertainty in the demands is considered in two dimensions: (i) the time between consumer orders (according to exponential distribution) and (ii) the order quantity (according to a log-normal distribution). An SAA approach is proposed. However, using a general-purpose solver, only small instances can be tackled. For this reason, the authors also presented a heuristic algorithm based on tabu search. More recently, insights related to different anticipation methods for hedging against uncertainty are discussed in [314].

Chapter 13
Territory Design

Abstract This chapter discusses territory design problems under uncertainty. This is a topic much related to discrete p-median problems under uncertainty. Thus, the contents of this chapter complement in some way part of the contents of Chap. 6. Initially, a classical setting that has been considered as basis for many districting problems is briefly reviewed. Afterwards, assuming uncertainty related to the activity level generated at the districts, a two-stage stochastic programming modeling framework is discussed with several extensions. The chapter proceeds by analyzing the use of chance constraints to replace customary hard balancing constraints. Afterwards, a different model set at the core of a robust optimization districting problem is analyzed. Budget uncertainty is assumed in this case. The chapter concludes with an overview of its contents, new research directions, and hints for further reading.

Keywords Districting · Re-districting · Compactness objectives · Equilibrium objectives · Chance constraints

13.1 Introduction

A districting problem (DP) consists of grouping a given set of territorial units (TUs) into a fixed number of larger geographical clusters called districts. Four features are desirable in these problems: integrity, balancing, compactness, and contiguity. The first imposes each TU to be assigned to one and only one district. The second stems from some activity measure originated at the TUs (e.g., demand for some commodity) and expresses the need for the districts to be of similar size concerning such measure. Compactness underlies the idea that the TUs within the same district should be close to each other (e.g., elongated districts should be avoided as much as possible). Finally, contiguity indicates that one should not cross other districts when moving between two TUs within a district; in other words, enclaves and exclaves should be avoided. In Fig. 13.1, we illustrate the design of five districts in a geographical area.

© The Author(s), under exclusive license to Springer Nature Switzerland AG 2024
F. Saldanha-da-Gama, S. Wang, *Facility Location Under Uncertainty*,
International Series in Operations Research & Management Science 356,
https://doi.org/10.1007/978-3-031-55927-3_13

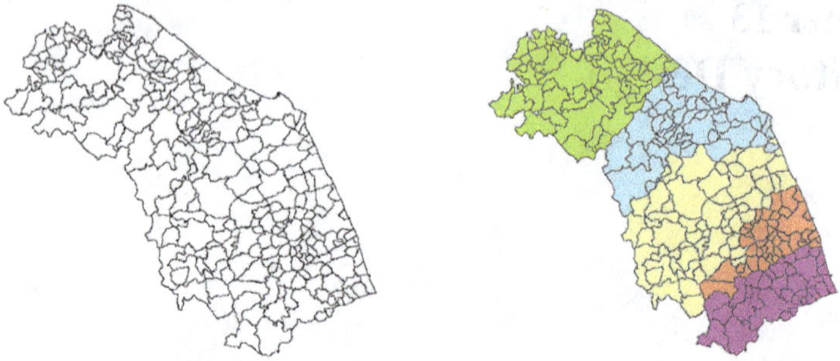

(a) (b)

Fig. 13.1 Dividing a geographical area into districts. (**a**) Original area. (**b**) Five districts created

Districting problems have applications in many areas as political districting [15, 341, 454], strategic service planning and management, for instance, in healthcare [61, 390, 522], school systems [90], energy and power distribution networks [555], police districts [149], waste collection [391, 457], transportation [91, 518], design of commercial areas [465], distribution logistics [318]. The interested reader can find more details in [455] and in the overview provided in [294].

Districting problems have played a central role in the context of distribution logistics. In practice, when uncertainty stems from the demands, several authors have resorted to districting approaches for dealing with the underlying stochastic vehicle routing problems (see, e.g., [102, 103, 262, 330, 331]). In this chapter, we ignore routing decisions, thus focusing on how to organize a territory into districts if uncertainty is to be hedged against. Nevertheless, the above works show the strong relation between this chapter and Chap. 10.

A seminal work in DPs is [267]. The authors introduce a mathematical model for districting. Nowadays, the model introduced in that work is still at the core of much work done. Next, we revisit it. The problem consists of creating a given number, say p, of districts according to three rules: integrity, balancing, and compactness.

Let I be the set of basic TUs, and d_i the demand of TU $i \in I$. This is a measure of the activity generated at the TUs. The activity level of a district is given by the sum of the activity generated at the TUs that are part of it. Thus, to account for balancing, we can compute a reference value, say μ, representing the activity level in each of the p districts to create in case they were perfectly balanced. We have $\mu = (1/p) \sum_{i \in I} d_i$. In practice, such a perfect balancing is not possible. Thus, we consider a tolerance $0 < \alpha < 1$ and assume a territory design to be balanced when the activity level at every district does not deviate from μ more than $100\alpha\%$.

In districting problems, we often consider that each created district has a TU representing it. In this way, we consider a cost that we denote by c_{ij} for assigning TU i to the district represented by j. We abuse the language by saying that TU i is assigned to TU j $(i, j \in I)$. Let us denote by δ_{ij} the distance between TUs i and j, $(i, j \in I)$. This distance is typically measured using the centroids of the TUs.

Compactness measures the closeness of the TUs units within each district. One way to accomplish that is to assume a cost $c_{ij} = \delta_{ij}$ for every pair of TUs ($c_{ii} = 0$ $\forall i \in I$). For each district, if k is its representative and D_k is the set of TUs in the district, a measure of dispersion is given by $\sum_{i \in D_k} c_{ik}$. Note that other distance-related costs could be considered to account for compactness, such as $c_{ij} = \delta_{ij}^2$ $\forall i, j \in I$.

Given the above ingredients, we can formulate the problem mathematically using a set of binary variables. For $i, j \in I$ x_{ij} equals one if TU i is assigned to TU j and zero, otherwise. For some $j \in I$, having $x_{jj} = 1$ indicates that the TU j represents its district. The problem can be formulated as follows:

$$\text{minimize} \quad \sum_{i \in I} \sum_{j \in I} c_{ij} x_{ij}, \tag{13.1}$$

$$\text{subject to} \quad \sum_{j \in I} x_{ij} = 1 \qquad \forall i \in I, \tag{13.2}$$

$$\sum_{j \in J} x_{jj} = p, \tag{13.3}$$

$$(1-\alpha)\mu x_{jj} \leq \sum_{i \in I} d_i x_{ij} \leq (1+\alpha)\mu x_{jj} \qquad \forall j \in I, \tag{13.4}$$

$$x_{ij} \leq x_{jj} \qquad \forall i, j \in I, \tag{13.5}$$

$$x_{ij} \in \{0, 1\} \qquad \forall i, j \in I. \tag{13.6}$$

In this model, the objective function (13.1) seeks to minimize the overall dispersion accounted for by the distances between TUs. Constraints in (13.2) ensure that all TUs are assigned to one district. Equality (13.3) impose the establishment of p districts. The double inequalities in (13.4) regard the balancing requirements. Constraints in (13.5) impose that TUs can only be assigned to representatives of the established districts. Finally, we have the domain constraints.

It is straightforward to conclude that inequalities in (13.5) are implied by (13.4). Nevertheless, they are relevant in some modeling frameworks for hedging against uncertainty, to be discussed later in the chapter.

A close look at the above model shows that we face a p-median problem with additional (balancing) constraints. This fact has established a strong link between facility location and districting. The model just presented captures three aspects of relevance in territory design: compactness (the objective function (13.1)), integrity (Constraints (13.2) and (13.6)), and balancing (Constraints (13.4)). Note that the model does not explicitly ensure contiguity. Empirically, it has been shown that compactness favors contiguity in the sense that the more compact a solution is, the more contiguous it tends to be. Nevertheless, this is not an implication. Thus, several authors have explicitly considered contiguity (or connectivity) constraints. Different possibilities exist for accomplishing this. One way consists of using a set of network flow constraints (see [504]). In this case, we look at each TU as generating one unit of supply and the districts' representatives acting as sinks. A district is contiguous if we can find a feasible flow from its TUs to its representative using only TUs in the district.

Let us assume that the problem has a graph $G = (V, A)$ underlying it, with vertices representing the TUs ($V = I$) and A containing all the direct links between TUs (a link exists for every pair of adjacent TUs). Giving this graph, for each $i \in I$ we can find the set I_i of nodes adjacent to i, i.e., $I_i = \{j \in I \mid (i, j) \in A \vee (j, i) \in A\}$.

Contiguity constraints can be included in the above model by considering one additional set of decision variables: for $i, j, m \in I$, and $(m, i) \in A$, w_{mij} represents the flow of district j sent from m to i.

The following constraints inspired by those proposed in [504] can be added to model (13.1)–(13.6):

$$\sum_{m \in I \mid (m,i) \in A} w_{mij} \leq (|I| - p)x_{ij} \qquad \forall i, j \in I, \tag{13.7}$$

$$\sum_{m \in I \mid (i,m) \in A} w_{imj} - \sum_{m \in I \mid (m,i) \in A} w_{mij} = x_{ij} \qquad \forall i, j, \in I,\ i \neq j. \tag{13.8}$$

Constraints in (13.7) ensure that flow corresponding to district j can be sent to a node i only if that node is assigned to that district. The right-hand side results from the fact that, in each district, we have at most $|I| - p$ TUs that are not representatives. Constraints in (13.8) guarantee that if a node is assigned to a district, it generates exactly one flow unit in the district if it is not the representative.

Alternatively, as proposed in [469], we can model the connectivity (contiguity) constraints in line with the sub-tour elimination constraints for the Traveling Salesman Problem:

$$\sum_{i \in S} x_{ij} - \sum_{i \in \cup_{v \in S}(I_v \setminus S)} x_{ij} \leq |S| - 1 \quad \forall j \in I,\ S \subset [I \setminus (I_j \cup \{j\})]. \tag{13.9}$$

These constraints are exponential in number. In [469], they are embedded in a cut generation approach (i.e., only those needed are generated). One advantage of such a representation of contiguity is that no additional decision variables are necessary.

Irrespective of the contiguity constraints selected, the model can be enhanced by including the following constraints also proposed in [469]:

$$\sum_{\ell \in I_i} x_{\ell j} \geq x_{ij} \quad \forall j \in I,\ i \in I \setminus \{\{j\} \cup I_j\}. \tag{13.10}$$

These constraints ensure that if TU i is assigned but is not adjacent to TU j, there is at least another TU ℓ adjacent to i and assigned to j.

In this chapter, we focus on districting solutions that hedge against uncertainty in the activity of the districts. In Sect. 13.2, base models are discussed. Stochastic programming is the paradigm used. In Sect. 13.3, balancing chance constraints are considered, and different algorithmic possibilities are presented. In Sect. 13.4, balancing becomes an objective, and dispersion is captured as a set of constraints. Robust optimization models are presented, considering a general graph underlying the problem and a specific topology. The chapter concludes with an overview and suggestions for further reading.

13.2 Coping with Uncertainty—First Modeling Aspects

In practice, the activity of the TUs may change. This may be due to the expansion of urban areas, migration flows, economic conditions, *et cetera*. Different possibilities exist for hedging against such changes, which depend on the specific problem we are solving. In particular, we may need to solve a redistricting problem, a multi-period districting problem, or a districting problem under uncertainty. In the first case, the demand changes have already occurred, and a reactive posture is assumed. A so-called redistricting problem needs to be solved. This is an optimization problem aiming at redesigning existing districts.

In the second case, the demand changes can be accurately predicted for some future—the planning horizon. In this case, a proactive posture can be assumed, and a multi-period districting problem can be considered in which a districting/redistricting plan is devised to cope with the varying demands.

In the third case, demand is uncertain and cannot be accurately forecast. Assuming a proactive posture, different paradigms can be considered: (i) online optimization (typically for short-term uncertainty), (ii) robust optimization (uncertainty sets describe the demand), (iii) stochastic programming (uncertainty described by some probability distribution), *et cetera*. The existing literature has focused on stochastic optimization models for hedging against uncertainty in districting problems. Accordingly, this is also the emphasis given in this chapter.

Let us consider a setting in which we seek to plan for organizing a geographical area into districts under uncertain demand. Assume that the latter is stochastic and represented by a random vector $\boldsymbol{\xi} = \{\boldsymbol{\xi}_1, \ldots, \boldsymbol{\xi}_{|I|}\}$ whose CDF is known.

A here-and-now decision must be made about the territory design. Looking for a solution feasible for all possible observations of the demand may be impossible, i.e., no matter the districts we consider, an unbalanced solution may be obtained for some demand future scenarios. Even if it was possible to find such a solution, it could easily lead to an overly conservative solution.

A good compromise is to make an initial territory design decision. Afterwards, when the demand becomes known, if necessary, implement some recourse action. Indeed, upon observing the demand we may conclude that the initial territory design is infeasible, i.e., the solution is unbalanced. In this case, we can consider extraordinary actions to be undertaken to overcome shortage or surplus (w.r.t. the established threshold) at some districts, e.g., downsizing workforce (in case of demand shortage) or outsourcing (in case of demand surplus).

The recourse action just mentioned can be accommodated in the model by considering two additional sets of (auxiliary) decision variables: ψ_j (φ_j) represents the demand surplus (shortage) in district $j \in J$ w.r.t. the maximum (minimum) threshold. Regarding the recourse costs, we consider g_j (h_j) standing for the unit cost incurred for demand above (below) the maximum (minimum) threshold at district $j \in I$. Considering the model proposed in [267] as a starting point and assuming a risk-neutral decision-maker, the problem can be formulated as follows:

$$\text{minimize} \quad \sum_{i\in I}\sum_{j\in I} c_{ij}x_{ij} + \mathcal{Q}(\boldsymbol{x}), \tag{13.11}$$

$$\text{subject to} \quad (13.2),\ (13.3),\ (13.5),\ (13.6),$$

with $\mathcal{Q}(\boldsymbol{x}) = \mathbb{E}_{\boldsymbol{\xi}}[Q(\boldsymbol{x}, \boldsymbol{\xi})]$ and, for each observation $\hat{\boldsymbol{\xi}}$ of $\boldsymbol{\xi}$,

$$Q(\boldsymbol{x}, \hat{\boldsymbol{\xi}}) = \min \quad \sum_{j\in I}\left(g_j\psi_j(\hat{\boldsymbol{\xi}}) + h_j\varphi(\hat{\boldsymbol{\xi}})\right), \tag{13.12}$$

$$\text{s. t.} \quad (1-\alpha)\mu x_{jj} \le \sum_{i\in I} d_i(\hat{\boldsymbol{\xi}})x_{ij} + \varphi_j(\hat{\boldsymbol{\xi}}) - \psi_j(\hat{\boldsymbol{\xi}}) \le (1+\alpha)\mu x_{jj}$$

$$\forall j \in I, \tag{13.13}$$

$$\psi_j(\hat{\boldsymbol{\xi}}), \varphi_j(\hat{\boldsymbol{\xi}}) \ge 0 \qquad \forall j \in I. \tag{13.14}$$

If the support of the random vector $\boldsymbol{\xi}$ is finite, then we can go further in terms of modeling because we can number the scenarios in a finite set Ω and index the demands and the second-stage decision variables in that set: for $i \in I$ and $\omega \in \Omega$, $d_{i\omega}$ is the demand of TU i under scenario ω, $\psi_{j\omega}$ ($\varphi_{j\omega}$) is the demand surplus (shortage) in district i under scenario ω. Additionally we take π_ω as the probability associated with scenario $\omega \in \Omega$ with $\pi_\omega \ge 0\ \forall \omega \in \Omega$ and $\sum_{\omega\in\Omega}\pi_\omega = 1$.

A relevant issue now is the reference value used for considering a district solution to be balanced. The initially defined value μ relied on the demands, which now are uncertain. A simple choice is to use the overall expected demand in the computation, thus obtaining

$$\hat{\mu} = \frac{1}{p}\sum_{\omega\in\Omega}\left(\pi_\omega \sum_{i\in I} d_{i\omega}\right).$$

We can now formulate the extensive form of the deterministic equivalent:

$$\text{minimize} \quad \sum_{i\in I}\sum_{j\in I} c_{ij}x_{ij} + \sum_{\omega\in\Omega}\left[\pi_\omega \sum_{j\in I}\left(g_j\psi_{j\omega} + h_j\varphi_{j\omega}\right)\right], \tag{13.15}$$

$$\text{subject to} \quad (13.2),\ (13.3),\ (13.5),\ (13.6),$$

$$(1-\alpha)\hat{\mu}x_{jj} \leq \sum_{i\in I} d_{i\omega}x_{ij} + \varphi_{j\omega} - \psi_{j\omega} \leq (1+\alpha)\hat{\mu}x_{jj} \qquad \forall j \in I,\ \omega\in\Omega, \tag{13.16}$$

$$\psi_{j\omega}, \varphi_{j\omega} \geq 0 \qquad \forall j \in I\ \omega \in \Omega. \tag{13.17}$$

We seek now to enrich the set of second-stage decisions. In addition to handling possible shortages or surpluses in districts, we can be more proactive. In particular, we can think of "moving" TUs across districts, thus incorporating redistricting into the discussion.

We keep assuming a finite set of scenarios capturing demand uncertainty. To embed redistricting as a recourse decision, we introduce the following notation: $r_{ij\omega}$ is the cost for reassigning TU $i \in I$ to TU $j \in I$ under scenario $\omega \in \Omega$. Regarding the decision variables, for $i, j \in I$ and $\omega \in \Omega$, we define $w_{ij\omega}$ as a binary variable equal to one if and only if TU i is assigned to district j under scenario ω, and $v_{ij\omega}$ a binary variable equal to one if ad only if $w_{ij\omega} = 1 \wedge x_{ij} = 0$. The extended problem can be formulated as follows:

$$\text{minimize} \quad \sum_{i\in I}\sum_{j\in I} c_{ij}x_{ij} + \sum_{\omega\in\Omega}\pi_\omega\left[\sum_{i\in I}\sum_{j\in I} r_{ij\omega}v_{ij\omega} + \sum_{j\in I}\left(g_j\psi_{j\omega} + h_j\varphi_{j\omega}\right)\right], \tag{13.18}$$

$$\text{subject to} \quad (13.2),\ (13.3),\ (13.5),\ (13.6),(13.17)$$

$$(1-\alpha)\hat{\mu}x_{jj} \leq \sum_{i\in I} d_{i\omega}w_{ij\omega} + \varphi_{j\omega} - \psi_{j\omega} \leq (1+\alpha)\hat{\mu}x_{jj} \qquad \forall j \in I,\ \omega\in\Omega, \tag{13.19}$$

$$\sum_{j\in I} w_{ij\omega} = 1 \qquad \forall i \in I,\ \omega\in\Omega, \tag{13.20}$$

$$\sum_{\ell\neq j} w_{j\ell\omega} \leq 1 - x_{jj} \qquad \forall j \in I,\ \omega\in\Omega, \tag{13.21}$$

$$v_{ij\omega} \geq w_{ij\omega} - x_{ij} \qquad \forall i, j \in I,\ \omega\in\Omega, \tag{13.22}$$

$$v_{ij\omega} \geq 0 \qquad \forall i, j \in I,\ \omega\in\Omega, \tag{13.23}$$

$$w_{ij\omega} \in \{0, 1\} \qquad \forall i, j \in I,\ \omega\in\Omega. \tag{13.24}$$

In this model, the double inequalities in (13.19) are the readjusted balancing constraints, with the second-stage assignments playing the major role. Equalities (13.20) ensure that in every scenario, the recourse decision is such that each TU is assigned to one and only one district. Inequalities in (13.21) prevent an initial district representative from being reassigned. This is important to ensure that the recourse territory design does not represent a complete disruption w.r.t. the initial one. Constraints in (13.22) and(13.23) guarantee that $v_{ij\omega} \geq \max\{0, w_{ij\omega} - x_{ij}\}$ $\forall i, j \in i, \omega \in \Omega$ which, together with the objective function, ensure that the optimal value of $v_{ij\omega}$ is $\max\{0, w_{ij\omega} - x_{ij}\}$ as desired.

The above redistricting recourse decision entails complete freedom to change the first-stage districting decision, which in practice may be worrisome. Indeed, we may significantly disrupt the initial districting, which is not the purpose. Moreover, "correcting" an initial solution according to each a scenario may disfavor compactness and contiguity. For this reason, we consider more realistic settings in which some "limits" are imposed on the redistricting decisions. This may emerge in different ways: (i) limitation on the number of reassignments in the second stage or (ii) an imposed similarity w.r.t. the initial districting plan.

If m is the maximum number of reassignments allowed in the second stage, then we can include the following additional constraint:

$$\sum_{i\in I}\sum_{j\in I} v_{ij\omega} \leq m \quad \forall \omega \in \Omega. \tag{13.25}$$

In some applications, it may also be relevant to guarantee a certain degree of similarity w.r.t. the territory design originally planned. We can write such similarity constraints as follows:

$$\sum_{i\in I} x_{ij} w_{ij\omega} \geq \gamma \sum_{i\in I} x_{ij} \quad \forall j \in I, \ \omega \in \Omega, \tag{13.26}$$

with $\gamma \in [0, 1]$ representing the minimum proportion of TUs to keep together in the same district according to the first-stage plan. Constraints in (13.26) can be easily linearized considering:

$$v_{ij\omega} \leq w_{ij\omega} \qquad \forall i, j \in I, \ \omega \in \Omega, \tag{13.27}$$

$$x_{ij} + v_{ij\omega} \leq 1 \qquad \forall i, j \in I, \ \omega\Omega, \tag{13.28}$$

$$\sum_{i\in I}(w_{ij\omega} - v_{ij\omega}) \geq \gamma \sum_{i\in I} x_{ij} \qquad \forall j \in I, \ \omega \in \Omega. \tag{13.29}$$

One aspect that can also be accounted for in the problem we are discussing is the existence of a maximum desirable dispersion, say $l_{\max}$ for the districting solutions being devised. Denoting by ℓ_{ij} the shortest path distance between TUs $i \in I$ and

$j \in I$ in the underlying graph, a maximum dispersion can be easily ensured for the first-stage solution:

$$x_{ij} \leftarrow 0 \qquad \forall i, j, \in I, \mid \ell_{ij} > l_{\max}. \tag{13.30}$$

A maximum value for the dispersion may also be imposed for the second-stage solution. For every $i, j \in I$ such that $\ell_{ij} > l_{\max}$ and for every $\omega \in \Omega$:

$$x_{ij} \leftarrow 0,$$
$$w_{ij\omega} \leftarrow 0,$$
$$v_{ij\omega} \leftarrow 0.$$

For every instance of the problem, there is a maximum absolute dispersion that makes sense to impose, which is obtained when solving the stochastic model.

Diglio et al. [171] consider the above modeling framework to show that capturing uncertainty in these problems may be relevant. The authors solved instances with up to 120 TUs using a general-purpose solver.

One particular aspect investigated by those authors regards the relevance of hedging against uncertainty (using the EVPI) and the possibility of using a simplified (single-scenario) model to approximate the stochastic problem (using the VSS). The application of the EVPI and the VSS is straightforward. Thus, we omit the details here (see Sect. 3.6). Nevertheless, one interesting aspect regarding the VSS is that its usage may blur the results. The objective function includes (re)assignment costs and the penalty costs for shortage/surplus demands w.r.t. the reference values. Since the penalty costs are high, they may blur a feature of great relevance to us: compactness. For this reason, in [171] the authors take the instances for which penalties are observed and focus only on the (re)assignment costs. As explained before, this induces a reliable measure of the solution compactness. It leads to a modified measure that the authors call the Compactness Value of the Stochastic Solution (CVSS). It is computed like the VSS but ignores the penalty costs (i.e., setting them to 0).

The above setting explicitly ignores contiguity. On the other hand, we are treating the balancing requirements as hard constraints. In practice, this may not be the case. The decision-maker may be willing to accept solutions that will violate the balancing requirements with a small probability. These are aspects we explore next.

13.3 Chance-Constrained Districting

Adopting the notation already presented, let $\mathbf{\Lambda} = \sum_{i \in I} \boldsymbol{\xi}_i$ be the random variable representing the total demand and let γ minimum threshold for the probability of a districting solution to turn out to be balanced. We can formulate a chance-constraint districting problem as follows:

$$\begin{aligned}
&\text{minimize} \quad (13.1),\\
&\text{subject to} \quad (13.2),\ (13.3),\ (13.6)-(13.8),\\
&\mathbb{P}\left[(1-\alpha)\frac{\Lambda}{p}x_{jj} \le \sum_{i\in I} d_i x_{ij} \le (1+\alpha)\frac{\Lambda}{p},\ j\in I\right] \ge \gamma, \qquad (13.31)\\
&w_{mij} \ge 0 \qquad \forall i,j,m \in I,\ (m,i)\in A. \qquad (13.32)
\end{aligned}$$

For the above model, a simheuristic (see [292]) is prosed in [172] taking advantage of the fact that by ignoring the chance constraints, the resulting problem is a p-median problem with connectivity requirements. A feasible solution is initially sought for the underlying p-median problem. Then, contiguity is checked. The solution is corrected if necessary to ensure that. An initial territory design is obtained. The probability of satisfying the chance constraints can now be estimated by simulation: multiple demand scenarios are generated. For each one, the balance is checked. If the proportion of balanced scenarios is greater than or equal to γ, we have a feasible solution to the chance-constrained problem; otherwise, the territory design must be adjusted. From the simulation, we can retrieve information about the balancing requirement for each district. This information can be used to update the solution by moving TUs from districts that are too often overbalanced to those that are too often underbalanced. When a new districting solution is obtained, the simulation procedure is repeated either to confirm that a feasible solution was obtained or to get hints for changing the solution to make it balanced with a higher probability.

An aspect of relevance in the procedure proposed in [172] regards the mechanism for recovering contiguity (connectivity) if it does not hold. After finding a feasible solution to the p-median problem, we need to check the contiguity constraints and correct the solution if necessary. With this purpose, we can make use of the adjacency matrix for the TUs and compute for every district the connected component that includes its representative (a procedure of $O(|I|^2)$). Then, look for the nodes not in the connected component of the district they belong to. These are the nodes contributing to non-contiguity.

Let $\overline{C}$ denote the set of TUs (if some exist) that are not contiguous to the corresponding representative, i.e., every path in the underlying graph between the TU and its representative crosses another district. Furthermore, for $i \in \overline{C}$, let C_i denote the set containing every TU adjacent to i that is either a representative or is contiguous to its representative. In [172] the authors prove that if the graph underlying the problem is connected and if $\overline{C} \neq \emptyset$, there is at least one $i \in \overline{C}$ such that $C_i \neq \emptyset$.

Based upon this result, if $\overline{C} \neq \emptyset$ after a feasible solution is obtained for the underlying uncapacitated p-median, then contiguity can be ensured by applying Algorithm 13.1. In this algorithm, for a TU $i \in I$, the representative of its district is denoted by j_i.

Algorithm 13.1: Ensure contiguity ($\overline{C} \neq \emptyset$)

1 **while** $\overline{C} \neq \emptyset$ **do**
2   **for** $i \in \overline{C}$ **do**
3   Build C_i
4   **if** $C_i \neq \emptyset$ **then**
5             $\ell^* \leftarrow \arg\min_{\ell \in C_i} \{c_{ij_\ell}\}$;
6             $x_{ij_i} \leftarrow 0; \quad x_{ij_{\ell^*}} \leftarrow 1$;
7             $\overline{C} \leftarrow \overline{C} \setminus \{i\}$;

The reasoning underlying the algorithm is simple: take a TU $i \in \overline{C}$ and build the list, say C_i, with all TUs ℓ adjacent to it, that are either representatives of a district or that are connected to the representatives of the districts they belong to. If $C_i \neq \emptyset$, assign i to the district represented in that set that induces the lowest-cost solution. This makes the solution "more contiguous" since node i is assigned to a district adjacent to it.

In Algorithm 13.1, when considering a node i in line 2, the corresponding set C_i may be empty. This means that there is no TU adjacent to i that is a district representative or is connected to its district representative. In such a case, we choose another TU in $\overline{C}$. Given the above result proved in [172], we know that there is at least one node in $\overline{C}$ whose contiguity to a representative is ensured. Hence, all $i \in \overline{C}$ will eventually be reassigned (in a "contiguous" way). Thus, the algorithm terminates in a finite number of steps and returns a contiguous solution.

Diglio et al. [173] derived another heuristic for the chance-constrained districting problem that outperforms the one above described. The authors consider the contiguity requirements imposed using connectivity Constraints in (13.9). Furthermore, they seek to replace the chance constraints with a deterministic counterpart. Next, we present the details.

We assume that uncertainty can be captured by a finite set of scenarios denoted by Ω. For each scenario $\omega \in \Omega$, let π_ω be its occurring probability that we also assume to know.

We can now index in Ω the demand vector and the reference thresholds for balancing constraints. In particular, for $\omega \in \Omega$, we denote by $d_{i\omega}$ the demand of TU $i \in I$ under scenario ω as by μ_ω the reference value for the demand to be served in each district under scenario $\omega \in \Omega$. A natural choice for μ_ω is D_ω/p, where D_ω is the total demand under scenario ω, i.e., $D_\omega = \sum_{i \in N} d_{i\omega}$.

Let us now consider an additional set of decision variables, say χ_ω, each of which is equal to one if the territory design is balanced under scenario $\omega \in \Omega$ and zero otherwise. The following scenario-indexed model emerges:

$$\text{minimize} \quad (13.1)$$

$$\text{subject to} \quad (13.2),\ (13.3),\ (13.6),\ (13.9),\ (13.10),$$

$$\sum_{i \in I} d_{i\omega} x_{ij} + U_\omega \chi_\omega \geq (1-\alpha)\mu_\omega x_{jj} \qquad \forall j \in I,\ \omega \in \Omega, \tag{13.33}$$

$$\sum_{i \in N} d_{i\omega} x_{ij} - L_\omega \chi_\omega \leq (1+\alpha)\mu_\omega x_{jj} \qquad \forall j \in N,\ \omega \in \Omega, \tag{13.34}$$

$$\sum_{\omega \in \Omega} \pi_\omega \chi_\omega \leq 1-\gamma, \tag{13.35}$$

$$\chi_\omega \in \{0, 1\} \qquad \forall \omega \in \Omega. \tag{13.36}$$

Inequalities in (13.33) and (13.34) are the "new" balancing constraints. Parameters U_ω and L_ω are large numbers that help to activate variables χ_ω for the scenarios in which the solution is not balanced. In other words, if, for a given scenario ω, at least one district is under(over)-served, then the corresponding χ_ω variable becomes equal to 1, thus rendering inequalities in (13.33) (respectively, inequalities in (13.34)) inactive. Constraint (13.35) guarantees that the joint probability of having solutions not balanced under the considered scenarios is at most equal to $1-\gamma$: we seek a solution that has a probability at least equal to γ of being balanced, as we want.

Note that this model is exact in case the support of the underlying random vector is finite and approximate when the scenarios in Ω are a sample of the actual set of scenarios.

When tackling real applications, the support of the random vector $\boldsymbol{\xi}$ may have a very large cardinality, possibly infinite. This is the case of demands reasonably described by continuous probability distributions. In this case, we can seek an approximate solution to the problem by drawing a random sample from the distribution underlying $\boldsymbol{\xi}$, thus obtaining a finite set Ω of demand scenarios.

For such a random sample, an estimate for the probability of a given districting solution to be balanced is the proportion of scenarios for which the solution is balanced. An approximate solution to our problem is a solution that is balanced for at least $\gamma|\Omega|$ of its elements. To guarantee this, we can replace constraint (13.35) with

$$\sum_{\omega \in \Omega} \chi_\omega \leq \lfloor (1-\gamma)|\Omega| \rfloor. \tag{13.37}$$

This inequality ensures that the number of unbalanced scenarios is at most $\lfloor(1-\gamma)|\Omega|\rfloor$, i.e., that the districting is balanced with the given probability γ. Hereafter, we consider (13.37) instead of (13.35) because facing a high number of scenarios (even infinite) is the most challenging and likely situation in practice. In this case, the model becomes approximate for the problem.

In synthesis, we focus on the model that consists of minimizing (13.1) subject to (13.2), (13.3), (13.6), (13.9), (13.10), (13.33), (13.34), (13.36), (13.37). The following result holds:

Proposition 13.1 *Constraints* (13.33) *and* (13.34) *can be re-written, respectively, as follows:*

$$\sum_{i\in I} d_{i\omega}x_{ij} + [(1-\alpha)\mu_\omega - d_{[1]\omega}]\chi_\omega \geq (1-\alpha)\mu_\omega x_{jj} \quad \forall j \in I,\ \omega \in \Omega, \tag{13.38}$$

$$\sum_{i\in I} d_{i\omega}x_{ij} - [(p-1-\alpha)\mu_\omega - \sum_{\ell=1}^{p-1} d_{[\ell]\omega}]\chi_\omega \leq (1+\alpha)\mu_\omega x_{jj} \quad \forall j \in I,\ \omega \in \Omega, \tag{13.39}$$

with $d_{[\ell]\omega}$ denoting the ℓ-th demand value when the demands under scenario ω are sorted non-decreasingly.

Proof See [173]. □

Despite this result, we are dealing with an NP-hard problem. For this reason, a heuristic is proposed in [173] based upon a popular scheme in location analysis. It consists of iterating between location and allocation decisions until no improvement is observed. A standard location-allocation scheme is revisited in Algorithm 13.2.

Algorithm 13.2: Location-allocation scheme

```
1 Select p initial centers and denote the resulting set by I_C;
2 repeat
3     I* ← I_C;                                  // best set of centers so far.
4     Solve the model fixing x_jj = 1 ∀j ∈ I_C;
5     Find the best center within each group;
6     I_C ← new set of centers
7 until I_C = I*;
8 return x*.                                     // feasible districting solution.
```

In Algorithm 13.2 we see the location phase in lines 1 and 5, and the allocation one in line 4. The center update in line 5 is accomplished by solving a 1-median problem within each district.

In districting problems, the above scheme starts by fixing p representatives (*Location phase*). Then, assign TUs to these representatives to obtain balanced and contiguous districts (*Allocation phase*). Afterwards, we update the representatives. We proceed until the representatives do not change in two consecutive iterations.

The main loop defines an *Improvement phase*. The algorithm returns a feasible districting solution, say $x^\star$, whose representatives are $I^\star$.

Given that the above location-allocation scheme defines a local search descent method, we should find a good set of initial centers. Not only can this impact the quality of the solution but also the computational effort required to execute the algorithm. Two algorithms are devised in [173] that can be seen as variants of the above search scheme—different mechanisms are proposed for finding an initial set of representative TUs.

One possibility consists of considering an initial restricted model induced by a random sample of scenarios Ω' drawn from set Ω. Of course, we target a sample such that $|\Omega'| < |\Omega|$. The restricted model is then used to obtain an initial promising set of representative TUs. After that, we recover the complete model and iterate the location-allocation scheme. This is formalized in Algorithm 13.3.

Algorithm 13.3: A sub-sampling scheme

1 Draw a random sample Ω' from Ω;
2 Consider the model restricted to Ω';
3 Select p centers for the restricted model denoting the resulting set by I_C;
4 Solve the restricted model setting $x_{jj} = 1, \forall j \in I_C$;
5 Find the best center within each district;
6 $I'_C \leftarrow$ new set of centers;
7 **repeat**
8     $I^* \leftarrow I'_C$; // best set of centers so far
9   Solve the full model setting $x_{jj} = 1, \forall j \in I^*$;
10   Find the best center within each district;
11     $I'_C \leftarrow$ new set of p centers;
12 **until** $I'_C = I^*$;
13 **return** $x^\star$; // a feasible districting solution

Solving the model restricted to the (sub)sample Ω' corresponds to the location phase in the location-allocation scheme for the problem at hand. Solving a "smaller problem" is a way to obtain a promising initial set of representative TUs for the original problem.

The selection of p centers in line 3 can be accomplished in different ways. One possibility is to use the competitive heuristic proposed in [449]. Alternatively, we can consider the well-known p-means algorithm as suggested in [456] for a similar problem. The resolution of the restricted model in line 4 and model in line 9 can be done using the row-generation algorithm proposed in [469]. Finally, the best center in each district (lines 3 and 10) reduces to solving the 1-median problem induced by the TUs in the district, minimizing the allocation cost (i.e., optimizing compactness).

The above procedure still relies on a multiple-scenario set. In some cases, it may be preferable to consider the extreme case of single-scenario instances. This can be done by sampling scenarios (randomly without repetition) from Ω. Each single-scenario problem can be tackled using a general-purpose solver imposing a time limit. To avoid spending too much time at this stage, such a limit should be small. Accordingly, we must be prepared to have some single-scenario instances not successfully solved. When planning for a specific scenario, we face a deterministic setting: the balancing constraints must be satisfied for that single scenario (which corresponds to directly set $\gamma = 1$).

Ideally, we would like to successfully solve a certain number m of single-scenario problems with m being a user-defined parameter. Thus, we must set the maximum number of single-scenario problems to consider larger than or equal to m. Taking the m (or less) sets of p centers found in the above process, we can define an *attractiveness* measure for each TU to become a representative of a district and also for a set of centers to be the set of representatives. The p most attractive TUs or the most attractive set of centers are selected and set as the initial set of centers to be fixed in the complete model. From here, the classical location-allocation scheme is executed. The procedure is formalized in Algorithm 13.4.

Algorithm 13.4: A single-scenario-based scheme

1 Draw a sample Ω' randomly without repetition from Ω;
2 counter $\leftarrow$ 0;
3 **for** $\omega \in \Omega'$ **do**
4 | Consider the restricted model; // single-scenario problem
5 | Select p centers for the restricted model and denote the resulting set by I_C;
6 | Attempt solving the restricted model fixing $x_{jj} = 1\ \forall j \in I_C$;
7 | **if** *succeeded* **then**
8 | | counter $\leftarrow$ counter + 1; // the restricted model was successfully solved
9 | | Find the best center within each district;
10 | | Update the attractiveness measure;
11 | | **if** *(counter $\geq m$)* **then**
12 | | | break; // exit the `for` loop
13 Select the most attractive p centers and denote the resulting set by I'_C;
14 **repeat**
15 | $I^* \leftarrow I'_C$; // best set of centers so far
16 | Solve full scenario-index model setting $x_{jj} = 1, \forall j \in I^*$;
17 | Find the best center within each district;
18 | $I'_C \leftarrow$ new set of p centers;
19 **until** $I'_C = I^*$;
20 **return** $x^\star$; // a feasible districting solution

In this algorithm, lines 1–10 correspond to the location phase in the overall location-allocation scheme. The selection of p centers in line 5 can be accomplished as above described (following [449] or [456]). The resolution of the single-scenario model (line 6) and the full scenario-indexed model in line 16 can be accomplished as in [469]. Finally, like above, the best center in each district (lines 9 and 17) is found solving the 1-median problem induced by the TUs in each district to optimize compactness.

The aspect remaining to discuss regards measuring the attractiveness of the TUs or set of TUs (line 13 in Algorithm 13.4).

Let us denote by m the number of single-scenario problems successfully solved (desirably the original targeted number). For $k = 1, \ldots, m$, let us denote by $\hat{\boldsymbol{x}}^k$ the solution obtained for the k-th single-scenario problem successfully solved, and by $z(\hat{\boldsymbol{x}}^k)$ its corresponding objective function value. Additionally, we consider $b(\hat{\boldsymbol{x}}^k)$ the number of balanced scenarios associated with solution $\hat{\boldsymbol{x}}^k$. This number can be easily computed by considering the demand vector associated with each scenario $\omega \in \Omega$ and checking whether the solution is balanced for that demand occurrence. Two attractiveness measures are proposed in [173]:

(i) Probabilistic attractiveness.
In this case, for each TU $j \in I$, we compute the probability, say P_j, of having j as a representative across the m scenarios:

$$P_j = \frac{1}{m} \sum_{k=1}^{m} \hat{x}_{jj}^k.$$

In Algorithm 13.4, line 13, we select the p centers with the highest probabilities.

(ii) Merit attractiveness.
In this case, for each solution $\hat{\boldsymbol{x}}^k$ $(k = 1, \ldots, m)$, we evaluate the attractiveness of the corresponding set of centers to be selected as the initial representatives. Following [456], we compute the value of a so-called merit function G as:

$$G(\hat{\boldsymbol{x}}^k) = \varsigma \left(\min_{\ell \in \{1,\ldots,m\}} z(\hat{\boldsymbol{x}}^\ell) \times \frac{1}{z(\hat{\boldsymbol{x}}^k)} \right) + (1 - \varsigma) \left(\frac{1}{\max_{\ell \in \{1,\ldots,m\}} b(\hat{\boldsymbol{x}}^\ell)} \times b(\hat{\boldsymbol{x}}^k) \right) \qquad \forall \varsigma \in [0, 1].$$

In the above expression, ς is a predefined weight that gives more relevance to a smaller objective function value or a higher number of balanced scenarios. When adopting this merit function, in line 13 of Algorithm 13.4, we select the set of p centers (among the m available from the solutions of the single-scenario problems) having the highest value of such measure.

Using the above procedures, the authors in [173] report results of computational tests for instances with up to 1000 TUs a 1000 scenarios.

A third procedure for the fully-indexed scenario model is proposed in [41]. The algorithm is based on cutting-plane generation. An initially restricted model is considered that results from drawing a sample Ω' randomly from Ω. Desirably, $|\Omega'| <<< |\Omega|$. One may even start with a single scenario, i.e., $|\Omega'| = 1$. In any case, the initial number of scenarios can be a user-defined parameter. For this initial set of scenarios, we solve the restricted model setting $\gamma = 1$, i.e., ensuring that a districting solution is balanced for all scenarios in set Ω'.

Denote by $\hat{x}(\Omega')$ the solution obtained by solving the model restricted to Ω' (and balanced for all scenarios in that set). This solution can be checked regarding balancing for every scenario in the original set Ω. Let $\Omega_B \subseteq \Omega$ be the set of scenarios for which the solution is balanced. Obviously, $\Omega' \subseteq \Omega_B$. If $|\Omega_B| > \gamma|\Omega|$, we stop. Otherwise, we look for an additional sample, say Ω'' in $\Omega \setminus \{\Omega_B \cup \Omega_T\}$, with Ω_T denoting a set of tabu scenarios. The cardinality of Ω'' can again be a user-defined parameter. The authors in [41] observed empirically that including new scenarios (and the corresponding balancing constraints) may result in fewer balanced scenarios associated with the resulting solution. Thus, one may wish to avoid considering the scenarios causing such a reduction. Set Ω_T accommodates this.

In a generic iteration of this procedure, we should check whether the number of balanced scenarios is smaller than that found in the previous iteration. If this is the case, the set Ω_T is enlarged by those additional scenarios that lead to such a reduction. Naturally, a maximum number of iterations should be defined in which scenarios in Ω_T are considered tabu.

Finally, the procedure stops when either (i) the desired number of balancing scenarios $\gamma|\Omega|$ is attained, (a feasible solution is found), (ii) a user-defined number of iterations is reached, (iii) a user-defined time limit is attained, or (iv) the set of "eligible" scenarios that can be added to the model is empty (i.e., it is not possible to add more scenarios to the restricted model because the size of the tabu list is large).

When a feasible solution is obtained, it can be worked out for further improvement in line with the location-allocation scheme already described. Other elements required by the procedure have already been described (the selection of p centers or the resolution of the restricted model).

When a solution is obtained, it induces p partitions of the underlying graph. We need to check if each partition contains more than one connected component. That check can be done in polynomial time (see [172]). If so, a connected component induces a violated contiguity constraint in (13.9). The (restricted) model must be solved again with such constraints embedded.

In this third procedure, what remains to be defined is a criterion for selecting the additional scenarios to include in the model. Baldassarre et al. [41] provide some ideas in addition to a straightforward random selection.

For each $\omega \in \Omega \setminus \Omega_T$, we can associate to each district j a relative deviation, say $\delta_{j\omega}$ concerning the thresholds for considering the district balanced. Let $D_{j\omega}$ be the total demand originated at the district under the scenario. We have

$$\delta_{j\omega} = \max\left\{0, \frac{D_{j\omega} - UB_{\omega}}{UB_{\omega}}, \frac{LB_{\omega} - D_{j\omega}}{LB_{\omega}}\right\} \quad \forall j \in J \mid x_{jj} = 1,\ \omega \in \Omega \setminus \Omega_T,$$

with $UB_\omega = (1+\alpha)\mu_\omega$, $LB_\omega = (1-\alpha)\mu_\omega$ and $\mu_\omega = (1/p)\sum_{i\in I} d_{i\omega}$. Now we can associate each scenario ω with different balancing violation measures:

Lowest deviation: $\delta_\omega^L = \min_{j\in I|x_{jj}=1} \{\delta_{j\omega}\}.$

Highest deviation: $\delta_\omega^U = \max_{j\in I|x_{jj}=1} \{\delta_{j\omega}\}.$

Average deviation: $\overline{\delta}_\omega = (1/p) \sum_{j\in I|x_{jj}=1} \delta_{j\omega}.$

The above measures can be used separately or combined to select the scenarios to include in Ω''.

13.4 Other Ways for Hedging Against Uncertainty in Districting

In this section, we change in different ways the flavor of the districting problems considered so far. First, we discuss a different type of mathematical model. Second, we cast uncertainty using robust optimization.

The districting problem we consider stems from applications where a maximum acceptable dispersion is given. Each TU $i \in I$ has a demand that we keep denoting by d_i. Now, it also has a service capacity given by q_i. The goal is to organize the territory into a given number p of districts to minimize the maximum difference between capacity and demand across districts.

Using the notation considered so far, we can formulate the problem as follows:

$$\text{minimize} \quad \max_{j,k\in I|j\neq k} \left\{ \sum_{i\in I} (q_i x_{ij} - d_i x_{ij}) - \sum_{i\in I} (q_i x_{ik} - d_i x_{ik}) \right\}, \tag{13.40}$$

$$\text{subject to} \quad \sum_{j\in I} x_{ij} = 1 \qquad \forall i \in I, \tag{13.2}$$

$$\sum_{j\in J} x_{jj} = p, \tag{13.3}$$

$$\sum_{i\in S} x_{ij} - \sum_{i\in \cup_{v\in S}(I_v\setminus S)} x_{ij} \le |S| - 1$$

$$\forall j \in I,\ S \subset [I \setminus (I_j \cup \{j\})], \tag{13.9}$$

$$x_{ik} + x_{jk} \le 1 \qquad \forall i, k, j \in J,\ \ell_{ij} > l_{\max}, \tag{13.41}$$

$$x_{ij} \in \{0, 1\} \qquad \forall i, j \in I. \tag{13.6}$$

Recall that $l_{\max}$ is the maximum allowed distance between TUs within the same district, and ℓ_{ij} is the shortest path distance between TUs $i \in I$ and $j \in J$ in the underlying network. I_v is the set of TUs adjacent to v in the underlying graph.

Compared to the models discussed so far, we now see the balancing accounted for as an objective function. On the other hand, dispersion is handled in the constraint set with a maximum threshold imposed.

As it becomes clear from the previous sections, contiguity in a general graph is challenging when seeking feasible districting solutions. In the context of health care systems planning, a specific topology for solving the problem is assumed in [390]. If the districting problem is to be solved on a spanning tree for the underlying graph, a sufficient condition for contiguity is that if two TUs i and j are in the same district, then all the TUs in the single path connecting those nodes should also be in the same district.

Let $\mathcal{T}$ be a spanning tree for the graph $G = (V, A)$ underlying our districting problem (see Sect. 13.1). For $i, j \in I$, let $S_{ij}^{\mathcal{T}}$ be the set of intermediate TUs in the unique path connecting i and j in $\mathcal{T}$ and let $\ell_{ij}^{\mathcal{T}}$ be the length of that path. A spanning tree-based model can now be formulated as follows (we also linearize the objective function):

$$\text{minimize} \quad \eta, \tag{13.42}$$

$$\text{subject to} \quad \sum_{i \in I}(q_i x_{ij} - d_i x_{ij}) - \sum_{i \in I}(q_i x_{ik} - d_i x_{ik}) \leq \eta \qquad \forall j, k \in I,\ j \neq k, \tag{13.43}$$

$$(13.2),\ (13.3),\ (13.6),\ (13.41),$$

$$\sum_{m \in S_{ik}^{\mathcal{T}}} x_{mj} \geq |S_{ik}^{\mathcal{T}}|(x_{ij} + x_{kj} - 1) \quad \forall i, j, k \in I. \tag{13.44}$$

Constraints in (13.44) ensure that if two nodes are part of the same district, all the nodes in the path connecting them in the tree are also in the same district. The above model can be simplified by observing that nominating district representatives is no longer necessary due to the adopted objective function. Given that the number of districts to create is equal to p, we can index the districts in set $P = \{1, \ldots, p\}$ and redefine the decision variables setting x_{ik} binary and equal to one if and only if TU $i \in I$ is part of district $k \in P$. The model can be reformulated as follows:

$$\text{minimize} \quad \eta, \tag{13.45}$$

$$\text{subject to} \quad \sum_{i \in I}(q_i - d_i)x_{ik'} - \sum_{i \in I}(q_i - d_i)x_{ik''} \leq \eta \qquad \forall k', k'' \in P,\ k' \neq k'', \tag{13.46}$$

$$\sum_{k \in P} x_{ik} = 1 \qquad \forall i \in I, \tag{13.47}$$

$$\sum_{m \in S_{ij}^{\mathcal{T}}} x_{mk} \geq |S_{ij}^{\mathcal{T}}|(x_{ik} + x_{jk} - 1) \qquad \forall k \in P, \forall i, j \in I, \tag{13.48}$$

$$x_{ik} + x_{jk} \leq 1 \qquad \forall k \in P, \ i, j \in I \mid \ell_{ij} > l_{\max}, \tag{13.49}$$

$$x_{ik} \in \{0, 1\} \qquad \forall l \in P, \ i \in I. \tag{13.50}$$

Based upon the above model, the authors in [390] propose finding a feasible solution to the districting problem by considering each TU at a time and finding a spanning tree underlying the tree of shortest paths rooted at the TU. The above model is solved for that spanning tree. A districting solution is obtained. This is repeated for all TUs. The final solution is that leading to the smallest value of the objective function.

Let us assume that demands are uncertain and captured using intervals. In Particular, we consider $d_i \in \mathcal{U}_i^D = [\overline{d}_i - \Delta_i, \overline{d}_i + \Delta_i]$, where $\overline{d}_i$ is the nominal value for the demand and Δ_i the maximum deviation from that nominal value. The robust version of constraints (13.46) is

$$\max_{d_i \in \mathcal{U}_i^D, \, i \in I} \left\{ \sum_{i \in I} (q_i - d_i) x_{ik'} - \sum_{i \in I} (q_i - d_i) x_{ik''} \right\} \leq \eta \qquad \forall k', k'' \in P, \ k' \neq k''. \tag{13.51}$$

To make the model easy-reading, we define $a_i = q_i - d_i \ \forall i \in I$. Accordingly, the above uncertainty sets induce $\mathcal{U}_i = [\overline{a}_i - \Delta_i, \overline{a}_i + \Delta_i]$, and $\overline{a}_i = q_i - \overline{d}_i, \forall i \in I$. Constraints in (13.51) can be written as follows:

$$\max_{a_i \in \mathcal{U}_i, \forall i \in I} \left\{ \sum_{i \in I} a_i x_{ik'} - \sum_{i \in I} a_i x_{ik''} \right\} \leq \eta \qquad \forall k', k'' \in P, \ k' \neq k''. \tag{13.52}$$

As discussed several times in this volume, in practice, it is not likely that all the parameters assume their worst values. This has led to alternative and more realistic uncertainty sets. Let us consider the scaled deviations (see Sect. 5.3): $\vartheta_i = \frac{a_i - \overline{a}_i}{\Delta_i}$ $\forall i \in I$. Accordingly, we have $a_i = \Delta_i \vartheta_i + \overline{a}_i \ \forall i \in I$. We can go even further. When the districts are created, they will end up operating independently. Thus, we can consider a scaling factor associated with a TU and the district to which it belongs. For $k \in P$ let us define a budget-uncertainty set [75]:

$$\mathcal{U}_k = \left\{ (\vartheta_{1k}, \ldots, \vartheta_{|I|k}) \in \mathbb{R}^{|I|} \mid \sum_{i \in I} \vartheta_{ik} \leq \Gamma \ \vee \ |\vartheta_{ik}| \leq 1, \ \forall i \in I \right\}.$$

Inequalities in (13.52) can be adapted as follows:

$$\max_{\substack{(\vartheta_{1k'},\ldots,\vartheta_{|I|k'})\in\mathcal{U}_{k'}\\(\vartheta_{1k''},\ldots,\vartheta_{|I|k''})\in\mathcal{U}_{k''}}} \left\{ \sum_{i\in I}(\Delta_i\vartheta_{ik'}+\overline{a}_i)x_{ik'} - \sum_{i\in I}(\Delta_i\vartheta_{ik''}+\overline{a}_i)x_{ik''} \right\} \leq \eta$$

$$\forall k', k'' \in P,\ k' \neq k'', \qquad (13.53)$$

which can be equivalently written as

$$\sum_{i\in I}\overline{a}_i x_{ik'} - \sum_{i\in I}\overline{a}_i x_{ik''} +$$

$$\max_{\substack{(\vartheta_{1k'},\ldots,\vartheta_{|I|k'})\in\mathcal{U}_{k'}\\(\vartheta_{1k''},\ldots,\vartheta_{|I|k''})\in\mathcal{U}_{k''}}} \left\{ \sum_{i\in I}\Delta_i\vartheta_{ik'}x_{ik'} - \sum_{i\in I}\Delta_i\vartheta_{ik''}x_{ik''} \right\} \leq \nu \quad \forall k', k'' \in P,\ k'\neq k''.$$

$$(13.54)$$

For a given districting decision, say $\hat{x}$, the scaled deviations $\vartheta_{1k'},\ldots,\vartheta_{|I|k'}$ will take positive values whereas the deviations $\vartheta_{1k''},\ldots,\vartheta_{|I|k''}$ take negative values. This allows writing the inner maximization problem as follows (for a fixed districting solution):

$$\underset{k',k''\in P,\ k'\neq k''}{\text{maximize}} \quad \sum_{i\in I}\Delta_i\vartheta_{ik'}\hat{x}_{ik'} + \sum_{i\in I}\Delta_i\vartheta_{ik''}\hat{x}_{ik''}, \qquad (13.55)$$

$$\text{subject to} \quad \sum_{i\in I}\vartheta_{ik} \leq \Gamma \qquad \forall k\in P, \qquad (13.56)$$

$$0 \leq \vartheta_{ik} \leq 1 \qquad \forall k\in P,\ i\in V. \qquad (13.57)$$

Dualizing the above linear programming problem allows replacing (13.54) by

$$\sum_{i\in I}\overline{a}_i x_{ik'} - \sum_{i\in I}\overline{a}_i x_{ik''} + \Gamma(\zeta_{k'}+\zeta_{k''}) + \sum_{i\in I}\varrho_{ik'} + \sum_{i\in I}\varrho_{ik''} \leq \eta$$

$$\forall k', k'' \in P,\ k' \neq k'',$$

$$\varrho_{ik} + \lambda_k \geq \Delta_i x_{ik} \qquad \forall k\in P,\ i\in I,$$

$$\lambda_k \geq 0 \qquad \forall k\in P,$$

$$\varrho_{ik} \geq 0 \qquad \forall k\in P,\ i\in I.$$

In the above constraints, λ_k ($k \in P$) are the dual variables associated with (13.56), and ϱ_{ik} are those associated with the right-hand side inequality in (13.57). These developments allow formulate our problem as an MILP. Note that $\Gamma = 0$ reduces the problem to the nominal formulation, whereas $\Gamma = |I|$ immunizes the solution against the simultaneous worst-case values for all parameters.

The above model may become too large in some realistic settings. A genetic algorithm is proposed in [390] for finding feasible solutions to the problems. The interested reader should refer to that work for the related algorithmic details.

13.5 Conclusions and Further Reading

This chapter focused on immunizing territory design decisions against uncertainty in the data, namely in the activity level generated at each territory unit. This topic has much potential growth due to its practical relevance and scarce work found in the literature. One aspect worth investigating regards the combination of uncertainty with a time-dependent decision. This calls for casting the problems in a multi-stage decision-making context. In the case of a stochastic programming paradigm, this leads to a multi-stage stochastic optimization problem. Another research trend to be pursued regards the inclusion of more comprehensive uncertainty, such as ambiguity in the underlying probability distributions and possible correlation between the underlying random variables.

When looking into a districting problem, one realizes that it can be seen as a clustering problem with additional features such as contiguity—a feature relevant in a geographical context—or balancing. Therefore, one may think to extend to clustering some developments related to districting and vice-versa.

Clustering uncertain data is not new (see [140]). The typical input consists of a set of probability distribution functions, each describing the locations of a point in a metric space (e.g., the n-th dimensional Euclidean space). The goal is to cluster the points around a subset of them to be found, each representing a cluster. Different policies may be considered. One is to assign each point to the same cluster irrespective of what its location turns out to be. The other is to keep the representatives fixed but adapt the allocations to the actual location of the points. Median or center (expected) costs can then be devised. Some further reading on the topic include [288, 484, 575].

Despite a close relation with clustering, districting calls for specific models and solution techniques to be devised due to the geographical context in which they emerge.

Part IV
Comprehensive Uncertainty with Applications

Chapter 14
Bilevel Recycling Facility Location

Abstract This chapter presents an application of distributional robust facility location to a resource recovery planning problem under a public-private partnership. The problem intertwines two subproblems: one of interest to the local authority and another sought by a private operator. The former seeks a waste sorting scheme and a cost-sharing budget to minimize the expected financial benefit. The economic feasibility of the private operator must be ensured. Given the budget and sorting scheme, the private operator seeks to locate and operate resource recovery facilities. The goal is to maximize the expected total profit. A salient feature of the problem is that the feedstock condition (composition and volume) is uncertain. Moreover, it can be influenced by the sorting-at-source scheme implemented. This feature drives the interactions between public and private partners. A bilevel resource recovery planning model is devised under a distributionally robust optimization framework. A decision-dependent ambiguity set is defined to model the influence of sorting schemes on the feedstock uncertainty. The problem can be formulated and solved as a mixed-integer linear program. Numerical experiments are reported demonstrating that the optimization framework provides a helpful decision-analysis tool for resource recovery planning under feedstock ambiguity.

Keywords Resource recovery system design · Robustness · Feedstock uncertainty · Economic feasibility

14.1 Introduction

In the wake of increasing pressures for hedging against global climate change, sustainable waste management continues to present challenges to regulators. As reported in [274], the annual waste generation is projected to increase to 2.2 billion tons (1.42 kg/capita/day) by 2025 (from 1.3 billion tons (1.2 kg/capita/day) in 2012). To mitigate the environmental impacts associated with waste disposal, many advances in the so-called waste-to-resource recovery technologies (e.g., Anaerobic Digestion (AD) and gasification for organic waste streams) have become com-

© The Author(s), under exclusive license to Springer Nature Switzerland AG 2024
F. Saldanha-da-Gama, S. Wang, *Facility Location Under Uncertainty*,
International Series in Operations Research & Management Science 356,
https://doi.org/10.1007/978-3-031-55927-3_14

mercially available in suitable configurations and scales to recover products, e.g., energy for power generation, fuel gas, and combustion ashes for road construction (see [307, 329, 395]). These technologies have been widely recognized to be much cleaner and more efficient than conventional mass incineration, creating exciting opportunities to improve *waste-to-energy economics* [483, 540].

Adopting new resource recovery technologies may not be sufficient to ensure successful waste management solutions. A case in point is the demise of a Singapore food waste recycling company. IUT Global, which shut down in 2011, operated only three years after its launch even though it was awarded "National ENERGY GLOBE Overall Winner" in 2008 for its sophistication in AD technology and recycling business model. In 2009, the company invested in building a Phase-II AD plant, doubling the capacity of the Phase-I plant. Unfortunately, the collected waste feedstock became only one-third of its treatment capacity.

A key lesson, in this case, is the importance of comprehensive planning for financially and environmentally sustainable waste management that takes into careful account the variability of operating conditions (e.g., the input feedstock composition and volume), the sharing responsibility between private and public stakeholders, the impact of waste *sorting-at-source* mechanisms, and also the goals and requirements for different stakeholders. In the case of IUT Global, as reported in [523], up to 40% of food waste collected by the company was contaminated with plastic bags and other inorganic trash, which significantly raised its labor costs in separating the impurities. With a poor feedstock condition projection and without contracting with the local authority in feedstock (waste) quality control (sorting) and fixed cost-sharing, IUT Global did not receive effective aid from the latter, exposing itself to the risks of the high variability of feedstock condition, and high upfront cost, which finally led to its liquidation.

In waste management, the formation of Public-Private Partnership (P3) between *local authorities* and *private operators* has become of increasing interest in recent years. Such partnerships, through different cost-sharing mechanisms, determine the manner and mode in which resource recovery operations are implemented [34, 35, 544].

We consider a resource recovery planning problem involving public and private stakeholders (in the P3 setting). The problem primarily involves two key phases: waste sorting-at-source and waste treatment. The waste generated at different sites is first sorted by applying the scheme implemented. It may include simple separation procedures or more sophisticated automated sorting. Next, the sorted waste streams—"feedstocks"—are allocated to various treatment and recovery facilities where valuable materials and components are extracted. These are sold off in the market or recycled. Finally, any remaining components or residues that cannot be further usefully recovered are given appropriate end-of-life treatment and disposal.

The investment cost of resource recovery facilities is shared between the local authority and the private operator. The local authority determines its budget for facility installation sharing costs. It also owns and manages all waste sorting-at-source operations at the waste collection sites. Hence, its problem seeks to decide the sorting schemes and the budget level to minimize the expected payout. The financial sustainability of the private operator must be ensured. Given the capital investment budget and sorting scheme, the private operator manages the waste collection and resource recovery operations. Its problem is to install the resource recovery facilities and conduct the recovery operations. The goal is to maximize the expected total recovery profit.

A salient feature of the problem is that the characteristics of input feedstock (composition and volume) are generally uncertain or only partially known. This can significantly impact the recourse recovery performance (see [472] and the example above). The input feedstock, in turn, can be directly influenced by the public sorting schemes implemented by the local authority. Consequently, the input feedstock condition is a natural coupling point between public and private partners in the waste management system. For instance, if the original feedstock condition is of good quality, such as the high carbohydrate content in food waste, the economic feasibility of the resource recovery operations can be easily achieved. Hence, the local authority may only need to invest in a sorting scheme with a low cost and allocate a smaller budget for cost-sharing. On the other hand, under poor feedstock quality and investment funding budget, the private operator may be unable to achieve an economically viable waste treatment solution. In such a case, the local authority must decide between allocating more facility investment budget or investing more in the sorting-at-source to improve the input feedstock condition. This is more challenging when the feedstock condition is highly uncertain.

From the modeling and optimization perspective, the above problem of resource recovery planning under P3 has a bilevel structure, where the local authority is the leader, and the private operator is the follower who solves a two-stage facility location and feedstock-flow allocation problem. More importantly, the feature of sorting-dependent feedstock condition in the problem leads to a mechanism of *proactive* optimization under decision-dependent uncertainty. It generalizes the more common *reactive* optimization problems under uncertainty, where all uncertainties are assumed to be exogenous to the problem [55, 76, 78]. To address this decision-dependent uncertainty in our resource recovery planning problem, we build a bilevel model under the framework of distributionally robust optimization.

The remainder of this chapter is organized as follows. Section 14.2 introduces the problem background, develops the distributionally robust bilevel model, and investigates several properties of the model. Section 14.3 looks into a sorting-dependent ambiguity set for the bilevel model. Section 14.4 investigates tractable reformulations for the problem. Section 14.5 presents computational results to assess the modeling frameworks discussed. The chapter concludes with an overview and some suggestions for further reading.

14.2 A Robust Bilevel Model for Resource Recovery Planning

We consider a resource recovery problem involving two entities: a Local Authority (LA) and a Private Operator (PO). Both envision participating in a P3 scheme for a large-scale program on waste recycling, which primarily involves two key phases: waste sorting-at-source and waste treatment by resource recovery. Figure 14.1 presents a simple example of such a program for household waste recycling. An issue of central interest in this problem is the efficiency of the resource recovery phase, which can be significantly influenced by the outcomes of the waste sorting phase upstream. Consequently, the successful implementation of the recycling program hinges on the effective design and management of these two phases in tandem.

In the P3 partnership, the LA determines the waste sorting schemes to be implemented (in all regions or communities). The LA is responsible for the cost of the waste sorting program implementation. Additionally, we assume a cost-sharing policy—a matching fund mechanism—where the LA sponsors a portion of the investment costs for installing the resource recovery facilities. Cost-sharing is a well-acknowledged P3 practice as a financial incentive for attracting qualified tenders [308, 467, 473, 494]. The LA budget for co-sharing investment costs, or what is referred to as the matching fund budget, affects the number and types of facilities to open, which consequently impacts the operating capacity of the PO. On the other hand, the sorting scheme influences the feedstock quality and thus the operational efficiency—hence the revenues—of the PO. Given the matching fund budget and the sorting schemes, the PO determines the location and capacity of the resource recovery facilities and manages the resource recovery operations to maximize its expected total profit.

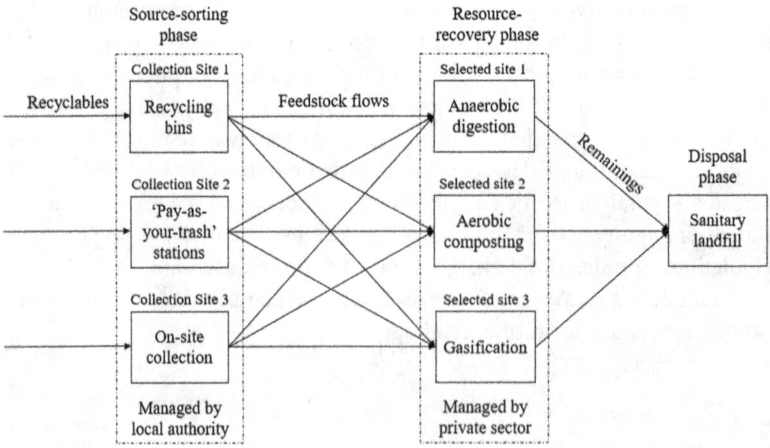

Fig. 14.1 The diagram of a 3 × 3 household waste recycling system under the P3 contract

For a P3 partnership to be viable, the recycling program must be economically feasible for both parties. Hence, the LA problem seeks to jointly determine the optimal waste sorting schemes and the matching fund budget to make the recycling program economically feasible for the PO (for instance, the optimal profit should be no smaller than some target level) while minimizing the costs. Both decisions to be made by the LA can be regarded as instruments that help the PO achieve economic feasibility. Furthermore, they must be carefully balanced to ensure the optimal system performance. Generally, higher investments in sorting lead to smaller matching fund budgets available, and *vice versa*.

A relevant parameter in the assumed matching fund mechanism is the cost-share ratio. Its value determines the proportion of resource recovery facility investment costs to be covered by the PO. The remaining runs for the LA in the context of the matching fund mechanism. Different values assumed for the cost-share ratio lead to distinct partnership arrangements in P3 contracts [34, 544]. This is discussed in detail in Sect. 14.2.3. Regarding model development, we assume that the cost-share ratio is a predetermined and exogenous parameter. Nevertheless, we show how it can be integrated as a decision variable without much technical difficulty.

Finally, we assume that the LA works with a single PO, who can be one among the multiple qualified bidders pursuing the P3 contract. Therefore, we do not consider the horizontal competition among the multiple POs. Instead, we focus on studying the dynamics between the LA and PO. Under this assumption, the model presented next can be applied to evaluate the optimized system design under the partnership between the LA and each PO, which is by the case-by-case bid evaluation approach in reality. Besides, we also assume that the sorted feedstock streams can be handled at all installed resource recovery facilities of the contract-awarded PO in the studied system. This assumption reflects the qualification in the bidding procedure of a public project in reality. In particular, when the LA contracts a PO to execute a resource recovery project, proficiency in treating target feedstock categories is usually one technical requirement to the public in bidding documents (see [473]).

We denote by H, I, J, and N the sets of alternative waste sorting schemes, alternative resource recovery facility sites, waste collection sites, and possible waste components in the sorted feedstock, respectively. We define v_j as the feedstock volume at the collection site $j \in J$ with the vector $\boldsymbol{v} = (v_j)_{j \in J}$ and define u_{jn} as the percentage of the waste component $n \in N$ in a unit amount of the sorted feedstock from the zone $j \in J$, with $\boldsymbol{u} = (u_{jn})_{j \in J, n \in N}$. In particular, we have $\sum_{n \in N} u_{jn} = 1 \; \forall j \in J$ with each $u_{jn} \geq 0$.

To formulate the problem, we consider several sets of decision variables. For $j \in J$ and $h \in H$, s_{jh} is a binary variable equal to one if the waste sorting scheme h is selected at the waste collection zone j and zero otherwise; $\boldsymbol{s} = (s_{jh})_{j \in J, h \in H}$. For $i \in I$, y_i is a binary variable equal to one if the resource recovery site i is selected for installing a facility and 0 otherwise; $\boldsymbol{y} = (y_i)_{i \in I}$. Finally, for $i \in I$ and $j \in J$, x_{ij} represents the (amount of) feedstock distributed from the collection zone j to the recovery site i.

The decision variables of the LA are the sorting scheme selection $\boldsymbol{s}$ and budget level ξ. The decision variables of the PO include the resource recovery site selection $\boldsymbol{y}$ and feedstock-flow allocation $\boldsymbol{x}$.

14.2.1 The Problem of the Private Operator

For a given resource recovery facility location solution $\boldsymbol{y}$, the sorted feedstock from collection zones is dispatched to open facilities for treatment and resource recovery. In practice, the private operator may charge service or "gate" fees for waste disposed of. Alternatively, they can pay some salvage value to waste collectors. In turn, treated and recovered products from the PO (e.g., electricity and recyclable materials) can be sold to the market. Specifically, the revenue accrued from the resource recovery unit at each treatment site $i \in I$ is given by

$$\sum_{j \in J} x_{ij} \varsigma_i + \sum_{j \in J} \sum_{n \in N} \mu_{in} u_{jn} x_{ij}, \tag{14.1}$$

where ς_i is the unit gate fee of handling per unit feedstock at the site i, and μ_{in} is the unit revenue obtained from one unit feedstock component n at the site i, i.e., the sales of recovered products.

The operational cost of resource recovery at $i \in I$ is given by

$$\epsilon \left(\sum_{j \in J} \vartheta_{ij} x_{ij} \right) + \sum_{j \in J} x_{ij} \iota_i + c_D \left(\sum_{j \in J} \sum_{n \in N} \sigma_{in} u_{jn} x_{ij} \right) + \sum_{j \in J} \sum_{n \in N} \eta_{in} u_{jn} x_{ij}. \tag{14.2}$$

This consists of transportation costs, waste processing costs, residual disposal costs, and possible environmental-related costs. In particular, the first cost term refers to the total transportation cost, where ϵ is the transportation cost per unit distance traveled for per unit amount of feedstock, and ϑ_{ij} is the distance between collection site j and resource recovery site i. The second term shows the total processing cost, where ι_i is the unit variable cost of incoming feedstock processed at the resource recovery site i. The third item is the residual disposal cost, where c_D is the unit residual disposal cost, and σ_{in} is the proportion of residual generated from treating per unit waste component n at the resource recovery site i. The last term refers to the total emission cost, where η_{in} refers to the possible emission cost from processing one unit of waste component n in the feedstock at the resource recovery site i.

Based on the above, the total profit for the PO (not considering fixed capacity repayments) accrued from treating $\boldsymbol{x}$ can be written as

$$\sum_{i \in I} \left[\sum_{j \in J} (\varsigma_i - \epsilon \vartheta_{ij} - \iota_i) x_{ij} + \sum_{j \in J} \sum_{n \in N} (\mu_{in} - c_D \sigma_{in} - \eta_{in}) u_{jn} x_{ij} \right]. \tag{14.3}$$

To simplify notation, we define

$$r_{ijn} = \varsigma_i + \mu_{in} - \epsilon\vartheta_{ij} - l_i - c_D\sigma_{in} - \eta_{in}$$

as the unit operational profit coefficient of treating component n in the feedstock sorted from the collection zone j at the resource recovery site i. The components of r_{ijn} reflect the preference of the PO to select a resource recovery site i that is nearer to the collection site j to reduce transportation cost (ϑ_{ij}). On the other hand, the PO would also like to receive the feedstock containing a higher concentration of the desired waste component n, i.e., the purer feedstock. Usually, the latter has a relatively higher recovery revenue (μ_{in}), and a relatively lower residual content (σ_{in}) and emission cost (η_{in}) (see [104]). Recalling that $\sum_{n\in N} u_{jn} = 1\ \forall j \in J$, we can re-write (14.3) as follows:

$$\sum_{i\in I}\sum_{j\in J}\sum_{n\in N} r_{ijn}u_{jn}x_{ij}. \tag{14.4}$$

For a given facility location design $\boldsymbol{y}$ and a feedstock condition $(\boldsymbol{v}, \boldsymbol{u})$, the PO maximizes its payoffs by optimizing the dispatching decisions $\boldsymbol{x}$, formulated as:

$$\phi(\boldsymbol{y}, \boldsymbol{u}, \boldsymbol{v}) = \max \quad \sum_{i\in I}\sum_{j\in J}\sum_{n\in N} r_{ijn}u_{jn}x_{ij} - c_D\sum_{j\in J}\left(v_j - \sum_{i\in I} x_{ij}\right), \tag{14.5}$$

$$\text{s. t.} \quad \sum_{i\in I} x_{ij} \le v_j \qquad \forall j \in J, \tag{14.6}$$

$$\sum_{j\in J} x_{ij} \le y_i\varpi_i \qquad \forall i \in I, \tag{14.7}$$

$$x_{ij} \ge 0 \qquad \forall i \in I,\ j \in J. \tag{14.8}$$

The objective function (14.5) includes the operational profit from the treated feedstock (14.4) and the disposal cost for the untreated feedstock. The latter emerges when there is insufficient treatment capacity. Constraints in (14.6) ensure that all feedstocks collected at each zone and processed by different resource recovery units cannot exceed the total feedstock quantity from that zone. Constraints in (14.7) impose the capacity limit ϖ_j on the total feedstock processed at each opened resource recovery unit.

Next, we consider a matching fund budget (cost-sharing budget) ξ and a sorting scheme $\boldsymbol{s}$ implemented by the LA. Recall that ξ is the maximum amount of funds that the LA will sponsor the PO for investments in resource recovery facilities and capacities.

We assume that the feedstock condition is uncertain, thus represented by a random vector $(\tilde{\boldsymbol{u}}, \tilde{\boldsymbol{v}})$. Moreover, the probability distribution of the random vector is also assumed uncertain and affected by the sorting scheme $\boldsymbol{s}$. In particular, we allow the feedstock condition distribution to be any member of a sorting-dependent family of distributions, denoted by $\mathbb{F}(s)$, i.e., $(\tilde{\boldsymbol{u}}, \tilde{\boldsymbol{v}}) \sim \mathbb{P} \in \mathbb{F}(s)$. We defer the detailed discussion on the modeling of $\mathbb{F}(s)$ to Sect. 14.3. Therefore, the problem that the PO seeks to solve consisting of optimizing resource recovery facility location design $\boldsymbol{y}$ given $(\xi, \boldsymbol{s})$ under uncertainty can then be written as follows:

$$Q^{\lambda}_{\mathbb{F}_s}(\xi) = \max \quad \inf_{\mathbb{P}\in\mathbb{F}(s)} \mathbb{E}_{\mathbb{P}}\Big[\phi\left(\boldsymbol{y}, \tilde{\boldsymbol{u}}, \tilde{\boldsymbol{v}}\right)\Big] - \lambda \boldsymbol{c}^\top \boldsymbol{y}, \tag{14.9}$$

$$\text{s. t.} \quad (1-\lambda)\boldsymbol{c}^\top \boldsymbol{y} \leq \xi, \tag{14.10}$$

$$\boldsymbol{y} \in \{0, 1\}^{|I|}, \tag{14.11}$$

where $\boldsymbol{c} = (c_i)_{i\in I}$ is the vector of the annualized upfront fixed cost of opening resource recovery units, and $\phi\,(\boldsymbol{x}, \boldsymbol{u}, \boldsymbol{v})$ is the optimal operational value given $\boldsymbol{y}$ and $(\boldsymbol{u}, \boldsymbol{v})$ by solving the operational problem (14.5)–(14.8).

In problem (14.9)–(14.11), $\lambda \in [0, 1]$ is the cost-share ratio parameter, whose value is the proportion of facility installation cost borne by the PO, while $(1 - \lambda)$ is the proportion to be paid by the LA. More specifically, the role of the parameter λ is to model the matching fund mechanism, similar to those in investment funds matching [308, 429]. In particular, for every monetary unit the PO spends installing new treatment capacity, the LA matches that with a co-fund of $\frac{1-\lambda}{\lambda}$ monetary units. Hence, if the PO pays $\lambda \boldsymbol{c}^\top \boldsymbol{y}$, the LA matching funds will be $(\frac{1-\lambda}{\lambda})\lambda \boldsymbol{c}^\top \boldsymbol{y} = (1-\lambda)\boldsymbol{c}^\top \boldsymbol{y}$. Since the limit on the matching funds available is ξ, we have $(1-\lambda)\boldsymbol{c}^\top \boldsymbol{y} \leq \xi$. Thus, the PO is sponsored by at least $(1-\lambda)\boldsymbol{c}^\top \boldsymbol{y}$ and at most ξ. The PO has this information, which explains the role of constraint in (14.10).

The objective of the PO in the problem (14.9)–(14.11) is to maximize its worst-case expected payoff over the family of distributions $\mathbb{F}(s)$. The problem is formulated using a two-stage distributionally robust mixed-integer optimization model (see [76]) with a decision-dependent ambiguity set. In this two-stage adaptive problem, the facility location $\boldsymbol{y}$ is the first-stage decision made before the observation of the feedstock outcomes $(\tilde{\boldsymbol{u}}, \tilde{\boldsymbol{v}})$, while the feedstock allocation $\boldsymbol{x}$ is the second-stage decision, which adapts to the realizations of $(\tilde{\boldsymbol{u}}, \tilde{\boldsymbol{v}})$. In particular, when $\mathbb{F}(s) = \{\mathbb{P}_s\}$, problem (14.9)–(14.11) reduces to a two-stage stochastic programming (see [78]) with a decision-dependent distribution.

14.2.2 The Problem of the Local Authority

The primary goal of the LA is to select and implement the appropriate waste sorting schemes in the recycling program.

To keep the exposition of the formulations cleaner, we assume that only one waste sorting scheme is selected at each waste collection zone. Hence, we define the feasibility set for the s variables as:

$$\mathcal{S} = \left\{ s \in \{0,1\}^{|J|\times|H|} \,\middle|\, \sum_{h\in H} s_{jh} = 1\,\forall j \in J \right\}.$$

Note that w.l.o.g. a waste sorting scheme may combine multiple sorting technologies (similar or different). Note also that this setting includes the case of not selecting any sorting scheme (e.g., by defining a "dummy" sorting scheme).

Given the volume v of feedstock (collected and sorted) and the sorting scheme s, the associated sorting cost, denoted by $f(s, v)$, is defined as

$$f(s, v) = \sum_{j\in J}\sum_{h\in H}(\varrho_{jh}v_j + \iota_{jh})s_{jh}.$$

In the above expression, ϱ_{jh} is the sorting cost per unit of waste, and ι_{jh} is the annualized setup cost by implementing the waste sorting scheme $h \in H$ at the collection site $j \in J$. Such sorting cost structure has been considered in [159], and [560], and it is chosen here also for tractability purposes.

In addition to the responsibility for the sorting scheme, the LA also determines the budget for sharing costs (resource recovery investment) with the PO. The overall problem of the LA, termed the Robust Bilevel Resource Recovery Planning (BRRP), is formulated as follows:

$$\text{minimize} \quad \xi + \sup_{\mathbb{P}\in\mathbb{F}(s)} \mathbb{E}_{\mathbb{P}}\left[f(s, \tilde{v})\right], \tag{14.12}$$

$$\text{subject to} \quad Q^{\lambda}_{\mathbb{F}(s)}(\xi) \geq \tau, \tag{14.13}$$

$$s \in \mathcal{S}, \tag{14.14}$$

$$\xi \in \mathbb{R}_+. \tag{14.15}$$

$Q^{\lambda}_{\mathbb{F}(s)}(\xi)$, defined in (14.9)–(14.11), represents the optimal worst-case expected payoff for the PO given the LA decisions (ξ, s) and under a given investment cost-share ratio λ. The objective function above minimizes the sum of the shared fixed cost and the worst-case expected sorting cost overall feedstock distributions in $\mathbb{F}(s)$. The economic feasibility constraint (14.13) requires that (in equilibrium) the PO achieves a target payoff level τ. Intuitively, if the LA sets a too low budget ξ for the cost-sharing, the PO may be unable to make the necessary capacity investments to achieve economic feasibility. On the other hand, if the LA selects a too inferior sorting scheme s, the resulting poor feedstock quality would inflate the expected operational costs of the PO, rendering economic feasibility unattainable. Hence, the issue of interest in (14.12)–(14.15) is the balance and mix of the two investment instruments of the LA.

Regarding uncertainty, we assume that the LA adopts an ambiguity-averse position. Thus, it seeks to optimize the worst-case expected cost. Such a position can improve the robustness of the planning solutions under uncertainty, which is very important to ensure a sustainable and economically feasible waste recycling system. It is interesting to note that the uncertainty faced by the LA has an "endogenous" characteristic since the feedstock distribution faced by the LA depends on the decision $\boldsymbol{s}$. Problem (14.12)–(14.15) can also be interpreted as an instance of a leader-follower (Stackelberg) game. The LA is the leader and decides the optimal budget ξ^* and sorting scheme $\boldsymbol{s}^*$. The PO is the follower, who determines the optimal solution $\boldsymbol{x}^*$ given the investment cost-sharing budget ξ^*, and performs operations under feedstock outcomes arising from sorting scheme $\boldsymbol{s}^*$.

14.2.3 *Model Analysis*

14.2.3.1 Analysis of Cost-Share Parameter λ

We first discuss the model (14.12)–(14.15) regarding the cost-share ratio λ. As mentioned above, different values of $\lambda \in [0, 1]$ lead to different types of P3 arrangements in practice. Next, we provide additional details.

1. $\lambda = 1$: The PO bears all costs related to resource recovery, including all his facility investments. This corresponds to the *Build-Operate-Transfer (BOT) contract*. In this case, the budget variable ξ in (14.12)–(14.15) can be set equal to zero because constraint (14.10) is redundant.
2. $\lambda = 0$: The LA provides a lump sum budget ξ for all facility asset purchases by the PO, thus bearing all capital investment costs. This corresponds to the *Central Management and Control (CMC)*, in which the LA has the sole proprietorship of all entities and resources for the end-to-end waste resource recovery activities. Meanwhile, by signing the management contract with the LA, the PO only takes over the operation and maintenance of the installed resource recovery unit within the contracting period. The overall problem of the LA.
3. $\lambda \in (0, 1)$: The LA provides a budget ξ to match capital investment expenditures made by the PO so that the LA effectively bears a proportion of $(1-\lambda)$ of the total resource recovery capital costs. This could correspond to the case of *Concession Agreement (CA)*—a negotiated contract that allows the PO to invest—manage, and rehabilitate the resource recovery units at his own commercial risk under an exclusive license from the LA.

We further consider the influence of the cost-share ratio λ on the optimal cost of the overall problem (14.12)–(14.15). It turns out that (not unexpectedly) the optimal value of that problem is, in general, not monotone in λ (this is illustrated in some results presented in Table 14.5, Sect. 14.5.1). Nevertheless, we can still derive qualitative insights about the relationship between λ and the optimal solutions.

Proposition 14.1 *Suppose* $0 \leq \lambda_1 < \lambda_2 < 1$, *and let* (s_1^*, ξ_1^*) *and* (s_2^*, ξ_2^*) *be the optimal solutions of problem* (14.12)–(14.15) *corresponding to* λ_1 *and* λ_2 *respectively. Then we have*

$$\sup_{\mathbb{P}\in\mathbb{F}(s_2^*)} \mathbb{E}_{\mathbb{P}}\left[f(s_2^*, \tilde{v})\right] \geq \sup_{\mathbb{P}\in\mathbb{F}(s_1^*)} \mathbb{E}_{\mathbb{P}}\left[f(s_1^*, \tilde{v})\right] + \left[\xi_1^* - \xi_2^*\left(\frac{1-\lambda_1}{1-\lambda_2}\right)\right]. \tag{14.16}$$

Proof See [550]. □

This result implies that when a higher cost-share ratio is shifted to the PO, e.g., from λ_1 to λ_2, and if the resulting optimal location design of the PO does not lead to a larger fixed cost investment than before, then the LA would need to pay more in the sorting investment. That is

$$c^\top y_1^* \geq c^\top y_2^*, \text{ or equivalently, } \xi_1^* - \xi_2^*\left(\frac{1-\lambda_1}{1-\lambda_2}\right) \geq 0.$$

In this case, the LA can lower the fixed cost-sharing (by shifting more portion to PO), but as a trade-off, it should invest more in the sorting scheme to ensure economic feasibility. This also implies that when a less attractive cost-sharing structure (to the PO) leads to scaling down on recovery facility investments of the LA, the latter is forced to improve useful feedstock conditions (quantity and quality) to help the PO improve the resource recovery performance (to the extent of satisfying the economic feasibility constraint). This can also be verified by some computational results that we present later (Table 14.5, Sect. 14.5.1).

14.2.3.2 A Central Control Counterpart

The structure of model (14.12)–(14.15) allows a reformulation as follows:

$$\text{minimize} \quad (1-\lambda)c^\top y + \sup_{\mathbb{P}\in\mathbb{F}(s)} \mathbb{E}_{\mathbb{P}}\left[f(s, \tilde{v})\right], \tag{14.17}$$

$$\text{subject to} \quad \mathbb{E}_{\mathbb{P}}\left[\phi\left(y, \tilde{u}, \tilde{v}\right)\right] - \lambda c^\top y \geq \tau \quad \forall \mathbb{P} \in \mathbb{F}(s), \tag{14.18}$$

$$s \in \mathcal{S}, \tag{14.19}$$

$$y \in \{0, 1\}^{|I|}, \tag{14.20}$$

which we term a "*central control counterpart*" for the problem (14.12)–(14.15), since it (literally) assumes a central controller who handles both the location and sorting of investment designs. Intuitively, the equivalence in the optimal solution values of the original problem (14.12)–(14.15) and the reformulation (14.17)–(14.20) is a simple consequence of the economic feasibility constraint in (14.13), which requires the optimal expected profits of the PO to achieve at least the level τ. When embedded in problem (14.12)–(14.15), it suffices to ensure that some feasible

$\boldsymbol{y}$ can be selected by the PO so that τ is achieved. To enable this, the LA only needs to allocate the necessary matching budget $\xi = (1 - \lambda)\boldsymbol{c}^\top \boldsymbol{y}$ corresponding to such $\boldsymbol{y}$ that also leads to minimizing her total costs, which explains the objective function in (14.17).

We now state more formally, in the following theorem, how the optimal matching budget ξ^* of the LA and the capacity-location decision $\boldsymbol{y}^{**}$ of the PO can be recovered from (14.17)–(14.20).

Theorem 14.1 *Let* $(\boldsymbol{s}^*, \boldsymbol{y}^*)$ *be the optimal solution of the problem* (14.17)–(14.20) *given* λ*. Then* $(\boldsymbol{s}^*, \xi^* = (1 - \lambda)\mathbf{c}^\top \boldsymbol{y}^*)$ *is the optimal solution of the LA problem* (14.12)–(14.15)*. Furthermore, as for the PO problem, we have*

1. *If* $\mathbf{c}^\top \boldsymbol{y} \neq \mathbf{c}^\top \boldsymbol{y}'$, $\forall \boldsymbol{y}, \boldsymbol{y}' \in \{0, 1\}^{|I|}$, $\boldsymbol{y} \neq \boldsymbol{y}'$*, then* $\boldsymbol{y}^{**} = \boldsymbol{y}^*$ *is also the optimal decision of the PO in the leader-follower model* (14.12)–(14.15)*.*
2. *In general, the optimal solution* $\boldsymbol{y}^{**}$ *of the PO can be recovered by solving the following adaptive distributionally robust problem:*

$$\boldsymbol{y}^{**} \in \operatorname*{argmax}_{\boldsymbol{y}} \left\{ -\lambda \boldsymbol{c}^\top \boldsymbol{y} + \inf_{\mathbb{P} \in \mathbb{F}(\boldsymbol{s}^*)} \mathbb{E}_{\mathbb{P}}\Big[\phi\left(\boldsymbol{y}, \tilde{\boldsymbol{u}}, \tilde{\boldsymbol{v}}\right)\Big] \;\middle|\; \begin{array}{l} \boldsymbol{c}^\top \boldsymbol{y} \leq \boldsymbol{c}^\top \boldsymbol{y}^* \\ \boldsymbol{y} \in \{0, 1\}^{|I|} \end{array} \right\}. \tag{14.21}$$

Proof See [550]. □

The reformulation (14.17)–(14.20) is in the form of distributionally robust optimization of $(\boldsymbol{s}, \boldsymbol{y})$ over a decision-dependent ambiguity set, which is constrained by another two-stage adaptive distributionally robust optimization subproblem. The attractiveness of this formulation (14.17)–(14.20) is that it enables a computationally amenable approach for solving the robust BRRP problem (14.12)–(14.15). The details are discussed in Sect. 14.4.

14.2.3.3 On the Optimal Cost-Share Ratio λ

Let us assume now that the cost-share parameter λ can also be optimized (by the LA). Let $\mathcal{L}(\xi, \boldsymbol{s}; \lambda)$ be the optimal value of problem (14.12)–(14.15). Recall that λ is embedded in the structure of $Q^{\lambda}_{\mathbb{F}(\boldsymbol{s})}(\xi)$. The resulting problem of interest is

$$\min_{\lambda \in [0,1]} \mathcal{L}(\xi, \boldsymbol{s}; \lambda). \tag{14.22}$$

The optimal cost ratio λ^* can be obtained by reformulating (14.22), which can be done as shown next. Define a new problem for the LA:

$$\text{minimize} \quad \xi_{\text{LA}} + \sup_{\mathbb{P} \in \mathbb{F}(\boldsymbol{s})} \mathbb{E}_{\mathbb{P}}\Big[f(\boldsymbol{s}, \tilde{\boldsymbol{v}})\Big], \tag{14.23}$$

$$\text{subject to} \quad Q_{\mathbb{F}(\boldsymbol{s})}(\xi_{\text{LA}}) \geq \tau, \tag{14.24}$$

$$s \in \mathcal{S}, \tag{14.25}$$

$$\xi_{\text{LA}} \in \mathbb{R}_+, \tag{14.26}$$

and a new problem for the PO given ξ_{LA}:

$$Q_{\mathbb{F}(s)}(\xi_{\text{LA}}) = \max \quad \inf_{\mathbb{P} \in \mathbb{F}(s)} \mathbb{E}_{\mathbb{P}}\Big[\phi\left(\boldsymbol{y}, \tilde{\boldsymbol{u}}, \tilde{\boldsymbol{v}}\right)\Big] - \xi_{\text{PO}}, \tag{14.27}$$

$$\text{s. t.} \quad \boldsymbol{c}^\top \boldsymbol{y} - \xi_{\text{PO}} \leq \xi_{\text{LA}}, \tag{14.28}$$

$$\boldsymbol{y} \in \{0, 1\}^{|I|}, \tag{14.29}$$

$$\xi_{\text{PO}} \in \mathbb{R}_+. \tag{14.30}$$

The variables ξ_{LA} and ξ_{PO} are the investment cost shares of the LA and PO, respectively. We can then recover the optimal λ^* of (14.22) via the solutions of the problem (14.23)–(14.26) with $Q_{\mathbb{F}(s)}(\xi_{\text{LA}})$ given according to (14.27)–(14.30).

Using similar reasoning as above, we can obtain the "central control counterpart" formulation of the problem (14.22) or equivalently of the problem (14.23)–(14.26) (with $Q_{\mathbb{F}(s)}(\xi_{\text{LA}})$ given according to (14.27)–(14.30)), i.e., when the cost-share ratio λ is to be optimized. In particular, we consider the following problem:

$$\text{minimize} \quad \xi_{\text{LA}} + \sup_{\mathbb{P} \in \mathbb{F}(s)} \mathbb{E}_{\mathbb{P}}\left[f(\boldsymbol{s}, \tilde{\boldsymbol{v}})\right], \tag{14.31}$$

$$\text{subject to} \quad \mathbb{E}_{\mathbb{P}}\left[\phi\left(\boldsymbol{y}, \tilde{\boldsymbol{u}}, \tilde{\boldsymbol{v}}\right)\right] - \xi_{\text{PO}} \geq \tau \quad \forall \mathbb{P} \in \mathbb{F}(s), \tag{14.32}$$

$$\boldsymbol{c}^\top \boldsymbol{y} \leq \xi_{\text{LA}} + \xi_{\text{PO}}, \tag{14.33}$$

$$\boldsymbol{s} \in \mathcal{S}, \tag{14.34}$$

$$\boldsymbol{y} \in \{0, 1\}^{|I|}, \tag{14.35}$$

$$\xi_{\text{PO}} \in \mathbb{R}_+. \tag{14.36}$$

This model is based on (14.23)–(14.26) with $Q_{\mathbb{F}(s)}(\xi_{\text{LA}})$ given according to (14.27)–(14.30).

The following result holds:

Proposition 14.2 *Let* $(\boldsymbol{s}^*, \boldsymbol{y}^*, \xi_{\text{LA}}^*, \xi_{\text{PO}}^*)$ *be the optimal solution of problem* (14.31)–(14.36). *Then*

$$\left(\boldsymbol{s}^*, \xi^* = \xi_{\text{LA}}^*, \lambda^* = \frac{\xi_{\text{PO}}^*}{\xi_{\text{LA}}^* + \xi_{\text{PO}}^*}\right)$$

is the optimal solution to the LA problem (14.22). *The optimal location* $\boldsymbol{y}^{**}$ *of the PO given* λ^* *can then be obtained following the rule in Theorem 14.1 accordingly.*

Proof See [550]. □

It is easy to see that the computation of (14.31)–(14.36) is not more challenging than problem (14.17)–(14.20) because both models share the same structure for optimization.

14.3 Modeling Feedstock Ambiguity

We now focus on the modeling of the distributional ambiguity of the feedstock condition $(\tilde{\boldsymbol{u}}, \tilde{\boldsymbol{v}})$. In practice, the waste sorting schemes implemented can influence the distributional characteristics of the feedstock condition $(\tilde{\boldsymbol{u}}, \tilde{\boldsymbol{v}})$. We begin by considering the dependence of the collected feedstock volume on the sorting scheme. Intuitively, the influence of the sorting scheme on the collected input feedstock volume can be related to the *participation rate* generated by the sorting scheme implemented. For instance, investments in advertisement and promotion campaigns can bring about greater awareness and interest of residents to participate in the waste separation-at-source activities [128, 417, 517]. Nevertheless, we remark that the feedstock volume is relatively less sensitive to the sorting schemes utilized, compared with the feedstock composition (to be introduced later), by noting that the main goal in sorting is to enhance the feedstock quality (in composition). Therefore, we focus on the dependence of the average feedstock volume on the sorting schemes, assuming that the support of all possible outcomes remains the same.

More specifically, the range of expected feedstock volume $\tilde{\boldsymbol{v}}$ given the sorting scheme $\boldsymbol{s}$ is modelled as:

$$\mathbb{E}_{\mathbb{P}}\left[\tilde{v}_j\right] \in \left[\boldsymbol{\alpha}_j^\top \boldsymbol{s}_j, \boldsymbol{\beta}_j^\top \boldsymbol{s}_j\right], \forall j \in J. \tag{14.37}$$

The parameters $\boldsymbol{\alpha}_j = (\alpha_{j1}, \ldots, \alpha_{jH})^\top$ and $\boldsymbol{\beta}_j = (\beta_{j1}, \ldots, \beta_{jH})^\top$, and each pair of α_{jh} and β_{jh} refer to the estimated lower and upper bounds of feedstock volume by using the waste sorting scheme h at the collection site j, i.e.,

$$\alpha_{jh} \leq \mathbb{E}_{\mathbb{P}}\left[\tilde{v}_j\right] \leq \beta_{jh} \quad \forall j \in J,\ h \in H.$$

The dependence stated in (14.37) is also motivated by several empirical studies in waste management [147, 219].

Furthermore, we construct the support $\mathcal{V}$ of (sorted) feedstock volume $\tilde{\boldsymbol{v}}$ as follows:

$$\mathcal{V} = \left\{\boldsymbol{v} \in \mathbb{R}^{|J|} \mid \boldsymbol{v} = \boldsymbol{v}^0 + \boldsymbol{\Delta}, \forall \boldsymbol{\Delta} \in \operatorname{conv}\left\{\boldsymbol{\Delta}^k = (\Delta_1^k, \ldots, \Delta_{|J|}^k), k \in [K]\right\}\right\}. \tag{14.38}$$

The above set uses historical observations at different periods and collection zones. In particular, $\boldsymbol{v}^0$ is the observation of the feedstock volumes in the latest period at all zones $j \in J$. Each Δ_j^k refers to a difference between the historical volume observations of adjacent periods at zone j. For instance, we can take $\Delta_j^k = v_j^t - v_j^{t-1}$, where $t = k$ refers to different historical periods. The set $[K] = \{1, \ldots, K\}$ contains the indices of the sample observations, with K standing for the sample size.

Note that we allow the uncertainty $\boldsymbol{\Delta}$ to vary within the convex hull of $\{\boldsymbol{\Delta}^k, k \in [K]\}$. This way of modeling the feedstock volume support (with the convex hull of scenarios) seeks the tractability of the resulting optimization models. In our model, general polyhedral support (without knowing the extreme points) would lead to a two-stage distributionally robust optimization problem with uncertainty in both the right-hand side (of the second-stage subproblem) and objective function. In general, this is intractable. The problem would become even more challenging if we let the support be dependent on the sorting decisions.

We also note that the use of historical data information on waste quantity has been commonly adopted in the literature (see [101, 537, 549]).

Regarding the feedstock composition $\boldsymbol{u}$, its dependence on the sorting scheme can be motivated by the separation purity level achieved by the sorting process. For instance, effective waste sorting may require investments in effective separation and disposal techniques and workforce training. The interested reader can refer to [148, 251, 363] for a deeper discussion.

In the models that follow, we consider the following representation of the uncertain feedstock composition:

$$\mathbb{E}_{\mathbb{P}}\left[\boldsymbol{G}\tilde{\boldsymbol{u}}_j\right] \geq \boldsymbol{F}\boldsymbol{s}_j \quad \forall j \in J, \tag{14.39}$$

$$\mathbb{P}\left[\tilde{\boldsymbol{u}} \in \mathcal{U}(\boldsymbol{s})\right] = 1, \tag{14.40}$$

where the support $\mathcal{U}(\boldsymbol{s})$, also affected by $\boldsymbol{s}$, is defined as

$$\mathcal{U}(\boldsymbol{s}) = \left\{ \boldsymbol{u} \in \mathbb{R}_+^{|J|\times|N|} \;\middle|\; \boldsymbol{A}\boldsymbol{u}_j \geq \boldsymbol{B}\boldsymbol{s}_j \;\; \forall j \in J; \quad \sum_{n\in N} u_{jn} = 1 \;\; \forall j \in J \right\}. \tag{14.41}$$

Matrices $\boldsymbol{G} \in \mathbb{R}_+^{L_1\times|N|}$, $\boldsymbol{F} \in \mathbb{R}_+^{L_1\times|H|}$, $\boldsymbol{A} \in \mathbb{R}_+^{L_2\times|N|}$ and $\boldsymbol{B} \in \mathbb{R}_+^{L_2\times|H|}$, with commensurate dimensions L_1, L_2, are introduced to model the feedstock quality improvement by the sorting scheme selected.

We illustrate the ambiguity model (14.39)–(14.41) with an example.

Example 14.1 Suppose anaerobic digestion (AD) units are implemented for biogas production by recycling the organic wastes. A preferred feedstock category C_ℓ (out of a set L of available categories) can be organic matter. Thus, the feedstock components $u_{jn}, \forall n \in C_\ell$, would refer to all types of organic waste components, e.g., meat, vegetable, leaves, fruit peels, *et cetera*. The purity level $b_{\ell h}$ refers to the content level of organic matter in the collected biomass that can be measured and is a central indicator for feedstock AD operations.

Suppose the sorting scheme $h \in H$ is selected for each waste collection site $j \in J$. The total fraction of all preferred waste components in category C_ℓ that is related to every required content $\ell \in L$ ($L_1 = L_2 = |L|$ in this example) can be guaranteed to be above a threshold level $b_{\ell h}$. Accordingly, the expected total fraction of preferred waste components in category C_ℓ also reaches level $\psi_{\ell h}$. Then, for the sorting scheme h, the above resulting (affected) feedstock composition characteristics (14.39) and (14.40) map to the following distributional constraints of feedstock composition:

$$\mathbb{E}_{\mathbb{P}}\left[\sum_{n\in C_\ell} \tilde{u}_{jn}\right] \geq \psi_{\ell h} \quad \forall j \in J,\ \ell \in L,$$

$$\mathbb{P}\begin{bmatrix} \sum_{n\in C_\ell} \tilde{u}_{jn} \geq b_{\ell h} & \forall j \in J,\ \ell \in L, \\ \sum_{n\in N} \tilde{u}_{jn} = 1 & \forall j \in J, \\ \tilde{u}_{jn} \geq 0 & \forall j \in J,\ n \in N, \end{bmatrix} = 1,$$

where the category C_ℓ can be mathematically represented by a proper data setting for matrix $\boldsymbol{G}$ and $\boldsymbol{A}$, i.e., setting $g_{\ell n} = a_{\ell n} = 1$ in this example if preferred waste component n belongs to category ℓ, and 0 otherwise, where $g_{\ell n}$ and $a_{\ell n}$ are defined as the element of matrix $\boldsymbol{G}$ and $\boldsymbol{A}$, respectively.

Gathering the distributional features (14.37)–(14.41) of feedstock condition $(\tilde{\boldsymbol{u}}, \tilde{\boldsymbol{v}})$ affected by the sorting scheme $\boldsymbol{s}$, we construct the following sorting-dependent feedstock condition ambiguity set:

$$\mathbb{F}(\boldsymbol{s}) = \left\{ \mathbb{P} \in \mathcal{P}\left(\mathbb{R}_+^{|J|\times|N|} \times \mathbb{R}_+^{|J|}\right) \;\middle|\; \begin{array}{ll} (\tilde{\boldsymbol{u}}, \tilde{\boldsymbol{v}}) \sim \mathbb{P} & \\ \mathbb{E}_{\mathbb{P}}\left[\tilde{v}_j\right] \in \left[\boldsymbol{\alpha}_j^\top \boldsymbol{s}_j, \boldsymbol{\beta}_j^\top \boldsymbol{s}_j\right] & \forall j \in J \\ \mathbb{E}_{\mathbb{P}}\left[\boldsymbol{G}\tilde{\boldsymbol{u}}_j\right] \geq \boldsymbol{F}\boldsymbol{s}_j & \forall j \in J \\ \mathbb{P}\left[\tilde{\boldsymbol{v}} \in \mathcal{V}, \tilde{\boldsymbol{u}} \in \mathcal{U}(\boldsymbol{s})\right] = 1 & \end{array} \right\}. \tag{14.42}$$

In this ambiguity set, it is clear that the feedstock condition has distributional characteristics affected by the sorting decision $\boldsymbol{s}$.

14.4 Cost/Revenue Evaluation and Optimization

Given the equivalent distributionally robust optimization problem (14.17)–(14.20), the key computational challenges can be identified as follows:

1. The evaluation of the objective (14.17). It contains the worst-case expected sorting cost of the local authority, namely

$$\sup_{\mathbb{P}\in\mathbb{F}(s)} \mathbb{E}_{\mathbb{P}}\left[f(s,\tilde{v})\right]. \tag{14.43}$$

2. The evaluation of the constraint in (14.18). It contains the worst-case expected resource recovery revenue of the private operator:

$$\begin{aligned}&\inf_{\mathbb{P}\in\mathbb{F}(s)} \mathbb{E}_{\mathbb{P}}\left[\phi\left(y,\tilde{u},\tilde{v}\right)\right]\\ &= \inf_{\mathbb{P}\in\mathbb{F}(s)} \mathbb{E}_{\mathbb{P}}\left[\max_{x:(14.6)-(14.8)} \sum_{i\in I}\sum_{j\in J}\sum_{n\in N} r_{ijn}\tilde{u}_{jn}x_{ij}\right.\\ &\qquad\left. - c_D\sum_{j\in J}\left(\tilde{v}_j - \sum_{i\in I} y_{ij}\right)\right],\end{aligned} \tag{14.44}$$

which evaluates a two-stage adaptive distributionally robust value given (s, y). Furthermore, by Theorem 14.1 (general case), given the solution (s^*, y^*) from (14.17)–(14.20) we also need to recover the optimal location decision y^{**} of the private operator by solving the following adaptive distributionally robust optimization problem

$$\max_{y\in\{0,1\}^{|I|}} \left\{\inf_{\mathbb{P}\in\mathbb{F}(s)} \mathbb{E}_{\mathbb{P}}\left[\phi\left(y,\tilde{u},\tilde{v}\right)\right] - \lambda c^\top y \;\middle|\; c^\top y \le c^\top y^*\right\}. \tag{14.45}$$

3. The optimization of the overall problem (14.17)–(14.20). Since the ambiguity set $\mathbb{F}(s)$ is affected by the sorting decision s, problem (14.17)–(14.20) is essentially a decision-dependent (adaptive) distributionally robust optimization problem. It is optimized by trading off the worst-case expected cost/revenue performance and the impact on the distributional conditions of feedstock by the decision (sorting scheme s). This is different from most existing studies on the (adaptive) distributionally robust optimization, which assumes a static ambiguity set (see, e.g., [76, 543]).

The computations of subproblem (14.43), subproblem (14.44), solution recovery problem (14.45), and the overall problem (14.17)–(14.20), are discussed in the forthcoming subsections, respectively.

14.4.1 The Worst-Case Expected Sorting Cost

Since the worst-case expected sorting cost is only associated with the uncertainty of feedstock volume $\tilde{\boldsymbol{v}}$, the subproblem (14.43) is equivalent to the worst-case expected sorting cost over the projection of the ambiguity set $\mathbb{F}(\boldsymbol{s})$ concerning the $(\tilde{\boldsymbol{u}}, \tilde{\boldsymbol{v}})$ onto the $\tilde{\boldsymbol{v}}$, which is denoted by $\mathbb{F}_{\tilde{\boldsymbol{v}}}(\boldsymbol{s})$ and can be expressed as

$$\mathbb{F}_{\tilde{\boldsymbol{v}}}(\boldsymbol{s}) = \left\{ \mathbb{P} \in \mathcal{P}\left(\mathbb{R}_{+}^{|J|}\right) \left| \begin{array}{ll} \tilde{\boldsymbol{v}} \sim \mathbb{P} & \\ \mathbb{E}_{\mathbb{P}}\left[\tilde{v}_j\right] \in \left[\boldsymbol{\alpha}_j^\top \boldsymbol{s}_j, \boldsymbol{\beta}_j^\top \boldsymbol{s}_j\right] & \forall j \in J \\ \mathbb{P}\left[\tilde{\boldsymbol{v}} \in \mathcal{V}\right] = 1 & \end{array} \right. \right\}.$$

That is, $\sup_{\mathbb{P}\in\mathbb{F}(\boldsymbol{s})} \mathbb{E}_{\mathbb{P}}\left[f(\boldsymbol{s}, \tilde{\boldsymbol{v}})\right] = \sup_{\mathbb{P}\in\mathbb{F}_{\tilde{\boldsymbol{v}}}(\boldsymbol{s})} \mathbb{E}_{\mathbb{P}}\left[f(\boldsymbol{s}, \tilde{\boldsymbol{v}})\right]$.

The following result states that subproblem (14.43) can be equivalently converted into a linear programming problem.

Proposition 14.3 *Given a waste sorting scheme* $\boldsymbol{s}$*, the evaluation of the worst-case expected sorting cost* (14.43) *solves the following linear optimization problem (in* $\boldsymbol{v}$*):*

$$\textit{maximize} \quad \sum_{j\in J}\sum_{h\in H} (\varrho_{jh} v_j + \iota_{jh}) s_{jh}, \tag{14.46}$$

$$\textit{subject to} \quad v_j \in \left[\boldsymbol{\alpha}_j^\top \boldsymbol{s}_j, \boldsymbol{\beta}_j^\top \boldsymbol{s}_j\right] \qquad \forall j \in J, \tag{14.47}$$

$$\boldsymbol{v} = \boldsymbol{v}^0 + \sum_{k=1}^{K} \gamma^k \boldsymbol{\Delta}^k, \tag{14.48}$$

$$\sum_{k=1}^{K} \gamma^k = 1, \tag{14.49}$$

$$\gamma^1, \ldots, \gamma^K \in \mathbb{R}_+. \tag{14.50}$$

Proof See [550]. □

14.4.2 The Worst-Case Expected Operational Revenue

Next, we deal with the computation of the worst-case expected resource recovery return (14.44). This is a two-stage distributionally robust problem containing uncertainties in both the objective (with $(\tilde{\boldsymbol{u}}, \tilde{\boldsymbol{v}})$) and the right-hand side of constraints (with $\tilde{\boldsymbol{v}}$) of the second-stage operational problem $\phi\left(\boldsymbol{y}, \tilde{\boldsymbol{u}}, \tilde{\boldsymbol{v}}\right)$ for the resource recovery revenue maximization:

$$\phi(\boldsymbol{y}, \tilde{\boldsymbol{u}}, \tilde{\boldsymbol{v}}) = \max \quad \sum_{i\in I}\sum_{j\in J}\sum_{n\in N} r_{ijn}\tilde{u}_{jn}x_{ij} - c_D \sum_{j\in J}\left(\tilde{v}_j - \sum_{i\in I} x_{ij}\right), \tag{14.51}$$

$$\text{s. t.} \quad \sum_{i\in i} x_{ij} \leq \tilde{v}_j \qquad \forall j \in J, \tag{14.52}$$

$$\sum_{j\in J} x_{ij} \leq y_i \varpi_i \qquad \forall i \in I, \tag{14.53}$$

$$x_{ij} \geq 0 \qquad \forall i \in I,\ j \in J. \tag{14.54}$$

A two-stage distributionally robust optimization problem with such a structure, even for the evaluation problem, is generally intractable (see [76]). In our case, leveraging on the support $\mathcal{V}$ of feedstock volume, which is defined with the convex hull of finite given scenarios as in (14.38), it is possible to show that problem (14.44) can also be reformulated as a linear program. This is formally stated in the following proposition.

Proposition 14.4 *Given the sorting scheme and facility location decision* $(\boldsymbol{s}, \boldsymbol{y})$, *the evaluation of the worst-case expected resource recovery revenue* (14.44) *can be accomplished by solving the following linear program:*

$$\textit{maximize} \quad \pi + \sum_{j\in J}\left[\boldsymbol{q}_j^\top \boldsymbol{F}\boldsymbol{s}_j + \overline{\delta}_j(\boldsymbol{\alpha}_j^\top \boldsymbol{s}_j) + \underline{\delta}_j(\boldsymbol{\beta}_j^\top \boldsymbol{s}_j)\right], \tag{14.55}$$

$$\textit{subject to} \quad \sum_{j\in J} c_D \boldsymbol{e}_{|I|}^\top \boldsymbol{v}_j^k - \sum_{j\in J}\left(\boldsymbol{B}\boldsymbol{s}_j\right)^\top \boldsymbol{p}_j^k - \boldsymbol{e}_{|J|}^\top \boldsymbol{\chi}^k$$

$$- c_D \boldsymbol{e}_{|J|}^\top \boldsymbol{v}^k - \left(\underline{\boldsymbol{\delta}} + \overline{\boldsymbol{\delta}}\right)^\top \boldsymbol{v}^k \geq \pi \qquad \forall k \in [K], \tag{14.56}$$

$$\mathbf{R}_j \boldsymbol{v}_j^k + \boldsymbol{A}^\top \boldsymbol{p}_j^k + \chi_j^k \boldsymbol{e}_{|N|} \geq \boldsymbol{G}^\top \boldsymbol{q}_j \qquad \forall j \in J,\ k \in [K], \tag{14.57}$$

$$\boldsymbol{v}^k \boldsymbol{e}_{|I|} \leq \boldsymbol{v}^k \qquad \forall k \in [K], \tag{14.58}$$

$$\left(\boldsymbol{v}^k\right)^\top \boldsymbol{e}_{|J|} \leq \mathrm{Diag}(\boldsymbol{\varpi}) \cdot \boldsymbol{y} \qquad \forall k \in [K], \tag{14.59}$$

$$\boldsymbol{v} \in \mathbb{R}_+^{|J|\times|I|\times K},\ \boldsymbol{p} \in \mathbb{R}_-^{|J|\times L_2\times K},$$

$$\boldsymbol{\chi} \in \mathbb{R}^{|J|\times K},\ \boldsymbol{q} \in \mathbb{R}_+^{L_1\times|J|},$$

$$\underline{\boldsymbol{\delta}} \in \mathbb{R}_-^{|J|},\ \overline{\boldsymbol{\delta}} \in \mathbb{R}_+^{|J|},\ \pi \in \mathbb{R}, \tag{14.60}$$

where $\boldsymbol{e}_{|I|} \in \mathbb{R}^{|I|}$, $\boldsymbol{e}_{|J|} \in \mathbb{R}^{|J|}$, *and* $\boldsymbol{e}_{|N|} \in \mathbb{R}^{|N|}$ *are the vector of ones,* $\mathrm{Diag}(\boldsymbol{\varpi})$ *is the* $\mathbb{R}^{|I|\times|I|}$ *diagonal matrix with vector* $\boldsymbol{\varpi}$ *setting the diagonal entries, and* $\boldsymbol{v}^k = \boldsymbol{v}^0 + \boldsymbol{\Delta}^k \ \forall k \in [K]$ *with* $\boldsymbol{\Delta}^k$ *being given by* (14.38).

Proof See [550]. □

From the above linear programming model, we can derive an MILP formulation for the facility location resource recovery problem (14.45).

Corollary 14.1 *Let* $(\boldsymbol{s}^*, \boldsymbol{y}^*)$ *be the solution of problem* (14.17)–(14.20)*. Then, the solution of the problem* (14.45) *is* $(\boldsymbol{s}^*, \boldsymbol{y}^{**})$*, where* $\boldsymbol{y}^{**}$ *can be obtained by solving the following MILP:*

$$\begin{aligned} \text{maximize} \quad & \pi + \sum_{j\in J}\left[\boldsymbol{q}_j^\top \boldsymbol{F}\boldsymbol{s}_j^* + \overline{\delta}_j(\boldsymbol{\alpha}_j^\top \boldsymbol{s}_j^*) + \underline{\delta}_j(\boldsymbol{\beta}_j^\top \boldsymbol{s}_j^*)\right] - \lambda \boldsymbol{c}^\top \boldsymbol{y} && (14.61)\\ \text{subject to} \quad & (14.56)\text{–}(14.60), \\ & \boldsymbol{c}^\top \boldsymbol{y} \le \boldsymbol{c}^\top \boldsymbol{y}^*, && (14.62)\\ & \boldsymbol{y} \in \{0,1\}^{|I|}. && (14.63) \end{aligned}$$

14.4.3 Solving the Overall Robust BRRP Problem

Using the tractable reformulations of the worst-case expected sorting cost and operational revenue stated in Propositions 14.3 and 14.4, we can reformulate the overall robust BRRP problem (14.17)–(14.20) as a mixed-integer program (MIP). Furthermore, utilizing some linearization techniques, we can further transform the problem into a linear MIP formulation. That linearization does not entail additional integer variables. This is formally presented in the following theorem.

Theorem 14.2 *The optimization problem* (14.17)–(14.20) *can be reformulated as the following mixed-integer linear problem with the original sorting scheme and location decisions* $(\boldsymbol{s}, \boldsymbol{x})$ *as the only integer variables:*

$$\text{minimize} \quad (1-\lambda)\boldsymbol{c}^\top \boldsymbol{y} + \sum_{j\in J}\left(v_j^0 \boldsymbol{\varrho}_j^\top + \boldsymbol{\iota}_j^\top\right)\boldsymbol{s}_j + \sum_{j\in J} t_j + \zeta - \left(\underline{\boldsymbol{\rho}} + \overline{\boldsymbol{\rho}}\right)^\top \boldsymbol{v}_0 \qquad (14.64)$$

$$\begin{aligned} \text{subject to} \quad & (14.57)\text{–}(14.59), \\ & (\underline{\boldsymbol{\rho}}_j + \overline{\boldsymbol{\rho}}_j)^\top \boldsymbol{\Delta}^k + \zeta \ge \sum_{j\in J} \boldsymbol{s}_j^\top \boldsymbol{\varrho}_j \Delta_j^k \qquad \forall k \in [K], && (14.65) \end{aligned}$$

$$\boldsymbol{b}_h^\top \boldsymbol{p}_j^k + \Theta(s_{jh} - 1) \leq \psi_j^k \qquad \forall j \in J,\ h \in H,\ k \in [K], \tag{14.66}$$

$$\sum_{j \in J} c_D \boldsymbol{e}_{|I|}^\top \boldsymbol{v}_j^k - \boldsymbol{e}_{|J|}^\top \boldsymbol{\chi}^k - c_D \boldsymbol{e}_{|J|}^\top \boldsymbol{v}^k$$

$$- (\overline{\boldsymbol{\delta}} + \underline{\boldsymbol{\delta}})^\top \boldsymbol{v}^k - \pi \geq \boldsymbol{e}_{|J|}^\top \boldsymbol{\psi}^k \qquad \forall k \in [K], \tag{14.67}$$

$$\boldsymbol{F}_h^\top \boldsymbol{q}_j + \overline{\delta}_j \alpha_{jh} + \underline{\delta}_j \beta_{jh} + \Theta(1 - s_{jh}) \geq \varphi_j \quad \forall j \in J,\ h \in H, \tag{14.68}$$

$$\boldsymbol{e}_{|J|}^\top \boldsymbol{\varphi} \geq \tau + \lambda \boldsymbol{c}^\top \boldsymbol{y} - \pi, \tag{14.69}$$

$$\alpha_{jh} \underline{\rho}_j + \beta_{jh} \overline{\rho}_j - \theta(1 - s_{jh}) \leq t_j \qquad \forall j \in J,\ h \in H, \tag{14.70}$$

$$t_j \leq \alpha_{jh} \underline{\rho}_j + \beta_{jh} \overline{\rho}_j + \Theta(1 - s_{jh}) \qquad \forall j \in J,\ h \in H, \tag{14.71}$$

$$\begin{aligned} &\boldsymbol{s} \in \mathcal{S},\ \boldsymbol{y} \in \{0, 1\}^{|I|}, \\ &\boldsymbol{v} \in \mathbb{R}_+^{|J| \times |I| \times K},\ \boldsymbol{p} \in \mathbb{R}_-^{|J| \times L_2 \times K},\ \boldsymbol{\chi} \in \mathbb{R}^{|J| \times K}, \\ &\boldsymbol{q} \in \mathbb{R}_+^{L_1 \times |J|},\ \underline{\boldsymbol{\delta}} \in \mathbb{R}_-^{|J|},\ \overline{\boldsymbol{\delta}} \in \mathbb{R}_+^{|J|},\ \pi \in \mathbb{R}, \\ &\underline{\boldsymbol{\rho}} \in \mathbb{R}_-^{|J|},\ \overline{\boldsymbol{\rho}} \in \mathbb{R}_+^{|J|},\ \zeta \in \mathbb{R}, \\ &\boldsymbol{\psi} \in \mathbb{R}^{|J| \times K},\ \boldsymbol{\varphi} \in \mathbb{R}^{|J|},\ \boldsymbol{t} \in \mathbb{R}^{|J|}, \end{aligned} \tag{14.72}$$

where $\boldsymbol{e}_{|I|} \in \mathbb{R}^{|I|}$ *and* $\boldsymbol{e}_{|J|} \in \mathbb{R}^{|J|}$ *are the vector of ones,* $\boldsymbol{\Delta}^k$, $k \in [K]$ *are given by* (14.38), $\boldsymbol{F}_h$, $h \in H$ *are the columns of matrix* $\boldsymbol{F}$ *given by* (14.42), *and* Θ *is a sufficiently large positive number.*

Proof See [550]. □

14.5 Numerical Results

We consider a food waste recycling planning problem under P3 in the land-scarce city-state Singapore. After the shutdown of IUT Global in 2008 (the largest food waste recycling company in Singapore at the time), the food waste recycling rate in the country stagnated below 20% until 2017 (see [394]). Meanwhile, as reported in [472], poor recycling practices of the residents caused up to 50% of food waste to be contaminated with plastic bags and other inorganic trash on bad days. The application we discuss emerges in this context.

We consider ten fixed collection zones for food waste source-sorting operations and ten alternative resource recovery facility installation sites spread in different

Table 14.1 Technical coefficient of AD units

Modular capacity (ton/year)	Recovered energy price ($SGD/kWh)	Variable processing cost ($SGD/ton)	Residual disposal cost ($SGD/ton)	Average annualized fixed cost ($SGD/year)
50,000	0.2365	42.46	77	1,558,589

Table 14.2 Resource recovery performance of different waste components in feedstock by AD unit

Waste type	Food waste (organics)	Horticultural waste (organics)	Other impurities (inorganics)
Energy recovery ratio	347.2 kWh/ton	198.0 kWh/ton	0 kWh/ton
Residual ratio	0.1 ton/ton	0.1 ton/ton	1 ton/ton

Table 14.3 Performance of different food waste sorting schemes

	Sorting scheme	Unit operational cost ($SGD/ton)	Fixed investment cost ($SGD/year)	Expected volume range per site (ton/tear)	Guaranteed organic content level (%)
1	Recycling bin	35.93	164,150	[12,957, 18,140]	50
2	On-site collection	44.91	196,980	[16,844, 21,427]	70
3	Incentive collection	53.06	264,691	[14,252, 22,002]	90

zones of Singapore. We assume that the recycling private operator only uses AD technology. Table 14.1 lists the modular capacity and economic data setting of the AD units based on a 10-year lifespan in the Singapore context [375, 523].

Table 14.2 shows the resource recovery performances on different waste components in collected food wastes (feedstock) by AD units [44].

Additionally, Table 14.3 provides some estimated data assumptions of the commonly-used food waste sorting scheme considered in the system design. Historical observations of food waste growth from 2003 to 2016 are used to construct the support $\mathcal{V}$ for sorted food waste volume and determine the expected volume range (with specific lower and upper bounds) corresponding to three different sorting schemes, for a 10-year planning horizon [164, 394]. Moreover, we consider the level of organic content (including food waste and horticultural waste) in the sorted feedstock as the sorting performance criterion in the recycling system. Hence, the guaranteed organic content level data reported in [206, 472] in Table 14.3 are used to construct the feedstock composition support $\mathcal{U}(\boldsymbol{s})$ and relevant constraints in our ambiguity set $\mathbb{F}(\boldsymbol{s})$ in (14.42), as illustrated in Example 14.1. All the linear programs and MIP models were solved using a general-purpose solver.

14.5.1 Analysis of (τ, λ) Combinations

In this section, we analyze the solution profile (ξ^*, s^*, y^*) of the robust BRRP model under different settings of parameters τ and λ.

We first conduct the experiment by fixing $\lambda = 0.5$ and varying $\tau \in [0, 4.0 \times 10^6]$ \$SGD/year. Specifically, for each scenario of τ, we obtain the solution of the robust BRRP problem by solving the equivalent MILP (14.64)–(14.72). The corresponding solution profiles for different values of τ are presented in Table 14.4. All monetary information is measured in 10^6 \$SGD/year. Also, in this table, $s^*_{j1} = 1, s^*_{j2} = 1$, and $s^*_{j3} = 1$ indicate that the collection zone j employs recycling bin, on-site, and incentive collection, respectively (Table 14.3). Finally, $y^*_i = 1$ indicates that the location i is selected for the resource recovery facility installation.

From the results in Table 14.4, the total payouts of the local authority (LA), i.e., the sorting cost plus the matching costs, increase as the target τ of the private operator (PO) target increases. It can be observed that the number of opened resource recovery facilities only increases one time (from 3 to 4) after τ surpasses 0.5×10^6 \$SGD/year and then remains unchanged. Likewise, the cost of facility installation follows the same trend. We remark that the capital investment costs of different types of food waste facilities are all relatively high in general (Table 14.1) and do not have significant differences (based on our data). Therefore, the investment cost (budget) can be well reflected by the number of facilities opened. On the other hand, the expense of implementing and operating the sorting schemes continues to increase with τ.

More specifically, as the target τ of the PO increases, the pattern of the optimal sorting policy of the LA can be summarized as follows:

(i) When τ is relatively low ($\tau \leq 0.5$), the LA can invest in the cheap yet relatively ineffective sorting scheme "recycling bin" with a small number of facilities opened.

(ii) As τ increases to an intermediate level ($1.0 \leq \tau \leq 2.5$), the LA must raise the budget for facility installation (allowing the private operator to invest in more capacity) and shift sorting investment from the "recycling bin" to the more effective yet more expensive scheme on-site collection.

(iii) When τ reaches a relatively high level ($3.0 \leq \tau \leq 4.0$), the optimal policy for the LA becomes to invest more into the sorting, i.e., replacing more recycling bin, and on-site collection schemes with the more effective sorting scheme incentive collection, albeit bearing a higher sorting cost, rather than to keep raising the matching budget for opening more facilities. This is reasonable and critical as the fixed installation cost is relatively high, which actually could significantly inflate the payoff of the PO under the P3 with budget sharing (e.g., $\lambda = 0.5$ in this case) while improving the feedstock condition via sound sorting schemes is relatively more cost-effective.

We next analyze the solution profiles of the robust BRRP across different λ with a fixed τ. For a more realistic analysis, we fix the payoff target τ of the PO as

Table 14.4 The robust BRRP solution profile across different resource recovery payoff targets τ with $\lambda = 0.5$

Payoff target (PO)	Worst-case total expense (LA)	Worst-case sorting cost (LA)	Worst-case total fixed cost $\mathbf{c}^\top \boldsymbol{y}^*$	Shared-fixed cost budget and sorting scheme (LA)				Location (PO)
				ξ^*	$s^*_{j1} = 1$	$s^*_{j2} = 1$	$s^*_{j3} = 1$	$y^*_i = 1$
0	12.94	10.60	4.68	2.34	$i = 3, 4, 7$	$j = 5, 8$	$j = 1, 2, 6, 9, 10$	$i = 4–6$
0.5	13.16	10.82	4.68	2.34	$j = 2, 5, 6$	$j = 10$	$j = 1, 3, 4, 7–9$	$i = 2, 4, 5$
1.0	13.48	10.36	6.23	3.12	$j = 7$	$j = 2–4, 6, 8–10$	$j = 1, 5$	$i = 2–5$
1.5	13.69	10.57	6.23	3.12	$j = 2$	$j = 3, 4, 6–9$	$j = 1, 5, 10$	$i = 2, 4–6$
2.0	14.13	11.01	6.23	3.12	$j = 8$	$j = 1, 3–5$	$j = 2, 6, 7, 9, 10$	$i = 4, 5, 7, 10$
2.5	14.34	11.22	6.23	3.12	$j = 9$	$j = 1–3$	$j = 4–8,10$	$i = 3–5,7$
3.0	14.56	11.44	6.23	3.12	$j = 6$	$j = 3, 5$	$j = 1, 2, 4, 7–10$	$i = 2, 4–6$
3.5	16.66	13.54	6.23	3.12	–	$j = 1, 7$	$j = 2–6, 8–10$	$i = 2–5$
4.0	16.90	13.78	6.23	3.12	–	$j = 1$	$j = 2–10$	$i = 2, 4–6$
≥4.5	Infeasible							

4.0×10^6 \$SGD/year based on the Singapore context. The results are presented in Table 14.5. The monetary information is measured in 10^6 \$SGD/year.

First of all, we see that under the current target level, the upper limit for λ is 50%, and the PO will not be able to achieve the target τ anyway if with a sharing ratio larger than this level. Furthermore, when the PO requires a relatively high-profit target, say 4.0×10^6 \$SGD/year, to support resource recovery operations, as the cost share λ of the PO increases, the sorting schemes "recycling bin" and "on-site collection" are gradually replaced by the more effective incentive collection scheme along with the sorting investment being increased. This implies that if the LA expects the PO to share more upfront investment, it should make more efforts in sorting schemes to ensure the feedstock quality to sustain the payoff target for the PO.

The insight and rationale behind the above observations are also intuitive. On the one hand, in waste management practice, installing resource recovery facilities can usually be more expensive compared to the sorting scheme investment. On the other hand, opening more facilities is advantageous if the feedstock quality is high and the feedstock volume is sufficient. Otherwise, it could not achieve break-even due to the poor conversion ratio and heavy capital investment cost. Moreover, investing in the sorting scheme can improve feedstock quality and thus enhance operational profit. Therefore, as our results in Table 14.5 suggest, when the PO and LA share the investment cost of resource recovery, the sorting scheme investment is effective by reducing the total LA cost while achieving the profit target of the PO.

Finally, we also analyze the cooperation efficiency between the LA and the PO by comparing with the following first-best approach of the completely centralized model (CCM):

$$v_{\text{CCM}}^{\tau_{\text{O}}} = \min \quad \boldsymbol{c}^{\top} \boldsymbol{y} + \sup_{\mathbb{P} \in \mathbb{F}(\boldsymbol{s})} \mathbb{E}_{\mathbb{P}}\Big[f(\boldsymbol{s}, \tilde{\boldsymbol{v}}) - \phi\left(\boldsymbol{y}, \tilde{\boldsymbol{u}}, \tilde{\boldsymbol{y}}\right) \Big], \tag{14.73}$$

$$\text{s. t,} \quad \mathbb{E}_{\mathbb{P}}\left[\phi\left(\boldsymbol{y}, \tilde{\boldsymbol{u}}, \tilde{\boldsymbol{v}}\right)\right] \geq \tau_{\text{O}} \quad \forall \mathbb{P} \in \mathbb{F}(\boldsymbol{s}), \tag{14.74}$$

$$\boldsymbol{s} \in \mathcal{S}, \tag{14.75}$$

$$\boldsymbol{y} \in \{0, 1\}^{|I|}. \tag{14.76}$$

In this case, the local authority decides the sorting schemes and recycling facility location and handles the recycling operations. Moreover, τ_{O} is the target for the recycling operational revenue used for a consistent comparison. The above first-best model is only for analysis purposes since, in practice, it is rare that the LA handles all operations in waste management systems.

To evaluate the cooperation efficiency (CE), we compare the total cost of the LA and PO in our optimized BRRP with that of the CCM (14.73)–(14.76), i.e., $v_{\text{CCM}}^{\tau_{\text{O}}}$. In particular, given a financial target τ and cost-share ratio λ, we can compute the optimal cost of the LA, denoted by $v_{\text{LA}}^{(\lambda,\tau)}$, and the optimal total revenue of the PO, denoted by $r_{\text{PO}}^{(\lambda,\tau)}$ under optimality using model (14.17)–(14.20). The total cost of

Table 14.5 The robust BRRP solution profile across different cost-share ratios λ with $\tau = 4.0 \times 10^6$ \$SGD/year

Sharing ratio (%) (PO)	Worst-case total expense (LA)	Worst-case sorting cost (LA)	Worst-case total fixed cost $\mathbf{c}^\top \boldsymbol{y}^*$	Matching fund budget and sorting scheme (LA)				Location (PO)
				ξ^*	$s_{j1}^* = 1$	$s_{j2}^* = 1$	$s_{j3}^* = 1$	$y_i^* = 1$
0	16.59	10.36	6.23	6.23	$j = 5$	$j = 1, 3, 6–10$	$j = 2, 4$	$i = 2–5$
10	16.39	10.78	6.23	5.61	$j = 7$	$j = 1, 3, 4, 8, 10$	$j = 2, 5, 6, 9$	$i = 2, 4–6$
20	16.00	11.01	6.23	4.99	$j = 10$	$j = 1, 5, 8, 9$	$j = 2–4, 6, 7$	$i = 2, 4–6$
30	15.79	11.43	6.23	4.36	$j = 4$	$j = 3, 7$	$j = 1, 2, 5, 6, 8–10$	$i = 2, 4–6$
40	15.39	11.65	6.23	3.74	$j = 3$	$j = 10$	$j = 1, 2, 4–9$	$i = 4–6, 10$
50	16.90	13.78	6.23	3.12	–	$j = 1$	$j = 2–10$	$i = 2, 4–6$
≥60	Infeasible							

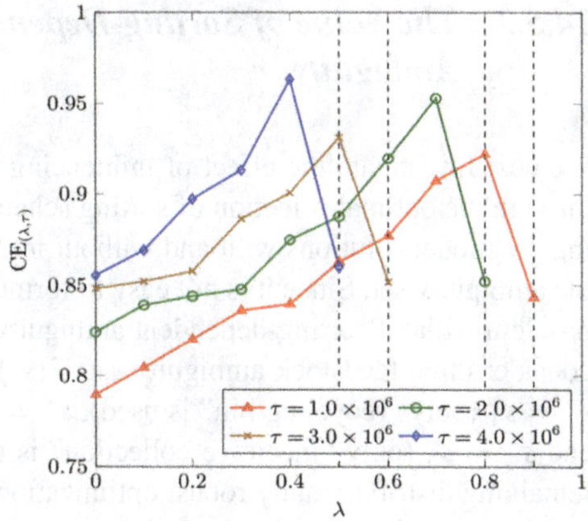

Fig. 14.2 Cooperation efficiency $CE_{(\lambda,\tau)}$ under different cost-share ratio λ and profit target τ of the PO

the LA and PO under (λ, τ) is then $v_{\text{LA}}^{(\lambda,\tau)} - r_{\text{PO}}^{(\lambda,\tau)}$. Furthermore, for a fair comparison with the consistent constraint, we set $\tau_{\text{O}} = \tau + \lambda c^{\top} y^*$, where y^* comes from the solution of the BRRP model, and solve for the optimal cost $v_{\text{CCM}}^{\tau_{\text{O}}}$ of the CCM. Finally, we calculate the cooperation efficiency (CE) concerning (λ, τ) as follows:

$$\text{CE}_{(\lambda,\tau)} = \frac{v_{\text{CCM}}^{\tau_{\text{O}}}}{v_{\text{LA}}^{(\lambda,\tau)} - r_{\text{PO}}^{(\lambda,\tau)}}.$$

In Fig. 14.2 we plot the cooperation efficiency $\text{CE}_{(\lambda,\tau)}$ for different combinations (λ, τ). The dashed line indicates the uppermost value of λ along each CE curve (when the PO becomes infeasible). Observing this figure, we can draw some conclusions. (i) Given a profit target, the cost share of the PO maximizes the CE at a relatively higher point within its feasible range (i.e., from zero until the point after which the PO cannot meet the target). (ii) for the relatively high-profit target (e.g., $\tau = 4.0 \times 10^6$), the CE-maximized cost-share ratio of the PO is relatively low (e.g., $\lambda = 0.8$). However, the CE-maximized ratio is high (e.g., $\lambda = 0.4$), when the profit target is relatively low (e.g., $\tau = 1.0 \times 10^6$).

Some interesting insights can be drawn from this analysis. First, effective sorting plays a pivotal role in supporting the financially sustainable operations for the PO, which deserves enough attention from the LA. Second, installing more resource recovery facilities (or expanding the existing capacity) should be handled carefully. This is due to its high upfront cost and expansion investment. Our analysis is consistent with the shutdown case of IUT Global [523] in practice, where one of the key reasons is that the firm with lower-than-expected operational revenue resulting from poor feedstock condition (purity) cannot afford upfront facility installation (expansion) cost. Third, when the profit target of the PO is relatively low, it is possible to assign a higher cost-share ratio to achieve a high cooperation efficiency. However, the sharing ratio should be low whenever the profit target of the PO is high.

14.5.2 The Value of Sorting-Dependency in Feedstock Ambiguity

We next investigate the effect of influencing the feedstock composition ambiguity through the optimal selection of sorting schemes. This is accomplished by comparing the model solutions with and without integrating sorting-at-source decisions in the ambiguity set. Since it is not easy to formulate the distributionally robust BRRP problem without sorting-dependent ambiguity in a single model, for simplicity, we consider three feedstock ambiguity sets $\mathbb{F}(s_1)$, $\mathbb{F}(s_2)$ and $\mathbb{F}(s_3)$, obtained by fixing $s^* = s_1$ (only "recycling bin" is used), $s^* = s_2$ (only "on-site collection" is used) and $s^* = s_3$ (only "incentive collection" is used) respectively. We then solve the remaining distributionally robust optimization models (without decision-dependent ambiguity) and denote R_1, R_2 and R_3 the obtained solutions. The outcome profiles for λ and τ are presented in Tables 14.6 and 14.7, respectively. The monetary information is given as 10^6 \$SGD/year.

Comparing the solutions R_1, R_2, and R_3 without sorting-dependent ambiguity in Tables 14.6 and 14.7, and that with sorting-dependent ambiguity in Tables 14.4 and 14.5 under the same parameter setting. Several conclusions can be drawn:

(i) There is no feasible R_1 solution neither for any payoff target τ with $\lambda = 0.5$ nor for any λ with $\tau = 4.0 \times 10^6$ \$SGD/year.

(ii) Feasible solutions for R_2 exist only for some parameter settings. Moreover, the feasible R_2 solutions result in more facilities opened (more fixed cost budget) and a higher sorting cost. For instance, when $\tau = 0.5 \times 10^6$ \$SGD/year and $\lambda = 0.5$, the R_2 solution opens four facilities with a total sorting cost of 11.60×10^6 \$SGD/year. However, the R-BRRP solution in Table 14.4 opens only three facilities with a total sorting cost of 10.82×10^6 \$SGD/year due to its flexibility in allocating cost-effectively the sorting schemes.

(iii) R_3 also induces higher sorting costs or opening more facilities. More importantly, we note that even R_3 solution is also infeasible for the case of $\tau = 4.0 \times 10^6$ \$SGD/year, $\lambda = 0.5$. This is counterintuitive because R_3 is generated assuming the implementation of the most expensive sorting scheme, whose cost is borne entirely by the local authority rather than the private operator. Upon closer inspection, this is due to the relatively high uncertainty level (more precisely, the higher upper bound of volume) of the feedstock volume (Table 14.3), resulting from the "incentive collection" scheme employed in R_3. The relatively high feedstock volume uncertainty with "incentive collection" could result from the significant differences among attitudes and sensitivities of residents toward the financial incentives [448, 557]. This volume uncertainty can only be handled by opening more facilities. This would inflate the fixed cost of the PO and make the target unachievable. In contrast, the sorting-integrated R-BRRP solution retains one sorting scheme as "on-site collection" to prevent opening more facilities to ensure feasibility (Table 14.4). This also implies the potential risk of rule-of-thumb decisions in sorting scheme implementation.

Table 14.6 Profiles of optimal solutions (with fixed sorting-at-source decision) for different feedstock ambiguity sets, across different τ with $\lambda = 0.5$

Model	Payoff target (PO)	Worst-case total expense (LA)	Worst-case sorting cost (LA)	Worst-case total fixed cost	Budget and sorting scheme (LA)				Location (PO)
				$\boldsymbol{c}^\top \boldsymbol{y}^*$	ξ^*	$s_{j1}^* = 1$	$s_{j2}^* = 1$	$s_{j3}^* = 1$	$y_i^* = 1$
R_1	≥ 0	Infeasible							
R_2	0–0.5	14.71	11.60	6.23	3.12	–	$j = 1$–10	–	$i = 2, 4$–6
	≥ 1.0	Infeasible							
R_3	0–2.5	17.44	14.32	6.23	3.12	–	–	$j = 1$–10	$i = 2, 4, 5, 10$
	3.0–3.5	17.44	14.32	6.23	3.12	–	–	$j = 1$–10	$i = 2, 4$–6
	≥ 4.0	Infeasible							

Table 14.7 Profiles of optimal solutions (with fixed sorting-at-source decision) for different feedstock ambiguity sets, across different λ with $\tau = 4.0 \times 10^6$ \$SGD/year

Model	Sharing ratio (%) (PO)	Worst-case total expense (LA)	Worst-case sorting cost (LA)	Worst-case total fixed cost $\mathbf{c}^\top \mathbf{y}^*$	Budget and sorting scheme (LA)				Location (PO)
					ξ^*	$s^*_{j1} = 1$	$s^*_{j2} = 1$	$s^*_{j3} = 1$	$y^*_i = 1$
R_1	≥0	Infeasible							
R_2	0	19.39	11.60	7.79	7.79	–	$j = 1$–10	–	$i = 2, 4, 5$–7
	≥10	Infeasible							
R_3	0	20.56	14.32	6.23	6.23	–	–	$j = 1$–10	$i = 2, 4$–6
	10	19.93	14.32	6.23	5.61	–	–	$j = 1$–10	$i = 2, 4$–6
	20	19.31	14.32	6.23	4.99	–	–	$j = 1$–10	$i = 2, 4$–6
	30	18.68	14.32	6.23	4.36	–	–	$j = 1$–10	$i = 2, 4$–6
	40	18.06	14.32	6.23	3.74	–	–	$j = 1$–10	$i = 2, 4$–6
	≥50	Infeasible							

The unit of all monetary terms is 10^6 SGD/yr

14.6 Conclusions and Further Reading

This chapter discussed the application of distributionally robust facility location in a resource recovery planning problem. A Public-Private Partnership (P3) was assumed. A distributionally robust bilevel recycling facility location model with a decision-dependent ambiguity set was devised. It was shown that it can be solved as a mixed-integer linear program.

The model was analyzed regarding the cost-share ratio. The value of sorting investment in supporting sustainable recovery operations was made clear. Numerical results were presented. They showed that the resource recovery solutions integrating sorting-at-source mechanisms to influence feedstock ambiguity perform significantly better than those that do not. The sorting-decision-integrated solutions can take advantage of the information of the private operator to more carefully and smartly influence the feedstock conditions with ambiguity through sorting-at-source investments. Simultaneously, the financial sustainability of the private operator is achieved while reducing the cost burdens of the local authority.

The topic discussed in this chapter is also related to distributionally robust game optimization. The interested readers can refer to [2, 347, 348, 440]. Other work related to the topic include [101, 222, 549] for stochastic resource recovery planning, and [333, 335, 536, 537, 552] for robust resource recovery planning.

Chapter 15
Multi-Period Hub Location with Time Series

Abstract This chapter discusses the application of distributionally robust optimization to an uncapacitated multi-period hub location. Uncertain periodic demand is assumed. A nested ambiguity set is constructed, incorporating a general multivariate time series model for uncertain periodic demands. To ensure stable cost flows, each expected periodic cost is limited by a budget while maximizing the robustness level in terms of the size of the ambiguity set. It is shown that the budget-driven uncapacitated model constructed optimizes a "Sharpe Ratio" type criterion over the worst case among all periods. The influence of the budgets on the optimal robustness level is also discussed. To efficiently solve the resulting model, a bisection search algorithm is presented that solves (in each iteration) a mixed-integer conic program. Numerical experiments support the attractiveness of the model.

Keywords Hub location · Financial budget · Time series · Robust optimization · Wasserstein distance

15.1 Introduction

Hub location was central in Chap. 11. As we explained, this is a widely studied area due to many applications in logistics, transportation, telecommunications, and computer networks. We recall that hub networks are commonly used to efficiently route commodities between many origin–destination (O–D) pairs (see [134, 137]).

The development of a hub network for operating in the long-term (e.g., decades) is often time-consuming and is usually accomplished over time. During that time, demands between O–D pairs also vary. Thinking of a time-discretized setting (e.g., monthly, yearly), a multi-period hub location (MHL) problem emerges. We seek to find where and when to locate a set of hubs over the planning horizon to better adapt to a time-evolving setting [144]. This chapter complements the analysis provided in Chap. 11. We consider a multi-period setting with demand uncertainty. At the beginning of the planning horizon, i.e., before any demand is observed, the decision-maker decides on the hub construction plan. Afterwards, in each period, a decision

© The Author(s), under exclusive license to Springer Nature Switzerland AG 2024
F. Saldanha-da-Gama, S. Wang, *Facility Location Under Uncertainty*,
International Series in Operations Research & Management Science 356,
https://doi.org/10.1007/978-3-031-55927-3_15

is made regarding flow routing. Two challenges are considered in this chapter: (i) the uncertain trend of the demand process over the panning horizon and (ii) the risk of operational cost fluctuations over the planning horizon.

Regarding the first, we stress again that MHL is usually a long-term strategic decision problem. Demand is dynamic and uncertain. Hardly can the underlying process be precisely specified. In such a setting, as the contents of this volume have shown, ignoring uncertainty when making decisions can lead to solutions that are far from good when uncertainty is revealed (see [145, 510, 527]). Adaptive two-stage hub location models deal mainly with cross-sectional uncertainty. Thus, it is difficult, or at least misleading, to extend it to a multi-period setting involving a path-dependent stochastic process (e.g., [468, 535]).

The second challenge emerges due to the heavy upfront investments typically required by a hub location problem. This requires relatively stable operational cash flows over the planning horizon. However, due to uncertainty in demand over time, the operational costs can also fluctuate significantly from one period to another. This may result in a heavy financial burden or even the financial failure of the project. Such a temporal risk in the operational cash flows has been well recognized in many multi-period planning problems (e.g., [115, 361, 492]). Therefore, in the MHL problem, it is desirable to effectively mitigate the risk of operational cost fluctuation over the planning horizon.

This chapter discusses a time series distributional robust optimization modeling framework seeking to address the above challenges. A nested ambiguity set is developed incorporating a general multivariate time series model (e.g., VAR, or VARMA) with Wasserstein balls for different periods. Furthermore, a budget-driven model is developed to mitigate the fluctuation of periodic operational costs. We follow closely the analysis presented in [278]. Nevertheless, we adapt some contents and much notation to this volume.

The remainder of this chapter is organized as follows. Section 15.2 introduces the budget-driven MHL model and presents the nested ambiguity set. Section 15.3 provides modeling insights and discusses a tractable reformulation for the problem. Section 15.4 presents several computational results to justify the value of the budget-driven MHL model. The chapter concludes with an overview and some suggestions for further reading.

15.2 Base Model and Time Series Ambiguity Set

In this section, we introduce the problem we set at the core of this chapter. We then briefly introduce the (multivariate) time series models for forecasting future demands and the Wasserstein ball for describing the ambiguity of the estimated random noises (ERN) and construct a Wasserstein-ball-and-Time-Series-model-based (WTS) ambiguity set to model the distributional ambiguity of the demand trend over time.

15.2.1 Base Uncapacitated Model

We consider a multi-period uncapacitated hub location problem with a complete hub-level network. The problem can be formulated by extending the model proposed in [256] and discussed in Chaps. 2 and 11. In particular, we start from the multicommodity version of it, given by (11.340)–(11.344).

We introduce the relevant notation. Regarding the underlying sets, we denote by K the set of potential locations for the hubs, by M the set of commodities (each associated with one O–D pair), and by T the planning horizon divided into $|T|$ periods (we assume $T = \{1, \ldots, |T|\}$).

The parameters involved in the problem are d_{mt}, the demand for commodity $m \in M$ in period $t \in T$, $c_{k\ell mt}$ the unit cost for shipping commodity $m \in M$ through hubs k and ℓ in period $t \in T$, $k, \ell \in K$, and f_{kt}, the fixed cost for building a hub at $k \in K$ in period $t \in T$.

The deterministic problem can be formulated considering two sets of decision variables. For $k \in K$ and $t \in T$, y_{kt} is one if a hub operates at k in period t and zero otherwise. For $m \in M$ and $t \in T$, $w_{k\ell mt}$ is the amount of commodity m shipped through hubs k and ℓ in period t, $k, \ell \in K$.

For every $t \in T$ we denote the location decision and demand in period t by $\boldsymbol{y}_t = (y_{kt})_{k \in K}$ and $\boldsymbol{d}_t = (d_{mt})_{m \in M}$. Collectively, we denote the set of location decisions up to period t by $\boldsymbol{Y}_t = (\boldsymbol{y}_1, \ldots, \boldsymbol{y}_t) = (y_{k\tau})_{k \in K, \tau \leq t}$.

The problem of finding the flow routing in period $t \in T$ minimizing the shipping cost can be formulated as follows:

$$\mathrm{Q}_t(\boldsymbol{Y}_t, \boldsymbol{d}_t) = \min \quad \sum_{m \in M} \sum_{k,\ell \in K} c_{k\ell mt} w_{k\ell mt}, \tag{15.1}$$

$$\text{s. t.} \quad \sum_{k,\ell \in K} w_{k\ell mt} = d_{mt} \quad \forall m \in M, \tag{15.2}$$

$$\sum_{\ell \in K} w_{k\ell mt} + \sum_{\ell \in K \setminus \{j\}} w_{\ell kmt} \leq d_{mt} \cdot \sum_{\tau \in [t]} y_{k\tau}, \quad \forall m \in M,\ k \in K, \tag{15.3}$$

$$w_{k\ell mt} \geq 0 \quad \forall m \in M,\ k, \ell \in K. \tag{15.4}$$

The first set of constraints ensures demand satisfaction, whereas the second guarantees that the commodities are routed using only selected hubs. In these constraints, $[t] = \{1, \ldots, t\}$.

An appealing structure of the above model is that the shipping cost $\mathrm{Q}_t(\boldsymbol{Y}_t, \boldsymbol{d}_t)$ can be represented as an affine function of demand $\boldsymbol{d}_t$ (see, e.g., [256]):

$$\mathrm{Q}_t(\boldsymbol{Y}_t, \boldsymbol{d}_t) = \sum_{m \in M} d_{mt} \cdot \Upsilon_{mt}(\boldsymbol{Y}_t) \quad \forall t \in T. \tag{15.5}$$

For each $m \in M$, $\Upsilon_{mt}(\boldsymbol{Y}_t)$, can be interpreted as the unit shipping cost of commodity m in period t given the location decision $\boldsymbol{Y}_t$. Let $x_{k\ell}$ be a variable representing the fraction of the commodity shipped through hubs k and ℓ ($k, \ell \in K$) and $\boldsymbol{X} = (x_{k\ell})_{k,\ell \in K}$. Then we have

$$\Upsilon_{mt}(\boldsymbol{Y}_t) = \min \quad \sum_{k,\ell \in K} c_{k\ell mt} x_{k\ell}, \tag{15.6}$$

$$\text{s. t.} \quad \sum_{k,\ell \in K} x_{k\ell} = 1, \tag{15.7}$$

$$\sum_{\ell \in K} x_{k\ell} + \sum_{\ell \in K \setminus \{k\}} x_{\ell k} \leq \sum_{\tau \in [t]} y_{k\tau} \quad \forall k \in K, \tag{15.8}$$

$$\boldsymbol{X} \in \mathbb{R}^{|K| \times |K|}. \tag{15.9}$$

Such an affine structure plays a role in facilitating our subsequent analysis and optimization when accounting for randomness in the demand process.

The deterministic uncapacitated MHL problem seeks the hub location plan and the flow routing throughout the planning horizon to minimize the total cost. It can be formulated as follows:

$$\text{minimize} \quad \sum_{t \in T} v_t, \tag{15.10}$$

$$\text{subject to} \quad \boldsymbol{f}_t^\top \boldsymbol{y}_t + \mathrm{Q}_t(\boldsymbol{Y}_t, \boldsymbol{d}_t) \leq v_t \quad \forall t \in T, \tag{15.11}$$

$$\boldsymbol{Y} \in \mathcal{Y}, \tag{15.12}$$

$$\boldsymbol{v} \geq \boldsymbol{0}. \tag{15.13}$$

In this model, $\boldsymbol{f}_t = (f_{kt})_{k \in K}$ for all $t \in T$, $\boldsymbol{v} = (v_1, \ldots, v_{|T|})$ with v_t being an epigraphical decision variable that corresponds to the total cost in period t. Additionally, we are considering $\boldsymbol{Y} = (\boldsymbol{y}_1, \ldots, \boldsymbol{y}_{|T|}) = (y_{kt})_{k \in K, t \in T}$, and $\mathcal{Y}$ is the feasible set of the location decision $\boldsymbol{Y}$. Throughout this chapter, we assume that a hub remains open once built, implying $\mathcal{Y} = \{\boldsymbol{y} \in \{0, 1\}^{|K| \times |K|} \mid \sum_{t \in T} y_{kt} \leq 1 \ \forall k \in K\}$. We note that additional requirements on $\boldsymbol{Y}$ can be imposed in $\mathcal{Y}$ straightforwardly.

15.2.2 Budget-Driven Uncapacitated Model

For financially sustainable operations management, in practice, we may need to control the cost level of each period. This motivates us to pre-specify for each $t \in T$, v_t as the cost target in period t. We consider demand uncertainty. We denote by $\tilde{\boldsymbol{d}}$ the random vector that represents it. For each $t \in T$, $\tilde{\boldsymbol{d}}$ is the random vector representing the demand for all commodities in period t. Furthermore, we assume we do not know the exact distribution underlying the demand. Instead, we consider that the correct demand distribution $\mathbb{P}$ is to be any member in an ambiguity set $\mathcal{F}(\theta)$ of size θ (we refer the reader to Sect. 4.5 and Chap. 8 for base related concepts).

Given the cost target, we seek to hedge against the distributional ambiguity as much as possible by maximizing the size of the ambiguity set. This leads to the following budget-driven model:

$$\text{maximize} \quad \theta, \tag{15.14}$$

$$\text{subject to} \quad \zeta_t(\boldsymbol{Y}, \theta) = \sup_{\mathbb{P} \in \mathcal{F}(\theta)} \mathbb{E}_{\mathbb{P}}[\boldsymbol{f}_t^\top \boldsymbol{y}_t + \mathrm{Q}_t(\boldsymbol{Y}_t, \tilde{\boldsymbol{d}}_t)] \leq v_t \quad \forall t \in T, \tag{15.15}$$

$$\boldsymbol{Y} \in \mathcal{Y}, \tag{15.16}$$

$$\theta \geq 0. \tag{15.17}$$

In this problem, $\zeta_t(\boldsymbol{Y}, \theta)$ is the so-termed, worst-case expected periodic cost under distributional ambiguity. In the above budget-driven model, periodic demands $\tilde{\boldsymbol{d}}_1, \ldots, \tilde{\boldsymbol{d}}_T$ are revealed sequentially. Hence, it is reasonable to view these demands as a time series subject to distributional ambiguity—an issue we discuss next.

15.2.3 Time Series-Model Based Wasserstein-Ball Ambiguity Set

Suppose that a collection of historical samples $\hat{\boldsymbol{d}}_{-1}, \ldots, \hat{\boldsymbol{d}}_{-N}$ is available, where for each $n \in [N] = \{1, \ldots, N\}$, $\hat{\boldsymbol{d}}_{-n} = (d_{-nm})_{m \in M}$ tracks the historical demands of commodities that was observed n periods before the current time. Based on estimation from historical data, the demand forecast function is a time series taking the following form:

$$\tilde{\boldsymbol{d}}_t = \boldsymbol{d}_t^0 + \boldsymbol{A}_t^1 \tilde{\boldsymbol{d}}_1 + \boldsymbol{A}_t^2 \tilde{\boldsymbol{d}}_2 + \cdots + \boldsymbol{A}_t^{t-1} \tilde{\boldsymbol{d}}_{t-1} + \tilde{\boldsymbol{\xi}}_t \quad \forall t \in T, \tag{15.18}$$

where $\boldsymbol{d}_t^0$ is the intercept, $\boldsymbol{A}_t^1, \ldots, \boldsymbol{A}_t^{t-1}$ are matrices of coefficients that capture the relation between $\tilde{\boldsymbol{d}}_t$ and previous demands, and $\tilde{\boldsymbol{\xi}}_t$ is some estimated random noise (ERN).

As we will elaborate in the following examples, our time series model (15.18) is rich enough to cover many popular time series approaches. Some existing time series models of uncertainty, such as the growth model in [537] and the additive demand model in [427], can be recast as a particular case of (15.18).

Example 15.1 (Vector Autoregression (VAR)) By multivariate least squares estimation, the pth-order VAR(p) model yields

$$\tilde{\boldsymbol{d}}_t = \boldsymbol{\kappa} + \boldsymbol{\Gamma}_1 \tilde{\boldsymbol{d}}_{t-1} + \cdots + \boldsymbol{\Gamma}_p \tilde{\boldsymbol{d}}_{t-p} + \tilde{\boldsymbol{\xi}}_t,$$

where $(\boldsymbol{\kappa}, \boldsymbol{\Gamma}_1, \ldots, \boldsymbol{\Gamma}_p) = \boldsymbol{D}\boldsymbol{\Lambda}^\top(\boldsymbol{\Lambda}\boldsymbol{\Lambda}^\top)^{-1}$ with $\boldsymbol{D} = (\hat{\boldsymbol{d}}_{-1}, \hat{\boldsymbol{d}}_{-2}, \ldots, \hat{\boldsymbol{d}}_{-N})$, $\boldsymbol{\Lambda} = (\boldsymbol{\Lambda}_{-1}, \ldots, \boldsymbol{\Lambda}_{-N})$, and $\boldsymbol{\Lambda}_{-n} = (1, \hat{\boldsymbol{d}}_{-n-1}, \ldots, \hat{\boldsymbol{d}}_{-n-p})^\top$ for all $n \in [N]$. Fitting into model (15.18), we have

$$\tilde{\boldsymbol{d}}_t = \boldsymbol{d}_t^0 + \boldsymbol{A}_t^1 \tilde{\boldsymbol{d}}_1 + \boldsymbol{A}_t^2 \tilde{\boldsymbol{d}}_2 + \cdots + \boldsymbol{A}_t^{t-1} \tilde{\boldsymbol{d}}_{t-1} + \tilde{\boldsymbol{\xi}}_t \quad \forall t \in T,$$

where $\boldsymbol{d}_t^0 = \boldsymbol{\kappa} + \sum_{s \in [p-t+1]} \boldsymbol{\Gamma}_{t+s} \hat{\boldsymbol{d}}_{-s}$, $\boldsymbol{A}_t^s = \mathbb{I}[s \geq t-p] \cdot \boldsymbol{\Gamma}_{t-s}$ for every $s \in [t-1]$, and $\mathbb{I}[\mathcal{E}]$ is an indicator function which takes a value of 1 if the logical expression $\mathcal{E}$ is true and 0 otherwise.

Example 15.2 (Vector Autoregressive Moving Average (VARMA)) With two parameters p and q, the VARMA(p, q) model yields

$$\tilde{\boldsymbol{d}}_t = \boldsymbol{\kappa} + \boldsymbol{\Gamma}_1 \tilde{\boldsymbol{d}}_{t-1} + \cdots + \boldsymbol{\Gamma}_p \tilde{\boldsymbol{d}}_{t-p} + \tilde{\boldsymbol{\xi}}_t + \boldsymbol{L}_1 \tilde{\boldsymbol{\xi}}_{t-1} + \cdots + \boldsymbol{L}_q \tilde{\boldsymbol{\xi}}_{t-q}.$$

Fitting into model (15.18), we have

$$\begin{aligned} \boldsymbol{d}_t^0 &= \boldsymbol{\kappa} + \sum_{s \in [p-t+1]} \boldsymbol{\Gamma}_{t+s} \hat{\boldsymbol{d}}_{-s} + \sum_{s \in [q-t+1]} \boldsymbol{L}_{t+s} \hat{\boldsymbol{\xi}}_{-s} - \sum_{s \in [t-1]} \boldsymbol{L}_s \boldsymbol{d}_{t-s}^0 \quad \text{and} \\ \boldsymbol{A}_t^s &= \sum_{\substack{j+i_1+i_2+\cdots=t-s \\ 1 \leq i_1 \leq i_2 \leq \cdots \leq q,\ 1 \leq j}} \boldsymbol{L}_{i_1} \boldsymbol{L}_{i_2} \cdots (\boldsymbol{\Gamma}_j + \boldsymbol{L}_j) \qquad \forall s = 1, \ldots, t, \end{aligned}$$

where the number of indices $i_1, i_2, \ldots$ can be zero, and the historical estimated noises $\hat{\boldsymbol{\xi}}_{-1}, \ldots, \hat{\boldsymbol{\xi}}_{-N}$ are calculated from historical demands.

Seeking computational tractability, we make the following assumption, which is also reasonable in a statistical sense because the ERNs tend to be mutually independent as the sample size increases (see [178]).

Assumption 15.1 The estimated random noises $\tilde{\boldsymbol{\xi}}_1, \ldots, \tilde{\boldsymbol{\xi}}_{|T|}$ are mutually independent.

Equipped with the time series model (15.18), we next model the distributional ambiguity of ERNs over the planning horizon. We assume that for each $t \in T$, the ERN $\tilde{\boldsymbol{\xi}}_t$ is governed by some distribution $\mathbb{Q}_t$, which is ambiguous due to difficulty in estimation but is known to be close to some reference $\hat{\mathbb{Q}}_t$ that is consistent with available information. We measure the closeness between $\mathbb{Q}_t$ and $\hat{\mathbb{Q}}_t$ by the following type-1 Wasserstein distance (as introduced in Sect. 4.5.2):

$$\Delta(\mathbb{Q}_t, \hat{\mathbb{Q}}_t) = \inf_{\Pi} \left\{ \mathbb{E}_\Pi[\|\tilde{\boldsymbol{\xi}}_t - \tilde{\boldsymbol{\xi}}_t'\|] : \begin{array}{l} \Pi \in \mathcal{P}(\mathbb{R}^{|M|} \times \mathbb{R}^{|M|}) \text{ with} \\ \text{marginals } \mathbb{Q}_t \text{ and } \hat{\mathbb{Q}}_t, \text{ respectively} \end{array} \right\},$$

where $\tilde{\boldsymbol{\xi}}_t \sim \mathbb{Q}_t$, $\tilde{\boldsymbol{\xi}}_t' \sim \hat{\mathbb{Q}}_t$, and $\mathcal{P}(\mathbb{R}^{|M|} \times \mathbb{R}^{|M|})$ represents the set of all joint distributions supported on $\mathbb{R}^K \times \mathbb{R}^K$, and $\|\cdot\|$ represents a general norm (e.g., ℓ-norm is denoted by $\|\cdot\|_\ell$, the Mahalanobis norm is denoted by $\|\cdot\|_{\mathrm{M}}$).

The distributional ambiguity of $\tilde{\boldsymbol{\xi}}_t$ is being captured by the fact that its distribution $\mathbb{Q}_t$ is only unknown to reside in a ball of radius θ centered at $\hat{\mathbb{Q}}_t$ with respect to the Wasserstein distance; that is, we have $\mathbb{Q}_t \in \mathcal{B}_t(\theta)$ with

$$\mathcal{B}_t(\theta) = \{\mathbb{Q} \in \mathcal{Q}(\mathbb{R}^K) \mid d(\mathbb{Q}, \hat{\mathbb{Q}}_t) \leq \theta\}.$$

Here, $\mathcal{B}_t(0) = \{\hat{\mathbb{Q}}_t\}$ because $\Delta(\mathbb{Q}_t, \hat{\mathbb{Q}}_t) = 0$ only if $\mathbb{Q}_t = \hat{\mathbb{Q}}_t$.

Next, we discuss a nested distributional ambiguity for the demands $\tilde{\boldsymbol{d}}_1, \ldots, \tilde{\boldsymbol{d}}_T$.

Recall the time series model (15.18). For $t = 1$, the distribution $\mathbb{P}_1$ of $\tilde{\boldsymbol{d}}_1$ is determined by the constant $\boldsymbol{d}_1^0$ and the ERN $\tilde{\boldsymbol{\xi}}_1$. Thus, the ambiguity set of $\tilde{\boldsymbol{u}}_1$ can be captured by

$$\mathcal{F}_1(\theta) = \left\{ \mathbb{P}_1 \in \mathcal{P}(\mathbb{R}_+^{|M|}) \;\middle|\; \begin{array}{l} \tilde{\boldsymbol{d}}_1 \sim \mathbb{P}_1,\ \tilde{\boldsymbol{\xi}}_1 \sim \mathbb{Q}_1 \\ \tilde{\boldsymbol{d}}_1 = \boldsymbol{d}_1^0 + \tilde{\boldsymbol{\xi}}_1 \\ \mathbb{Q}_1 \in \mathcal{B}_1(\theta) \end{array} \right\}.$$

Then for $t = 2, \ldots, T$, the ambiguity set of $\tilde{\boldsymbol{d}}_t$ conditioning on the realization $\boldsymbol{D}_{t-1} = (\boldsymbol{d}_1, \ldots, \boldsymbol{d}_{t-1})$ of $(\tilde{\boldsymbol{d}}_1, \ldots, \tilde{\boldsymbol{d}}_{t-1})$, denoted by $\mathcal{F}_{t|\boldsymbol{D}_{t-1}}$, is

$$\mathcal{F}_{t|\boldsymbol{D}_{t-1}}(\theta) = \left\{ \mathbb{P}_{t|\boldsymbol{D}_{t-1}} \in \mathcal{P}(\mathbb{R}_+^{|M|}) \;\middle|\; \begin{array}{l} \tilde{\boldsymbol{d}}_t \sim \mathbb{P}_{t|\boldsymbol{D}_{t-1}},\ \tilde{\boldsymbol{\xi}}_t \sim \mathbb{Q}_t \\ \tilde{\boldsymbol{d}}_t = \boldsymbol{d}_t^0 + \boldsymbol{A}_t^1 \boldsymbol{d}_1 + \cdots + \boldsymbol{A}_t^{t-1} \boldsymbol{d}_{t-1} + \tilde{\boldsymbol{\xi}}_t \\ \mathbb{Q}_t \in \mathcal{B}_t(\theta) \end{array} \right\},$$

where $\mathbb{P}_{t|\boldsymbol{D}_{t-1}}$ is the conditional distribution of $\tilde{\boldsymbol{d}}_t$ given $\boldsymbol{D}_{t-1}$.

Finally, we formulate the ambiguity set for the demand trend $(\tilde{\boldsymbol{d}}_1, \ldots, \tilde{\boldsymbol{d}}_T)$ as follows:

$$\mathcal{F}(\theta) = \left\{ \mathbb{P} \in \mathcal{P}\left(\mathbb{R}_+^{|M|\times|T|}\right) \;\middle|\; \begin{array}{ll} (\tilde{\boldsymbol{d}}_1, \ldots, \tilde{\boldsymbol{d}}_T) \sim \mathbb{P} & \\ \tilde{\boldsymbol{d}}_1 \sim \mathbb{P}_1,\ \tilde{\boldsymbol{d}}_{t|\boldsymbol{D}_{t-1}} \sim \mathbb{P}_{t|\boldsymbol{D}_{t-1}} & \forall t = 2, \ldots, |T| \\ \mathbb{P}_1 \in \mathcal{F}_1(\theta), & \\ \mathbb{P}_{t|\boldsymbol{D}_{t-1}} \in \mathcal{F}_{t|\boldsymbol{D}_{t-1}}(\theta) & \forall t = 2, \ldots, |T| \end{array} \right\}. \tag{15.19}$$

Before ending this section, we present some possible candidates for the reference distribution of the ERNs.

Example 15.3 (Gaussian Distribution) In the VARMA model (which covers VAR as a special case), it is usually assumed that the ERN follows a Gaussian distribution of zero mean and an estimated covariance matrix $\boldsymbol{\Sigma}$. That is, we can set $\tilde{\boldsymbol{\xi}}_t \sim \hat{\mathbb{Q}}_t = \mathcal{N}(\mathbf{0}, \boldsymbol{\Sigma})$.

Example 15.4 (Surrogate Empirical Distribution) Suppose that the ERNs are i.i.d. random vectors. We can calculate from historical demands the historical estimated noises $\hat{\boldsymbol{\xi}}_{-1}, \ldots, \hat{\boldsymbol{\xi}}_{-N}$, which form a surrogate empirical distribution such that $\tilde{\boldsymbol{\xi}}_t \sim \hat{\mathbb{Q}} = \frac{1}{N}\sum_{n\in[N]} \delta_{\hat{\boldsymbol{\xi}}_{-n}}$ for every $t \in T$, where $\delta_{\hat{\boldsymbol{\xi}}_{-n}}$ is a Dirac measure that places a unit mass at $\hat{\boldsymbol{\xi}}_{-n}$.

Example 15.5 (General Distribution) If ERNs are not identical, we can consider a general noise distribution $\tilde{\boldsymbol{\xi}}_t \sim \hat{\mathbb{Q}}_t$ as the reference distribution of $\mathcal{B}_t(\theta)$ for every $t \in T$. In this case, the uncapacitated model admits a tractable reformulation as long as the mean of $\hat{\mathbb{Q}}_t$ is known.

15.3 Budget-Driven Robust Model over a WTS Ambiguity Set

We consider uncertain demand captured by the Wasserstein-ball-and-Time-Series-model-based (WTS) defined according to (15.19). First, we introduce a periodic budget-driven uncapacitated MHL model to maximize the robustness level while ensuring that, in each period, the worst-case expected cost is not higher than the corresponding budget. We then point out an insightful interpretation of the model and derive a tractable reformulation for it. Finally, we analyze the impact of the periodic budgets on the optimal robustness level.

15.3.1 Model, Insights, and Tractable Reformulation

Exploring the nested ambiguity set $\mathcal{F}(\theta)$ in (15.19), we can re-express the worst-case expected periodic costs in problem (15.14)–(15.17) with a nested form as well.

Lemma 15.1 *Given* $\mathcal{F}(\theta)$ *defined as in* (15.19), *for all* $t \in T$ *we have*

$$\zeta_t(\boldsymbol{Y}, \theta) = \sup_{\mathbb{P}_1\in\mathcal{F}_1(\theta)} \mathbb{E}_{\mathbb{P}_1}\left[\sup_{\mathbb{P}_2\in\mathcal{F}_{2|\tilde{\boldsymbol{D}}_1}(\theta)} \mathbb{E}_{\mathbb{P}_2}\left[\cdots \sup_{\mathbb{P}_t\in\mathcal{F}_{t|\tilde{\boldsymbol{D}}_{t-1}}(\theta)} \mathbb{E}_{\mathbb{P}_t}\left[\boldsymbol{f}_t^\top \boldsymbol{y}_t + \mathrm{Q}_t(\boldsymbol{Y}_t, \tilde{\boldsymbol{d}}_t)\middle|\, \tilde{\boldsymbol{D}}_{t-1}\right]\cdots\middle|\,\tilde{\boldsymbol{D}}_1\right]\right]. \tag{15.20}$$

Proof See Appendix B in [278]. □

This nested representation involves the propagation of demand uncertainty over the different periods of the planning horizon. It seems inevitably affected by the notorious "curse of dimensionality" in the multi-stage optimization problems (see [492]). Fortunately, due to the affine structure of the objective function of the uncapacitated MHL given by (15.5) for all $t \in T$ and taking advantage of the structure of the proposed ambiguity set of independent ERNs, we can rearrange the terms in (15.20) based on each $\tilde{\boldsymbol{\xi}}_t$ and derive a regular distributionally robust optimization reformulation of $\zeta_t(\boldsymbol{Y}, \theta)$. The following result holds.

Proposition 15.1 *Given $\mathcal{F}(\theta)$ as defined in* (15.19), *for all $t \in T$ we have*

$$\zeta_t(\boldsymbol{Y}, \theta) = \sup_{\mathbb{Q}_{[t]} \in \mathcal{B}_{[t]}(\theta)} \mathbb{E}_{\mathbb{Q}_{[t]}} \left[\boldsymbol{f}_t^\top \boldsymbol{y}_t + \boldsymbol{\eta}_t^\top \boldsymbol{\Upsilon}_t(\boldsymbol{Y}_t) + \sum_{s \in [t]} [\boldsymbol{B}_t^s \tilde{\boldsymbol{\xi}}_s]^\top \boldsymbol{\Upsilon}_t(\boldsymbol{Y}_t) \right]. \tag{15.21}$$

Here, $\boldsymbol{\Upsilon}_t(\cdot) = (\Upsilon_{mt}(\cdot))_{m \in M}$, $\mathbb{Q}_{[t]} = \bigotimes_{s \in [t]} \mathbb{Q}_s$, $\mathcal{B}_{[t]}(\theta) = \bigotimes_{s \in [t]} \mathcal{B}_s(\theta)$, and deterministic vector and matrices from the estimated time series model (15.18) *are*

$$\boldsymbol{\eta}_t = \boldsymbol{d}_t^0 + \sum_{s \in [t-1]} \boldsymbol{A}_t^s \boldsymbol{\eta}_s \text{ and } \boldsymbol{B}_t^s = \sum_{s < i_1 < i_2 < \cdots < i_l < t-1} \boldsymbol{A}_{i_1}^s \boldsymbol{A}_{i_2}^{i_1} \cdots \boldsymbol{A}_t^{i_l} \quad \forall s \in [t-1],$$

and $\boldsymbol{B}_t^t$ is a $|M|$-dimensional identity matrix.

Note that for all $t \in T$, the term in the above worst-case expectation (15.21) is affine concerning $\tilde{\boldsymbol{\xi}}_t$. This enables employing the regularization (Theorem 6.3, [385]) to derive an insightful closed-form representation we formalize as follows.

Theorem 15.1 *The budget-driven uncapacitated MHL model* (15.14)–(15.17) *is equivalent to*

$$\theta^\star(\boldsymbol{v}) = \max_{\boldsymbol{Y} \in \mathcal{Y}} \min_{t \in T} \gamma_t(\boldsymbol{Y}) \tag{15.22}$$

with

$$\gamma_t(\boldsymbol{Y}) = \frac{\left(v_t - \mathbb{E}_{\hat{\mathbb{Q}}_{[t]}} \left[\boldsymbol{f}_t^\top \boldsymbol{y}_t + \tilde{\boldsymbol{d}}_t^\top \boldsymbol{\Upsilon}_t(\boldsymbol{Y}_t) \right]\right)^+}{\nabla_\theta \zeta_t(\boldsymbol{Y}, \theta)}, \tag{15.23}$$

where $\hat{\mathbb{Q}}_{[t]} = \bigtimes_{s \in [t]} \hat{\mathbb{Q}}_s$ and $\nabla_\theta \zeta_t(\boldsymbol{Y}, \theta) = \sum_{s \in [t]} \left\| [\boldsymbol{B}_t^s]^\top \boldsymbol{\Upsilon}_t(\boldsymbol{Y}_t) \right\|_*$ is the gradient of $\zeta_t(\boldsymbol{Y}, \theta)$ at θ.

Proof See [278]. □

The implications of Theorem 15.1 are two-fold. First, reformulation (15.22) reveals that for each period t, the corresponding component $\gamma_t(\boldsymbol{Y})$ is a "Sharpe Ratio"-type criterion on "the expected cost saving under the reference distribution $\hat{\mathbb{Q}}_{[t]}$" over "the sensitivity of the worst-case expected cost $\zeta_t(\boldsymbol{Y}, \theta)$ to the robustness level θ." Accordingly, the optimal robustness level $\theta^\star$ optimizes such a criterion over the worst case among all $|T|$ periods. The "Shape Ratio"-type criterion generalizes the classical Sharpe Ratio stemming from finance (see [495]).

Corollary 15.1 *Suppose that for all $t \in T$, $\hat{\mathbb{Q}}_t = \mathcal{N}(\boldsymbol{0}, \boldsymbol{\Sigma})$ and $\|\cdot\|$ in $\mathcal{B}_t(\theta)$ is the Mahalanobis norm based on the estimated covariance matrix $\boldsymbol{\Sigma}$ such that $\|\boldsymbol{b}\| = \|\boldsymbol{\Sigma}^{-1/2}\boldsymbol{b}\|_2$. Then for all $t \in T$, we have*

$$\gamma_t(\boldsymbol{Y}) = \frac{\left(v_t - \mathbb{E}_{\hat{\mathbb{Q}}_{[t]}}\left[\boldsymbol{f}_t^\top \boldsymbol{y}_t + \tilde{\boldsymbol{d}}_t^\top \boldsymbol{\Upsilon}_t(\boldsymbol{Y}_t)\right]\right)^+}{\sum_{s\in[t]} \sqrt{\mathbb{V}_{\hat{\mathbb{Q}}_s}[\boldsymbol{B}_t^s \tilde{\boldsymbol{\xi}}_s]^\top \boldsymbol{\Upsilon}_t(\boldsymbol{Y}_t)]}},$$

where $\mathbb{V}_{\hat{\mathbb{Q}}_t}[\cdot]$ is the variance under distribution $\hat{\mathbb{Q}}_t$. In particular, in the single-period case (i.e., $|T| = 1$), we have

$$\gamma_1(\boldsymbol{Y}) = \frac{\left(v_1 - \mathbb{E}_{\hat{\mathbb{Q}}_1}\left[\boldsymbol{f}_1^\top \boldsymbol{y}_1 + \tilde{\boldsymbol{d}}_1^\top \boldsymbol{\Upsilon}_1(\boldsymbol{y}_1)\right]\right)^+}{\sqrt{\mathbb{V}_{\hat{\mathbb{Q}}_1}[\boldsymbol{f}_1^\top \boldsymbol{y}_1 + \tilde{\boldsymbol{d}}_1^\top \boldsymbol{\Upsilon}_1(\boldsymbol{y}_1)]}}.$$

Proof See [278]. □

The second implication of representation (15.22) is computationally appealing. Observing that the robustness level of a location design $\boldsymbol{Y}$ can be evaluated via the following formula:

$$\min_{t\in T} \left\{ \frac{\left(v_t - \mathbb{E}_{\hat{\mathbb{Q}}_{[t]}}\left[\boldsymbol{f}_t^\top \boldsymbol{y}_t + \tilde{\boldsymbol{d}}_t^\top \boldsymbol{\Upsilon}_t(\boldsymbol{Y}_t)\right]\right)^+}{\sum_{s\in[t]} \left\|[\boldsymbol{B}_t^s]^\top \boldsymbol{\Upsilon}_t(\boldsymbol{Y}_t)\right\|_*} \right\}.$$

We conclude that optimizing the robustness level over $\boldsymbol{Y} \in \mathcal{Y}$ can be achieved by a bisection search wherein each iteration examines the feasibility of a mixed-integer conic-constrained system. This is formalized in the subsequent theorem and algorithm.

Theorem 15.2 *Given $\mathcal{F}(\theta)$ according to* (15.19), *the feasible set of model* (15.14)–(15.17) *is equivalent to the following mixed-integer conic-constrained system:*

$$
\begin{cases}
\boldsymbol{f}_t^\top \boldsymbol{y}_t + \boldsymbol{\eta}_t^\top \boldsymbol{h}_t + \sum\limits_{s\in[t]} \boldsymbol{h}_t^\top \boldsymbol{B}_t^s \boldsymbol{\mu}_s + \theta\lambda_t^s \le v_t & \forall t \in T \\
\|[\boldsymbol{B}_t^s]^\top \boldsymbol{h}_t\|_* \le \lambda_t^s & \forall t \in T,\ s \in [t] \\
\sum\limits_{k,\ell\in K} c_{k\ell mt} x_{k\ell mt} \le h_{mt} & \forall t \in T,\ m \in M \\
\sum\limits_{k,\ell\in K} x_{k\ell mt} = 1 & \forall t \in T,\ m \in M \\
\sum\limits_{\ell\in K} x_{k\ell mt} + \sum\limits_{\ell\in K\setminus\{j\}} x_{\ell kmt} \le \sum\limits_{s\in[t]} y_{ks} & \forall t \in T,\ m \in M,\ k \in K \\
\lambda_t^s \in \mathbb{R}_+,\ x_{k\ell mt} \in \mathbb{R}_+,\ \boldsymbol{h}_t \in \mathbb{R}_+^{|M|} & \forall t \in T,\ m \in M,\ k,\ell \in K,\ s \in [t] \\
Y \in \mathcal{Y},
\end{cases}
\tag{15.24}
$$

where $\boldsymbol{\mu}_t = \mathbb{E}_{\hat{\mathbb{Q}}_t}[\tilde{\boldsymbol{\xi}}_t],\ t \in T$.

The Pseudo code of the bisection search is summarized in Algorithm 15.1.

Algorithm 15.1: Bisection search algorithm for the optimal robustness level θ^*

Input: $\underline{\theta} = 0$, some suitably large $\bar{\theta}$, and error tolerance ε.
repeat
  $\theta = (\underline{\theta} + \bar{\theta})/2$.
  Check the feasibility of the mixed-integer conic constraint system (15.24).
  If (15.24) is feasible, $\underline{\theta} = \theta$; else, $\bar{\theta} = \theta$.
until $\bar{\theta} - \underline{\theta} \le \varepsilon$.;
Output: $\theta^\star = (\underline{\theta} + \bar{\theta})/2$.

15.3.2 Impact of the Cost Targets onto the Optimal Robustness Level

The periodic budgets can significantly affect the robustness level, which is of practical relevance for applying the budget-driven uncapacitated MHL model (15.14)–(15.17). Hence, in this section, we explore the sensitivity of the optimal robustness level $\theta^\star(\boldsymbol{v})$ to $\boldsymbol{v}$. Our analysis is closely related to the following subset of periods that correspond to the worst-case robustness level:

$$\mathcal{T}^\star(\boldsymbol{v}, \boldsymbol{Y}) = \underset{t \in T}{\operatorname{argmax}} \frac{\left(v_t - \mathbb{E}_{\hat{\mathbb{Q}}_{[t]}}\left[\boldsymbol{f}_t^\top \boldsymbol{y}_t + \tilde{\boldsymbol{d}}_t^\top \boldsymbol{\Upsilon}_t(\boldsymbol{Y}_t)\right]\right)^+}{\sum_{s \in [t]} \left\|[\boldsymbol{B}_t^s]^\top \boldsymbol{\Upsilon}]_t(\boldsymbol{Y}_t)\right\|_*}.$$

In particular, our argument breaks into two cases (*i*) $|\mathcal{T}^\star(\boldsymbol{v}, \boldsymbol{Y})| = 1$ and (*ii*) $|\mathcal{T}^\star(\boldsymbol{v}, \boldsymbol{Y})| \geq 2$, indicating whether there is a single period that corresponds to the worst-case robustness level.

Theorem 15.3 *Suppose that model* (15.14)–(15.17) *has a unique optimal location design* $\boldsymbol{Y}^\star$ *under the periodic budget* $\boldsymbol{v}$.

1. *If* $\mathcal{T}^\star(\boldsymbol{v}, \boldsymbol{Y}) = \{t^\star\}$, *then*

$$\nabla_{v_{t^\star}} \theta^\star(\boldsymbol{v}) = \frac{1}{\sum_{s \in [t^\star]} \|[\boldsymbol{B}_{t^\star}^s]^\top \boldsymbol{\Upsilon}_{t^\star}(\boldsymbol{Y}_{t^\star}^\star)\|_*};$$

while $\nabla_{v_t} \theta^\star(\boldsymbol{v}|\boldsymbol{Y}^\star) = 0$ *for all* $t \in T \setminus \{t^\star\}$.

2. *If* $|\mathcal{T}^\star(\boldsymbol{v}, \boldsymbol{Y})| \geq 2$, *then*

$$\nabla_{v_t}^- \theta^\star(\boldsymbol{v}) = \frac{1}{\sum_{s \in [t]} \|[\boldsymbol{B}_t^s]^\top \boldsymbol{\Upsilon}_t(\boldsymbol{Y}_t^\star)\|_*} \quad \text{and} \quad \nabla_{v_t}^+ \theta^\star(\boldsymbol{v}) = 0 \quad \forall t \in \mathcal{T}^\star(\boldsymbol{v}, \boldsymbol{Y});$$

while $\nabla_{v_t} \theta^\star(\boldsymbol{v}) = 0$ *for all* $t \in T \setminus \mathcal{T}^\star(\boldsymbol{v}, \boldsymbol{Y})$.

Proof See [278]. □

Theorem 15.3 implies that if there is only a single period corresponding to the optimal robustness level $\theta^\star(\boldsymbol{v})$, then $\theta^\star(\boldsymbol{v})$ is sensitive *only* to both increasing and decreasing ends of the budget in that period. If there are multiple periods corresponding to the optimal robustness level $\theta^\star(\boldsymbol{v})$, then $\theta^\star(\boldsymbol{v})$ is sensitive only to the decreasing end of the budgets in those periods. Recall that for all $t \in T$,

$$\nabla_\theta \zeta_t(\boldsymbol{Y}, \theta) = \sum_{s \in [t]} \left\|[\boldsymbol{B}_t^s]^\top \boldsymbol{\Upsilon}_t(\boldsymbol{Y}_t)\right\|_*.$$

The sensitivity of the optimal robustness level concerning the period budgets and that of the worst-case expected cost regarding the robustness level can be related:

$$(\nabla_\theta \zeta_1(\boldsymbol{Y}, \theta)), \ldots, \nabla_\theta \zeta_T(\boldsymbol{Y}, \theta))^\top \cdot \nabla_{\boldsymbol{v}}^- \theta^\star(\boldsymbol{v}) = |\mathcal{T}^\star(\boldsymbol{v}, \boldsymbol{Y})|.$$

15.4 Numerical Results

In this section, we present some computational results to assess the value of the budget-driven uncapacitated MHL model (15.14)–(15.17).

Instances of the problem were generated from the Civil Aeronautics Board (CAB) dataset introduced in [416]. We consider the pairwise directed demands between the first 15 cities as commodities (that is, $|M| = \binom{15}{2} = 105$). Additionally, we take the following $I = 15$ cities as potential hubs to be selected. Concerning the planning horizon, $|T|$ is selected from $\{5, 10, 15\}$. All optimization problems were solved using the algorithms available in a general-purpose solver (Gurobi 9.5.0).

We set the initial demand to the value in the CAB dataset, generate historical samples $\hat{\boldsymbol{d}}_{-1}, \hat{\boldsymbol{d}}_{-2}, \ldots, \hat{\boldsymbol{d}}_{-N}$ from the VAR(1) process:

$$\tilde{\boldsymbol{d}}_t = \boldsymbol{\kappa}^\dagger + \boldsymbol{\Gamma}_1^\dagger \tilde{\boldsymbol{d}}_{t-1} + \tilde{\boldsymbol{\varepsilon}}_t,$$

where $\boldsymbol{\kappa}^\dagger$ and $\boldsymbol{\Gamma}_1^\dagger$ are randomly generated intercept and coefficient matrix, respectively. The historical noises are generated from a Gaussian distribution with zero mean and covariance matrix $\boldsymbol{\Sigma}^\dagger = [\boldsymbol{\Gamma}_1^\dagger]^\top \boldsymbol{\Gamma}_1^\dagger$.

Based on the historical samples, we can apply techniques from time series analysis to obtain the estimated time series model (15.18).

In Fig. 15.1, we can observe that the temporal dependence of the (first commodity) demand can be decomposed by independent ERNs over periods. This is consistent with Assumption 15.1. In this Figure, the left panel shows the autocorrelation function of the historical demands of the first commodity. The middle panel depicts the autocorrelation function of the corresponding ERNs. Finally, the right panel shows the forecast resulting from the estimated time series model. The solid lines are the forecasted demand trends of the first two commodities. The shadow areas in the three panels correspond to the 95% confidence intervals.

Based on the estimated model, we can forecast the future demand trend. The future demand fluctuations of the first two commodities increase along the future 10 periods, showing the propagation of uncertainty over periods. We incorporate in the subsequent experiments the estimated time series model into the ambiguity set (15.19).

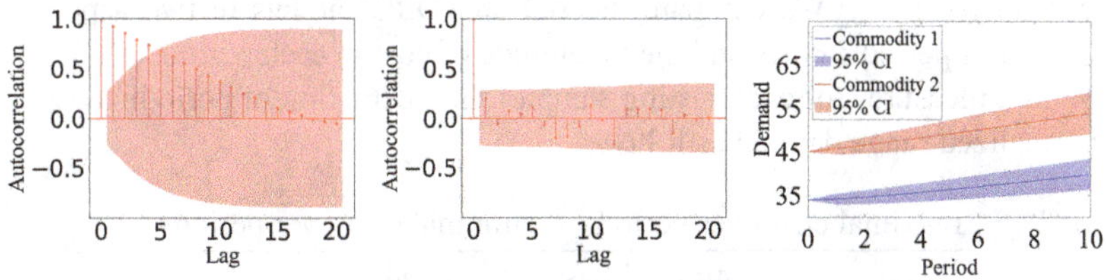

Fig. 15.1 Forecasting results of the time series model

The computational analysis consists of two parts: (i) analyzing the impacts of periodic budgets on the optimal location design (Sect. 15.4.1); (ii) showing the value of considering the time series model in the ambiguity set (Sect. 15.4.2);

15.4.1 Location Design under Different Target Settings

We study the performance of the budget-driven uncapacitated model over the ambiguity set (15.19) (denoted as "BD") under different budget settings. We compare that model against the distributionally robust optimization model (denoted as "DRO") that seeks the optimal location design with the least worst-case expected total cost over the same ambiguity set.

Seeking a fair comparison, we set the total cost target, i.e., $\sum_{t \in T} v_t$, in the BD model equal to the worst-case expected total cost in the DRO model over the ambiguity set (15.19) with the size $\theta_{\text{DRO}} \in \{\theta_0, 2\theta_0, 3\theta_0\}$ for some baseline $\theta_0 = 5$. Furthermore, to test the effect of different target settings, we let the budget increase at a growth rate r over the planning horizon, where $r \in \{-2\%, 0\%, 2\%\}$. That is, the budget in each period follows from

$$v_t = (1+r)^{t-1} v_1 \quad \forall t \in T.$$

In Table 15.1, we present results regarding the location design profiles resulting from using models DRO and BD models under different target settings. In this table, $\hat{\theta}$ stands for the size of ambiguity set in the DRO model, corresponding to the total cost target.

Observing the table, we conclude that in both models, the optimal solution prefers installing more hubs in the first period. This is understandable because the hubs located in the first period can serve longer. When the decision-maker is optimistic about the future with a positive growth rate for the periodic budget to guarantee that the cost is less than the budget in the first period, the BD model tends to postpone the construction plan to subsequent periods.

Once the optimal location design is obtained, it is interesting to analyze the out-of-sample performance. From the actual demand process, we generate 1000 testing samples of the demand trend $(\tilde{\boldsymbol{d}}_1, \ldots, \tilde{\boldsymbol{d}}_{|T|})$ over 10 periods. W.l.o.g. we focus on the case $\theta_{\text{DRO}} = \theta_0$. We compare the BD and DRO models in two aspects: the fluctuation of periodic costs and the magnitude of cost in each period.

We first measure (for each testing sample) the fluctuation of periodic cost using the normalized range defined as follows:

$$\frac{\text{maximal cost over periods} - \text{minimal cost over periods}}{\text{minimal cost over periods}}.$$

Table 15.1 Location design profiles of DRO and BD models under different target settings

			Number if installed hubs										
			Period t										
θ_{DRO}	Model	r	1	2	3	4	5	6	7	8	9	10	Total
θ_0	DRO	–	4	–	–	–	–	–	–	–	–	–	4
	BD	−2%	1	1	–	–	–	–	–	–	–	–	2
		0%	1	1	1	–	–	1	–	1	1	1	7
		2%	1	–	1	–	–	1	1	1	1	1	7
$2\theta_0$	DRO	–	5	–	–	–	–	–	–	–	–	–	5
	BD	−2%	2	1	1	–	–	–	–	–	–	–	4
		0%	1	1	–	–	–	–	–	–	–	–	2
		2%	1	1	–	–	–	–	1	1	1	1	6
$3\theta_0$	DRO	–	6	–	–	–	–	–	–	–	–	–	6
	BD	−2%	3	2	1	1	–	–	–	–	–	–	7
		0%	2	1	–	–	–	–	–	–	–	–	3
		2%	2	1	1	–	1	–	1	–	1	–	7

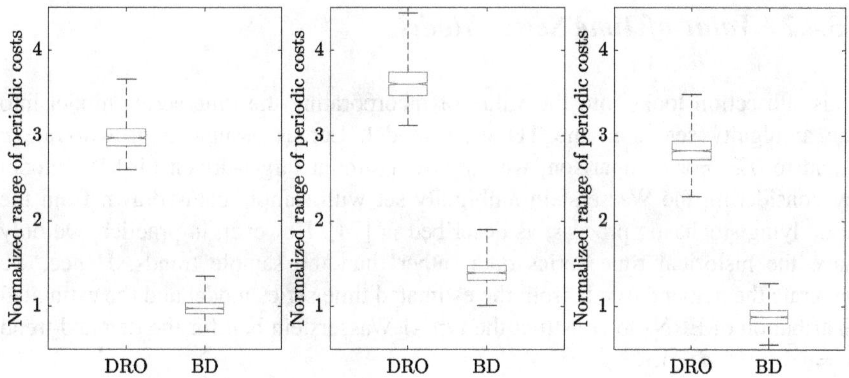

Fig. 15.2 Boxplots of the normalized range of periodic costs over 1000 testing demand trends drawn from the true process with $r = -2\%$ (left), $r = 0\%$ (middle) and $r = 2\%$ (right)

Figure 15.2 depicts several boxplots summarizing the variation of the normalized range over different testing samples. Under every budget setting we have considered, the BD model leads to a lower fluctuation of periodic costs compared to the DRO model. This shows that the BD model can yield an optimal location design that would lead to a more stable operational cash flow over the planning horizon.

We next focus on the magnitude of periodic costs. Figure 15.3 shows the related information. For each testing sample, the normalized cost in each period is the realized cost divided by the corresponding budget. The charts depict the mean (in solid line) and the 10% and 90% quantiles (shaded areas) of the (normalized) periodic costs. We also observe that the first periodic cost of the BD model is significantly smaller than that of the DRO model, which is much larger than the

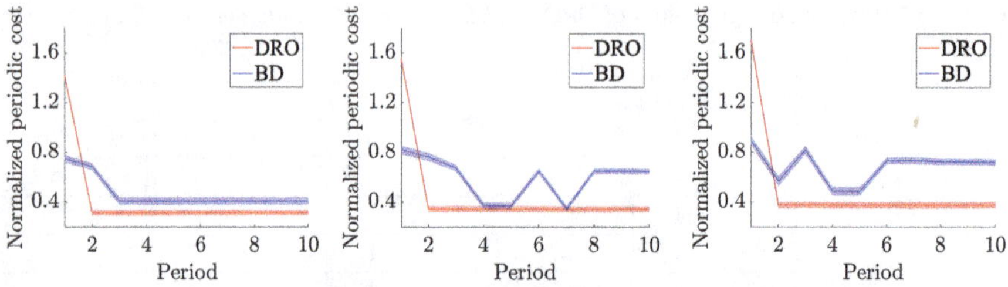

Fig. 15.3 Magnitude of the normalized periodic costs of the BD and DRO models over 10 periods with $r = -2\%$ (left), $r = 0\%$ (middle) and $r = 2\%$ (right)

corresponding budget and is thus undesirable. Although the periodic costs of the BD model in other periods are higher than those of the DRO model, the BD model yields more stable results, so it better ensures operational sustainability.

15.4.2 Value of Time Series Model

This subsection looks into the value of incorporating the time series model into the ambiguity set as in the TD-WTS model. Let us assume a growth rate r equal to 0%. For comparison, we employ a similar target-driven (TD-W) model by considering the Wasserstein ambiguity set with sample paths drawn from the underlying stochastic process, as described in [74]. However, in practice, we only have the historical time series data rather than the sample trends. Hence, we generate the demand trends from the estimated time series model and the estimated distribution of ERNs to construct the type-1 Wasserstein ball for the demand trend $\tilde{\boldsymbol{d}} = (\tilde{\boldsymbol{d}}_1, \ldots, \tilde{\boldsymbol{d}}_{|T|})$:

$$\mathcal{F}_{\mathrm{W}}^{\theta} = \left\{ \mathbb{P} \in \mathcal{P}(\mathbb{R}_{+}^{|M|\times|T|}) \mid \Delta_{\mathrm{W}}(\mathbb{P}, \hat{\mathbb{P}}) \leq \theta \right\},$$

where the empirical distribution $\hat{\mathbb{P}}$ assigns equal weight to each of 100 independent sample paths $\hat{\boldsymbol{d}}^n$, $n \in [100]$ drawn from the estimated time series process, i.e.,

$$\hat{\mathbb{P}} = \frac{1}{100} \sum_{n\in[100]} \delta_{\hat{\boldsymbol{d}}^n}.$$

In this case, the only difference between the two models regards using the estimated time series model in the modeling process. By exploiting the structure of the uncapacitated MHL problem, a tractable reformulation can be devised for the constraints in the TD-W model. We can calculate the optimal solution of two models via a bisection search algorithm (given in Sect. 15.3.1), where each iteration solves a mixed-integer conic linear program.

Table 15.2 Summary statistics of periodic costs over 1000 testing samples drawn from the true process

	$T=5$		$T=10$		$T=15$	
Model	BD	BD-W	BD	BD-W	BD	BD-W
Average ($\times 10^7$)	5.89	6.51	3.95	4.45	3.67	4.41
Range ($\times 10^7$)	0.99	1.03	3.26	3.32	3.24	4.08
Standard deviation ($\times 10^7$)	0.17	0.18	1.23	1.49	0.89	1.60

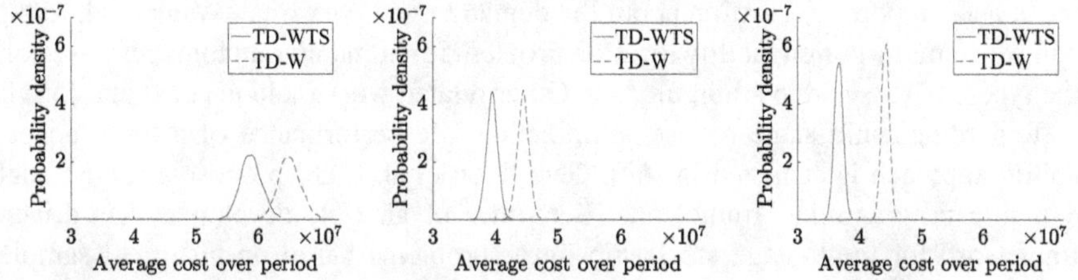

Fig. 15.4 The probability density distribution of the average periodic cost of TD-WTS and TD-W models over 1000 test samples drawn from the true demand process, for $T = 5$ (left), $T = 10$ (middle), $T = 15$ (right)

Table 15.2 summarizes the out-of-sample performance of the two models based on 1000 testing samples. For each period, we calculate the expected cost for these testing samples. Then, we obtain the average, range, and standard deviation of the expected cost for the planning horizon. Lower average, range, and standard deviation values reveal the advantage of considering the time series model for mitigating the magnitude and fluctuation of periodic costs. Such an advantage is more notable as the number of periods increases. This is clear in Fig. 15.4, which depicts the probability density distribution of the average periodic cost of TD-WTS and TD-W models. Intuitively, there are two reasons behind this: (i) the time series model sufficiently extrapolates the path-dependent uncertainty process by harnessing the historical data, and (ii) it decomposes the fluctuation of the demand by ERNs, which significantly decreases the magnitude of uncertainty.

15.5 Conclusions and Further Reading

In this chapter, we presented an application of a distributionally robust model for an uncapacitated MHL problem with uncertain demands. A nested ambiguity set was discussed, incorporating a general multivariate time series model with Wasserstein balls for different periods to characterize the time-dependent uncertainty on future periodic demands. To mitigate the fluctuation of periodic operational costs, a budget-driven model was developed that restricts the expected cost in each period of the planning horizon. Managerial insights of the budget-driven model were also provided. The budget-driven uncapacitated MHL model can be reformulated as a set of mixed-integer conic programs solved via a bisection search algorithm. Numerical results were discussed, showing the value of considering the time series in the ambiguity set.

Several streams of literature are related to the topic of this chapter. These include (i) time-dependent facility location under uncertainty, (ii) multi-stage robust optimization, and (iii) residual-based distributionally robust optimization. For the first, we refer to [144, 559], who develop stochastic programming models. Wang et al. [538] propose a target-oriented location-transportation problem based on the mean and support information about the demand trend over time. Wang et al. [539] consider a multi-period facility location problem for humanitarian logistics based on the type-∞ Wasserstein ambiguity set. Other related work include [278] and [561].

Regarding multi-stage robust optimization, the performance of a finite adaptability approach is analyzed in [69]. Georghiou et al. [228] propose a robust dual dynamic programming framework. Bertsimas et al. [74] develop a data-driven framework for multi-stage stochastic linear problems based on historical sample paths.

Finally, concerning residual-based distributionally robust optimization (DRO), the authors in [178] propose a (single-stage) DRO model with the data coming from a VAR(1) process. Kannan et al. [297] introduce a residual-based Wasserstein DRO framework to incorporate covariance information. Perakis et al. [427] propose an information-partitioning ambiguity set to characterize the random residuals in an additive demand model. Qi et al. [439] investigate a distributionally robust conditional quantile regression model with fixed-design of the covariates.

Chapter 16
Data-Driven Robust Production Planning

Abstract This chapter discusses a two-stage production planning problem with demand ambiguity, which can be seen as a distributionally robust facility location. The first stage regards a product service to offer. The second stage involves production planning. The problem is cast in a data-driven decision-making setting. Historical information is considered for demand and related covariates. An econometric model is developed to predict the demand, leveraging seemingly unrelated regression estimated with feasible generalized least squares. A predictive ambiguity set is constructed to harness the prediction model with the empirical covariates and residuals. A decision-dependent two-stage distributionally robust optimization (DRO) model is built, taking advantage of the econometric model and predictive ambiguity set. The problem is reformulated as an empirical counterpart under the predicted demand distribution regularized by a perceived shortage cost as the shadow price for ambiguity aversion. Exploiting this structure, ambiguity-averse operational properties are analyzed, including risk exposure. Numerical results demonstrate the effectiveness of the modeling framework proposed.

Keywords Production planning · Decision-dependent uncertainty · Distributionally robust optimization

16.1 Introduction

In the era of Industry 4.0 and with the ground being laid for Industry 5.0, disruptive technologies, such as the Internet-of-Things (IoT), 5G, and digital twins, have significantly enhanced data accessibility for production planning and control, which enables several statistical and machine learning methods to harness the side information or covariates for predicting and estimating the demand ([43, 70, 72]). Nevertheless, along with the production efficiency being constantly enhanced, the ever-increasing number of products and their ever-shorter life cycles in the market [447] further challenge demand prediction and thus production planning. On the other hand, "Servitization" in manufacturing (industries using their products to sell

© The Author(s), under exclusive license to Springer Nature Switzerland AG 2024
F. Saldanha-da-Gama, S. Wang, *Facility Location Under Uncertainty*,
International Series in Operations Research & Management Science 356,
https://doi.org/10.1007/978-3-031-55927-3_16

"outcome as a service" instead of a one-off sale) has grown immensely over the past decade, with an increasing number of products and services being bundled to provide to improve the value captured (see e.g., [443]). According to the European Commission, around 70% of the small and medium manufacturing enterprises (SMEs) in Denmark, France, Germany, and The Netherlands are providing product-service combinations (see [202]). Also, China, the USA, and the UK present a significant percentage of servitized companies, 38%, 53%, and 56%, respectively (see [374]). In servitized systems, a salient feature is that the service, as a covariate of demand, becomes a decision to be optimized, which can impact the demand distribution to be estimated in the planning and challenge the decision-making.

The joint production and service planning problem is under a subscription setting that is closely related to the servitization, where the product-service provider (e.g., manufacturer) charges the product-service recipient (e.g., retailers or customers) a fixed amount in advance. Then, it fulfills all demand realized over a specific period bearing the associated demand-fulfillment costs. The provider aims to determine the portfolio of products with the associated service levels and the capacity to maximize the expected total profit. A central assumption is that the historical information on demand and the associated covariates is available in which the service decision serves as a covariate for demand. Two challenges stemming from a related data-driven production-service planning problem include: (i) demand ambiguity with correlation and variance-heterogeneity and (ii) service dependency to be captured in the demand prediction and production-service planning.

Dealing with demand uncertainty has been a challenge for long in production planning and supply chain management. Much work on production planning employs a paradigm that endows the decision-maker with perfect distributional information about the underlying random demand (e.g., [1, 568]). However, this is a restrictive assumption that might not adhere to reality. A manufacturer often provides multiple products, resulting in a high-dimensional demand (e.g. [360]), which in general involves a variety of determinants (factors) and therefore is difficult to accurately calibrate ([205, 372]). More importantly, the possible correlation and variance-heterogeneity characterize a critical feature of the high-dimensional demand distribution, which should be captured for a more accurate demand description and prediction. For instance, numerous studies in the literature have illustrated that demand correlation affects operations as well as profit in production planning. This is the case in [127, 246, 399].

The second challenge mentioned regards the service-dependency effect. As mentioned previously, in servitization, the service attributes often impact demand. For example, it has been found that the effect on consumer utility of an 8% increase in warranty length (after-sales service quality) is equivalent, all else being equal, to decrease the price by 1%, and is, in turn, equivalent to increasing product quality by 2.2%, for a car with medium characteristics (see [247]). It has also been estimated that the impact of location-based mobile promotion, which is a new popular service channel, can represent more than half of the total sales (see [203]). Therefore, in a production-service planning problem context, the demand is affected by the possible exogenous features or factors, inter alia, weather, economic outlook, public health events, and the endogenous covariate of service decision to be optimized. Hence, it is vital to embed service dependency in production-service planning under demand ambiguity.

The remainder of this chapter is organized as follows. Section 16.2 introduces a decision-dependent two-stage DRO model that incorporates a demand prediction model leveraging feasible generalized least squares to capture cross-product correlation. Section 16.3 presents some insights with a tractable reformulation. Section 16.4 presents computational results supporting the value of the modeling frameworks discussed. The chapter concludes with a summary and suggestions for further reading.

16.2 Modeling Aspects

We first introduce the setting underlying the production planning problem we are analyzing. Then, we design a so-called predictive ambiguity set that depends on the service decision and captures the cross-product correlation. Afterward, we formulate a decision-dependent two-stage DRO model.

16.2.1 Background and Assumptions

We consider a manufacturer who produces a set P of products used to supply a retailer. Each product offers one service level from a set L of possibilities. The manufacturer has a set I of production facilities. The unit capacity cost for each facility $i \in I$ is f_i. The upfront production cost for product $p \in P$ is g_p^F, and its associated service investment at level $\ell \in L$ is $g_{p\ell}^S$. Both are fixed costs.

We assume that each product $p \in P$ endowed with a service level $\ell \in L$ costs the retailer a fixed subscription amount of $g_{p\ell}^R$ (revenue for the manufacturer). This value is paid in advance to secure (buy out) the (ownership of the) production of the manufacturer to fulfill all future demand. The retailer gets the revenue from the product sales. On the other hand, the manufacturer commits to produce the selected products to fulfill the demand realized in the selling period.

The service level usually affects the demand. For instance, a higher service level may increase sales. In our setting, this benefits the retailer but leads to a higher (production) cost for the manufacturer. As a trade-off, the manufacturer would thus charge a higher fixed payment to the retailer for the higher level of service. To simplify notation, we denote by $g_{p\ell} = g_{p\ell}^R - g_p^F - g_{p\ell}^S$ the total net fixed profit (for the manufacturer) resulting from providing product $p \in P$ at service level $\ell \in L$.

In our problem, the manufacturer must decide (i) the portfolio of products and the associated service levels to offer and (ii) the production capacity level for each facility.

For $i \in I$, let $q_i \in \mathbb{R}_+$ be the capacity decision for facility i. We represent the product-service bundle decision by binary variables. For $p \in P$, $\ell \in L$, $y_{p\ell}$ equals one if and only if the manufacturer produces product p at service level ℓ. Furthermore, we take $\boldsymbol{y} = (y_{p\ell})_{p\in P,\ell\in L}$. Additionally, $(\boldsymbol{q}, \boldsymbol{y})$ is the production-service decision that includes the endogenous capacities $\boldsymbol{q} = (q_i)_{i\in I}$.

The production planning problem we are considering can be cast as a two-stage adaptive optimization. In the first stage, the cost of a production-service decision $(\boldsymbol{q}, \boldsymbol{y})$ is accounted for

$$\sum_{p\in P}\sum_{\ell\in L} g_{p\ell} y_{p\ell} - \sum_{i\in I} f_i q_i. \tag{16.1}$$

To make the problem meaningful, we impose that at least one product is produced, and each product is bundled with at most one service level. Using our decision variables, this can be ensured considering the following $\boldsymbol{y}$-feasibility set:

$$\mathcal{Y} = \left\{ \boldsymbol{y} \in \{0,1\}^{|P|\times|L|} \,\middle|\, \sum_{p\in P}\sum_{\ell\in L} y_{p\ell} \geq 1; \sum_{\ell\in L} y_{p\ell} \leq 1 \,\forall p \in P \right\}. \tag{16.2}$$

In particular, $\sum_{\ell\in L} y_{p\ell} = 1$ indicates that product i is selected.

Denote by $\boldsymbol{\xi} = (\xi_p)_{p\in P}$ the random vector representing the demand of all products. In the second stage—the production stage—given the production-service decision $(\boldsymbol{q}, \boldsymbol{y})$, and a realization $\hat{\boldsymbol{\xi}}$ of the demand $\boldsymbol{\xi}$, the manufacturer solves a production problem, which determines production levels to minimize the production cost. The latter is denoted by $\mathrm{Q}(\boldsymbol{q}, \boldsymbol{y}, \hat{\boldsymbol{\xi}})$ and can be formulated as follows:

$$\mathrm{Q}(\boldsymbol{q}, \boldsymbol{y}, \hat{\boldsymbol{\xi}}) = \min \quad \sum_{p\in P}\sum_{i\in I_0} c_{pi} x_{pi}, \tag{16.3}$$

$$\text{s.t.} \quad \sum_{p\in P} x_{pi} \leq q_i \qquad \forall i \in I, \tag{16.4}$$

$$\sum_{i\in I_0} x_{pi} = \hat{\xi}_p \sum_{\ell\in L} y_{p\ell} \qquad \forall p \in P, \tag{16.5}$$

$$\boldsymbol{x} \in \mathbb{R}_+^{|P|(|I|+1)}. \tag{16.6}$$

In this model, $I_0 = I \cup \{0\}$ is an extended set of facilities with a dummy for accommodating surplus demand (w.r.t. the operating capacity). For $p \in P$ and $i \in I$, x_{pi} is a decision variable representing the amount of product p to produce at facility i, and c_{pi} is the corresponding unit production cost. As mentioned above, we set up a dummy facility with a large enough capacity for each product to indicate outsourcing the product or some other recourse action when the regular production facilities fall short of capacity. In particular, we use c_{p0} with $c_{p0} > c_{pi}$ $\forall i \in I$ to represent the unit outsourcing cost, also called shortage cost, of each product $p \in P$. The production variable x_{p0} represents the amount of product $p \in P$ outsourced (shortage).

The second-stage problem seeks to minimize the total production cost subject to capacity and demand-fulfillment constraints. Specifically, Constraints (16.4) ensure that the total amount produced at a facility does not exceed its capacity. Equalities (16.5) impose that the demand for each product is satisfied by regular production or outsourcing. Finally, we have the domain constraints.

Given the demand distribution $\boldsymbol{\xi} \sim \mathbb{P}^{\boldsymbol{y}}$ that is dependent on the production decision $\boldsymbol{y}$, the overall problem seeks to maximize the expected total profit. It can be formulated as a two-stage stochastic programming problem:

$$\max_{\substack{\boldsymbol{q} \in \mathbb{R}_+^{|I|} \\ \boldsymbol{y} \in \mathcal{Y}}} \left\{ \sum_{p \in P} \sum_{\ell \in L} g_{p\ell} y_{p\ell} - \sum_{i \in I} f_i q_i - \mathbb{E}_{\mathbb{P}^{\boldsymbol{y}}} \big[\mathrm{Q}(\boldsymbol{q}, \boldsymbol{y}, \boldsymbol{\xi}) \big] \right\}, \tag{16.7}$$

with $\mathcal{Y}$ given by (16.2).

16.2.2 The Ambiguity Set

In the production planning problem we are analyzing, demand can be affected by the service decision. We focus on the situation in which the service types are those commonly observable for most products.

We also consider the case in which the demand also depends on (is explained by) M exogenous covariates $\boldsymbol{\chi} = (\chi_1, \ldots, \chi_M)^\top \in \mathbb{R}^M$, such as resident income, socio-economic state, and seasonality, to mention a few.

Given the above, we consider a demand model capturing the explanatory effects of endogenous service dependency and exogenous covariates:

$$\xi_p = a_{p0} + \boldsymbol{a}_p^\top \boldsymbol{y}_p + \boldsymbol{b}_p^\top \boldsymbol{\chi} + \varepsilon_p, \quad \forall p \in P, \tag{16.8}$$

where a_{p0} is the intercept, $\boldsymbol{a}_p = (a_{p1}, \ldots, a_{p|L|})^\top$, and $\boldsymbol{y}_p = (y_{p1}, \ldots, y_{p|L|})^\top$. Each element $a_{p\ell}$ represents the coefficient of the explanatory variable $y_{p\ell}$, which is also the decision of providing product p at service level ℓ. The vector $\boldsymbol{b}_p = (b_{p1}, \ldots, b_{pM})^\top$ contains the coefficients for the explanatory covariates $\boldsymbol{\chi}$ for product p. Finally, ε_p is a random error.

We extend the notation as follows:

$$\boldsymbol{a}_0 = \begin{bmatrix} a_{10} \\ a_{20} \\ \vdots \\ a_{|P|0} \end{bmatrix}, \ \mathbf{A} = \begin{bmatrix} \boldsymbol{a}_1^\top & & & \\ & \boldsymbol{a}_2^\top & & \\ & & \ddots & \\ & & & \boldsymbol{a}_{|P|}^\top \end{bmatrix}, \mathbf{B} = \begin{bmatrix} \boldsymbol{b}_1^\top \\ \boldsymbol{b}_2^\top \\ \vdots \\ \boldsymbol{b}_{|P|}^\top \end{bmatrix}, \text{ and } \boldsymbol{\varepsilon} = \begin{bmatrix} \tilde{\varepsilon}_1 \\ \tilde{\varepsilon}_2 \\ \vdots \\ \tilde{\varepsilon}_{|P|} \end{bmatrix}.$$

The demand model (16.8) can be represented in a more compact way as follows:

$$\boldsymbol{\xi} = \boldsymbol{a}_0 + \mathbf{A}\boldsymbol{y} + \mathbf{B}\boldsymbol{\chi} + \boldsymbol{\varepsilon}. \tag{16.9}$$

Given a set of historical observations $\{\hat{\boldsymbol{\xi}}^1, \ldots, \hat{\boldsymbol{\xi}}^N\}$, $\{\boldsymbol{y}^1, \ldots, \boldsymbol{y}^N\}$, and $\{\hat{\boldsymbol{\chi}}^1, \ldots, \hat{\boldsymbol{\chi}}^N\}$, we are able to obtain estimation $(\hat{\boldsymbol{a}}_0, \widehat{\mathbf{A}}, \widehat{\mathbf{B}})$ of the multivariate regression model of demand (16.9). To capture the correlation and variance-heterogeneity and thus achieve a more accurate description of the demand (uncertainty), we consider a generalized multivariate linear regression approach to describe the explained and explanatory variables in our demand model. We can estimate the model using a two-stage estimation process, namely, the Feasible Generalized Least Squares (FGLS) method.

Since the production planning is a long-run decision problem, the covariates $\boldsymbol{\chi}$ are not necessarily to be given. Therefore, they are treated as random variables over the planning horizon, while the production design (decision) $\boldsymbol{y}$ once decided is fixed over time.

Using the estimated coefficients $(\hat{\boldsymbol{a}}_0, \widehat{\mathbf{A}}, \widehat{\mathbf{B}})$, we can re-write the demand given a production design $\boldsymbol{y}$ as

$$\boldsymbol{\xi}|\boldsymbol{y} = \hat{\boldsymbol{a}}_0 + \widehat{\mathbf{A}}\boldsymbol{y} + \widehat{\mathbf{B}}\boldsymbol{\chi} + \boldsymbol{\varepsilon}_{\mathrm{F}}, \tag{16.10}$$

where $\boldsymbol{\varepsilon}_{\mathrm{F}} = \boldsymbol{\varepsilon} + (\boldsymbol{a}_0 - \hat{\boldsymbol{a}}_0) + (\mathbf{A} - \widehat{\mathbf{A}})\boldsymbol{y} + (\mathbf{B} - \widehat{\mathbf{B}})\boldsymbol{\chi}$ is the residual and the subscript "F" indicates the FGLS estimation. Formally, we denote this predicted demand (distribution) given $\boldsymbol{y}$ as

$$\hat{\boldsymbol{\xi}}(\boldsymbol{y}, \boldsymbol{\chi}, \boldsymbol{\varepsilon}_{\mathrm{F}}) = \hat{\boldsymbol{a}}_0 + \widehat{\mathbf{A}}\boldsymbol{y} + \widehat{\mathbf{B}}\boldsymbol{\chi} + \boldsymbol{\varepsilon}_{\mathrm{F}}. \tag{16.11}$$

Not only does the distribution of the predicted demand $\hat{\boldsymbol{\xi}}(\boldsymbol{y}, \boldsymbol{\chi}, \boldsymbol{\varepsilon}_{\mathrm{F}})$ depend on design $\boldsymbol{y}$, but also it depends on the random covariates $\boldsymbol{\chi}$ and residuals $\boldsymbol{\varepsilon}_{\mathrm{F}}$. If the true distribution of $(\boldsymbol{\chi}, \boldsymbol{\varepsilon}_{\mathrm{F}})$ is known, then $\hat{\boldsymbol{\xi}}(\boldsymbol{y}, \boldsymbol{\chi}, \boldsymbol{\varepsilon}_{\mathrm{F}})$ becomes the true demand given $\boldsymbol{y}$. This, however, is nearly impossible. The true model $(\boldsymbol{a}_0, \mathbf{A}, \mathbf{B})$ that determines the $\boldsymbol{\varepsilon}_{\mathrm{F}}$ is never known.

Being aware of this ambiguity, we assume the manufacturer is ambiguity-averse and wishes to hedge against the distributional uncertainty reasonably. To this end, we construct the ambiguity set by incorporating the above demand prediction (16.11). First, we denote by $\mathbb{Q}^{\boldsymbol{y}}$ the joint distribution of $(\boldsymbol{\xi}, \boldsymbol{\chi}, \boldsymbol{\varepsilon}_{\mathrm{F}})$

conditional on the decision $\boldsymbol{y}$. Since the manufacturer only cares about the demand prediction of products that have been selected to provide, we define

$$\boldsymbol{\Lambda}_{\boldsymbol{y}} = \text{Diag}\Big[(\boldsymbol{y}_p^\top \boldsymbol{e}_{|L|})_{p\in P}\Big]. \tag{16.12}$$

This is the product-selection matrix, whose p-th diagonal zero-one element $\boldsymbol{y}_p^\top \boldsymbol{e}_{|L|}$ equals to one if and only if the manufacturer selects product p, where $\boldsymbol{e}_{|L|} \in \mathbb{R}^{|L|}$ is the all-one vector.

Furthermore, since it is impossible to know perfectly the distribution of $(\boldsymbol{\chi}, \boldsymbol{\varepsilon}_{\text{F}})$ in practice, we propose to use the available historical observations $\{\hat{\boldsymbol{\chi}}^1, \ldots, \hat{\boldsymbol{\chi}}^N\}$ and the observed residuals

$$\hat{\boldsymbol{\varepsilon}}_{\text{F}}^n = \hat{\boldsymbol{\xi}}^n - (\hat{\boldsymbol{a}}_0 + \widehat{\mathbf{A}}\boldsymbol{y}^n + \widehat{\mathbf{B}}\hat{\boldsymbol{\chi}}^n) \quad \forall n \in [N] \tag{16.13}$$

to approximate the distribution of $(\boldsymbol{\chi}, \boldsymbol{\varepsilon}_{\text{F}})$. More specifically, we construct the following empirical distribution with observations for $(\boldsymbol{\chi}, \boldsymbol{\varepsilon}_{\text{F}})$:

$$\widehat{\mathbb{P}}_{(\boldsymbol{\chi},\boldsymbol{\varepsilon}_{\text{F}})} = \frac{1}{N}\sum_{n\in[N]} \delta_{(\hat{\boldsymbol{\chi}}^n, \hat{\boldsymbol{\varepsilon}}_{\text{F}}^n)}, \tag{16.14}$$

where δ_X denotes the *Dirac delta* distribution with mass one on X.

Finally, we construct the ambiguity set of $\mathbb{Q}^y$ as follows:

$$\mathcal{F}_\theta(\boldsymbol{y}) = \left\{ \mathbb{Q} \in \mathcal{P}(\mathbb{R}^{|P|}\times\mathbb{R}^M\times\mathbb{R}^{|P|}) \;\middle|\; \begin{array}{l} (\boldsymbol{\xi}, \boldsymbol{\chi}, \boldsymbol{\varepsilon}_{\text{F}}) \sim \mathbb{Q};\ (\boldsymbol{\chi}, \boldsymbol{\varepsilon}_{\text{F}}) \sim \widehat{\mathbb{P}}_{(\boldsymbol{\chi},\boldsymbol{\varepsilon}_{\text{F}})} \\ \mathbb{E}_{\mathbb{Q}}\Big[\Big\|\boldsymbol{\Lambda}_{\boldsymbol{y}}(\boldsymbol{\xi} - \hat{\boldsymbol{\xi}}(\boldsymbol{y}, \boldsymbol{\chi}, \boldsymbol{\varepsilon}_{\text{F}}))\Big\|_l\Big] \le \theta \sqrt[l]{K(\boldsymbol{y})} \\ \mathbb{Q}\Big[(\boldsymbol{\xi}, \boldsymbol{\chi}, \boldsymbol{\varepsilon}_{\text{F}}) \in \mathbb{R}_+^{|P|}\times\mathbb{R}^M\times\mathbb{R}^{|P|}\Big] = 1 \end{array} \right\}. \tag{16.15}$$

It is noted that the above ambiguity set $\mathcal{F}_\theta(\boldsymbol{y})$ includes (i) a constraint on the expected prediction error to be no larger than $\theta\sqrt[l]{K(\boldsymbol{y})}$. θ is the ambiguity-aversion level and $K(\boldsymbol{y})$ represents the total number of products selected, (ii) an estimated empirical distribution $\widehat{\mathbb{P}}_{(\boldsymbol{\chi},\boldsymbol{\varepsilon}_{\text{F}})}$ for the covariates and residuals $(\boldsymbol{\chi}, \boldsymbol{\varepsilon}_{\text{F}})$ in the prediction $\hat{\boldsymbol{\xi}}(\boldsymbol{y}, \boldsymbol{\chi}, \boldsymbol{\varepsilon}_{\text{F}})$, and (iii) a support constraint.

16.2.3 A Decision-Dependent Two-Stage DRO Model

Taking the predictive ambiguity set $\mathcal{F}_\theta(\boldsymbol{y})$ given by (16.15), we consider the worst-case expected total profit, given a production-service decision $(\boldsymbol{q}, \boldsymbol{y})$ and ambiguity-aversion level θ. We denote such profit by $\vartheta(\boldsymbol{q}, \boldsymbol{y}; \theta)$:

$$\vartheta(\boldsymbol{q}, \boldsymbol{y}; \theta) = \boldsymbol{g}^\top \boldsymbol{y} - \boldsymbol{f}^\top \boldsymbol{q} - \sup_{\mathbb{Q}\in\mathcal{F}_\theta(\boldsymbol{y})} \mathbb{E}_{\mathbb{Q}}\big[\mathrm{Q}(\boldsymbol{q}, \boldsymbol{y}, \boldsymbol{\xi})\big]. \tag{16.16}$$

For an observation $\hat{\boldsymbol{\xi}}$ of $\boldsymbol{\xi}$, $\mathrm{Q}(\boldsymbol{q}, \boldsymbol{y}, \hat{\boldsymbol{\xi}})$ is the second-stage production cost function given by (16.3)–(16.6), and it is also noted that the expectation under $\mathbb{Q}$ coincides with that under its marginal on $\boldsymbol{\xi}$.

Finally, the ambiguity-averse manufacturer or provider employs the resultant model of robust predictive production-service planning (RP-PSP), which maximizes the worst-case expected total profit:

$$\vartheta_N^{\star}(\theta) = \max_{\substack{\boldsymbol{q} \in \mathbb{R}_+^{|I|} \\ \boldsymbol{y} \in \mathcal{Y}}} \vartheta(\boldsymbol{q}, \boldsymbol{y}; \theta), \tag{RP-PSP}$$

with $\mathcal{Y}$ being the production feasible set (16.2).

16.3 Model Reformulation and Optimization

For a given production decision $(\boldsymbol{q}, \boldsymbol{y})$ and ambiguity-aversion level θ, we denote by

$$\mathcal{Q}(\boldsymbol{q}, \boldsymbol{y}; \theta) = \sup_{\mathbb{Q} \in \mathcal{F}_\theta(\boldsymbol{y})} \mathbb{E}_{\mathbb{Q}}\big[\mathrm{Q}(\boldsymbol{q}, \boldsymbol{y}, \boldsymbol{\xi})\big]$$

the worst-case expected production cost of $(\boldsymbol{q}, \boldsymbol{y})$ over the predictive ambiguity set $\mathcal{F}_\theta(\boldsymbol{y})$. We point out that $\mathcal{Q}(\boldsymbol{q}, \boldsymbol{y}; \theta)$ possesses an insightful equivalent reformulation that consists of an empirical expected production cost and a perceived shortage cost. This result leverages a reformulation of the second-stage production cost function $\mathrm{Q}(\boldsymbol{q}, \boldsymbol{y}, \hat{\boldsymbol{\xi}})$ and the duality argument for the regularization of Wasserstein DRO (the interested reader should refer to [80, 223, 487]).

In particular, given a decision $(\boldsymbol{q}, \boldsymbol{y})$ and ambiguity-aversion level θ, let $\mathcal{K}(\boldsymbol{y})$ be the index set of products selected by $\boldsymbol{y}$. We have

$$\mathcal{Q}(\boldsymbol{q}, \boldsymbol{y}; \theta) = \mathbb{E}_{\widehat{\mathbb{P}}_N^{\boldsymbol{y}}}\big[\mathrm{Q}(\boldsymbol{q}, \boldsymbol{y}, \boldsymbol{\xi})\big] + \theta \left(\sqrt[l]{K(\boldsymbol{y})} \sqrt[l^*]{\sum_{p \in \mathcal{K}(\boldsymbol{y})} c_{p0}^{l^*}} \right), \tag{16.17}$$

where l^* is the Hölder conjugate number of l, i.e., the number such that $1/l + 1/l^* = 1$, and

$$\widehat{\mathbb{P}}_N^{\boldsymbol{y}} = \frac{1}{N} \sum_{n \in [N]} \delta_{\hat{\mathbf{a}}_0 + \widehat{\mathbf{A}}\boldsymbol{y} + \widehat{\mathbf{B}}\hat{\boldsymbol{\chi}}^n + \hat{\boldsymbol{\varepsilon}}_{\mathrm{F}}^n}.$$

The above reformulation has several important implications. First, from the reformulation (16.17), when $\theta = 0$, the worst-case expected (second-stage) production cost $\mathcal{Q}(\boldsymbol{q}, \boldsymbol{y}; \theta)$ reduces to an empirical counterpart $\mathbb{E}_{\widehat{\mathbb{P}}_N^{\boldsymbol{y}}}[\mathrm{Q}(\boldsymbol{q}, \boldsymbol{y}, \boldsymbol{\xi})]$, where $\widehat{\mathbb{P}}_N^y = \frac{1}{N}\sum_{n\in[N]} \delta_{\hat{\mathbf{a}}_0+\widehat{\mathbf{A}}y+\widehat{\mathbf{B}}\hat{\chi}^n+\hat{\varepsilon}_F^n}$ is exactly the empirical distribution of the prediction model $\hat{\boldsymbol{\xi}}(\boldsymbol{y}, \boldsymbol{\chi}, \boldsymbol{\varepsilon}_F)$ with $(\boldsymbol{\chi}, \boldsymbol{\varepsilon}_F) \sim \widehat{\mathbb{P}}_{(\boldsymbol{\chi},\boldsymbol{\varepsilon}_F)}$. Accordingly, the decision-dependent DRO model (RP-PSP) degenerates to the decision-dependent empirical (stochastic) program (16.7) under the predicted demand distribution $\widehat{\mathbb{P}}_N^y$, i.e.,

$$\vartheta_N^\star(0) = \max_{\substack{\boldsymbol{q}\in\mathbb{R}_+^{|I|}\\ \boldsymbol{y}\in\mathcal{Y}}} \left\{\boldsymbol{g}^\top\boldsymbol{y} - \boldsymbol{f}^\top\boldsymbol{q} - \mathbb{E}_{\widehat{\mathbb{P}}_N^y}[\mathrm{Q}(\boldsymbol{q}, \boldsymbol{y}, \boldsymbol{\xi})]\right\}.$$

Second, exploiting this reformulation for the worst-case expected production cost $Q(\boldsymbol{q}, \boldsymbol{y}; \theta)$, we can show that the problem can be reformulated as a mixed-integer conic linear program that is computationally attractive. More specifically, the problem (RP-PSP) can be reformulated as the following optimization problem:

$$\text{maximize} \quad \sum_{p\in P}\sum_{\ell\in L} g_{p\ell}y_{p\ell} - \sum_{i\in I} f_i q_i - \frac{1}{N}\sum_{n\in[N]}\sum_{p\in P}\sum_{i\in I_0} c_{pi}x_{pi}^n - \theta v, \tag{16.18}$$

$$\text{subject to} \quad \left\|(\boldsymbol{w}_p^\top \boldsymbol{e}_{|L|})_{p\in P}\right\|_l \le v, \tag{16.19}$$

$$\left\|\boldsymbol{\Lambda}_y \boldsymbol{c}_0\right\|_{l^*} \le u, \tag{16.20}$$

$$u - \|\boldsymbol{c}_0\|_{l^*}(1 - y_{p\ell}) \le w_{p\ell} \le u \qquad \forall p \in P,\ \ell \in L, \tag{16.21}$$

$$0 \le w_{p\ell} \le \|\boldsymbol{c}_0\|_{l^*} y_{p\ell} \qquad \forall p \in P, \ell \in L, \tag{16.22}$$

$$\sum_{p\in P} x_{pi}^n \le q_i \qquad \forall i \in I,\ n \in [N], \tag{16.23}$$

$$\sum_{i\in I_0} x_{pi}^n = \sum_{\ell\in L} \xi_p^{\ell,n} y_{p\ell} \qquad \forall p \in P,\ n \in [N], \tag{16.24}$$

$$\boldsymbol{q} \in \mathbb{R}_+^{|I|},\ u \in \mathbb{R},\ v \in \mathbb{R}, \tag{16.25}$$

$$\boldsymbol{w} \in \mathbb{R}^{|P|\times|L|},\ \boldsymbol{x} \in \mathbb{R}_+^{|P|(|I|+1)N}, \tag{16.26}$$

$$\boldsymbol{y} \in \mathcal{Y}, \tag{16.27}$$

where $\boldsymbol{\Lambda}_y$ is the diagonal product-selection matrix given by (16.12), and each x_{pi}^n is a decision variable representing the amount of product $p \in P$ to produce at facility $i \in I$ in the scenario $n \in [N]$. Furthermore, $\|\cdot\|_{l^*}$ stands for the dual norm of $\|\cdot\|_l$ with l^* being the Hölder conjugate of l.

Reformulation (16.18)–(16.27) has a computationally appealing structure. First, it is a mixed-integer conic linear program that reduces to a mixed-integer linear program (MILP) and a mixed-integer second-order cone program (MISOCP) when norm $\|\cdot\|_1$ (or $\|\cdot\|_\infty$) and $\|\cdot\|_2$ norms are chosen for defining the ambiguity set $\mathcal{F}_\theta(\boldsymbol{y})$. Second, compared with the corresponding stochastic programming counterpart with $\theta = 0$, the problem (RP-PSP) with reformulation (16.18)–(16.27) only induces additional variables and constraints of size $O(|P| \times |L|)$ that is independent of the sample size N. Finally, after the linearization of the (nonlinear) term $\sqrt[l^*]{K(y)} \sqrt[l^*]{\sum_{p \in \mathcal{K}(y)} c_{p0}^{l^*}}$, the resulting reformulation (16.18)–(16.27) does not require additional integer variables, and the induced constraints (16.21) and (16.22) involving the original binary $\boldsymbol{y}$ variables are also independent of the sample size N. These appealing scalability features reflect well the computational advantage of the model (RP-PSP) as a decision-dependent two-stage DRO.

16.4 Numerical Experiments

In this section, we demonstrate the performance of model (RP-PSP) via numerical experiments with real-life retailing sales data. The experiments consist of three parts: the value of robustness, the value of capturing heteroskedasticity, and the value of capturing decision dependency.

We consider a production-service planning problem of ten potential products ($|P| = 10$) with a single production facility ($|I| = 1$). Each product is associated with two different service types ($|L| = 2$): I and II, referring to a high- and low-quality service level, respectively. The raw dataset for demand prediction consists of daily sales data for more than 100 products of different categories over two years. In particular, the type-I services correspond to those that effectively promote product sales, while the type-II do not. We use the data from the first year as the training set. The data from the second year are used as the testing set.

We consider the prediction model with a single exogenous covariate season. Accordingly, the ambiguity set $\mathcal{F}_\theta(\boldsymbol{y})$ with the prediction $\hat{\boldsymbol{\xi}}(\boldsymbol{y}^n, \hat{\boldsymbol{\chi}}^n, \hat{\boldsymbol{\varepsilon}}_{\mathrm{F}}^n)$ is constructed as given by (16.15) in Sect. 16.2.2. In particular, for a more intuitive exposition of model performance, we let the ambiguity-aversion level θ be determined using the following rule:

$$\theta = \frac{1}{|P| \times N} \sum_{p \in P} \sum_{n \in [N]} \hat{\xi}_p(\boldsymbol{y}^n, \hat{\boldsymbol{\chi}}^n, \hat{\boldsymbol{\varepsilon}}_{\mathrm{F}}^n) \times \rho, \tag{16.28}$$

where ρ represents an average prediction-error ratio across different products and scenarios, and $\hat{\xi}_p(\boldsymbol{y}^n, \hat{\boldsymbol{\chi}}^n, \hat{\boldsymbol{\varepsilon}}_{\mathrm{F}}^n)$, $n \in [N]$, $p \in P$ are predicted values. We can vary the prediction-error ratio $\rho \geq 0$ for different analysis purposes.

Given the ambiguity-aversion level θ, we solve model (RP-PSP) and obtain the optimal solution with the associated (in-sample) worst-case expected total profit. For different evaluation purposes, we also solve (i) the empirical stochastic counterpart model (denoted as "SAA") that also corresponds to model (RP-PSP) with $\theta = 0$, (ii) the DRO model with prediction model estimated by OLS that ignores the heteroskedasticity (denoted as "DRO-OLS"), and (iii) the DRO model over an ambiguity set that ignores the decision dependency (called "DRO-Naïve").

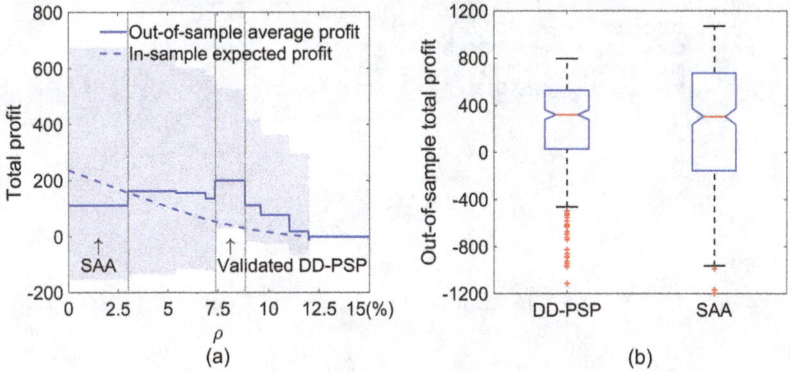

Fig. 16.1 (**a**) In-sample and out-of-sample profits of RP-PSP under different values of ρ, (**b**) Box plot of out-of-sample performance comparison between SAA ($\rho = 0$) and validated RP-PSP ($\rho = 7.5\%$)

Considering the optimal solution for the different optimization models under different values of prediction-error ratio ρ, we compare the out-of-sample (total) profits. The comparison is made using the average value and empirical distribution of out-of-sample profits. We also investigate the out-of-sample performances of the models with perspective best values of ρ that we term the *validated* models.

We first investigate the value of robustness by comparing the out-of-sample performance of model (RP-PSP) with the SAA model. Figure 16.1a plots the in-sample and out-of-sample profits of the RP-PSP solution as the prediction-error ratio ρ varies from 0 to 15%. The shaded area represents the 25%–75% confidence interval of the out-of-sample total profit. Observing the figure, we conclude the following: (i) The SAA solution ($\rho \leq 3.0\%$) incurs significant performance disappointments (i.e., out-of-sample performance is worse than the in-sample performance). Meanwhile, the robust solution of RP-PSP ($\rho > 3.0\%$) can successfully avoid the disappointments. (ii) The RP-PSP solution outperforms the SAA solution in the average out-of-sample profit when ρ is properly selected (e.g., $\rho \in [3.0\%, 8.8\%]$). In particular, the validated RP-PSP solution (e.g., $\rho = 7.5\%$) outperforms the SAA solution in distributions of out-of-sample profit, for instance, with better location and dispersion statistics (Fig. 16.1a and b). *These observations well justify the value of robustness in the RP-PSP model.*

We then evaluate the value of model (RP-PSP) in capturing heteroskedasticity by incorporating the prediction model in the ambiguity set $\mathcal{F}_\theta(\mathbf{y})$. For comparison, as previously mentioned, we construct an OLS prediction model counterpart incorporated in the ambiguity set, thus obtaining the DRO-OLS model. We compare the out-of-sample performances of both models. We first observe Fig. 16.2a. The blue (red) shaded area represents the 25%–75% confidence values of the out-of-sample total profit of RP-PSP (DRO-OLS). This figure indicates three things. (i) The RP-PSP solution outperforms the DRO-OLS solution in average out-of-sample performance for most values of ρ. (ii) The RP-PSP solution can also mitigate the

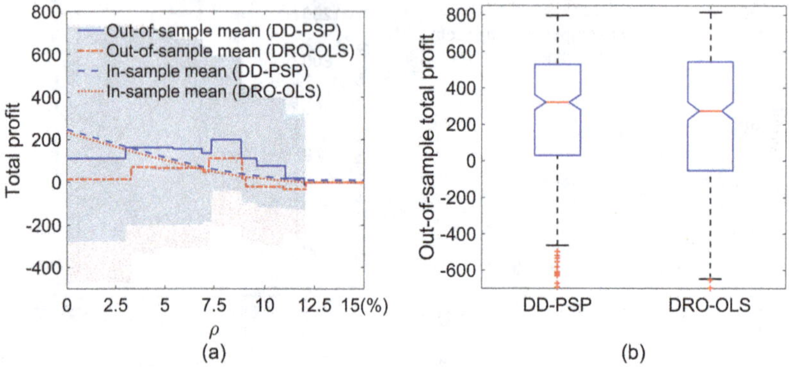

Fig. 16.2 (**a**) In-sample and out-of-sample profits of RP-PSP and DRO-OLS under different ρ; (**b**) Box plot of out-of-sample performance comparison between validated RP-PSP ($\rho = 7.5\%$) and validated DRO-OLS ($\rho = 7.5\%$)

performance disappointment more effectively than the DRO-OLS solution for most values of ρ. (iii) The validated RP-PSP solution (e.g., $\rho = 7.5\%$) outperforms the validated DRO-OLS solution (e.g., $\rho = 7.5\%$), in terms of a significantly better out-of-sample total profit distribution (Fig. 16.2b).

Finally, to evaluate the value of RP-PSP in capturing the decision dependency, we consider a predictive ambiguity set $\widehat{\mathcal{F}}_\theta$ that incorporates the prediction model, yet disregarding the service dependency. In other words, we allow the service as a part of covariates to be included in the prediction model but ignore the impact of service decision onto the prediction in $\widehat{\mathcal{F}}_\theta$. In particular, we consider

$$\widehat{\mathcal{F}}_\theta = \left\{ \mathbb{Q} \in \mathcal{P}(\mathbb{R}^{|P|} \times \mathbb{R}^M \times \mathbb{R}^{|P|}) \left| \begin{array}{l} (\boldsymbol{\xi}, \boldsymbol{\chi}_\mathrm{S}, \boldsymbol{\varepsilon}_\mathrm{F}) \sim \mathbb{Q};\ (\boldsymbol{\chi}_\mathrm{S}, \boldsymbol{\varepsilon}_\mathrm{F}) \sim \widehat{\mathbb{P}}_{(\boldsymbol{\chi}_\mathrm{S}, \boldsymbol{\varepsilon}_\mathrm{F})} \\ \mathbb{E}_\mathbb{Q}\left[\left\| \boldsymbol{\xi} - \hat{\boldsymbol{\xi}}(\boldsymbol{\chi}_\mathrm{S}, \boldsymbol{\varepsilon}_\mathrm{F}) \right\| \right] \leq \theta |P| \\ \mathbb{Q}\left[(\boldsymbol{\xi}, \boldsymbol{\chi}_\mathrm{S}, \boldsymbol{\varepsilon}_\mathrm{F}) \in \mathbb{R}_+^{|P|} \times \mathbb{R}^M \times \mathbb{R}^{|P|} \right] = 1 \end{array} \right. \right\},$$

where $\boldsymbol{\chi}_\mathrm{S}$ includes covariates of service, the prediction model now has the form of $\hat{\boldsymbol{\xi}}(\boldsymbol{\chi}_\mathrm{S}, \boldsymbol{\varepsilon}_\mathrm{F}) = \hat{\boldsymbol{a}}_0 + \widehat{\mathbf{B}}\boldsymbol{\chi}_\mathrm{S} + \boldsymbol{\varepsilon}_\mathrm{F}$, and $\widehat{\mathbb{P}}_{(\boldsymbol{\chi}_\mathrm{S}, \boldsymbol{\varepsilon}_\mathrm{F})}$ now includes the observed service levels that spread the service effect over different scenarios. We then solve the RP-PSP model with this (naïve) ambiguity set $\widehat{\mathcal{F}}_\theta$, thus adopting the term DRO-Naïve model as mentioned previously.

Figure 16.3 presents the in-sample and out-of-sample total profits for models of RP-PSP and DRO-Naïve under different values of $\rho \in [0\%, 30\%]$. The blue (red) shaded area represents the 25%–75% confidence interval of the out-of-sample total profit of RP-PSP (DRO-Naïve); observing the figure, we conclude the following. (i) The DRO-Naïve model loses off (to the RP-PSP model) in mitigating the performance disappointment: The former incurs a positive disappointment for almost all scenarios of ρ. The latter effectively controls the disappointment to be non-positive for 90% of ρ (i.e., $\rho \geq 3.0\%$). (ii) The validated RP-PSP solution

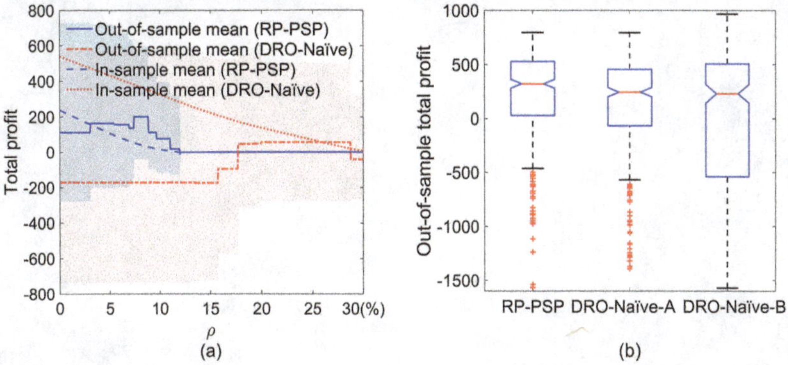

Fig. 16.3 (**a**) In-sample and out-of-sample profits of RP-PSP and DRO-Naïve under different ρ. (**b**) Box plot of out-of-sample performance comparison for validated RP-PSP ($\rho = 7.5\%$), DRO-Naïve-A (validated with $\rho = 20.0\%$), and DRO-Naïve-B (with $\rho = 7.5\%$, same as validated RP-PSP)

($\rho = 7.5\%$) can produce better out-of-sample profits than both the validated DRO-Naïve solution ($\rho = 20.0\%$) and the DRO-Naïve solution with the same $\rho = 7.5\%$ as the validated RP-PSP. These observations justify the value of RP-PSP in capturing the decision dependency.

16.5 Conclusions and Further Reading

In this chapter, we discussed an application of distributionally robust optimization to a joint production and service planning problem. The effect of demand ambiguity with correlation and variance-heterogeneity (cross-product heteroskedasticity) and the effect of service dependency were considered in the data-driven model. A predictive ambiguity set based on a prediction model was constructed to capture demand correlation. Service dependency was also captured, and a decision-dependent two-stage DRO model was developed. The model can be reformulated as a mixed-integer conic linear program that is computationally attractive. Finally, the numerical results reported demonstrate the effectiveness of the model.

The interested reader can find additional insights on the topic focused in this chapter in different research streams. Production planning under uncertainty is considered in [230, 239, 407, 427, 534]. Decision-dependent two-stage DRO is analyzed [175, 359, 409, 558]. Finally, insights on residuals-based DRO/stochastic optimization can be found in [178, 278, 297, 298, 439].

Reference

1. Adulyasak, Y., Cordeau, J.-F., & Jans, R. (2015). Benders decomposition for production routing under demand uncertainty. *Operations Research, 63,* 851–867.
2. Ahipasaoglu, S. D., Meskarian, R., Magnanti, T., & Natarajan, K. (2015). Beyond normality: A distributionally robust stochastic user equilibrium model. *Transportation Research Part B: Methodological, 81,* 333–354.
3. Ahmadi-Javid, A., & Seddighi, A. H. (2013). A location-routing problem with disruption risk. *Transportation Research Part E: Logistics and Transportation Review, 53,* 63–82.
4. Ahmed, M., Seraj, R., & Islam, S. M. S. (2020). The k-means algorithm: A comprehensive survey and performance evaluation. *Electronics, 9,* 1295.
5. Ahmed, S., Luedtke, J., Song, Y., & Xie, W. (2017). Nonanticipative duality, relaxations, and formulations for chance-constrained stochastic programs. *Mathematical Programming, 162,* 51–81.
6. Aksen, D., Piyade, N., & Aras, N. (2010). The budget constrained r-interdiction median problem with capacity expansion. *Central European Journal of Operations Research, 18,* 269–291.
7. Albareda-Sambola, M., Alonso-Ayuso, A., Escudero, L. F., Fernández, E., & Pizarro, C. (2013). Fix-and-relax-coordination for a multi-period location-allocation problem under uncertainty. *Computers & Operations Research, 40,* 2878–2892.
8. Albareda-Sambola, M., Fernández, E., & Laporte, G. (2007). Heuristic and lower bound for a stochastic location-routing problem. *European Journal of Operational Research, 179,* 940–955.
9. Albareda-Sambola, M., Fernández, E., & Saldanha-da-Gama, F. (2011). The facility location problem with Bernoulli demands. *Omega, 39,* 335–345.
10. Albareda-Sambola, M., Fernández, E., & Saldanha-da-Gama, F. (2017). Heuristic solutions to the facility location problem with general Bernoulli demands. *INFORMS Journal on Computing, 29,* 737–753.
11. Albareda-Sambola, M., Fernández, E., & Saldanha-da-Gama, F. (2023). Some heuristic methods for discrete facility location with uncertain demands. In H. A. Eiselt & V. Marianov (Eds.), *Uncertainty in facility location models* (pp. 391–431). Springer Nature Switzerland AG.
12. Albareda-Sambola, M., & Rodríguez-Pereira, J. (2019). Location-routing and location-Arc routing. In G. Laporte, S. Nickel, & F. Saldanha-da-Gama (Eds.), Location Science (Ch. 15, 2nd ed., pp. 431–451). Springer.

© The Author(s), under exclusive license to Springer Nature Switzerland AG 2024
F. Saldanha-da-Gama, S. Wang, *Facility Location Under Uncertainty*,
International Series in Operations Research & Management Science 356,
https://doi.org/10.1007/978-3-031-55927-3

13. Ales, Z., & Elloumi, S. (2018). Compact MILP formulations for the *p*-center problem. In J. Lee, G. Rinaldi, & A. Mahjoub (Eds.), *Lecture Notes in Computer Science 10856, International Symposium on Combinatorial Optimization, ISCO 2018* (pp. 14–25). Springer.
14. Alibeyg, A., Contreras, I., & Fernández, E. (2018). Exact solution of hub network design problems with profits. *European Journal of Operational Research, 266,* 57–71.
15. Almeida, T., & Manquinho, V. (2022). Constraint-based electoral districting using a new compactness measure: An application to Portugal. *Computers & Operations Research, 146,* 105892.
16. Alonso-Ayuso, A., Escudero, L. F., & Teresa Ortuño, M. (2003). BFC, a branch-and-fix coordination algorithmic framework for solving some types of stochastic pure and mixed 0–1 programs. *European Journal of Operational Research, 151,* 503–519.
17. Altın, A., Yaman, H., & Pınar, M. Ç. (2011). The robust network loading problem under hose demand uncertainty: Formulation, polyhedral analysis, and computations. *INFORMS Journal on Computing, 23,* 75–89.
18. Alumur, S., & Kara, B. Y. (2008). Network hub location problems: The state of the art. *European Journal of Operational Research, 190,* 1–21.
19. Alumur, S. A., Nickel, S., & Saldanha-da-Gama, F. (2012). Hub location under uncertainty. *Transportation Research Part B: Methodological, 46,* 529–543.
20. Alumur, S. A., Nickel, S., Saldanha-da-Gama, F., & Verter, V. (2012). Multi-period reverse logistics network design. *European Journal of Operational Research, 220,* 67–78.
21. Álvarez-Miranda, E., Fernández, E., & Ljubić, I. (2015). The recoverable robust facility location problem. *Transportation Research Part B: Methodological, 79,* 93–120.
22. Aly, A., & White, J. (1978). Probabilistic formulation of the emergency service location problem. *Journal of the Operational Research Society, 29,* 1167–1179.
23. Amini, A., Tavakkoli-Moghaddam, R., & Ebrahimnejad, S. (2020). A robust location-arc routing problem under uncertainty: Mathematical model with lower and upper bounds. *Computational and Applied Mathematics, 39,* 318.
24. An, Y., Zeng, B., Zhang, Y., & Zhao, L. (2014). Reliable p-median facility location problem: Two-stage robust models and algorithms. *Transportation Research Part B: Methodological, 64,* 54–72.
25. Anand, P. (1987). Are the preference axioms really rational? *Theory and Decision, 23,* 189–214.
26. Andaryan, A. Z., Mousighichi, K., & Ghaffarinasab, N. (2024). A heuristic approach to the stochastic capacitated single allocation hub location problem with Bernoulli demands. *European Journal of Operational Research, 312,* 954–968.
27. Ang, E., Iancu, D. A., & Swinney, R. (2017). Disruption risk and optimal sourcing in multitier supply networks. *Management Science, 63,* 2397–2771.
28. Ang, M., Lim, Y. F., & Sim, M. (2012). Robust storage assignment in unit-load warehouses. *Management Science, 58,* 2114–2130.
29. Angulo, G., Ahmed, S., & Dey, S. S. (2016). Improving the integer L-shaped method. *INFORMS Journal on Computing, 28,* 483–499.
30. Anily, S., & Federgruen, A. (1991). Structured partitioning problems. *Operations Research, 39,* 130–149.
31. Ardestani-Jaafari, A., & Delage, E. (2018). The value of flexibility in robust location-transportation problems. *Transportation Science, 52,* 189–209.
32. Arora, R., & Gao, R. (2022). *Data-driven multistage distributionally robust optimization with nested distance: Time consistency and tractable dynamic reformulations.* Optimization Online: https://optimization-online.org/?p=20641.
33. Artzner, P., Delbaen, F., Eber, J.-M., & Heath, D. (1999). Coherent measures of risk. *Mathematical Finance, 9,* 203–228.
34. Asian Development Bank (2010). *Municipal solid waste treatment: Case study of PPPs in Wenzhou.* https://www.adb.org/publications/municipal-solid-waste-treatment-case-study-ppps-wenzhou

35. Asian Development Bank (2017). *Case studies in private sector participation: Solid waste management*. https://www.adb.org/publications/private-sector-participation-solid-waste-management
36. Atamtürk, A., Berenguer, G., & Shen, Z.-J.M. (2012). A conic integer programming approach to stochastic joint location-inventory problems. *Operations Research, 60,* 366–381.
37. Atamtürk, A., & Narayanan, V. (2009). The submodular knapsack polytope. *Discrete Optimization, 6,* 333–344.
38. Averbakh, I., & Berman, O. (1997). Minimax regret *p*-center location on a network with demand uncertainty. *Location Science, 5,* 247–254.
39. Averbakh, I., & Berman, O. (2000). Algorithms for the robust 1-center problem on a tree. *European Journal of Operational Research, 123,* 292–302.
40. Azizi, N., & Salhi, S. (2022). Reliable hub-and-spoke systems with multiple capacity levels and flow dependent discount factor. *European Journal of Operational Research, 298,* 834–854.
41. Baldassarre, S., Bruno, G., Diglio, A., & Piccolo, C. (2023). A sample approximation solution procedure for chance-constrained districting problems. *Computers & Operations Research, 160,* 106376.
42. Ball, M. O., & Lin, F. L. (1993). A reliability model applied to emergency service vehicle location. *Operations Research, 41,* 18–36.
43. Ban, G.-Y., Gallien, J., & Mersereau, A. J. (2019). Dynamic procurement of new products with covariate information: The residual tree method. *Manufacturing & Service Operations Management, 21,* 798–815.
44. Banks, C. J., & Heaven, S. (2013). Optimisation of biogas yields from anaerobic digestion by feedstock type. In A. Wellinger, J. D. Murphy, & D. Baxter (Eds.), *The biogas handbook: Science, production and applications* (pp. 131–165). Elsevier.
45. Barahona, F., & Jensen, D. (1998). Plant location with minimum inventory. *Mathematical Programming, 83,* 101–111.
46. Baron, O., Milner, J., & Naseraldin, H. (2011). Facility location: A robust optimization approach. *Production and Operations Management, 20,* 772–785.
47. Basciftci, B., Ahmed, S., & Shen, S. (2021). Distributionally robust facility location problem under decision-dependent stochastic demand. *European Journal of Operational Research, 292,* 548–561.
48. Basole, R. C., & Bellamy, M. A. (2014). Supply network structure, visibility, and risk diffusion: A computational approach. *Decision Sciences, 45,* 753–789.
49. Batta, R., Dolan, J. M., & Krishnamurthy, N. N. (1989). The maximal expected covering location problem: Revisited. *Transportation Science, 23,* 277–287.
50. Bayram, V., Kara, B. Y., Saldanha-da-Gama, F., & Yaman, H. (2023). Humanitarian logistics under uncertainty: Planning for sheltering and evacuation. In H. A. Eiselt & V. Marianov (Eds.), *Uncertainty in facility location problems* (pp. 81–105). Springer.
51. Bayram, V., Yıldız, B., & Farham, M. S. (2023). Hub network design problem with capacity, congestion, and stochastic demand considerations. *Transportation Science, 57,* 1276–1295.
52. Ben-Tal, A., Brekelmans, R., Den Hertog, D., & Vial, J.-P. (2017). Globalized robust optimization for nonlinear uncertain inequalities. *INFORMS Journal on Computing, 29,* 350–366.
53. Ben-Tal, A., Den Hertog, D., & Vial, J.-P. (2015). Deriving robust counterparts of nonlinear uncertain inequalities. *Mathematical Programming, 149*(1–2), 265–299.
54. Ben-Tal, A., Den Hertog, D., Waegenaere, A. D., Melenberg, B., & Rennen, G. (2013). Robust solutions of optimization problems affected by uncertain probabilities. *Management Science, 59,* 341–357.
55. Ben-Tal, A., El Ghaoui, L., & Nemirovski, A. (2009). *Robust optimization*. Princeton University Press.
56. Ben-Tal, A., Golany, B., Nemirovski, A., & Vial, J.-P. (2005). Retailer-supplier flexible commitments contracts: A robust optimization approach. *Manufacturing & Service Operations Management, 7,* 248–271.

57. Ben-Tal, A., Goryashko, A., Guslitzer, E., & Nemirovski, A. (2004). Adjustable robust solutions of uncertain linear programs. *Mathematical Programming, 99,* 351–376.
58. Ben-Tal, A., & Nemirovski, A. (1998). Robust convex optimization. *Mathematics of Operations Research, 23,* 769–805.
59. Ben-Tal, A., & Nemirovski, A. (1999). Robust solutions of uncertain linear programs. *Operations Research Letters, 25,* 1–13.
60. Ben-Tal, A., & Nemirovski, A. (2002). Robust optimization–methodology and applications. *Mathematical Programming, 92,* 453–480.
61. Benzarti, E., Sahin, E., & Dallery, Y. (2013). Operations management applied to home care services: Analysis of the districting problem. *Decision Support Systems, 55,* 587–598.
62. Beraldi, P., & Bruni, M. E. (2009). A probabilistic model applied to emergency service vehicle location. *European Journal of Operational Research, 196,* 323–331.
63. Beraldi, P., Bruni, M. E., & Conforti, D. (2004). Designing robust emergency medical service via stochastic programming. *European Journal of Operational Research, 158,* 183–193.
64. Beraldi, P., & Ruszczyński, A. (2002). The probabilistic set-covering problem. *Operations Research, 50,* 956–967.
65. Berman, O., & Drezner, Z. (2008). The p-median problem under uncertainty. *European Journal of Operational Research, 189,* 19–30.
66. Berman, O., & Krass, D. (2019). Stochastic location models under congestion. In G. Laporte, S. Nickel, & F. Saldanha-da-Gama (Eds.), *Location Science* (Ch. 17, 2nd ed., pp. 477–535). Springer.
67. Berman, O., Krass, D., & Tajbakhsh, M. M. (2012). A coordinated location-inventory model. *European Journal of Operational Research, 217,* 500–508.
68. Bertsimas, D., Brown, D. B., & Caramanis, C. (2011). Theory and applications of robust optimization. *SIAM Review, 53,* 464–501.
69. Bertsimas, D., Goyal, V., & Sun, X. A. (2011). A geometric characterization of the power of finite adaptability in multistage stochastic and adaptive optimization. *Mathematics of Operations Research, 36,* 24–54.
70. Bertsimas, D., & Kallus, N. (2020). From predictive to prescriptive analytics. *Management Science, 66,* 1025–1044.
71. Bertsimas, D., & Kim, C. W. (2023). *A machine learning approach to two-stage adaptive robust optimization*. arXiv:2307.12409 [cs.LG]. Available at https://doi.org/10.48550/arXiv.2307.12409.
72. Bertsimas, D., & Koduri, N. (2022). Data-driven optimization: A reproducing kernel Hilbert space approach. *Operations Research, 70,* 454–471.
73. Bertsimas, D., & Shtern, S. (2018). A scalable algorithm for two-stage adaptive linear optimization. *arXiv:1807.02812* [math.OC]. Available at: https://doi.org/10.48550/arXiv.1807.02812
74. Bertsimas, D., Shtern, S., & Sturt, B. (2023). A data-driven approach to multistage stochastic linear optimization. *Management Science, 69,* 51–74.
75. Bertsimas, D., & Sim, M. (2004). The price of robustness. *Operations Research, 52,* 35–53.
76. Bertsimas, D., Sim, M., & Zhang, M. (2019). Adaptive distributionally robust optimization. *Management Science, 65*(2), 604–618.
77. Bieniek, M. (2015). A note on the facility location problem with stochastic demands. *Omega, 55,* 53–60.
78. Birge, J., & Louveaux, F. V. (2011). *Introduction to stochastic programming* (2nd ed.). Springer.
79. Biuki, M., Kazemi, A., & Alinezhad, A. (2020). An integrated location-routing-inventory model for sustainable design of a perishable products supply chain network. *Journal of Cleaner Production, 260,* 120842.
80. Blanchet, J., Murthy, K., & Nguyen, V. A. (2021). Statistical analysis of Wasserstein distributionally robust estimators. In *Tutorials in Operations Research: Emerging optimization methods and modeling techniques with applications* (pp. 227–254). INFORMS.

81. Blanco, V., Fernández, E., & Hinojosa, Y. (2023). Hub location with protection under interhub link failures. *INFORMS Journal on Computing*. https://doi.org/10.1287/ijoc.2023.1296
82. Bock, H.-H. (2007). Clustering methods: A history of k-means algorithms. In P. Brito, G. Cucumel, P. Bertrand, & F. de Carvalho (Eds.), *Selected contributions in data analysis and classification* (pp. 161–172). Springer.
83. Bodur, M., Dash, S., Günlük, O., & Luedtke, J. (2016). Strengthened Benders cuts for stochastic integer programs with continuous recourse. *INFORMS Journal on Computing, 29,* 77–91.
84. Borges Lopes, R., Plastria, F., Ferreira, C., & Sousa Santos, B. (2014). Location-arc routing problem: Heuristic approaches and test instances. *Computers & Operations Research, 43,* 309–317.
85. Borrás, F., & Pastor, J. T. (2002). The ex-post evaluation of the minimum local reliability level: An enhanced probabilistic location set covering model. *Annals of Operations Research, 111,* 51–74.
86. Boyd, S., & Vandenberghe, L. (2004). *Convex optimization*. Cambridge University Press.
87. Brown, D. B., de Giorgi, E., & Sim, M. (2012). Aspirational preferences and their representation by risk measures. *Management Science, 58,* 2095–2113.
88. Brown, D. B., & Sim, M. (2009). Satisficing measures for analysis of risky positions. *Management Science, 55,* 71–84.
89. Bruni, M. E., Pugliese, L. D. P., Beraldi, P., & Guerriero, F. (2017). An adjustable robust optimization model for the resource-constrained project scheduling problem with uncertain activity durations. *Omega, 71,* 66–84.
90. Bruno, G., Esposito, E., Genovese, A., & Piccolo, C. (2016). Institutions and facility mergers in the Italian education system: Models and case studies. *Socio-Economic Planning Sciences, 53,* 23–32.
91. Bruno, G., & Laporte, G. (2002). An interactive decision support system for the design of rapid public transit networks. *INFOR: Information Systems and Operational Research, 40,* 111–118.
92. Bütün, C., Petrovic, S., & Muyldermans, L. (2021). The capacitated directed cycle hub location and routing problem under congestion. *European Journal of Operational Research, 292,* 714–734.
93. Cacchiani, V., Iori, M., Locatelli, A., & Martello, S. (2022). Knapsack problems—an overview of recent advances. Part I: Single knapsack problems. *Computers & Operations Research, 143,* 105692.
94. Cadarso, L., Escudero, L. F., & Marín, A. (2018). On strategic multistage operational two-stage stochastic 0–1 optimization for the rapid transit network design problem. *European Journal of Operational Research, 271,* 577–593.
95. Calafiore, G., & Dabbene, F. (2006). *Probabilistic and randomized methods for design under uncertainty*. Springer.
96. Calafiore, G. C., & Campi, M. C. (2006). The scenario approach to robust control design. *IEEE Transactions on Automatic Control, 51,* 742–753.
97. Calafiore, G. C., & El Ghaoui, L. (2006). On distributionally robust chance-constrained linear programs. *Journal of Optimization Theory and Applications, 130,* 1–22.
98. Çalik, H., Labbé, M., & Yaman, H. (2019). p-center problems. In G. Laporte, S. Nickel, & F. Saldanha-da-Gama (Eds.), *Location Science* (Ch. 3, 2nd ed., pp. 51–65). Springer.
99. Çalik, H., & Tansel, B. Ç. (2013). Double bound method for solving the p-center location problem. *Computers & Operations Research, 40,* 2991–2999.
100. Campbell, J. F., & O'Kelly, M. E. (2012). Twenty-five years of hub location research. *Transportation Science, 46,* 153–169.
101. Cardin, M.-A., Xie, Q., Ng, T. S., Wang, S., & Hu, J. (2017). An approach for analyzing and managing flexibility in engineering systems design based on decision rules and multistage stochastic programming. *IISE Transactions, 49,* 1–12.
102. Carlsson, J. G. (2012). Dividing a territory among several vehicles. *INFORMS Journal on Computing, 24,* 565–577.

103. Carlsson, J. G., & Delage, E. (2013). Robust partitioning for stochastic multivehicle routing. *Operations Research, 61*(3), 727–744.
104. Carrere, H., Antonopoulou, G., Affes, R., Passos, F., Battimelli, A., Lyberatos, G., & Ferrer, I. (2016). Review of feedstock pretreatment strategies for improved anaerobic digestion: From lab-scale research to full-scale application. *Bioresource Technology, 199,* 386–397.
105. Carrizosa, E., & Nickel, S. (2003). Robust facility location. *Mathematical Methods of Operations Research, 58,* 331–349.
106. Carrøe, C. C., & Schultz, R. (1999). Dual decomposition in stochastic integer programming. *Operations Research Letters, 24,* 37–45.
107. Casement, C. J., & Kahle, D. J. (2018). Graphical prior elicitation in univariate models. *Communications in Statistics-Simulation and Computation, 47,* 2906–2924.
108. Caunhye, A. M., Zhang, Y., Li, M., & Nie, X. (2016). A location-routing model for prepositioning and distributing emergency supplies. *Transportation Research Part E: Logistics and Transportation Review, 90,* 161–176.
109. Chakravarty, A. K., Orlin, J., & Rothblum, U. G. (1985). Consecutive optimizers for a partitioning problem with applications to optimal inventory groupings for joint replenishment. *Operations Research, 33,* 820–834.
110. Chan, Y., Carter, W. B., & Burnes, M. D. (2001). A multiple-depot, multiple-vehicle, location-routing problem with stochastically processed demands. *Computers & Operations Research, 28,* 803–826.
111. Chapman, S., & White, J. (1974). Probabilistic formulations of emergency service facilities location problems. In *ORSA/TIMS Conference, San Juan, Puerto Rico*.
112. Charnes, A., & Cooper, W. W. (1959). Chance-constrained programming. *Management Science, 6,* 73–79.
113. Chavez, M. M. M., Costa, Y., & Sarache, W. (2021). A three-objective stochastic location-inventory-routing model for agricultural waste-based biofuel supply chain. *Computers & Industrial Engineering, 162,* 107759.
114. Chen, G., Daskin, M. S., Shen, Z.-J.M., & Uryasev, S. (2006). The α-reliable mean-excess regret model for stochastic facility location modeling. *Naval Research Logistics, 53,* 617–626.
115. Chen, L. G., Long, D. Z., & Sim, M. (2015). On dynamic decision making to meet consumption targets. *Operations Research, 63,* 1117–1130.
116. Chen, Q., Fu, C., Zhu, N., Ma, S., & He, Q.-C. (2023). A target-based optimization model for bike-sharing systems: From the perspective of service efficiency and equity. *Transportation Research Part B: Methodological, 167,* 235–260.
117. Chen, W., & Sim, M. (2009). Goal-driven optimization. *Operations Research, 57,* 342–357.
118. Chen, X., Sim, M., & Sun, P. (2007). A robust optimization perspective on stochastic programming. *Operations Research, 55,* 1058–1071.
119. Chen, Z., Kuhn, D., & Wiesemann, W. (2024). Data-driven chance constrained programs over Wasserstein balls. *Operations Research, 72,* 410–424.
120. Chen, Z., Kuhn, D., & Wiesemann, W. (2023). On approximations of data-driven chance constrained programs over Wasserstein balls. *Operations Research Letters, 51,* 226–233.
121. Chen, Z., Sim, M., & Xiong, P. (2020). Robust stochastic optimization made easy with RSOME. *Management Science, 66,* 3329–3339.
122. Chen, Z., Sim, M., & Xu, H. (2019). Distributionally robust optimization with infinitely constrained ambiguity sets. *Operations Research, 67,* 1328–1344.
123. Cheng, C., Adulyasak, Y., & Rousseau, L.-M. (2021). Robust facility location under demand uncertainty and facility disruptions. *Omega, 103,* 102429.
124. Cheng, C., Adulyasak, Y., & Rousseau, L.-M. (2021). Robust facility location under disruptions. *INFORMS Journal on Optimization, 3,* 298–314.
125. Cheng, C., Yu, Q., Adulyasak, Y., & Rousseau, L.-M. (2024). Distributionally robust facility location with uncertain facility capacity and customer demand. *Omega, 122,* 102959.
126. Cheng, J., Gicquel, C., & Lisser, A. (2019). Partial sample average approximation method for chance constrained problems. *Optimization Letters, 13,* 657–672.

127. Chod, J., Pyke, D., & Rudi, N. (2010). The value of flexibility in make-to-order systems: The effect of demand correlation. *Operations Research, 58,* 834–848.
128. Choon, C. M. (2016). *Culture shock over South Korea's mandatory recycling of food waste*. Available at: https://www.straitstimes.com/asia/east-asia/culture-shock-over-south-koreas-mandatory-recycling-of-food-waste
129. Church, R., & ReVelle, C. (1974). The maximal covering location problem. *Papers of the Regional Science Association, 32,* 101–118.
130. Church, R. L., & Drezner, Z. (2022). Review of obnoxious facilities location problems. *Computers & Operations Research, 138,* 105468.
131. Church, R. L., & Murray, A. (2018). *Location covering models: History, applications and advancements*. Springer International Publishing AG.
132. Coelho, I. M. (2023). *A multi-period stochastic location-inventory problem.* University of Lisbon, MSc Thesis.
133. Contardo, C., Iori, M., & Kramer, R. (2019). A scalable exact algorithm for the vertex *p*-center problem. *Computers & Operations Research, 103,* 211–220.
134. Contreras, I. (2021). Hub network design. In T. G. Crainic, M. Gendreau, & B. Gendron (Eds.), *Network design with applications to transportation and logistics* (Ch. 18, pp. 567–598). Springer.
135. Contreras, I., Cordeau, J.-F., & Laporte, G. (2011). Benders decomposition for large-scale uncapacitated hub location. *Operations Research, 59,* 1477–1490.
136. Contreras, I., Cordeau, J.-F., & Laporte, G. (2011). Stochastic uncapacitated hub location. *European Journal of Operational Research, 212*(3), 518–528.
137. Contreras, I., & O'Kelly, M. (2019). Hub location problems. In G. Laporte, S. Nickel, & F. Saldanha-da-Gama (Eds.), *Location Science* (Ch. 12, 2nd ed., pp. 327–363). Springer.
138. Cordeau, J.-F., Furini, F., & Ljubić, I. (2019). Benders decomposition for very large scale partial set covering and maximal covering location problems. *European Journal of Operational Research, 275,* 882–896.
139. Cordeau, J.-F., Klibi, W., & Nickel, S. (2021). Logistics network design. In T. G. Crainic, M. Gendreau, & B. Gendron (Eds.), *Network design with applications to transportation and logistics* (Ch. 19, pp. 599–625). Springer.
140. Cormode, G., & McGregor, A. (2008). Approximation algorithms for clustering uncertain data. In *Proceedings of PODS'08, The twenty-seventh ACM SIGMOD-SIGACT-SIGART Symposium on Principles of Database Systems, June 9–12, Vancouver, BC, Canada* (pp. 191–199).
141. Correia, I., & Melo, T. (2017). A multi-period facility location problem with modular capacity adjustments and flexible demand fulfillment. *Computers & Industrial Engineering, 118,* 307–321.
142. Correia, I., Melo, T., & Saldanha-da-Gama, F. (2013). Comparing classical performance measures for a multi-period, two-echelon supply chain network design problem with sizing decisions. *Computers & Industrial Engineering, 64,* 366–380.
143. Correia, I., Nickel, S., & Saldanha-da-Gama, F. (2010). The capacitated single-allocation hub location problem revisited: A note on a classical formulation. *European Journal of Operational Research, 207,* 92–96.
144. Correia, I., Nickel, S., & Saldanha-da-Gama, F. (2018). A stochastic multi-period capacitated multiple allocation hub location problem: Formulation and inequalities. *Omega, 74,* 122–134.
145. Correia, I., & Saldanha-da-Gama, F. (2019). Facility location under uncertainty. In G. Laporte, S. Nickel, & F. Saldanha-da-Gama (Eds.), *Location Science* (Ch. 8, 2nd ed., pp. 185–213). Springer.
146. Current, J., Ratick, S., & ReVelle, C. (1997). Dynamic facility location when the total number of facilities is uncertain: A decision analysis approach. *European Journal of Operational Research, 110,* 597–609.
147. Dahlén, L., & Lagerkvist, A. (2008). Methods for household waste composition studies. *Waste Management, 28,* 1100–1112.

148. Dahlén, L., Vukicevic, S., Meijer, J.-E., & Lagerkvist, A. (2007). Comparison of different collection systems for sorted household waste in Sweden. *Waste Management, 27,* 1298–1305.
149. D'Amico, S. J., Wang, S. J., Batta, R., & Rump, C. M. (2002). A simulated annealing approach to police district design. *Computers & Operations Research, 29,* 667–684.
150. Danach, K., Gelareh, S., & Neamatian Monemi, R. (2019). The capacitated single-allocation p-hub location routing problem: A Lagrangian relaxation and a hyper-heuristic approach. *EURO Journal on Transportation and Logistics, 8,* 597–631.
151. Dashti, H., Conejo, A. J., Jiang, R., & Wang, J. (2016). Weekly two-stage robust generation scheduling for hydrothermal power systems. *IEEE Transactions on Power Systems, 31,* 4554–4564.
152. Daskin, M. S. (1982). Application of an expected covering model to emergency medical service system design. *Decision Sciences, 13,* 416–439.
153. Daskin, M. S. (1983). A maximum expected covering location model: Formulation, properties and heuristic solution. *Transportation Science, 17,* 48–70.
154. Daskin, M. S. (2013). *Network and discrete location: Models, algorithms, and applications* (2nd ed.). Wiley.
155. Daskin, M. S., Coullard, C. R., & Shen, Z.-J.M. (2002). An inventory location model: Formulation, solution algorithm and computational results. *Annals of Operations Research, 110,* 83–106.
156. Daskin, M. S., Hesse, S. M., & ReVelle, C. S. (1997). α-reliable p-minimax regret: A new model for strategic facility location modeling. *Location Science, 5,* 227–246.
157. De Maesschalck, R., Jouan-Rimbaud, D., & Massart, D. L. (2000). The Mahalanobis distance. *Chemometrics and Intelligent Laboratory Systems, 50,* 1–18.
158. De Maio, A., Laganà, D., Musmanno, R., & Vocaturo, F. (2021). Arc routing under uncertainty: Introduction and literature review. *Computers & Operations Research, 135,* 105442.
159. De Oliveira Simonetto, E., & Borenstein, D. (2007). A decision support system for the operational planning of solid waste collection. *Waste Management, 27,* 1286–1297.
160. Delage, E., & Ye, Y. (2010). Distributionally robust optimization under moment uncertainty with application to data-driven problems. *Operations Research, 58,* 596–612.
161. Delbaen, F. (2002). Coherent risk measures on general probability spaces. In K. Sandmann & P. J. Schönbucher (Eds.), *Advances in finance and stochastics*. Springer.
162. Demange, M., Gabrel, V., Haddad, M. A., & Murat, C. (2020). A robust p-center problem under pressure to locate shelters in wildfire context. *EURO Journal on Computational Optimization, 8,* 103–139.
163. Dentcheva, D., Prékopa, A., & Ruszczyński, A. (2000). Concavity and efficient points of discrete distributions in probabilistic programming. *Mathematical Programming, 89,* 55–77.
164. Department of Statistics Singapore (2009). *General household survey 2015*. Available at: https://www.singstat.gov.sg/find-data/search-by-theme/population/geographic-distribution/latest-data
165. Desaulniers, G., Desrosiers, J., Erdmann, A., Solomon, M. M., & Soumis, F. (2002). VRP with pickup and delivery. In P. Toth & D. Vigo (Eds.), *The vehicle routing problem* (Ch. 9, pp. 225–242). SIAM.
166. Di Domenica, N., Mitra, G., Valente, P., & Birbilis, G. (2007). Stochastic programming and scenario generation within a simulation framework: An information systems perspective. *Decision Support Systems, 42,* 2197–2218.
167. Diabat, A. (2016). A capacitated facility location and inventory management problem with single sourcing. *Optimization Letters, 10,* 1577–1592.
168. Diabat, A., Richard, J. P., & Codrington, C. W. (2013). A Lagrangian relaxation approach to simultaneous strategic and tactical planning in supply chain design. *Annals of Operations Research, 203,* 55–80.

169. Días, J., & Fernández, E. (2002). A branch-and-price algorithm for the single source capacitated plant location problem. *Journal of the Operational Research Society, 53,* 728–740.
170. Diecidue, E., & Van de Ven, J. (2008). Aspiration level, probability of success and failure, and expected utility. *International Economic Review, 49,* 683–700.
171. Diglio, A., Nickel, S., & Saldanha-da-Gama, F. (2020). Towards a stochastic programming modeling framework for districting. *Annals of Operations Research, 292,* 248–285.
172. Diglio, A., Peiró, J., Piccolo, C., & Saldanha-da-Gama, F. (2021). Solutions for districting problems with chance-constrained balancing requirements. *Omega, 103,* 102430.
173. Diglio, A., Peiró, J., Piccolo, C., & Saldanha-da-Gama, F. (2023). Approximation schemes for districting problems with probabilistic constraints. *European Journal of Operational Research, 307,* 233–248.
174. Dillenberger, C., Escudero, L. F., Wollensak, A., & Zang, W. (1994). On practical resource allocation for production planning and scheduling with period overlapping. *European Journal of Operational Research, 75,* 275–286.
175. Doan, X. V. (2022). Distributionally robust optimization under endogenous uncertainty with an application in retrofitting planning. *European Journal of Operational Research, 300,* 73–84.
176. Dolgui, A., & Ivanov, D. (2021). Ripple effect and supply chain disruption management: New trends and research directions. *International Journal of Production Research, 59,* 102–109.
177. Dönmez, Z., Kara, B. Y., Karsu, Ö., & Saldanha-da-Gama, F. (2021). Humanitarian facility location under uncertainty: Critical review and future prospects. *Omega, 102,* 102393.
178. Dou, X., & Anitescu, M. (2019). Distributionally robust optimization with correlated data from vector autoregressive processes. *Operations Research Letters, 47,* 294–299.
179. Drezner, Z. (Ed.) (1995). *Facility location: A survey of applications and methods*. Springer.
180. Drezner, Z., & Hamacher, H. (Eds.) (2002). *Facility location: Applications and theory*. Springer.
181. Dunke, F., Heckmann, I., Nickel, S., & Saldanha-da-Gama, F. (2018). Time traps in supply chains: Is optimal still good enough? *European Journal of Operational Research, 264,* 813–829.
182. Dupačová, J., Gröwe-Kuska, N., & Römisch, W. (2003). Scenario reduction in stochastic programming. *Mathematical Programming, 95,* 493–511.
183. Duran-Mateluna, C., Ales, Z., Elloumi, S., & Jorquera-Bravo, N. (2023). Robust MILP formulations for the two-stage weighted vertex p-center problem. *Computers & Operations Research, 159,* 106334.
184. Ebery, J., Krishnamoorthy, M., Ernst, A., & Boland, N. (2000). The capacitated multiple allocation hub location problem: Formulations and algorithms. *European Journal of Operational Research, 120,* 614–631.
185. Eiselt, H. A., & Marianov, V. (Eds.) (2011). *Foundations of location analysis*. Springer.
186. Eiselt, H. A., & Marianov, V. (Eds.) (2023). *Uncertainty in facility location models*. Springer Nature Switzerland AG.
187. Elhedhli, S., & Wu, H. (2010). A Lagrangean heuristic for hub-and-spoke system design with capacity selection and congestion. *INFORMS Journal on Computing, 22,* 282–296.
188. Elloumi, S., Labbé, M., & Pochet, Y. (2004). A new formulation and resolution method for the p-center problem. *INFORMS Journal on Computing, 16,* 84–94.
189. Eppen, G. (1979). Effects of centralization on expected costs in a multi-location newsboy problem. *Management Science, 25,* 498–501.
190. Erlebacher, S. J., & Meller, R. D. (2000). The interaction of location and inventory in designing distribution systems. *IIE Transactions, 32,* 155–166.
191. Erlenkotter, D. (1978). A dual-based procedure for uncapacitated facility location. *Operations Research, 26,* 992–1009.
192. Ernst, A. T., & Krishnamoorthy, M. (1996). Efficient algorithms for the uncapacitated single allocation p-hub median problem. *Location Science, 4,* 139–154.

193. Ernst, A. T., & Krishnamoorthy, M. (1998). Exact and heuristic algorithms for the uncapacitated multiple allocation p-hub median problem. *European Journal of Operational Research, 104,* 100–112.
194. Ernst, A. T., & Krishnamoorthy, M. (1999). Solution algorithms for the capacitated single allocation hub location problem. *Annals of Operations Research, 86,* 141–159.
195. Escudero, L. F., Araceli-Garín, M., Merino, M., & Pérez, G. (2007). The value of the stochastic solution in multistage problems. *TOP, 15,* 48–64.
196. Escudero, L. F., Araceli-Garín, M., Merino, M., & Pérez, G. (2010). On BFC-MSMIP strategies for scenario cluster partitioning, and twin node family branching selection and bounding for multistage stochastic mixed integer programming. *Computers & Operations Research, 37,* 738–753.
197. Escudero, L. F., Araceli-Garín, M., Pérez, G., & Unzueta, A. (2013). Scenario cluster decomposition of the Lagrangian dual in two-stage stochastic mixed 0–1 optimization. *Computers & Operations Research, 40,* 362–377.
198. Escudero, L. F., Araceli-Garín, M., Pizarro, C., & Unzueta, A. (2018). On efficient matheuristic algorithms for multi-period stochastic facility location-assignment problems. *Computational Optimization and Applications, 70,* 865–888.
199. Escudero, L. F., & Monge, J. F. (2023). On risk management of multistage multiscale FLP under uncertainty. In H. A. Eiselt & V. Marianov (Eds.), *Uncertainty in facility location models* (pp. 355–390). Springer Nature Switzerland AG.
200. Escudero, L. F., & Pizarro-Romero, C. (2017). On solving a large-scale problem on facility location and customer assignment with interaction costs along a time horizon. *TOP, 25,* 601–622.
201. Escudero, L. F., & Salmeron, J. (2005). On a fix-and-relax framework for a class of project scheduling problems. *Annals of Operations Research, 140,* 163–188.
202. European Commission (2018). *Study on the potential of servitisation and other forms of product-service provision for EU SMEs: Final report. Part A, main report.* Publications Office. Available at: https://data.europa.eu/doi/10.2826/558915
203. Fang, Z., Gu, B., Luo, X., & Xu, Y. (2015). Contemporaneous and delayed sales impact of location-based mobile promotions. *Information Systems Research, 26,* 552–564.
204. Fazayeli, S., Eydi, A., & Kamalabadi, I. N. (2018). Location-routing problem in multimodal transportation network with time windows and fuzzy demands: Presenting a two-part genetic algorithm. *Computers & Industrial Engineering, 119,* 233–246.
205. Feng, Q., Luo, S., & Zhang, D. (2014). Dynamic inventory-pricing control under backorder: Demand estimation and policy optimization. *Manufacturing & Service Operations Management, 16,* 149–160.
206. Feng, Z. (2014). *Recycling gets easier with a bin at every HDB block.* Available at: https://www.straitstimes.com/singapore/recycling-gets-easier-with-a-bin-at-every-hdb-block
207. Fernandes, L. J., & Saldanha-da-Gama, F. (2008). Contingency planning – a literature review. In *Proceedings of SCMCC-08 Supply Chain Management and Competitiveness*.
208. Fernández, E., Hinojosa, Y., Puerto, J., & Saldanha-da-Gama, F. (2019). New algorithmic framework for conditional value at risk: Application to stochastic fixed-charge transportation. *European Journal of Operational Research, 277,* 215–226.
209. Fernández, E., & Landete, M. (2019). Fixed-charge facility location problems. In G. Laporte, S. Nickel, & F. Saldanha-da-Gama (Eds.), *Location Science* (Ch. 4, 2nd ed., pp. 67–98). Springer.
210. Fischetti, M., Laporte, G., & Martello, S. (1993). The delivery man problem and cumulative matroids. *Operations Research, 41,* 1055–1064.
211. Fischetti, M., Ljubić, I., & Sinnl, M. (2017). Redesigning Benders decomposition for large-scale facility location. *Management Science, 63,* 2146–2162.
212. Fonseca, M. C., García-Sánchez, Á., Ortega-Mier, M., & Saldanha-da-Gama, F. (2010). A stochastic bi-objective location model for strategic reverse logistics. *TOP, 18,* 158–184.
213. França, P. M., & Luna, H. P. L. (1982). Solving stochastic transportation-location problems by generalized Benders decomposition. *Transportation Science, 16,* 113–126.

214. Freimer, M. B., Linderoth, J. T., & Thomas, D. J. (2012). The impact of sampling methods on bias and variance in stochastic linear programs. *Computational Optimization and Applications, 51,* 51–75.
215. Gaar, E., & Sinnl, M. (2022). A scaleable projection-based branch-and-cut algorithm for the p-center problem. *European Journal of Operational Research, 303,* 78–98.
216. Gabrel, V., Lacroix, M., Murat, C., & Remli, N. (2014). Robust location transportation problems under uncertain demands. *Discrete Applied Mathematics, 164,* 100–111.
217. Gabrel, V., Murat, C., & Thiele, A. (2014). Recent advances in robust optimization: An overview. *European Journal of Operational Research, 235,* 471–483.
218. Gal, S., & Klots, B. (1995). Optimal partitioning which maximizes the sum of the weighted averages. *Operations Research, 43,* 500–508.
219. Gallardo, A., Bovea, M. D., Colomer, F. J., Prades, M., & Carlos, M. (2010). Comparison of different collection systems for sorted household waste in Spain. *Waste Management, 30,* 2430–2439.
220. Gallego, G. (1998). New bounds and heuristics for (Q, r) policies. *Management Science, 44,* 219–233.
221. Gambella, C., Ghaddar, B., & Naoum-Sawaya, J. (2021). Optimization problems for machine learning: A survey. *European Journal of Operational Research, 290,* 807–828.
222. Gambella, C., Maggioni, F., & Vigo, D. (2019). A stochastic programming model for a tactical solid waste management problem. *European Journal of Operational Research, 273,* 684–694.
223. Gao, R., Chen, X., & Kleywegt, A. J. (2022). Wasserstein distributionally robust optimization and variation regularization. *Operations Research.* https://doi.org/10.1287/opre.2022.2383
224. Gao, R., & Kleywegt, A. (2023). Distributionally robust stochastic optimization with Wasserstein distance. *Mathematics of Operations Research, 48,* 603–655.
225. García, S., & Marín, A. (2019). Covering location problems. In G. Laporte, S. Nickel, & F. Saldanha-da-Gama (Eds.), *Location Science* (Ch. 5, 2nd ed., pp. 99–119). Springer.
226. Gendreau, M., Jabali, O., & Rei, W. (2014). Stochastic vehicle routing problems. In P. Toth & D. Vigo (Eds.), *Vehicle routing: Problems, methods, and applications* (Ch. 11, 2nd ed., pp. 213–239). SIAM.
227. Gendron, B., Scutellà, M. G., Garroppo, R. G., Nencioni, G., & Tavanti, L. (2016). A branch-and-Benders-cut method for nonlinear power design in green wireless local area networks. *European Journal of Operational Research, 255,* 151–162.
228. Georghiou, A., Tsoukalas, A., & Wiesemann, W. (2019). Robust dual dynamic programming. *Operations Research, 67,* 813–830.
229. Ghadge, A., Er, M., Ivanov, D., & Chaudhuri, A. (2022). Visualisation of ripple effect in supply chains under long-term, simultaneous disruptions: A system dynamics approach. *International Journal of Production Research, 60,* 6173–6186.
230. Ghadimi, F., Aouam, T., & Uzsoy, R. (2023). Safety stock placement with market selection under load-dependent lead times. *IISE Transactions, 55,* 314–328.
231. Ghaffarinasab, N. (2022). Stochastic hub location problems with Bernoulli demands. *Computers & Operations Research, 145,* 105851.
232. Ghaffarinasab, N., Andaryan, A. Z., & Torkayesh, A. E. (2020). Robust single allocation p-hub median problem under hose and hybrid demand uncertainties: Models and algorithms. *International Journal of Management Science and Engineering Management, 15,* 184–195.
233. Ghaffarinasab, N., Çavuş, Ö., & Kara, B. Y. (2023). A mean-CVaR approach to the risk-averse single allocation hub location problem with flow-dependent economies of scale. *Transportation Research Part B: Methodological, 167,* 32–53.
234. Ghaffarinasab, N., & Kara, B. Y. (2022). A conditional β-mean approach to risk-averse stochastic multiple allocation hub location problems. *Transportation Research Part E: Logistics and Transportation Review, 158,* 102602.
235. Givens, C., & Shortt, R. (1984). A class of Wasserstein metrics for probability distributions. *The Michigan Mathematical Journal, 31,* 231–240.
236. Glover, F., & Laguna, M. (1997). *Tabu search.* Kluwer.

237. Goh, J., & Hall, N. G. (2013). Total cost control in project management via satisficing. *Management Science, 59,* 1354–1372.
238. Goh, J., & Sim, M. (2010). Distributionally robust optimization and its tractable approximations. *Operations Research, 58,* 902–917.
239. Gökalp, E., Gülpınar, N., & Doan, X. V. (2020). Capacity planning for networks of stem-cell donation centers under uncertainty. *Production and Operations Management, 29,* 281–297.
240. Golden, B. L., & Wong, R. T. (1981). Capacitated Arc routing problems. *Networks, 11,* 305–315.
241. Gonçalves, J. N. C., Sameiro Carvalho, M., & Cortez, P. (2020). Operations research models and methods for safety stock determination: A review. *Operations Research Perspectives, 7,* 100164.
242. Gorissen, B. L., Yanıkoğlu, İ, & Den Hertog, D. (2015). A practical guide to robust optimization. *Omega, 53,* 124–137.
243. Goryashko, A. P., & Nemirovski, A. S. (2014). Robust energy cost optimization of water distribution system with uncertain demand. *Automation and Remote Control, 75,* 1754–1769.
244. Gourtani, A., Nguyen, T.-D., & Xu, H. (2020). A distributionally robust optimization approach for two-stage facility location problems. *EURO Journal on Computational Optimization, 8,* 141–172.
245. Gouveia, L., & Saldanha-da-Gama, F. (2006). On the capacitated concentrator location problem: A reformulation by discretization. *Computers & Operations Research, 33,* 1242–1258.
246. Goyal, M., & Netessine, S. (2011). Volume flexibility, product flexibility, or both: The role of demand correlation and product substitution. *Manufacturing & Service Operations Management, 13,* 180–193.
247. Guajardo, J. A., Cohen, M. A., & Netessine, S. (2016). Service competition and product quality in the US automobile industry. *Management Science, 62,* 1860–1877.
248. Guan, Y., Ahmed, S., & Nemhauser, G. L. (2009). Cutting planes for multistage stochastic integer programs. *Operations Research, 57,* 287–298.
249. Gülpınar, N., Pachamanova, D., & Çanakoğlu, E. (2013). Robust strategies for facility location under uncertainty. *European Journal of Operational Research, 225,* 21–35.
250. Gunawardane, G. (1982). Dynamic versions of set covering type public facility location problems. *European Journal of Operational Research, 10,* 190–195.
251. Gundupalli, S. P., Hait, S., & Thakur, A. (2017). A review on automated sorting of source-separated municipal solid waste for recycling. *Waste Management, 60,* 56–74.
252. Guo, H., Li, Y., Shang, J., Gu, M., Huang, Y., & Gong, B. (2017). Learning from class-imbalanced data: Review of methods and applications. *Expert Systems with Applications, 73,* 220–239.
253. Habibi, M. K. K., Allaoui, H., & Goncalves, G. (2018). Collaborative hub location problem under cost uncertainty. *Computers & Industrial Engineering, 124,* 393–410.
254. Hakimi, S. L. (1964). Optimum locations of switching centers and the absolute centers and medians of a graph. *Operations Research, 12,* 450–459.
255. Hakimi, S. L. (1965). Optimum distribution of switching centers in a communication network and some related graph theoretic problems. *Operations Research, 13,* 462–475.
256. Hamacher, H. W., Labbé, M., Nickel, S., & Sonneborn, T. (2004). Adapting polyhedral properties from facility to hub location problems. *Discrete Applied Mathematics, 145,* 104–116.
257. Hanasusanto, G. A., & Kuhn, D. (2018). Conic programming reformulations of two-stage distributionally robust linear programs over Wasserstein balls. *Operations Research, 66,* 849–869.
258. Hanasusanto, G. A., Kuhn, D., Wallace, S. W., & Zymler, S. (2015). Distributionally robust multi-item newsvendor problems with multimodal demand distributions. *Mathematical Programming, 152,* 1–32.
259. Haneveld, W. K., van der Vlerk, M. H., & Romeijnders, W. (2020). *Stochastic programming: Modeling decision problems under uncertainty*. Springer.

260. Hao, Z., He, L., Hu, Z., & Jiang, J. (2020). Robust vehicle pre-allocation with uncertain covariates. *Production and Operations Management, 29,* 955–972.
261. Hassanpour, S. T., Ke, G. Y., Zhao, J., & Tulett, D. M. (2023). Infectious waste management during a pandemic: A stochastic location-routing problem with chance-constrained time windows. *Computers & Industrial Engineering, 177,* 109066.
262. Haugland, D., Ho, S. C., & Laporte, G. (2007). Designing delivery districts for the vehicle routing problem with stochastic demands. *European Journal of Operational Research, 180,* 997–1010.
263. Heckmann, I., Comes, T., & Nickel, S. (2015). A critical review on supply chain risk–definition, measure and modeling. *Omega, 52,* 119–132.
264. Heckmann, I., & Nickel, S. (2019). Location logistics in supply chain management. In G. Laporte, S. Nickel, & F. Saldanha-da-Gama (Eds.), *Location Science* (Ch. 16, 2nd ed., pp. 453–476). Springer.
265. Heckmann, I., Nickel, S., & Saldanha-da-Gama, F. (2023). Facility location and supply chain risk analytics. In H. A. Eiselt & V. Marianov (Eds.), *Uncertainty in facility location models* (pp. 155–181). Springer Nature Switzerland AG.
266. Heitsch, H., & Römisch, W. (2009). Scenario tree modeling for multistage stochastic programs. *Mathematical Programming, 118,* 371–406.
267. Hess, S. W., Weaver, J. B., Siegfeldt, H. J., Whelan, J. N., & Zitlau, P. A. (1965). Nonpartisan political redistricting by computer. *Operations Research, 13,* 998–1006.
268. Hewitt, M., Ortmann, J., & Rei, W. (2022). Decision-based scenario clustering for decision-making under uncertainty. *Annals of Operations Research, 315,* 747–771.
269. Hinojosa, Y., Puerto, J., & Saldanha-da-Gama, F. (2014). A two-stage stochastic transportation problem with fixed handling costs and a priori selection of the distribution channels. *TOP, 22,* 1123–1147.
270. Holguín-Veras, J., Pérez, N., Jaller, M., van Wassenhove, L., & Aros-Vera, F. (2013). On the appropriate objective function for post-disaster humanitarian logistics models. *Journal of Operations Management, 31,* 262–280.
271. Holmberg, K., Rönnqvist, M., & Yuan, D. (1999). An exact algorithm for the capacitated facility location problems with single sourcing. *European Journal of Operational Research, 113,* 544–559.
272. Holmberg, K., & Tuy, H. (1999). A production-transportation problem with stochastic demand and concave production costs. *Mathematical Programming, 85,* 157–179.
273. Homem-de-Mello, T., & Pagnoncelli, B. K. (2016). Risk aversion in multistage stochastic programming: A modeling and algorithmic perspective. *European Journal of Operational Research, 249,* 188–199.
274. Hoornweg, D., & Bhada-Tata, P. (2012). What a waste: A global review of solid waste management. *World Bank Open Repository*. Available at: http://hdl.handle.net/10986/17388
275. Hosseini, S., & Ivanov, D. (2022). A new resilience measure for supply networks with the ripple effect considerations: A Bayesian network approach. *Annals of Operations Research, 319,* 581–607.
276. Hosseini, S., Ivanov, D., & Dolgui, A. (2019). Review of quantitative methods for supply chain resilience analysis. *Transportation Research Part E: Logistics and Transportation Review, 125,* 285–307.
277. Høyland, K., & Wallace, S. W. (2001). Generating scenario trees for multistage decision problems. *Management Science, 47,* 295–307.
278. Hu, J., Chen, Z., & Wang, S. (2022). *Budget-driven multi-period hub location: A robust time series approach.* SSRN 4221971. Available at: https://papers.ssrn.com/sol3/papers.cfm?abstract_id=4221971
279. Hu, Q.-M., Hu, S., Wang, J., & Li, X. (2021). Stochastic single allocation hub location problems with balanced utilization of hub capacities. *Transportation Research Part B: Methodological, 153,* 204–227.

280. Huang, S.-H. (2015). Solving the multi-compartment capacitated location routing problem with pickup-delivery routes and stochastic demands. *Computers & Industrial Engineering, 87,* 104–113.
281. Isii, K. (1963). On the sharpness of chebyshev-type inequalities. *Annals of the Institute of Statistical Mathematics, 14,* 185–197.
282. Ivanov, D., Dolgui, A., Sokolov, B., & Ivanova, M. (2017). Literature review on disruption recovery in the supply chain. *International Journal of Production Research, 55,* 6158–6174.
283. Jabbarzadeh, A., Fahimnia, B., & Sabouhi, F. (2018). Resilient and sustainable supply chain design: Sustainability analysis under disruption risks. *International Journal of Production Research, 56,* 5945–5968.
284. Janschekowitz, M., Taherkhani, G., Alumur, S. A., & Nickel, S. (2023). An alternative approach to address uncertainty in hub location. *OR Spectrum, 45,* 359–393.
285. Javid, A. A., & Azad, N. (2010). Incorporating location, routing and inventory decisions in supply chain network design. *Transportation Research Part E: Logistics and Transportation Review, 46,* 582–597.
286. Jena, S. D., Cordeau, J.-F., & Gendron, B. (2016). Solving a dynamic facility location problem with partial closing and reopening. *Computers & Operations Research, 67,* 143–154.
287. Jena, S. D., Cordeau, J.-F., & Gendron, B. (2017). Lagrangian heuristics for large-scale dynamic facility location with generalized modular capacities. *INFORMS Journal on Computing, 29,* 388–404.
288. Jiang, B., Pei, J., Tao, Y., & Lin, X. (2013). Clustering unertain data based on probability distribution similarity. *IEEE Transactions on Knowledge and Data Engineering, 25,* 751–763.
289. Jiang, N., & Xie, W. (2022). ALSO-X and ALSO-X+: Better convex approximations for chance constrained programs. *Operations Research, 70,* 3581–3600.
290. Jiang, N., & Xie, W. (2023). ALSO-X#: Better convex approximations for distributionally robust chance constrained programs. *arXiv:2302.01737* [math.OC]. Available at: https://doi.org/10.48550/arXiv.2302.01737
291. Joncour, C., Kritter, J., Michel, S., & Schepler, X. (2023). Generalized relax-and-fix heuristic. *Computers & Operations Research, 149,* 106038.
292. Juan, A., Faulin, J., Grasman, S., Rabe, M., & Figueira, G. (2015). A review of simheuristics: Extending metaheuristics to deal with stochastic combinatorial optimization problems. *Operations Research Perspectives, 2,* 62–72.
293. Jucker, J. V., & Carlson, R. C. (1976). The simple plant-location problem under uncertainty. *Operations Research, 24,* 1045–1055.
294. Kalcsics, J., & Ríos-Mercado, R. Z. (2019). Districting problems. In G. Laporte, S. Nickel, & F. Saldanha-da-Gama (Eds.), *Location Science* (Ch. 25, 2nd ed., pp. 705–743). Springer.
295. Kall, P., & Mayer, J. (2005). *Stochastic linear programming*. Springer.
296. Kamalahmadi, M., & Parast, M. (2016). A review of the literature on the principles of enterprise and supply chain resilience: Major findings and directions for future research. *International Journal of Production Economics, 171,* 116–133.
297. Kannan, R., Bayraksan, G., & Luedtke, J. R. (2020). Data-driven sample average approximation with covariate information. *Optimization Online*. Available at: http://www.optimization-online.org/DBHTML/2020/07/7932.html
298. Kannan, R., Bayraksan, G., & Luedtke, J. R. (2020). Residuals-based distributionally robust optimization with covariate information. *arXiv:2012.01088* [math.OC]. Available at: https://doi.org/10.48550/arXiv.2012.01088
299. Kantorovich, L., & Rubinstein, G. (1958). On a space of totally additive functions. *Vestnik Leningrad. Univ, 13,* 52–59.
300. Kara, B., & Rancourt, M.-È. (2019). Location problems in humanitarian supply chains. In G. Laporte, S. Nickel, & F. Saldanha-da-Gama (Eds.), *Location Science* (Ch. 21, 2nd ed.). Springer.
301. Kara, B. Y., & Tansel, B. Ç. (2000). On the single-assignment p-hub center problem. *European Journal of Operational Research, 125,* 648–655.

302. Kargar, K., & Irfan Mahmutoğulları, A. (2022). Risk-averse hub location: Formulation and solution approach. *Computers & Operations Research, 143,* 105760.
303. Kaufman, L., & Rousseeuw, P. J. (2005). *Finding groups in data: An introduction to cluster analysis*. Wiley.
304. Kaya, O., & Ozkok, D. (2020). A blood bank network design problem with integrated facility location, inventory and routing decisions. *Networks and Spatial Economics, 20,* 757–783.
305. Kellerer, H., Pferschy, U., & Pisinger, D. (2004). *The multiple-choice knapsack problem* (pp. 317–347). Springer.
306. Keutchayan, J., Ortmann, J., & Rei, W. (2021). Problem-driven scenario clustering in stochastic optimization. *arXiv:2106.11717* [math.OC]. Available at: https://doi.org/10.48550/arXiv.2106.11717
307. Khoo, H. H., Lim, T. Z., & Tan, R. B. (2010). Food waste conversion options in Singapore: Environmental impacts based on an LCA perspective. *Science of the Total Environment, 408,* 1367–1373.
308. Kim, Y., Oh, I., & Lee, J.-D. (2015). Economic impact assessment of public-private matching fund programs using firm-level data. *The Singapore Economic Review, 60,* 1550060.
309. Kınay, Ö. B., Kara, B. Y., Saldanha-da-Gama, F., & Correia, I. (2018). Modeling the shelter site location problem using chance constraints: A case study for Istanbul. *European Journal of Operational Research, 270,* 132–145.
310. Kınay, Ö. B., Saldanha-da-Gama, F., & Kara, B. Y. (2019). On multi-criteria chance-constrained capacitated single-source discrete facility location problems. *Omega, 83,* 107–122.
311. Kiya, F., & Davoudpour, H. (2012). Stochastic programming approach to re-designing a warehouse network under uncertainty. *Transportation Research Part E: Logistics and Transportation Review, 48,* 919–936.
312. Kleywegt, A. J., Shapiro, A., & Homem-de-Mello, T. (2002). The sample average approximation method for stochastic discrete optimization. *SIAM Journal on Optimization, 12,* 479–502.
313. Klibi, W., Lasalle, F., Martel, A., & Ichoua, S. (2010). The stochastic multiperiod location transportation problem. *Transportation Science, 44,* 221–237.
314. Klibi, W., Martel, A., & Guitouni, A. (2016). The impact of operations anticipations on the quality of stochastic location-allocation models. *Omega, 62,* 19–33.
315. Klose, A., & Drexl, A. (2005). Facility location models for distribution system design. *European Journal of Operational Research, 162,* 4–29.
316. Kochenderfer, M. J. (2015). *Decision making under uncertainty: Theory and application*. The MIT Press.
317. Kochenderfer, M. J., Wheeler, T. A., & Wray, K. H. (2022). *Algorithms for decision making*. The MIT Press.
318. Konur, D., & Geunes, J. (2019). Integrated districting, fleet composition, and inventory planning for a multi-retailer distribution system. *Annals of Operations Research, 273,* 527–559.
319. Kouvelis, P., & Yu, G. (1997). *Robust discrete optimization*. Kluwer.
320. Krokhmal, P. A. (2007). Higher moment coherent risk measures. *Quantitative Finance, 7,* 373–387.
321. Küçükyavuz, S., & Jiang, R. (2022). Chance-constrained optimization under limited distributional information: A review of reformulations based on sampling and distributional robustness. *EURO Journal on Computational Optimization, 10,* 100030.
322. Kuhn, D., Esfahani, P. M., Nguyen, V. A., & Shafieezadeh-Abadeh, S. (2019). Wasserstein distributionally robust optimization: Theory and applications in machine learning. In *Operations research & management science in the age of analytics* (pp. 130–166). INFORMS.
323. Laporte, G., Louveaux, F., & Mercure, H. (1989). Models and exact solutions for a class of stochastic location-routing problems. *European Journal of Operational Research, 39,* 71–78.
324. Laporte, G., & Louveaux, F. V. (1993). The integer L-shaped method for stochastic integer programs with complete recourse. *Operations Research Letters, 13,* 133–142.

325. Laporte, G., Louveaux, F. V., & van Hamme, L. (1994). Exact solution to a location problem with stochastic demands. *Transportation Science, 28,* 95–103.
326. Laporte, G., Nickel, S., & Saldanha-da-Gama, F. (Eds.) (2019). *Location Science* (2nd ed.). Springer.
327. Laporte, G., Nobert, Y., & Taillefer, S. (1988). Solving a family of multi-depot vehicle routing and location-routing problems. *Transportation Science, 22,* 161–172.
328. Larson, R. C. (1975). Approximating the performance of urban emergency service systems. *Operations Research, 23,* 845–868.
329. Lee, S.-H., Choi, K.-I., Osako, M., & Dong, J.-I. (2007). Evaluation of environmental burdens caused by changes of food waste management systems in Seoul, Korea. *Science of the Total Environment, 387,* 42–53.
330. Lei, H., Laporte, G., & Guo, B. (2012). Districting for routing with stochastic customers. *EURO Journal on Transportation and Logistics, 1,* 67–85.
331. Lei, H., Wang, R., & Laporte, G. (2016). Solving a multi-objective dynamic stochastic districting and routing problem with a co-evolutionary algorithm. *Computers & Operations Research, 67,* 12–24.
332. Li, Y., Chen, K., Collignon, S., & Ivanov, D. (2021). Ripple effect in the supply chain network: Forward and backward disruption propagation, network health and firm vulnerability. *European Journal of Operational Research, 291,* 1117–1131.
333. Li, Y., & Huang, G. (2012). Robust interval quadratic programming and its application to waste management under uncertainty. *Environmental Systems Research, 1,* 1–16.
334. Li, Y., & Zobel, C. W. (2020). Exploring supply chain network resilience in the presence of the ripple effect. *International Journal of Production Economics, 228,* 107693.
335. Li, Y. P., Huang, G. H., Nie, X. H., & Nie, S.-L. (2008). A two-stage fuzzy robust integer programming approach for capacity planning of environmental management systems. *European Journal of Operational Research, 189,* 399–420.
336. Li, Z. C., Bing, X., & Fu, X. (2023). A hierarchical hub location model for the integrated design of urban and rural logistics networks under demand uncertainty. *Annals of Operations Research*. https://doi.org/10.1007/s10479-023-05189-6
337. Liebchen, C., Lübbecke, M., Möhring, R., & Stiller,. (2009). The concept of recoverable robustness, linear programming recovery, and railway applications, lncs 5868. In R. Ahuja, R. Mohring, & C. Zaroliagis (Eds.), *Robust and online large-scale optimization* (pp. 1–27). Springer.
338. Lin, C. K. Y. (2009). Stochastic single-source capacitated facility location model with service level requirements. *International Journal of Production Economics, 117,* 439–451.
339. Liu, B. (2015). *Uncertainty theory* (4th ed.). Springer.
340. Liu, F., Chen, Z., & Wang, S. (2023). Globalized distributionally robust counterpart. *INFORMS Journal on Computing, 35,* 1120–1142.
341. Liu, H., Erdogan, A., Lin, R., & Tsao, H.-S.J. (2020). Mathematical models of political districting for more representative governments. *Computers & Industrial Engineering, 140,* 106265.
342. Liu, K., Li, Q., & Zhang, Z.-H. (2019). Distributionally robust optimization of an emergency medical service station location and sizing problem with joint chance constraints. *Transportation Research Part B: Methodological, 119,* 79–101.
343. Liu, M., Grigas, P., Liu, H., & Shen, Z.-J. M. (2023). Active learning in the predict-then-optimize framework: A margin-based approach. *arXiv:2305.06584* [cs.LG]. Available at: https://doi.org/10.48550/arXiv.2305.06584
344. Liu, T., & Li, G. (2021). Robust recycling facility location with clustering. *Computers & Operations Research, 136,* 105466.
345. Liu, T., Saldanha-da-Gama, F., Wang, S., & Mao, Y. (2022). Robust stochastic facility location: Sensitivity analysis and exact solution. *INFORMS Journal on Computing, 34,* 2776–2803.

346. Liu, Y., Li, Z., Liu, J., & Patel, H. (2016). A double standard model for allocating limited emergency medical service vehicle resources ensuring service reliability. *Transportation Research Part C: Emerging Technologies, 69,* 120–133.
347. Liu, Y., Xu, H., Yang, S.-J.S., & Zhang, J. (2018). Distributionally robust equilibrium for continuous games: Nash and stackelberg models. *European Journal of Operational Research, 265,* 631–643.
348. Loizou, N. (2015). Distributionally robust game theory. *arXiv:1512.03253* [cs.GT]. Available at: https://doi.org/10.48550/arXiv.1512.03253
349. Loke, G. G., Tang, Q., & Xiao, Y. (2022). *Decision-driven regularization: A blended model for predict-then-optimize.* SSRN 3623006. Available at: https://papers.ssrn.com/sol3/papers.cfm?abstract_id=3623006
350. Long, D. Z., Sim, M., & Zhou, M. (2023). Robust satisficing. *Operations Research, 71,* 61–82.
351. Longo, H., de Aragão, M. P., & Uchoa, E. (2006). Solving capacitated arc routing problems using a transformation to the CVRP. *Computers & Operations Research, 33,* 1823–1837.
352. Louveaux, F. V. (1986). Discrete stochastic location models. *Annals of Operations Research, 6,* 23–34.
353. Louveaux, F. V. (1993). Stochastic location analysis. *Location Science, 1,* 127–154.
354. Louveaux, F. V., & Peeters, D. (1992). A dual-based procedure for stochastic facility location. *Operations Research, 40,* 564–573.
355. Lu, C.-C. (2013). Robust weighted vertex *p*-center model considering uncertain data: An application to emergency management. *European Journal of Operational Research, 230,* 113–121.
356. Lu, C.-C., & Sheu, J.-B. (2013). Robust vertex *p*-center model for locating urgent relief distribution centers. *Computers & Operations Research, 40,* 2128–2137.
357. Luedtke, J. (2016). Stochastic integer programming. *Course Presented at the 14th International Conference on Stochastic Programming*. Available at: https://www.stoprog.org/sites/default/files/tutorials/SP16/jimluedtkestochip2016.pdf
358. Luedtke, J., & Ahmed, S. (2008). A sample approximation approach for optimization with probabilistic constraints. *SIAM Journal on Optimization, 19,* 674–699.
359. Luo, F. (2021). A distributionally-robust service center location problem with decision dependent demand induced from a maximum attraction principle. *arXiv:2011.12514* [math.OC]. Available at: https://doi.org/10.48550/arXiv.2011.12514
360. Ma, S., Fildes, R., & Huang, T. (2016). Demand forecasting with high dimensional data: The case of SKU retail sales forecasting with intra-and inter-category promotional information. *European Journal of Operational Research, 249,* 245–257.
361. Mahmoud, O. (2017). The temporal dimension of risk. *The Journal of Risk, 19,* 57–83.
362. MahmoumGonbadi, A., Genovese, A., & Sgalambro, A. (2021). Closed-loop supply chain design for the transition towards a circular economy: A systematic literature review of methods, applications and current gaps. *Journal of Cleaner Production, 323,* 129101.
363. Malik, N. K. A., Abdullah, S. H., & Abd Manaf, L. (2015). Community participation on solid waste segregation through recycling programmes in putrajaya. *Procedia Environmental Sciences, 30,* 10–14.
364. Mara, S. T. W., Kuo, R. J., & Asih, A. M. S. (2021). Location-routing problem: a classification of recent research. *International Transactions in Operational Research, 28,* 2941–2983.
365. Marianov, V., & Méndez-Vogel, G. (2023). Customer-related uncertainties in facility location problems. In H. A. Eiselt & V. Marianov (Eds.), *Uncertainty in facility location models* (pp. 53–77). Springer Nature Switzerland AG.
366. Marianov, V., & ReVelle, C. S. (1994). The queuing probabilistic location set covering problemd and some extensions. *Socio-Economic Planning Sciences, 28,* 167–178.
367. Marín, A., Martínez-Merino, L. I., Rodríguez-Chía, A. M., & Saldanha-da-Gama, F. (2018). Multi-period stochastic covering location problems: Modeling framework and solution approach. *European Journal of Operational Research, 268,* 432–449.

368. Marín, A., & Pelegrín, B. (1998). The return plant location problem: Modelling and resolution. *European Journal of Operational Research, 104,* 375–392.
369. Marinakis, Y. (2015). An improved particle swarm optimization algorithm for the capacitated location routing problem and for the location routing problem with stochastic demands. *Applied Soft Computing, 37,* 680–701.
370. Markowitz, H. (1952). Portfolio selection. *The Journal of Finance, 7,* 77–91.
371. Marques, M. C., & Dias, J. M. (2018). Dynamic location problem under uncertainty with a regret-based measure of robustness. *International Transactions in Operational Research, 25,* 1361–1381.
372. Martínez-de-Albéniz, V., & Belkaid, A. (2021). Here comes the sun: Fashion goods retailing under weather fluctuations. *European Journal of Operational Research, 294,* 820–830.
373. Marín, A., & Pelegrín, M. (2019). p-median problems. In G. Laporte, S. Nickel, & F. Saldanha-da-Gama (Eds.), *Location Science* (Ch. 2, 2nd ed., pp. 25–50). Springer.
374. Mastrogiacomo, L., Barravecchia, F., & Franceschini, F. (2019). A worldwide survey on manufacturing servitization. *The International Journal of Advanced Manufacturing Technology, 103,* 3927–3942.
375. McCrea, M., Tan, T. K., Ting, H.-H., & Zuo, X. (2009). *A cost-benefit analysis of different waste-to-energy technologies for the management of municipal solid waste in Singapore*. The University of Chicago. Available at: https://bigproblems.uchicago.edu/team6.pdf
376. Mehrjerdi, Y. Z., & Nadizadeh, A. (2013). Using greedy clustering method to solve capacitated location-routing problem with fuzzy demands. *European Journal of Operational Research, 229,* 75–84.
377. Melo, M. T., Nickel, S., & Saldanha da Gama, F. (2006). Dynamic multi-commodity capacitated facility location: a mathematical modeling framework for strategic supply chain planning. *Computers & Operations Research, 33,* 181–208.
378. Melo, M. T., Nickel, S., & Saldanha-da-Gama, F. (2009). Facility location and supply chain management– A review. *European Journal of Operational Research, 196,* 401–412.
379. Meraklı, M., & Yaman, H. (2016). Robust intermodal hub location under polyhedral demand uncertainty. *Transportation Research Part B: Methodological, 86,* 66–85.
380. Meraklı, M., & Yaman, H. (2017). A capacitated hub location problem under hose demand uncertainty. *Computers & Operations Research, 88,* 58–70.
381. Miranda, P. A., & Garrido, R. A. (2006). A simultaneous inventory control and facility location model with stochastic capacity constraints. *Networks and Spatial Economics, 6,* 39–53.
382. Mirchandani, P. B., & Francis, R. L. (Eds.) (1990). *Discrete location theory*. Wiley.
383. Mirchandani, P. B., Oudjit, A., & Wong, R. T. (1985). 'Multidimensional' extensions and a nested dual approach for the m-median problem. *European Journal of Operational Research, 21,* 121–137.
384. Mirzaei-khafri, S., Bashiri, M., Soltani, R., & Khalilzadeh, M. (2020). A robust optimization model for a location-arc routing problem with demand uncertainty. *International Journal of Industrial Engineering, 27,* 288–307.
385. Mohajerin-Esfahani, P., & Kuhn, D. (2017). Data-driven distributionally robust optimization using the Wasserstein metric: Performance guarantees and tractable reformulations. *Mathematical Programming, 171,* 11–166.
386. Mohamed, I. B., Klibi, W., Sadykov, R., Şen, H., & Vanderbeck, F. (2023). The two-echelon stochastic multi-period capacitated location-routing problem. *European Journal of Operational Research, 306,* 645–667.
387. Mohammadi, M., Torabi, S., & Tavakkoli-Moghaddam, R. (2014). Sustainable hub location under mixed uncertainty. *Transportation Research Part E: Logistics and Transportation Review, 62,* 89–115.
388. Mohammadi, S., Darestani, S. A., Vahdani, B., & Alinezhad, A. (2020). A robust neutrosophic fuzzy-based approach to integrate reliable facility location and routing decisions for disaster relief under fairness and aftershocks concerns. *Computers & Industrial Engineering, 148,* 106734.

389. Moshref-Javadi, M., & Lee, S. (2016). The latency location-routing problem. *European Journal of Operational Research, 255,* 604–619.
390. Mostafayi Darmian, S., Fattahi, M., & Keyvanshokooh, E. (2021). An optimization-based approach for the healthcare districting under uncertainty. *Computers & Operations Research, 135,* 105425.
391. Mourão, M. C., Nunes, A. C., & Prins, C. (2009). Heuristic methods for the sectoring arc routing problem. *European Journal of Operational Research, 196,* 856–868.
392. Murray, A. (2023). Sources of uncertainty in location analysis. In H. A. Eiselt & V. Marianov (Eds.), *Uncertainty in facility location problems* (pp. 3–24). Springer Nature Switzerland.
393. Nadizadeh, A., & Nasab, H. H. (2014). Solving the dynamic capacitated location-routing problem with fuzzy demands by hybrid heuristic algorithm. *European Journal of Operational Research, 238,* 458–470.
394. National Environment Agency (2018). *Waste and recycling statistics for 2017.* Available at: http://www.nea.gov.sg/energy-waste/waste-management/waste-statistics-and-overall-recycling
395. National Environment Agency (2019). *3R programmes and resources.* Available at: https://www.nea.gov.sg/our-services/waste-management/3r-programmes-and-resources/types-of-recyclables-and-recycling-processes
396. Nekooghadirli, N., Tavakkoli-Moghaddam, R., Ghezavati, V. R., & Javanmard, S. (2014). Solving a new bi-objective location-routing-inventory problem in a distribution network by meta-heuristics. *Computers & Industrial Engineering, 76,* 204–221.
397. Nemirovski, A., & Shapiro, A. (2006). Scenario approximations of chance constraints. In G. Calafiore & F. Dabbene (Eds.), *Probabilistic and randomized methods for design under uncertainty* (pp. 3–47). Springer.
398. Nemirovski, A., & Shapiro, A. (2007). Convex approximations of chance constrained programs. *SIAM Journal on Optimization, 17,* 969–996.
399. Netessine, S., Dobson, G., & Shumsky, R. A. (2002). Flexible service capacity: Optimal investment and the impact of demand correlation. *Operations Research, 50,* 375–388.
400. Neyshabouri, S., & Berg, B. P. (2017). Two-stage robust optimization approach to elective surgery and downstream capacity planning. *European Journal of Operational Research, 260,* 21–40.
401. Ngueveu, S. U., Prins, C., & Wolfler Calvo, R. (2010). An effective memetic algorithm for the cumulative capacitated vehicle routing problem. *Computers & Operations Research, 37,* 1877–1885.
402. Nickel, S., & Puerto, J. (2005). *Location theory: A unified approach.* Springer.
403. Nickel, S., & Saldanha-da-Gama, F. (2019). Multi-period facility location. In G. Laporte, S. Nickel, & F. Saldanha-da-Gama (Eds.), *Location Science* (Ch. 11, 2nd ed., pp. 303–326). Springer.
404. Nickel, S., Saldanha-da-Gama, F., & Ziegler, H.-P. (2010). *Stochastic programming approaches for risk aware supply chain network design problems.* Technical Report, Fraunhofer ITWM, Kaiserslautern, Germany. Fraunhofer ITWM, Report 181. Available at: https://kluedo.ub.rptu.de/frontdoor/deliver/index/docId/2204/file/bericht_181.pdf
405. Nickel, S., Saldanha-da-Gama, F., & Ziegler, H.-P. (2012). A multi-stage stochastic supply network design problem with financial decisions and risk management. *Omega, 40,* 511–524.
406. Nikoofal, M. E., & Sadjadi, S. J. (2010). A robust optimization model for p-median problem with uncertain edge lengths. *The International Journal of Advanced Manufacturing Technology, 50,* 391–397.
407. Notz, P. M., & Pibernik, R. (2022). Prescriptive analytics for flexible capacity management. *Management Science, 68,* 1756–1775.
408. Noyan, N. (2012). Risk-averse two-stage stochastic programming with an application to disaster management. *Computers & Operations Research, 39,* 541–559.
409. Noyan, N., Rudolf, G., & Lejeune, M. (2022). Distributionally robust optimization under a decision-dependent ambiguity set with applications to machine scheduling and humanitarian logistics. *INFORMS Journal on Computing, 32,* 729–751.

410. Nozick, L. K., & Turnquist, M. A. (2001). Inventory, transportation, service quality and the location of distribution centers. *European Journal of Operational Research, 129,* 362–371.
411. Nozick, L. K., & Turnquist, M. A. (2001). A two-echelon inventory allocation and distribution center location analysis. *Transportation Research Part E: Logistics and Transportation Review, 37,* 425–441.
412. Nucamendi-Guillén, S., Martínez-Salazar, I., Khodaparasti, S., & Bruni, M. E. (2022). New formulations and solution approaches for the latency location routing problem. *Computers & Operations Research, 143,* 105767.
413. Oakley, J. (2017). *SHELF: Tools to support the sheffield elicitation framework (R package).* Available at: https://github.com/OakleyJ/SHELF
414. O'Hagan, A. (1998). Eliciting expert beliefs in substantial practical applications. *Journal of the Royal Statistical Society: Series D (The Statistician), 47,* 21–35.
415. O'Kelly, M. E. (1986). The location of interacting hub facilities. *Transportation Science, 20,* 92–106.
416. O'Kelly, M. E. (1987). A quadratic integer program for the location of interacting hub facilities. *European Journal of Operational Research, 32,* 393–404.
417. Omran, A., Mahmoud, A., Abdul, A. H., & Robinson, G. M. (2009). Investigating households attitude toward recycling of solid waste in malaysia: A case study. *International Journal of Environmental Research, 3,* 275–288.
418. Oyola, J., Arntzen, H., & Woodruff, D. L. (2017). The stochastic vehicle routing problem, a literature review, part II: Solution methods. *EURO Journal on Transportation and Logistics, 6,* 349–388.
419. Oyola, J., Arntzen, H., & Woodruff, D. L. (2018). The stochastic vehicle routing problem, a literature review, part I: Models. *EURO Journal on Transportation and Logistics, 7,* 193–221.
420. Ozsen, L., Coullard, C. R., & Daskin, M. S. (2008). Capacitated warehouse location model with risk pooling. *Naval Research Logistics, 55,* 295–312.
421. Pagnoncelli, B. K., Ahmed, S., & Shapiro, A. (2009). Sample average approximation method for chance constrained programming: Theory and applications. *Journal of Optimization Theory and Applications, 142,* 399–416.
422. Pantuza, G., Jr., & de Souza, M. C. (2022). Formulations and a Lagrangian relaxation approach for the prize collecting traveling salesman problem. *International Transactions in Operational Research, 29,* 729–759.
423. Parragh, S. N., Tricoire, F., & Gutjahr, W. J. (2022). A branch-and-Benders-cut algorithm for a bi-objective stochastic facility location problem. *OR Spectrum, 44,* 419–459.
424. Paul, J. A., & Wang, X. (2019). Robust location-allocation network design for earthquake preparedness. *Transportation Research Part B: Methodological, 119,* 139–155.
425. Peiró, J., Corberán, A., Martí, R., & Saldanha-da Gama, F. (2019). Heuristic solutions for a class of stochastic uncapacitated p-hub median problems. *Transportation Science, 53,* 1126–1149.
426. Pekel, E., & Kara, S. S. (2019). Solving fuzzy capacitated location routing problem using hybrid variable neighborhood search and evolutionary local search. *Applied Soft Computing, 83,* 105665.
427. Perakis, G., Sim, M., Tang, Q., & Xiong, P. (2022). Robust pricing and production with information partitioning and adaptation. *Management Science*. https://doi.org/10.1287/mnsc.2022.4446
428. Pflug, G. C., & Pichler, A. (2014). *Multistage stochastic optimization*. Springer.
429. Phillips, D. A. (2000). Implementing the market approach to enterprise support: A comparative evaluation of matching grant schemes. *The World Bank, Policy Research Working Paper Series 2589*. Available at: https://ideas.repec.org/p/wbk/wbrwps/2589.html
430. Pichler, A., & Shapiro, A. (2021). Mathematical foundations of distributionally robust multistage optimization. *SIAM Journal on Optimization, 31,* 3044–3067.
431. Pires Ribeiro, J., & Barbosa-Póvoa, A. (2018). Supply chain resilience: Definitions and quantitative modelling approaches–A literature review. *Computers & Industrial Engineering, 115,* 109–122.

432. Pomes, A. (2022). Multi-period stochastic districting: An optimization model and heuristic algorithm. EURO Summer Institute on Location Science, 11–24 June 2022, Edinburgh, Scotland, UK.
433. Postek, K., Den Hertog, D., Kind, J., & Pustjens, C. (2019). Adjustable robust strategies for flood protection. *Omega, 82,* 142–154.
434. Powell, W. B. (2022). *Reinforcement learning and stochastic optimization: A unified framework for sequential decisions*. Princeton University Press.
435. Prékopa, A. (1990). A dual method for the solution of a one-stage stochastic programming problem with random RHS obeying a discrete probability distribution. *ZOR—Methods and Models of Operations Research*, 34:441–461.
436. Prékopa, A. (2003). Probabilistic programming. In *Stochastic programming*. Handbooks in Operations Research and Management Science (Vol. 10, pp. 267–351). Elsevier.
437. Prins, C., Prodhon, C., Ruiz, A., Soriano, P., & Wolfler Calvo, R. (2006). Solving the capacitated location-routing problem by a GRASP complemented by a learning process and a path relinking. *4OR*, 4:221–238.
438. Prins, C., Prodhon, C., Ruiz, A., Soriano, P., & Wolfler Calvo, R. (2007). Solving the capacitated location-routing problem by a cooperative lagrangean relaxation-granular tabu search heuristic. *Transportation Science, 41,* 470–483.
439. Qi, M., Cao, Y., & Shen, Z.-J.M. (2022). Distributionally robust conditional quantile prediction with fixed design. *Management Science, 68,* 1639–1658.
440. Qu, S., Meng, D., Zhou, Y., & Dai, Y. (2017). Distributionally robust games with an application to supply chain. *Journal of Intelligent & Fuzzy Systems, 33,* 2749–2762.
441. Quintero-Araujo, C. L., Guimarans, D., & Juan, A. A. (2021). A simheuristic algorithm for the capacitated location routing problem with stochastic demands. *Journal of Simulation, 15,* 217–234.
442. Rabbani, M., Heidari, R., & Yazdanparast, R. (2019). A stochastic multi-period industrial hazardous waste location-routing problem: Integrating NSGA-II and Monte Carlo simulation. *European Journal of Operational Research, 272,* 945–961.
443. Raddats, C., Baines, T., Burton, J., Story, V. M., & Zolkiewski, J. (2016). Motivations for servitization: The impact of product complexity. *International Journal of Operations & Production Management, 36,* 572–591.
444. Rahimian, H., & Mehrotra, S. (2019). Distributionally robust optimization: A review. *arXiv:1908.05659* [math.OC]. Available at: https://doi.org/10.48550/arXiv.1908.05659
445. Rajesh Chauhan, D., Unnikrishnan, A., Figliozzi, M. A., & Boyles, S. D. (2023). Robust multi-period maximum coverage drone facility location problem considering coverage reliability. *Transportation Research Record, 2677,* 98–114.
446. Ramachandra, A., Rujeerapaiboon, N., & Sim, M. (2021). Robust conic satisficing. *arXiv:2107.06714* [math.OC]. Available at: https://doi.org/10.48550/arXiv.2107.06714
447. Ren, S., Chan, H.-L., & Siqin, T. (2020). Demand forecasting in retail operations for fashionable products: methods, practices, and real case study. *Annals of Operations Research, 291,* 761–777.
448. Reschovsky, J. D., & Stone, S. E. (2017). Market incentives to encourage household waste recycling: Paying for what you throw away. In T. C. Kinnaman (Ed.), *The economics of residential solid waste management* (pp. 233–252). Routledge.
449. Resende, M. G., & Werneck, R. F. (2004). A hybrid heuristic for the p-median problem. *Journal of Heuristics, 10,* 59–88.
450. ReVelle, C., & Marianov, V. (1991). A probabilistic fleet model with individual vehicle reliability requirements. *European Journal of Operational Research, 53,* 93–105.
451. ReVelle, C. S., & Hogan, K. (1988). A reliability-constrained siting model with local estimates of busy fractions. *Environment and Planning B: Planning and Design, 15,* 143–152.
452. ReVelle, C. S., & Hogan, K. (1989). The maximum availability location problem. *Transportation Science, 23,* 192–200.

453. ReVelle, C. S., & Hogan, K. (1989). The maximum reliability location problem and α-reliable p-center problem: Derivatives of the probabilistic location set covering problem. *Annals of Operations Research, 18,* 155–174.
454. Ricca, F., Scozzari, A., & Simeone, B. (2013). Political districting: from classical models to recent approaches. *Annals of Operations Research, 204,* 271–299.
455. Ríos-Mercado, R. Z. (Ed.) (2019). *Optimal districting and territory design*. Springer.
456. Ríos-Mercado, R. Z., Álvarez-Socarrás, A. M., Castrillón, A., & López-Locés, M. C. (2021). A location-allocation-improvement heuristic for districting with multiple-activity balancing constraints and p-median-based dispersion minimization. *Computers & Operations Research, 126,* 105106.
457. Ríos-Mercado, R. Z., & Bard, J. F. (2019). An exact algorithm for designing optimal districts in the collection of waste electric and electronic equipment through an improved reformulation. *European Journal of Operational Research, 276,* 259–271.
458. Rockafellar, R. T. (1997). *Convex analysis*. Princeton University Press.
459. Rockafellar, R. T., & Uryasev, S. (2002). Conditional value-at-risk for general loss functions. *Journal of Banking & Finance, 26,* 1443–1471.
460. Rockafellar, R. T., & Wets, R.J.-B. (1991). Scenario and policy aggregation in optimisation under uncertainty. *Mathematics of Operations Research, 16,* 119–147.
461. Rossi, R. (2021). *Inventory analytics*. Open Book Publishers.
462. Rostami, B., Errico, F., & Lodi, A. (2023). A convex reformulation and an outer approximation for a large class of binary quadratic programs. *Operations Research, 71,* 471–486.
463. Rostami, B., Kämmerling, N., Naoum-Sawaya, J., Buchheim, C., & Clausen, U. (2021). Stochastic single-allocation hub location. *European Journal of Operational Research, 289,* 1087–1106.
464. Roundy, R. (1985). 98%-effective integer-ratio lot-sizing for one-warehouse multi-retailer systems. *Management Science, 31,* 1416–1430.
465. Ríos-Mercado, R. Z., & López-Pérez, J. F. (2013). Commercial territory design planning with realignment and disjoint assignment requirements. *Omega, 41,* 525–535.
466. Sadeghi, M., Jolai, F., Tavakkoli-Moghaddam, R., & Rahimi, Y. (2015). A new stochastic approach for a reliable p-hub covering location problem. *Computers & Industrial Engineering, 90,* 371–380.
467. Sadka, E. (2007). Public-private partnerships—a public economics perspective. *CESifo Economic Studies, 53,* 466–490.
468. Saif, A., & Delage, E. (2021). Data-driven distributionally robust capacitated facility location problem. *European Journal of Operational Research, 291,* 995–1007.
469. Salazar-Aguilar, M. A., Ríos-Mercado, R. Z., & Cabrera-Ríos, M. (2011). New models for commercial territory design. *Networks and Spatial Economics, 11,* 487–507.
470. Saldanha-da-Gama, F. (2022). Facility location in logistics and transportation: An enduring relationship. *Transportation Research Part E: Logistics and Transportation Review, 160,* 102903.
471. Samanta, M., & Welsh, A. H. (2013). Bootstrapping for highly unbalanced clustered data. *Computational Statistics & Data Analysis, 59,* 70–81.
472. Samantha, B. (2016). *Food waste raises a stink for recycling*. Available at: https://www.asiaone.com/food-waste-raises-stink-recycling
473. Sandra, C.-L. (1994). *Private sector participation in municipal solid waste services in developing countries*. The World Bank, Urban Management Programme Discussion Paper 13. Available at: https://elibrary.worldbank.org/doi/abs/10.1596/0-8213-2825-5
474. Sarykalin, S., Serraino, G., & Uryasev, S. (2014). Value-at-risk vs. conditional value-at-risk in risk management and optimization. *INFORMS TutORials in operations research* (pp. 270–294).
475. Sauvey, C., Melo, T., & Correia, I. (2020). Heuristics for a multi-period facility location problem with delayed demand satisfaction. *Computers & Industrial Engineering, 139,* 106171.

476. Saxena, A., Goyal, V., & Lejeune, M. A. (2010). MIP reformulations of the probabilistic set covering problem. *Mathematical Programming, 121,* 1–31.
477. Sazvar, Z., Tafakkori, K., Oladzad, N., & Nayeri, S. (2021). A capacity planning approach for sustainable-resilient supply chain network design under uncertainty: A case study of vaccine supply chain. *Computers & Industrial Engineering, 159,* 107406.
478. Scaparra, M. P., & Church, R. (2012). Protecting supply systems to mitigate potential disaster: A model to fortify capacitated facilities. *International Regional Science Review, 35,* 188–210.
479. Scaparra, M. P., & Church, R. L. (2019). Location problems under disaster events. In G. Laporte, S. Nickel, & F. Saldanha-da-Gama (Eds.), *Location Science* (Ch. 22, 2nd ed., pp. 631–656). Springer.
480. Scarf, H. (1958). A min-max solution of an inventory problem. In K. J. Arrow, S. Karlin, & H. Scarf (Eds.), *Studies in the mathematical theory of inventory and production* (pp. 201–209). Stanford University Press.
481. Schiffer, M., & Walther, G. (2018). Strategic planning of electric logistics fleet networks: A robust location-routing approach. *Omega, 80,* 31–42.
482. Schilling, D., Elzinga, D. J., Cohon, J., Church, R., & ReVelle, C. (1979). The team/fleet models for simultaneous facility and equipment siting. *Transportation Science, 13,* 163–175.
483. Schmidt, J. H., Holm, P., Merrild, A., & Christensen, P. (2007). Life cycle assessment of the waste hierarchy-a Danish case study on waste paper. *Waste Management, 27,* 1519–1530.
484. Schubert, E., Koos, A., & Emrich, T. (2015). A framework for clustering uncertain data. *In Proceedings of the VLDB Endowment, 8,* 1976–1979.
485. Serra, D., & Marianov, V. (1998). The *p*-median problem in a changing network: The case of Barcelona. *Location Science, 6,* 383–394.
486. Serra, D., Ratick, S., & ReVelle, C. S. (1996). The maximum capture problem with uncertainty. *Environment and Planning B: Urban Analytics and City Science, 23,* 49–59.
487. Shafieezadeh-Abadeh, S., Kuhn, D., & Esfahani, P. M. (2019). Regularization via mass transportation. *Journal of Machine Learning Research, 20,* 1–68.
488. Shahabi, M., & Unnikrishnan, A. (2014). Robust hub network design problem. *Transportation Research Part E: Logistics and Transportation Review, 70,* 356–373.
489. Shang, X., Zhang, G., Jia, B., & Almanaseer, M. (2022). The healthcare supply location-inventory-routing problem: A robust approach. *Transportation Research Part E: Logistics and Transportation Review, 158,* 102588.
490. Shapiro, A. (2001). On duality theory of conic linear problems. In M. A. Goberna & M. A. López (Eds.), *Semi-infinite programming. Nonconvex optimization and its applications* (Vol. 57, pp. 135–165). Springer.
491. Shapiro, A. (2021). Tutorial on risk neutral, distributionally robust and risk averse multistage stochastic programming. *European Journal of Operational Research, 288,* 1–13.
492. Shapiro, A., Dentcheva, D., & Ruszczyński, A. (2021). *Lectures on stochastic programming: Modeling and theory.* SIAM.
493. Shapiro, J. F. (2004). Challenges of strategic supply chain planning and modeling. *Computers & Chemical Engineering*, 28:855–861. FOCAPO 2003 Special Issue.
494. Sharafi, A., Taleizadeh, A. A., & Amalnick, M. S. (2018). Fair allocation in financial disputes between public-private partnership stakeholders using game theory. *Service Science, 10,* 1–11.
495. Sharpe, W. F. (1994). The Sharpe ratio. *Journal of Portfolio Management, 21,* 49–48.
496. Shehadeh, K. S. (2023). Distributionally robust optimization approaches for a stochastic mobile facility fleet sizing, routing, and scheduling problem. *Transportation Science, 57,* 197–229.
497. Shehadeh, K. S., & Sanci, E. (2021). Distributionally robust facility location with bimodal random demand. *Computers & Operations Research, 134,* 105257.
498. Shen, H., & Jiang, R. (2021). Convex chance-constrained programs with Wasserstein ambiguity. *arXiv:2111.02486* [math.OC]. Available at: https://doi.org/10.48550/arXiv.2111.02486

499. Shen, H., Xue, M., & Shen, Z.-J. M. (2023). *Data-driven reliable facility location design*. SSRN 4525872. Available at: https://papers.ssrn.com/sol3/papers.cfm?abstract_id=4525872
500. Shen, S. (2013). Optimizing designs and operations of a single network or multiple interdependent infrastructures under stochastic arc disruption. *Computers & Operations Research, 40,* 2677–2688.
501. Shen, Z.-J.M., Coullard, C. R., & Daskin, M. S. (2003). A joint location inventory model. *Transportation Science, 37,* 40–55.
502. Shen, Z.-J.M., & Qi, L. (2007). Incorporating inventory and routing costs in strategic location models. *European Journal of Operational Research, 179,* 372–389.
503. Shen, Z.-J.M., Zhan, R. L., & Zhang, J. (2011). The reliable facility location problem: Formulations, heuristics, and approximation algorithms. *INFORMS Journal on Computing, 23,* 470–482.
504. Shirabe, T. (2009). Districting modeling with exact contiguity constraints. *Environment and Planning B: Planning and Design, 36,* 1053–1066.
505. Shu, J., Teo, C.-P., & Shen, Z.-J.M. (2005). Stochastic transportation-inventory network design problem. *Operations Research, 53,* 48–60.
506. Sim, M., Tang, Q., Zhou, M., & Zhu, T. (2021). *The analytics of robust satisficing: Predict, optimize, satisfice, then fortify*. SSRN 3829562. Available at: https://papers.ssrn.com/sol3/papers.cfm?abstract_id=3829562
507. Sim, T., Lowe, T. J., & Thomas, B. W. (2009). The stochastic p-hub center problem with service-level constraints. *Computers & Operations Research, 36,* 3166–3177.
508. Skorin-Kapov, D., Skorin-Kapov, J., & O'Kelly, M. (1996). Tight linear programming relaxations of uncapacitated p-hub median problems. *European Journal of Operational Research, 73,* 501–508.
509. Snyder, L., & Daskin, M. S. (2006). Stochastic p-robust location problems. *IIE Transactions, 38,* 971–985.
510. Snyder, L. V. (2006). Facility location under uncertainty. *IIE Transactions, 38,* 537–554.
511. Snyder, L. V., & Shen, Z.-J.M. (2019). *Fundamentals of supply chain theory*. Wiley.
512. Solomon, M. M. (1987). Algorithms for the vehicle routing and scheduling problems with time window constraints. *Operations Research, 35,* 254–265.
513. Taha, H. A. (1973). Concave minimization over a convex polyhedron. *Naval Research Logistics Quarterly, 20,* 533–548.
514. Taherkhani, G., & Alumur, S. A. (2023). Hub location models under uncertainty. In H. A. Eiselt & V. Marianov (Eds.), *Uncertainty in facility location models* (pp. 337–354). Springer Nature Switzerland AG.
515. Taherkhani, G., Alumur, S. A., & Hosseini, M. (2020). Benders decomposition for the profit maximizing capacitated hub location problem with multiple demand classes. *Transportation Science, 54,* 1446–1470.
516. Taherkhani, G., Alumur, S. A., & Hosseini, M. (2021). Robust stochastic models for profit-maximizing hub location problems. *Transportation Science, 55,* 1322–1350.
517. Tai, J., Zhang, W., Che, Y., & Feng, D. (2011). Municipal solid waste source-separated collection in China: A comparative analysis. *Waste Management, 31,* 1673–1682.
518. Tavares-Pereira, F., Figueira, J. R., Mousseau, V., & Roy, B. (2007). Multiple criteria districting problems. *Annals of Operations Research, 154,* 69–92.
519. Teitz, M. B., & Bart, P. (1968). Heuristic methods for estimating the generalized vertex median of a weighted graph. *Operations Research, 16,* 955–961.
520. Teo, C.-P., Ou, J., & Goh, M. (2001). Impact on inventory costs with consolidation of distribution centers. *IIE Transactions, 33,* 99–110.
521. Teo, C.-P., & Shu, J. (2004). Warehouse-retailer network design problem. *Operations Research, 52,* 396–408.
522. Teymourifar, A. (2022). Simulation-based optimization for resectorization in healthcare systems. *Journal of Simulation*. https://doi.org/10.1080/17477778.2022.2152395
523. The Straits Times (2011). *Food waste recycler folds*. Available at: http://www.eco-business.com/news/food-waste-recycler-folds/

524. Thomas, A., & Kopczak, L. R. (2005). From logistics to supply chain management: the path forward in the humanitarian sector. *Fritz Institute, White paper*. Available at: http://www.fritzinstitute.org/PDFs/WhitePaper/EnablingDisasterResponse.pdf
525. Toregas, C., Swain, R., ReVelle, C. S., & Bergman, L. (1971). The location of emergency service facilities. *Operations Research, 19,* 1363–1373.
526. Tran, T. H., O'Hanley, J. R., & Scaparra, M. P. (2017). Reliable hub network design: Formulation and solution techniques. *Transportation Science, 51,* 358–375.
527. Unnikrishnan, A., Valsaraj, V., & Waller, S. T. (2009). Stochastic and dynamic shipper carrier network design problem. *Networks and Spatial Economics, 9,* 525–550.
528. Uryasev, S. (2000). *Probabilistic constrained optimization: Methodology and applications* (Vol. 49). Springer Science & Business Media.
529. Vajda, S. (1972). *Probabilistic programming*. Academic.
530. Van Engeland, J., Beliën, J., De Boeck, L., & De Jaeger, S. (2020). Literature review: Strategic network optimization models in waste reverse supply chains. *Omega, 91,* 102012.
531. Vatsa, A. K., & Jayaswal, S. (2016). A new formulation and Benders decomposition for the multi-period maximal covering facility location problem with server uncertainty. *European Journal of Operational Research, 251,* 404–418.
532. Vatsa, A. K., & Jayaswal, S. (2021). Capacitated multi-period maximal covering location problem with server uncertainty. *European Journal of Operational Research, 289,* 1107–1126.
533. Wagner, M. R., Bhadury, J., & Peng, S. (2009). Risk management in uncapacitated facility location models with random demands. *Computers & Operations Research, 36,* 1002–1011.
534. Wang, K., Chen, S., Jiang, Z., Zhou, W., & Geng, N. (2021). Capacity allocation of an integrated production and service system. *Production and Operations Management, 30,* 2765–2781.
535. Wang, S., Chen, Z., & Liu, T. (2020). Distributionally robust hub location. *Transportation Science, 54,* 1189–1210.
536. Wang, S., & Ng, T. S. (2019). Robustness of resource recovery systems under feedstock uncertainty. *Production and Operations Management, 28,* 628–649.
537. Wang, S., Ng, T. S., & Wong, M. (2016). Expansion planning for waste-to-energy systems using waste forecast prediction sets. *Naval Research Logistics, 63,* 47–70.
538. Wang, X., Kuo, Y.-H., Shen, H., & Zhang, L. (2021). Target-oriented robust location-transportation problem with service-level measure. *Transportation Research Part B: Methodological, 153,* 1–20.
539. Wang, Z., You, K., Wang, Z., & Liu, K. (2021). Multi-period facility location and capacity planning under ∞-Wasserstein joint chance constraints in humanitarian logistics. *arXiv:2111.15057* [math.OC]. Available at: https://doi.org/10.48550/arXiv.2111.15057
540. Waste Management World (2016). *Go-ahead for waste to energy gasification plant in Nottinghamshire, UK*. Available at: https://waste-management-world.com/artikel/go-ahead-for-waste-to-energy-gasification-plant-in-nottinghamshire-uk/
541. Watson, J.-P., & Woodruff, D. L. (2011). Progressive hedging innovations for a class of stochastic mixed-integer resource allocation problems. *Computational Management Science, 8,* 355–370.
542. Weaver, J. R., & Church, R. L. (1983). Computational procedures for location problems on stochastic networks. *Transportation Science, 17,* 168–180.
543. Wiesemann, W., Kuhn, D., & Sim, M. (2014). Distributionally robust convex optimization. *Operations Research, 62,* 1358–1376.
544. World Bank Group (2016). *BOT DBO and concession agreements for municipal solid waste disposal, recycling, sanitary landfill and waste to energy*. Available at: https://ppp.worldbank.org/ppp/sector/solid-waste/sample-contracts-waste-disposal-treatment-recycling
545. Xie, W. (2020). Tractable reformulations of two-stage distributionally robust linear programs over the type-∞ Wasserstein ball. *Operations Research Letters, 48,* 513–523.
546. Xie, W. (2021). On distributionally robust chance constrained programs with Wasserstein distance. *Mathematical Programming, 186,* 115–155.

547. Xie, W., & Ahmed, S. (2018). On deterministic reformulations of distributionally robust joint chance constrained optimization problems. *SIAM Journal on Optimization, 28,* 1151–1182.
548. Xie, W., Ouyang, Y., & Wong, S. C. (2016). Reliable location-routing design under probabilistic facility disruptions. *Transportation Science, 50,* 1128–1138.
549. Xiong, J., Ng, T. S. A., & Wang, S. (2016). An optimization model for economic feasibility analysis and design of decentralized waste-to-energy systems. *Energy, 101,* 239–251.
550. Xiong, J., Wang, S., & Ng, T. S. (2021). Robust bilevel resource recovery planning. *Production and Operations Management, 30,* 2962–2992.
551. Xu, N., & Nozick, L. (2009). Modeling supplier selection and the use of option contracts for global supply chain design. *Computers & Operations Research, 36,* 2786–2800.
552. Xu, Y., Huang, G. H., Qin, X. S., Cao, M. F., & Sun, Y. (2010). An interval-parameter stochastic robust optimization model for supporting municipal solid waste management under uncertainty. *Waste Management, 30,* 316–327.
553. Yaman, H. (2011). Allocation strategies in hub networks. *European Journal of Operational Research, 211,* 442–451.
554. Yang, X., Bostel, N., & Dejax, P. (2019). A MILP model and memetic algorithm for the hub location and routing problem with distinct collection and delivery tours. *Computers & Industrial Engineering, 135,* 105–119.
555. Yanık, S., Sürer, Ö., & Öztayşi, B. (2016). Designing sustainable energy regions using genetic algorithms and location-allocation approach. *Energy, 97,* 161–172.
556. Yao, Z., Lee, L. H., Jaruphongsa, W., Tan, V., & Hui, C. F. (2010). Multi-source facility location-allocation and inventory problem. *European Journal of Operational Research, 207,* 750–762.
557. Yau, Y. (2010). Domestic waste recycling, collective action and economic incentive: The case in Hong Kong. *Waste Wanagement, 30,* 2440–2447.
558. Yu, X., & Shen, S. (2022). Multistage distributionally robust mixed-integer programming with decision-dependent moment-based ambiguity sets. *Mathematical Programming, 196,* 1025–1064.
559. Yu, X., Shen, S., & Ahmed, S. (2021). On the value of multistage stochastic facility location with risk aversion. *arXiv:2105.11005* [math.OC]. Available at: https://doi.org/10.48550/arXiv.2105.11005
560. Yuan, H. P., Shen, L., Hao, J. J. L., & Lu, W. S. (2011). A model for cost-benefit analysis of construction and demolition waste management throughout the waste chain. *Resources, Conservation and Recycling, 55,* 604–612.
561. Yuan, Y., Song, Q., & Zhou, B. (2023). A multi-period emergency medical service location problem based on Wasserstein-metric approach using generalised Benders decomposition method. *International Journal of Systems Science, 54,* 1173–1185.
562. Zanjirani Farahani, R., Rashidi Bajgan, H., Fahimnia, B., & Kaviani, M. (2015). Location-inventory problem in supply chains: A modelling review. *International Journal of Production Research, 53,* 3769–3788.
563. Zarandi, M. H. F., Hemmati, A., Davari, S., & Turksen, I. B. (2013). Capacitated location-routing problem with time windows under uncertainty. *Knowledge-Based Systems, 37,* 480–489.
564. Zeng, B., & Zhao, L. (2013). Solving two-stage robust optimization problems using a column-and-constraint generation method. *Operations Research Letters, 41,* 457–461.
565. Zetina, C. A., Contreras, I., Cordeau, J.-F., & Nikbakhsh, E. (2017). Robust uncapacitated hub location. *Transportation Research Part B: Methodological, 106,* 393–410.
566. Zhalechian, M., Tavakkoli-Moghaddam, R., Zahiri, B., & Mohammadi, M. (2016). Sustainable design of a closed-loop location-routing-inventory supply chain network under mixed uncertainty. *Transportation Research Part E: Logistics and Transportation Review, 89,* 182–214.
567. Zhang, B., Peng, J., & Li, S. (2017). Covering location problem of emergency service facilities in an uncertain environment. *Applied Mathematical Modelling, 51,* 429–447.

568. Zhang, G., Shang, J., & Li, W. (2011). Collaborative production planning of supply chain under price and demand uncertainty. *European Journal of Operational Research, 215,* 590–603.
569. Zhang, S., Chen, M., & Zhang, W. (2019). A novel location-routing problem in electric vehicle transportation with stochastic demands. *Journal of Cleaner Production, 221,* 567–581.
570. Zhang, X., Zheng, Z., Zhang, S., & Du, W. (2016). Partial interdiction median models for multi-sourcing supply systems. *The International Journal of Advanced Manufacturing Technology, 84,* 165–181.
571. Zhang, X., Zheng, Z., Zhu, Y., & Cai, K.-Y. (2014). Protection issues for supply systems involving random attacks. *Computers & Operations Research, 43,* 137–156.
572. Zhang, Y., Qi, M., Lin, W.-H., & Miao, L. (2015). A metaheuristic approach to the reliable location routing problem under disruptions. *Transportation Research Part E: Logistics and Transportation Review, 83,* 90–110.
573. Zhao, Y., Chen, Z., & Zhang, Z. (2023). Distributionally robust chance-constrained p-hub center problem. *INFORMS Journal on Computing*. https://doi.org/10.1287/ijoc.2022.0113
574. Zhong, S., Cheng, R., Jiang, Y., Wang, Z., Larsen, A., & Nielsen, O. A. (2020). Risk-averse optimization of disaster relief facility location and vehicle routing under stochastic demand. *Transportation Research Part E: Logistics and Transportation Review, 141,* 102015.
575. Zhou, K., Pan, Y., Chen, C. L. P., Wang, D., & Han, S. (2016). k-medoids method based on divergence for uncertain data clustering. In *Proceedings of SMC'16, IEEEE International Conference on Systems Manufacturing and Cibernetics, June 9–12, Budapest, Hungary* (pp. 2671–2674).
576. Zhu, N., Fu, C., & Ma, S. (2018). Data-driven distributionally robust optimization approach for reliable travel-time-information-gain-oriented traffic sensor location model. *Transportation Research Part B: Methodological, 113,* 91–120.
577. Zhu, T., Boyles, S. D., & Unnikrishnan, A. (2022). Two-stage robust facility location problem with drones. *Transportation Research Part C: Emerging Technologies, 137,* 103563.
578. Ziegler, H.-P. (2012). *Algorithms for Linear Stochastic Programs and their Application in Supply Chain Network Design Problems.* PhD Thesis, Karlsruhe Institute of Technology, KIT, Karlsruhe.
579. Zipkin, P. H. (2000). *Foundations of inventory management*. McGraw Hill.
580. Zsidisin, G. A., & Henke, M. (Eds.) (2019). *Revisiting supply chain risk.* Springer Nature Switzerland.
581. Zymler, S., Kuhn, D., & Rustem, B. (2013). Distributionally robust joint chance constraints with second-order moment information. *Mathematical Programming, 137,* 167–198.

Index

© The Author(s), under exclusive license to Springer Nature Switzerland AG 2024
F. Saldanha-da-Gama, S. Wang, *Facility Location Under Uncertainty*,
International Series in Operations Research & Management Science 356,
https://doi.org/10.1007/978-3-031-55927-3

GPSR Compliance
The European Union's (EU) General Product Safety Regulation (GPSR) is a set of rules that requires consumer products to be safe and our obligations to ensure this.

If you have any concerns about our products, you can contact us on

ProductSafety@springernature.com

In case Publisher is established outside the EU, the EU authorized representative is:

Springer Nature Customer Service Center GmbH
Europaplatz 3
69115 Heidelberg, Germany

www.ingramcontent.com/pod-product-compliance
Lightning Source LLC
Chambersburg PA
CBHW051611131025
33954CB00003B/160
* 9 7 8 3 0 3 1 5 5 9 2 6 6 *